standard catalog of AMERICAN MOTORS

Edited by John A. Gunnell

FIRST EDITION

All rights reserved. No part of this book may be reproduced or transmitted in any form or by any means, electronic or mechanical, including photocopying, recording, or by an information storage and retrieval system without the permission in writing from the Publisher.

© 1993

Published by Krause Publications, Inc.
700 E. State St.
Iola, WI 54990
Telephone: 715-445-2214

ISBN: 0-87341-232-X
LIBRARY of CONGRESS NUMBER: 92-74794

Printed in the United States of America

FOREWORD

The concept behind Krause Publication's "standard catalogs" is to compile massive amounts of information about motor vehicles and present it in a standard format which the hobbyist, collector or professional dealer can use to answer some commonly asked questions.

Those questions include: What year, make and model is the vehicle? What did it sell for new? How rare is it? What is special about it? In our general automotive catalogs, some answers are provided by photos and others by the fact-filled text. In one-marque catalogs like this one, articles are also included up front.

Chester L. Krause of Krause Publications is responsible for the basic concept of creating the standard catalogs covering American cars. David V. Brownell, of Special-Interest Autos, undertook preliminary work on the concept while editing Old Cars Weekly in the 1970s. John A. Gunnell continued the standard catalog project after 1978. The *Standard Catalog of American Cars 1946-1975* was first published in 1982. Meanwhile, Beverly Rae-Kimes and Henry Austin Clark, Jr. continued writing and researching the *Standard Catalog of American Cars 1805-1942*, which was published in 1985. In 1987, The *Standard Catalog of Light-Duty American Trucks 1900-1986*, was compiled by John Gunnell and a second edition of the 1946-1975 volume was printed. In 1988, the 1805-1942 volume by Kimes and Clark appeared in its second edition. Also in 1988, James M. Flammang authored the *Standard Catalog of American Cars 1976-1986*, which went into its second edition in 1990. More recently, the 1946-1975 book was re-edited during 1992.

While the four-volume set of standard catalogs enjoyed high popularity as all-inclusive guides for car collectors, there seemed to be many auto enthusiasts who focused their energies on only one make of car or multiple makes made by closely related manufacturers. This led to creation of standard catalogs about Chrysler Corp., FoMoCo and GM makes, which in turn brought requests for catalogs about independent automakers. The *Standard Catalog of AMC* is our first venture into this niche. It covers a network of firms that led to, joined in or followed what enthusiasts know as the AMC "family" of automobiles. Rambler, Jeffery, Ajax, Nash, Lafayette, Hudson, Essex, Terraplane, Dover, Metropolitan, Nash-Healey, AMC, Jeep...even Renault...are among the companies included.

The *Standard Catalog of AMC* was compiled by an experienced editorial team consisting of the automotive staff of Krause Publications and numerous experts on a certain marque or specific area of AMC history. A major benefit of this "teamwork" has been the gathering of more significant facts about each model than a single author might find.

No claims are made that these catalogs are infallible history texts or encyclopedias. Nor are they repair manuals or "bibles" for motor vehicle enthusiasts. They are meant as a contribution to the pursuit of greater knowledge about many wonderful vehicles. They are also much larger in size, broader in scope and more deluxe in format than most other collector guides, buyer guides or price guides.

The long-range goal of Krause Publications is to make all of the catalogs as nearly perfect as possible. At the same time, we expect they will always raise new questions and bring forth new facts that were not previously unearthed. All our contributors maintain an ongoing file of new research, corrections and additional photos that are used, regularly, to refine and expand future editions.

Should you have knowledge you wish to see in future editions, please don't hesitate to contact the editors at *Standard Catalog of AMC*, editorial department, 700 East State Street, Iola, WI 54990.

ABBREVIATIONS

AAA	American Automobile Assoc.
Aero	Aerodynamic
Air	Air Conditioning
AM	AM Radio
AM	American Motors Corp.
Amp	Amperes
AMX	American Motors Experimental
B-Pillar	Second Roof Pillar
Brghm/Brgm	Brougham
Br2	Breezeway
BSW	Black Sidewall Tires
Bus	Business
Cabr	Cabriolet
Cam	Camshaft
CC	Cubic Centimeters
cid	cubic inch displacement
clb	Club
Co.	Company
Conv.	Convertible
Corp	Corporation
Cpe	Coupe
C-Pillar	Third Roof Pillar
cr	Cross (AMC)
cty/ctry	Country
cus	Custom
cyl	Cylinder
del	Deluxe
Div	Division
d/l	Deluxe Level
dly	Delivery
dpl	Diplomat (AMC)
dr.	Door(s)
DSO	District Sales Office
EFI	Electronic Fuel Injection
Eight	Eight cylinder auto
ESA	Economic Stabilization Agency
Exec	Executive
Fam	Family (Family Sedan)
FLO	Flow
FM	FM Radio
Fml	Formal
FsBk	Fastback
FT	Formal Top
GI	WWII Soldiers
Go-Pack	AMC performance options
GT	Gran Turismo
Hatch	Hatchback
HO	High-output
Holly	Hollywood (Hudson)
Hp	Horsepower
HT	Hardtop
I	Inline
I.O.	Identification
L4	L-head four cylinder
LAN	Landau
LAPD	Los Angeles Police Dept.
L-head	Valve in Block Motor
Limo	Limousine
LWB	Long Wheel Base
MM	Millimeters
mpg	Miles Per Gallon
mph	Miler Per Hour
MTL	Metal
NASCA	Nat. Assoc. Stock Car Racing
NHP	Net horsepower
NHRA	Nat. Hot Rod Assoc.
NPA	National Price Administration
Notch	Notchback
OHV	Overhead Valve(s)
Ont.	Ontario, Canada
OSRV	Outside Rearview (mirror)
O/W	Opera Window
P	Passenger (GP=6-passenger)
Phae	Phaeton
PRM	Revolutions Per Minute
RWL	Raised White Letter
SAE	Society Automotive Engineer
SC/360	AMC Model Name
SCCA	Sports Car Club of Amer.
Sed	Sedan
Six	Six-cylinder Model
SOHC	Single Overhead Camshaft
Spec.	Special
Spl.	Special
Spt.	Sport(s)
SST	Super Sport Touring AMC Model
Sta.	Station
Tach	Tachometer
Taxi	Taxicab
Trans AM	Trans American Race
Tu-Tone	Two-tone
Twn	Town
US	United States
USA	United States of America
USAC	U.S. Auto Racing Club
UTL	Utility
Veed	V-shaped
V-8	V-8 cylinder engine
V-Shaped	Shaped like a V
Wag.	Wagon
Wis	Wisconsin
WL	White Letter
w/o	Without

INTRODUCTION

The catalog you hold in your hands, focusing on cars and trucks from the American Motors "family," is a compilation of photos, articles, specifications tables and current values for a variety of vehicles including Rambler, Nash, Hudson, AMC, and Jeep models.

Realizing that many car hobbyists prefer a specific AMC-related brand or model, it seemed reasonable to think that a catalog formatted along marque lines would be appealing. The editor was asked to organize such a catalog from material existing here at Krause Publications.

The catalog begins with informative histories and articles which have appeared in *Old Cars* since 1971. The stories were organized to provide a look at the histories and products of the companies which preceded or joined American Motors Corp.

Inside the catalog you'll find dozens of articles which were first published so long ago that they are now, themselves, considered "classics." They highlight milestones from the times of the 1902 Rambler Runabout to the 50th anniversary of the Jeep in 1991.

This material has been obtained from the contributors whose work is familiar to hobbyists throughout the world. Masque experts who wrote articles or compiled data include Arch Brown, Alex Burr, John Code, Larry Daum, Charles Liskow, Jack Miller, Larry Mitchell, John O'Halloran, Vince Ruffolo and Darryl A. Salisbury. Writers and historians whose work appears are Robert C. Ackerson, Terry Boyce, Ned Comstock, Betty Cott, John Gunnell, Phil Hall, Bill Mason, Robert F. Mehl, Jr., Gerald Perschbacher, Ed Robinson, Shelden Rody, Bill Siuru, Wally Wyss and R. Perry Zavitz.

The second section of this catalog is the lengthy and detailed specifications tables which present, in standardized format, styling and engineering features of virtually all AMC-related models. You'll find engine sizes, horsepower ratings, wheelbases, measurements and tire sizes...just to name a few things.

Additional thanks for the gathering of this standardized catalog data goes to Beverly Rae Kimes and Henry Austin Clark Jr. (authors of Krause Publication's *Standard Catalog of American Cars 1805-1942*) and James Flammang (author of the *Standard Catalog of American Cars 1976-1986*).

The primary source of photos of American Motors-related automobiles was the *Old Cars* photo archives, a vast collection containing over 16,000 pictures, advertisements and illustrations of vehicles. This archives includes automakers' publicity stills, photos obtained from specialized vendors such as Applegate & Applegate, pictures taken at hobby shows by the *Old Cars* staff and news photos snapped at thousands of hobby events.

Photos were also obtained through American Motors, Chrysler-Jeep-Eagle and from the collections of Jack Miller and Vince Ruffolo. In addition, where photographs were unavailable, illustrations from sales and technical literature and advertisements have been used to show what the cars looked like when they were new.

The final element of this catalog is the presentation of current "ballpark prices" for all AMC-related models from 1902 to 1984. These values came from the *Old Cars Price Guide*, which is compiled as a bimonthly magazine edited by Ken Buttolph and James T. Lenzke. The prices are formatted according to Krause Publications' time-tested 1-6 condition scale so that collectors who use the catalog can determine their cars' values in a variety of conditions.

MEET THE EDITOR

John Gunnell has been involved in the car collecting hobby since 1972 and began writing about vintage cars in 1975. Since then, he has authored thousands of stories and dozens of books about collectible cars and the hobby of collecting them. Gunnell's experience with the ownership of AMC models is limited to past ownership of a 1978 Pacer D/L station wagon. On many occasions, the reliable little bubble-shaped "woodie" carried him from his adopted home in Iola, Wis. to his home town of Staten Island, N.Y., with two adults, four kids and the family dog inside. For the future, Gunnell dreams of someday buying a 1953 Hudson Hornet Hollywood hardtop to go cruising in.

BODY STYLES

Body style designations describe the shape and character of an automobile. In the early days, automakers exhibited great imagination in coining words to name their products. This led to descriptions that were not totally accurate. Many of the car words were taken from other fields: mythology, carriage building, architecture, railroading, and so on. Therefore, they have no "correct" automotive meanings; only those brought about through actual use. Some seeming inconsistencies have persisted to recent years, while other imaginative terms of past eras have faded away. One manufacturer's sedan might resemble another's coupe. Some automakers have persisted in describing a model by a word different from common usage, such as Ford calling the Mustang a sedan. Following the demise of the pillarless hardtop in the mid-1970s, various manufacturers continued using the term hardtop to describe their sedans. Some used the descriptions pillared hardtop or thin-pillared hardtop to label what most call a sedan. Descriptions in this catalog generally follow the manufacturers' terms, unless they conflict strongly with accepted usage.

One specific AMC example is worth noting. Some AMC hatchbacks are called three- or five-door models, even though the extra door is not an entrance. And the Hornet version of a "Sportabout" model is quite different from the roadster-like Wayfarer Sportabout that Dodge marketed in early postwar years.

TWO-DOOR (CLUB) COUPE: The Club Coupe designation seems to come from club car, describing the lounge or parlor car in a railroad train. The early postwar Club Coupe combined a shorter-than-sedan body structure with the convenience of a full back seat, unlike the single-seat business coupe. Club Coupe has been used less frequently after World War II, as most two-door models have been referred to simply as coupes. The distinction between two-door coupes and two-door sedans has grown fuzzy, too. Hudson used the term Club Coupe until 1954, the year the company merged with Nash to form AMC.

TWO-DOOR SEDAN: The term sedan originally described a conveyance seen only in movies today: a wheel-less vehicle for one person, borne on poles by two men...one ahead and one behind. Automakers pirated the word and applied it to cars with a permanent top that seated four to seven people (including driver) in a single compartment. The two-door sedan of recent times has sometimes been called a pillared coupe or just plain coupe, depending on the manufacturer's whims. On the other hand, some cars commonly referred to as coupes carry the sedan designation on factory literature. One of AMC's most unusual two-door sedans was the Pacer.

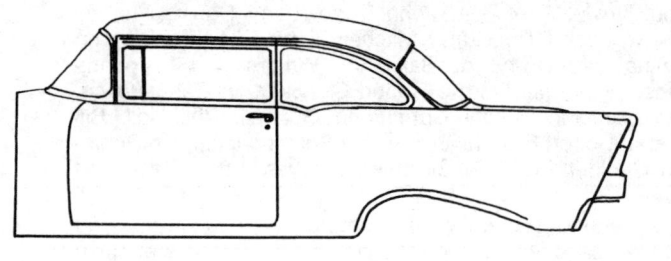

TWO-DOOR (THREE-DOOR) LIFTBACK COUPE: Originally a small opening in the deck of a sailing ship, the term hatch was later applied to airplane doors and to passenger cars with rear liftgates. Most automakers called these cars hatchbacks, but AMC used the term liftback. Various models appeared in the early 1950s, but weather tightness was a problem. The concept emerged again in the early 1970s, when fuel economy factors began to signal the trend toward compact cars. Technology had remedied the sealing difficulties. By the 1980s, most manufacturers produced one or more hatchback models, though the question of whether to call them two-doors or three-doors never was resolved. Their main common feature was the lack of a separate trunk. Liftback coupes may have had a different rear end shape, but the two terms often described essentially the same vehicle. The Gremlin was an interesting hatchback coupe that AMC created.

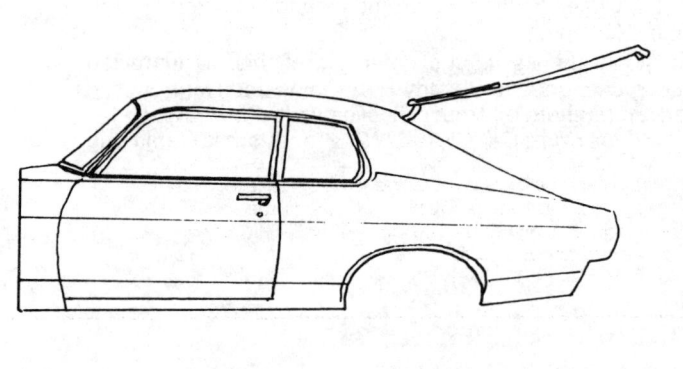

TWO-DOOR FASTBACK: By definition, a fastback is any automobile with a long, moderately curving, downward slope to the rear of the roof. This body style relates to an interest in streamlining and aerodynamics and has gone in and out of fashion at various times. Some fastbacks (Mustangs for one) have grown quite popular. Others have tended to turn customers off. Certain fastbacks are really two-door sedans or pillared coupes. Four-door fastbacks have also been produced. Many of these (such as Buick's late 1970s four-door Century sedan) lacked sales appeal. Fastbacks may or may not have a rear-opening hatch. The Hudson Hornet's fastback styling helped it in stock car racing and the AMX's aerodynamic lines helped its dragstrip and road racing performance.

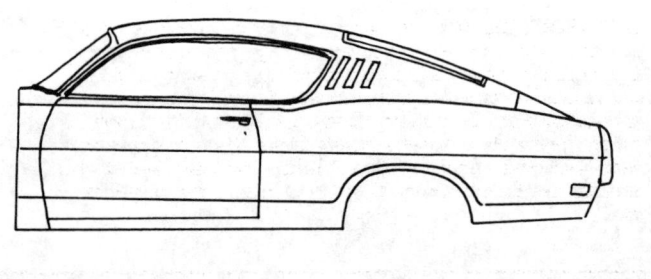

TWO-DOOR HARDTOP: The term hardtop, as used for postwar cars up to the mid-1970s, describes an automobile styled to resemble a convertible, but with a rigid metal or fiberglass top. In a production sense, this body style evolved after World War II. It was first called hardtop convertible. Other generic names have included sports coupe, hardtop coupe or pillarless coupe. In the face of proposed federal government rollover standards, nearly all automakers turned away from pillarless cars by 1976 or 1977. **FORMAL HARDTOP:** The hardtop roofline was a long-lasting fashion hit of the postwar era. The word formal can be applied to things that are stiffly conservative and follow the established rule. The limousine, being the popular choice of conservative buyers who belonged to the establishment, was looked upon as a formal motor car. When designers combined the lines of these two body styles, the result was the Formal Hardtop. This style has been marketed with two- or four-doors, canopy or vinyl roofs (full or partial) and conventional or opera-type windows, under various trade names. The distinction between a formal hardtop and plain pillared-hardtop coupe (see above) hasn't always followed strict rules. AMC did not offer this body style.

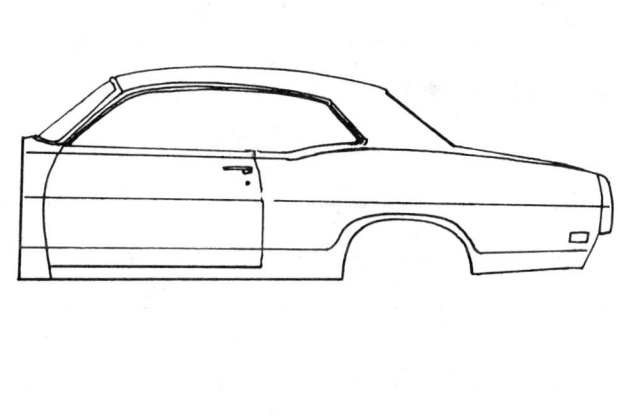

CONVERTIBLE: To depression-era buyers, a convertible was a car with a fixed-position windshield and folding. When raised, the top displayed the lines of a coupe. Buyers in the postwar period expected a convertible to have roll-up windows, too. Yet the definition of the word includes no such qualifications. It states only that such a car should have a lowerable or removable top. American convertibles became almost extinct by 1976, except for Cadillac's Eldorado. In 1982, though, Chrysler brought out a LeBaron ragtop; Dodge a 400; and several other companies followed it a year or two later. Today, many other cars are available in the convertible format. The last AMC ragtop was the Alliance convertible, offered in 1985 and 1986.

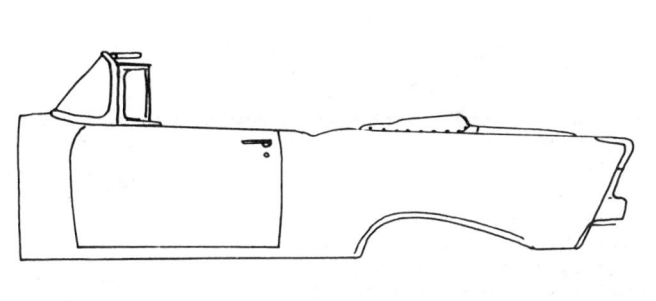

ROADSTER: This term derives from equestrian vocabulary, where it was applied to a horse used for riding on the roads. Old dictionaries define the roadster as an open-type car designed for use on ordinary roads, with a single seat for two persons and, often, a rumbleseat as well. Hobbyists associate folding windshields and side curtains (rather than roll-up windows) with roadsters, although such qualifications stem from usage, not definition of term. Most recent roadsters are either sports cars, small alternative-type vehicles or replicas of early models. Hudson built its last roadster in 1931. Nash ended production of true roadsters in 1930, although a "convertible-roadster" (with roll-up windows) was cataloged through 1932.

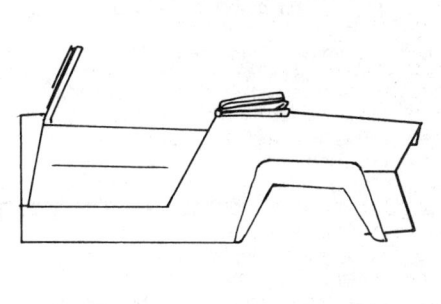

RUNABOUT: By definition, a runabout is the equivalent of a roadster. The term was used by carriage makers and has been applied in the past to light, open cars on which a top is unavailable or totally an add-on option. None of this explains its use by Ford on certain Pintos. Other than this usage, recent runabouts are found mainly in the alternative vehicle field, including certain electric-powered models. The most famous runabout in the AMC family of cars is the first Rambler.

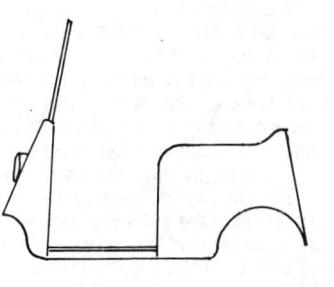

Standard Catalog of AMC

FOUR-DOOR SEDAN: If you took the wheels off a car, mounted it on poles and hired two weight lifters (one in front and one in back) to carry you around in it, you'd have a true sedan. Since this idea isn't very practical, it's better to use the term for an automobile with a permanent top (affixed by solid pillars) that seats four or more persons, including the driver, on two full-width seats.

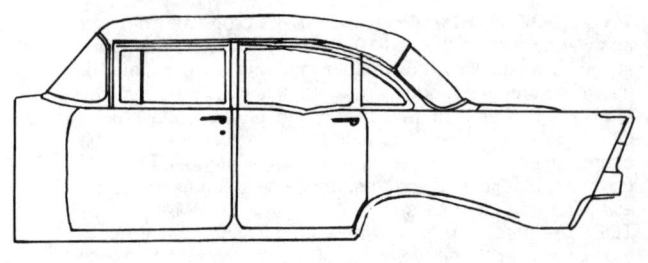

FOUR-DOOR HARDTOP: This is a four-door car styled to resemble a convertible, but having a rigid top of metal or fiberglass. Buick introduced a totally pillarless design in 1955. A year later most automakers offered equivalent bodies. Four-door hardtops have also been labeled sports sedans and hardtop sedans. By 1976, proposed federal rollover standards and waning popularity had taken their toll of four-door hardtop output. Only a few makes still produced a four-door hardtop. They disappeared soon thereafter. AMC's 1957 Rambler Rebel is probably the company's most famous four-door hardtop.

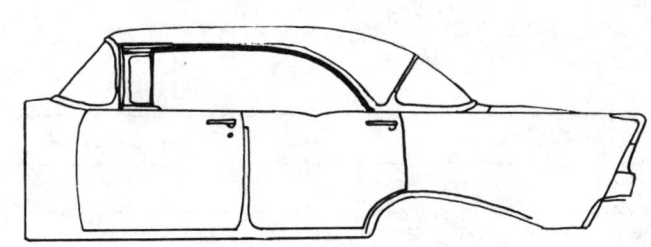

FOUR-DOOR PILLARED HARDTOP: Once the true four-door hardtop began to fade away, manufacturers needed another name for their luxury four-doors. Many were styled to look almost like the former pillarless models, with thin or unobtrusive pillars between the doors. Some, in fact, were called "thin-pillar hardtops." The distinction between certain pillared hardtops and ordinary (presumably humdrum) sedans occasionally grew hazy.

FOUR-DOOR (FIVE-DOOR) LIFTBACK: Essentially unknown among domestic models in the mid-1970s, the four-door liftback or hatchback became a popular model as cars grew smaller and front-wheel drive caught on. Styling was similar to the original two-door hatchback, except for the extra doors. Luggage was carried in the back of the car. It was loaded through the hatch opening, not in a separate trunk. AMC's first hatchback sedan, a five-door model, appeared in the 1984 Encore line.

LIMOUSINE: This word's literal meaning is 'a cloak.' In France, limousine means any passenger vehicle. An early dictionary defined limousine as an auto with a permanently enclosed compartment for 3-5, with a roof projecting over a front driver's seat. However, modern dictionaries drop the separate compartment idea and refer to limousines as large luxury autos, often chauffeur-driven. Some have a movable division window between the driver and passenger compartments, but that isn't a requirement.

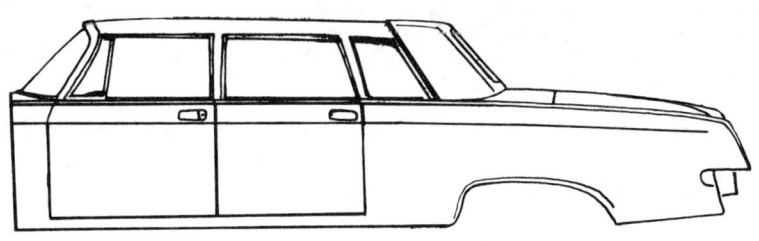

TWO-DOOR STATION WAGON: Originally defined as a car with an enclosed wooden body of paneled design (with several rows of folding or removable seats behind the driver), the station wagon became a different and much more popular type of vehicle in the postwar years. A recent dictionary states that such models have a larger interior than sedans of the line and seats that can be readily lifted out, or folded down, to facilitate light trucking. In addition, there's usually a tailgate, but no separate luggage compartment. The two-door wagon often has sliding or flip-out rear side windows.

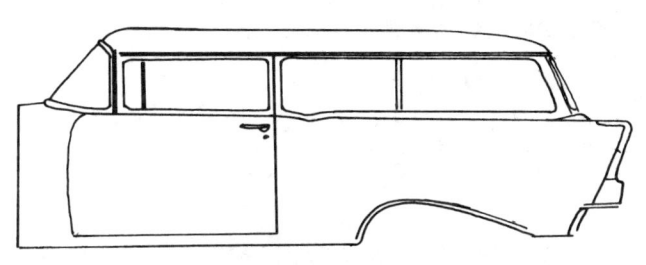

FOUR-DOOR STATION WAGON: Since functionality and adaptability are advantages of station wagons, four-door versions have traditionally been sales leaders. At least they were until cars began to grow smaller. This style usually has lowerable windows in all four doors and fixed rear side glass. The term "suburban" was almost synonymous with station wagon at one time, but is now more commonly applied to light trucks with similar styling. Station wagons have had many trade names, such as Country Squire (Ford) and Sport Suburban (Plymouth). Quite a few have retained simulated wood paneling, keeping alive the wagon's origin as a wood-bodied vehicle. AMC was famous for introducing the four-door hardtop (pillarless) station wagon in the mid-1950s.

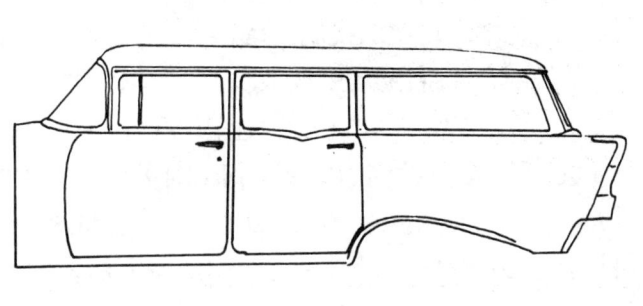

LIFTBACK STATION WAGON: Small cars came in station wagon form too. The idea was the same as bigger versions, but the conventional tailgate was replaced by a single lift-up hatch. For obvious reasons, compact and subcompact wagons had only two seats, instead of the three that had been available in many full-sized models. The Hornet Sportabout station wagon should become a collectible example of this body style, along with early American Eagle station wagons.

INDEX TO ARTICLES

Meet the Family 13

AMC Had History of Growth, Various Name Changes 14
Prophet of 1901 Vindicated as AMC Caps 75th Year 19
Rambler American ... The First "Compact" 22
From Bikes to Buggies: Nash Lineage Traced to Mid-19th Century 27

The Early Years 28

The Quad - Nash's Four-Wheeled Wonder 32
When Jeffery Made Trucks 34

Kin From Kenosha — Nash 36

The Bargain-Priced Ajax 37
1933 Nash Advanced Eight 38
Lafayette Has Arrived 39
Good-Bye Nash: The Hardest Words in the English Language 41
Last of the Nash-Lafayettes: 1940 "All-Purpose" Coupe 44
The '42 Nash Was Hardly a Smash 46
Cars That Are Trucks...and Vice Versa 47
Was Gyro Gearloose a Nash Enthusiast? 48
Nash in the Late '40s and Early '50s 49
Rambler Luxury in a Compact Package 53
Nash - The End of the Line 55

Car City Cousins — Hudson 57

Hudson is Handy Hobby Car 58
Dover Commercial Trucks 60
"Hot Diggety Dog ... That's Terraplaning!" 61
Terraplane in Review 63
1933 Terraplane Eight Was a Performance Compact 68
Hudson 112 - Homely But Endearing 70
Hudson After World War II: 1946 and 1947 72

Cars in Plaster 74
The Enigmatic Hudson.. 76
Hudson's Postwar Record: From Go-Go to Long Gone............................ 78
Topless Hudsons... 80
Was the Jet Hudson's Edsel? 83
A '52 Hudson Visits the Old "Filling Station".. 85

Rapid Relatives: Hudson Race Cars 88

When Hudson Went Racing... 89
Hornet's Buzz Meant the End of Queen Bee Rocket's Reign 91
Hudson Led "Big 3" in Backing Stockers on National Circuit 96

Foreign Relations... 99

Nash Went Abroad to Satisfy U.S. Hunger for Sports Cars 100
AMC's First Sports Car 102
Cars of AMC: The Metropolitan... 104
Once Spurned by Collectors, Metropolitan Now Enjoying
    a Revival in Popularity 106
Project Italia's Goal: Enhancing Hudson.. 109

Identity Crisis: Nash/Hudson Merger 111

History of AMC: Part I 112
History of AMC: Part II 114

Family Gathering: AMC Style... 116

Rambler - The Second Time Around.. 117
Hardtop Wagon Was an American Motors' First 119
The '63 Ramblers: They Still Look Good Today..................................... 120
Recycled AMC Model Names 122
AMX Led AMC Into Youth Market: Now It's a Hot 'Street Classic' 124
SC/Rambler Hurst was Hot "Tot"... 127
AMC Sought Success With Innovative Designs 129

Renagade Relatives: AMC Jeep... 133

Jeep Celebrated 50 Years of Leadership in 1991 134

ACKNOWLEDGEMENTS

Since this catalog combines articles published in *OLD CARS WEEKLY* with standard catalog material from five different books, there is a long list of sources to thank for their artistic, photographic and written contributions.

Some of these contributors wrote articles, supplied photos and did standard catalog research. Some created drawings and illustrations. Others were primary compilers of complete standard catalog data bases. Inside the catalog, you will find credit lines on articles, catalog entries and artwork recognizing the major individual contributors.

On this page, we'd like to generally acknowledge all the people and organizations that helped to make this catalog a reality. Their names are listed alphabetically. Following each name, in parentheses, are codes indicating their role in the creative process. The codes used are: A=article(s); C=catalog compilation, primary; D=drawings; E=editorial staff; P=photo(s) contributor; R=research for marques/models in catalog data base; V=values researched. You will notice that many contributors wear more than one hat.

Robert C. Ackerson (A,C,P,R)
American Motors Corp. (P)
Antique Truck Club (P)
William L. Bailey (P)
Brad Bowling (E)
Terry V. Boyce (A,P)
Arch Brown (A,C,P,R)
Alex Burr (A)
Kenneth Buttolph (E,V)
Henry Austin Clark, Jr. (C,P)
Linda Clark (A)
Ned Comstock (A)
John A. Conde (A,C,P)
Betty Cott (A)
Larry R. Daum (A,C)
Wayne R. Graefen (P,R)
Phil Hall (A,P)
Harrah's Collection (P)
Bob Hovorka (A,D)
Hudson Motor Car Co. (P)
Indianapolis Motor Speedway Corp. (P)
Jeep Division, Chrysler Corp. (A,P)
Kaiser-Jeep Corp. (P)
Beverly Rae Kimes (A,C,P,R)
Ron Kowalke (E)
Tom LaMarre (A)
John Lee (P)
Jim Lenzke (E,V)
Library of Congress (P)
Charles Liskow (R)
Jack L. Martin (P)
Bill Mason (A,P)
Mary Mason (A,P)
Jack Miller (P,R)
Robert F. Mehl (A)
Larry G. Mitchell (A,P,R)
Steve Mostowa (P)
Nash Motors (P)
John O'Halloran (P,R)
Old Cars Weekly (P)
Gerald Perschbacher (A,P)
Ramsey Winch Corp. (P)
Ed Robinson (A)
Sheldon Rody (A)
Vincent Ruffolo (A,P,R)
Dennis Sagvold (P)
Darryl Salisbury (R)
Mary Sieber (E)
Bill Siuru (A)
Charles Webb (P)
Donald F. Wood (P)
Wally Wyss (A)
R. Perry Zavitz (A,C,P)

Of course, general thanks also go out to the hundreds of Hudson, Nash, AMC enthusiasts who provided much of the inspiration for this catalog, including (but not limited to) Grover Cleveland, John A. Conde, Larry R. Daum, George Doughtie, Jack Miller, Larry G. Mitchell, Vince Ruffolo, Darryl Salisbury, Bob Stevens and the late Richard A. "Dick" Teague.

DEDICATION

Krause Publications would like to dedicate this catalog to the memory of Richard A. Teague, car designer, creator and collector. Known and loved by many marque enthusiasts who recognized him as "Mr. AMC," Mr. Teague added flavor and flair to the "Pride of Kenosha" and gave the marque an honored thread in the tapestry of automotive history.

Meet the Family

AMC's roots trace back to the 1902 Rambler, such as this example that two old car lovers used to venture out into modern traffic during an antique auto tour.

AMC had history of growth, various name changes

American Motors Corp., one of the oldest pioneers in the U.S. automobile industry, had a long history of automotive milestones established by a variety of great marques and different nameplates.

On March 1, 1902, the first one-cylinder Rambler automobile was sold at the Chicago Automobile Show. The car, which had a retail price of $750, was built in Kenosha, Wis., by the Thomas B. Jeffery Co. This firm would become Nash Motors in 1916, Nash-Kelvinator in 1937 and part of American Motors in 1954.

Eventually, American Motors would market its AMC passenger cars and Jeep vehicles in every state of the U.S. and more than 140 countries. AMC administrative functions were headquartered in American Center, Southfield, Mich. Research, engineering, styling and purchasing departments were located in Detroit. Passenger-car manufacturing was carried out in Wisconsin, with plants at Kenosha, Milwaukee and Coleman, plus a proving grounds near Burlington. The company's four-wheel drive Jeep vehicles continued to be built in Toledo, Ohio, where Willys-Overland launched the famous freedom fighter. AM General Corp., which concentrated on military and government business as well as transit-bus production, had "AMC family" plants operating in South Bend, Mishawaka and Indianapolis, Ind., Livonia, Mich., and in Marshall, Texas.

Other factories were operated by AMC to make automotive components. They included modern plastics plants at Evart and Mt. Clemens Michigan, and Evansville, Ind. An electrical components plant was located in Iron River, Mich. Engines were produced at a plant in Richmond, Indiana. Some of these plants are now producing Jeep-Eagle components under Chrysler's ownership of AMC.

American Motors inherited decades of manufacturing experience and vast facilities when it came into being, on May 1, 1954, through the merger of Nash-Kelvinator Corp. and Hudson Motor Car Co. It broadened its heritage with the acquisition in February, 1970, of Jeep Corp.

Nash-Kelvinator had been created January 4, 1937, by the merger of Kelvinator Corp. and Nash Motors Co. Its first chairman was Charles W. Nash. After running a carnage and wagon factory in Flint, Mich., Nash became head of Buick in 1910 and president of General Motors in 1912. He resigned from GM in 1916 to purchase the Thomas B. Jeffery Co. of Kenosha, Wis.

Founded by an English-born inventor who first made bicycles, this firm had produced Rambler automobiles from 1902 to 1914, when, the name was changed to Jeffery. The Jeffery Company also had developed the four-wheel-drive "Quad" truck, made famous in World War I. Nash reorganized the firm as the Nash Motors Co. on July 29, 1916, and greatly increased production of both cars and trucks. Nash Motor's built 11,000 trucks in 1918, more than any company previously had produced in a single year.

As it expanded, Nash bought a half interest (1919) in the Seaman Body Corp. of Milwaukee. In 1936, Nash took Seaman over completely. Seaman started in 1846 as the A.D. Seaman Co., makers of fine furniture. The firm was among the first to reinforce automobile bodies with steel.

Nash bought the Racine, Wis. plant of the defunct Mitchell Motor Car Co. in 1924 and acquired the name and equipment of LaFayette Motors Corp. of Indianapolis, Ind. While nearly all cars bore the Nash name, LaFayette automobiles were made from 1934 to 1939 and the Ajax in 1925 and 1926. Peak prewar Nash production was 138,169 in 1928.

Charles Nash, who was 72 years old in 1936, began to look for a successor. His search led to George W. Mason, a former Chrysler works manager who was then president of the Kelvinator Corp. of Detroit. The Kelvinator enterprise dated from 1914, the year it began experiments on the first successful household electric refrigeration system. Mason declined to leave Kelvinator, and the companies merged on Jan. 4, 1937, as Nash-Kelvinator Corp. Mason became president and Nash was elected chairman of the board. Headquarters of the new company were in Detroit.

No stranger to the automobile business, Mason had demonstrated and sold motorcycles and automobiles as a boy in North Dakota. After learning mechanical engineering and business administration at the University of Michigan, he worked for Studebaker and Dodge. In 1921, then only 30, he became chief of all Chrysler manufacturing at the Maxwell-Chalmers Co.

As president of Nash-Kelvinator, Mason put Nash engineers to work on a new lightweight, economy car. Working with engineers of the Edward G. Budd Manufacturing Co., which had pioneered the all-steel automobile body, he devoted four years to the development of the Nash 600. Introduced in the fall of 1940 as a 1941 model, the 600 marked the first application in a mass-produced U.S. automobile of the single-unit construction principle. All Nash cars beginning with 1949 models were of single-unit construction.

Mason's enthusiasm for small cars was shared by George Romney, who had joined Nash-Kelvinator in 1948 as assistant to the president after a successful career as managing director of the Automobile Council for War Production and the Automobile Manufacturers Assoc.

One result was an 84 inch wheelbase prototype car, the Nash Experimental International. Originally called the NXI, it was shown to special "surview" audiences in a dozen cities from New York to San Francisco in the winter of 1949-1950. U.S. motorists for the first time were invited to give their opinions of a car not yet produced. The NXI was manufactured later (starting in 1954) as the Metropolitan. It was built by Austin, in England, and sold by Nash and Hudson dealers in the U.S. and Canada. Production of Metropolitans was discontinued in 1962, with deliveries from 1954-1962 totaling 94,986.

In March 1950, Nash-Kelvinator introduced the first modern compact car. The initial model was a convertible on a wheelbase of 100 inches. A station wagon was added in May 1950, and a hardtop sedan followed the next spring. For the new car, the nostalgic name Rambler was revived.

These four gents out for a winter ride in their 1909 Hudson touring car had little idea that Hudson would become part of the AMC automotive network some 45 years after the photo was taken.

Nash executives had been small car oriented for 15 years when the Rambler compact of 1950 appeared. The first models, released in spring, were all convertibles. A wagon and hardtop soon followed.

The Rambler was given only modest promotion at first and standard-sized Nash cars continued in production. In 1952, Nash offered fun-sized cars styled by Pinin Farina, the renowned Italian designer. Rambler production had increased from 20,782 in 1950 to 57,555 in 1951. While total Nash sales dropped during the next few years, the Rambler's percentage of industry sales increased. A 108 inch wheelbase four-door Rambler sedan and station wagon were added in 1954.

One of Mason's postwar dreams was to explore the possibility of a merger of independent automobile manufacturers to utilize facilities more efficiently and to spread tooling costs. Early in 1954, executives of the Hudson Motor Car Co., one of the firms approached earlier by Mason, proposed a merger with Nash-Kelvinator. Thus, American Motors Corp. was created on May 1, 1954.

Hudson was named for the founder of Detroit's noted department store, the J.L. Hudson Company. Joseph L. Hudson backed his niece's husband,, Roscoe B. Jackson, Roy D. Chapin and others in launching the company in 1909 and was briefly its president. Chapin was its leading spirit. As a young man in 1901, he had driven a Curved Dash Oldsmobile from Detroit to New York. He later served as Secretary of Commerce under President Hoover. Chapin was president of Hudson from 1910 to 1923 and from 1933 until his death in 1936.

The company's initial offering, the four-cylinder Hudson Twenty, was the first low-cost automobile using a selective sliding gear transmission with three forward speeds and reverse. The roadster had a guaranteed speed of 50 mph. Hudson cars became known for speed, ruggedness and mechanical innovations. Hudson engineers solved the problem of six-cylinder engine vibration with a fully-balanced crankshaft, introduced in the 1916 Hudson Super Six.

Perhaps Hudson's greatest contribution was bringing the closed car within financial reach of average car buyers. In 1919, Hudson set up a separate company (later dissolved into the parent firm) to build the low-priced Essex. In the first year, a touring car, roadster and sedan were offered. Total 1919 sales more than tripled the previous year's total, with Essex out-selling Hudson. In 1922, Essex introduced the coach body style. For the first time, the closed car was offered at a price competitive with open models.

In 1929, Hudson shipped 300,962 cars, two-thirds bearing the Essex name. Hudson ranked third in the industry in registrations - the highest place any independent make would achieve until the record-breaking performance of the modern Rambler, more than a quarter century later.

The Essex car was dropped with the introduction, in mid-1932 of the low-priced Essex-Terraplane, (soon to be plain Terraplane). Christened by aviatrix Amelia Earhart, Terraplane cars were built through 1937.

Hudson was one of the first to resume car production after the World War II. For several years, it was among the leading independents in total sales. Hudson Hornet models participated in and won numerous stock car racing events in the postwar period. Sales, however, were dwindling and President A.E. Barit sought a merger with another independent. The consolidation with Nash-Kelvinator to form American Motors Corp. resulted.

American Motors began operations in the toughest year for independents in the history of the industry. With Ford and Chevrolet waging a battle for sales leadership, sales of all other companies suffered. An unexpected blow came with the passing of George Mason at 63. He died in Detroit on Oct. 8, 1954, following an illness of only five days. Four days later, George Romney was elected to succeed Mason as chairman and president.

At American Motors, Romney assembled a young management team. Nash and, Hudson sales and dealer organizations were consolidated. Shortly after the merger, Roy Abernethy was brought into the company as vice president in charge of Nash sales. In 1953, Nash-Kelvinator and Hudson together had 3.56 per cent of industry sales. In 1954, the year of the merger, it dropped to 2.14 per cent and in 1955, a year of record total sales in the industry, it fell to 1.91 percent. In 1956, the percentage was 1.93 per cent, but fewer cars were sold overall in the industry.

The bright spot was the Rambler. Its sales had risen steadily from 1954 to 1956. A Rambler won the Mobilgas Economy Run that year (to be followed in subsequent years by many other economy run championships) and Ramblers began to show higher resale values on used car lots.

American Motors elected to stake everything on the Rambler. Beginning with the 1958 model year, American Motors stopped making full-sized cars and dropped the Nash and Hudson names. The company chose to fight its way out of heavy losses by concentrating on compact Ramblers. New models of other companies, meanwhile, were continuing the trend of longer, heavier and more powerful cars each year.

A revolution in automobile preference and design was under way. American Motors' cars sales rose from 114,084 and 1.94 per cent of industry sales in 1957 to 217,332 cars (all Ramblers) and 4.66 per cent of the industry in 1958. The company also earned its first profits. It was the only U.S. automobile company to sell more cars than in the previous year.

American Motors built 401,446 cars in 1959, setting an all-time production and sales record for an independent automobile company. Of the more than 5,000 companies that had appeared on the American scene in more than-half a century none, other than those of the Big Three, had ever built more than 307,000 cars in a single year. In 1960, all companies offered compact models, most of single-unit construction, but American Motors' sales and share of the greatly expanded market continued to increase.

Surveys showed motorists preferring Rambler cars for more reasons than their convenient size. To meet the increased demand for Ramblers, more than $43 million was invested in an expansion program to boost capacity. In 1960, Rambler built 485,746 cars, still another all-time high for an independent producer. Then, in 1963, Rambler's worldwide production surpassed the half-million mark for the first time.

In February 1962, George Romney resigned as president and chairman of American Motors to campaign successfully for governor of Michigan. The board of directors elected him vice-chairman and granted him a leave of absence. Richard E. Cross was named board chairman and Roy Abernethy became president. Following his election to the governorship in November of that year, Romney resigned as vice-chairman and director.

During the next few years, the company invested $300 million in new advanced design engines, bodies and plant facilities. On June 6, 1966, Cross resigned as chairman of the board to devote more time to his Detroit law fired. He continued as a director. The board of directors elected industrialist Robert B. Evans as chairman. An entirely new line of larger cars was introduced in the 1967 model year, with the Classic series becoming the Rebel (renamed the Matador in 1971) and new luxury models, including convertibles, making their bow. Roy D. Chapin, Jr., was named executive vice-president and general manager of the automotive division.

On Jan. 9, 1967, Abernethy retired and was succeeded by William V. Luneburg as president and chief operating officer. Chapin was elected chairman and chief executive officer. Evans, who stepped down as chairman, continued as director. Chapin staked out a new and bold direction for American Motors. Among the many forward steps taken to strengthen confidence in the company was a sharp reduction in delivered prices of the low-priced Rambler American models to make them more competitive with imports. Imported car sales were again were on the rise. A two-passenger sports car, the AMX, bowed in February, 1968.

In July, 1968, American Motors sold its Kelvinator appliance business to White Consolidated Industries, of Cleveland, to devote its full and complete energies to the automobile business. The compact, Hornet line of cars was introduced in the fall of 1969, and the new subcompact Gremlin bowed, in April, 1970, when the Rambler name was discontinued, As part of a planned expansion and acquisition program American Motors, on Feb. 5, 1970. acquired Kaiser-Jeep Corp. in a transaction in volving cash debentures and American Motors stock. Jeep is the leading worldwide manufacturer of four-wheel drive commercial recreational vehicles.

Both Jeep and AMC are pioneers in the automobile industry, tracing their beginnings to the turn of the century. Jeep Corp. had its origin with the successful road testing of the Overland runabout in February, 1903. The first Overland was built by the Overland Automobile Div. of Standard Wheel Co. at Terre Haute, Ind.

At that time, John North Willys, an energetic auto salesman from Elmira, N.Y. had contracted for the entire year's output of Overland automobiles. In 1908, he reorganized the firm as the Willys-Overland Co. and became its president, treasurer, sales manager and purchasing agent at facilities in Indianapollis. In need of additional production space, Willys purchased the Pope-Toledo plant at Toledo, Ohio. In 1911, he moved all of the company's operations to Toledo. By 1915 Willys-Overland was the second largest automobile company in the U.S.

In 1929, Willys sold his interest in the company and took the post of Ambassador to Poland. He soon returned to Toledo as active head of the company.

In 1940, experiments were started on what was to become the famed Jeep four-wheel drive military vehicle of World War II. The U.S. Army selected the Willys design over competitive models, and the famed four-wheel drive quarter-ton truck was placed in production in 1941.

The company's postwar production was programmed to relate the manufacture of the wartime Jeep vehicle to peacetime utilitarian needs at home and abroad, and the company began development of a line of vehicles stemming from the military model. It registered the Jeep trademark. In 1946, it introduced the industry's first all-steel station wagon.

The name Willys Motors, Inc. was adopted in April 1953, when Willys-Overland facilities, plants and other physical assets were purchased by the Henry J. Kaiser interests. In March, 1963, the name was changed to Kaiser-Jeep Corp., and to Jeep Corp. with its acquisition, early in 1970, by American Motors.

In March, 1971, a new, wholly-owned subsidiary, AM General Corp., was created by American Motors. The new company assumed the assets and government contracts of the former General Products Div., whose plants were in South Bend, Mishawaka and Indianapolis, Ind. Cruse W. Moss was president of AM General, which produced tactical wheeled vehicle's for the military and delivery vehicles for the U.S. Postal Service. In 1972, the company announced plans to enter the urban transit bus field, with production quartered in the AM General plant at Mishawaka. In 1972, AM General received a contract from the Urban Mass Transit Administration to develop, prototypes of the "Transbus," a transit bus of the future.

In 1973, AM General successfully bid on a contract to build 620 transit buses for the Washington Area Mass Transit Association. It subsequently received bids to build thousands of buses for many other U.S. cities. In 1975, AM General produced 350 electric-powered delivery vehicles for the U.S. Postal Service. Early in 1977 construction began on a plant addition at Marshall, Texas, where articulated buses would be built for several U.S. transit authorities.

American Motors definitely was a worldwide company. From 1958 on, it expanded not only its domestic markets, but its rapidly growing overseas markets as well. Its International Division directed all overseas operations. American Motors (Canada) Ltd., located at Brampton, Ontario, operated as a separate subsidiary. In total, at the peak, AMC cars and Jeep vehicles were produced in 29 countries.

A new Canadian Rambler plant was opened, in January, 1961, in Brampton. In January, 1969, Canadian Fabricated Products Ltd., of Stratford, Ontario, was acquired. An all-new Stratford plant, formally opened in the summer, of 1971, produced soft trim for AMC passenger cars. Holmes Foundry, Ltd., of Sarnia, Ontario, was added to the list of wholly-owned subsidiaries in 1971.

In the spring of 1972, a plant was acquired in Coleman,

In 1975, AMC purchased a new factory in Richmond, Ind. to build Gremlin engines. This is the collectible "Levis" edition Gremlin X of that year parked outside an 1890s style ice cream parlor.

This 1975 Pacer X was featured in Goodyear Tire Co.'s tribute to the 100th year of the automobile in 1986. It would certainly be appropriate to recognize this American Motors model again, in 1993, since it marks the 100th anniversary of the American automobile.

17

Wis., where a new AM subsidiary, Coleman Products Co., produced wiring harnesses for AMC cars and Jeep vehicles. In 1974, the company established a stamping plant for passenger cars and Jeep vehicles at South Charleston, W.V. In 1974, Coleman Products built another plant at Iron River, Mich., where electrical components were produced. In July 1974, American Motors also reached agreement to acquire Wheel Horse Products, Inc., of South Bend, Ind., a long-time manufacturer of lawn, and garden tractors. In 1975, American Motors purchased the tooling and rights to an all-new water-cooled four-cylinder engine from Volkswagenwerk of West Germany. AMC also purchased a modern plant in Richmond, Ind., where the company built engines for Gremlins. Three subsidiary companies produced plastic components for AMC and other companies, too.

(This article was originally published by OLD CARS WEEKLY in honor of AMC's "Diamond Jubilee" 75th anniversary celebration. It is reprinted here because it gives a good outline of the company's history through the mid-1970s. Later, AMC would go through a series of additional changes, such as its partnership with Renault, of France, the development of the Eagle and the Chrysler takeover in August 1987. As you can see, the AMC's later history continued to bear out the title of this article.)

Willys-Overland would become an AMC family member in 1970, when the Jeep deal was sealed. Seen here is the original pilot model of the famed military vehicle, which Willy-Overland delivered to the U.S. Army on Nov. 11, 1940.

The Ambassador Suburban was a limited-production wood-paneled sedan, similar in concept to the Chrysler Town & Country. (Arch Brown photo)

Prophet of 1901 vindicated as AMC caps 75th year

By R. Perry Zavitz

"When the fizz and fireworks cease to sizz and sputter in the automobile business, and the trade settles down to the normal level, it is fairly safe to say that the firm of T.B. Jeffery and Co. of Kenosha, Wis., will be found well up on the list of the fittest who survived." That statement was made way back in 1901 and appeared in *Motor World*. It was quoted by John B. Rae in his book *American Automobile Manufacturers* (Chilton Co. 1959).

No doubt equally optimistic predictions were made about other car makers. However, this forecast was extremely accurate. American Motors, the fourth largest car builder in the largest car building country of the world, traces its origin to the T.B. Jeffery Company. As reported in *Old Cars*, March 22, 1977, the firm observed the 75th anniversary of their first sale on March 1, 1977.

Thomas Jeffery, with his partner Philip Gormully, started making bicycles, bearing the trade name Rambler, nearly 115 years ago. Their vision later encompassed the automobile, when few people had ever seen such a contraption. Production began in early 1902. The car, like the bicycle, was called Rambler. The following year, Rambler ranked fifth in popularity among automakers, with a total of 1,350 cars built.

After Charles W. Nash resigned from the General Motors presidency in 1916, he bought the Thomas B. Jeffery Company. The car's name had been changed to Jeffery after 1913. The 1917 models were the first to carry the Nash name.

Before leaving the early history of Rambler, Jeffery and Nash, we cannot overlook one fascinating incident which typifies the company. It is generally assumed that during the legal battle Henry Ford waged against the Selden patent, automakers were divided into two camps. The majority belonged to the Association of Licensed Automobile Manufacturers (ALAM) and paid royalties to the Selden patent holders on each car they sold. A small group, headed by Ford, refused to pay the royalties and fought against the patent's legality. However, there was a third group, of which Rambler was the largest. They simply ignored the whole Selden situation, neither paying royalties, nor fighting the regime.

But, getting down to my main area of interest, a postwar review of Nash, we find that the prophecy made in 1901 was vindicated by years of success and interesting products. Postwar production began on Oct. 27, 1945, after a 44-month hiatus courtesy Hitler, Hirohito and their henchmen. The 1946 model Nashes were practically repeats of the '42s.

Styling was slightly altered. Model lines were resumed as the 600 and Ambassador. However, there was no Ambassador Eight. Only six-cylinder engines were available. A 172.6 cid flathead powered the 600, and a 234.8 cid ohv six drove the Ambassador. Advertised horsepower was 82 and 112, respectively, up five in each case from prewar ratings.

Body styles were changed somewhat for '46. No two-door sedans were made, just four-door sedans and club coupes (called Broughams). Sedans were offered, however, in trunk-back or fastback (slipstream) styling. Trunk models were about $40 more.

Prices ranged from $1,293 for the 600 Brougham to $1,929 for the Ambassador Suburban. The latter was a limited edition wood-panelled sedan, similar in concept to Chrysler's Town & Country sedan, but some $460 to $650 less expensive.

During 1946, Nash captured nearly five percent of the new

car market and took eighth in sales. It did better than any other Independent.

Slightly longer upper grille bars were featured on 1947 Nashes. Model availability remained the same.

For 1948, sub-series versions were offered. Both the 600 and the Ambassador came in Super or Custom trim. Two additional body types were added. The 600 had a bargain basement DeLuxe coupe. It was the first postwar three-passenger Nash. At the top of the range was another two-door; the first Nash convertible in seven years.

Nash sales for '48 were the highest since the depression, even though Nash was the last of the Independents still running around with its prewar body.

When the restyled Nash appeared for 1949, it made a great impression, resulting in even more sales. But, competition was stiffer. The styling was not like that of other cars. Featuring enclosed front wheels that were often seen on designers' drawingboards, this was the first example of such enclosures on a production car.

Two series were continued with the same engines as in 1948. However, each series now came in three versions: Super, Super Special, and Custom. Body type availabilities were revised. Sedans were now offered in two- and four-door versions. There were no coupes, but a Brougham was offered. It used the two-door sedan body. Inside, it featured two-passenger rear seating. A permanent armrest separated two lounge chair styled seats, which were angled slightly toward the center. Nash's infamous "bed" (folding seat) was redesigned for '49 to make it much more practical. It was not available in the Broughams.

There were no three-passenger models in '49. Dropped permanently from the Nash roster were the convertible (in full-size form), and the wood-veneered Suburban.

Very little exterior styling change was made on 1950 Nashes. The 600 name was replaced by Statesman. There was a three-passenger Statesman DeLuxe business coupe. The Super Special editions were deleted in both series. All this was rather unexciting news. The big story for '50 came on April 15, when the Nash Rambler was unveiled. The Rambler was a compact convertible on a 100-inch wheelbase (176 inches overall), using the former 600 engine. The Statesman, in contrast, had an 85 hp version of that engine, which had a 1/4-inch longer stroke for 184 cid.)

The Rambler was highly regarded for the type of car it was, though some buyers were scared away by its $1,808 price tag. They just could not conceive luxury in a small package. On June 23, a station wagon was added to the Rambler line. Its price was the same as the convertible. The convertible featured an unusual top. There were three solid frames around the side windows. The canvas top raised and lowered on those frames. The contemporary Crosley convertible used almost the same idea (as had 1931-1932 Ford convertible sedans), but its top was not power-operated like the Rambler's.

A hardtop, Nash's first, was included in the Rambler range as of June 28, 1951. It was called the Country Club. Full-size 1951 Nashes had a restyled grille and protruding taillights. The same model line was repeated for '51 with no full-size hardtops.

A second new car in as many years was introduced by Nash in February, 1951. The Nash-Healey was another innovative product. Looking towards the growing popularity of European sports cars, Nash announced a British-American hybrid. It was basically a Healey with the grille of a Nash and the motor from the Ambassador (no pun intended). The motor was souped up to 125 horsepower.

For its sophomore year, the Nash-Healey underwent complete restyling. Such a move is almost incredible, at least in recent car history. But, there was a reason. Nash had engaged the services of famous Italian designer Pinin Farina (who later changed his name to Pininfarina) to restyle their 1952 full-size cars. This was done to commemorate the company's 50th anniversary. Farina's work included not only the styling of the Nash and Nash-Healey, but the construction of the latter, as well.

The full-size Nashes lost their fastback shape for a crispier, leaner look. Statesman and Ambassador were still made in Super and Custom versions each offering two- and four-door sedans. There were no Broughams, but the Custom editions each included a two-door hardtop, known, like the Rambler, as the Country Club.

Ramblers, unchanged for '52, got a slight facelift for 1953 from Pinin Farina. A Nash-Healey hardtop was added. It was made slightly longer than the ragtop in order to seat four people. Only very minor changes marked '53 Statesman and Ambassador styling.

The 1954 models were likewise given minor appearance changes. The two-door sedans were available only in Super trim...not Custom. In what was to be its final year, the Nash-Healey came in only the hardtop model.

Rambler, growing in popularity, added new four-door sedans and station wagons to its lineup. Eight inches was added to the length of the four-door models. These Ramblers, and the shorter ones with Hydra-Matic, both used the Statesman engine. For 1952, it was enlarged to 195.6 cid. and rated at 110 hp. Two-door Ramblers with manual transmissions used the old 184 cid Statesman motor.

The Ambassador engine was enlarged for the '52 models to 252.6 cid and produced 120 hp. In mid '53, this larger engine was used in the Nash-Healey, in which it developed 140 hp. This same engine was optional in the Ambassador during late 1953-1954. Nash was trying to keep pace in the horsepower race with its old sixes.

A new era began on May 1, 1954, when Nash merged with Hudson. Both companies were struggling to compete with the "Big Three." Uniting to form American Motors Corp. did not solve all their problems by any means, but obviously it eased their burdens to a point where they could survive.

The Nash (and Hudson) Rambler was another car aimed at an untapped market niche. Back in early 1950, before the Rambler bowed, Nash displayed one of the first postwar experimental cars. It was a tiny two-seater, with an 84-inch wheelbase. Its four-cylinder motor was expected to come from either Fiat or (British) Standard and be of not more than 36 horsepower. Dubbed NXI, for Nash Experimental International, it was used in a survey of public reaction to the idea of producing such a car

The Metropolitan, introduced on March 18, 1954, was essentially the same car as the NXI. Some obvious changes to the production version included standard style bumpers (instead of a grille-encircling front and a spare tire-encircling rear bumper. The engine was from the 42 hp Austin A40. The Metropolitan was built in Britain and shipped to the United

When the restyled Nash appeared in 1949, it made a great impression, resulting in higher sales. This 1949 Nash Super Statesman four-door sedan characterizes the all-new postwar appearance. (John Lee photo)

States and Canada.

During production from 1953 to 1962, 83,442 Metropolitans were shipped to the U.S. Almost one-fourth arrived in 1959. Another 11,544 were shipped to Canada. The biggest year for Canadian sales was 1955, when some 20 percent of the year's run went there. In addition to these nearly 95,000 Metropolitans sent to America, some remained in Britain for sale there.

A revised Metropolitan was announced on April 9th, 1956. Called the 1500, it had a larger engine rated at 52 horsepower. A restyled grille, and zig-zag two-tone separating chrome side trim were some of the obvious changes on the new model. Some convenience items were added as well.

Years ahead of its time, the Metropolitan was American Motors' approach to the economy car market. Despite its origin, the "Met" never seemed to suffer from the foreign car image that most of its contemporaries did. The Metropolitan was sold as both a Nash and a Hudson. The only difference was the marque name and insignia. After the establishment of American Motors Corp., the same marketing strategy was used for the Rambler.

The 1955 Rambler received a new grille, similar to that of the 1949-1950 Nash. A different model mix was offered. The two-door sedan introduced in '54 (made and sold only in Canada in 1953) was available in DeLuxe, Super, and Custom versions. The four-door sedan was also available in all three trims. The two-door wagon came only in Super trim and the four-door wagon in Custom.

An all-new body was introduced on 1956 Ramblers. The wheelbase was 108 inches, same as the previous four-door models. Overall length grew to 191-1/4 inches. If that is compact, then so was the '49 Plymouth! The shortest '49 Plymouth, with a 111-inch wheelbase, was the first real, mass-production postwar compact. Its overall length was 1/16-in. more than that of the 1954-1955 four-door Ramblers. Though the Rambler did not initiate the compact car class, it most certainly made it popular.

All the 1956 Ramblers were four-door models. Sedans were offered in each of the three trim-lines. Station wagons came in either Super or Custom form. The most interesting Ramblers were the hardtops in a Custom sub-series.

General Motors introduced four-door hardtops in mid-1955. They came as Buick Specials and Centurys and as Oldsmobile Super 88s and 98s. Coming in a close second was Rambler's 1956 four-door hardtop. Rambler, however, scored an industry first with its 1956 four-door hardtop station wagon.

Although they had few styling changes, the 1957 Ramblers were noteworthy because of their 250 cid/190 hp V-8 engine. It was optional in all models, except the DeLuxe sedan and the hardtop wagon, in which it was standard.

Full-size Nash and Hudson cars were discontinued at the end of '57 production. American Motors Corp. then concentrated on the Ramblers, making that the marque name.

Ramblers were basically unchanged for 1952. The Greenbrier station wagon was an upgraded model with two-tone paint, woodgrained window frames and richer trim. This name would later surface on Chevrolet's Corvair station wagon.

Nash used the name "Country Club" to identify its 1956 Ambassador two-door hardtop model. A continental tire was standard on Custom trim models, but could be ordered as an option for models in other car-lines.

An English magazine, *The Motor*, found this 1951 Rambler Country Club hardtop "quite attractive" and liked its roominess and performance.

Rambler American...the first "compact"

By Robert C. Ackerson

When the Nash Rambler was first introduced in April 1950, *Business Week* (April 15, 1950) regarded it as an automobile whose primary reason for existence was to serve as a sales stimulant for larger Nash models. Demonstrating a lack of appreciation for George Mason's innovative spirit, *Business Week* claimed that "Nash is counting on its $13 million Rambler investment to be a sales leader for its standard models. Customers, lured into the showroom with a small car price, can be upgraded to more expensive models. That's the way Nash sales strategists have figured it out."

The Rambler, both in its design philosophy and production life, was far more than a stalking horse for larger Nash automobiles. In its original form it was anything but a low-priced economy car. Although its sales in the early '50s were relatively modest, the steady growth of its popularity gave AMC president George Romney the key to American Motors' survival. For 1956-1957, the original Rambler was benched. It gave way to a slightly larger, slightly more modern version. But, in 1958, the timing was perfect for the 100-inch Rambler to make a comeback. In the years of AMC's greatest success, the Rambler American, as it was renamed in 1958, was an important contributor to American Motors' prosperity.

In 1950, Nash-Kelvinator (which four years later absorbed Hudson to become American Motors) was regarded by *The Motor* "as among the most enterprising of American automobile manufacturing groups." The Rambler's small size and far from austere appointments created some uncertainty among journalists as to its precise role in the automobile market. But, with an eye towards the future, *The Motor* predicted, "the Nash Rambler is obviously destined to have a wide influence on automobile sales in many parts of the world." Griff Borgeson concluded his evaluation of the new Rambler on a similar note, reporting (*Motor Trend*, July 1950) "whatever you call it, it's perhaps the most interesting and refreshing addition to the U.S. automotive market since the war's end."

The Rambler's appearance gave it instant identity as a running mate to the existing Nash Statesman and Ambassador models. But, in an age when size was a virtue, the Rambler's dimensions were startling. Compared to the Nash Statesman, its 100-inch wheelbase was a foot less, its length shorter by 25 inches. With a 6.5-inch narrower rear tread it was slimmer by four inches. From an engineering and technical viewpoint, the Rambler reflected the state-of-the-art at Nash-Kelvinator. It had a unitized body/frame structure and a flathead six derived from the prewar Nash 600 engine. With 172.6 cid and a 7.25:1 compression ratio, it developed 82 horsepower at 3800 rpm. This engine did not have intake and exhaust manifolds as such. Instead, the single-barrel Carter downdraft carburetor delivered its mixture to a pseudo-manifold which was part of the cylinder block casting. The exhaust pipe was clamped directly to the engine block and holes drilled in the pipe were matched against openings in the exhaust ports.

The Rambler's suspension system differed from those on the Statesman and Ambassador, which used coil springs at all four wheels. Instead, the Rambler was fitted, at the rear, with six semi-elliptic leaf springs. Its front suspension used unequal length wishbones and coil springs mounted high in the body. The standard transmission for the Rambler was a Warner three-speed gearbox.

Overdrive with a dash-mounted control button that allowed instant shifts from overdrive to direct drive was optional.

In its original Airflyte Custom Convertible Landau form the $1,800 Nash Rambler was priced, it appeared, uncompetitively against existing convertibles from Ford, Chevrolet and Plymouth. However, the Rambler's standard equipment features were impressive: radio, electric wipers, custom steering wheel, Weather-Eye heating system, directional signals, electric clock, courtesy lights, chrome wheel discs and a power-operated top. Years before the Japanese were associated with selling automobiles that possessed a long list of features as standard equipment, Nash was doing the very same thing with the Rambler.

Mechanix Illustrated (May 1950) described the Nash Rambler as "America's first luxury small car" and Tom McCahill, who rated the "cute as a cupcake" Rambler's top speed as between 84 and 88 mph, regarded it as possessing "excellent riding qualities and quite a bit of snap and punch." Praise for the Nash Rambler was by no means limited to American publications. *The Motor* tested a Rambler in 1951 and found it quite attractive. Said *The Motor* (April 8, 1951) "the degree of fuel economy offered by this model is altogether outstanding in relation to its roominess and performance. For example, its fuel economy figures at steady speeds of 40 mph or 60 mph have, in our road tests since 1946, only been equalled by one model of comparable roominess (but less performance)."

Probably the most controversial feature of the Rambler, aside from its status as a luxury compact, was its handling. Ted Koopman, (*Speed Age*, January 1952) claimed the Rambler "may be classified as a semi-sports car and the nearest approach to the MG produced in America today. The Rambler corners and accelerates nearly as well as the best MG." Tom McCahill of *Mechanix Illustrated* (May 1950), while finding his Rambler praiseworthy on many points, wasn't quite that enthusiastic. He wrote "steering was quick and extremely short for an American car, though it still won't corner as well as an MG or any imported sports job."

In essence, the Nash Rambler was not a sports car, nor did Nash intend it to be. It provided a fairly smooth ride on rough roads, but when its driver began to suffer illusions of grandeur and attempt to corner with gusto, the Rambler exhibited lots of body roll and a strong oversteering tendency. The real virtues of the Rambler were its quality construction, fuel economy (with a 20 gallon tank and 25 mpg capability the Rambler went many miles between refills) and its unique status among American automobiles. Wilbur Shaw (*Popular Science*, May 1951) conducted a three-way comparison test between the Rambler, the newly introduced Henry J and the Austin A40. At that time the A40 was the most popular foreign car sold in the United States. Shaw concluded that the Rambler "belongs in the class of the Henry J and Austin only as to size . . . in aesthetics, ride and interior appointments, the Rambler obviously was the class of the three little cars."

Just two months after the introduction of the convertible model, Nash added a station wagon version to the Rambler lineup. This strategy of gradually expanding the Rambler model offerings was repeated in June 1951 when the Country Club two-door hardtop was introduced. Total Nash Rambler production for 1952 totalled 53,055 (in 1950 and 1951 it was 20,782 and 57,555 respectively). *Motor Trend*, (September 1952) noted that "the Nash Rambler remains rather low in car sales.... It is felt by many people that the reason the Nash Rambler remains low on the sales list is not that it is a small car, but that its price is too close to the established cars in the low-priced field."

Nash was, however, gradually refining the Rambler's position in the marketplace to yield a higher sales harvest. The introduction of the station wagon version had been a step in the right direction since it captured 22 percent of the station wagon sales during 1951. Also, helping to broaden the Rambler's appeal was the introduction of a low-priced Deliveryman station wagon. Without a rear seat, available only in a single exterior color, and devoid of some body brightwork, the Deliveryman retailed in 1952 for $1,900. The Greenbrier station wagon and convertible models were each priced at $2,132.20 and the Country Club hardtop delivered for $2,112.20.

Following the lead of the 1952 Golden Anniversary Ambassador and Statesman models, the 1953 Ramblers received new styling with strong overtones of Pinin Farina's influence. The result went a long way towards relieving the Rambler of its slightly squeezed look. The lowering of the hood, by nearly four inches, plus a larger (by 165 square inches) windshield, provided a substantial improvement in forward vision. Other changes included a new grille, reshaped door handles, and a higher rear fender line. Whereas a continental tire mount had been optional on all models except the station wagon in 1952, it became part of the standard equipment for all 1953 Rambler convertibles and hardtops. Mechanically, the most important development was the availability, for the first time, of Hydra-Matic on all Rambler models. Ramblers so-equipped were powered by flathead sixes displacing 195.6 cid and developing 90 horsepower. Those with either the three-speed manual or overdrive transmission used smaller, 184 cid versions rated at 85 horsepower.

Although output declined in 1953 to 41,885, the revamped Rambler received good reviews by its road testers. *Auto Age* (August 1953) reported the 1953 version cornered "remarkably well, better even than last year's model." Tom McCahill, in *Mechanix Illustrated* (December 1953) regarded the Nash Rambler as "the best businessman's car of the year, for economy and service." Almost unnoticed in the Rambler lineup was a simple, unadorned, two-door sedan. It received little attention at that time. However, it was this model which, four years later, would be resurrected as the Rambler American.

Their styling remained almost unchanged, but the addition of a four-door sedan and station wagon with a longer, 108-

The 1953 Nash Ramblers, such as this Country Club hardtop, received new styling with strong overtones of Italian designer touches.

All of the 1953 Nash Ramblers, including this station wagon, sported a redesigned grille. Another new-for-1953 feature was Hydra-Matic transmission.

inch wheelbase broadened the 1954 Nash Rambler's appeal. In the case of the sedan, an additional two inches of rear legroom, plus 14 inches more hip room, were important results. So was a larger trunk. In addition, *Motor Trend* (January 1954) said "the longer chassis allows this newcomer to wear the virtually unchanged Nash styling well, perhaps better than its big brothers..."

The opening of the Rambler's front wheelwells and the availability of Ramblers either as Nash or Hudson models in 1955 tends to overshadow the 1954 models Still it's worthwhile to spend a few moments taking stock of the Rambler's strengths and weaknesses at that time. First of all, American Motors was in financial deep water with a calendar-year loss of over $11 million and a total output of only 67,192 cars. Rambler's proportion of this production was a sizeable 37,779.

This was not enough to reverse the tide of red ink then flowing freely in Kenosha, Wisconsin. But, whereas Nash controlled 1.6 percent of the total market, Rambler wagon sales represented 2.1 percent of the station wagon market. As a new model in 1954, the 108-inch wheelbase Cross Country four-door station wagon represented 95 percent of Rambler wagon output. Another important plus for the Rambler was its wide price spread. This enabled it to appeal to economy car buyers with a $1,695, two-door sedan. It maintained its reputation as a quality small car with a custom series of sedans, covertibles and wagons priced between $2,095 and $2,175.

Yet, after five years of production, real sales success still eluded the Rambler. *Motor Trend*'s intentions were good, however, its praise was inadvertantly damning. "It seems to us" said the magazine (January 1954) "that the Rambler has a separate category: that of a 'specialty car.' In other words it's in a class by itself on the American market in what it offers the purchaser. It's a small, yet fairly roomy and definitely luxurious car with good performance and comfort." So, in spite of its many positive points, the Rambler still was falling short of its sales target. The concept was sound, but its execution needed some fine tuning.

Thanks to George Romney, changes took place in 1956. The Rambler received all-new styling, an engineering update and four-door hardtop station wagon models. American Motors still lost nearly $20 million dollars during 1956, but of 104,189 cars it built, 79,166 were Ramblers. With the decision to drop the old Nash and Hudsons after the 1957 models came greater public acceptance of the Rambler. Over 114,000 were built in 1957 and the next year AMC reported its first profit, a healthy $26 million.

With market momentum finally on its side, American Motors moved rapidly to exploit its position as the U.S. auto manufacturer most closely associated by the public with quality, compact-sized automobiles. George Romney's "dinosaurs in the driveway" speeches set the theme for an effective advertising campaign. It cast AMC in the role as the only U.S. manufacturer of cars that were really practical.

But, AMC's really unique move was the decision to dust off the original Rambler dies and resume production of the old, 100-inch wheelbase model in 1958. In order to build prototypes for the "new" Rambler American, AMC purchased a number of 1955 models and reconstructed them to suit engineering and design requirements. The actual changes made to upgrade the Rambler for its second reincarnation were minor, but they were effective in giving the American a fresh, contemporary outlook. A new mesh grille with an "R" medallion, plus a smooth hood without its old air scoop helped. There was a new roofline that lowered overall height by approximately two inches and allowed a larger (by 100 square inches) rear window to be used. Re-shaped rear quarter and deck panels were the major revisions. All that was needed to give the Rambler's taillights a new look without changing any components was to invert their chromed bezels!

The Rambler's mechanical structure was closely patterned after that of the earlier models. Its front and rear tread was slightly increased and minor changes were made in its front end suspension to improve its alignment. Larger brakes with 139.4 sq. in. of effective area (up from 104 sq. in.) were also fitted. The Rambler's 195.6 cid, 90 hp "Flying Scot" engine now had an 8.0:1 (rather than 7.3:1) compression ratio. Its water pump, which formerly had operated off the generator, now was mounted in a conventional front of the block position.

The response to the born-again Rambler was excellent. Even though a two-door, five-passenger sedan was the only American model offered, sales totaled 42,196. When the model year ended, AMC still had a considerable number of unfilled orders from their dealers on hand. Almost without exception the Rambler received praise from motoring journalists. *Motor Trend* (June 1958) reported, "In many departments the car reminded us of the Swedish Volvo because of its excellent performance, ride and handling." *The Autocar* (Jan. 16, 1959) reached a similar conclusion, telling its readers, "Sound and sensible in concept, the Rambler American has many European characteristics." *The Autocar* approached the Rambler with some misgivings admitting "since the Rambler specification is thoroughly orthodox, even dated it some respects, it was frankly not expected to inspire much enthusiasm. Yet, during a test distance of over 1,000 miles, it has proven thoroughly acceptable in almost all important aspects."

Very quickly the car of yesterday had become the car of the day. Its acceleration with the optional Flash-O-Matic automatic transmission was acceptable: 0-to-60 mph in 20 seconds. When equipped with overdrive (also an extra-cost item) it was lively, requiring only 16.4 seconds to reach 60 mph. Its brakes provided excellent stopping power and each American was delivered with a comprehensive do-it-yourself service manual which included "adjustment and repair" procedures for the carburetor, distributor, wheel bearings, ignition and lubrication.

The base price for the DeLuxe two-door was $1,789. The Super versions, which had standard front and rear armrests, dual sun visors, cigarette lighter and roll-down rear windows was $1,874.

George Romney attributed the Rambler's success in 1958 to AMC's response to "the large segment of consumers who want safe, sound, comfortable automobiles of a sensible size... ." As applied to the 1959 Rambler American, this theory meant more of the same, plus a little extra, in the form of a two-door station wagon. Romney said the new wagon was the "result of a write-in vote from owners." The lack of any styling changes of consequence in 1959 did little to dampen enthusiasm for the American. "The high resale value of the Rambler American" said AMC, "is indicative of the buyer's preference for simplicity and economy." *Motor Life* (April 1959) reported, "appreciation of the Rambler American station wagon grows in direct ratio to the number of miles covered in this eminently sensible vehicle." *Road & Track* (March 1959), whose test car with the standard three-speed gearbox recorded a 0-to-60 mph time of 16 seconds and a top speed of 88 mph, reached the conclusion that "the imported family sedans will find the American very rough competition. Very few, if any, can match its combination of roominess, performance and price.... We liked it very much, for the roll in a turn is negligible and the general handling qualities are excellent." The Rambler also was doing pretty well in the feather foot league, turning in a 25.29 mpg average with an automatic transmission in the 1959 Mobilgas Economy Run. A year earlier, an overdrive equipped Rambler under NASCAR supervision, traveled the 2,837 miles from Los Angeles to Miami at an average 35.4 mpg fuel consumption.

When introducing AMC's cars for 1960, company president

George Romney said, "American Motors makes year-to-year changes only if we are convinced the changes improve the economy, appearance, quality comfort, convenience, safety or performance of our products. Sometimes such changes are made during a model run. We do not change for change's sake."

Consistent with this philosophy, the Rambler Americans were only moderately changed for 1960. A rooftop luggage rack was now standard on all station wagons. Bonded brake linings replaced the old rivet type. The American's gas tank capacity was increased by two gallons, to 22 gallons, and the doors on all models opened to a wider 75-degree angle. New options included self-adjusting brakes and power steering.

Griff Borgeson, writing in *Sports Cars Illustrated* (October 1959) offered a fresh perspective on the American. "The most salient impression made upon us by the American" he wrote, "is the integrity of its package ... as equipped with standard transmission and overdrive. In this most popular version everything fits and harmonizes. The car looks fine, even when mingling with cars costing several times its price. It has style, without contrived gimmickry. The compatability of body, engine, transmission and chassis is thoroughly satisfying. To drive it is a genuine pleasure." Ed Anderson, then American Motors styling chief, reflected on the American's success by noting, "People want to take pride in their cars. We want to give them a car which is economical, but which has no air of cheapness about it. We could make a cheap car by making it chintzy ... and we'd fail in the market. The American must be a car in which its owner can take pride of ownership. This raises the dickens with costs, but in the long view it's the only route."

The American had come a long way from its 1950 introduction and 1958 rebirth. But, in spite of Romney's public professions in support of design continuity, no one really objected to spending some money on a redesign for the 1961 American. What was remarkable about the result was not its appearance (which was trim crisp original and up-to-date), but the willingness of AMC to make the American more compact than ever.

The American's new body, at 173.1 inches long, was 5.2 inches shorter and three inches narrower than before, yet more luggage space was provided. The American lineup was expanded to include a convertible and four-door station wagon for 1961. The top-of-the-line Custom models were powered by a 125 hp ohv version of AMC's 195.6 cid six. This engine debuted in mid-1960 in the larger Rambler Classic models. It was also offered as an option for the less expensive Super and DeLuxe series.

It was a great year for AMC, which expanded its production capacity in response to a new calendar year production mark of 372,485 automobiles. The American, with its sales bouyed by its fresh new look, accounted for nearly 25 percent of AMC sales. *Car Life* (May 1961) said it achieved a "smashing defeat" of the other compacts in the 1960 Mobilgas economy run with 28.35 mpg average. As before, the American's construction and assembly impressed road testers. Said *Car Life* (May 1961) "the way the Rambler was put together was not only acceptable, but actually admirable, in view of the way many higher-priced makes are cutting corners these days." Small wonder that *Car Life* (May 1961) described the American as "'a perfect example of a very desirable automobile." Perhaps the only blemish on the American's 1961 scorecard could be traced to the small number of Custom models equipped with aluminum block versions of the 195.6 cid, ohv six. Approximately 80 pounds was saved by adopting this design, but many owners were less than happy with its durability.

Fortunately the American's public image escaped serious harm as a result of this misadventure. Although only slight styling changes were made for 1962, a number of worthwhile engineering modifications further enhanced the American's public appeal. Self-adjusting brakes were now standard and an extended service mileage schedule was placed in effect providing for front suspension lubrication at 2,000-mile instead of 1,000-mile intervals. AMC now recommended oil changes every 4,000 miles, rather than 2,000 miles. Receiving a good deal of attention in the media was the adoption, by all AMC models, of a "Double Safety" dual braking arrangement. It provided separate hydraulic systems for the front and rear brakes. The optional Flash-O-Matic transmission was redesigned for lighter weight and smoother, more positive shifting. An interesting, low-cost ($59.20) option for the 1962 American was the Borg-Warner E-Stick semi-automatic transmission. E-Stick was available with either of the three-speed overdrive manual transmissions. Combined engine oil and intake manifold pressure engaged the clutch, leaving the driver to operate the shift lever in the usual manner.

The ability of the Rambler American to effectively compete with the newer American compacts was underscored by its first place compact car class finish in the 1962 Pure Oil performance and 1962 Mobilgas Economy Run contests. A 125 hp American with a standard transmission maintained a 28.73 mpg at a 40.89 mph speed to win the first event and a Rambler American with the 125 hp engine and standard transmission had the best average (31.11 mpg) in the latter. Not surprisingly, *Car Life* (February 1962) reported, "The smaller Rambler started out as a good car but, today, it's been seasoned into a superior one that needs no apology in the face of its many more recent, more sophisticated competitors."

With new styling just a year away, the American was only moderately changed for 1963. Its grille had vertical, rather

A rooftop luggage rack was added to the American's standard equipment list in 1960, along with brake system upgrades and a larger fuel tank.

Surprisingly, AMC's 1961 redesign of the American made it even smaller (by 5.2 inches) than before. The convertible was an all-new body style.

25

than horizontal, bars. An interesting addition to the Rambler lineup was a hardtop coupe in the top-of-the-line 440 series. Its ridged roofline and nearly vertical rear window gave it a convertible top look. The standard engine for this 440-H American was the 195.6 cid ohv six with a two-barrel carburetor and a rating of 138 hp. This engine was optional in the other 440 models whose standard power plant was the 125 hp overhead valve six. The remaining DeLuxe and Custom models still used the veteran 90 hp flathead six.

An option available only in the 440 models, which also included a station wagon and convertible, was Twin-Stick overdrive. This feature consisted of two, floor-mounted shift levers, one for the three-speed gearbox, the second operating the overdrive. This arrangement provided the American with the equivalent of a five-speed gearbox. When the overdrive lever was in its rear position, overdrive could be engaged at any speed above 25 to 28 mph by the driver lifting his foot momentarily from the accelerator. Similarly the car would drop out of overdrive below the same speed level. An amber light on the dash indicated when overdrive was in operation. When the driver pressed a button on the gearshift lever, the car would instantly shift out of overdrive as long as some accelerator pressure existed.

Car Life (January 1964) headlined its road test of the 1964 American with this caption: The original plain Jane compact car just got back from the beauty parlor. Indeed it had. Dick Teague, who credits the 1964 American with getting him his job as an AMC vice-president in charge of styling, gave the American " a look of quality and luxury of design heretofore little exploited in the field of basic transportation." the magazine said. The new American's 106 inch wheelbase not only helped give it a balanced appearance, but provided a real bonus in the form of 12 inches of additional rear seat hip room, although the car was 1.44 inches narrower. While its length was increased from 173.1 inches to 177.25 inches, the American possessed the shortest wheelbase and overall length of any American sedan. Mechanically, the 1964 American differed only moderately from its predecessors. The same trio of engines and the various transmission offering were continued.

Unfortunately, the new American's debut coincided with a shift in sentiment among the U.S. car buying public. While Ford was introducing the Mustang and enjoying the wonderful feeling of being in the right place at the right time, AMC had expended a good deal of effort developing an automobile whose potential market was drying up. Thus, the Rambler American's final years were less than outstanding ones. Nonetheless between 1964 and 1969, when Rambler production (the American portion of its name was deleted after the 1968 models) ended, several interesting versions were marketed by AMC.

In late 1964, after evaluating a number of designs, including V-6s with both 60-degree and 90-degree configurations, AMC introduced a new 232 cid Typhoon six in the Rambler Classic series. Its designers had as their goal an engine capable of performing on a par with competitive mid-displacement V-8s, while still delivering fuel economy compatible to that of contemporary AMC sixes. Thanks to its thin wall castings, the Typhoon six weighed (with flywheel) just 441 pounds, or some 15 percent less than the older AMC six. With a single-barrel carburetor it developed 145 hp at 4300 rpm and its fairly flat torque curve peaked at 215 lb./ft.

The following year, this engine with a single dual-barrel carburetor and ratings of 155 hp at 4400 rpm and 222 lb./ft. of torque at 1600 rpm became an American option. Nineteen sixty-five was the year of the "Sensible Spectaculars" at AMC. The sensible side of the American's personality showed itself in the Mobilgas Economy Run where it averaged 25.65 mpg. In the Pure Oil Performance Trials, at Daytona, it scored a victory with a 27.54 mpg score.

Spectacular was not a way to describe the American's acceleration, but with the 155 hp six and Shift-Command (which allowed the driver to control the shift points of the Flash-O-Matic transmission) it could accelerate from 0 to 60 mph in 10.9 seconds.

AMC attempted to change the American's well-earned and deserved reputation for economy in 1966. Economy simply was not a very salesworthy feature in the '60s, as opposed to a high-performance image. The 440-H hardtop was replaced with a new Rogue model. It had special identification trim and a two-tone paint scheme in which the roof and rear deck were painted the same color. This cautious testing of the "image car" market was followed by the more dramatic step, in mid-1966, of providing the Rouge with AMC's new 290 cid V-8.

The first 1,700 Rogue V-8s were given a glossy black and metallic "Sun Gold" color combination, special rocker panel moldings, blacked-out front grille, plus a black vinyl interior. With 200 hp, a four-speed gearbox, front disc brakes and a standard handling package, The Rogue, said *Car Life* (August 1966) is "an enthusiast's Rambler." *Motor Trend* (May 1966) viewed the Rogue in similar fashion noting, "Here at last is an enthusiast's car - and a pretty good one - from American Motors."

In 1967, American Motors made the 290 cid V-8 with either 200 hp or 225 hp an option for any Rambler American. But, this move was overshadowed, shortly after the 1967 models were unveiled, by addition of a larger 343 cid V-8/280 hp V-8 to the options list. Paralleling these moves was the development of an AMC high-performance kit, part number 3208580 (priced at $160), suitable for use in either the 290 or 343 V-8s. The major components of this over-the-counter package included a high-lift (.295 inch), long-duration cam, anti-pump-up hydraulic lifters and stronger valve springs. For an additional $4, intake manifold gaskets that blocked off the heat riser were available, as was (for another $60) a 4.44:1 rear axle gear set. A 225 hp, four-speed Rogue with these components and open headers ran the quarter-mile in 15.30 seconds and 93 mph. This particular Rambler was a *Motor Trend* project car. Eventually its drivers coaxed it to a 13.7 second 103 mph quarter-mile run.

The ultimate performance Rambler was yet to come. After it arrived, the world of super cars was never quite the same. There was nothing improper, of course, with AMC marketing a super car in 1969. Both the Javelin and AMX had been successful image changers for AMC and the idea of transforming a mundane family sedan into a quarter-mile terror wasn't exactly new. But, to convert a Rambler into a blatant factory hot rod with an outrageous red, white and blue paint job, replete with a huge hood-mounted cold air induction system, NASCAR type hood tie-downs and a raucous exhaust tone, was a development the world wasn't ready for.

AMC wasn't bothered by such matters. With assistance from George Hurst, the AMC Hurst SC/Rambler debuted with a $2,995 sticker price. AMC claimed 14.3 second time for the quarter-mile. Although the Scrambler's suspension included the Rouge handling package (stiffer springs and shocks, plus a larger-diameter front stabilizer bar), its handling was not exceptional.

Also unexceptional was the car's interior. Except for headrests that were colored-keyed to the Scrambler's exterior and a Sun tach mounted on the steering column, it was identical to that of a modest 128 hp Rambler.

How that car could go and stop, though! Both *Car Life* and *Car and Driver* reported identical 0-to-60 mph times of 6.3 seconds. *Car Life* achieved a highly respectable quarter-mile run of 14.20 seconds and 100.8. mph. The Scrambler demonstrated its braking ability to *Car and Driver* with an 80-to-0 mph stopping distance of 244 feet immediately after completing an identical run.

In terms of performance, the Scrambler represented the American's finest hour. But, in the more practical sense of

26

keeping AMC in business, the time came in 1967 for a price cut. Depending on the model, from $154 to $234 was slashed off the stickers to stimulate sales and help restore the public's confidence in AMC's future.

All good things have to come to an end and 1969 was the Rambler's final production year. In essence, the Rambler was a far more significant car than is generally recognized. Its role as the first modern American luxury compact (a theme the Concord traded fairly successfully on) certainly was an important one. Add to this its return to production in 1958, its various high-performance versions, the many economy runs it paced (even a record run down the Baja in 1967) and a rather extraordinary automobile emerges. Some will say, "Ramblers are Ramblers."

On the other hand a rose by any other name is still a rose!

American Motors went the total redesign route again, in 1964, thanks to Dick Teague. One magazine said "it just got back from the beauty parlor."

One of AMC's "sensible spectaculars" for model-year 1965 was the handsome and sporty American 440 convertible.

The Early Years

Demand for the new Rambler was so great that 1,500 were sold in 1902, the first year of mass production. This is the Model C runabout.

From bikes to buggies
Nash lineage traced to mid-19th century

By Vincent Ruffolo
(Associate editor NASH TIMES magazine)

Behind the Nash lies an unusual and romantic story. It encompasses three changes of name, one-half century of steady progress and a tradition, the source of which lies over one hundred years in the past.

In May 1847, Alonzo Duretto Seaman of New York set up as a furniture maker in the brand new city of Milwaukee. He delivered his wares in an ox-driven cart. Before dying in 1868, he had made a fortune. His sons William and Harry carried on the business. William had pioneered the idea of a phone booth and was consequently under contract to the Western Electric Company.

The Seaman plant was destroyed by fire in 1906. Their new building had more space than they needed. This allowed them to begin building automobile bodies for the Petrel Friction Driven Car. Soon, they had a dozen other automobile manufacturers as their clients. Included was the Rambler. Built by Jeffrey, it was one of the most popular buggy-like runabouts of its day.

In 1879, English born Thomas Buckland Jeffrey offered for sale, in the U.S., a bicycle assembled from parts made in England. Also named the Rambler, it was an immediate success and, in 1881, led to the formation of a partnership named the G & J Manufacturing Co. The partners were Thomas Jeffrey and Phillip Gormully.

Before turning to bicycle manufacturing, Jeffery had been a struggling inventor credited with developing a railroad velocipede and a gasoline arc lamp. He also invented the pneumatic rubber bicycle tire, which revolutionized the bicycle industry and became the forerunner of the automobile clincher tire.

In 1895, Jeffery and his son Charles witnessed the *Times-Herald* Automobile race held in Chicago. This was the first event of its kind ever run in America. J. Frank Duryea piloted the winning Duryea car at an average speed of 5.05 mph over the 52 mile course. The Jefferys decided to give up bicycles and manufacture automobiles instead.

Chancing future success, father, son and partner Gormully sold out their bicycle business to the American Bicycle Co. They then hand-built a prototype car, designed by Charles Jeffery. Experiments on the new car continued until two test cars were ready in 1901. Called Models A and B, they had two major innovations. First, the engine was under the hood in the front. Second, the steering mechanism was on the left.

While the cars were under development, Gormully died and Jeffery and his son went forward alone. They bought a plant in Kenosha, Wisconsin where they quietly set about manufacturing automobiles.

The scene is the Chicago Colosseum Automobile Exposition. The date, March, 1902. Jeffery presented to a curious and soon interested public the first two in a long series of Rambler automobiles. The two models were priced at $750 and $825 and referred to as the Runabout and Stanhope. So great was the demand that 1,500 Ramblers were built and sold during that year, making Rambler the world's second mass-produced automobile. It was a year behind Olds, but a year ahead of Ford.

The first Ramblers had a 72-inch wheelbase, a weight of 1,200 pounds and an eight horsepower engine with a top speed of 25 mph. They had green and red finish colors.

On June 14, 1902 a Rambler set the pace in a speed test at Minnehaha, Minn. At the time of the announcement of the race, a car was selected from the stock on hand and shipped. Handicapped by a bad track, the race was highlighted by a climb up a 35 percent grade. The Rambler made a record of three miles in seven minutes and eight seconds. Racing to a close finish against cars built and equipped specially for the race, the Rambler's victory was a significant win for the Kenosha firm.

As a result of the outstanding performance of the Rambler in the race, the car was sent back to the factory and used as a show piece. Orders for cars and inquiries from jobbers began to swamp the Kenosha office. However, the plant ran into difficulties because of the lack of labor throughout the city. The critical problem was also encountered by Simons Co., Allen Tannery Co., Badger Brass Co., Chicago Brass Co. and Mieselbach, Co., all Kenosha manufacturers.

The Jeffery's new product was an immediate success. Typical of a satisfied Rambler customer was a Lima, Ohio owner who wrote, "It is a truly wonderful piece of mechanism. It starts immediately, runs like a jack rabbit and stops only at our will."

By 1905, the factory site in Kenosha occupied 15 acres with an additional 33-1/2 acres available for expansion and testing. The Rambler was steadily improved by the Jeffery Company. Larger cars were introduced with a novel optional feature... a spare wheel and tire.

Thomas B. Jeffery died in March, 1910 after 35 years of valuable pioneer work in the automotive industry. His heirs carried on the business and the Rambler continued to bounce and trudge over the nation's roads, giving satisfaction to its growing list of customers in the dawning of the automotive era.

Soon, Rambler moved out of the low-priced bracket for good. Under the leadership of Jeffery heirs, a Rambler five-passenger limousine was marketed for $3,500. The catalog described it as having 36 x 5 tires, a Bedford cord interior, electrical lighting, mahogany ceiling and sides, speaking tube, mirror, clock, cigar case and broom holder.

For 1914, the heirs decided to drop the Rambler marque and perpetuate the family name on their automobile. Therefore, after the close of 1913 production, a new model bearing a Jeffery emblem was introduced. Even though the name had been changed, the public went for this car with equal enthusiasm, knowing that the quality of the company's products could be counted on to remain the same.

In 1915, an advertising milestone was reached. A combined Jeffery-Chesterfield advertisement, which appeared in the Saturday Evening Post, was the first accepted by a magazine openly showing a woman smoking a cigarette. The Jeffery car, however, retained its new identity for a short time. On July 13, 1916 the Kenosha Evening News carried a headling consisting of three words: Jeffery Plant Sold!

In 1897, Thomas B. Jeffery posed for this photo in the first experimental Rambler automobile, which he constructed in his Chicago bicycle factory.

This 1902 Rambler was owned by Chester L. Krause, founder of *Old Cars Weekly*. It can now be seen on display at the Kissel Museum, in Hartford, Wis.

After its successful beginning, Rambler moved out of the low-priced bracket and began marketing larger, more luxurious models like this 1908 Model 36 limousine. Jeffery gave silver watch fobs to owners who drove 15,000 miles.

This Rambler model was offered a year after Thomas B. Jeffery's death in 1910. His heirs continued making cars and eventually put the family name on them.

Classic car dealer Dr. Art Burrichter gave members of our editorial staff a ride in the 1914 Jeffery he had for sale. That old car could really fly!

After the close of production in 1913, a new model bearing a Jeffery nameplate was launched. The Rambler name was to reappear again, but not until the 1950s.

The Jeffery Quad was a 1-1/2-ton truck introduced in February, 1912 by Thomas B. Jeffery's heirs and was the first vehicle to carry the family name.

The Quad - Nash's four-wheel wonder

By John A. Conde

One of the most unusual and versatile motor vehicles ever built was the four-wheel drive Quad truck, manufactured in Kenosha, Wis., from 1913 to 1928, by the Thomas B. Jeffery Co. (builder of the original Rambler car). Jeffery subsequently became Nash Motors and American Motors Corporation.

"It drives, steers and brakes on all four wheels," was the fitting slogan for the Quad, which in 1918 was the largest selling truck in the world.

Well before World War I began, U.S. Army engineers announced they were looking for a motor truck to replace the famous four-mule escort wagon. Since the ordinary rear-drive truck of commerce did not meet the requirements, an Army officer set out to discover "a truck that would travel 24 hours a day if necessary but which would also go anywhere a four-mule team would go."

Early in 1913, the officer visited the Jeffery factory in Kenosha. Within a matter of months, engineers developed a 1-1/2-ton Jeffery Quad truck which was turned over to the Army Quartermaster Corps for testing.

The company announced the revolutionary vehicle with measured pride: "The Jeffery Quadruple-drive truck was designed to meet the requirements of the U.S. Army, the specifications calling for a 1-1/2-ton truck with 20 per cent overload capacity, that under its own power would go anywhere a four-mule team could haul a load. The truck now in the active service of the Quartermaster's Department of the U.S. Army has not only conclusively demonstrated its ability to follow a four-mule team, but under some conditions has been able to handle its load where the muledrawn equipment was helpless."

The new truck was hailed by automobile trade journals. *The Automobile*'s Sept. 25, 1913 issue reported: "In braking, the truck can be stopped in eight feet when traveling 15 miles per hour. In spite of its 125-inch wheelbase, it can be turned in a circle 45 feet in diameter, a result made possible by mounting all four road wheels so they can be used for steering. The pulling ability of the truck should come well up to requirements in that, on the fourth or high speed, the gear ratio between the motor and rear wheels is 8 to 1 and low speed 32 to 1, this latter reduction making it possible to meet road difficulties such as deep sand, mud and steep hills."

The four-wheel drive principle was new, and required detailed explanation: "The power is applied to all four wheels simultaneously and through the use of patented differentials any one wheel will drive the truck regardless of the traction afforded by the other three. This makes it almost impossible to stall it. All four wheels are used for steering, which makes it possible to turn this truck completely around in a 45-foot circle. All four wheels are also equipped with brakes, and the danger of skidding is almost entirely eliminated, while the truck loaded to its full capacity and driven at maximum speed can be stopped within its own length."

An outstanding feature of the new truck was four-wheel brakes (which did not appear on U.S. production automobiles until early in the 1920s.) Army engineers applauded the application of interchangeability of key components. Three differentials used in the driving system were interchangeable, as were the four driving wheels (with hard-rubber tires), the four sets of brakes, two propeller shafts and four driveshafts.

Almost overnight, the Jeffery company was inundated with orders for the new four-wheel drive vehicles. In 1913, Jeffery turned out 5,578-trucks (more than its total of passenger cars for the year). Most of them, but not all were Quads.

Ambulances for the governments of Belgium were built. France and Serbia - obviously worried over the imminence of war - constructed them on standard Jeffery six-cylinder passenger car chassis. Most of the trucks produced in 1913 were Quads delivered to the Russian and French governments.

Severe tests were given the first Quad units. "It plowed through hub-deep mud and sand that were absolutely impassable to any rear-drive truck," according to a company engineering report. "It lifted itself bodily over a bulk of lumber 16 inches high. It ran up and down seemingly impossible grades with apparently ridiculous ease. It forded a stream in which only the tops of the tires showed above the water."

Sales of the versatile vehicles were not limited to government agencies. Great demand for the Quad developed immediately among domestic truck users in the United States. Before World War I, more than 1,800,000 of the nation's 2,000,000 miles of roads were unpaved - and the Quad's ability to negotiate, mud and mire was a key factor in quick public acceptance.

In 1915, three Jeffery Quads were in service in Death Valley, California, hauling supplies and ore between Zabriskie, the Ibex Mine and other properties of the Death Valley Trucking Company. In its operations, the company's longest haul was 52 miles one way, in which the trucks encountered a number of steep grades from two to 28 percent over a distance of 19 miles, and six miles of the haul was through deep sand. In addition to their load of two tons, each truck hauled from three to five extra tons on trailers.

The Marston-Gass Lumber Co. of Donors, Pennsylvania, wrote Jeffery in November 1915. "To say that our Jeffery Quad is doing work that no other truck could do would not be emphasizing it at all," said the letter. "In fact, since the weather has been bad and the dirt roads are soft, we are sending this truck where teams cannot pull a load. Several times it has pulled our rear-drive truck, with its load, out of the mud."

In 1915, a total of 7,600 Quads were sold. Six standard bodies were offered for domestic sale - the express or box body with or without flare boards, stake platform body, compartment oil tank body, hand-dump body with Lally hoist, power-dump body with wood hydraulic hoist, and a combination hose and chemical body.

In August, 1916, the Jeffery company was purchased by Charles W. Nash, who had resigned as president of General Motors to build a car under his own name. One of his most important newly acquired assets was the Quad. During 1917, it carried both the Jeffery and Nash names, with production totaling 3,801. In 1918, all trucks made were Nash Quads and 11,490 were produced. Nearly all went to the U.S. Government. It was the highest number of trucks ever built in a single year by any truck company up to that time.

Production in 1919 dropped to 4,090; in 1920 to 3,697. Only a few more than a hundred a year were built from then on. Total Quad Production in the 15-year life of the famous vehicle was 41,674.

During the World War I campaign in France, many Quad trucks were converted to ambulances and saw front line service. The Signal Corps of the U.S. Army found the vehicles invaluable as they moved troops, heavy equipment and food supplies up to the front.

After the war, many of the trucks were given without charge to county and state governments. Nash continued to service the Quads for many years.

While the Quad made history in World War I, it was the famed four-wheel drive Jeep, developed by Willys-Overland, that became famous in the annals of military vehicle history. The Jeep served as the American GI's right arm in World War II. Thus, the four-wheel drive principle, which proved to be so vital in two major world conflicts, received a good deal of development from two separate predecessor companies of AMC.

(John A. Conde, who penned this article for *Old Cars Weekly* shortly after his retirement from American Motors Corporation, received the "1992 Friends of Automotive History Award" from the Society of Automotive Historians.)

The Highland Brewing Co. used a 1914 Jeffery Quad to carry heavy beverage kegs and crates over the unimproved roads of the era. The brewing company was located in Highland, Illinois. Note the fabric weather enclosure and through-the-windshield spotlight.

The 1915 Jeffery Quad Model 4010-B Motor Lorry belonged to the French Government, which sold it to the owners of the LeMans race course in 1920. It was used as a maintenance vehicle there through 1952, due to the failure of a differential. In 1979, the vehicle was restored in civilian format.

The 1914 Jeffery flareboard express sold for $1,325.

The 1914 Jeffery flareboard with stakes model was $1,575.

When Jeffery made trucks

By Larry R. Daum

The Thomas B. Jeffery Co. of Kenosha, Wis. first offered automobiles for sale to the public in 1902. These cars were called Ramblers. Trucks would also be produced by Jeffery under this name, as well as several other names. This is a story about Jeffery-made trucks.

By 1904, the company had made their first delivery wagon. It was based on a Rambler automobile chassis. An early forerunner of today's van, it was set up so that the delivery top could easily be detached from the body. This created a single-seat runabout to which a tonneau could be attached, instead of the delivery top. This vehicle was known as the 1904 Rambler Delivery Wagon, Type I and had a carrying capacity of 500 pounds of merchandise. This model was painted in Rambler's traditional carmine red and came with brass trim.

The Rambler Delivery Wagon grew in size. It gradually evolved into a distinct type of Rambler truck sharing fewer and fewer components of the automobile line. Some were used by the Thomas B. Jeffery Co. to deliver parts.

In 1911, a Model 63 Rambler roadster body was converted into a custom-built Rambler three-quarter-ton truck of a type that is now known as a pickup. This style of truck turned out to be such a good idea that it was available from the Thomas B. Jeffery Co. for a few years starting in 1913.

In 1912, the truck line of the Thomas B. Jeffery Co. was reorganized into a separate division of the company and a new heavier line of trucks was introduced in one-half, three-quarter, and one-ton capacities. These were the first vehicles to bear the Jeffery name, in honor of Thomas B. Jeffery, founder of the company. The new 1912 line of Jeffery Trucks were chain-drive units, similar to the famous chain-drive Mack trucks of the period.

By 1914, Jeffery was building 10,417 automobiles and 3,096 trucks annually. In 1915, automobile production dropped to only 3,100 units. Truck production reached 7,600 and herein lies an interesting story of how the famous Jeffery Quad Truck came to be.

In July of 1914, World War I began in Europe. This was to be the first mechanized war in the history of the world and a great number of trucks and other supplies would be needed by the allied nations of Europe.

The U. S. Army field tested a Four Wheel Drive Co. (FWD) truck in 1912, with several British Army observers at the test. The British observers were impressed with the FWD vehicle and ordered two trucks to be taken to England for tests. Shortly afterwards, an order for 50 trucks for the British Army was received at FWD Co. headquarters in Clintonville, Wis. This was followed by still larger orders after World War I started.

In 1913, a Captain Cranston of the U.S. Army Quartermaster Corps came to see his good friend Charles T. Jeffery, who ran the Thomas B. Jeffery Co. after his father died in 1910. Captain Cranston pleaded with Jeffery, on behalf of the U.S. Army, for the development of a four-wheel drive truck. Jeffery sent Martin P. Winther, head of his experimental department, to the FWD factory in Clintonville to purchase a truck to study. Winther's group looked the truck over and came up with an improved four-wheel drive system with better parts interchangeability. The three differentials were swappable, along with the four wheels, four sets of brakes, two propeller shafts and four driveshafts.

The Jeffery Quad drove, steered and braked on all four wheels, and had a rated carrying capacity of 1-1/2 tons, with a 20 percent overload capacity. These trucks would use a variety of Jeffery, Rambler, Buda and Nash four-cylinder engines.

The U.S. Army's first use of the Quad truck in battle came in 1916. Pancho Villa, the infamous Mexican bandit, had raided Columbus, N.M. on March 9, 1916. Quad trucks that had been destined for England were instead hurriedly sent to the Southwestern part of the U.S. There, General John J. Pershing made use of them when he led 12,000 troops after Pancho Villa into Mexico. By 1916, the U.S. Army was so impressed with the Quad, that they asked Jeffery to make two armored cars based on the vehicle.

Jeffery Quads on their way to the front lines in France during World War I.

One armored car was taken to the Mexican border by General Pershing in 1916. The other was built in 1914 by Jerry Decou, who was factory superintendent for Jeffery. All of the Quad Trucks had four-speed transmissions that allowed them to go forward and reverse in all four gears. This was a handy feature for fast retreats on a battlefield, as no turning around was necessary.

The military did not get all of the trucks that Jeffery made during that period of time. Autos and trucks were still available to the civilian population during World War I. Many Quads were used in construction work and for the pickup and delivery of goods in rural areas.

Charles Jeffery sold the Thomas B. Jeffery Co. to Charles Nash in 1917. Nash had been president of General Motors. By 1918, Nash Motor Co. was the world's largest truck manufacturer, with production of 11,490 units (compared to just 10,283 automobiles the same year. Most, but not all of these trucks were Quads and most went to the American government.

During World War I, Jeffery also built two-wheel drive trucks, but due to the need for the Quad by the allied military forces, not many conventional trucks were produced. This changed after World War I. Charles Nash reorganized the company.

Part of the change in the post World War I truck market was due to the sudden end of fighting. During the conflict, many trucks were sold to the U.S. government. When the war ended, these were sold at low cost, to get rid of the inventory. This caused sales in the post World War I truck market to decline. This policy in general, also caused a general economic recession which hurt all manufacturing.

Nash trucks failed to sell after 1921, due to the market glut and their expensive prices. The Quad sold for $3,250, while standard two-wheel drive trucks were $1,895 for a one-ton and $2,550 for a two-ton truck. It was $700 more for a Quad with smaller carrying capacity.

Nash two-wheel drive trucks also failed to sell after 1921, mostly due to their old-fashioned technology. They had an internal-gear rear axle. It made a great howling noise as it wore and may have discouraged buyers.

Nash truck production decreased rapidly in the 1920s. The Nash Quads and two-wheel drive trucks were used mostly for construction work. A few Quads were used by county and state governments for road maintenance, but many were war surplus units.

The last Quads were made in 1928 and the last prewar two-wheel drive trucks in 1929. The two-wheel drive Nash truck had never really sold well, in part because it lacked any major outstanding feature against the competition. Only one other type of Nash truck was made during the prewar period. A Nash dealer had the Nash factory convert a 1927 Nash Special Six into a pickup. It was used by the Nash dealer, as a tow truck, for many years. This truck still exists and was the only one of this type made.

Nash truck production wouldn't resume until 17 years later, when the Nash factory used Nash car sheetmetal to build a line of post World War II trucks. Production of these commercial vehicles was also limited.

(Thanks to Chuck Rizzo, Bob Aaron and Don Oakley of the Nash Car Club and Jim Harris and Dennis Kerperschimid of the American Motors Owners Association for their help with this story.)

Kin From Kenosha - Nash

The 1925 Ajax touring car was a 40 horsepower six on a 109 inch wheelbase with a list factory price of $865. It could hit 60 mph with its 170 cid L-head engine.

The bargain-priced Ajax

By Tom LaMarre

"To the best of our knowledge, there is no other manufacturer ...who builds so large a proportion of his products as does Nash," boasted Charles W. Nash - "the modest millionaire" - in the 1920s. One of those products was the short-lived Ajax, which was described as an "entirely new idea of motor car purchasing power in the $1,000 price field."

Before starting Ajax, Nash had been involved with Lafayette Motors Co., an Indiana firm backed by several General Motors finance executives. Nash was formerly the president of General Motors and saw this as the opportunity to establish a separate luxury car. When sales failed to thrive in 1922, Nash moved Lafayette to Milwaukee, Wis., near his own auto company.

In August 1923, Lafayette stockholders received a letter asking them to agree to the sale of Lafayette to Ajax Motors Co. for $225,000. Ajax was another Nash venture, organized to develop an economy car. The stockholders approved the transaction.

On April 2, 1924, Charles W. Nash bought the old Mitchell factory in Racine, Wis. Engineer Earl G. Gunn got the assignment to create the new car. Gunn, a former Packard engineer, had designed the Lafayette for Nash. On June 27, 1924, Ajax Motors Co. was officially established.

Charles Nash picked the name for the new car himself. In mythology, Ajax was the bravest of the Greek warriors next to Achilles. In the automotive world, however, it was no secret that the Ajax was a Nash in everything, but name. Charlie Nash even posed with the car for publicity photos.

The Ajax models were shown to dealers and distributors during a bash held at a Racine baseball field on May 6, 1925. The public got a closer look at the cars at the National Automobile Show. What they saw were small, quality-built cars that sold for under $1,000.

There were only two body styles in the Ajax line: the model 221 tourer ($925) and model 223 sedan ($995). Both of the boxy cars were designed by Earl G. Gunn and featured steel disc wheels, painted drum headlights with bright metal rims, a rear mounted spare and a seven main bearing crankshaft. One of the selling points for the sedan was its stylish exterior visor. The standard Ajax color was Mallard green.

Most historians claim that the Ajax was a failure. In reality, the car was a success; its name was a failure. While Nash sales were increasing, calendar year production of the Ajax was only 10,693 cars in 1925. About 20,000 Ajaxes were sold before the car was renamed the Nash Light Six on May 1, 1926. In an early example of badge engineering, dealers were sent Nash nameplates and hubcaps to put on any unsold Ajaxes.

The name change was all that was needed for sales to take off. Before long, a two-passenger coupe was even added to the line. In the automotive sales war, at least, Nash proved to be a better warrior than Ajax.

Priced at a modest $995, the Ajax four-door sedan was offered in 1925 and 1926. This same car became the Nash Light Six on May 1, 1926. Owners were provided with parts to convert their Ajax into a Nash at that time.

Nash Advanced Eight close-coupled coupe has twin horns, wipers and covered sidemounts. The streamlined parking lamps were a new feature for 1933. (Bill Mason photo)

1933 Nash Advanced Eight

By Bill & Mary Mason

Nash was hard hit by the depression. In 1930, the Kenosha, Wis. company was America's 10th ranked automaker and built nearly 55,000 cars. By 1932, production fell to 17,696 units and Nash found itself 13th in sales.

New styling and a reshuffling of car lines came the following season. Anxious to pick up sales volume, Nash adopted several popular design features such as V-shaped radiators, hood doors (instead of louvers) and streamlining for its 1933 models.

The nine 1932 series became five; the Special Eight was renamed the Advanced Eight. The car featured in this story is a 1933 Nash Advanced Eight close-coupled coupe.

This car is the only one of its kind listed in the Nash Car Club of America roster. It belongs to Ron Goularte, an attorney in Santa Clara, Calif. It was purchased at the Harris Collection Auction, in Cape Girardeau, Mo. on Aug. 6, 1983.

Advanced Eights were in the second highest 1933 Nash series, just one step under the Ambassador line. These cars were roughly comparable to a 60 series Buick, Chrysler Imperial CQ or Studebaker Commander. They were relatively large and powerful machines.

Standard equipment on the Advanced Eight models included an Autolite twin ignition system, mechanical brakes, synchromesh transmission (with freewheeling), Bijur lubrication system, airplane type instrument panel, safety glass windshield and coincidental transmission lock. Nash coachwork, by the Seaman Body Co., was carried on an X-type, dual double-drop frame.

An underslung worm-drive rear axle was used because of the car's low profile and ground hugging posture. A Ride-Control suspension system could be regulated from the dash. Nash introduced the floor-mounted headlight dimmer switch in 1933.

Styling and engineering improvements for 1933 were not conducive to increased sales. Calendar year output dropped to 14,973 cars, but Nash managed to hold onto 13th rank in industry.

When purchased, Goularte's close-coupled coupe had an older restoration. It seemed to be in generally good shape. To get it running, Ron's father, Tony, rebuilt the fuel pump and replaced the lines. After flushing the cooling system, the car was operating smoothly again.

A check of vehicle identification numbers came out perfect. Advanced Eight serial numbers for 1933 were B70021 to B70770 and engine numbers were B82671 to B83420. The serial number stamped on the car's right rear frame member was B70203, indicating that it was the 183rd unit made. The engine number, on the right side of the crankcase, was B82853, also 183rd in the production sequence. In short, all numbers seemed to match.

Several weeks after taking delivery, Goularte entered the car in the Modesto Concours, where it was photographed for The Standard Catalog of AMC. A Nash enthusiast from Santa Cruz saw the photo and called to point out that the hood mascot was wrong. He offered to send Ron Goularte the correct one. It arrived, as promised, a short time later.

In overall appearance, the Lafayette looked like the Nash of the same year, although it was smoother and had less elaborate body trim. This 1935 coupe from Ontario, Canada, features deluxe style plated headlamp shells.

Lafayette has arrived

By Gerald Perschbacher

In the dawning days of the American Revolutionary War, volunteers from several European nations, holding the same ideals as many American colonists, joined General George Washington in the fight for freedom. Among them was a French statesman, nobleman, and general named Marie Joseph Paul Yves Roch Gilbert deu Motier. His official title was "Marquis de Lafayette."

Lafayette fought valiantly for the American cause, distinguishing himself in battle and advancing to a divisional command. He experienced Valley Forge, was wounded at Brandywine, and was in the front line at Yorktown. Established as a hero in the colonies, he later returned to France and participated in the French Revolution.

It's natural that the name Lafayette should apply to an American automobile. But, it could not be just any auto. It had to be one of unquestionable character, of proven stock and pedigree. The car had to be stately and beyond reproach.

These thoughts ran through the minds of the executives in Indianapolis, Indiana. who were forming the Lafayette Motor Car Company.

You might have thought the Nash was the first company to use the name on an automobile, but from 1920 through 1924, the name Lafayette was carried by a car that was every inch a hero like the man it was named after.

The original Lafayette automobile carried a cameo of the famous Revolutionary general as its insignia. The car was built for the well-to-do. This is exemplified in the beautifully smooth tone of its rare advertisements: "One mile is enough to demonstrate the superb power of the Lafayette, its flexibility and its comfort. It takes 10,000 miles or ten times that, however, to show the Lafayette infinitely, to sound its fineness, to measure fully the benefits to be derived from the skill and the care that are spent upon it. Old owners and new alike, tell us they never have known the Lafayette's equal - and they have driven the best of American and foreign cars. Steadily the conviction that the Lafayette is one of the world's finest motor cars is finding wider and wider acceptance as the experience of Lafayette owners becomes known."

In the early 1920s, the economic waters into which American automobiles were sailing became turbulent with recession and the depression, an aftermath of World War I. Lafayette, as an automobile, became history, as did the man after which it was named.

But, good ideas often are reborn. That rebirth came in 1934, when Nash Motors, which had bought the assets of the Lafayette company, chose to dive into the risky lower-priced field with their new Lafayette. Perhaps the leaders at

Nash saw a golden opportunity to enter that market as the world watched Chevrolet pull ahead of Ford in calendar year production from 1931 through 1934. If there was a new king on the throne of American car mass production and sales, perhaps there could be an heir-apparent, a future king, in the new Lafayette.

The people at Nash were wise. The company's calendar year production reached a high of just over 138,000 in 1928. But, from there it was a downhill slide to the just under 15,000 units produced in 1933. The depression had hit, and it seemed as though Nash itself was in jeopardy!

While the first Lafayette automobile was a victim of bad times, the new Lafayette hit the depression sales scene and saved Nash. Calendar year production for the company grew from 1934, when the Lafayette was introduced by Nash, up to 1937. It reached a high level of production, just short of 86,000 units. But, 1938 was another bad economic year, and production for the year dropped to around 32,000. By 1940, the Lafayette had contributed to Nash's production mark, which broke 63,600 units.

Although the new Lafayette by Nash was not an expensive car (prices ranged from $585 up to nearly $1,000), it was a snappy job. Nash advertised it as the "big car in the low-priced field." On a 113-inch wheelbase, powered by an L-head engine rated at 75 hp, the car attracted a good number of buyers, but not in numbers sufficient to make the new Lafayette the apparent heir to the throne as the American sales leader.

The car proved the wisdom of the leadership at Nash. It not only helped push the company through the depression, but it showed solid foresight for success. This was something most luxury-oriented American auto companies failed to see in 1934.

While expensive cars got more expensive and more lavish in hopes of gleaning a greater dollar volume from a diminishing clientele, Nash went the other way; it entered the low-priced field and amassed extra sales. Nash made other wise moves, too. It merged with Kelvinator Corp. in 1936, which also helped keep it afloat.

By 1940, the Lafayette was doing so well in production that it was about three times greater in sales over the standard Nash offerings for that year. Perhaps the people at Nash saw danger in continuing the name Lafayette.

In the fall of 1940, the new Nash 600 was launched, replacing the Lafayette as the company's new, low-priced flagship. The new car was an immediate success.

It had to be successful. It had been built upon the tradition of low-priced excellence achieved by the Lafayette...that old soldier who helped revolutionize the market and fight to make Nash a winner.

Companies in Detroit, Mich. (1903); Lafayette, Ind. (1911); and Easton, Pa. (1912) used the name Lafayette before Lafayette Motors Co. of Mars Hill, Ind. and Milwaukee, Wis. launched its upscale Lafayette Model 134 in 1921 (1923 model shown). Started by a group of ex-GM execs including Charles Nash, the firm lasted until 1924. It was separate from Nash at the time. Ten years later, Charles Nash resurrected the name for his new low-priced model.

By 1938, the Lafayette business coupe was up to $770 in price. Lafayettes still had the same general design as their Nash relatives, except for bullet-shaped headlamps and shorter front end sheet metal.

Ed Robinson did not buy a wrecked 1939 Nash, but later picked up a '35 Nash Lafayette business coupe, such as the one seen here. It was a good performer that could outrun most of the other cars in his town.

Good-bye Nash: the hardest words in the English language

By Ed Robinson

In golden olden days, every neighborhood had its resident genius. These geniuses were almost always non-conformist mechanical wizards. In my hometown, Old Les was it. About once each year, usually in the spring, he would arrive driving a creation which invariably started great waves of admiration in all the younger car nuts. Les's skills were such that admiration rarely resulted in emulation. We admired his work, but we could rarely, duplicate it. I remember the time he arrived driving a 1932 Plymouth roadster with the frame installed upside down, "underslung" if you will.

Up to that point, the four-cylinder Plymouth had been an object of disdain. I studied that one for days. The steering still worked. The springs still worked. The hydraulic brakes still functioned. The car appeared to be flying, even while sitting still. After several weeks of study, I reluctantly concluded that I could not reverse the frame on my 1927 Chevy coupe.

The next year Les showed up in a 1936 Ford roadster. He had made solid, non-louvered hood sides. The grille was from a 1940 LaSalle. All of this was accomplished without the use of pop rivets or Bondo, which had yet to be invented (I think).

That year I happened to be driving a 1936 Ford three-window coupe. Again I envied. Again I studied. Again I wished to duplicate. This time I decided it was possible. I made one change. I decided to use a 1939 Nash grille rather than the 1940 LaSalle type. This decision was based upon expedience, rather than aesthetics. My favorite junkyard contained a 1939 Nash, but not a 1940 LaSalle.

In those golden olden days, junkyards were not recycling centers. They had no offices, they had no catalogs, they had no parts retrievers. You walked into the yard and located the pieces that you wanted. You removed said pieces yourself and then you went to the shack to negotiate price.

In the process of removing the grille, I made several interesting discoveries about 1939 Nashes. At that time dual exhausts and dual carbs were all the rage. That '39 Nash had no exhaust manifold. A header pipe simply bolted to the side of the block. Even more fascinating was the apparent absence of any intake manifold. The carburetor simply bolted to what appeared to be a cast-in channel. Furthermore, the thing had 12 spark plugs in a six-cylinder block.

I wasn't sure just what dual ignition would contribute to my project, but I reasoned that if dual points helped, dual plugs would really make some sort of a difference. The exhaust and intake arrangement I fully understood. I had visions of individual header pipes from each cylinder and I could mentally picture at least six Stromberg 97s bolted to the top of the head. Unfortunately, the particular Nash I studied had been hit so hard in the rear that resurrection was impossible. The grille and the motor were about all that was left. However, this junkyard introduction to a novel snout and a demolished rear began, for me, a long association with the "Prides of Kenosha." I'm still not sure whether I am glad I got into Nashes or not. With a Nash there is no middle ground. It is either a great car or a gutless wonder. You either love it or hate it. I have had both. I have done both.

I did not acquire the wrecked '39, but a year later I convinced my father to buy a new 1940 Nash Lafayette sedan. It was a black four-door with red pinstriping. It had six cylinders, 12 spark plugs and an overdrive behind the three-speed standard transmission. I never had the opportunity to try any of my dreamed of modifications on that one. All Dad ever permitted me to do to his cars was to apply Simonize. Rubbing hard wax was not my idea of what one should do with a new car, but "Dear Old Dad" had his own ideas of what you did with both cars and sons. Wax caused horseflies to skid off and break their necks before dad ever let me behind the wheel of his Nash.

My long planned modifications were not needed. Pappy's Lafayette turned out to be the fastest car in town. In our Clandestine "sneak out dad's car and make a timed run to the next town" sweepstakes, I held undisputed second place. I defeated a Pontiac Eight, an Olds Eight, a Buick, and all the Chevys in the county. First place, belonged to a Chrysler

New Yorker. I think I could have beaten it, but we were both afraid to ask for a rematch. The Nash was the first car that I ever drove past the magic 100 mph mark. Of course, that took a long, downhill run in overdrive.

Another favorite sport of the era was called "chase the couple." If one half of the couple being chased was your older sister, the fun was even greater. All older sisters' boy friends were obliged to accept the challenge. They had to try to ditch little brother. Some of the resulting chases made the scenes in "The French Connection" look like a Sunday afternoon drive.

My sister's "steady" considered himself a good, fast driver. He also thought that his sharknose supercharged Graham was unbeatable. The Nash and I gave him a real education in close-order drafting. He never did lose us. He was so upset and so determined that he traded the Graham for a new 1940 Pontiac "Silver Streak" eight. The contest was repeated. The results were the same. The Nash stuck to his Pontiac as stink sticks to a dog.

Johnny was not one to give up easily, nor was I. He traded the Pontiac for a Nash Ambassador, which to us seemed like the same thing Dad owned, except the Ambassador had eight cylinders, of course. Our Lafayette had only six, but I still was not outrun. Dad's Nash Lafayette lost a couple of headlight covers from thrown gravel on country roads, but we stayed right in there. Johnny finally solved his problem by keeping his Nash and getting a different girlfriend. Needless to say, my sister was not particularly happy, but she never told Dad what I had been doing with his pride and joy.

I was so impressed with the performance of dad's car that I traded my 1936 Chevy Convertible for a 1935 Nash Lafayette coupe. (I'm not sure whether Nash Lafayette is the correct terminology or not; most people simply called the '35 a Lafayette.)

I don't know how many Lafayette coupes were made in 1935, but the production figures must have been fairly low. I have to this day not seen another coupe like the one I had. I was not disappointed with my trade. Even without the later modifications to intake and exhaust systems, the '35 Lafayette was capable of outrunning most of the Fords and Chevys owned by my friends. World War II separated me from the '35. I was sorry to see both it and me go. I wish I had it back.

Late in 1941, having observed the performance of dad's '40 and my '35, my sister decided to buy a Nash. She purchased a new 1941 Nash 600. She was pleased with the economical purchase price and the cheap operating costs. Fortunately, she didn't drive hard enough to ever discover what a real dog she had. She never turned a corner fast enough to realize that the new design on those bodies let the doors fly open during stress. That Nash, I'm glad I never owned. I'm almost sorry I ever even drove it.

In the postwar years our family continued its relationship with Kenosha. Dad drove his 1940 Nash until 1949 when he purchased a new one. His 1949 upside down bathtub was painted a "dead grass brown." It had to be the ugliest car ever to grace our driveway. As far as I was concerned, it had only one redeeming feature: the front seats were fully reclining.

The idea of a portable bed disguised as a family sedan was philosophically intriguing. However, since I was newly married to a thoroughly housebroken bride, I never tried the bathtub/bedroom on wheels. Dad liked it. He made several cross country trips in it and bragged endlessly about the fantastic mileage he got. Since gas cost only 23-cents per gallon, I was not impressed. I never bothered to borrow the brown bathtub. It must have been a fairly solid vehicle. Dad kept it until 1954.

Through a series of unplanned trades, in 1951 I became the owner of a 1949 Rambler convertible landau. I'm glad the trades were unplanned. I would be eternally shamed and frustrated had I acquired such a headache intentionally. Two words "springy" and "mushy" describe the driving sensations inherent in the Rambler convert. The steering never indicated exactly which way you were heading. Steering wheel free-play changed to over-control with no appreciable difference in sensation. The rear end was equally ambivalent. You never knew in advance whether the rear wheels were going to follow the front ones, skid to one side, or simply roll under the tire casings.

The gearshift lever had the same indefinite springy sensation. You wriggled the stalk into what you assumed might be the approximate position of the gear which you hoped to select. Next you released the clutch pedal and waited to see if the car moved. The clutch pedal had the same disconcerting lack of feel as did the shift lever. In case you had missed the shift linkage, you could always depress the brake pedal and start over. Here again, the sensations were consistently inconsistent. Despite re-ground drums and new wheel cylinders, the Rambler brake pedal never lost its mushiness. Doors, too, were springy. The latches never quite caught. Endless adjustments never succeeded in locating the doors with enough precision to permit seeing anything other than generalized blurs in the door-mounted rearview mirrors.

All of the above descriptions of springiness and mushiness are simply generalized driving impressions, empirically derived, and not supported by facts. Firm statistical data is available regarding the frequency of replacement of the rubber connector between the generator tailshaft and the isolated water pump. Frequently hung-up exhaust valves and cracked valve seats could also be verified numerically.

The convertible top was another abomination unto the Lord. The doors had side rails and door posts the same as sedans. The visual impact and side ventilation of a convertible were, therefore, lost. The cost of canvas replacement, ever-present cold drafts, constant rattles, plus regular leaks around the edges were always evident.

The Nash Rambler convertible carefully retained all the bad features of a closed car, plus all the disadvantages of a convertible. It had a few features uniquely its own, which could be attributed to its being neither a closed or open car. When the top was fully retracted, it was impossible to avoid pinching the canvas and wearing instant holes. This problem was not too significant however, because one rarely got the top fully retracted.

The sliding top bows usually cocked themselves in their respective slots refusing to move either forward or backward. When that happened you had neither a convert nor a sedan. You had a two-man, two-hour headache. With a little ingenuity the top raising and lowering mechanism could probably have been converted into a garage door closer providing, of course, that you had a canvas garage door. And providing that you didn't really care whether it was fully closed or not. You could not have made such use, of course, unless your house had 12-volt wiring.

The demise of the Rambler fits well within the "springiness" concept. The rear spring hangers and the frame rusted in two and the body fell into the street. Glutton for punishment that I am, I then purchased a 1950 Rambler convertible. The experience was repeated. The 1950 had all the problems that the 1949 displayed.

By 1954, almost our entire family had somehow or other become involved with FoMoCo products. My sister was driving a 1951 Mercury. I had a 1950 Ford. Dad, who tended to buy a new car every six or seven years whether he needed it or not, decided that year to try the new 1954 overhead valve Ford. Three thousand miles later, he discovered a soft camshaft on his new Ford. That year Dad bought two new cars! The 3,000 mile new 1954 Ford was traded for a 1954 Nash Ambassador.

The 1954 Nash fairly closely resembled the 1949 bathtub. A few notches and square corners had been added and the rear end sported a mismatched continental spare tire. The long-stroke, high-torque, overhead valve six attached to an overdrive still provided operating economy. However, a fast resale attrition rate gobbled up any real savings or financial benefits.

I didn't own another Nash until two years ago.

In 1978, I bought a 1940 Nash Lafayette four-door sedan. The 1-to-6 condition categories used by *Old Cars Price Guide* hardly fit this particular vehicle. I think it is what they call "all there." In other words, all that's there sat out in the weather for at least 20 years before I purchased it.

I don't know whether the motor stuck before or after it was retired from active service. I don't know whether to call it a challenging restoration possibility or simply a parts car. I'm beginning to think I should describe it as one of those which I'm sorry I bought, I'll be glad to sell and I won't miss it after it's gone.

On the other hand, the spacious back seat is just as large as they were when I was in high school. If fixed up, it probably would still do a 100 mph downhill. Every old barn should have its resident Queen awaiting restoration. So, unless it begins to crowd the pieces of the 1960 Metropolitan...or unless somebody offers me a profit...I probably will leave it as part of the auction to be held after I'm gone. I have a hard time saying "Good-by Nash."

Ed Robinson dreamed of adding a '39 Nash grille to his '36 Ford three-window coupe. The handsome grille was only one of Nash's advanced styling features, as characterized by this 1939 sedan seen at Wally Rank's 1992 auto show.

Steve Gregori's 1940 Nash Lafayette all-purpose coupe is perched on a 117-inch wheelbase. It cost $860 when new. (Bill Mason photo)

Last of the Nash-Lafayettes: 1940 "All-Purpose" Coupe

By Arch Brown

By 1940, the Lafayette, as old Charlie Nash had originally envisioned it, was long gone. The name remained for one final year in 1940 but, since 1938, the Lafayette had been a Nash series, rather than an independent marque.

When it was introduced, back in 1934, the Lafayette had been Nash Motor's depression-fighter priced just $20 higher than the Master series Chevrolet. For in those days, the low-priced three...Chevrolet, Ford and Plymouth...held more than two-thirds of the new car market.

Times were tough and the depression's effect upon the independent automobile manufacturers was catastrohic. A number of the smaller companies had already packed it in. They included some time-honored marques that had been around nearly as long as the industry itself. Moon, Kissel, Elcar and Jordon were all gone. Peerless had converted its factory into, of all things, a brewery. Franklin, Stutz and Marmon were just then closing up shop. Other makers would follow.

The message was clear enough and the larger independents took appropriate steps. In 1932, Studebaker introduced the Rockne and Hudson metamorphosed its Essex into the high-stepping little Terraplane. But, Studebaker was encountering "heavy weather" and soon found itself in receivership. The Rockne, a fine little automobile that deserved a better fate, lasted only a couple of years. The Terraplane did better than that, but it wasn't setting any sales records either.

That left Nash alone. By far the strongest of the independents, in terms of financial structure, Nash also had the most efficient manufacturing facilities of the group. If anyone could successfully mount an assault on the mass market, surely it would be Nash Motors.

Accordingly, early in 1934 the Lafayette model made its debut. A handsome car it was, too. It was designed by no less a personage than Count Alexis de Sakhnoffsky. Under its hood lay the seven main bearing, 75 hp flathead that had previously powered the Nash "Big Six." Charlie Nash must have had high hopes for the Lafayette.

But, it bombed! Sales that first year came to just over 9,300 cars; less than one-half of one percent of the U.S. market. Nor did 1935 bring much improvement. Lafayette sales that year were nearly double the 1934 figure, but the market was a little better by then and the Lafayette's market penetration didn't quite reach two-thirds of one percent. Furthermore, there were indications that even this limited success was largely at the expense of Nash itself, rather than the lower-priced three.

And so it was that, in 1937, Nash abandoned its attempt to invade the low-priced field. That year the Nash-Lafayette 400 replaced both the Lafayette and the Nash 400 of the previous season. Priced, as the 400 had been, to compete with Pontiac and Dodge, it was larger, heavier and more powerful (as well as more expensive) than earlier Lafayettes. It was also, perversely, far more popular. Combined Nash and Lafayette sales increased by an impressive 64 percent.

From that time on, the Lafayette name was progressively downplayed. For 1938, it was placed on the side of the hood, with Nash appearing on the radiator badge. A year later, one had to virtually do a headstand to find the tiny body plate which announced "Lafayette Series." And, in 1941, the Lafayette was phased out altogether. Its place was taken by the smaller, cheaper, lighter and far less powerful Nash 600.

Pictured here is a trophy-winning 1940 Nash-Lafayette all-purpose coupe belonging to Steve and Karen Gregori, of Fresno, Calif. Restored by Steve from a one-owner original, it serves the Gregori family as a tour car with its twin jump seats providing accomodations for the two Gregori children. Its competent, 99 hp L-head engine is easily capable of sustaining highway speeds well in excess of the legal limit, if the driver is inclined to let the car stretch its long legs.

Comfortable and stylish, this particular model was designed by George Walker, who later gained fame with Ford Motor Co. At $860, the handsome Lafayette all-purpose coupe was a stellar value in its day.

To many motorists, the 1942 Nash Ambassador looked a lot like a Packard as it moved down the road. Up close, there were differences between the two marques, but the overall appearance was very similar.

The '42 Nash was hardly a smash

By Gerald Perschbacher

When the 1942 Nash hit the market, it was in a world marked by several armed conflicts and one world war. Today, 50 years later, we sometimes forget that our modern perspective cannot explain why a 1942 car found success...or failure.

Back in 1942, U.S. President Franklin D. Roosevelt was calling for the defeat of Hitler's "insane violence. He declared that under no circumstance would he, as President of the United States, betray the cause of freedom with a negotiated peace. Berlin, the Nazi capital, boasted that its troops had entered the suburbs of Leningrad. United States airplane production had reached a record high of 1,854 in anticipation of armed conflict. Eleven Allied governments pledged their committment to the new Roosevelt-Churchill Atlantic Charter. And 17,000 union steel workers ended their strike at three large plants in the Birmingham, Alabama area.

It was into these unsettled international and domestic waters that the new Nash set sail in late 1941. Nash registrations for all of 1941 broke 77,000 new cars, but actual sales of 1942 models amounted to just 31,700 units, before auto production ground to a halt in January of 1942.

In the fall of 1940, the new, low-priced Nash 600 jumped into the market. Clearly an economy model (the designation 600 was based on its 20-gallon fuel tank providing 30 mpg), the Nash got the attention of many economy-minded motorists who were recovering from the depression and facing the uncertain future of armed conflict and gas rationing.

Also remaining available in 1942 were Ambassador sixes and eights. The six utilized the venerable 234.8 cid/105 hp Nash overhead valve motor, instead of the 172.6 cid/75 hp flathead found in 600s. The top-of-the-line Ambassador powerplant was a 260.8 cid/115 hp overhead valve straight eight.

The Nash 600 was the first mass-produced U.S. car to carry unitized construction. It came on a 112-inch wheelbase and sold for between $700 and $900, depending on the particular body style ordered. Ambassador sixes and eights shared a 121-inch wheelbase. Sixes sold in the $994 to $1,069 range, while eights ran from $1,044 to $1,119.

The new 1942s, like most other U.S. autos at the time, were not purchased in large quantities or long remembered. The war saw to that. But, the basic 1942 Nash designs were carried-over to the immediate postwar years, when sales bounced back to near the 100,000 mark.

Some differences between 1942 and 1946 Nashes include the relocation of parking lights. They were moved from above the headlights, atop the front fenders, to lower side positions between the small center grille section and the headlights. The Nash "badge" at the front of the hood was redesigned and numerous other minor changes were implemented. For example, the Nash name, formerly spelled out, in chrome-plated block letters right under the center grille section, was deleted.

At a glance, more than a few people viewed the 1942 Nashes as "small Packards" as they came barreling their way. There were definite styling similarities between the two. But, as the Nash came closer, it was apparent that the front grille treatment was not quite identical to the Packard design.

In 1946, Nash would broaden the center grille section to project a "bigger is better" impression. There was also more chrome trim to make the postwar cars look richer.

However, the 1942 Nash models didn't get as much attention. With the short sales year, they puttered around just a few months to the beat of such Hit Parade tunes as "Der Fuehrer's Face" and "Praise the Lord and Pass the Ammunition." Then, as Americans put on ammo belts and suited-up for global conflict, the 1942 Nash line-up became one of the year's early casualties.

This beautiful 1946-1948 Nash stake bed truck illustrates the passenger-car-inspired styling found on these rare commercial vehicles. Most of the trucks were equipped with dual rear wheels.

Cars that are trucks...and vice versa

By R. Perry Zavitz

The idea of building a truck on a passenger car frame is an old one, and continues today with Ford's Ranchero and Chevrolet's El Camino. However, the greatest credit for this innovative body style rests with the independents, especially Hudson.

Hudson introduced pickups and delivery vans based on the Terraplane chassis in the mid-'30s and revived this style, after World War II, in pickup form only. These coupe-express models used a passenger car chassis, which had its wheelbase extended to 128 inches.

Available in six-cylinder form only, the Hudson pickup was produced from late 1945-1947, with styling virtually unchanged during this period. Total production for the postwar pickups was 6,021. Hudson's pickup/passenger car era came to an end with the introduction of the revolutionary 1948 "Step-Down" Hudson, which was not conducive to pickup truck conversions.

While the 1947 model year was the last for Hudson trucks, it was the first for postwar Nash trucks. Nash was not a total stranger to the truck building field, having produced thousands of Jeffery and Nash Quad trucks in the teens. However, the company's postwar effort was their first fling at truck-making in almost 30 years.

Nash truck styling was inspired by (though not identical to) the 1946-1948 Nash passenger cars. Built only in cab-and-chassis form, many pf the postwar Nash trucks were equipped with dual rear wheels. They all used the Ambassador's big overhead valve six-cylinder engine for power.

Nash produced these truck for nine model years without a styling facelift. Production totals were never earth shaking, with a total of 4,998 of the Kenosha products produced during their entire life span. The best year was 1948 when 1,052 trucks came off the line. In 1955, the last model year, only 16 trucks were produced by the newly-formed American Motors Corporation.

Rare in their time, the Nash trucks are very unusual to find in the U.S. today since almost the entire production run was made for export. A few were sold to domestic Nash dealers to use as service trucks, but your chances of finding a Nash truck in 1992 are better if you're looking in South America or Europe.

Although the bulk of production went to the overseas market, many Nash dealers in the U.S. had tow trucks bearing Nash nameplates, such as this example which survived in Houston, Texas.

Walt Disney cartoon character Gyro Gearloose would look right at home behind the wheel of Vince Ruffolo's 1951 Nash Ambassador Custom two-door sedan.

Was Gyro Gearloose a Nash enthusiast?

By Terry Boyce

Remember Gyro Gearloose, the wacky inventor in those old Disney comic books? I always figured that if he drove a car, it would be a bathtub Nash.

It is still difficult to keep from chuckling about the "turtleback" Nash of 1949-1951. But, I feel guilty about the unbridled decision I heaped on those old cars as they limped down the streets of my callow adolescent years. Then, I didn't know about the sincerity of their design nor that they were really comfortable, pleasing automobiles. Later, I owned a beautiful, low-mileage 1951 Ambassador Custom sedan and I saw the light.

The Nash Airflyte was the car of the future when it was introduced in 1949; the future as it had been perceived during World War II. This streamlined car was the cumulative statement of a brilliant designer, Nils Erik Wahlberg. He had joined Charley Nash at the beginning, in 1916, and by World War II was the firm's vice president in charge of design. The Airflyte was the summation of his experience. When Nash dropped it for 1952, he quit.

Like most car designers, Wahlberg and his team believed the bulbous, softly rounded 1941 Chrysler Newports and Thunderbolts and Packard's similar Phantom were forecasting the future of automotive design. During the war years they carried the streamlining theme to its radical conclusion, with fully skirted wheel openings, hidden headlamps and uncluttered, aerodynamic bodies prominent in their design proposals.

However, except for Studebaker, most immediate prewar cars were hardly changed from 1942 models. When the new designs began to appear in 1949, Ford had a fresh, slab-sided approach and General Motors remained cautiously evolutionary instead of revolutionary. Chrysler was even more restrained. Only Nash brought the wartime "car of tomorrow" to the showroom. It was not met with great enthusiasm.

The Nash Airflytes were Buck Rogers futuristic inside and out.

All the instruments on the 1949-1950 Nashes were grouped into a streamlined pod on the steering column called the Uniscope. Only the radio, hidden behind a sliding screen, remained on the dash panel.

Exterior styling of the Airflyte was truly aerodynamic. It was more than a gimmick. Long hours of wind-tunnel testing during the war years had lead Wahlberg into a study of surface cracks and orifices and the results were incorporated into the Airflytes. A 1949 Nash cut through the atmosphere more cleanly and with less drag than any other American production car. A series of wind-tunnel tests at Wichita University, in early 1949, confirmed the Airflyte's design efficiency. A new Ambassador generated 113 pounds of drag at 60 mph. Only the Studebaker could come close to this figure and the scale ran all the way from there, to Packard's 171 pounds of drag.

Still, buyers were not impressed. They wanted postwar performance. The hot Oldsmobile 88 defined what that meant. By comparison, the Nash 600 (renamed Statesman in 1950) was a true slug. It took 25 seconds for its little 85 hp L-head six to find the end of a standing quarter-mile run and the Uniscope would only indicate 55 mph at that time. Top speed was less than 80 mph.

Gas mileage was 18-27 mpg, but who cared? The larger Ambassador line was considerably more capable, but the image was cast.

The Airflyte design, with its unit-body construction, restricted body types to two. There was a four-door sedan and a two-door sedan. An unusual Brougham interior package was listed for the two-door style, expanding the model availability without adding a true body style. Two large, armchair style seats were placed in the cavernous rear body when the Brougham was ordered. They were slightly turned in toward the center and were separated by a high triangular center armrest. The Brougham interior added $22 to the base two-door sedan's price. It was available in standard and deluxe models of both series.

The name of the 1946 Nash 600 indicated how far it could travel on a full tank of gasoline.

Nash in the late '40s and early '50s

By Robert C. Ackerson

Nash began the 1950s with a proud heritage, a veteran management team, a respected reputation and an automobile with styling unlike that of any other American competitor.

George Mason, the president of Nash-Kelvinator since its conception in 1937, was a man whose rotund physique belied an outlook and philosophy characterized by dynamism and energy. Mason was an early proponent both of the American small car and the necessity, if they were going to survive, of mergers between the independent automobile manufacturers. Yet, beyond doubt, it was his persuading of George Romney to join Nash-Kelvinator in 1948, that stands as George Mason's most far reaching decision. Without Romney's missionary zeal and unremitting faith in the sales viability of the small Rambler, Mason's success in bringing Nash and Hudson together, in 1954, as American Motors, would have been nothing more than a respite before what would have ranked as one of the greatest business failures of American history.

Nothing of that sort happened. Instead, American Motors, after sustaining losses for the first years of its existence, earned a profit of $26 million in 1958. After that turnabout, there was no stopping AMC. In 1959, it produced 374,240 cars, yielding a profit of $60,341,823. George Mason would have been very proud and very happy.

Nash hadn't been the first American manufacturer to adopt a unitized body-chassis construction for its automobiles. Both the Chrysler Airflow and Lincoln Zephyr were earlier examples. But, casting modesty to one side, Nash announced the 1941 Ambassador 600 in a grand fashion, declaring "in automotive history, the year 1941 will probably be noted principally for the introduction of a new kind of automobile body construction." Nash did have a point, since the 600's monocoque design was both more advanced than that used by either Chrysler or Lincoln. It also offered a real-world bonus; the ability to travel 600 miles on a 20-gallon tankful of fuel.

Like that of other new cars offered by most American companies in 1942, the 600's production ended a few months after the bombing of Pearl Harbor. When the war ended, Nash was quick to get back into the automobile business. It meant little when Mason depicted the postwar 600 (of which 6,148 were built in 1945) as "the forerunner of cars to come." The assumption was widespread that every manufacturer was planning vehicles far more interesting than their initial postwar offerings.

The new Ambassador and 600 models introduced on Oct. 1, 1948 were described by Nash as having "that streamlined look...that big, broad, bold look...that clean, unbroken fender sweep...that Airflyte look..!" To many wags, the Nash had the appearance of an inverted bathtub missing its legs. But aerodynamically, at least, it was entitled to the graceful Airflyte label. In wind tunnel tests conducted at the University of

Wichita, in Kansas, the Nash's drag co-efficient had been calculated at 0.43, an impressive figure for the time.

George Mason had favored enclosing the Nash's front wheelwells and this did contribute to its wind cheating form. On the other hand, their adoption resulted in a larger than necessary turning circle and made front tire changing an unpleasant and inconvenient task. England's The Autocar charitably avoided passing judgement on Nash's appearance, instead describing them as "very interesting technically" because of their unit-body construction. In truth the Airflytes were, aside from this feature and their appearance, conventional. Both the 121-inch wheelbase, 210-inch overall length Ambassador and the smaller 600 (112-inch wheelbase, 200-inch overall length) used torque tube drive with coil springs at all four wheels as had the prewar 600. Similarly, their engines were familiar to earlier generations of Nash owners. The 600's L-head six, with a 3-1/8-inch stroke and 4-inch bore, had 184 cid. With a 7.0:1 compression ratio, it developed 82 hp at 3600 rpm. The Ambassador's 234.6 cid overhead valve six, with a 7.3:1 compression ratio had a 112 hp rating.

The Nash interior, while not as startling in appearance as its exterior, still possessed a bit of the Buck Roger's look with its "Uniscope" instrument pod. It placed the speedometer and major gauges directly in front of the driver. The Uniscope's life span was mercifully short, as it was discontinued after 1951. Also destined for a limited existence was the Brougham model, a favorite of George Mason. He thought customers would be eager to own an automobile with two large rear seats, slightly angled toward each other, with a wide armrest between them. Just the thing, thought Mason, to provide a comfortable setting for a friendly game of cards on those long trips to the country. Nash offered this feature on both the Ambassador and 600, but it was quietly withdrawn after the 1950 model year.

The accomodations for both the Nash's driver and occupants were quite appealing without the Brougham's unorthodox seating arrangement. Since 1938, its "Weather-Eye" ventilating and heating system had been a Nash accessory that delivered what it promised: window-up ventilation when desired and a thermostatically-controlled passenger compartment temperature.

Equally long-lived was Nash's unique twin convertible-bed feature. The front seatback, which adjusted to five different positions, fully reclined to provide a reasonably comfortable bed with a 64.5-inch length.

Obviously, being tucked-in for the night inside a Nash wasn't quite the same as settling down on the old mattress back home, but it wasn't all that bad either. At nominal cost, Nash dealers also offered air mattresses with built in pillows and tight-fitting nylon screens for the window openings.

With postwar demand for new automobiles still far from satisfied in 1949, Nash had little difficulty in setting a new production record of 142,592. This strong momentum was carried over into 1950. Statesman replaced the 600 as the designation for the Ambassador's smaller and less expensive companion series. Except for a larger rear window (which only marginally improved rearview vision) the 1950 models were virtually undistinguishable from their year-old counterparts. Via an agreement which provided General Motors with manufacturing rights to Nash's Weather-Eye system, the 1950 Ambassador and Statesman were available with GM's Hydra-Matic transmission. Tom McCahill who considered the Ambassador "America's most versatile car" reported a 0-to-60 mph time of 16.19 seconds with a Hydra-Matic Ambassador.

The Nash lost none of its spacious interior accomodations in 1951, but it did acquire a new rear fenderline that lessened its roly-poly look. Few owners treated their Nashes as sports cars but nonetheless Griff Borgeson (MOTOR TREND, December 1951) felt impelled to caution that the Nash "body heels more than the passengers realize and it's important not to underestimate your speed when cornering."

Although Nash engineers succeeded in coaxing an additional three horsepower from the Statesman's 184 cid six, it was still an 85 hp weakling for 1951. With Hydra-Matic it accelerated in a most leisurely fashion, arriving at 30 mph from rest in 7.8 seconds and eventually reaching 62 mph in another 19.5 seconds. Not surprisingly, the Statesman had the dubious distinction of being both the slowest car in standing start quarter-mile acceleration (24.47 seconds) and the car with the lowest (77.17 mph) top speed of all American automobiles tested by Motor Trend during 1951. The Ambassador, also slightly more powerful (115 hp to 112 hp) for 1951, did substantially better. With overdrive, it needed just over 15 seconds to reach 60 mph from rest and in the 1951 speed trials at Daytona Beach, Johnny Mantz's Ambassador reached a respectable maximum speed of 94.29 mph. Later, in the NASCAR 160-mile Grand National race, Ambassadors finished in seventh and ninth positions.

During 1951, Oldsmobile and Hudson battled for supremacy in NASCAR racing. However, on April 1, in a 112-mile event at Charlotte, NC, Curtis Turner gave the Ambassador its one and only race victory of the season. A total of 14 Ambassadors took part in NASCAR competition during 1951, most, if not all of them equipped with twin carburetors and a special intake manifold.

For partisans who appreciated Nash's solid construction and roomy interior, the NASCAR version held little appeal. But, it was yet another example of George Mason's leadership style. He was a continual practitioner of the fine art of innovation. He knew, perhaps better than any of his contem-

Indianapolis Motor Speedway selected the 1947 Nash Ambassador as "Official Pace Car" for that year's Indy 500 race on May 30.

Rarely seen postwar model from Nash is this 1948 Ambassador Suburban four-door sedan with wood body paneling.

poraries at the other independent manufacturers, that standing pat was just a step away from rigor mortis. Although his enthusiasm for Nash's enclosed front fenderwells was aesthetically flawed, Mason appreciated the finer points of automotive styling. He was also aware that the American manufacturers by no means possessed a monopoly on design talent and after attending the 1949 Paris Salon, he returned to the United States full of enthusiasm for the efforts of Pinin Farina.

George Romney's itinerary for a planned European trip was expanded to include a visit to Farina. Two days after they met, Pinin Farina and George Romney had reached an agreement. In explaining the suddenness of this decision, Farina expressed an admiration for the zeal and enthusiasm of Nash's leaders, their willingness to consider new ideas and the overall quality of Nash automobiles. Two years later, after the introduction of the first Farina-designed Nash, Farina elaborated on his perception of domestic automobiles. "Personally," he said, "I like the American cars and their large comfortable interiors very much. In the midst of their panorama of the prevailing conditions, circumstances and scenery of U.S. life, they are beautiful; and our European cars look poor in comparison."

Although Nash publicly regarded the Farina "as the Rembrandt of automotive design," its committment to his proposals was qualified. While he labored on a design for the 1952 models, in-house Nash stylists were doing likewise. In essence, Farina's design won this trans-Atlantic styling contest, but it was not a total victory. Explained Nash: "The design he gave us was not approved entirely by our designers, but basically it's his design with some modifications." These included a reshaping of the sloping front and rear deck of the Farina prototype, the removal of rear quarter windows and some superfluous, but expensive to manufacture grille bars. Regarding the appearance of the final product, Tom McCahill reported (*Mechanix Illustrated*, September 1952), "some anonymous wag on the West Coast is reputed to have remarked that there was 'more Wheatena than Farina in the new Nash line.' When I rudely asked one of the Nash brass just how much of the design was actually Farina, he admitted that there was, through necessity, a bit of the Midwest mixed in with Sunny, Italy."

Some critics faulted Nash for fumbling the opportunity to produce what they regarded as one of the handsomest cars on the road. (The company exercised final veto power over some of Farina's ideas). However, the end result was an attractive automobile. From its British perspective, *The Motor* passed favorably on its appearance, concluding, "Farina has achieved a design of classic simplicity." Farina never relied upon excessive amounts of chrome trim for effect and the "Golden Anniversary" Nash reflected a similar restraint. Evidence of Farina's preference 'for smooth body curves that gently flowed into each other was also apparent in the new Nash.

Regardless of how onlookers regarded the new Nash, the view from inside it was like that of no other Detroit car. Said Tom Mc Cahill, "To get behind the wheel of the Ambassador is like being on the bridge of the Queen Mary." Its larger windshield, 40 percent thinner window posts, optional tinted Solex glass and "Road-Guide" front fenders all contributed to easing the driver's task in maneuvering the 78 inch wide Nash through traffic. Said McCahill, "it still gives the sensation of steering a three-acre lot or going through the Tunnel of Love in a coal barge."

Weighing in at 3,480 pounds and stretching 209.25 inches on a 121.3 inch wheelbase, the Ambassador was one of America's largest automobiles. Not far behind was the Statesman. Its wheelbase now was 114.25 inches and its length was just seven inches short of the Ambassador's. Although Tom McCahill thought it odd that a Nash official "bragged that this new job was two inches wider than a Cadillac," the payoff came for drivers who enjoyed commodious accomodations. No other American car had a front seat exceeding the width of the Ambassador's nor did any competitor, at any price, surpass its 37.5 inches of front headroom.

Having disposed of the unpopular Uniscope instrument pod, Nash reverted to a more conventional dash arrangement in 1952. It placed all instruments, including orange-colored lights to indicate low oil pressure or battery discharge, in a recessed sun shield. Although the dash's top edge remained exposed, its center area was covered with Ensolite, a shock absorbing material and given a leather-like vinyl finish. At each end of the dash were radio speakers providing "Duo Coustic" sound. Operation of the Weather-Eye heating-cooling system was the model of simplicity. Lateral movement of a single control arm regulated the temperature and its rotation controlled the speed of the combination heater-defroster fan.

Along with its new look, the 1952 Nash also had a new "Airflex" front suspension. In truth, it wasn't actually very new at all, being an up-sized version of the 1950 Rambler's. Highlighted by long coil springs mounted directly above the steering knuckles, it eliminated the need for the usual front crossmember. Nash claimed this arrangement also reduced unsprung weight.

There was little change apparent in the Nash's basic road personality, but a pleased Tom McCahill (*Mechanix Illustrated*, September 1952) reported that, "...the new Ambassador has, in my book, the finest shock-proof ride in the world today... . A short time ago I reported that the Buick Roadmaster had the finest rough road ride of any car made in

Nash executive George Mason is said to have favored the enclosed front wheel look that debuted on the company's 1948 "bathtub" models.

A new hood ornament was one of the changes seen for 1953, when this Ambassador Custom four-door sedan was issued.

51

America, but that was before I tested the Nash Ambassador."

The Nash's fresh look and design innovations, plus the attention paid by its designers to small operating details, brought it many other plaudits during 1952.

"Nash, it would appear," wrote *The Motor* (March 19, 1952), "is definitely the most international minded of the American manufacturers. The Farina designs on its production cars are the most dramatic proof of that." Automobile critic Barney Clark told readers of *Auto Sport Review* that most of the independents didn't take advantage of the flexibility their relatively small size made possible. "They seem," he said, "as hidebound as a rhinoceros; (the) one exception to this has been Nash."

Neither the Ambassador or Statesman were being prepped by Nash for entry into the horsepower race, but both cars did receive a few additional horses for 1952. The Statesman's engine, with a longer 4-1/4-inch bore, grew to 195.6 cid and an output of 88 hp at 3800 rpm. Its compression ratio remained at 7.0:1, but a new horizontal carburetor replaced the older downdraft type. The bore of the Ambassador's overhead valve six was increased 1/8 inch to 3.50 inches, which with its 4.38-inch stroke provided 252.6 cid. With overdrive, the Ambassador was a good performer capable of 0-to-60 mph in just under 16 seconds with a top speed of 97 mph. Of all cars driven by Tom McCahill up his special "back-breaker" test hill, only the Ambassador with overdrive and Cadillac 61 with a three-speed manual transmission reached the top in third gear.

Bob Christie's neither-here-not-there 30th place (21st in class) finish in the third annual Mexican Road Race provided Nash's public relations department with little to crow about. A more impressive demonstration of Nash Ambassador stamina was provided by two Swiss drivers who traveled the 8,078 miles from Calcutta to Paris in less than 16 days at an average speed of 42.38 mph.

Nash was the apparent exception to the economic malaise that was rapidly draining the independents of their life's blood. Hudson reported a $10.4 million loss for 1953 and Kaiser-Frazer lost $4.7 million in 1952. Packard was still barely in the black for 1953, with a $5.4 million profit. Studebaker, in spite of its attractive 1953 coupe, reported razor-thin earnings of $2.7 million.

Nash looked much better. A profit of just over $14 million seemed healthy enough. But, a quick comparison with earlier years was sobering. If not for the sales strength of the Kelvinator division and the modest popularity of the Rambler, the situation at Nash would have been considerably bleaker. Nash's sales decline was due to many factors but not among them was the crime of producing an inferior product.

Nash styling remained almost unchanged for 1953, but new Country Club, two-door hardtop models joined the Statesman and Ambassador lines. Both series also received fairly significant horsepower boosts. For an additional $192.50, Ambassadors could be powered by the "LeMans Dual Jetfire" engine, derived from the successful Nash-Healey racing venture at LeMans. With twin downdraft Carter YH carburetors and an 8.0:1 compression aluminum head, the Ambassador's horsepower was boosted from its normal 120 at 3700 rpm to 140 at 4000 rpm.

With a 55-5/8-inch front tread, five inches narrower than the rear tread, the Ambassador's handling remained its weakest performance point. With overdrive and the LeMans engine, its top speed was just over 100 mph.

A pair of Ambassadors with the standard, single-carb, 120 hp engine averaged 22.55 and 21.12 mpg in the 1953 Mobilgas Economy Run. That gave Nash a 1-2 finish in the Upper-Medium Class division. No one would mistake the Statesman for an Oldsmobile 88 by virtue of its acceleration, but 100 hp (resulting from a higher 7.45:1 compression ratio, new twin-barrel carburetor, larger intake manifold passages, higher lift cam and an improved exhaust system with less back pressure), instead of 88, did at least bring its performance to the fringe of respectability.

| Early postwar Nash stats | | | |
Year	Net Sales	Net Earnings	Production (*)
1947	$250,262,581	$18,097,697	113,315
1948	302,860,264	20,132,954	118,621
1949	364,193,360	26,229,930	142,592
1950	427,203,107	28,836,326	168,652
1951	401,148,293	16,220,173	103,585
1952	358,400,502	12,603,701	99,086
1953	478,697,891	14,123,026	93,504

(*) Production totals exclude Rambler

This young woman found the 1951 Nash Ambassador Custom four-door sedan just right for a trip to a quiet lakeside setting. The "tubs" had a 168,652 unit production run for calendar year 1951.

Italian designer Pinin Farina restyled the 1952 Nash line. However, company designers at Nash's Midwest headquarters added their touches, prompting some critics to say the new appearance was "more Wheatena than Farina."

A surprisingly new Nash product for 1950 was the Rambler convertible. It was smaller than most cars of its era, but long on luxury features. Women buyers found it an attractive offering at $1,808.

Rambler luxury in a compact package

By R. Perry Zavitz

Nash was just about the first company to introduce its 1950 cars. New car introduction used to be a very exciting time, but Sept. 23, 1949 was probably not a date that stuck long in the minds of those who saw the 1950 models in the Nash showrooms. In fact, it was actually a disappointment to anyone looking for a change. Good as Nash was, it had nothing really new or different for 1950.

Another disappointment could have been the lack of a convertible in the Nash lineup. One appeared for 1948, but failed to make it for 1949, and now again for 1950. Yet Nash still had a big surprise up its Wisconsin sleeve.

On April 14, 1950, when the 1950 Nashes had been on the market nearly seven months, a totally new model was introduced. Being a new model, some people thought it was the production version of the little experimental NXI that had been widely displayed to test public reaction. The new Nash was small, but not as tiny as the NXI (which finally did reach the showrooms early in 1954 as the Metropolitan).

The surprising new Nash was called the Rambler, which was an appropriate name. Nash was the direct descendant of a company that built bicycles before the turn of the century and cars right after the turn of the century. Both the bikes and cars bore the Rambler name.

The new Rambler was of smaller than normal size. Before World War II, this size of car was often referred to as a "light car," but Nash called it a compact car. That term has remained in use throughout the auto industry for the last three-and-a-half decades.

It is significant that the term describing Rambler has stuck, because the Rambler was an industry leader. It was not the first postwar compact. That honor goes to the 111 inch wheelbase Plymouth DeLuxe that debuted in 1949. But, Rambler popularized the concept of the compact car. Nash, and its successor American Motors, has always produced a small car since the 1950 Rambler was introduced. For that matter, so have the Big Three, since the end of 1959.

But, the Rambler was more than one in a series of several compact cars of the 1950s. It was also a luxury car. This was a calculated risk that Nash made, which could have brought that name to an early end if it had not succeeded. Fortunately it did succeed. It actually put American Motors on the road to survival, something that another new model might not have done.

Rambler rode on a 100 inch wheelbase. Its total length was

53

176 inches. That meant it was just a little more than an inch longer overall than the Jeepster. The next longest 1950 car was the Plymouth, which was nearly a foot and a half longer than the Rambler.

In the weight department, Rambler tipped the scales at 2,515 pounds. It was the heaviest of the smaller cars. Next up the scale of lightweights was the Studebaker Champion, but its convertible was nearly 400 pounds heftier than the Rambler.

The bottom line in getting potential buyers to sign on the dotted line was price. The Rambler's basic list price was $1,808. That placed it right between the Dodge Wayfarer roadster and the Chevrolet convertible. Since those cars were larger than the Rambler, a lot of otherwise interested people thought the Rambler was over-priced. They could not understand a price that high for a car that small. They failed to understand that the Rambler was a luxury car and that luxury need not be tied to bigness.

There is another way to analyze price and size. That is to check the cost of a car per inch of overall length. Rambler cost $10.27 per inch of length. That was nine cents less than the Plymouth convertible.

Price in relation to weight is another good way of determining the degree of luxury in a car. (No matter what its length, all the extra goodies aboard add weight to a car.) Rambler's cost per pound was 71.9 cents. That is very comparable to the 1950 Chrysler Windsor convertible's 70.7 cents per pound.

So what made the Rambler a luxury car? The Rambler came with such items as wheelcovers, courtesy lights, two-tone upholstery, foam cushioned seats front and rear, custom steering wheel, Nash's excellent heater and defroster, radio and antenna, cigar lighter, and direction signals. All these features were included in the $1,808 base price, but each of them cost extra in many other 1950 cars.

Then, too, the Rambler was a convertible; the type of car many people longed for, but usually did not buy because it was more expensive than their budget would permit.

Another unusual feature of the Rambler was the way the top opened and closed. The side window and door frames were a fixed part of the body. A reversible electric motor controlled nylon-coated steel cables on rails atop these side frames. Thus, the top was pulled, in one direction or the other, to open or close the car. It was not only a simple arrangement, but also offered better protection in case of a rollover than conventional convertibles. Nash boasted that their Rambler offered "all the thrill of an open car with the comfort and safety of a sedan."

Just to put this in proper perspective, it must be pointed out that this type of top was also being offered by the Crosley at this time. They introduced it around mid-1946. Looking overseas, as Nash was doing in the early postwar years, we find that Fiat's tiny 500C, introduced in 1948, was a two-seater coupe with a fold-back canvas roof. But, the Rambler was hardly the kind of car the Crosley and Fiat runabouts were.

Rambler's motor developed 82 hp, which was three to five times the power those little puddle-jumpers could muster. It was the former L-head six Nash 600 motor. (For 1950, the 600 was replaced by the Statesman and received a longer stroke version of its previous motor.) Rambler was comparable in power to the Statesman and also the Studebaker Champion. Both Jeepster four- and six-cylinder engines were smaller and less potent than the Rambler.

Rambler ranked as the fifth least powerful convertible of 1950. On the surface, that seems to contradict the luxury character it was supposed to have. (An increase in luxury is usually accompanied by an increase in power.) Its $22.05 cost per pound of weight compares very favorably with the $22.84 price per pound of the Cadillac convertible. The industry average for the convertibles of the year was $22.60.

Horsepower has to be considered along with the car's weight. The Rambler weighed 30.67 pounds per maximum horsepower developed, which was right between the Dodge Wayfarer roadster (30.63) and the Olds 98 convertible (30.74). Convertibles of 1950 averaged 32.98 pounds per horsepower developed.

So much for the mathematical manipulations. With its 82 hp, 172.6 cid (2.8 liter) six-cylinder engine, how did the Rambler perform in practical terms? The Mobilgas Economy Run of 1951 is quite revealing.

In normal driving, Rambler fuel consumption was as high as 20 mpg, using the $100 optional overdrive. But, it achieved an average of 31.053 mpg in the Los Angeles to Grand Canyon jaunt. That was the best of all the 32 participating cars and even better than the four-cylinder Willys and Henry J.

Ton miles per gallon was a figure often boasted when the economy run results were published. It was a calculation of fuel consumption in relation to the car's weight. In the latter years it was not used because it gave an advantage to the heavier cars. Nevertheless, in the 1951 Economy Run, Rambler scored a ton miles per gallon figure of 53.489. That was very close to the 53.785 figure chalked up by the big Hudson Hornet.

Acceleration from 0-to-60 was around 17 seconds, which was about the same as a contemporary Mercury with Merc-0-Matic. Top speed was about 85 mph or comparable to Chevrolet. So, the Rambler could get around without creating a traffic jam behind.

During calendar 1950, there were 14,639 Nash Rambler convertibles built. That catapulted Nash from low production in 1949 and made it the fifth largest convertible manufacturer in 1950. More than seven of every 100 convertibles built was a Rambler, which was more than either Oldsmobile or Plymouth averaged. Obviously, most of the initial resistance to the Rambler's price had disappeared.

Production fell to 9,594 during 1951. Overall convertible production dropped that year, too. As a result, Rambler virtually maintained its share of the total ragtop market. A much greater production drop occurred in 1952 when there were just 3,108 Rambler convertibles made. The total increased somewhat for 1953, to 39,501. Add the four cars made in calendar 1954 and total Rambler convertible production amounts to 30,846. This was not a bad figure for a four-year production run of a specialty model by an independent car maker.

One reason for the continuous drop in Rambler convertible production was that other body types were periodically introduced. On June 23, 1950, Nash introduced the Rambler station wagon. Then, on June 28 of the following year, the Rambler Country Club hardtop went on display.

In mid-1952, a two-door Rambler sedan was introduced to the Canadian market. After its successful launch, it was offered to the American buyer on Nov. 29, 1953, as a 1954 model. Along with it, a four-door sedan and four-door station wagon were also made available on a wheelbase stretched to 108 inches.

The last Rambler convertible, using the permanent side rails, was the 1954 model. By this time the price had risen somewhat, though the overall increase over four years was not out of line. Retailing for $1,980 in 1954, the base price had risen only about two-thirds as fast as the cost of living had climbed since 1950. Meanwhile, Ford (the nation's largest convertible builder at the time) let its prices rise consistent with the rate of inflation.

The Nash Rambler convertible was perhaps the most influential car of 1950s. Its influence is still very strong today.

This 1954 Nash Ambassador Custom Country Club two-door hardtop represents the type of cars the Kenosha, Wis. automaker was producing when the merger with Hudson to form AMC took place on May 1, 1954.

Nash - the end of the line

By Robert C. Ackerson

The 1954 Nashes possessed little that was startling or new. The Ambassador's standard engine's output was boosted to 130 hp and the Statesman's 195.6 cid flathead six reached its ultimate stage of development with twin carburetors and an aluminum cylinder head. This 110 hp, "Dual Powerflyte" engine's 8.5:1 compression ratio equalled that of Buick's V-8 and was exceeded only by the 8.7:1 of Packard's 212 hp straight eight.

But, the sales appeal of Nashes...those cars with vast expanses of interior space, styling that was familiar to the point of monotony, and performance inferior to their peer's...-was limited. Statesman production for the year totalled only 20,225. Output of 21,428 Ambassadors was equally dismal. The introduction of the Metropolitan in March had brought Nash-Kelvinator some welcome favorable press coverage.

Of far greater importance was the event that took place on May 1, 1954. In the largest merger in automobile history, Nash-Kelvinator Corp. and the Hudson Motor Car Co. joined together to form American Motors. Such a union of forces had long been a goal of Nash's George Mason, who as early as 1946, had courted both Hudson and Packard.

In 1948, Packard turned down a formal proposal of merger with Nash. Mason and Packard chairman Alvan Macauley couldn't see eye-to-eye. That same year, Macauley offered George Romney, a former executive with the Automobile Manufacturer's Assoc., a chance to join Packard as executive vice-president at an annual salary of $50,000.

George Mason had succeeded Macauley as AMA president, becoming Romney's boss. He was vacationing in Bermuda when he received word Romney was leaving the AMA for Packard. Mason persuaded Romney not to make his decision final until Nash had an opportunity to counter Packard's bid. If Romney came to Nash-Kelvinator, the counter-offer would be a $30,000 per year and a role as Mason's assistant.

Romney knew of Mason's dreams concerning a merger but, it was Nash's economic strength and innovative product design that lead him to join Nash on April 1, 1948. Romney did not become Mason's heir-apparent overnight, but, by 1950, he was vice-president.

In a report to Mason in 1949, Romney included "some observations on merger" in which he correctly forecast the conditions in which most of the independent automakers would find themselves by 1953. They had to act decisively, said Romney, while they still had financial strength. Once the losses began to mount up and sales had declined, it would be too late.

Hudson allowed this to happen before president A.E. Barit arranged to meet with George Mason on June 16, 1953. It only took a few hours and a handshake to bring Nash and Hudson together. By mid-January 1954, the boards of both companies had given approval to a merger of Hudson into Nash-Kelvinator. The resulting partnership, on May 1, 1954, became American Motors.

The optimism surrounding this event was short lived. Romney had suggested that independent producers could benefit by buying components from each other. American Motors agreed to purchase Packard V-8 engines from Studebaker-Packard in 1955, with the expectation that S-P would reciprocate and favor American Motors with parts contracts. This arrangement soon fizzled in an acrimonious exchange between Romney and S-P president James Nance.

Of great consequence to American Motors' present and future role in the automobile industry was the death of George Mason in October, 1954. *Fortune* magazine had said of Mason, "he appears to be the only executive among the independents who foresaw clearly what could happen to them in a fully competitive market." With his passing, American Motors' future seemed dim. As its losses steadily increased, speculation about its survival seemed to grow in a geometric proportion. On Oct. 12, 1954 George Romney succeeded Mason as chairman, president and general manager of American Motors.

Sensing that Judgement Day for American Motors was at hand, Romney first undertook a major effort to reduce expenses via a "Let's Be Competitive" program within the company. Tom Mahoney, George Romney's biographer, placed the savings resulting from a host of cost-cutting moves, such as executive salary reductions (including Romney's), leaving offices unpainted, and cleaning offices every-other-day, at millions of dollars. But, these were strictly finger-in-the-dike efforts to buy time.

Only new products, geared to reach a market overlooked by the Big Three, would reverse the tide for American Motors. Such a reorientation of its resources forced AMC's leaders to assess the merits of continuing production of the large Nashes and their post-1954 "kissing cousins," the Hudson "Hashes." The issue was decided at an emotion-packed directors meeting in 1956; both cars were to be dropped after the 1957 model run and the company was to tie its future to the Rambler.

When the 1955 Nashes were redesigned, their continued existence was not yet in question. A strong effort was made to improve their saleability with the grafting of a wraparound windshield onto the four-year old body shell, the use of re-tooled cowls, new front and rear quarter panels, doors and roof dies. They also made available a 208 hp, 320 cid Packard V-8 as an alternative to the Ambassador ohv six.

If the final product still fell short of being an artistic triumph, it has to be recognized as the product of a near-Herculean effort, performed during trying times and in far less than perfect conditions. The contract with Packard for V-8 engines wasn't signed until July 1 and the decision to adopt a wrap-around windshield came very late in the Nash's styling development. In point of time, it was one of George Mason's last official acts as leader of American Motors.

Nash claimed that by moving the headlights on its 1955 model closer together, in a fashion seen earlier on the 1953 Nash-Healey, more light was directed onto the road surface. It said this made for safer driving, especially in foggy or rainy conditions. To prevent any potential miscalculation of the Nash's width by drivers approaching at night, its parking lights were positioned in the outer edges of the front fenders. They operated whenever the headlights were in use.

With the Packard V-8 and Twin-Ultramatic transmission, the Ambassador V-8's performance was on a par with its competitors. It was good for 0-to-60 mph in 13.7 seconds, a 102 mph top speed and an 18.6 second time in the quarter-mile. But, a modern V-8 added to Nash's traditional six-cylinder engine and its interior comforts weren't really enough to assure its survival. Production of Ambassadors and Statesmans did rise modestly to 57,619, but Nash's appeal was limited mostly to its small band of marque loyalists. There was no improvement in its market strength the following year.

What Nash described as "Speed-line styling" included higher rear fenders with reshaped "clinched fist" taillights and "lightening bolt" side trim. None of these changes could disguise what was now a very outdated design.

American Motors began the year offering the Ambassador with a 220 hp version of the Packard 352 cid V-8. This engine had been rated at 275 hp by Packard, in 1955, when it powered Clipper Custom models. George Romney would have been quite happy not to have purchased any Packard engines in 1956. Shortly after the Nash-Hudson merger, Romney had been hopeful that a high rate of component reciprocity could be reached with Studebaker-Packard. However, the understanding floundered almost immediately as American Motors' willingness to purchase Packard parts was not reciprocated by Studebaker-Packard's use of AMC parts.

After a terse exchange of telegrams and letters between Romney and Packard's James Nance, American Motors began development of its own V-8 engine during 1954. Two years later, in April 1956, it appeared with a 250 cid displacement, single two-barrel carburetor 8.0:1 compression ratio and 190 hp. It became the power plant for the Ambassador Special, of which 4,145 were built.

Although its manufacturer was now known as American Motors Corp. (AMC) the 1955 Nash Ambassador Custom Country Club two-door hardtop reflects only minor detail changes from the previous edition. How many can you spot?

A continental rear tire and bumper extension was standard equipment on Custom level models, such as this 1955 four-door sedan. This feature could also be obtained for less expensive models at extra cost.

Car City Cousins - Hudson

The strong, powerful Hudson "Super Six" chassis and running gear was popular with aftermarket firms that made early campers. This Hudson based camping car was constructed in Tomahawk, Wisconsin in 1927.

Hudson is handy hobby car

"For everything, there is a season; and a time for every purpose, under heaven."

By John Gunnell

The wonderful, warm, golden days of summer are a precious commodity to the hearty people of northeastern Wisconsin. This is a season to be soaked up and savored; a time for pressing flowers and packing tasty fruits into Mason jars. The cold and snow will come, as they always do, before the cranberry crop reaches the Thanksgiving table.

A ride from Iola to Green Bay, on a hot afternoon, reveals some of the ways Badger State residents increase their enjoyment of the shortest, but sweetest season. Passing through Symco, you see the home of the Union Threshermen and imagine the sights and sounds of their August Thresheree. In Bear Creek you almost taste, and maybe smell, the kraut boiling for the summertime Sauerkraut Festival.

But, what you see more than signs of fairs and festivals are the machines of summer. Machines are a big part of life in Wisconsin, and there are machines for every season, too: snowmobiles and sleds and plows for winter. And, there are different machines for summer: warm weather tractors and horse trailers, but most importantly, the campers and motorhomes.

Camping is what summer is all about in Wisconsin. Camping is not a new thing here. Wisconsinites have been enjoying the pleasures of traveling and camping for many, many years. In fact, the purpose of our trip to Green Bay was to visit Clete Van Erem and photograph a unique car he bought last year...a Hudson camper built in 1927.

Van Erem's camper left the Hudson factory as a 1927 Hudson Super Six Victoria Brougham that was delivered to the original customer in partially complete form. It wasn't a chassis-only job. It came with front and rear fenders, radiator grille, hood, cowl and the "cab" section of the passenger compartment. This vehicle was then sent to the Tomahawk Boat Works, in Tomahawk, Wisconsin, where the custom made camping body was constructed over wood framing.

The camper is in excellent, unrestored condition since it was purchased in 1978 with only 42,700 actual miles showing on the odometer. Only a small amount of detailing was required to put the Hudson into its current condition. Van Erem added new carpeting throughout the vehicle and made new curtains to replace the worn, pull-down window shades.

They were still present, but in poor shape. New canvas was installed on the two bunks inside, which were part of the original conversion. The body was repainted in a bright blue color and trimmed in white. The horsehair insulated ice chest, made back in 1927, was used as a model to construct a replica, which was lined with styrofoam. Almost everything else about the vehicle is exactly the same as the day it left the Tomahawk Boat Works.

The engine was in very good condition and required little more than a good cleaning. To be on the safe side, Van Erem dropped the oil pan to check the condition of the rods and bearings and, finding everything in good order, replaced the pan using new gaskets. The only missing item, a screw-on Hudson hubcap, was recently found at an old car swap meet.

Special features added to the Hudson, in addition to the wooden camping body, included the two iron-frame cots, overhead cabinets in the rear, window screens, the ice chest and a fold-down table. The right rear wheel hub is fitted with pegs that bolt to a generator. The system can be used to provide electricity at a camping site. A dome light is provided for interior illumination, but it is not working now.

The 1927 Super Six powerplant is an F-head design with 3-1/2 X 5 in. bore and stroke. The engine has no provisions for rocker arm lubrication, making it necessary to manually inject oil into flip-up cups on the rocker cover every 250 miles. An oil can is carried in a special holder, mounted on the firewall for this purpose. Van Erem recently completed a 1,300 mile trip in the Hudson without incident. He oiled the rocker arms each time he stopped for gas and got about 12 mpg on the trip. The Hudson drove strongly at highway speeds and used no oil, other than that consumed by normal lubrication of the rockers.

The wooden body is built over the chassis and fender assembly of a normal Hudson Victoria Brougham and extends about three or four feet longer in back. By lifting the carpets below the rear of the bunk, you can see the Brougham's gas tank in its original position, complete with what should have been mountings for a suitcase-style trunk. The original gas filler pipe can also be seen, but a second pipe that passes to the outside of the wooden body is now used to fill the tank.

The body itself is built from plywood bolted and nailed to a sturdy wood frame. At the front, the width of the body is about the same as the original sedan body. It extends outward, behind the doors, about 8 to 10 inches. In the rear, it curves up and over the stock rear fenders, but has distinctively-shaped wheelwells. Two screened windows are placed high on either side, towards the rear. On the right-hand side, the window is centered in a door panel also made of plywood. Two smaller windows are seen in the back. The roof curves both lengthwise and widthwise, to conform to the curvature of the bow-shaped top frame. The upholstery in the driver's compartment is totally original. The taillamp assembly mounts in the center of the rear panel.

Van Erem bought the car from the family that originally commissioned the camper conversion by Tomahawk, but records of the cost of construction seem to be lost forever to time. However, this beautiful Hudson-based camping car is rolling proof that people in northeast Wisconsin have been using campers to get the most out of summer for over 50 years. As such, it stands as a unique part of the state's automotive history.

Dover commercial vehicles were a product of the Hudson Motor Car Co. They were introduced to the public during the summer of 1929, but the marque lasted for just two years.

Dover commercial trucks

By Sheldon Rody

By 1929, Hudson Motor Car Co. had an enviable reputation as a builder of six-cylinder automobiles. The key to their success was the Hudson Super Six and their fast selling lower-priced line, the Essex Super Six.

The Essex was introduced as a four-cylinder car in 1919. By 1924, it was a six-cylinder car. The Essex is credited with being the first auto to offer a closed car in the low-priced range.

Hudson management wanted to cash in on the commercial vehicle market. The first experiment was a 1922 Essex pickup truck. Only one was built and it was used at the factory as a company vehicle.

Business times were very good in 1928 and 1929 when the Dover...a commercial vehicle based on the Super Six was planned. The Dover was announced to the world in July, 1929. Hudson was proud of the Dover and advertised it as "built by the Hudson Motor Car Company." This is directly the opposite approach used when the Essex was introduced in 1919.

To launch the British-sounding Essex, Hudson set up a separate company to build and sell the car. Hudson's advertising did not connect the two marques directly. It is also interesting to note an English name was used on the truck, just as the name Essex was of English origin.

The Dover came to the marketplace with the following models, and prices: Screen Side Express - $885, Canopy Express - $870, Open Flareboard Express - $835, Panel Delivery - $895 and cab and chassis - $595.

The Screen Side, Canopy, and Flareboard Express models all shared the same basic pick-up box. The Dover brochure points out that the Flareboard could later be converted to either of the other Express models.

The bodies were built for Hudson, in Evansville, Indiana, by Hercules Products Co. The Dover Catalog stated the paint used was a dark blue pyroxlin finish identical to that used for Hudson and Essex Cars.

Equipment that was standard on the Dover included a special delivery slip holder, a light in the body for reading delivery slips at night, and an overhead toolbox with a spring-hinged cover. Also standard were adjustable leather-covered seats.

Mechanically, the engine was adapted from Essex passenger cars and the truck chassis was specially designed for commercial use. The Dover was rated to carry a 3/4-ton load.

The U.S. Post Office was the buyer of a large fleet of Dovers with special mail truck bodies. These bodies featured sliding side doors. The durability of these vehicles was proven by the U.S. Post Office. The agency reported that some of the Dovers were still in use in the late 1940s and early 1950s.

The Dover truck was only built in 1929 and 1930. The model's production total and the date of final production are unclear. Two different Hudson parts books indicate there may have been Dovers produced in 1931. Other sources suggest 1930 was the last year.

The name Dover may have disappeared, but the Hudson commercial cars did not. Commercial cars were continued under Terraplane and Hudson names through 1947.

No more than a handful of the Dover commercial vehicles exist today. One that was reportedly being restored in Michigan was a former mail truck.

Hudson president Roy D. Chapin was another fan of fast driving who enjoyed slipping behind the controls of the 1933 Essex Terraplane Eight convertible coupe.

"Hot diggety dog...that's Terraplaning!"

By Ned Comstock

I don't remember the "hot diggety" part of the old radio commercial I ran across the other day in *Automobile Quarterly's* (Summer, 1971) Hudson piece, but the rest of it comes through so clear, just as if it were 1933 again. "In the air it's aeroplaning, on the water it's hydroplaning, on the ground, hot diggety dog, that's *Terraplaning!*" the announcer would say.

But, more than the radio program, the memory of the car itself came flooding back, as if it were still parked outside. I don't think anybody who owned a Terraplane Eight convertible coupe in 1933 ever forgot it.

In those drab depression days, most cars came and went in muted colors, but Terraplane with its bright paintwork stood out like a pretty girl.

When most cars were still unconscious copies of covered wagons, the 1933 Terraplane Eight convertible was styled like a down-sized Deusenberg.

While contemporary automobiles were lugging around 40 pounds for every horsepower...43 that year for Cadillac...the 1933 Terraplane Eight convertible coupe ran with only 28. And it ran like a deer.

1933 was the year FDR inspired Americans with bright new hope and Terraplane Eight, with its fresh good looks and eager power, translated the spirit of the times into flashing action. It seemed like a part of the future.

Battered by four years of the hardest hard times, Americans considered any car a privilege in 1933. Driving a new car was a joy. But, traveling in a 1933 Terraplane Eight convertible coupe with the top down, well, that was like flying...and flying was for heroes, then.

Breaking performance patterns had long been stock in trade at Hudson. What Ford did for the poor man and Cadillac for the rich, Hudson did for the middle class. By making better use of technology, instead of simply increasing size and weight, these leaders improved the breed. A good case could be made the Hudson's contribution at the middle level was not the least.

In 1916, a year when the great P-cars (Packard, Peerless and Pierce) turned to oversize, Hudson took a road less traveled and that made all the difference. By refining the old (1914) model 6-40 design, Hudson turned out the famous Super Six. It jumped horsepower from 40 to 76, a whopping 90 percent, without increasing engine size. At the same time, they reduced the price.

This magic was still working in the brilliant 1927 F-head, maybe the high point of the Hudson years. Here was a $1,095 automobile with performance equal to many great and famous makes selling at three times as much.

Again in 1933, in the most difficult times, this old technique was put to test. The Terraplane Eight came out in 1933 at

$725. Combining the Hudson Eight engine (bored out to 244 inches and hyped up to 94 horses) with the short 113-inch Essex chassis, the net result weighed in at 2,640 pounds.

Nothing I ever saw on the road that year could out-perform 1933 Terraplane Eight convertible coupe...not even Ford. It was not until the next year that Ford stepped up the V-8 to equal, or nearly equal, the Terraplane Eight. And Ford was the only one!

Hudson held dozens of stock car records in 1933. I don't know the Terraplane Eight's official Flying Mile. I do know I took mine to a special section of new concrete highway running along the Utica, N.Y. airport. There were no intersections and I could watch for state police. Long ago, on that summer day, I wiped the clock on that Terraplane's Eight. The speed needle bounced on the peg full circle and held steady at 100 mph.

Exactly like Alfred P. Sloan, I do not count "fancy-price" cars. There were too few of them. I'm talking about cars you and I could own...the kind most anybody could afford. Of these, I believe the 1933 Terraplane Eight was the fastest stock car in America that year.

It had another performance characteristic not seen before in the way that it was geared. Always, before even the Model A (notably fast in second gear), you down-shifted for heavy going. When the sand was too deep or the road too steep to make it on high, you slogged along in a smaller gear. That is what it was for.

Not so in Terraplane Eight. Second gear catapulted you ahead like a thrown stone. Second gear would press your back against your seat...make, you crouch down and hold on tight. Second gear was for sport. Others, more stately, might make progress and build speed slowly. In Terraplane, you rocketed out in front of them.

To their everlasting credit, Hudson put up this performance prodigy in a beautiful package. Like all good design, it was a model of simplicity. Mine was set off with lovely coffee-with-cream paintwork, emphasized by bright tomato red wire wheels. I have seen silver-and-black, too. It's just as effective.

But, the real beauty was built-in. Because people are basically the same size usually, small car bodies bulk bigger and spoil proportions. Not so of the Terraplane Eight. The passenger compartment was low and small. You crouched in it, as you would in an old-time airplane cockpit. The hood was high and long, as if it accomodated a long powerful engine, which in fact it did. And the afterdeck was understated. In a picture without scale or reference, you could not tell the size of the car. It might well have been a new English super-sport of 143 inches, or 113.

It was a clean design. The rear fenders swooped gracefully and naturally alongside the body, with modest skirts new that year. The single spare wheel was slung at the rear in a rakish angle, adding to the impression of low length. The top was plain, without any outside hardware. You could tell it was a Terraplane Eight because the hood was ventilated with louver doors. The Terraplane Six had simple slits.

When you sat on the flat, low seat, you looked straight ahead through a windshield not much taller than the span of your hand. Instead of the usual old-style dashboard about halfway to the floor, where you might have put your feet in a horse drawn buggy, instruments were mounted on a panel just below your forward line of vision. There was a speedometer (ground speed indicator?) as big as a jack-o-lantern...the first warning lights on any car I ever saw...and an oil level gauge I haven't seen since.

In 1933, the Terraplane Eight convertible's style matched performance. You got the feeling of flying that you may never get on the ground again.

In 1932, the Essex-Terraplane proved to be quite a performance car when Chet Miller took the Penrose Trophy at the Pikes Peak Hillclimb in a new Terraplane convertible. This established the series' high-speed reputation.

Terraplane in review

By Alex Burr

Unlike its predecessor Essex, which was originally incorporated as a separate division of the Hudson Motor Car Co. in September 1917, the Terraplane was introduced as a model within the Hudson corporate structure. The reasons for dropping Essex are shrouded in the mists of time, as are the reasons for dropping Terraplane in 1938. There are unconfirmed rumors of a power struggle between Hudson and Essex management during the late 1920s, but there were undoubtedly other reasons as well.

Numerous interesting comparisons between Essex and Terraplane are possible. Essex was introduced during a business recession following World War I. Terraplane was introduced during a depression. Both cars were instrumental in bringing Hudson into a strong sales position during the time they were in production. And both were utilized to bring advances to the automotive industry that are still in use today.

Exactly where the name Terraplane originated is not clear. One Hudson historian says the name came from a Harry Miller designed sports car featuring Kirchoff designed bodies that was promoted under the name Terraplane. At the time, Frank Spring, later working in Hudson's design department, was Kirchoff's neighbor and supposedly remembered "Terraplane" when Hudson was looking for a name for the new car in 1932. The name also happened to fit the increasing interest in aviation during the early 1930s.

Lindbergh's New York-to-Paris flight was still fresh in many minds and the great aviation routes were beginning to appear. In keeping with this aeronautical theme, Amelia Earhart was invited to the factory to introduce the new car. And, the first Essex-Terraplane off the line was presented to Orville Wright.

The new Essex-Terraplane Model K, designed by Stuart G. Baits and the Hudson engineering staff, featured several innovations not generally found on cars of the time. Among these were an adjustable steering column and generator and oil pressure warning lights in place of gauges. Weight saving was an important criteria in the new car.

A fairly rigid unit was produced by extending the bottom body pan over nearly the whole frame length and joining the frame and body structure together. This procedure had first been used on the 1930 Hudson and Essex cars under the name "unit engineering." A new type wire wheel, also designed by Hudson engineers, featured a large diameter hub allowing brake drums to be placed within the hub. This produced shorter axle shafts giving a somewhat narrower tread design.

The same motor used in Essex models, featuring a bore and stroke of 2-15/16 inches by 4-3/4 inches (193 cid), was used in the new car. Several changes had been made. Through use of a redesigned combustion chamber and a composite aluminum-cast-iron cylinder head, a compression ratio of 5.8:1 was achieved. Cooling was improved by using a water pump in place of the previously used thermo-syphon system. More weight was saved by skeletonizing the bellhousing, using a light stamping for the clutch cover, and a transmission of a new design featuring compactness and light weight.

In keeping weight to a minimum, Hudson was taking aim at the new Ford V-8 in the performance arena. Ford's new motor, at 221 cid, was rated at 65 hp. The Essex-Terraplane was rated at 70 hp. By keeping weight down to nearly 100 pounds below Ford in comparable body styles, Hudson had a car with a power-to-weight ratio definitely in favor of the Essex-Terraplane.

To say the new car could scat was an understatement. Essex-Terraplane owners used to say, with tongue in cheek, that the new Ford V-8 had to be fast to stay ahead of their car. Its performance capabilities were quickly proven at the 1932 Pikes Peak Hillclimb, when Chet Miller brought home the Penrose trophy with a new record time of just over 21 minutes.

Hudson really took a gamble with their new offering. During a business recession or depression, the intelligent thing to do is to draw back a bit. Not Hudson. Not only did they introduce a new car in the middle of a depression...the new line offered 12 different body configurations. To their credit, they did aim at the low-price market.

Prices started at $425, f.o.b. Detroit, for a roadster...$20 below Chevrolet and $35 below Ford...and went on up to $610 for a convertible coupe. In between, the prospective customer could opt for a business coupe at $470, a coach at $475, phaeton for $495, special business coupe for $510, as well as the four-passenger coupe special coach $515, sport roadster $525, sedan $550, special coupe (also at $550) or a special sedan at $590.

Production, shipment and registration figures for 1932 provide an indication of how the company was doing during the depression years. The peak year for shipments was 1928, with 229,887 Essex cars being shipped. Interestingly enough, Essex was in production for 15 years, or one-third of the life of the Hudson company, and produced one-third of total Hudson Motor Car Co. output. According to *Moody's Analysis of Investments*, combined Essex and Essex-Terraplane production in 1932 totaled 34,389 units. Hudson shipping records show that 17,425 Essex cars and 16,581 Essex-Terraplanes left the factory.

However, it appears that some people could afford a new car. The *Automotive News 1968 Almanac* shows combined registrations of 28,778 cars. Net sales totaled $25,861,671 with a net loss of $8,459,982. The company managed to come out in production behind Chevrolet, Ford, Willys-Overland and Whippet.

If Hudson had appeared to be gambling in 1932, they seemed to be headed for the casino with a vengeance in 1933. The Hudson line itself included both sixes and eights, whereas only an eight was available in 1930-1932. The Essex-Terraplane line also featured both sixes and eights in no less than six different model lines.

The 193 cid six was to be found in the Model K and Model K commercial line on a 106-inch wheelbase and in the Model KU line on a 113-inch wheelbase brought over from the 1929 through 1932 Essex cars. The Model KT, offered on the same 113-inch wheelbase, featured a straight eight motor. The Model K was available in Special and Standard lines, the Model KU in Special and DeLuxe, and the Model KT in Standard and DeLuxe.

Prices in the Model K Standard lineup started at $425 for a roadster, while in the Special series the price for a sport roadster was $525. Coupes could be had at $530 or $570, coaches for $505 or $545, and sedans topped the listings at either $535 or $595. In addition, the Standard series also offered a phaeton at $515 and a business coupe at $484, while a convertible coupe at $595 was available in the Special series.

Slightly higher in the price range, for those still gainfully employed, was the Model KU. Prices started here at $505 for a Special business coupe or $585 for the DeLuxe. On up the line were coaches at $525 or $605, coupes either $555 or $635, and sedans at $575 or $655.

At the top of the Essex-Terraplane line, for those who still had their money stuffed in the family mattress, was the Model KT eight. The only year for a production eight in the Terraplane line was 1933. Here, roadsters were offered at $565 for the Standard or $635 for the DeLuxe, followed by either business coupes or coaches at $615 or $685 depending on the model. It was $625 or $695 for a four-passenger roadster, $655 or $725 for a coupe, $675 or $745 for a sedan, and $695 or $765 for a convertible coupe.

The 193 cid six banger used in 1933 had started life in the 1924 Essex. Coupled with a 5.4:1 rear end ratio it was, at the time, a near disaster. Since the motor was turning too high, it had a disturbing tendency to come unglued at speeds above 40 mph. Hudson dealers did spend some amount of time replacing motors. As time went by, this engine starting out at 129 cid, was improved and turned out to be a pretty good piece of Hudson engineering. In any event, it stuck around until 1947, giving it almost as much life as Chevy's famous stovebolt six.

The eight was a different matter. Instead of using the Hudson 254 cid eight already in production, the Essex-Terraplane eight was the 1932-1933 six with two cylinders added. This gave 244 cid. Several components such as connecting rods, pistons and the oil pump were readily interchangeable between either the six or Hudson eight.

In 1931, Briggs and Murray replaced Biddle & Smart, of

Double-deck hood louvers identified the 1933 Terraplane Six. The five-window coupe went for $555 in standard trim and $635 as a deluxe. This factory photograph calls it a Terraplane, although it was still an Essex-Terraplane.

More streamlined styling was a characteristic of the 1934 Terraplane. The Essex part of the model's name was dropped officially this season. Also gone was the eight-cylinder car-line.

Amesbury, Massachusetts, as the suppliers of roadster, sedan, and phaeton bodies. Briggs supplied Hudson and Essex with phaeton bodies in 1931 and added convertible coupe bodies in 1932, replacing Murray. Murray had supplied the roadster and certain limited production, seven-passenger and club sedan bodies used in 1931-1932. Briggs remained with Hudson until 1935.

If anything, the 1933 Essex-Terraplane proved even more formidable in racing events than the 1932 model. Chet Miller had taken the 1932 Penrose trophy at Pikes Peak. At the 1933 Pikes Peak event, his brother Al kept it in the family with an Essex-Terraplane eight, tearing up the record book in the process. Not only that, but the little cars took second, third and fourth places, as if to prove that winning first wasn't a fluke.

By the end of 1933, more than 50 hillclimbing records had fallen, including a trek up Mt. Washington, in New Hampshire, with the transmission sealed in third gear. At Daytona Beach, several Class C records had been set under AAA supervision. It should be noted that Class C included production cars of 185-300 cid, or nearly every make in the country except monsters like Packard, Cadillac and Marmon.

An interesting note closes out the 1933 model year. A few Essex-Terraplanes ended up registered as Essexes for one reason or other. Occasionally one finds a 1933 Essex for sale in antique car ads. Undoubtedly, this is one of the misregistered cars. Or is it? Hudson records show one Essex shipped in 1933.

Despite the incredible array of offerings for 1933, and the fact that the car was a performer with a reasonable price tag, Hudson slipped into seventh place behind Chevrolet, Ford, Plymouth, Dodge, Pontiac and Studebaker. Total production was 40,982 (including 38,494 Essex-Terraplanes). Hudson shipment records show 38,150 cars were sent out to dealers, while the 1968 *Automotive News Almanac* says 35,831 were registered. The financial picture was a little better, showing a loss of $4,409,930 on net sales of $23,521,458.

Along with just about everybody else, the Hudson Motor Car Co. floated into 1934 on a sea of red ink. There were more than a few firms, however, that didn't stay alive long enough to make it into the new year. Hudson was down, but definitely still in the ballgame. Since the deluge of models in 1933 hadn't produced the desired effect, there was a reduction to only two 1933 lines initially.

The name Essex-Terraplane was shortened to Terraplane and the eight was dropped. Introduced on Jan. 1, 1934, were the Terraplane Model K Special on a 112-inch wheelbase and the Terraplane Model KU on a 116-inch wheelbase. Apparently, there was a feeling that a third, lower-priced model was needed and the Terraplane Model KS Challenger was introduced on May 16, 1934.

The 1934 Hudsons and Terraplanes were characterized by all-new styling. The square boxy lines of the '20s and early '30s were giving way to smooth lines that flowed from bumper to bumper producing an outstanding and pleasing effect. Hudson ads proclaimed "dynamic streamlining" and sales brochures stressed the new "Hudson principal of Unit-Engineering."

Carryover features included the adjustable steering column, adjustable seats, warning lights and other features. Under the hood the venerable small-block six was stretched to its final configuration of three inches by five inches, giving a piston displacement of 212 cid.

Outside a new feature was introduced on "compartment models" - a trunk compartment. As the sales brochures told it, "a capacious checkroom for baggage actually built in." Other Terraplane models featured only a luggage vestibule for the spare tire, but sidemount spare tires were available, so the space could be used for packages or luggage. In either case luggage, packages, and other items were finally to be safe "from dirt, moisture, and possible theft."

Higher prices for the 1934 models indicated an improving economy. The Model K Special Six offered a two-passenger coupe for $600, a two/four-passenger coupe for $635, a convertible coupe for $695, the compartment victoria at $655, a coach for $615, a $675 sedan, and the compartment sedan at $715. Not shown in the price lists were a phaeton and a roadster, although these styles do appear in both the 1930-1936 parts books and 1934 sales brochures. These two, along with the convertible coupe, are listed as having Briggs bodies.

As in previous years, a commercial line was part of the lower-priced lineup. The Model K commercial line carried a utility coach, sedan delivery, cab pickup and cab chassis - all with Hudson-built bodies. There was also a station wagon produced by J.T. Cantrell & Co. of Long Island, New York, but this appears to have been a private operation, with no direct factory connection to Hudson.

Stepping up a bit, one would find the Terraplane Major Model KU. As with all models in both Hudson and Terraplane lines, the styling was the same. About the only way to differentiate models was by serial numbers on the data plates. This was true for 1932-1938 models, but more so on the 1934-1938 cars. Some minor variations show differences in that some models have a temperature gauge and outside mounted horns, while lower-priced models do not.

Model KU prices began at $665 for the two-passenger coupe. It was $710 for the two/four-passenger coupe, $750 for a convertible coupe, $720 for the compartment victoria, $680 for a coach, $740 for a sedan, and $780 for the compartment sedan.

Apparently, sales were good enough to lead Hudson to believe that a somewhat lower-priced model would be a good idea. Or maybe sales were slipping. Whatever the reason, the Model KS Challenger line was introduced in May featuring only four offerings. The two-passenger coupe went for $565, the coach $575, the two/four-passenger coupe $610 and a sedan $635.

While changes in the 1934 models were to be found mainly in styling, there were some mechanical changes. Front axle design was changed to provide smoother operation and a smoother ride. Known as Axle-Flex, it consisted of two end pieces to which the hubs were attached pivoting on a two-piece center section. Springs were lengthened to help smooth out the ride and mechanical brakes were retained. (These changes were made while Hudson engineers were attempting to overcome management resistance to hydraulics.)

That the new styling was acceptable to the buying public was proven by the fact that of 58,536, cars produced 56,804 were shipped and 40,510 were registered. As its predecessor Essex did before, Terraplane was proving to be Hudson salvation. The company managed to climb back into fifth place in production with a total Hudson-Terraplane production of 85,835 cars.

Although the financial picture remained bleak, things improved slightly with a net loss of only $3,239,202. Net sales, however, increased dramatically to $52,567,561. It is a measure of the company's integrity that Hudson president Roy D. Chapin's persistent efforts were able to secure outside financing between 1930 and 1934 and would continue to do so until 1936.

"You feel safer...You are safer in America's Only Body All of Steel" and "Performance You Want ... Ruggedness and Economy You Demand!" proclaimed ads promoting the 1935 Terraplane line. "How about bodies?" another ad asked. It then proceeded to inform reader's that the new line featured bodies that were bigger inside and out. In an attempt to get the jump on competition, Hudson introduced the new models on Nov. 21, 1934, three weeks before Chevrolet and Ford and six weeks before makers such as Dodge, Olds and Plymouth.

Placed before the buying public were the Model G Special, Series 51, with its attendant commercial lineup, and the Model GU DeLuxe, Series 52. Model identification was simplified to a number system, with the first number (in this case 5) being the last digit of the year of production, and the second number denoting the series. For 1935, serial-numbers on the data plates therefore began at either 51-101 or 52-101 for Terraplanes. To denote Canadian built cars the letter C preceded the numbers.

Generally speaking, most features found on the 1934 cars remained on 1935 models. There was one new feature that made its appearance. Called "Electric-Hand" this was a vacuum-operated semi-automatic gearshift of the pre-selector type. It was controlled by a device mounted on the steering column with a gear shift pattern built in. One merely selected the required gear range, and by depressing the clutch pedal, gear change was accomplished automatically.

In the event of failure of the "Electric-Hand," Hudson was considerate enough to include a gearshift lever placed in the right side kick panel. This could be installed in a socket on the floor, in the normal floor shift position, allowing normal gearshifting. Horsepower was increased to 88 and 100 hp was available with an optional high compression cylinder head. Easier riding qualities were provided through longer springs and improved oil-cushioned shock absorbers.

Both 1935 lines featured a coach, touring brougham, sedan, suburban sedan, two-passenger coupe and two/four-passenger coupe. In addition, the Series 52 DeLuxe line offered a convertible coupe with a Briggs built body. In the Series 51 commercial lineup, a utility coach and cab chassis with Hudson built bodies, a panel delivery with a York-Hoover body, and a cab pickup with a Detweiler body were available.

Prices overall were reduced some $15 to $40 from 1934. This put the Terraplane Series 51 prices at $585 for a two-passenger coupe, $625 for either the two/four-passenger coupe or touring brougham, $595 for a coach, $655 for the sedan, and $685 for the suburban sedan. In the Series 52 line, the two-passenger coupe was listed at $635, coach $645, two/four-passenger coupe and Touring Brougham $675, sedan $705, convertible $725 and Suburban Sedan $735.

Despite price reductions and a fairly intensive ad campaign, Hudson and Terraplane slipped into eighth place in production behind higher-priced cars such as Oldsmobile, Buick, and Pontiac. This didn't mean Hudson was really in trouble; rather, it showed that the nation's economy was becoming healthier and people could afford more expensive, cars. Of course, General Motors also had more extensive production facilities and could, thus, manufacture more cars in a given time.

The state of the economy did show in Hudson's overall financial picture. Net sales came to $63,077,415 and, for the first time in several years, there were net earnings. They amounted to $584,749. The cars did sell, as was shown by Terraplane production of 71,604 cars of which 70,323 were shipped and 53,838 showed up as registered. In this production total, there was at least one Terraplane eight special. It was ordered for a California police department, according to a former Hudson-Terraplane dealer in that state.

The Hudson Motor Car Co. entered 1936 on a note of tragedy, with the death of Roy D. Chapin on Feb. 16, 1936. Chapin, one of the original founders of the company, started out with Oldsmobile in 1901. In 1906 he was with E.R. Thomas, which led to the founding of Hudson in 1909. In 1916, he headed the Highways Transport Committee of the Council of National Defense.

He was instrumental, in 1917, in the formation of Essex Motors, a subsidiary of Hudson during the 1920s. Between 1926 and 1928, he was president of the prestigious Automobile Manufacturer's Association and was President Hoover's Secretary of Commerce in 1932-1933. In 1933, Chapin returned to Detroit and took over the reigns of a faltering Hudson Motor Car Co. Upon his death, he was succeeded by A.E. Barit.

Barit had begun his career with Hudson, in 1910, as a stenographer in the purchasing department. Over the years he had worked his way up as treasurer, vice-president of purchasing and general manager. He was destined to become Hudson's last president, resigning in 1956 after an acrimonious debate over dropping the Hudson name (which American Motors decided to do, since no one named Hudson had been associated with the company for years). American Motors Corp. was the result of the merger between Hudson and Nash in 1954.

The 1936 Hudsons and Terraplanes finally brought the company into the world of streamlined and semi-streamlined cars. The 1934-1935 cars had, in effect, been a transition between the boxy cars of the '20s and early '30s. The 1936 lines, to remain generally the same through 1939, showed the 1934 Chrysler Airflow influence. The effect was more rounded and somewhat bulbous, but quite pleasing overall. Again, as in 1935, only two model lines were offered in the Terraplane line the DeLuxe Model 61 and Custom Model 62.

Wheelbase length was increased to 115 inches and the same 3 x 5 inch bore and stroke six-banger was used, with some improvements over 1934-1935. Hydraulic brakes were the big internal news for 1936. In typical Hudson fashion, they were different. Since Hudson management still wasn't convinced of the safety of hydraulics, the boys in engineering devised a back-up system. Using a bell-crank arrangement

John Bischoff, of Rocky Hill, Connecticut, was the owner of this handsome 1936 Terraplane rumbleseat coupe. Hudson founder Roy D. Chapin passed away during February of 1936.

Even greater emphasis was placed on streamlining in the appearance of the 1936 models. This two-door sedan featured "suicide" type doors that opened from the front of the body.

hooked into the brake pedal, they connected that to the rear brakes through the emergency brake cables. If the hydraulics failed, the driver still had braking through the rear shoes. The only weak point in the system is cable stretch due to age. In any event, it was good enough to remain in use through the rest of Hudson's life.

Body styles remained about the same as in 1935, with some expansion in the Model 61 DeLuxe lineup. Here was offered a two-passenger coupe for $595, a two/four-passenger coupe for $640, a convertible coupe for $715, a brougham for $615, the touring brougham for $635, a sedan for $670, a touring sedan for $690 and a-new-for-1936 station wagon at $750. Model 61 commercial offerings were a cab chassis, cab pickup, sedan delivery and utility coach.

The Custom Model 62 line offered the same body styles as the Model 61. Major differences between the two were twin outside horns and a temperature gauge on the Model 62, while the Model 61's horn was mounted under the hood and it had no temperature gauge. Prices began at $650 for the two-passenger coupe. It was $690 for the two/four-passenger coupe, $760 for the convertible coupe, $655 for a brougham, $685 for the touring brougham, $720 for the sedan and $740 for the touring sedan.

Since fuel economy runs were in the news, it was only natural to find both Hudsons and Terraplanes entered. In previous years, Hudson-built cars had made a shambles of most record books. In the 1936 Yosemite trials, a Terraplane entered in Class 3-A against a Pontiac six, was only able to tie it at 23.95 mpg, but placed second due to a ton miles/gallon rating of 45.37. Interestingly enough, a Hudson eight in Class 5-A placed fourth behind a Graham, DeSoto, and a Chrysler...all sixes... with an average of 22.85 mpg.

Despite ever increasing production and new styling, Hudson-Terraplane remained in eighth place with a reported combined production of 123,266 units. According to *Moody's Industrials* production of Terraplanes totaled 97,857 with 78,471 registered. The company had an astounding net earnings of $3,305,616 on net sales of $77,150,680, their best year since 1925. Back then, Hudson-Essex had net earnings of $21,378,504, the highest ever recorded in the company's history.

In 1937, Hudson and Terraplane cars were introduced in October 1936, as had been done with the 1936 models in 1935. In doing so, they led Chevrolet, Ford, Pontiac and Oldsmobile to market by about a month. Styling remained unchanged from 1936, with a change in grille treatment to differentiate the two years. Two models were again featured: the Model 71 DeLuxe and the Model 72 Super. Wheelbase on all Terraplanes was increased to 117 inches, except for the Model 78 "Big Boy" commercials, which were set on a 124-inch wheelbase. There were two commercial lines in 1937: the Model 70 and the Model 78.

The biggest change came in prices. The Model 71 DeLuxe line, now offering 10 different body configurations, started out with a business coupe at $740. Following were a three-passenger coupe at $755, a three-passenger Brougham $780, three-passenger Victoria Coupe and Touring Brougham at $800, a sedan at $830, a touring sedan at $850, a convertible coupe at $875, the station wagon for $905, and four-passenger Convertible Brougham at $955.

The Model 72 Super Terraplane finally brought Terraplane prices over the $1,000 mark. A three-passenger coupe was offered at $835, a two-door brougham at $855, a three-passenger victoria coupe or touring brougham at $875, a sedan at $905, a touring sedan at $925, a convertible coupe at $945, and a four-passenger convertible brougham at $1,025.

Hudson and Terraplanes naturally entered the 1937 Yosemite Economy Run. This year brought things to the usual Hudson conclusion. In Class C competition, Terraplane placed first with an overall 22 mpg and 48.4 ton miles per gallon. In Class E, a Hudson eight came in second with 22.71 mpg and 52.23 ton miles per gallon. It ended up in third place overall.

Early introduction and performance not withstanding, the company remained in eighth place in production and Terraplane production dropped to 91,490 cars with 83,436 shipped. Registration figures are unclear, since Hudson and Terraplane were combined at 90,027. The financial picture told even worse news. Net sales dropped $2 million to $74,502,130, while net earnings dropped to $670,716.

The world was 23 months away from the beginning of the holocaust of World War II in October, 1937. But, it was business as usual for the nation and Hudson; introduction month for the new models. (For 1940, the company would move back to August). Despite the fact that Terraplane was on its way out, four models were introduced. The Model 80 commercial or utility, on a 117-inch wheelbase, offered a utility coupe, utility coach, utility touring coach and station wagon. The Model 88, on a 124-inch wheelbase, offered only six-passenger sedans and touring sedans.

As in previous years, Terraplane led off with the DeLuxe Model 81 and Super Model 82. Terraplane's replacement, the Hudson 112, made its debut as the 89 model-line's "112," so named for its 112-inch wheelbase. Within this framework were the Standard, DeLuxe and Utility series. The public wasn't too happy over Terraplane's demise and showed their displeasure by ignoring the new offering. The only two years it was offered were 1938 and 1939.

The Model 81 DeLuxe series offered a three-passenger coupe at $789, three/five-passenger Victoria Coupe at $835, three-passenger convertible coupe at $926, six-passenger brougham at $822, six-passenger touring brougham at $843, six-passenger sedan at $864, six-passenger touring sedan at $884 and a six-passenger convertible brougham at $990.

Next up the line was the Model 82 Super Series. Prices began at $845 for a three-passenger coupe, $886 for a three/five-passenger victoria coupe, $878 for the six-passenger brougham, $599 for a six-passenger touring brougham, $915 for a six-passenger sedan, $935 for a touring sedan, $971 for the three-passenger convertible coupe and $1,034 for a six-passenger convertible brougham.

With all those offerings, one would expect more production from Terraplane than was actually the case. Only 7,065 cars were produced, with a mere 6,588 shipped. Combined Hudson-Terraplane registrations were 40,889. Rather a dismal ending for the car that helped Hudson survive the depression years.

The financial picture was also a minor disaster. Net sales dropped to $38,845,238 with a net loss of $4,670,004. Financial losses were to continue through 1939 and 1940, with 1941 showing a profit.

The company was to show a profit for each of the following 10 years, until 1951. By then, the signs were beginning to point to merger, which still was not to be enough to keep the company alive. The last car to bear the Hudson nameplate came in 1957.

With a $750 price tag, the 1936 Terraplane station wagon was a new model aimed at the town-and-country set.

In only one year, 1933, did Hudson Motor Car Co. offer an eight-cylinder model of its Terraplane "companion car." It was quite a good-looking automobile, as well as a champion performer.

1933 Terraplane Eight was a performance compact

By Robert F. Mehl

Anyone remember the Terraplane? It had a short life in the automotive world. It was introduced, in 1932, as the Essex-Terraplane. For 1934, it was simply called the Terraplane. In its last year, 1938, it was named the Hudson-Terraplane.

In the 1920s, Hudson was a company to contend with in the marketplace. After a successful start in 1909, Hudson (the name of the car came from a department store owner and investor) introduced a lower-priced companion car in 1919. It was called the Essex.

The Essex (the name was derived from England) represented a completely new car. It was the sensation of the Detroit Automobile Show that year. Before long, a big selling feature of the Essex line evolved...the inexpensive closed bodied car.

The touring car, priced at $1,395, was still the best selling Essex. However, a model that would ultimately become especially popular with buyers was the basic coach (two-door sedan). It cost only $100 more than the lowest-priced open model. Thousands of Americans could...and did...buy a comfortable, closed bodied car for the first time. Until this point, the enclosed automobile had generally been considered a luxury.

Some 21,879 examples of the new nameplate were sold in the first year of production. By 1929, Essex was the number three best selling American car behind Ford and Chevrolet. When Essex finally died with the 1932 models, 1,331,107 had been sold. Few other makes of that era equaled its record.

The Essex evolved into what was essentially a small, less expensive Hudson. The two even shared components. Somehow, the company felt, in the early 1930s, that it needed a new gimmick. Sales had sagged terribly and the Essex suddenly lost its former great popularity.

A new car and a new concept were under development, perhaps motivated by the depression. By the summer of 1932, the new car bowed as the Essex-Terraplane. It was one of the first compact cars on the American scene.

The wheelbase was 106 inches and the engine was the Essex flathead six. Bodies were scaled down, yet they still retained a clear identity with big brother Hudson. The car was, in appearance, a miniature Hudson with similar styling by designer Frank Spring.

Coming to Hudson as chief stylist, Spring had left Murphy of California, the custom-body building concern. Most car buffs remember him as the designer of the step-down Hudson of 1948, but he made numerous other important contributions to Hudson history.

The company, after kicking around several ideas for the new name, came up with Terraplane. Terra was Latin for earth, and plane, as in airplane, implied aviation. The name was copyrighted and registered. Said the sales promotion, "the plane that skims the earth...today, the earthbound automobile takes on phantom wings, and you have the Terraplane."

A full line of models was offered, including roadster and phaeton and sales came to 16,581, almost as many as the 1932 Essex, which sold for a longer period of time. The bigger Hudson sold only 7,777 cars.

The Essex Terraplane made its mark immediately by winning the Pike's Peak Hill Climb with a run that broke all stock car class records. The winning car's time was 21 minutes and 21 seconds. Terraplanes had a good power-to-pound ratio, as they weighed just a little more than a ton.

Then in 1933, Hudson introduced the Essex Terraplane Eight. The six was still offered in a full line of models, but the eight was the exciting car. The wheelbase of the new eight was 113 inches; the six remained at the original 106 inches.

Both cars shared a standard track. The new inline eight-cylinder engine was masterfully created by merely adding two more cylinders to the 70 hp six-cylinder engine. This resulted in a 94 hp engine and an even better power to weight ratio than the Pike's Peak winning six.

One key Terraplane production advantage was the interchangeability of parts, as well as manufacturing simplification. Both the six and eight were L-head engines. The eight developed its 94 hp at 3,200 rpm with a compression ratio of 5.8:1. The engine was mounted on "rubber and air cushions" to absorb vibrations. Pistons were silicon aluminum, cam ground. Lubrication was by "positive duo-flo automatic oiling" (splash lubrication), meaning that the connecting rods sort of scooped oil out of the pan. It sounds scary, but it worked for Hudson for years. Other features included a mechanically operated fuel pump, Servo Bendix mechanical brakes, synchromesh transmission and a "triple-sealed, oil-cushioned, cork insert single-plate" clutch. Hudson used a wet clutch successfully for years, too. A downdraft carburetor also replaced the 1932 updraft system. Buyers could, at extra cost, order an automatic clutch.

The Terraplane Eight was the talk of the local grocery store loafers. At Daytona Beach, Florida, in February 1933, one of the cars broke 12 stock car records for speed and acceleration. It also set a new record for the five-mile flying start run at 85.431 mph.

For 1933, the Terraplane was a better looking car than the 1932 model. It had a handsomer front end and was better proportioned altogether. The fenders were skirted, just like those used on Hudsons. Wire wheels, always painted aluminum, were standard. While the radiator nameplate labeled the car "Essex Terraplane," the hubcaps displayed an eight surrounded by the name Terraplane on two sides.

For 1933 only, a marvelously sculpted and unique Griffin hood ornament graced the Terraplane's radiator shell. Hudson used the same ornament. Both before and after, the typical Hudson ornament was a stylized bird.

Most spares were rear-mounted, but dual sidemounted spares were a Terraplane option. They had metal covers, while rear-mounted spares usually had fabric covers on the 6:00x16 tires. Parking lights that repeated the headlights in miniature, adorned each front fender. The Terraplane Eight had four doors in each side of the hood for ventilation, instead of the louvers used on the six.

Dual taillights were standard, as was a set of curved horns, one mounted under each headlight. The doors were hinged at the rear in all models. They opened from the front to rear in a style that became known as "suicide" doors.

A full choice of color options seems to have been available, with the exception of red. Cars could be either a single color or have black fenders with color on the body. A wide range of opalescent colors was available. These were metallic finishes, which Hudson claimed to have pioneered in 1932, although there is evidence of metallics used on automobiles many years earlier. Hudson was, indeed, among the first to mass produce a popular-priced car offering metallic finish as a regular factory option.

Bodies were of all steel construction, except for a fabric top panel. A claim was also made as to originating unitized body construction. "The chassis and body are of solid steel construction and are coupled together as a unit to make one complete whole," said the promotional copy writers.

Inside, the Terraplane driver found an adjustable front seat, pedals and steering column. On the dash were "idiot lights" to register oil pressure and generator performance. This dubious monitoring system appeared first on the 1932 Hudson line.

The Terraplane's choke was combined with its throttle. Starting was accomplished by a button on the dash. There was an interior light, assist cords and rear quarter and back window shades (for closed models). Safety glass was optional. Dials were huge and poised in the center of the dash. Although there was no dash provision for radio controls, aerials were built into all closed models.

Interesting Terraplane options included: wind wings, trunk racks for six-wheel models, and special trunk installations for five-wheel models. These trunks fitted in the space between the body and the spare. They were useful for increasing storage room, but were very awkward looking and required an extra bumper section to protect the spare.

There were five Terraplane Eights for 1933 including a two-door coach (sedan); a four-door sedan; a coupe with either rumble seat or baggage compartment; a convertible coupe with rumble seat; and a roadster with rumble seat. There was no eight-cylinder phaeton, but one was available on the six-cylinder running gear.

Prices began at $585 for the eight-cylinder coupe to the $745 for the deluxe four-door sedan. Sales total for the 1933 Terraplane six and eight combined was 38,150. How many of these were eights is not definitely known, but certainly not as many as the cheaper sixes.

Hudson lost over $2 million in 1933. Few Terraplanes for the year 1933 ever appear at shows today. They are really rare and the eight is the rarer of the two.

For some strange reason, Hudson dropped the Terraplane Eight for 1934. This decision came in spite of its wide acclaim for high performance. Possibly it competed with the senior Hudson Eight. Perhaps Hudson was economizing and did not feel able to produce two distinct body sizes.

Whatever the reason, Hudson abandoned this spectacular performance car. Also abandoned was the compact car concept. What would a few more years of the compact performance package have done for the company in the long run? No one knows. Unfortunately, the idea was not given sufficient time to gain a foothold in the industry before it was killed.

For 1934, the Terraplane (the Essex prefix was dropped) became a full-sized six sharing the same bodies and appointments as the Hudson Six and Hudson Eight. For one brief, shining moment, the Terraplane Eight blazed a trail not to be followed until years later...and never to be forgotten.

A pair of late-1930s "Yuppies" looking over the 1938 Hudson 112 coach. Chances are, the young, upscale urban professionals of that era purchased something else. The 112 was an economy compact without a real market niche.

Hudson 112 - homely but endearing

By Robert C. Ackerson

Just about everyone remembers Hudson's ill-fated Jet, of which only 35,367 were produced for the 1953 and 1954 model years. A little less precise in most memories is the Hudson 112 of 1938 and 1939. Like the Jet, the Hudson 112 was a bit on the homely side. But, since it was every inch a Hudson, the 112 endeared itself to owners who appreciated its solid virtues of sound design, good workmanship and, in spite of its small engine, excellent performance and durability.

The year 1938 wasn't a good year for the American economy, as it fell back into depression. In such an environment, Hudson received plaudits for investing in the production of an automobile whose success could mean thousands of jobs for American workers.

In reality, the 112 was far from a new car. Wherever possible, existing body panels from the Hudson Terraplane were used. The Terraplane was phased out as the 112 became available.

The Terraplane represented a small, but snappy Hudson product. Its engine dated to 1924. Back then, with 130 cid and 28 hp, it had raised a furor as the replacement for the outstanding Essex four-cylinder F-head engine. But, as the great Hudson historian "Doc" Daugherty said in the *White Triangle News*, it was destined to go from "Somewhat Sick to Super Six."

For use in the Hudson 112, this L-head six was de-stroked to 4-1/2 inches (from five inches) and had the same three-inch bore. This provided a displacement of 175 cid. A maximum of 83 hp was developed at 4000 rpm, with a compression ratio of 6.5:1.

In 1936, Hudson had introduced Radial Control. It consisted of radius arms attached to the front axle and pivoted to the frame side members. This allowed softer springs to be used, resulting in a softer ride with no less of roadability. The 112 lacked this feature. However, with semi-elliptic front and rear leaf springs and direct-acting hydraulic shock absorbers, its suspension was patterned after that of larger Hudsons.

Since most of its body components had been seen before, the 112's styling was familiar and presented no surprises. A high grille with a center divider separating a series of horizontal bars and high positioned headlights mounted in each side of the 112's hood, provided a somewhat somber and conservative front end appearance. The 112's forward hinged "alligator" hood design would be adopted en masse by the Hudson lineup the following year.

In spite of various cost cutting efforts like reduction of the Terraplane wheelbase (by five inches) for use with the 112, its price still exceeded that of Ford, Chevrolet and Plymouth by $25-$45. It even had just a single windshield wiper and sun visor.

Even compared to the Hudson Terraplane, the 112's price advantage was not impressive. The least expensive four-door sedan retailed for $755. A Hudson Terraplane DeLuxe four-door sedan listed for $846. Making a car smaller didn't automatically bring about big cuts in its manufacturing costs.

Although not everyone was thrilled with the 112's small engine, it proved well-matched to its relatively light weight of approximately 2,600 pounds. It was capable of upholding the Hudson performance reputation. In runs supervised by the AAA at Bonneville, a standard 112 sedan set new distance records for Class D cars (engines of 112 to 183 cid) from the standing kilometer to 30,000 kilometers. In addition, the 112 also established new endurance marks from one hour to 12 hours. To cite two results: the Hudson averaged 80.54 mph for 25 hours and after 12 days it had traveled 20,327 miles at an average speed of 70.58 mph.

Although the 112 was the least powerful car ever to pace the Indianapolis 500, there's no doubt that it was an honor well earned.

Hudson's output for 1938 model year totalled 51,078 cars, a big drop from the 111,342 level of 1937. Just over 50 percent (25,769) were 112s. The following year, only moderate changes were made. Tire size increased to 6.00 x 16 from 5.50 x 16 and engine horsepower was boosted to 86, from 83, due to improved carburetion and a larger 1-1/4 inch instead of 1-1/8 inch carburetor venturi.

A smoother front end with a nicely curved grille sporting graceful horizontal bars made it easy to identify the new model. Just before production began, Hudson management decreed that the 112 would appear with "catwalk" grilles, just like the other Hudsons for 1939.

History often swings on small hinges. In the case of the Hudson 112, there are at least two "what if" questions that arise. What if Hudson had used the larger 212 cid Terraplane engine for the original 112? (With up to 101 hp available the answer, if not all the consequences, is obvious.) Second, what if Hudson had not dropped the 112 after the 1939 model year? (Carried over into the postwar years, when anything sold, the excellent little 112 as a smaller step-down model, could have reshaped thinking about small cars.)

At the very least, the Hudson Jet would have been viewed as a successor to a great down-sized Hudson. Unfortunately, the 112 vanished and the Jet was seen as a car that was both too little and too late.

As I said, small hinges can really mess up history.

The 1947 Hudson held its ground like a good soldier till it was relieved by fresh troops. All the while, it maintained the high standards of Hudson tradition: with performance, good service and value.

Hudson after World War II: 1946 and 1947

By Gerald Perschbacher

"Amazing is the only word for this new Hudson Six, the car to see with the 'other three.' Brilliant new design! New riding and driving ease! Amazing in performance, economy, endurance, safety and beauty," said the ads.

Trumpet blasts from the pen of the copywriter would have led you to believe that the Hudson for 1941 was just short of a heavenly creation.

Selling for $695, the car was a bargain. It offered a lot to the buyer. But, by the beginning of World War II, the styling and innovations on the Hudson were becoming common. Many other auto manufacturers shared in the two-tone color combinations and new interior cloth selections which began to dress up the steel soldiers of the road.

That prewar styling marched before the public through several years of world conflict...until America's fighting men had secured peace for the country. It was then, amid a new seller's market, that Hudson premiered its new 1946 model.

It really wasn't all-new. Only the Kaiser and Frazer were totally fresh cars, but that was mostly because they had not been made before the war. The 1947 Hudson, much like its allies and enemies in the American automobile market, was a tired soldier with new buttons and frills on an old, dusted-off uniform. Much like Studebaker, the Hudson people were looking forward to a revolutionary new design. However, every last mile was coaxed out of the old one.

That design was good and postwar sales among car-hungry Americans were over 93,000 for 1946. They broke 100,000 for 1947. These were good years, but not memorable.

The 1946 Hudson Super was available as a six- or eight-cylinder vehicle, with 212 cid for the former and 254 cid for the latter. That placed horsepower at 103 and 128. The Hudson Commodore carried the same power plants. There was only one wheelbase, 121 inches, and an overall length of slightly above 207 inches.

Minor styling changes included a new grille, with a recessed center section, and different outer trim. Only a Hudson salesman or historian could tell the difference.

In 1947, Hudsons came as four-door sedans, Broughams, Club Coupes (for six passengers), coupes (for three passengers) and convertibles. But for those who wanted Hudson styling for commercial purposes, the Carrier Six was available for the modest sum of $1,675.

This cross between a car and truck was lighter than its brothers, weighing in at 3,000 pounds. It was lighter in looks, too, as the attractive passenger car lines and contours mated with the all-steel hauling bed. Special caravan tops were optional. A good bit of weight was reduced by implementing an aluminum frame and the coupe-pickup was a hard-working vehicle; one willing to take punishment in the line of duty.

Of the 95,000 1947 Hudsons produced, the pickup was the rarest of the bunch. Only 2,917 were made. Even by the standards of an independent automobile manufacturer in the postwar days, that was low production. It was one reason the pickup was not continued after '47. But the main reason was Hudson's new postwar styling and engineering coup d'etat - the step-down Hudson for 1948. That new, low, styling made a pickup version impractical.

Rear view of a "Carrier Six" owned by Hudson historian Jack Miller shows the large dimensions of the dispatch box. As you can tell, Miller makes good use of his original condition truck around his circa-1957 Hudson dealership in Ypsilanti, Mich. He sells collector cars and parts in the old facility.

Model of the sleek Commodore Eight is finished in dark blue and features wide white sidewall tires. The models show the sleek, aerodynamic lines of early '50s Hudsons, which made them successful in stock car racing.

Cars in plaster

By John Gunnell

Plaster, as in Plaster-of-Paris, is a white powder made from gypsum and used as a quick-setting paste with water for casts and molds. Chances are you've had your dentist stick some plaster in one of those painful holes in your teeth at one time or another.

To some people, "getting plastered" means a painful case of over indulgence in one of life's more temporary pleasures. In ancient Egypt, the great kings got coated with plaster after the pain was over and the bandages applied on top of the plaster weren't there to ease the pain. But, pain and plaster don't always go together and plaster isn't always connected to temporary things either.

Plaster has been used to create both grotesque and beautiful objects. Flamboyant flamingo lawn ornaments are a good example of the former. The Hudson Motor Car Co. once provided its dealers with a form of the second.

Large, Plaster-of-Paris dealer showroom models that Hudson issued in the '30s, '40s and '50s are popular items with Hudson enthusiasts and old car lovers in general. When such models make one of their rare appearances at swap meets today, they are often snapped up quickly, regardless of price.

Depending on which Hudson model the plaster casts depict, the models range in size from approximately 30 inches to about 45 inches. They were used in much the same way dealer promotional models are today...to show prospective buyers what new cars looked like. They were painted in available colors or color combinations and detailed down to grilles, hubcaps, trim and insignia. One big difference between modern plastic models and this vintage type is that the windshield and side windows are part of the plaster cast and are painted silver to represent glass. Today's models use clear plastic inserts, which look much more realistic.

The Hudson models are hollow on the inside, the same way most modern replicas are, but the wheels are part of the casting itself. Modern dealer promos usually have a screwed-on chassis and molded rubber tires that look more like the real things. Another departure from reality is in the finish of the chrome trim and brightwork. The vintage models utilized silver enamel paint to highlight bumpers, grilles, ornaments and hubcaps. Today, we have chrome-plastic plating to make the trim on models look more realistic.

Passenger side of the Hudson Jet model has four doors. Plaster on and above the rear fender skirt has deteriorated a little. The roof looks a little flatter than that of the full-sized Jet.

Even though these old dealer models suffer a little in the realism department, they have a nice character of their own. This makes them highly sought after by hobbyists. It's that magic touch of deja vu, blended with old-fashioned styling, that pumps-up their appeal. Where else can you find a miniature representation of your 1937 Hudson, your Commodore Eight or your energy-efficient 1953 Hudson Jet?

The models shown here come from the collection of Jack Miller. They are located in the circa-1957 Hudson dealership that doubles as headquarters for the Hudson-Essex-Terraplane Club, Inc. It's located at 100 E. Cross St., Ypsilanti, MI 48197.

One unusual feature of the models is that they are cast as four-door sedans on one side and two-door sedans on the other. That way, Hudson dealers only had to have one model to illustrate both body styles. That was probably considered a good idea at the time, but unfortunately, it means that less of these beautiful models were made.

Miller has these fine examples in his large collection of all types of miniature Hudsons and marque memorabilia. In addition, he has at least one damaged plaster model, which he hopes to restore through a process he's developing with a friend, who's a dentist, of course. But, both men insist they have no plans to move onto lawn flamingoes or Egyptian mummies when the project is done.

Driver's side of the Hudson Jet model has the styling of a two-door sedan. It is finished in two-tone green. The miniatures gave the potential Hudson buyer a hint of what models looked like in different colors.

This appears to be a model of a 1937 Hudson, but the fender-mounted headlamps are missing. This model was painted a pea-soup green color. It features four-door styling on the opposite side of the body.

Part of the postwar Hudson enigma is illustrated by this 1946 Commodore Six four-door sedan, which sold at a record pace, while the all-new postwar cars encountered market share drops.

The enigmatic Hudson

By R. Perry Zavitz

Like nearly all other cars, Hudson resumed postwar production with warmed-over 1942 models. The 1946 Hudson looked like a '42, but with grille kicked in. The resulting cavity became a favorite place for mounting fog fights.

There were two Hudson lines for 1946, Super and Commodore. Both used 121-inch wheelbase chassis. The latter was the top series. Unlike the Commodore, the Super did not have lights above the headlights and on either side of the hood. Each series offered a choice of six- or eight-cylinder engines, so in effect, there were four series.

The six-cylinder engine displaced 212 cid and produced 102 hp. The straight eight was of 254 cid and 128 hp.

Body types offered in both series with a choice of eight- or six-cylinder engines included the four-door sedan, two-door sedan (called Brougham), Club Coupe and convertible. The models with two doors had a convenient front seat feature. When the seatback was pushed forward, the whole bench pivoted forward as well. This allowed easier passage in and out of the car. These "swivel seats," though of different design than those on later Chrysler products, preceded Chrysler's by some 13 years.

New Hudson registrations for calendar 1946 totaled 72,414. That was enough for ninth place, Hudson's highest standing since 1925. It was Hudson's biggest share of the market ever at 3.99 percent.

The 1947 Hudsons were nearly identical to '46 models. Registrations rose to 83,344, but Hudson slipped to 12th place.

For 1948, Hudson introduced a totally and dramatically new body. Of all the early postwar styling, Hudson's was the most unique because of its step-down design and unit construction. The floor was placed below the frame instead of on top of it. This technique reduced overall height, yet Hudson had more headroom than any of its contemporaries. Ground clearance was above average, even though the car looked as low-slung as a pregnant dachshund. A bonus benefit of the step-down design was enhanced safety, due to a lower center of gravity. Little wonder this layout has become common today.

Super and Commodore models continued for 1948 with the addition of a three-passenger coupe in the Super line, but the Super convertible was deleted. The two-door Brougham was omitted in the Commodore series and the convertible got the Brougham name.

Six- and eight-cylinder engines were offered in both series. The in-line eight underwent little change, but the six was substantially bigger. Its displacement was increased to 262 cid, upping the output to 121 hp. That outdid the eight. In fact, it was bigger and more powerful than any six available in a current American car. Hudsons with the "Big Six" suddenly became the cars to try to beat in stock car racing.

With new styling and the Big Six, it is not surprising that Hudson registrations shot up to 109,497. This brought it up to 10th place in popularity.

No noticeable changes marked the 1949 Hudson models. Despite new styling by Nash and most of the "Big Three," Hudson held 10th spot in 1949 with a record 137,907 registrations.

To broaden its market, Hudson brought out a smaller car, called the Pacemaker, for 1950. Styled like other Hudsons, it had 119-inch wheelbase (other Hudsons increased to 124 inches in 1948) and a 232 cid/112 hp six-cylinder motor. The Pacemaker competed directly in the Dodge/Mercury/Pontiac price, size and power ranges. It was offered in five body types, four-door sedan, two-door sedan, three-passenger Club Coupe and convertible. The two larger lines retained the same models as before, except for the Super three-passenger coupe.

The grille on the large and small 1950 Hudsons shared some components and changes from 1949. Compression was raised a bit, with the six up to 123 hp, but the eight's rating was unchanged.

In spite of the addition of the Pacemaker series, Hudson's 1950 registrations dropped to 134,719 cars and 12th place.

A restyled oval grille and optional Hydra-Matic transmission were features of the 1951 Hudsons. Another car-line was offered with introduction of the famous Hornet. It was basically a Commodore with a block-busting 308 cid six-cylinder motor. That was the biggest six made for any American passenger car after World War II. It was even larger than the celebrated V-8 that Oldsmobile introduced in 1949. It was more powerful too, with 145 hp versus the Oldsmobile's 135 hp. On a power/displacement basis, the Hornet six was about six percent more efficient than the Olds V-8, even with its slightly higher compression.

The Hornet was offered in the same models and at the same prices as the Commodore Eight. All other Hudson models were continued. In September, a hardtop called the Hollywood was added to the senior lines.

Another dip in popularity occurred in 1951, and only 96,847 registrations were recorded. Hudson had dropped to 14th place.

A bigger rear window and slight, but noticeable, trim changes marked 1952 Hudsons. The Super was discontinued. In its place was the Wasp. It used the Pacemaker chassis, but the 262 cid motor powered it. With higher compression, it was rated at 127 hp...just one puny pony short of the eight.

The Hornet stung the Olds by winning 27 out of 34 NASCAR events in 1952. It has been said, with lots of supporting data, that racing moves cars out of the showroom faster. Hudson must have been the exception to the rule. During 1952, its registrations dropped nearly 20 percent to 78,509.

For 1953, Hudson introduced another new model called the Jet. It was a small car...what we now call a compact. Its styling was unlike that of the other Hudsons. It had a 105-inch wheelbase and was powered by a 202 cid/104 hp six-cylinder motor.

The Pacemaker name was changed to Wasp. The former Wasp became the Super Wasp. The Commodore series was gone. So was eight-cylinder engines. The 262 cid/127 hp six was optional in the Wasp. A power package was optional on Hudson engines. It consisted of a high-compression head (aluminum head optional), dual carburetors and a few other performance goodies. Called Twin H-Power, it raised output to 114 hp on the Hudson Jet.

Again sales fell. In 1953 Hudson registrations totaled 66,797, leaving it in 15th place. For 1954, yet another new model was introduced by Hudson...the fifth in as many years. This was called the Italia. It was styled and built in Italy by an Italian coachbuilding firm called Touring. The two-passenger, aluminum-bodied hardtop was mounted on 105 inch wheelbase like the Jet. The Jet motor, with a "Twin H" power package installed, drove the Italia. Provisions were made to use other Hudson engines in Italias, but only the 202 cid/104 hp six was used in the 26 cars that were actually built. Had the Hornet engine been used, Hudson engineers predicted a top speed of close to 150 mph.

For 1954, the Jet received minor trim changes. Other models were restyled front and back. Up front, they resembled the Jet. At the rear, new taillights were mounted higher and more protruding. Registrations tumbled to 35,824. In reality, that was the last year for Hudson. In January 1954, Hudson and Nash merged to form American Motors Corporation. Just prior to that agreement, it was reported that a Hudson official declared they would not merge if Hudson could not keep its identity. In the following three model years, its name was about all that actually remained of that identity. The 1955-1957 Hudsons were thinly disguised Nashes, some making use of Packard V-8 engines.

Statistics recorded a number of years ago show that a smaller percentage of Hudson production was registered at the time than any other comparable orphan. This indicates that Hudson experienced the least loyalty. Why this would be, in view of their long history of excellent engineering, leadership in safety features, enviable performance records, and (towards the end) increasing model choice, is the Hudson engima.

Mysteriously, Hudson's percent of total industry sales started "stepping down" after the step-down models, like this Commodore sedan, appeared for the 1948 model year. These unit-body cars were dramatically new in overall concept.

Since an overhead valve V-8 was what Hudson needed to spur postwar sales, why did it bring out the Italia sports car with a small six-cylinder engine. That's another of the enigmas surrounding the company's history.

Another enigmatic aspect of Hudson history is the question of why the marque's 1951-1953 successes in stock car racing didn't increase sales of showroom cars like this 1951 Super Six four-door sedan.

Interesting, but again enigmatic, was the thinking behind the badge-engineered 1955-1957 Hudsons, which were basically Nash products with Hudson badges. Apparently, Hudson demanded that its identity survive after the AMC merger.

The 1946 Hudson Commodore Six used a prewar body shell with new front end styling. Many of these cars were two-toned, an unusual feature for the day.

Hudson's postwar record: from go-go to long gone

By R. Perry Zavitz

It was in 1909 when the first Hudsons were made. Had it not been for Olds, there just might not have been a Hudson at all. Founders of Hudson were Roy D. Chapin, Howard E. Coffin, Frederick O. Benzer, and James J. Brady. All were former Olds employees. Other Hudson pioneers included Roscoe B. Jackson and George W. Dunham. The Hudson name came from one of the new company backers. An uncle of Mr. Jackson's wife, the well known Detroit department store owner J. L. Hudson, invested $90,000.

The Hudson Motor Car Co. was initially a subsidiary of the automaker Chalmers-Detroit. Before the end of the first year, Chapin, Coffin and Bezner bought out from Hugh Chalmers. Hudson became an independent company.

The cars built by Hudson soon gained a reliable reputation. By the end of prewar car production, in February 1942, some two-thirds of a million were in use in the United States.

On Aug. 30, 1945, Hudson resumed car production. By the end of 1945, it had already built 4,735 cars and stood fifth in production, although standings in that hectic start-up year are rather meaningless.

The cars themselves were little changed from the '42s. The grille looked like it met head on with a timber. However, the cavity became a favorite location for fog lights.

Super and Commodore lines were offered with a choice of one 212 cid/102 hp six or a 254 cid/128 hp straight eight engine. A club coupe and four-door sedan were available with either motor in either series. In addition, there was a business coupe, two-door sedan, and convertible offered in the Super Six, and a convertible in the Commodore Eight series.

The same models and engines were continued in the 1947 models. In fact, there was practically no change for '47. Car production for 1946 totalled 91,215, and for 1947 it increased to 100,862. Among them was the 3,000,000th Hudson. A record 3,000 dealers and distributors had no problem selling all the new cars being made.

So, in a seller's market, it came as a bit of a surprise when the 1948 models were introduced in December, 1947. They had completely different unitized bodies with postwar styling. Sensational as these cars were, the most unusual feature was the "step-down" floor structure. It was placed beneath the frame, rather than on top of it. That allowed the whole car to be much lower than the others. The Hudson was an even five feet in height, which was notably lower than even the new Kaiser and restyled Studebaker.

The six-cylinder engine was given a bigger bore, but shorter stroke. The result was a larger displacement than the eight. At 262 cid, it was rated at 121 hp and not quite as powerful as the eight. The same model availability continued as before, with the addition of a Commodore Six convertible. Prices ran from $2,069 to $3,138 basic.

Hudson deemed it unnecessary to make any changes, superficial or otherwise, on their cars for 1949. A two-door sedan was made available as a Super Eight.

Production during calendar 1948 was 143,697. For 1949, it dropped slightly to 142,462. It was an excellent showing, however, because all makes which had not been restyled before got all-new bodies for '49. Even Kaiser and Frazer were facelifted for 1949.

Slight grille changes marked the 1950 Hudsons, but a

smaller, lower-priced series was introduced. The Pacemaker, a name resurrected from Hudson's past, was aimed directly at the Mercury, full-size Dodge, and Pontiac six in regard to power, size, and price.

The Pacemaker, 201 1/2 inches long, had a 232 cid/112 hp six-cylinder motor, and was priced from $1,806 to $2,444 depending on the model. It was available in all the body types formerly offered in the Super Six series. The Pacemaker DeLuxe sub-series did not have a three-passenger coupe. It was discontinued from the Super Six as well. The other series offered the same body types as in 1949.

Front end styling for 1951 was changed to an oval-shaped grille. For the second year running, a new line was added. This time it was in the upper price ranges. Hudson Hornets were offered in the same body types at prices identical to Commodore Eights. The Hornet was a six...but what a six!

Its displacement of 308 cid made it the indisputable giant of the six-cylinder car motors, at least since World War II. Rated at 145 hp, it was bigger and more powerful than the contemporary straight eight in the Packard 200 or the V-8 in the Oldsmobile.

This engine powered Hudson to checkered flags all over the country. Seldom considered a slouch in official or impromptu races, Hudson piled up 79 NASCAR wins in the Grand National division from 1951 through 1955. In 1952, 1953 and 1954, Hudson accounted for substantially more victories than any other make. For instance, of the 34 races in '52, Hudson drove home first in 27. Nearly 80 percent of all races were first-place wins for Hudson.

With that kind of performance from six cylinders, combined with Hudson's unique body engineering, little wonder the 1951-1954 Hornets are certified Milestone cars and recognized by the Milestone Car Society.

Automobile Quarterly has compiled the NASCAR and AAA wins by the Hornet. An incredible total of 131 firsts are listed for the 1951-1954 seasons. In addition, there were 67 second and 36 third place finishes.

Unfortunately, the same pace was not maintained in the showrooms. Hudson was losing the sales race. It stood 10th in 1948 and 1949. By 1951, it had tumbled to 14th. The decline came despite the hot Hornet and the introduction of hardtops in all but the Pacemaker lines.

No styling changes were made for 1952, except for less chrome on the grille. There was more trim around the rear window and along the roof lips, though. No new models were introduced, but Hudson tried to hide that fact by replacing the Super Six name with Wasp. That name capitalized on the widespread reputation the Hornet had earned for itself in its introductory year. There was no eight-cylinder engine available in the Wasp, only in the Commodore.

Slight changes in body type availability were seen. The convertible was deleted from the Pacemaker car-line, though it was offered in the Wasp series.

The 1953 model year saw the elimination of several models, but the introduction of yet another series. What might be considered a postwar counterpart to the Essex appeared as the Jet. It was a small car, the last of five compacts introduced by the industry in as many years.

The Jet was approximately the same size as the smaller 1949 Plymouth Deluxe, 1950 Nash Rambler, 1951 Henry J and 1952 Willys-Aero. The market for such cars was very small then, but extremely competitive.

The Jet had a 202 cid/104 hp six-cylinder motor, which made it the most powerful standard equipped compact of its day. Two- and four-door sedans were offered in the Jet and Super Jet series. Prices started at $1,858.

The Wasp continued, but not as a convertible. Its engine was the six, formerly used in the Commodore. That name was eliminated, but there was a Super Wasp line. With the same six-cylinder engine, it offered two and four-door sedans, and club, hardtop, and convertible coupes. The Hornet continued with the same models as before. Gone was Hudson's straight eight engine.

Limited styling changes, which were applied to the 1954 full-sized Hudsons, made the greatest appearance change since the '48 models. The grille was revised, and the familiar, slightly notched back was given a boxier shape - both in the trunk lid and with the use of Cadillac-inspired taillights.

By 1954, Twin H-Power had become a performance option on many Hudson models. It consisted of dual carburetors and higher compression heads.

On the Jet, Twin H-Power increased the output to 114 hp and on the Hornet up to 170 hp. The standard Hornet engine was rated at 160 hp for '54.

The Wasp reverted to the '52 Pacemaker's 127 hp motor, while the Super Wasp retained the 262 cid six-cylinder engine. Its rating was increased to 140 hp. Two new sub-series were added to Hudson's 1954 lineup. A super deluxe Jet Liner made its appearance, as did the bargain-priced Hornet Special.

Let's not forget the 1954 custom-built Italia. With a sensational looking aluminum body, this two-passenger coupe used the Jet chassis and motor with Twin H-Power. Just 26 were built in Italy.

In 1954, Hudson Motor Car Co. merged with Nash-Kelvinator to become a division of American Motors Corp. The independent automakers were in a financial bind and sought consolidation for strength.

The last Hudson built in Detroit left the assembly line on Oct. 2, 1954. On Dec. 28, 1954 of that year, the first Wisconsin-built Hudson came off the assembly line.

When Hudson car production was finally terminated, in June 1957, there were almost 600,000 Hudsons still in use in America. The total number of Hudsons registered had been steadily dropping from a high of 866,000 in 1953. What these figures tell us is the sad fact that, in its last four years, the factory was not building Hudsons as fast as the public was scrapping them.

A smaller Hudson Pacemaker was introduced after World War II. Shown is the 1951 four-door sedan model.

This 1952 Hudson convertible was part of the new-that-year Wasp series, which was derived from the Pacemaker Super Custom Six car-line of one year earlier.

Prewar look characterized 1947 Hudson convertibles.

Topless Hudsons

By Bill Siuru

In the early postwar years, if you wanted a sporty car you bought a convertible; and,if you wanted good performance and excellent handling, (at least as measured by the standards of American cars of the era) you probably considered purchasing a Hudson.

In every year from 1946-1954, Hudson offered several convertibles. Unlike some manufacturers who limited the ragtop to top-of-the-line series, Hudson had convertibles at both ends of the scale. The lowest-priced Hudson ragtops were usually the best sellers. Hudson convertibles were called Brougham convertibles, in parallel with Hudson's policy of calling its two-door sedans Broughams.

While Hudson offered both six- and eight-cylinder engines in its convertibles until 1953, it was the Hudson six that saw the greatest development. The inline eight stayed pretty much the same through its years of production, 254 cid and 121-128 hp. However, it can be said Hudson pushed the L-head six to its limit. Starting out at 212 cid and 103 hp in 1946, the engine was increased to 262 cid and 121 hp for 1948, and was raised to 308 cid and 145 hp for the legendary Hornet six in 1951. By 1953, this was up to 160 hp in standard form and the "severe usage," (read stock car racing) 7-X version was rumored to have put out as much as 200 horses, when all the available performance goodies were installed.

In the early 1950s, Hudsons were the cars to beat on the stock car circuit. Hudson capitalized on this in its advertising. The Hudson catalogs let the buyer build quite a "street machine" for its day, with such options as Twin-H Power, Power Dome cylinder heads, "export" suspension systems, a variety of rear end ratios and heavy-duty clutches and brakes for "police and taxi" duty. Unfortunately, the high-performance youth market would not be around for a few more years, so this emphasis on performance was in vain. Like so many other ideas from the independents, Hudson's performance image was ahead of its time.

If you were after all out performance, you probably would not buy a convertible. Ragtops weighed in 200-300 pounds above your basic two-door sedan or coupe. The convertibles were the most luxurious of the Hudsons and were priced as much as $600 more than the lowest-priced sedans. The Hudson ragtops were fitted with beautiful leather interiors. Top-of-the-line models like the Commodore and Hornet were equipped with hydraulically-operated power windows as standard equipment. (They were often an option on lower-priced versions.)

Hudsons were noted for their extensive array of standard equipment and they had a list of options that allowed a person to make his car as personal as he wanted. This was another area in which Hudson kept apace of the mainstream of American-built cars.

Because of the 1948 step-down design with its low stance, the Hudsons were considered one of the safest cars of the era. Adding to this reputation were such safety features as dual braking (hydraulic, backed up by mechanical brakes), unitized construction, and direct-to-frame bumper attachments. The design made for a very roomy car. The convertibles were full six-passenger vehicles. The 1948-1954

Hudsons used basically the same body, with facelifts.

The failure to change was probably the biggest reason for the marque's demise. It surely was not poor engineering, lack of quality, or a shortage of available equipment that brought on Hudson's decline. Rather, it was the fact that young buyers felt that the cars looked old-fashioned and had an overall dated character, even though they were excellent machines.

The following gives a rundown of the Hudson convertible story during the postwar years:

[1946-1947]: The warmed-over prewar style Hudsons included two convertible models in the Super Six and Commodore Eight series. Production of the six was approximately 1,037 for 1948 and 1,462 for 1947. Commodore Eight convertibles were sold at a rate of 140 (1946) and 361 (1947).

[1948-1949]: Production of Brougham convertibles based on Hudson's famous step-down design got off to a slow start in 1948, with only 86 Super Six, 49 Commodore Six and 65 Commodore Eight convertibles built. This makes the 1948 Hudson ragtops the rarest of the lot. By 1949, production was up to 1,868, 656 and 596 for the three models in the same respective order.

[1950]: When Hudson's low-priced Pacemaker series appeared, convertibles were offered in both the base 500 and the slightly upgraded Deluxe version. The Pacemaker 500 series accounted for an estimated 1,865 ragtop assemblies, while approximately 660 open-bodied Pacemaker Deluxes were built. These are about the most popular Hudson convertibles made after World War II. In addition, some 464 Super Six; 700 Custom Commodore Six and 426 Custom Commodore Eight Brougham Convertibles were made. This was definitely the best year for topless Hudsons.

[1951]: This year only the Pacemaker Custom Six line had a convertible. Its sales dropped to 425 units. Convertibles were also available in the Super Custom Six, Custom Commodore Six and Custom Commodore Eight (181) car-lines, with approximate production of 282, 211 and 181, respectively. From the collector's standpoint, a most desirable Hudson ragtop is one of 551 Hornet convertibles, with big 308 cid L-head sixes, made in 1951. This was the first year Hudson offered a fully automatic transmission (GM's Hydra-Matic) on the Hornet and Commodore. Hudson's own Drive-Master and Super-matic semi-automatic transmissions were still available, as they had been since 1946 and 1949 respectively, but were not offered on all models.

[1952]: This year saw even a further decline in Hudson convertible sales, as well as rearrangement of series names. The upscale Pacemaker was now the Wasp Six and about 220 convertibles were made in this series. The Super nameplate was gone, but about 20 convertibles were made in the Commodore Six line and the Commodore Eight series added about 30 more. The hottest-selling convertible was the Hornet Six, with some 360 made. 1952 was the first year for Hudson's Hollywood hardtop in Wasp, Commodore and Hornet formats.

[1953]: The number of convertible models shrunk in 1953 with ragtops being available only in Super Wasp and Hornet form. Only 50 Super Wasps were estimated to have been built, and the number of Hornet versions is not known.

[1954]: This was Hudson's last year as a distinctive marque. In 1955, Hudsons would become based on the Nash. Also, 1954 was the last year for a convertible to wear the Hudson emblem. Sales of the facelifted 1954 Super Wasp Six were 222 for the calendar year. Hornet convertibles were also offered, but the number built is not available.

From a collectible standpoint, Hudson convertibles are very desirable. They are rare, historically important, and feature outstanding engineering and performance.

The 1948 Hudson Super Six convertible.

The 1951 Commodore Six convertible had distinct side trim.

Few styling changes marked the '49 ragtops.

Convertibles had top level interior trim.

Hot performance was a feature of this 1951 Hornet ragtop.

Side trim on this topless 1951 Hudson reads "Commodore Eight."

Smallest, most popular drop top Hudson was the 1951 Pacemaker.

Rear view of the 1954 Hudson Hornet convertible.

Front view of the 1954 Hudson Hornet ragtop.

This Hornet ragtop paced a Milwaukee stock car race in 1954.

The 1954 Hudson Jet Liner was anything but a cash cow for Hudson. Its high retail price kept sales low, so that production start-up costs for the compact Hudson were not recouped.

Was the Jet Hudson's Edsel?

The Hudson Jet was introduced on Jan. 2, 1953, and automotive historians generally agree that the Jet was Hudson's Edsel.

The tooling and start-up costs for the Jet were $12 million. It cost the Ford Motor Co. $100 million to tool their line of cars, but the company survived and prospered later. Hudson paid for its mistake with its life.

Hudson dealers were clamoring for a small car to sell. The Nash Rambler was selling reasonably well and the Henry J was helping Kaiser-Frazer get extra sales.

What Hudson dealers saw in a much-distorted drawing of the proposed Jet compact and what they got "in the flesh" were two entirely different things.

Some Hudson historians say that the '52 Ford carried a lot of influence at Hudson. Company designers and company President A.E. Barit liked the Ford's styling. But, probably having more influence on the Jet's appearance was Barit's use of the Fiat 1400 as a guide.

Chair-high seats were insisted upon by Barit, who also dictated the compact car's overall dimensions. The resulting package became the production version of the Jet.

The Hudson Hornet was one of the most respected automobiles on the AAA and NASCAR stock car circuits. It won more races than any previous make and did so in the span of just a few years. The engine in the Hornet was a huge L-head six-cylinder motor of 308 cid.

Dealers tried to claim that the Jet's 202 cid L-head six was a "scaled-down" version of the Hornet's engine. Not so. One look at the bore and stroke specifications of the old Commodore Eight engine reveals that it and the Jet have the same bore (three inches) and stroke (4-3/4 inches). Hudson simply chose to save money and built a six from the tooling used to make the old inline eight.

The Jet was built with the same "armored car" engineering that the large Hudson step-downs featured. It tipped the scales at a not-so-compact sounding 2,700 pounds. When one considers that Ford and Chevrolet full-sized cars weighed only 400 pounds more, it is hard to imagine Hudson calling the Jet a "light six," although that is what such a car would have been tagged prior to World War II.

A high list price proved to be another stumbling block in sending the Jet down the road to success. For instance, let's compare the factory price of a Jetliner four-door sedan with its competition:

Jetliner four-door	$2057
Ford Mainline Six four-door sedan	$1701
Ford Customline four-door sedan	$1845
Ford Crestline four-door sedan	$1960
Chevy 150 four-door sedan	$1680
Chevy 210 four-door sedan	$1771
Chevy Bel Air four-door sedan	$1884
Plymouth Plaza four-door sedan	$1745
Plymouth Savoy four-door sedan	$1853
Plymouth Belvedere four-door sedan	$1933
Willys Aero Lark four-door sedan	$1823
Willys Aero Ace four-door sedan	$2023
Rambler Super four-door	$1795

Rambler Custom four-door sedan..........................$1965
Studebaker Custom four-door sedan.....................$1801
Studebaker DeLuxe four-door sedan.....................$1918
Dodge Meadowbrook four-door sedan $2000

As you can tell from the above list, the Jet didn't have a chance in the marketplace. The Jet was more expensive than the top of the line Fords, Chevys and Plymouths.

Magazine test drivers found the Jet one of the hottest cars in its class. It had a top speed of over 95 mph and would accelerate from 0-to-60 mph in 14.5 seconds,

Production of the 1953 Jet was 21,143 units. In 1954, that number slipped to 14,224. The merger of Hudson and Nash was complete by 1954 and the Jet didn't figure into the new American Motors production plan. The Jet was dropped in favor of the Nash Rambler. For 1955, there was even a Hudson Rambler and a Hudson Metropolitan. With the introduction of the 1958 model cars, AMC dropped both the Nash and the Hudson name to concentrate on the Rambler.

It is impossible to blame the demise of a large industrial concern, such as Hudson, on one factor. However, the drain of funds used to bring the Jet to market robbed Hudson of cash to update the full-size Hudson line. This cash flow problem eventually contributed to the death of Hudson.

Automotive artist Bob Hovorka made this sketch of the '52 Hudson Hornet to illustrate his column "Notes from the Filling Station" for *Old Cars Weekly*. Today, Hovorka's famous drawings and stories appear regularly in *Special-interest Autos* magazine.

A '52 Hudson visits the old "Filling Station"

By Bob Hovorka

The softly rounded sedan would be considered huge by today's down-sized standards. But back in 1962, the '52 Hudson was just another 10-year old used car. Once a proud, independent manufacturer, Hudson had joined the growing ranks of postwar orphans. Like Kaiser, Nash, Packard and Willys...it deserved better.

Since its introduction in 1909, Hudson had garnered a reputation for building quality cars that were a trifle unconventional. Not content to be a follower, Hudson Motor Co. brandished an enviable list of automotive firsts. Throughout its history, Hudson's triangular emblem graced a long line of rapid rebels that defied convention, and amassed numerous speed and stamina records. Like most auto companies, it struggled through the infamous depression of the '30s. Unlike many, Hudson survived.

After World War II, the company again staked its reputation on something out of the ordinary...the Monobuilt body-and-frame on the step-down Hudsons. While recessing the body between the frame rails is common today, back in 1948 it was truly an exciting idea.

Why the company failed has been, and will be, discussed as long as there are Hudson devotees. Suffice to say, that during the early '50s, yearly body changes...to make last year's models look "all-new"...and overhead valve V-8 engines, were mandatory advertising features. And while the step-down Hudsons received many safety awards, including *Safety Engineering's*..."America's Safest Car" trophy, the Monobuilt body-and-frame construction was nearly impossible to change on a yearly basis. Too much of the basic shape was necessary to maintain structural integrity; making changes very costly. As for a V-8 engine, Hudson did not offer a V-8 till after the Nash-Hudson merger. By then, the roadability inherent in the step-down design was gone.

To be fair: although the '55 through '57 Nashes and Hudsons shared the same body, there were more differences between them than most of the so-called A, J, or K cars built today. Both Nash and Hudson carried extensive front end sheet metal variations; both offered their own six-cylinder engines (Nash retaining its overhead valve configuration, while Hudson continued its tried and true flathead). Hudson even modified its exclusive "Triple-Safe Braking System" to fit the new body. It was all quite different from today, when one vehicle can wear three or four different nameplates by changing plastic grilles and taillights while engines and underpinnings remain amazingly alike. Yet, while those '55 through '57 Hudsons were comfortable highway cruisers, the precise handling and racing prowess that had made Hudson undisputed stock car champion, was gone. No longer would the obsolete flathead six do battle with Oldsmobile's Rocket V-8, or Chrysler's mystical Hemi. No longer would Hudson owners point with pride to those bulbous beauties that showed Chrysler, Ford and General Motors the short way around the oval track. No longer would the uniquely shaped step-downs cradle their passengers in "box-section steel frames" that encircled occupants with roll cage-like protection.

Back in 1957, the last automobile to wear a Hudson triangle trundled down the American Motors' Kenosha assembly line. But, back in 1962, the old '52 Hornet that limped into the filling station was truly an orphan. Its humpback fuselage was considered as antiquated as the long stroke, flathead six that powered it. The new 1962 cars carried fresh crisp styling. More importantly, they offered the advertising magic of "short stroke V-8s." Oh, sixes were still available, but almost all were overhead valve engines built for economy, not performance.

Slowly, the biggest six-cylinder engine to grace an American postwar production car labored up the driveway. Filling station owner Mr. Miller looked up from the car he was working on. He took a shop cloth from his pocket and wiped his hands. It had been a quiet day, and the scent of pipe tobacco wafted from the familar old briar. His grey eyebrows formed a quizzical arch. He didn't recognize the car. The skinny lad who pushed the step-down's massive door open was also unknown. Long gangly legs stretched over Hudson's infamous ledge and headed towards the filling station.

"Hi, Mr. Miller. I'm Bill Cauldwell." He pushed a sandy lock of hair from his eyes and continued, "We've never met, but mom and I moved into the old Turner house last summer. Mom's having trouble with the car and some of my friends told me you were the man to see."

"Cauldwell?" The squint that settled in the old mechanic's eye was not from the sun, but from the name he was trying to place. "Oh yes. I think I know. Your mother works at the school."

"Yeh, that's her," echoed the cheery voice.

"What seems to be the problem with the Hudson?"

A grin broke across the youngster's face. "At least you know what it is." He laughed out loud. "My mom's brother gave it to her when we moved up here. He kept it pretty clean. Guess he kind of felt sorry for her after dad died. Said he was looking at a new car, and nobody wanted the Hudson for a trade-in because it was an orphan. So he gave it to mom...told her he had some trouble going a long distance, but for short trips it was a reliable car. Boy, did we have a time getting it up here. It's got some problem after it warms up. Doesn't seem to have much pep...but it's only a six. Sometimes it just dies! And if you have to start it right away when it's hot, forget it!"

"He was pretty disgusted, because he had the carburetors rebuilt. It's got two of 'em you know...put on a new fuel pump, and a new starter. But it didn't help. Didn't matter much up here though, because mom only used it up and down to work. It's only got to go about a mile, then it sits all day till she comes home. It was alright last summer, but it's getting pretty bad now."

Mr. Miller leaned back, puffed on his pipe, nodded occasionally, but didn't say a word. "You know, I even put a set of points and plugs in it, but that didn't seem to make any difference. Then one of the guys at school told me to have you take a look at it. I know they don't make 'em anymore, but if you could maybe get it to run a little better, we could still use it for a while. Mom says if it's not too serious, well...I mean we'd like to find out for sure what's wrong with it. But if it's too serious, well, after all it's a 10-year old car that's not even made anymore, so it would be foolish to stick too much money into it."

Mr. Miller took the pipe from his mouth. "You're sort of the businessman of the family, I take it."

"Yep, mom says a man should take care of the car problems; not a woman. I'm not looking for something for nothing, but I mean to get a fair deal. The guys at school all said you were a square shooter, so here I am. I'll pay a fair price for a fair assessment."

The old mechanic chuckled to himself as he scraped the tobacco from his pipe. Billy Cauldwell rambled on. After he finished cleaning his pipe, Mr. Miller repacked it with fresh tobacco. Then, leaning forward, he successfully interrupted the young man's conversation, "Mind if I look at the automobile?"

"I don't mind, but I will need to know your hourly rate."

"Well son, a look won't cost you anything. If I turn something up, we'll discuss payment."

Bill felt pretty smug. He had gotten a free estimate without even trying.

Mr. Miller slid behind the wheel of the old Hudson; its massive steering wheel sat atop a completely chrome-plated steering column. His eyes caught the mileage. "You say your uncle owned the automobile?"

"Yes sir. He bought it new."

The old mechanic turned towards the rear seat, noting the high polish on the burl-toned dashboard, and the firmness of the thick, ribbed upholstery. After surveying the rear passenger area, he turned back towards the front. Once again a quizzical arch raised his eyebrows, as he tried to recall whether the twin carburetors were a factory or dealer-installed option in 1952. He reached for the crank to open the window. It was mounted inside a deeply recessed door panel. The recessed panel added several inches to the already wide interior. He recalled an early Hudson advertisement that showed a woman standing beside a Hudson seat, which had been set vertically alongside the car. It graphically illustrated Hudson's claim that the car was actually lower than it was wide. Hudson certainly pushed the low overall height and low center of gravity, but its torpedo styling was beginning to look dated even in 1952. It carried many traits of an earlier era: little details like the light in the ignition switch, so it could be easily found at night, and the knob over the windshield that raised and lowered the roof-mounted radio antenna. Its narrow, slit-like windows were also showing their age. Still, Hudson's exclusive "Triple-Safe" Brakes, which provided a mechanical hook-up to the rear wheels should the hydraulic system fail, had not been adopted by any other automobile manufacturer, even in 1962. He smiled. The old car was in excellent shape.

Mr. Miller inserted the round ignition key and turned it. The six-volt battery struggled to push pie plate-size pistons up and down sewer pipe-diameter cylinders. They moved slowly. He released the key.

"Hey, you knew it was the round key. Everybody always tries the square one." The young man grinned at his private joke.

Once again, the old mechanic turned the key. The noise of the laboring starter filled the cloth-wrapped interior.

Leaning his head through the narrow window opening, Billy shouted, "Sometimes after it's hot, it just won't kick over!"

Mr. Miller nodded. He jiggled the gas pedal judiciously, and finally coaxed the big six to life. He stepped on the accelerator a few times to smooth it out, then reached down and pulled the hood release. Stopping in front of the high-domed hood, he found the safety latch and opened it. The metallic churning that greeted his ears was typical Hudson Hornet. He leaned over the ironing board-size engine, listening carefully for several minutes.

"All right Bill, you can turn her off now."

The youngster slid behind the wheel and turned the key. The silence was broken by the chirping of birds.

Carefully, the old mechanic unsnapped the distributor cap, grasped the rotor, and turned. Satisfied, he replaced the cap, walked around to the other side of the engine and touched the ram-like intake manifold. He took a shop cloth from his coveralls and pushed the heat riser up and down. It was free.

Moving to the front of the car, he pressed the fan belt with one hand, and moved the fan blades with the other. He then reached over the fender and removed the breather cap. "Start her up Bill."

Again the big six was slowly coaxed to life. Watching for a moment, Mr. Miller replaced the oil breather cap, then walked around to look at the breather pipe.

"OK. You can turn her off now."

Wiping his hands in the shop cloth, he headed towards the driver's side. He removed the unlit pipe that dangled from his lips. "Well son, by all rights we should take a compression reading. But, based on the fact that the mileage seems to be original and the automobile was well cared for, I'd say there are probably two things that could be causing the problem. First, I believe the automobile needs a timing chain. That long-stroke six was well known for its ability to make rubber bands from timing chains, even when it was new. Your uncle ever mention replacing one?" The wily old mechanic never waited for an answer. "No matter, it sure needs one now."

"Secondly, I'd advise removing the twin carburetors from the vehicle. Your mother surely won't miss the extra power and they have a reputation for inviting vapor lock problems. Here again, if everything was working properly, they technically should cause no trouble. However, we have a 10-year old automobile that's not driven very far. I also believe that when the manifold is removed, we'll find out that there's a hole or crack in it that's letting the exhaust gases get back into the intake manifold. It can happen on this type of setup because the manifolds come in contact with each other. This upsets not only the mixture, by pouring burnt exhaust gases back into the engine, but also warms the intake manifold more than normal. It sure doesn't need a any extra heat."

"Depending on how bad the crack is, the extra heat might not have a big effect while the engine is running. But, trying to restart a warm engine is another story. The extra heat can overheat the fuel, compounding the problem. Those twin carburetors just multiply everything by two. If this is the case, and based on how hot that intake manifold felt when I touched it, I think it is, it might be wise to replace those twin carburetors with a single carburetor. Back in the early '50s, Hudson was competing in the mid-price field against automobiles like Oldsmobile and Mercury. They both had V-8 engines. Hudson had a six and the twin carburetion. They called it Twin-H-Power...gave Hudson owners something to talk about when others were bragging about their V-8 engines. Hudson wasn't the first automaker to offer compound carburetion, but the words Twin-H-Power caught on like magic. Sort of like Chrysler's Hemi...it had a nice ring to it."

"Actually, there were some internal modifications that became standard in 1954; they improved performance over the earlier models. That was the year Hudson featured a new functional hood scoop. They talked about Instant-Action engines and Super-Induction and a lot of people felt the hood scoop rammed air into the engine. I always felt that hood scoop did more good after the engine was turned off. Then, it let the hot air that built up under the hood escape through the scoop. Well, no matter, this isn't a '54. More important, your mother needs a good reliable automobile, not a racing machine. I think that replacing those two pieces will cure the problem.

The young lad looked up quizzically, "Are a timing chain and manifold usually responsible for hard starting troubles?"

"Not really. A weak coil, defective starter motor, poor battery, points, carburetion...there's just a whole list of things that can cause starting difficulties. But, from your description, it sounds like most of the obvious solutions have been tried. The starting problem, combined with the automobile's loss of power, especially when hot made me think the manifold was the culprit."

"But what made it, crack?"

"Oh, it doesn't have to be a physical crack. It could be burned through. After all, there's a lot of heat going out a manifold and this is an older vehicle. The casting could have been a little thin at one spot, who knows? Regardless, like I said, a timing chain and a manifold should give that old Hornet back some of its sting. It was quite an automobile in its day."

Mr. Miller struck the match he was holding and puffed on the old briar. "Yes sir, quite an automobile." He looked towards the Hudson and smiled. "A lot of people didn't like its shape, but personally, I still like its streamlined lines over those boxy new models. Always seemed more aerodynamic. Who knows? Maybe someday that sloping roofline will come back in style. Tastes change. Did you notice how they chopped off a lot of tailfins this year? Maybe there will come a time when they start building automobiles to do what they were originally intended to do...to move people from one place to another reliably and safely. That old Hudson had a good record of doing just that."

"Well, what do you say Bill. Should we give it a try?"

Rapid Relatives: Hudson Race Cars

The Hudson Eight sedan at speed during the 1935 Muroc Lake speed runs in California.

When Hudson went racing

By Robert Ackerson

A half-century ago, the Hudson Motor Car Co. ranked third in output among all American automobile companies. Total production stood at 300,962, a level that Hudson was destined never to be reached again. In 1930, the first full year of the depression, Hudson-Essex production fell drastically to 113,898. The year 1931 was even worse, with only 40,338 Essex and 17,487 Hudson cars sold. During the next three years Hudson losses totalled in excess of $7.5 million dollars, in spite of a doubling of sales of Hudsons and Terraplanes in 1933, when both cars were extensively restyled.

A good deal of Hudson's sales strength was based upon its reputation for stamina (in 1932 Hudson claimed that 73.3 percent of all Hudson products built since 1909 were still in operation) and performance. In an attempt to get back black ink back in the ledgers, Hudson emphasized both of these virtues in 1935.

Steel bodies were in vogue in the mid-1930s and Hudson boasted that it was the only American manufacturer building cars with all-steel bodies. Unlike some manufacturers who, said Hudson "use steel only where it shows," Hudson used steel "all around." There was not a splinter of wood in any structural part of any Hudson body. Hudson may have been exaggerating when it called its 1935 models a "new kind of automobile," but this advancement in automotive design was certainly one of which Hudson could rightfully be proud.

Hudson also waxed eloquently in its advertising for 1935 regarding its new Electric Hand transmission. It was standard on the Hudson Custom models and optional on the lower-priced Hudsons and Terraplanes. The Electric Hand Hudsons had a conventional clutch and transmission, but the normal floor-mounted gearshift was replaced by a small, steering column-located lever that allowed the driver to pre-select, via electric and vacuum power, his next change of gear. The amount of electricity this system used was minimal. Hudson claimed that running through the gears 48 times or making 144 different shifts used no more electricity than burning the headlights for an hour. Incidentally this system, which was built by Bendix, was the same unit later used on the Cord 810.

For a large automobile powered by a 113 hp engine, the Hudson Eight was a fairly economical automobile. In economy tests held around the country, Hudson Eights averaged 20.1 mpg.

The 1935 Hudson's all-steel body, Electric Hand and full economy were certainly catalysts to its sales, but clearly its most noteworthy attribute was performance. In April, 1935, a Hudson Eight was driven around two five-mile circular courses on the Muroc Dry Lake, in California, to set 36 new AAA records. Among the records set by the trio of Wilbur Shaw, Babe Stapp and Al Gordon was a speed in the Flying Kilometer of 92.84 mph. Most of the new Hudson records were for Class C American cars with engines of 185 to 305 cid. However, the Hudson average for 1,000 miles at 85.84 mph established a new record for closed cars, regardless of engine size or price.

The Hudson's impressive performance, plus that of its Terraplane running mate (which held hillclimb records for stock cars from Lookout Mountain, in Tennessee, to Mt. Baldy in California) certainly made the proper impression upon the American public. A total of 101,080 cars were retailed by Hudson-Terraplane dealers, yielding a profit to the Hudson Motor Car Co. of $584,749.

Both in Europe, as well as America, the Hudson and Terraplane were highly regarded for their performance.

The Autocar (Sept. 29, 1933) tested a British-built Terraplane (Hudson had an assembly plant in London) fitted with a four-seater touring body. Zero-to-60 mph was reported to require only 18.5 seconds. In 1934, Hudson claimed that a Terraplane could accelerate from rest to 75 mph in just 21 seconds. During the same year a contest, including hillclimbing and acceleration tests, was held in Buffalo, New York. It pitted a Ford V-8, a Terraplane 8 and a Hudson 8 against each other. The Terraplane won the economy competition, but the Hudson emerged the victor both in the hillclimbing and acceleration tests.

Hudsons continued to set performance records right up to the eve of the American entry into World War II. In 1940, a Hudson sedan traveled 20,000 miles under AAA supervision, at a speed of 70.58 mph, giving Hudson a total of nearly 150 stock car performance records.

When the postwar Hudson Hornet started winning stock car races in the early '50s, a lot of young enthusiasts hadn't known any other performance cars except an Olds Super 88 or a Mercury. They couldn't quite understand just how that "whale" of a big Hudson could do what it did on a stock car track. Old-time Hudson fans didn't have any such problems. All they had to do was to recall 1935 and note that the Hornet was just a "chip off the old block." (And it was still a chrome alloy block at that!)

A cloud of dust rises into the sky behind the hot-performing 1935 Hudson Eight.

This is the radio timer control used during the 1935 speed run made in a Hudson Eight four-door sedan on April 12, 1935.

Driver Al Gordon relieved A.C. Pillsbury (viewable over gasoline attendant's shoulder), when the "Hudson Stock 8 Record Breaker" took on a tankful of Red Lion gas at Muroc Lake.

Buddy Marr checks below the Hudson's hood, as (l. to r.) E.J. Saunders discusses record run with drivers Wilbur Shaw, Babe Strapp and Al Gordon. Saunders worked in Gilmore Oil Co.'s engineering department.

Hudson highlighted racing in its ads. This one includes a chart recording the 31 victories of 1952 and shows Dick Rathman taking the checkered flag (in Walt Chapman's Hornet) on May 4, 1952. Small pictures show scene from Darlington 100-miler on May 10 and Tim Flock in his Hudson.

Hornet's buzz meant the end of queen bee Rocket's reign

By Phil Hall

In February, 1951, pioneer automotive scribe Tom McCahill was test driving a 1951 Chrysler New Yorker in Florida for *Mechanix Illustrated*. He stopped by Daytona Beach to look over the NASCAR speed trials about to be run there. McCahill, not familiar with rapid motoring, was more or less challenged by Bill France, NASCAR president, to run the big

FirePower in the trials.

Accepting the challenge, McCahill had the big sedan tuned at a local dealer. On Feb. 8, 1951, he turned in a two-way average of 100.13 mph over a less than ideal beach, to be declared the winner in the strictly stock passenger car class.

Three days, later the 160-mile Grand National event on the beach-road course oval found a newcomer in the winner's circle, a 1951 Hudson Hornet. It was driven by former hot rodder Marshall Teague, a Daytona Beach garage owner.

Within three days, the once invincible Oldsmobile 88 had been solidly shot out of the box by a pair of makes that weren't even considered challengers the year before.

Looking at the cars themselves, the two brash newcomers for 1951 each offered something that the Olds didn't. And, since strictly stock was the name of the game, there was little Olds drivers could do about it.

Chrysler offered raw power. Though the chassis was still the basic 1949 model (known more for strength than handling) and the transmission was the rather inefficient Fluid-Matic, the engine was something else.

The 331.1 cid, hemispherical head V-8 was advertised at 180 hp at 4000 rpm, making it the most powerful production passenger car in the country at a time when Americans were starting to take notice of such things. The potential of the power plant seemed unlimited, with the big valves and low-compression ratio of the early model, not to mention the restrictive-two barrel carb.

Meanwhile, the 1951 Olds 88 continued with the 135 hp/ 303.7 cid engine it started with in 1949.

The Hornet's long suit was handling. The step-down chassis, introduced on the 1948 models, had always made for a good handling car. Contributing to this was the wide tread, low center of gravity, center point steering and the combination of front and rear anti-roll bars.

Hudson, however, was not popular among stock car entrants because of the lack of power. That was remedied with the introduction of the 1951 Hornet, which featured a bored and stroked version of the 262 cid Super Six. The new 308 cid flathead six turned out an honest 145 hp...10 more than Olds.

McCahill's run of 100.13 mph did not break the 100.28 mph run of Joe Littlejohn that was set in 1950 in an Olds. Beach conditions were responsible for the drop. They become more apparent when it is considered that the fastest Olds in 1951, driven by Bill Trent, turned in a time of 93.95 mph.

In the beach-road event, Tim Flock finished second in a 1950 Lincoln. The first Olds was driven by brother Fonty Flock, who took third.

Oldsmobile did not roll over and die in 1951. After Daytona, it returned to its traditional king of the hill role, but the king had a very potent challenger in NASCAR racing...the Hornet.

Teague took a second in the 112-mile Grand National at Charlotte, North Carolina, then won the event on the half-mile at Gardena, California. It was time for a visit to the Hudson factory in Detroit.

The combination of his two victories and his background of Hudson hot rodding made him a welcome visitor to the company, which was looking for a new way to promote sales. Hudson had an extensive background in the world of speed, so the idea of helping out stock car racing was a bit more palatable to them than the other manufacturers.

At Daytona, Teague used the heavy-duty or export suspension, which was right out of the catalog. He also utilized another cataloged item, the 3.58 rear axle. His talks with Hudson pointed out the advantages for race car builders, emphasizing the extensive option list for chassis and engine parts.

The rapport Teague established with the factory was the harbinger of a new necessity in auto racing, both overt and covert, for the next two decades.

After his return from the factory, Teague set up a second Hornet for NASCAR racing. He persuaded Herb Thomas to drive it. Thomas started the season in a Plymouth and switched to an 88 before he joined Teague.

One of the most important events on the 1951 NASCAR Grand National schedule was the 250-mile event at Detroit. The track was nothing to get excited about, unless you were a thrill seeker and liked to see cars bounce around. More important was the fact that the auto company big wigs would be there to see what this stock car racing thing was all about.

Packard supplied the pace car; Hudson and Olds backers were sure their favorites would show who was boss, but a Chrysler won.

The victory was the only one of the season for Chrysler but it couldn't have come at a more opportune moment. After the event, Chrysler engineers filled the pit of winner Tommy Thompson. They offered to rebuild the car for him, at least acknowledging that the sport existed.

Thompson's earliest taste of "factory support" wasn't especially beneficial, however. When it was delivered, a day before the race at Langhorne, Pa., it looked nice enough. However, the engineers had replaced the export springs and special reinforced parts with stock ones and, as a result, Thompson was unable to race it.

Executives of the other companies were amazed to see their creations dump broken parts all over the track, especially wheels. This didn't look good to a crowd of potential buyers. Other than Chrysler's token involvement and Hudson's cooperation with Teague, little evolved from the Detroit show on the surface.

The 1/4-mile, banked and paved oval at Darlington, S.C., was still the stock car superspeedway in 1951 and Hudson's first product of Teague engineering was ready for its premiere....Twin-H carburetion.

Instead of the two-barrel carb, Teague and Thomas drove Hornets sporting two one-barrel carbs spaced to give better fuel distribution to the big six.

Yes, it was in the parts bulletin Hudson issued before the race!

A big-time winner for Hudson was Marshall Teague, who became the first driver to take the Daytona Beach Classic twice in a row, in 1951 and 1952.

Hudson smoothly rounds a curve at N. Wilkesboro, Pa., to win again!

Tim Flock, Bud Shuman and Herb Thomas (l. to r.) swept many races in their step-down Hudsons. This picture was taken after a one-two-three victory in a 250-mile race at Detroit on June 29, 1952.

Teague set the fastest time of the day and Thomas won the race.

"Although my qualifying time at Darlington was the fastest, it was not the maximum speed of the car, as was later proved in the race," Teague said in a letter to *Speed Age*. "After running the track enough to learn the groove, I was able to obtain a top speed of over 110 mph at the end of the straightaway.

Even with all these tremendous speeds, I never once ran at maximum speed, as tires are the governing factor at Darlington. Hudsons are able to run at greater speeds, for longer periods with less tire wear, due to their superior roadability, aided by their low center of gravity," Teague noted.

In addition to being a good driver and builder, Teague was also a good public relations man for Hudson.

Plymouth continued to pester the more powerful makes on the short tracks. Lee Petty notched a couple of victories in his venerable 1949 business coupe, but the other 39 checkered flags went to other makes.

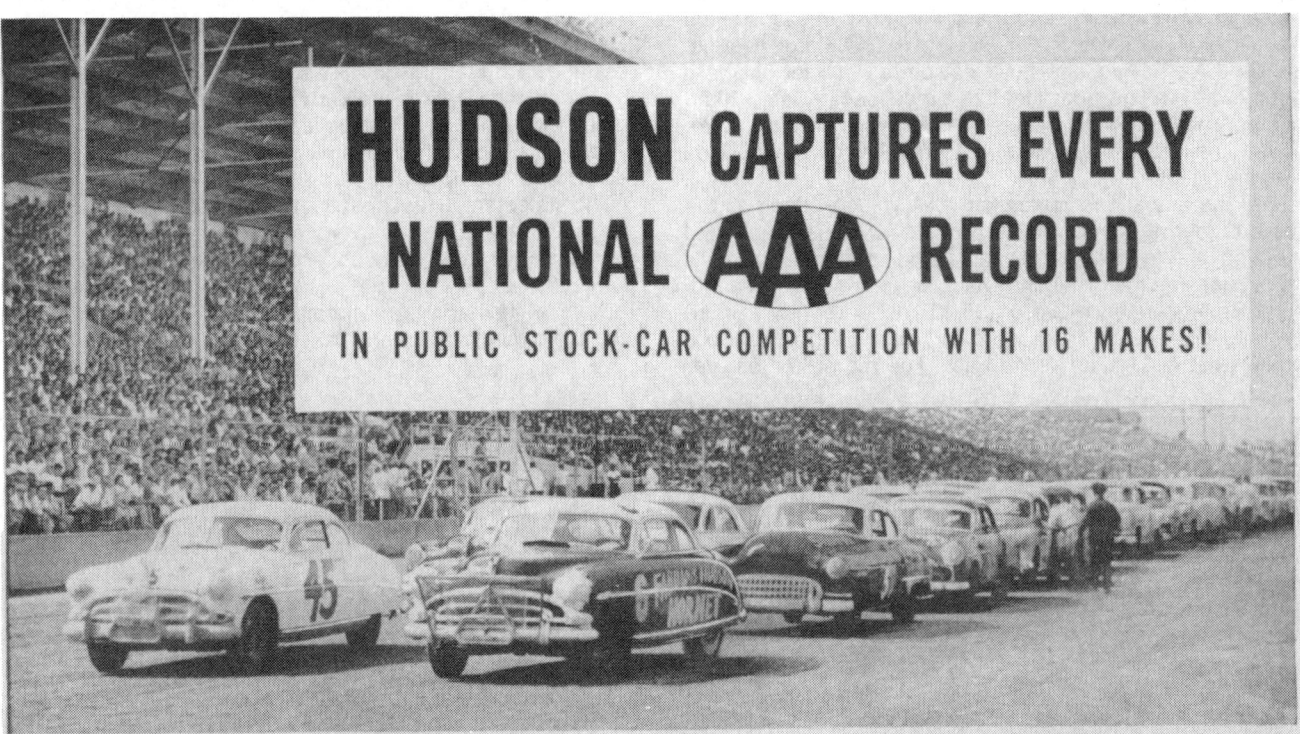

Three Hudsons and one Oldsmobile held the first four starting berths in this AAA stock car race shown in the Hudson ad. Although the company earned a solid reputation for performance, racing never translated into showroom sales.

Oldsmobile again accounted for the majority of Grand National victories with 20. Hudson was next with 12. Studebaker was good for three, Mercury two and Nash joined Chrysler in the one-shot category. Olds clinched the manufacturer's championship in NASCAR with 163 points to 98 for Plymouth and 92 for Hudson. Part of the reason for Hudson's poor showing was the lack of numbers. There were a lot of 88s and Plymouths around from the previous season and relatively few of the new Hudsons, but that was changing fast.

When the drivers' points were tallied, Thomas was declared the champion...a gold star for Hudson. Meanwhile, up north, the American Automobile Association (AAA) contest board stock car racing circuit was in trouble. Only three events were held in 1951 and all were in Milwaukee on the one-mile dirt oval.

The rules were very prohibitive, both in requirements for the race tracks and requirements for the cars. Only strictly stock passenger cars were allowed, with no reinforcement for safety and no options.

Rodger Ward won the first event, July 8, in an Olds 88. On August 23, Norm Nelson finished first in an 88, but was disqualified for having non-stock equipment; runner-up Ward got the win. The next day featured a race for cars with 120 hp or less, and Nelson got the win in a Mercury, fair and square.

Ward, needless to say, was the 1951 AAA champ and Olds was the top car.

After the embarrassing season, AAA voted to open up the rules to permit safety modifications to wheels, hubs and steering, and to let the cars compete on shorter tracks.

Chrysler had the initial and final glory of the 1951 season. Tony Bettenhausen drove a 1951 Saratoga coupe to third overall and first for American cars in the La Carrera Panamericana, or Mexican Road Race. The red no. 7 was prepared by the Kiekhafer Corp., builders of outboard motors. The car was fully setup for the 1,933 mile run from Tuxtia Gutierrez to Cuidad Juarez, through the Mexican countryside.

The car was reportedly producing 30 percent more power than stock and the record 114.33 mph average (breaking the old mark set by an Olds by 13.9 mph) would seem to verify that something had been done. The Kiekhafer-Chrysler combination would be heard from again in the racing world.

Looking over the 1951 season, Chrysler won two major non-oval speed events, (Daytona Speed Trials and La Carrera); Oldsmobile took the most wins and manufacturer's championship in NASCAR and also the best record in the limited AAA competition. Yet, it was Hudson that made the most of their accomplishments in stock car racing.

There was nary a mention of racing accomplishments from Chrysler and Olds.

In June, 1951, Hudson issued a folder entitled "Meet the Winner, the Fabulous Hudson Hornet." It touted Hudson victories at Daytona; Gardena, California; Canfield, Ohio and Phoenix, Arizona.

More important, the brochure tied in the wins with the car itself. "The superiority of Hudson's new and better way to build passenger cars proved itself over all competition," it noted over a photo of Teague leading the pack at Daytona.

Teague also gave some of his biased testimonial: "My Hudson Hornet is strictly standard!"

The statement, "Hudson rules the road and America knows it," was part of a plan to "psyche-out" the competition in racing. Unfortunately, Hudson sales in 1951 dropped from 143,586 to 92,859.

Despite the lack of performance in the showroom, the stage was set for Hudson's takeover in stock car racing.

HUDSON HORNET
Scores Sweeping Stock Car Successes Using Dependable
CHAMPION SPARK PLUGS!

(1952 Record through July 13th)

26 VICTORIES IN 30 STARTS

Meet Mr. M. H. Toncray, Chief Engineer, HUDSON MOTOR CAR CO.

"*CHAMPION SPARK PLUGS have been standard equipment on Hudson Motor Cars for over twenty years. We know we can depend on Champions for top performance, economy of operation and long life, both on the highway and on the track.*"

HERB THOMAS, 1951 Stock Car Racing Champion, Nascar Grand National Circuit

"*The Hudson Hornet and CHAMPION SPARK PLUGS have teamed up to make a winning combination in stock car racing. I know, for that's my team! I use stock Champions, the same as those sold by your Hudson dealer or service station.*"

The record-breaking success of the Hudson Hornet in strictly stock car competition in 1952—as well as in 1951—is a tribute to the handling qualities, ruggedness, dependability and safety in-built in Hudson's step-down design.

Each car must be certified as strictly stock and identical with one you can purchase from your local dealer.

Here, as in racing of all types, Champion Spark Plugs are unchallenged for top performance and dependability. For not only have they been in the winning Hudsons, but in other victorious stock cars in many other events.

All of these cars, regardless of make or year, have used the standard type Champions recommended for that car. Here's solid proof that the Champions for your *own* car are tops in performance and dependability.

CHAMPION SPARK PLUG COMPANY
TOLEDO 1, OHIO

FOLLOW THE EXPERTS
USE THE SPARK PLUG CHAMPIONS USE

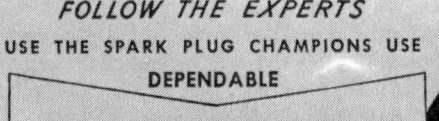

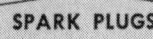

Champion Spark Plug, Co. highlighted the Hornet's stock car racing victories of 1951-1952 in this ad featuring the company's chief engineer M.H. Toncry and 1951 NASCAR Grand National champion Herb Thomas.

Herb Thomas campaigned the no. 92 Fabulous Hudson Hornet all over the country. With Twin-H power, the "old-fashioned" Hudson sixes were invincible in stock car races throughout the 1953 season, usually beating the "modern" V-8s.

Hudson led "Big 3" in backing stockers on national circuit

By Phil Hall

The year 1953 represented the dawn of the entrance of the "Big 3" automakers into national championship stock car racing. Oldsmobiles, Cadillacs, Fords, Lincolns, Plymouths and Chryslers had been racing since 1948, backed by the skill of mechanics and courage of the drivers. Then, something happened to spark the interest of General Motors, Ford and Chrysler; the Hudson Hornet got direct race backing from the Hudson factory.

Late-model stock car racing was growing in popularity by leaps and bounds, but Hudson, an outsider, was getting all the gravy. "Something has to be done to stop it," thought GM, Ford and Chrysler. So, they began following Hudson's example, by getting involved in stock car racing and using their involvement to sell cars in the showrooms.

On paper, the Hudson Hornet should not have been dominating the sport. It was a big, heavy car powered by an outdated flathead six.

In reality, a combination of excellent handling, full cooperation from the factory's parts division and a large number of top drivers in Hornets made Hudson the king of stock car racing. Proof was found in the company's 1952 record of 40 wins in 48 NASCAR (National Association for Stock Car Racing) and AAA (American Automobile Association Contest Board) events.

Hudson's challengers for that year were:

(1) Oldsmobile...with its modern design V-8. The Olds also featured a twin stabilizer suspension, but still had inferior handling, when compared to the Hudson.

(2) Chrysler...featuring a very powerful Hemi V-8, but dated chassis.

(3) Plymouth...a small car, with a 97 horsepower engine, that couldn't move as fast as the big cars, but which ran almost forever and was very easy on fuel and tires.

(4) Nash, a car with soft springs, but with twin SU sidedraft carbs on its Ambassador six.

(5) Ford, whose drivers had a choice of the outdated flathead V-8 or an overhead valve six that was almost as powerful.

Theoretically, for 1953, Hudson was in for a rough time. Potent new challengers from Dodge and Lincoln were announced. Also premiered was a new lightweight V-8 in the Buick Roadmaster. This engine would be very potent the next year when added to the lighter Century.

Dodge for 1953 was built on a smaller chassis with reductions in length, wheelbase and most important, weight (down about 200 pounds). Under the hood was the Red Ram V-8, a scaled down "semi-Hemi" displacing 241.2 cid and putting out 140 horses when breathing through the stock two-barrel carb.

Lincoln unleashed a four-barrel carb and some performance oriented options including solid lifters. The unoptioned 317.5 cid V-8 produced an advertised 205 hp.

Lincoln premiered its 1953 models in the La Carerra Panamericana (Mexican Road Race) late in 1952. With a bit of tweaking, they took the first four positions in their class.

Hudson, on the other hand, did relatively little to the 1953 Hornet. The standard two-barrel, 308 cid six still put out an advertised 145 hp as when it was introduced in 1951. The "police" (or 7-X) engine, with two-one barrels...called "Twin-H" carburetion...produced more power than the standard version, but the horses were never advertised. Many guessed the output at over 200 hp. In 1952, race driver Marshall Teague figured Twin-H was good for 10 honest extra horses throughout the rpm range. But, other than minor cosmetic revisions, the 1953 Hornet was little changed from the 1952 model.

Though late model stock car racing during the year is mainly contested on ovals, the first attention-getting event of 1953 was the Speed Trials over the sands of Daytona Beach, in February.

The NASCAR supervised events included the Flying Mile and standing-start acceleration tests (also over one mile.) The modern-day Speed Trials started in 1950 as a contest between men to see who had the fastest car. It developed into a contest between manufacturers to see who had the fastest product.

Proof of the more than casual interest by Big 3 manufacturers was the Oldsmobile contingent at the 1953 Speed-Weeks. Six cars were entered. Four ran in the 160-mile Grand National race on the beach road course. The other two were meant specifically for the Speed Trials, but some wound up doing double duty.

All six entrants were equipped with "packages" of optional equipment listed by Oldsmobile. The packages, six in all, covered modifications to the engine, differential, axle shafts, radiator, suspension and fuel tank. All of these packages carried Olds parts numbers and were considered legal by NASCAR, as long as they were available to the public. Olds said they were, although the claim that they were authorized for installation on all 1951-1952 Oldsmobiles was controversial. Those just happened to be the models that were legal for NASCAR and AAA racing. Shades of Hudson!

The bulletin covering the Olds packages also listed five items not manufactured by Lansing or listed on Olds' authorized parts lists. They were a heavier steering linkage, Air-Lift spring boosters, reinforced wheels, harder rear engine mounts and straight-through dual exhaust pipes. Tom McCahill complained of these modifications in *Mechanix Illustrated*. "This sounds to me like saying, 'pick, your own speed shop, It's okay with us,'" wrote Uncle Tom. "Now I maintain that this is hardly 'factory available optional equipment.' If it is, all George Mason of Nash has to do is 'authorize' any Rambler owner to install a 270 Offenhauser in his little gem and stock car records would be blown sky-high overnight."

The Olds packages did their thing and as a result, Oldsmobiles took all three events they entered. Bob Pronger of Blue Island, Ill., drove his 88 two-door sedan to a two-way Flying Mile average of 113.38 mph, well above the old mark of 100.28 mph set by Joe Littlejohn, in 1950, also in an Olds. In the acceleration trials, Pronger also swept honors with a 74.41 mph average through the traps.

In the Grand National race, Pronger rolled, but Bill Blair went without a pit stop and snapped Hudson's winning streak in the event. Hudson driver Marshall Teague had won the last two years in a row. Fonty Flock, in another 'factory' Olds took second, giving the package cars a 1-2 sweep. Tommy Thompson was third in a 1953 Lincoln and Herb Thomas brought the first Hornet across the finish line in fourth place.

Making things even more uncomfortable for Hudson in NASCAR country was the fact that long-time nemesis Lee Petty had switched from his 97 horse Plymouth short tracker to a Red Ram Hemi-powered Dodge. (Petty continued to use Chryslers on larger tracks.) Also in the Mopar ranks, Chrysler stayed with its two-barrel carburetion system and 180 advertised horsepower for 1953, despite getting a new body.

Though not as direct as Olds, Dodge issued a parts bulletin "in response to occasional requests from Dodge dealers and car owners for a maximum output package for their cars when under certain types of special service." That "special service" referred to stock car racing, there was little doubt.

The package contained some Mopar parts and some authorized Dodge parts. Included were special cylinder heads, manifolds, valve springs, dual exhausts, suspension parts, differential gears and heavy-duty wheels and hubs. These weren't for the flathead sixes by any stretch of the imagination. They were to give more thrust to the "crimson ram."

The performance kits helped both Oldsmobile and Dodge in NASCAR. Olds tripled its 1953 victory total from the previous year, scoring nine wins. This included Daytona and Buck Baker's big win, at Darlington, N.C., in the Southern 500. Dodge, thanks to Petty, took six wins. This wasn't bad, considering a Grand National had never been won by a Dodge before.

While the Big 3 race cars were starting to win, Hudson was far from dead in the water. There was a total of 37 events on the 1953 schedule. With Oldsmobile and Dodge sharing a total of 15, that left 22 other races. Hudson took all 22 checkered flags. Once again, this made the Hornet 1953's undisputed circuit king.

Herb Thomas won the championship driver title, repeating his performance of 1951. Lee Petty finished second, but Hudson driver Jim Rathman took third, giving the Hornet pilots two-out-of-three...and "two-out-of-three ain't bad," as they say. Baker, the first Olds driver to hit in the standings, came in fourth.

Up north, in AAA racing, Hudson was also as tough as ever. It took 13 wins in 16 starts. Frank Mundy (Francisco Edwardo Menendez) of Atlanta, Ga., a former stunt driver, won five events and took the championship. Other Hudsonites taking victories were Marshall Teague (3), Jack McGrath (3) and Sam Hanks (2). Clarence LaRue of Akron, Ohio, scored Oldsmobile's lone victory. One surprise was the 1951 Packard driven by Don O'Dell of Blue Island, Illinois.

A 1953 Hudson Hornet slides around the outside of the track, while a '52 Hornet takes the inside line through the same curve. Driving the '53 is Dick Rathman, while the pilot of the older car is Tim Flock.

With absolutely no factory help, O'Dell finished second in the point standings and even picked up a pair of wins on the hot Milwaukee Mile, where the biggest purses were paid. O'Dell's final win came in the last race of the season, a September 20, 200 miler.

Though it will never be recorded in the win column, one of Hudson's victories, of sorts, came at Milwaukee on August 23, 1953. Don Dunfee's 1953 Dodge had stalled just off the south turn. Dunfee got out of the car, but left it on the track. McGrath didn't see the car until it was too late. He crashed into it while going full bore in his Hornet. The Hornet was demolished, but the structure around McGrath held up. He escaped with a cut on his wrist, which seemed truly amazing after viewing the wreck and the car.

The "paper threat" at the beginning of 1953 turned out to do little to change Hudson's winning ways. Even when the Big 3 started challenging, these AMC predecessors proved capable of putting up an excellent fight.

Through the dust you can see several Hudsons and Oldsmobiles taking the starter's flag in this early stock car race. The two marques were very competitive through the end of the 1953 season.

Foreign Relations

The sport of polo was used as a symbol of the 1954 Nash-Healey's sports car appeal. Coupes with this reverse-curve style back window are known as Farina Coupes and 90 of these cars were built.

Nash went abroad to satisfy U.S. hunger for sports cars

By Wally Wyss

America in the early 1950s was a nation emerging from the automotive equivalent of the Ice Age. G.I.s returning from Europe had brought over some spindley wheeled shoeboxes called M.G.s and started a craze for sports cars. The only trouble was, nobody in Detroit was too sure what in "Sam Hill" a sports car was. GM, in its rolling Motorama shows, kept displaying cars like the Buick Wildcat. It had wire wheels. That made it a sports car, right?

But, little old Nash figured out that if you really wanted a sports car, the thing to do was go to Europe and buy the damn thing off of someone overseas who could knock them together for nine and six pence. So that's how the Nash-Healey came to be.

The name Healey may sound familiar. Yes, it's the same Healey as in the Jensen-Healey and the Nash was one of his earlier numbers. Donald Healey, managing director of the Donald Healey Motor Co. of Warwick, England, had built a special sports car using a Nash Ambassador engine and driveline. He entered it in the 24 hour LeMans endurance race, in July, 1950. So well did the sports car perform in the race (finishing fourth), that Nash decided to contract for a limited number of the sports models.

For the new production Nash-Healey, the high-compression, six-cylinder Nash Ambassador engine was fitted with an aluminum head and dual carburetors. Overdrive was standard. The prototype, which had an aluminum body built by the Healey company, was shown publicly for the first time at the Paris Automobile Show in early fall of 1950.

Production began in December, 1950. In that month, 36 models were built. An additional 68 were produced in the months of January, February and March of 1951. All were two-door convertibles.

General specifications of the initial 1951 Nash-Healey included, as standard equipment: leather upholstery, adjustable steering wheel, directional signals, chrome wheel discs, foam rubber cushions and five 4-ply whitewall tires. Standard colors were Champagne Ivory and Sunset Maroon. (No other colors were available.)

The six-cylinder engine, of 234.8 cid (3847 cc), was rated at 125 hp. It had 8:1 compression ratio, seven main bearing crankshaft; intake manifold sealed-in-head and two S.U. sidedraft carburetors. Other details included: torque tube drive; rear coil springs; 6.40 x 15 tires; 20 gallon fuel tank; and plexiglas side windows. Dimensions were: overall length 170 inches, width 60 inches, wheelbase 102 inches, tread 53 inches front and rear, turning radius 17 feet 6 inches, road clearance seven inches, weight 2,400 pounds.

No Nash-Healeys were made from April 1951 until January 1952. Then, an entirely new roadster body was created by Pinin Farina of Turin, Italy. A total 150 of these 1952 convertible models were produced.

By this time, the Nash-Healey was truly an international car. The engine and main parts were manufactured by Nash in Kenosha, Wisconsin. They were then shipped to England, where the chassis with "trailing link" front suspension was added by the Donald Healey's company. The chassis with engines were then shipped to Italy, where the custom Farina body was built by hand. The new Farina designed Nash-Healey was shown for the first time at the Chicago Automobile Show in February, 1952.

A Nash-Healey took first place in its class behind a Ferrari and a Talbot (and third among all entries) in the 1952 French Grand Prix at LeMans. Fifty-eight cars had started and only 17 finished.

In January, 1953, a Farina designed hardtop model was added to the Nash-Healey series. A total of 162 roadsters and hardtops were built in 1953.

Nash-Healeys haven't exactly vaulted to the top scale in postwar car values, but suffer the same as the Dual Ghia and other hybrid cars in that their perverse pedigree tends to give them the status of an orphan.

Someday, their value will be recognized...if only because of the magic of the Pinin Farina name.

A second example of the Farina coupe shows off the Pinin Farina coachbuilder's tag a bit clearer against its dark colored body. This car also has a Nash Healey script on the front fendersides.

The Pinin Farina body tag also graces the flanks of this Nash-Healey coupe, but note the difference in its rear window treatment. Cars with this style roofline are known as LeMans coupes. Only 60 were made.

Originally, the Nash-Healey came out as a roadster with a folding fabric top and side curtains for weather protection. The twin-carb engine used in the hybrid sports cars was also known as the "LeMans Six."

George Mason (l.) and Battista Pinin Farina look over the cockpit of a 1952 Nash-Healey roadster.

AMC's first sports car

By Wally Wyss

When American Motors was bought out by the French automaker Renault, some wags said "it won't be long before Renault thinking begins to influence what's left of the AMC designs." However, if You look at American Motors history, you'll see that tie-ins with foreign automakers were common.

One of the most memorable tie-ins was the three-way deal they worked out with Donald Healey, of England, to supply chassis, and Battistas Pininfarina, of Italy, to supply coachwork for the Nash-Healey sports car.

The car came about as a result of Healey's efforts. In 1950, he built a "Special" (a one-off car) around a Nash engine. He competed with it at LeMans in '50, '51 and '52, taking fourth, third and sixth in his class, respectively.

Healey had obtained the Nash engine from George W. Mason, president of the Nash-Kelvinator Corp. He commenced production on a Nash-Healey roadster as early as 1950. Those were the days when you could easily sell a replica of your LeMans cars to eager buyers. Healey built the frame and suspension and designed the body. It was a dud, as far as styling went. The engine, transmission and clutch came from America.

All Nash-Healeys were shipped to the U.S. to be sold at Nash dealerships. But, Nash had been talking with Battista Pininfarina about their sports car. Farina (he didn't change his name to "Pininfarina" until 1961) was eager to design a sports car, as well as sedans for Nash. After all, he had done many sports cars for Ferrari, Lancia and other sports car makers.

Unlike the sedans, Pininfarina was able to be more creative with the sports car, evolving a long hood/short rear deck look that still looks svelte, except for the grille, which was a '50s Nash trademark.

One reference source this author consulted says Healey built the Pininfarina-designed bodies, but it may be that Pininfarina also built bodies...at least for some of them.

The engine was an inline six rated at 140 hp at 4,000 rpm. Redline was at 4,500 rpm. Road testers at the time reported first gear peaked out at 30 mph, second gear at 52 mph and third at just over 100 mph. Nash built the blocks and heads in the United States and sent the chassis and blocks to England. There, Healey added two British S.U. carburetors, an 8.1:1 aluminum head and a high-lift cam to the stock Ambassador "Dual Jetfire" engine.

The 0-to-60 mph performance was not exhilarating. Being on the order of 14-16 seconds (one-third longer than it took a Jaguar XK120), it was snappy, but not speedy.

The 2,500 pound, American specification, Nash-Healeys were never to attain the success that Donald Healey had with the lightweight originals. That may have been because, in those early postwar days of racing, things were becoming more competitive each month. A car that won LeMans in '50 couldn't even qualify for the '54 event.

The Nash-Healey as a convertible was quite stylish, but a bit impractical. Severe wind buffeting set in when you got above 50 mph. The side curtains were inconvenient and may be one reason that the '55 T-bird, with roll-up windows, was welcomed by sports car buffs.

A coupe version was also offered in 1953, in fact two different styles. One had a wraparound rear window with two chrome rear window trim bars, like a '50s Chevy. The other had a smaller rear window of more conventional oblong shape.

The Nash-Healeys didn't stay around long. Corvettes were starting to catch on. By 1955, the Corvette V-8 and the V-8 powered, roll-up-windows, plush-trimmed T-bird soaked up what would have been the Nash-Healey's market. This contributed to Nash and Hudson merging to form American Motors.

Three decades later, critics of the Renault deal joked that the company ought to be called "Franco-American Motors." Few of them recalled that AMC's predecessors had flirted with English and Italian automakers years before.

Nash expert Vince Ruffolo is the owner of this LeMans coupe with the non-wraparound back window.

Causing quite a stir in this photo taken in Italy is a 1954 Nash-Healey Pinin Farina hardtop. Although a product of American/British/Italian linkage, the Nash sports car was built exclusively for the U.S. domestic market.

The NXI prototype as shown by Nash in 1950.

Cars of AMC: the Metropolitan

By Robert C. Ackerson

Although it was introduced in March, 1954, some two months prior to the creation of American Motors, it seems appropriate to regard the Metropolitan as a true AMC automobile since it remained available until early 1962.

Prior to World War II, Nash-Kelvinator management, particularly George Mason, had desired to expand the price range of its models as a means to increase sales volume. Little attention was paid to any moves into the upper price field, since the high tooling costs involved could not be justified in light of the relatively limited size of the market. Thus, Nash-Kelvinator directed its efforts to the development of less expensive automobiles that were not currently produced by American companies.

Although not every one there agreed with this program, it had the full backing of George Mason. He firmly believed that "many people will pick something new and different if given half a chance." Thus, after the war ended, Nash began an extensive study into the field of "personal transportation." One of the key factors affecting this research was the awareness that the average passenger load of an automobile was just 1.8 persons and that the driver was, most of the time, the car's only occupant.

Mason's Northern Michigan estate became a transportation laboratory where various automobiles, mainly of European origin, were tested. In addition, a number of motorcycles and motor scooters were evaluated, along with engine designs with one-, two-, four-, six- and eight-cylinders placed in both inline and V-type configurations.

In 1948, George Mason showed designs for a three-wheeler to George Romney, but by 1949, he abandoned this idea in favor of a conventional four-wheel layout. Mason, like many would-be manufacturers of inexpensive automobiles, hoped to keep the price of his dream car under the magical $1,000 level. He soon realized that it would be impossible to domestically produce the type of car he desired at a price anywhere near that point.

Existing European economy cars were found lacking in such areas as styling, comfort and prestige value. This lead to the rather original position that Nash should consider designing an automobile for the American market that could use existing components of a European manufacturer and be assembled abroad.

Initially Italian, French, German and British firms were in the running for receiving this contract if and when it was offered, but the public display of the NXI (Nash Experimental International) in 1950 tipped the scales heavily in favor of Austin, in England, rather than an automobile based upon the Fiat Topolino chassis and running gear.

The first NXI, a maroon convertible, was introduced at the Waldorf Astoria Hotel in New York City. Then, it embarked on a cross-country showing. At each exhibition a questionnaire was available. In excess of 235,000 were distributed. Tabulations of this early sampling of public opinion indicated a preference for a single-cushion front seat, rather than the divided seats of the prototype. The public also preferred a wider tread and longer wheelbase, plus auxiliary seating for two passengers at the rear. Not surprisingly, few Americans were terribly excited about the prospect of purchasing a NXI with a two-cylinder, 500 cc. Fiat engine. The overwhelming preference was for a more powerful four-cylinder engine.

Nash-Kelvinator introduced a new, small-car in 1950. But, when measured against this compact Rambler's 100-inch wheelbase, the NXI was (in the American view) ultra-small. It weighed 1,000 pounds less than a Rambler, had a wheelbase 16 inches shorter and possessed an overall length nearly three feet smaller.

Still, the development of the Metropolitan continued. The next step came on Oct. 5, 1952 when George Mason announced that Austin and the body firm of Fisher & Ludlow had been selected to produce the NIX, which soon was being referred to as the NKI (Nash-Kelvinator International).

By the following summer, preliminary models of the Nash mini were being evaluated by Nash-Kelvinator's engineering section. In March 1954, it debuted as the Metropolitan. In its convertible form, the Metropolitan listed at $1,469. The coupe version was priced slightly lower at $1,445.

From a styling viewpoint, the Metropolitan was pure Nash. Both front and rear fenders had built-in wheel skirts. Also shared with its big brothers was the Metropolitan's oval grille opening, high fender lines, indented lower windshield edge and low-mounted side body rails. This format, on a car with an 85-inch wheelbase and 149.5-inch overall length, wasn't a styling triumph, but it gave the Metropolitan friendly looks to go with a good ride.

The Metropolitan's suspension was also a design. Although all Metropolitan components (except for the sealed beam headlights and windshield wipers) were of British manufacture, the entire car was engineered in Detroit. Thus, the front springs were mounted between the wheelwell structure and upper control arms, as on all Nash cars since 1952. A Hotchkiss-type rear suspension was used. It had rubber-mounted, semi-elliptic leaf springs. Body construction was of single-unit type.

Powering the Metropolitan was a modified Austin A-40, four-cylinder, overhead valve engine displacing 73.17 cid and developing 42 hp at 4500 rpm. Peak torque was 62 lbs.-ft. at 2,400 rpm and the compression ratio was 7.2:1. Combined with a weight of 1,784 pounds (convertible) or 1,825 pounds (hardtop), this output enabled the Metropolitan to reach a top speed of just over 70 mph. It could achieve 0-to-60 mph in approximately 21 seconds. For anyone tuned into foreign cars, the Austin A-40 sport engine's dual Zenith carburetors could easily replace the Metropolitan's single Zenith for a quick and easy power boost.

Nash was apparently concerned that a four-speed manual transmission would be unsettling to prospective Metropolitan buyers. Thus, the A-40s bottom gear was blocked out to provide the Metropolitan with three forward speeds. As was the case with larger Nash models, the Metropolitan's shift lever was dash-mounted. This made skinned knuckles a common trait among Metropolitan drivers.

The Metropolitan's interior was simple, but certainly not spartan. The front bench seat was upholstered in leather, nylon and Bedford cord and all gauges were gathered in a single circular housing. Warning lights were provided for oil pressure and ammeter, but neither a warning light nor a gauge was provided for water temperature monitoring.

This shortcoming was remedied in 1958, when a water temperature gauge was optional. Standard equipment included directionals, electric wipers, cigarette lighter, dual sun visors, foam rubber front seat cushions, rear seat upholstery and a continental tire mount and cover. Optional equipment included the Nash Weather-Eye heating cooling system, custom radio, white sidewall tires and a partial flow oil filter.

Nash described the Metropolitan as "a practical car for small families, a sensible second for any family." Within four months, some 8,000 had been sold. The Metropolitan's sales undoubtedly benefited from positive comments made by both domestic and European automotive journalists about its design. Floyd Clymer (*Popular Mechanics*, November, 1954) reported: "It may well be that Nash has started a new trend in American motoring." *Motor Trend* (May, 1954) concluded: "If this car doesn't crack the American small car market, there is no such market." While the Metropolitan was by no means a sports car it could, said *The Motor* (Sept. 1, 1954) "be cornered quite as fast as other small cars on both wet and dry roads."

Although over 13,000 Metropolitans were sold in 1954, only 6,096 were retailed during 1955, when the Metropolitan was available either as a Nash or Hudson model. Nonetheless, it outsold all imported automobiles, with the exception of Volkswagen.

For 1956 a revamped Metropolitan 1500 model was introduced. It had a new mesh grille and side trim. This allowed two-tone paint schemes. These changes, plus the removal of the false hood scoop, were accompanied by the use of the 90 cid Austin A-50 engine with 52 hp at 4500 rpm and 77 lbs.-ft. at 2500 rpm. Since this was the basic BMC "B" engine, as used in the MGA, enterprising Metropolitan owners could easily modify their cars into mini Q-ships. Actually AMC wasn't ignorant of this potential and constructed at least one extra-performance Metropolitan with a high-compression head, dual S.U. carburetors, overdrive transmission and stiffer suspension.

The Metropolitan's best sales year was 1959 when 22,209 were sold. By that point, the Metropolitan had an outside trunk lid, larger 5.60 x 13 tires, and side window vents. However, American Motors regarded the revived Rambler as possessing far greater sales potential and Metropolitan production ended in mid-1960. Additional factors contributing to the Metropolitan's demise were an indifferent service attitude by many AMC dealers, the questionable reliability of some British components, and a limited advertising program.

Metropolitans were still available as late as 1962, when 412 were sold. The previous year, 856 had been purchased.

While Metropolitans came in only two body styles (the convertible was most popular), AMC did construct three station wagon prototypes that could have been available for the 1960 model year. The wagon, it's very likely, could have become very popular. But AMC, sensing big profits with its domestic compact, regarded the Metropolitan as more of a nuisance than an asset in its great comeback drive.

The 1960 Metro station wagon prototype. A clever design that might have caught on.

Though better known as Nash-Metropolitans, some of the early minicars, like this one, were Hudson-Metropolitans. The only real differences were the use of a Hudson grille center medallion and Hudson hubcaps.

Once spurned by collectors, Metropolitan now enjoying a revival in popularity

By Betty Cott

Twenty years ago, most collectors wouldn't have given the Metropolitan a second glance. The majority of Mets were tired, tatty little cars; too young to interest restorers and too old for the used car lot.

They occupied an automotive limbo and many a Met ended its days in a junkyard. However, as interest in collecting, preserving and restoring postwar cars gathered momentum, Nash's tiny car slowly, but surely, began to enjoy a revival. Today, it has all the appearances of turning into a full-blown cult car.

The Metropolitan, brainchild of former Nash Prexy George Mason, carries a number of automotive firsts to its credit. It was the first U.S. car for the American market that was built in England. British components, right down to its Pye radio, were used. It would have to be credited with becoming the first successful American sub-compact. This was no small feat at a time when the average U.S. automobile was suffering from ever-increasing bloat, fins, and a relentless horsepower race. It also was in tune with the times, as the overseas manufacturers, such as VW and Renault, were launching little cars with a vengeance and carving out an increasing market share for themselves.

Americans had been offered small cars before. Cyclecars, for example. But, the cyclecar craze of 1912-1914 began on a dubious premise and ended in dismal failure. Later, the American Austin and Bantam provided grist for the comedian's mill, but little money for company stockholders. And the Crosley, while nearly achieving a breakthrough, was just a tad too small for U.S. buying tastes.

Prior to start up, the Metro went through many changes. To begin with, the original design had been accepted in haste to get the ball rolling. By March, 1954, the cars had acquired severely simple bumpers, a new grille, and a sporty hood improvement in the shape of a stamped, chrome-faced air scoop. For the first two years, chrome-on-brass high quality hubcaps sported the "Nash" script. Their design was then changed over to "M" for Metropolitan. Coupe and convertible coupe were delivered in plain, solid colors, unmarked by any side trim, whatever.

In 1956, complaints of sluggishness of the Austin A-40 engine (chosen for its absolute economy) brought the decision to switch to the A-50 engine. This changed horsepower from 42 to 52. Austin, of course, could take this upgrading without missing a stroke. A mesh grille replaced the original bar type, with an oval chrome enclosure around it and center medallions (which read "M" or "H," depending on whether the car was bought from a Nash or Hudson dealership.)

The 1956 upholstery was changed over to a smart combination of striped broadcloth and vinyl in grey and white. The English-type open glovebox on the dash received an American-type locking door. The most visible change (accompanied by re-scaling of prices by more than $100) was two-tone paint. The contrasting colors were separated, horizontally, by chromed striping.

The car's creators were happy to invest in whatever changes the public wanted, primarily because of startling sales from the beginning. While the sales division had prayed for 10,000 car sales in the first 12 months, no less than 8,100 cars were ordered in the first quarter of the model year! The record year was 1959, when 22,200 Metropolitans were sold.

At no time was there a "typical" buyer. Even though the sales theme was always impressive economy, far more cars were sold to higher-income buyers, than to budget-wary purchasers. Often, they became second cars for wealthy families. Many were bought, by indulgent husbands, for their wives. Nowhere were they more in evidence than on college campuses, particularly in the East where Metropolitan sales were always strongest.

A black convertible with gold fittings went to Princess Margaret of England with AMC's compliments. Convertibles turned up regularly in movies...and on TV shows...with tops down. Tales of 40 mpg helped to keep prospects streaming in. As the car convinced the public that it was a relatively maintenance-free, dependable vehicle that was equally satisfying in city traffic or on the freeway, skeptical buyers accepted them.

Not that the Metro was perfection personified. Thousands of owners burst brake lines by pouring mineral-based American brake fluid into the Girling brake system. It was designed only for vegetable-oil brake fluids. Long-haul drivers complained that the wheels, 13 inchers, were too small, which promoted heavy bearing wear. Like their full-sized Nash brothers, the Metropolitans developed an annoying habit of sticking in gear. This situation could only be overcome by raising the hood and lining up the two short arms which operated the transmission levers by hand.

Few owners liked the original access to the luggage compartment, which was through the hinged rear seatback. To move anything heavy or awkward in and out of the trunk from inside meant constant knee damage to the front seat cushion, twisted backs, and torn upholstery. In 1959, AMC corrected this universal complaint by adding a trunk lid.

Door locks were prone to disintegration, as were strikers and pins. As a result, doors often flew open while the cars were rounding turns, or locked owners out without warning. The hand-operated convertible tops were difficult to raise and lower.

Long and loud were complaints over two errors, in the Metro, which involved the use of plastic. Owners were angered over the bright chrome-finished plastic bezels around the taillights. They crumbled easily, if hit by even a slight blow, or weathered swiftly into cracked segments that discolored badly and spoiled the car's appearance. Similarly, inside the car, the plastic drum-shaped ignition and light switch also cracked and crumbled, or melted whenever the slightest short occurred. After a year or two of use, few Mets had the original light switch in operation. Although plastic was used only sparingly in the Metro body, the results were disastrous. There was also an annoying drumming caused by sheet metal. It could be noticed at various speeds, usually in the trunk area.

On the plus side, however, the little Nash produced much respect from its owners. With the A-50 engine replacing the A-40, the car was fast enough for freeway use, nimble on the acceleration lanes, exhilarating to drive, and more than reasonably dependable. Even though 90,000 were sold in the U.S. (the remainder went to England and Canada) there were never enough mixed in with U.S. traffic to make them commonplace. Probably, no new marque was less understood by the general public.

Most owners liked their cars. Few, if any, became hot rods or funny cars. Few were repainted in anything, but factory-spec colors. The factory was aware of whatever shortcomings existed and, fearing the competition of VW on the horizon, kept pace by coming out with improvements (particularly in the 1959 Metropolitan 1500). Enhancements included quarter vent windows, more comfortable seats, better door locks, and window riser mechanisms, as the years went by,

With its 85 inch wheelbase, over 30 mpg economy and 78 mph top speed, the Metropolitan was undoubtedly the first successful sub-compact in American automotive history (although detractors will point out that it was 99 percent an English car).

The Metropolitan benefitted, as did the dowdy Model A Ford, from a unique combination of fortunate circumstances. The choice of Austin Motor Co., with its long experience, eliminated the inevitable bugs which might have littered the way if Nash had attempted to develop the powertrain and running gear on its own. Ludlow & Fisher, the body suppliers, compounded the quality of every Metro by building in far greater integrity than was specified in the original contracts. The firm applied huge quantities of hand-work to welding, fitting and shaping the bodies. Salted roads inevitably took their toll and finding a rust-free Met became the exception,

There was no external lid on this 1954 Nash-Metropolitan convertible's luggage compartment, so it made perfect sense to mount the spare tire outside on the rear of the car.

Many of the cars were purchased by relatively well-to-do families for service as a second family car, usually used by the lady of the house, while the "big car" (a Nash in this case) stayed home in the garage.

rather than the rule. Rust problems aside, though, many Metros posted more than 100,000 miles with far less repair and maintenance than their American-built contemporaries.

By 1959, however, the handwriting was on the wall. Suddenly, the Metro production schedule was cancelled. The reason was simple: the impressive success of the new AMC Rambler, which by then had gone through several model-years, saw it emerge as a much more attractive family car than the stark 1950 Rambler. When the Rambler division sold more than 50,000 cars in 1959 (just under half of total AMC sales) President George Romney had a decision to make.

The Metropolitan, popular as it was, was a two-seater. It showed less unit profit per sale and, unquestionably, appealed to a somewhat limited segment of the market. In contrast, the Rambler proved the salvation of AMC and sold in base form for only $105 more than the Metro. Obviously, the future was with the four- or five-passenger car.

AMC issued the shutdown order at the turn of the year. The last Metro rolled off the Austin lines in June 1960. Since then, the little Metro has gone largely the way of the Model A. For some years, the car seemed to vanish, turning up now and then in junkyards. Many were scrapped between 1965 and 1970. Others were kept in service by owners who simply admired "the little beast." Around 1970, collectors spotted the Metropolitan as a definite sleeper. The cars, which had been selling for $25-$50 out of backyards and salvage points, suddenly became desirable. Like the Model A of yesteryear (thousands changing hands for $10-$20 in running condition 35 years ago) collectors have recognized the Metropolitan as a doubly valuable acquisition. Not only does it easily lend itself to restoration, with almost all parts still available, but the car is likewise useful day-in/day-out transportation. It fits comfortably into today's 55 mph traffic.

Restored Metros, particularly later ones with trunks, are hitting price levels of $5,700 to $6,500. Metropolitan Owners Clubs were formed in the budding years of the marque. The Metropolitan Owners Club of North America, 5009 Barton Rd., Madison, WI 53711 can be very helpful to new owners or veteran collectors. This organization can be reached by phone at (608) 271-0457. Another source of help and parts is a West Coast business known as The Metropolitan Pit Stop, 5324-30 Laurel Canyon Rd., North Hollywood, CA

Though the bulk of Metros were sold in the U.S., some were marketed in England (like this 1957 convertible) or Canada.

The Metropolitan has become a very popular collector car in the last 20 years. This 1959-1962 style convertible was featured in the American Motors Owners (AMO) national meet a few years ago.

"Metropolitan Owners Club of North America" reads the license plate frame on this 1959 Metropolitan ragtop at Spud City Nats in Plover, Wisconsin. The plate itself advises that the car's owner belongs to the national club's Lemon Bay Chapter.

Even though it was a pioneer in the postwar small-car field, the Metropolitan was not a stripped-down model. Even a radio was standard equipment, along with twin visors. The "fuzzy dice" are not original, of course.

The appearance of the aluminum-bodied car was unusual and unique, with an oval eggcrate grille and functional air scoops over the front brakes. In order to maintain the sleek appearance of the Italia and still make it possible to get inside, the doors were recessed 14 inches into the roof.

Project Italia's goal: enhancing Hudson

By Robert C. Ackerson

In 1952, the Hudson Motor Car Co. reported to its stockholders that the production of 79,117 cars had netted the Company a tidy profit of $8,307,000. On the surface, this was reasonably cheery news since the previous year a loss of $1,125,000 had been sustained. While much of this could be traced to labor problems, a shortage of steel and government price control policies, the fact that Hudson sales had started a long downhill slide was painfully evident.

When the new step-down Hudsons had made their debut as 1948 models, the reception of a car-hungry nation was as expected; they were bought up with ravenous enthusiasm.

In excess of 140,000 Hudsons were shipped to dealers in 1948 and in each of the following two years. But in 1951, in spite of the introduction of the Hornet and its great success in stock car racing, production fell to 92,859. Even with continued victories in competition and the addition of the lower-priced Wasp series in 1951, sales failed to rebound to anything near the levels of 1948-1950. Instead they were dropping perilously close to Hudson's 70,000 break-even point.

By 1952, the situation was sufficiently serious to be called critical, but not yet of proportions to be considered of a terminal nature. In other words, Hudson could still afford a minor adventure or two into the world of dream cars and automobile racing. Both served as jumping off points for the Hudson Italia project.

In 1952, Hudson's long-time styling boss, Frank Spring, was given the go-ahead to create a limited production automobile based on Hudson Jet components. It was to be a car that could score an outright victory on the Mexican Road Race. Whether or not such an accomplishment would have had the effect of improving Hudson sales is highly questionable. Hard-core Hudson owners were a vociferous lot, highly prone to following the successes of such Hudson drivers as Marshall Teague and Herb Thomas with fanatic loyalty. It can, of course, be argued that without the favorable publicity generated by its racing accomplishments, Hudson would have been in even more serious trouble than it was. But even if racing helped sell some Hudsons, it had nowhere near the impact a new, good looking automobile sitting in Hudson showrooms would have had upon sales.

This obvious point was not overlooked by Hudson resident, A.E. Barit. Nor was the high level of interest the public displayed in the dream cars from GM, Ford and Chrysler. Hudson couldn't quite match GM's Motorama effort, but an automobile put into limited production that could compete successfully in racing would enhance Hudson's image as a forward looking company and sound out reaction to new styling concepts.

The mechanical elements of the Italia came from the Hudson Jet. This 105-inch wheelbase car had been introduced in February, 1953. It was quite a decent little sedan powered, in

109

"Super Jet," by a 202 cid flathead six with dual carburetors and a maximum output of 114 bhp. planned. When set up for racing, the 7X gave some 200 hp and would have made this automobile a competitor to be reckoned with.

An early 1953 Jet was sent to Carrozzeria Touring in Milan, Italy to be used as the prototype for the Italia. All of the 25 or 26 (sources differ as to the exact number completed) Italias built were constructed in Milan. The price of labor costs made it less costly to make limited production autos.

When the Carrozzeria Touring people finished getting the stock Jet sedan ready for transformation, all that remained was the chassis platform (the Jet had unit construction), the cowl section and some rear support braces. The appearance of the aluminum body car was unusual and unique. But the mixture of early '50s Italian themes such as the oval, egg-crate grille with garish American styling (ie. a stop/directional back-up light design impersonating a triple exhaust system), was not entirely successful. Yet, as the source of a potential design for a new production Hudson sedan, the Italia offered much of interest. Overall, its design was not hampered with excessive chrome and its rear deck, if one ignored the intrusive rear lighting system, was an extremely clean design. It was slightly suggestive of the 1963 Buick Riviera.

Above each single headlight were functional scoops that directed air to the front brakes. A similar favor was done for the rear brakes, via intakes set into the rear fenders. Production Hudsons, with their step-down design, were pretty low with a height of five feet. The Jet, which was less of a looker than Frank Spring had wanted it to be, stood 60-7/8 inches above the ground. In contrast, the Italia was 54-inches high.

In order to maintain the sleek appearance of the Italia and still make it possible for people to get inside, its doors were recessed 14 inches into its roof. All Italias were painted Italian Cream and had identical red and white interiors. The dash, covered with a non-reflective red crinkle and non-reflecting red finish, used stock Hudson Jet instruments. They were rearranged to give the Italia's interior an additional distinction of its own. Also adding to the Italia's classy interior was deep-pile Italian red carpeting.

Accommodations for the driver and his passenger were well thought out, functional and attractive. Each anatomical seat had a reclining back and was constructed of three different sections. Each contained foam rubber of a different density. This design, along with a shape that followed the contours of the occupants' shoulders, back and hips, enabled the passenger and driver, in Hudson's words, to "sit relaxed, without 'holding on,' even when the car corners at high speeds."

Certainly, the Italia was capable of such behavior with only a 1,900 lb. weight and good power. The excellent 202 cid Super Jet flathead six, with twin single-throat carburetors, developed 114 hp at 4000 rpm. The Italia could put more than a few of its V-8 contemporaries to shame in a drag race.

When Hudson announced the Italia to the press on Aug. 25, 1953, it described it as a "car of the future." President Barit added, "The Italia is a preview of many exciting projects now underway in Hudson's engineering laboratories. It introduces for the first time many new features and innovations which without question, will set a pattern of things to come."

The Italia's $4,350 New York port-of-entry price included Borrani racing-type chrome wire wheels (which Hudson claimed were designed especially for the Italia), directional signals, radio, heater, backup lights, and white sidewall tires.

Overall the automotive press received the Italia well. *The Autocar*, in general, approved its appearance, though describing some of its styling features as "frivolous." John R. Bond, writing in *Road & Track* (November, 1953) concluded that "of all the current crop of dream cars, this one seems the most practical." In his eyes, the blend of Italian styling and 'American space-luxury requirements' was seen as eminently successful."

On the heels of the Italia's press debut, Hudson dealers received a letter from the home office advising them that orders would be accepted from them for production model Italias. Just how strong and firm their response was is questionable. *Special Interest Autos* (Nov./Dec. 1971) suggested that Hudson received only 19 orders in response to this Sept. 23, 1953 letter. Yet a press release from Hudson Director of Public Relations, Thomas P. Rhoades, with a January 1954 dateline claimed that "orders from dealers far exceed planned production of the Italia. In an earlier press sheet for the Chicago Auto Show, Hudson's vice-president of sales, N.K. VanDerzee called reports on the Italia from dealers 'highly favorable.' The January 1954 press release quoting VanDerzee, added that "allocation to dealers will be on a regional basis to permit nationwide public demonstration of the Italia's performance." It's also interesting to note that whereas the Sept. 23, 1953, dealer letter mentions a $4,800 f.o.b. Detroit price, the January 1954 press release noted that while it "would be comparable to that of leading European sports-type cars," the delivered price of the Italia had not yet been determined.

Most accounts of the Italia conclude with the assertion that the creation of American Motors in May, 1954, brought the Italia venture to an end.

Essentially this is correct although its demise was not immediate. For example, on Aug. 18, 1954, AMC announced that the Italia was still in limited production. The date that Italia production actually ended remains vague.

Nine Italias (with serial-numbers ranging from IT-10001 to IT-10009) were designated as 1954 models. The remaining Italias, with serial-numbers IT-10010 and up, were given 1955 model year status. However, whether these latter cars were actually produced in 1955 is uncertain.

Another interesting bit of Italia folklore is that, according to AMC, "no owner's manual, catalog or other literature was printed by the company." However, Italia owners were provided with a mimeographed, four-page document simply titled "*Hudson Italia: Supplementary Operating Instructions: for Use with Hudson Jet Owner Manual.*" This little pamphlet contained information concerning the operation of such items as the Italia's hood lock, dome light and back-up lights.

A four-door derivative of the Italia, known as the X-161, was completed in early 1954 by Carrozzeria Touring. It's doubtful, however, that Hudson could have held out until 1957, when an X-161 inspired automobile would have been ready for production.

In any case, the formation of AMC and George Romney's strategy of putting all the corporate eggs into a compact car basket left no room for such a move anyway. After 1957 Hudson was no more.

In order to maintain the sleek appearance of the Italia and still make it possible to get inside, the doors were recessed 14 inches into the roof.

Identity Crisis: Nash/Hudson Merger

Ironically, the first major redesign of the step-down Hudson took place just before the company's merger with Nash-Kelvinator. This is the 1954 Hudson Hornet Special Club Sedan. (Jack Miller photo)

History of AMC: Part I

By Robert C. Ackerson

The merger of the Hudson Motor Car Co. and Nash-Kelvinator into American Motors, in May 1954, brought under one corporate umbrella a mixed bag of automobiles with widely divergent origins and characteristics.

The Hudson line consisted of the Hudson Hornet, Wasp and Super Wasp, plus the compact Hudson Jet, Super Jet and Jet Liner. These models were destined for an early demise as AMC products. It would be the Nash models, particularly those of compact size, that ultimately justified the merger.

Ironically, the big step-down Hudsons had just received the most extensive restyling since their 1948 introduction. In the eyes of many critics it was an excellent effort. Albrecht Goertz, designer of the BMW 507, remarked (*Motor Sport*, June 1954), "Hudson's rear-quarter treatment adds, so much to the design that it seems a shame that it was not done sooner." A similar analysis could have been made of the raised fender line, wider and squared off grille form and side trim. Hudson called it "Flight Line Styling."

Hudson had always produced great engines. The "Instant Action" Hornet engine of 1954 delivered 160 hp with a single two-barrel carburetor and 170 hp with Twin-H-Power. The less expensive Wasp and Super Wasp Hudsons had standard cast-iron 7.0:1 compression ratio cylinder heads. The 232 cid Wasp engine developed 126 hp in this form, while the larger 262 cid Super Wasp engine was rated at 140 hp. With the optional 7.5:1 compression ratio aluminum head (standard in the 308 cid Hornet engine) their respective maximum horsepower levels were 129 and 143.

On March 19, 1954 Hudson added a lower-priced Hornet Special four-door sedan that combined the 124-inch wheelbase Hornet body and engine with Super Wasp interior trim. For exterior identification Hudson Special front fender script was added.

A few weeks later, on April 11, 1954, Hudson attempted to stimulate sales of the Jet by announcing a Family Club Sedan model with a list price of $1,465. This represented a significant price drop of $200 from previous Hudson models and made the Jet less expensive than any product offered by the "Big 3." Hudson's sales vice-president, N. K. VanDerzee, sounding like a voice from the past noted "there are many millions of American families without an automobile, and many more in need of an economical and dependable second family car. By careful planning, we were able to effect new economies in the production of a Jet Family Club Sedan, which will bring Hudson quality and performance within the reach of many more families."

Although this was a laudable move, the Jet simply wasn't attracting enough attention in the first place to get any mid-season sales promotion off the ground. This isn't to say that the Jet was an automobile unworthy of consideration by new car buyers. All Hudsons had redesigned combustion chambers, a longer dwell cam, improved porting and the traditional chrome alloy cylinder block. These attributes, plus the Jet engine's oversize bearings, made for a tough, rugged engine of simple L-head design. With its standard 7.5:1 cast-iron cylinder head, the Instant Action Jet engine developed 104 hp. With the optional 8.0:1 aluminum head and Twin-H-power this 202 cid 6-cylinder's output rose to 114 hp.

At the opposite end of the Jet's price range from the Family Club Sedan was the Jet Liner, whose interior featured pleated, foam rubber seats, two-tone door panels and an antique-white Plasti-Hide headliner. All Jets had redesigned interior layouts that added over two inches to the rear legroom.

Rounding out the Hudsons inherited by AMC was the Italia. In many ways this automobile was the final manifestation of many elements of Hudson's glorious past. A relationship with Railton and Brough, in the 1930s, lead to high-performance cars powered by Hudson engines and fitted with European styled bodies. This provided a natural background for the Italia, with its body designed and produced by Carrozzeria Touring in Italy. Hudson had serious plans to enter the Italia in the 1954 La Carrera Panamericana. Hudsons had done well in the previous "Mexican Road Races."

In an early memo from Hudson vice-president Stuart G. Baits to president A. E. Barit, Baits noted that "we are planning to use the Hornet power plant, axles, wheels, tires, etc." This wasn't carried out. The Italia, in its short production run, utilized the Jet's engine and running gear. However, the Italia's importance transcends this obviously limited impact. A four-door derivative of the Italia, the X-161, very likely would have served as the basis for a new generation of Hudsons, if the company could have survived as a separate entity.

Although Nash-Kelvinator had followed a different path of historical development than Hudson, its model lineup in 1954 was remarkably similar. Three years before the Hudson Jet, Nash had launched its Rambler. By 1954, it evolved into a line of four-door sedans, station wagons, convertibles and two-door sedans. When it was first introduced, in 1950, the Nash Rambler had attracted attention as a compact car whose fairly high list price reflected a product with quality appointments and many standard features. Nash didn't retreat from this format, but by 1954, divided the Ramblers into DeLuxe, Super and Custom models. This allowed a basic two-door Club Sedan to be sold for $1,695. Also expanding the Nash Rambler's market was a longer, 108-inch wheelbase for the 1954 four-door sedan. Standard on this model was a more powerful, 195.6 cid/90 hp "Super Flying Scot" engine. The other Ramblers had slightly smaller (184 cid) and less powerful (85 hp) versions of this L-head six.

The full-size Nash Statesman and Ambassador models were reskinned in 1952. Their styling was influenced by Italian designer Pinin Farina. All Nash products were holding firm to skirted front wheels, first introduced in 1949. Both the Stateman and Ambassador were large automobiles. Tom McCahill in *Popular Mechanix*, (September 1952) once remarked that "to get behind the wheel of the Ambassador is like being on the bridge of the Queen Mary." Their 78 inch width and 209-1/4 inch overall length (202-1/4 inches on Statesman) certainly justified this description.

As early as 1951, Nash had been marketing the Nash-Healey sports car. In its 1954 form, this was an extremely handsome Pinin Farina-styled two-seater available either in roadster or coupe form. Although its production was terminated in August 1954, after a total of 506 cars had been built, the final form of the Nash-Healey coupe body was used by Farina to create a prototype for an all-new full-sized Nash sedan.

It's not surprising, in light of the overall sales appeal of these cars inherited by AMC, that prospects for its survival seemed bleak. Hudson had lost nearly $10.5 million in 1953 and equally dismal was the production tally for 1954...only 50,660 Hudsons had been assembled. Nash was in a slightly stronger financial position, having earned just over $14 million in 1953. However, the big Statesman and Ambassador had clearly seen their glory days and their respective output was down to 20,225 and 21,425 cars. Even the Rambler's production for the 1954 model year was a modest 36,321 units.

Casting a darker shadow over the early days of American Motors was the death of its chairman and president, George W. Mason on Oct. 8, 1954.

Mason was the leader of Nash-Kelvinator since the union of Nash Motors and Kelvinator in 1937. He had vigorously pursued three goals he regarded vital to company survival: marketing of a quality, lightweight automobile; merger with another independent automobile manufacturer; and the grooming of a successor.

The 1940 Nash 600 was a successful manifestation of the first goal and the Nash Rambler was its offspring. The merger with Hudson followed the second goal, though it had fallen short of Mason's desire to include Packard. (To some, the acquistion of Hudson's outdated plants and products seemed hardly a step forward.) As for goal number three, at the helm of this new company with a proud heritage and a very uncertain future was one of the most remarkable individuals ever to head an American automobile company ... George Romney.

It would be Romney's leadership and intuitive sense that America really was ready to embrace a domestic small car that would enable American Motors not only to survive, but to prosper.

Nash Motors was in somewhat better financial shape than its new partner, but cars like this 1954 Nash Ambassador four-door sedan had already seen their glory days in the marketplace and sales were tapering off.

AMC's future would lie in its smaller models, like this 1954 Rambler four-door sedan. It featured Pinin Farina styling and a longer wheelbase than the early two-door Rambler wagons and convertibles.

Advertisement for the 1955 Nash Ambassador Country Club hardtop highlights its modern "Airliner" styling and high resale value. The 208 hp V-8 was shown, with no indication that the engine was sourced from Studebaker-Packard.

History of AMC: Part II

By Robert C. Ackerson

The fourth year for the Nash "Golden Anniversary" styling of 1952 was 1955. In the '50s, even a three-year styling cycle seemed long if sales were to be maintained. Any car with an appearance as dated as the Nash's seemed headed for trouble. However, 1955 was a banner year for the industry. Nash output saw a substantial increase to 57,535. (In fairness to Nash, at least some of the improvement was due to changes made in its 1955 format.)

Few new-car buyers who looked over Ford, Chrysler and GM offerings were likely to opt for a big Nash, but at least it had enough new features to stimulate a longtime Airflyte owner to trade in his faithful old inverted bath tub for a new model.

The transfer of Hudson production to Kenosha, Wis., plus problems inherent in bringing two corporate operations under single command, put the development of a substantially changed Nash on the level of a minor miracle. A contract to purchase V-8 engines from Packard was dated July 1, 1955 and it wasn't until early October that management approved a wraparound windshield for 1955 models.

The new Nashes also had retooled cowls, front and rear quarter panels, doors, pillars and roof dies. Coming directly from the discontinued Nash-Healey were its inboard headlights set in an oval grille with vertical-concave bars. Nash claimed placing the headlights closer together shed more light on the road, especially under rainy and foggy conditions. To alert oncoming drivers of the Nash's width, fender-mounted parking lights operated with the headlights on.

Two-tone paint schemes were in vogue. The curved forward front fender edges, plus a similarly shaped side character line, encouraged a rather pleasing dual color format. It couldn't disguise the Nash's age, but along with a larger front fender wheel cutout (that also reduced its turning radius by 1/2 foot) it gave the Nash a more modern appearance. Credit is also due a new wraparound windshield. Nash claimed it was the industry's largest.

Both four-doors and hardtops were available in Ambassador and Statesman lines, in either Super or Custom trim, with a choice of six-cylinder or V-8 engines. Standard equipment for the more expensive Custom versions included such traditional Nash features as tinted sun visors, a six inch wide retractable rear armrest, visors with storage netting, and a continental tire kit, as well as the "twin bed" seats. For the latter, Nash offered optional air mattresses and side window screening.

A 320 cid/208 hp Packard engine was used in Ambassador V-8s. They were $290 to $300 more expensive than sixes and carried small emblems on their rear fenders for identification. Early models had a 7.8:1 compression ratio, but a midyear change boosted this to 8.25:1. With the standard Twin-Ultramatic transmission, the V-8 Nash was a good performer with a 0-to-60 time of 13 seconds, a quarter-mile mark of 18.6 seconds, and a top speed of 108 mph.

Custom model Ambassadors were also available with the ohv, 252.6 cid six-cylinder engine in its twin-carburetor, 140 hp "LeMans Dual Jetfire" form. Super versions had a single-barrel carburetor and a 7.6:1. compression ratio head in place of the Custom's 8.0:1 version. The smaller Statesmans were offered only with two versions of a 196 cid, L-head six. The standard engine carried a 100 hp rating and a 7.5:1 compression ratio. The installation of a dual carburetor setup and an 8.5:1 aluminum cylinder head increased horsepower to the 110 level.

The successors to these cars debuted on Nov. 17, 1955. A 2-3/4 inch higher rear fender line ended in a new "clenched fist" taillight, with new "Speedline" side trim providing the

Ambassadors with a revamped two-tone paint arrangement. Only one Statesman model, a four-door "Super" sedan was offered. Its engine was updated with overhead valves. The 195.6 cid displacement of the L-head was retained, but with 130 hp and a 7.4:1 compression ratio.

AMC claimed the Typhoon six had the highest horsepower to displacement ratio of any six-cylinder engine. As before, all Ambassadors and Statemans with six-cylinder engines could have dual-range Hydra-Matic or overdrive in place of the standard three-speed manual transmission.

The Ambassador continued to be available in four models. Only one, the Super four-door sedan, was offered in six-cylinder form. This engine, in single carburetor form, developed 135 hp. With dual Jetfire carburetion, its output rose to 145 hp. A Super four-door sedan, along with a Custom four-door sedan and two-door hardtop, had a Packard V-8 standard. For 1956, it displaced a hefty 352 cid. With a 9.55:1 compression ratio and single Carter two-barrel, it rated for 220 hp at 4600 rpm. As in 1955, Ultramatic was the only transmission available.

American Motors was not happy with the V-8s from Studebaker-Packard. Shortly after assuming control of AMC in 1954, George Romney conducted a press conference at which he had reported that no merger talks were planned with Studebaker-Packard. However, he expressed hopes there would be more product reciprocity between the two companies. When this failed to materialize, AMC began development of its own V-8. It debuted in the midyear Nash Ambassador Special model. This 3-1/2 x 3-1/4 inch bore and stroke engine first displaced 250 cid. With an 8.0:1 compression ratio and two-barrel Carter carburetor, it's peak horsepower was 190 at 4900 rpm.

After losing $6.9 million in 1955, AMC reported a nearly 20 million loss for 1956. Although 4,145 Special V-8s had been sold, the appeal of the dated full-sized Nash format had long since peaked. Of AMC's 104,189 unit output for 1954, a total of 79,166 were Ramblers. Total production of Ambassador and Statesman models was represented by just 17,842 cars. Thus, at a November 1956 meeting of AMC's Board of Directors, the decision was made to drop both Nash and Hudson after the 1957 model run.

The result was a severly curtailed Nash lineup for 1957. All Statesman models were dropped, along with all six-cylinder engine options. Ironically, the final form of the Nash Ambassador became the first American automobile to offer dual headlights as standard equipment. Their use in a vertical form, plus an eggcrate mesh, oval-shaped grille with a concentric outer chrome ring, was well suited to the large Nash body. A flatter and lower (by nearly two inches) roofline, plus reshaped side trim and full front wheel cutouts, gave the Ambassador Country Club two-door hardtop its best appearance in years.

With AMC's V-8 bored out an additional 1/4 inch and displacing 327 cid, there was no need to purchase engines from Studebaker-Packard, even if the Packard V-8 was available.

Thanks to a higher 9.0:1 compression ratio and a four-barrel carburetor, the AMC V-8 developed 255 hp at 4700 rpm and 345 lbs.-ft. of torque at 2600 rpm. With Hydra-Matic, this engine could propel the Ambassador from 0-to-60 mph in under 10 seconds.

Total production of the last examples of the Nash Ambassadors totalled only 3,562 cars.

"Too Hot to Hold for '57," was American Motor's theme for the midyear 1956 Nash Ambassador Special. This two-door hardtop was the first to use the Nash-built Torque-Flo V-8 shown in the ad noting "Product of American Motors."

For 1957, the new in-house V-8 was bored to 327 cid and advertised as "255 hp strong." The copywriters credited it with blazing acceleration and amazing economy. It made the Ambassador Custom sedan the "world's finest travel car."

Family Gathering: AMC Style

The controversial Marlin was the first real fastback in the mold of today's 2+2 models. Sales were not exactly awe-inspiring and the car is slowly becoming a collector's item today.

Rambler - the second time around

By R. Perry Zavitz

It was the Rambler bicycle which begat the Rambler car, which begat the Jeffery, which begat the Nash, which begat the Rambler, which begat the Hornet and the Matador, the Ambassador, the Marlin, the Gremlin, the Javelin, the American, the AMX, ad infinitum.

This review concerns the second coming of the Rambler ... the 1958-1967 models.

Their history goes back to 1950 when Nash offered a compact car and revived the Rambler name for it. The Nash Rambler enjoyed moderate success until the recession began in the latter part of the 1950s. By that time, however, it had grown noticeably bigger than its 1950 predecessor.

In 1958, American Motors Corp., formed as a result of the merger of Nash and Hudson decided to discontinue their full-size cars, as well as the Nash and Hudson names. They concentrated on the Rambler, instead. Rambler became the marque name.

The Rambler line was expanded in two directions. A larger Rambler was introduced. It continued the top model name Ambassador, begun in 1932 by Nash. The Rambler Ambassador was nine inches longer in wheelbase and overall length than the standard Rambler. A 327 cid 270 hp V-8 engine was used.

A 250 cid/215 hp version was optional in a regular-size Rambler called the Rebel. Some have erroneously assumed that AMC's 327 cid V-8 was really the Chevrolet 327 power plant. There is no connection between these two engines. If there were, the relationship would be the reverse...AMC had its 327 V-8 five years before Chevrolet had one.

Smaller than the regular size Rambler of 1958 was the Rambler American. It was a reintroduction of the basic 1954-1955 Nash Rambler two-door sedan on a 100-inch wheelbase. Offhand, we can't recall any other car maker reviving one of its own models after such a short interval.

The product and marketing change begun for '58 at AMC, soon paid off. The company was the only U.S. car maker to increase sales in the recession year of 1958. Every other make of car was down substantially that year, except the new Edsel. The consumer trend was towards compact cars. Rambler was the only U. S. compact then. AMC was at the right place at the right time.

For 1959, the two-door station wagon was revived for the Rambler American line. Slight styling changes were made, as expected, to the Rambler six, Rebel and Ambassador. The introduction of the compact Studebaker Lark for 1959 went virtually unnoticed at Kenosha, Wisconsin Rambler registrations doubled from the 1958 figure.

However, 1960 was to be a bit different. Rambler had to brace itself for new compact competition from Ford, Plymouth and Chevrolet. So, a four-door American was offered.

At this time, all companies were emphasizing the station wagon. AMC was right in there, too. For 1960, they offered 14 such models. These included three-seat models in all but the American series. The three-seat models were also called "five-doors," because the rear door was side hinged and did not have the usual tailgate arrangement.

The last four-door hardtop station wagon made by AMC was in the 1960 Ambassador lineup. It had been in production since 1957, when it made history as a first. Oldsmobile, Buick, Dodge and Chrysler were the only others to offer four-door hardtop station wagons.

It was around this time that the Ambassador was being touted as "the world's most luxurious compact." This claim was really pure Madison Avenue hype. With an overall length of 198-1/2 inches, this Ambassador was hardly compact. Certainly it was less compact than Britain's Mark IX Jaguar, France's Citroen-DS, or Germany's Mercedes-Benz 220...and less luxurious, too.

117

Rambler held its own in 1960, despite all the competition from the Big 3 compacts. Even more competition came in 1961. Pontiac, Oldsmobile, Buick and Dodge introduced cars in the mid-Rambler size. Rambler now called this model the Classic. Perhaps the mad Madison Avenue boys got carried away again.

The most exciting news from Rambler that year was made by the American series. It received a new body and several new models. A four-door station wagon and a convertible were added. This convertible was conventional. It was not based on the 1950 Landau convertible.

The American and Classic series came in three degrees of trim: Deluxe, Super, and Custom. The Custom Classic six had a different engine from the others. Instead of the old 195.6 cid 90 hp L-head, it had an ohv engine of the same displacement. This was rated at 127 hp (138 hp with the optional two-barrel carburetor). This engine had an aluminum block with cast-iron cylinder liners.

Two things disappeared from the Ramblers of 1962. One was the optional 250 cid V-8 engine for the Classic. The other was the 117-inch Ambassador wheelbase. It became the same size as the Classic, using the 108-inch wheelbase.

A realignment of some trim designations marked the 1962 model year, too. Super and Custom names were replaced by Custom and 400, respectively. Two-door sedans were added to the Classic line, a first for this size Rambler. The ohv six engine was optional on the Deluxe and Custom American.

There was no Classic V-8 until mid season. When introduced, it featured an engine with 1/4-inch greater bore than before. It had a 287 cid and was rated at 198 hp.

A two-door hardtop was added to the American series for 1963. The Classic and Ambassador got fresh styling that year. It included curved side windows for the first time, except in luxury cars. The wheelbase for the new models was increased four inches to 112, but overall length was actually, shortened 1-1/4 inches.

A complete revamping of trim names took place. "Renumbering" would be more correct. The American came in 220, 330, and 440 versions. The Classic offered 550, 660, and 770 models. The Ambassador came in 880 and 990 variations. The latter two titles were reminiscent of the Nash 8-80 and 9-90 of 1931-1932. So often numbers used as model names have no significance to the buyer. This applied to the Rambler as well, but at least the sequence made sense.

A new six-cylinder engine of 232 cid was standard on the Classic 660 and 770, optional on the 550. Optional on all Classics were the 287 and 327 cid V-8s.

The 1965 American was little changed, but the Classic and Ambassador had crisper lines with a greater difference in front end design between them than before. The Ambassador grew four inches in wheelbase and five inches overall.

Convertibles were added to the Classic 770 and Ambassador 990 models.

A new 232 cid six-cylinder engine was introduced. Rated at 145 hp with one-barrel carburetor, it was standard in the 660 and 770 Classics. In its 155 hp two-barrel carburetor version, it was optional in the American (with automatic transmission) and Classic, but standard in the Ambassador.

Much publicity was devoted to a new model AMC introduced after the emergence late in the previous season of the Ford Mustang and Plymouth Barracuda. It was called the Marlin.

Perhaps the name is appropriate. There was something fishy about the way the automotive press panned that car. If they had a point, they also should have panned the Ford Torino and Mercury Monterey two-door fastbacks with equal venom. The later cars appeared after the Marlin died. What all these cars had in common was a fastback roof grafted to a standard lower body.

The Marlin was basically a Classic with an attractive gentle sloping roof. A choice of seven solid or 26 two-tone colors were offered. The two-tones used the second color on the roof and narrowing down the back. This application of two colors looked very good in some combinations, but rather hideous in others.

The Marlin's standard engine was the 155 hp six, with either of the AMC V-8s optional. Although the Marlin and the Ambassador continued in 1966, the Rambler name was not applied to them. They were distinct marques.

The Rambler name remained in the American and Classic car-lines. A juggling of sub-series names occurred again. The American was offered in 220, 440, and Rogue versions, while the Classic came in 550, 770 and Rogue trims. The same varieties were continued by the American for 1967. However, the Classic name was dropped, thank goodness. Its newly styled replacement was called Rebel. The Rebel's top version was the SST. AMC denied that SST stood for anything, but the similarity to the designation of the futuristic supersonic transport plane was obvious and no doubt deliberate.

Little change to these cars was made for 1968, but the name Rambler was omitted. American and Rebel now became marque names. Nevertheless, the spirit of Rambler, if not the name, lives forever.

For 1959, a two-door station wagon was revived for the Rambler American line. These cars represented a mild update of the early postwar Rambler. Of course, the Rambler name itself went back to bicycles and horseless carriages.

Station wagons sold well in Rambler-land. Side-hinged rear doors replaced conventional tailgates on "five-door" models with three seats, so this 1962 Rambler Ambassador with tailgate is a four-door with just two seats.

American Motors debuted the auto industry's first four-door hardtop station wagon with its introduction of the 1956 Nash Rambler Custom Cross Country model.

Hardtop wagon was an American Motors' first

By R. Perry Zavitz

It is the exception, rather than the rule among European cars. The Japanese auto industry only adopted it to a significant degree when they invaded our market. So, by and large, hardtop styling is a North American phenomenon.

Chrysler introduced the first modern hardtop: the 1946 Town & Country Custom Club Coupe. It was an experimental model, of which only seven were made. Not realizing its great potential, Chrysler did not bother putting the hardtop into production until the 1950 model year.

By 1950, every division of General Motors offered a hardtop, some for nearly a year. Those were two-door models. In 1949 and 1950, Kaiser produced the first four-door hardtop. Frazer followed suit in 1951. It took GM until 1956 to prepare for full line four-door hardtop production. GM put out midyear four-door hardtops in the small Buick series and in the Olds Super 88 and 98 model lines.

In 1956, the introduction of a unique new hardtop model came not from GM, Chrysler or Ford, but from American Motors. Included in the redesigned Rambler range that year was a four-door hardtop to parallel GM's corporate-wide expansion of that body type. However, AMC also added an all-new four-door hardtop station wagon. It was an industry first.

The Rambler, then being sold under both Nash and Hudson badges, offered three station wagons called Cross Country models. All were four-doors, because Rambler built nothing else in 1956 and 1957. One Cross Country was in the Super series, and the other two were in the top Custom range. The Super and the lower-priced Custom models were orthodox wagons. It was the high-priced wagon that combined the popular hardtop styling with the utility of a station wagon.

Listing for $2,494, the Rambler Cross Country hardtop station wagon was powered by the 195.6 cid/120 hp ohv six-cylinder motor. At extra cost, overdrive, HydraMatic, power steering and air conditioning were available.

In 1957, the Cross Country hardtop was again offered, but with Rambler's new 250 cid/190 hp V-8 engine. The base price was increased to $2,715, mainly due to the V-8.

That was the final year of Nash and Hudson. For 1958, American Motors concentrated on the Rambler, which became a new marque name. The name Ambassador, which had been used by Nash off and on for 30 years, was then applied to the senior Rambler series.

The Rambler Ambassador Custom included a Cross Country hardtop in its line-up. It had the new engine used in all 1958 Ambassadors, which was a 327 cid/270 hp V-8.

In 1959, the Ambassador Cross Country hardtop was still offered, in addition to a similar hardtop wagon in the Rambler Rebel Custom line. The Ambassador's price had climbed to $3,116 in 1958 and 1959, but the Rebel brought the hardtop station wagon's price down to $2,588...within $100 of its 1956 introductory price. That lower price included a 250 cid V-8, now rated at 215 hp.

The hardtop Rebel Cross Country was available just during the one season. For 1960, the Ambassador line included the only Rambler hardtop wagon available. Its price remained the same, as did engine size, though advertised horsepower was reduced to 250. That was the last year that AMC offered a hardtop station wagon.

The 1957 version of the Rambler Cross-Country station wagon featured many changes to its trim and decorative embellishments, although the body was basically unchanged. Badge on the rear pillar indicates a new 250 cid V-8 under the hood.

With different engines and trim levels, the 1963 Ramblers came in 30 or more variations. 660s, such as this four-door station wagon, had double bodyside molding at the front and a single molding towards the rear.

The '63 Ramblers: They still look great today

By Bill Siuru

Today, people are paying big five-digit prices for some highly styled, but conservative sedans with nameplates like BMW, Mercedes-Benz and Audi. Over two decades ago, American Motors produced some cars that compare quite well with these contemporary classics, at least from a styling standpoint. These were the nicely restyled Rambler Classic and Ambassador models introduced for 1963. OLD CARS WEEKLY subscriber Rick Myers of New Brighton, Pa. brought these "forgotten classies" to my attention.

The 1963 AMC products were the first cars from this marque that styling genius Richard Teague had a hand in designing and the results were extremely elegant. The entire design was well integrated from the concave grille to the companion tail section. The formal roofline and cutaway wheel openings looked just right. Incidently, Rambler was the first non-luxury make to use curved side glass. Trim was used sparingly, and then mostly to accent the smooth, basic lines. Indeed, the '63 Ramblers bear a faint resemblance to the equally attractive 1961 Lincoln Continental four-door hardtops.

While there were only three basic body styles used by the Classic and Ambassador in 1963, a two-door sedan, four-door sedan and four-door station wagon, there were lots of models based on these. I counted some 30 variations, when you consider the Classic came in both six-cylinder and V-8 series with 550, 660, and 770 trim levels in each. In the

The 1963 Rambler Ambassador 990 four-door sedan looked longer, but really wasn't. It came with a choice of 250 hp standard or 270 hp optional. The 990 Ambassador trim was the fanciest you could order.

Ambassadors, which all came with V-8s, there were the 800, 880 and 990 trim levels. A hardtop would not appear until 1964, even though the Ambassador and Classic's baby sister, the American, had both a hardtop and convertible in its 1963 lineup.

A stylish hardtop-like roofline was used on 1963 Rambler four-door sedans, such as the 660 model shown here. Note that the thinner whitewalls indicate this car has later vintage tires than the original style.

The 1963 Ramblers won the coveted *Motor Trend* "Car of the Year Award," and the car was relatively popular with the public. 1963 was the best year on record for the Rambler marque.

Besides its styling, the 1963 Ramblers had features that would fit right into current era of small, luxury models. For example, Rambler offered an aluminum block six and a "Twin Stick" transmission that consisted of a floor-mounted four-speed, plus overdrive. Naturally, the Ramblers in keeping with AMC tradition, used Unibody construction.

A top-of-the-line Ambassador 990 sedan with its 327 cid V-8 made a great highway cruiser, even by today's definition of the term. It also gave reasonable fuel economy. The Ramblers did win the 1963 Mobilgas Economy Run. A 1963 Rambler, especially a 990 model, would make a good "low buck classic" that would be noticed at car shows. However, they are rare. Only 14,019 four-doors were made and most of them have already probably met the crusher. One with lots of options would be especially rare considering only 0.50 percent of the '63 Ramblers were equipped with power windows, eight percent had power brakes and another eight percent got air conditioning.

A hidden storage compartment in the floor was standard in the 1963 Rambler Classic 770 four-door Cross-Country station wagon. Another selling feature was a ceramic armored muffler with lifetime guarantee to the car's original owner.

Motor Trend awards, Twin-Stick shifter, six-passenger seating and the 198 hp V-8 were highlighted in this 1963 advertisement for the Rambler Classic 770 four-door sedan. The 770s had full-length double bodyside moldings.

The Hornet name was used by Hudson, in the early '50s, to identify its hot-in-racing "Big Six" series. (Courtesy Jack Miller Collection)

Recycled AMC model names

By R. Perry Zavitz

In the last 15 years, Detroit introduced many new cars. Each has a name. Quite a few have gone to the past, taking names the same as forgotten cars. It is not unusual for a car to revive a name that had been a part of that company's past, but what we are about to look at, specifically, are names dug out of other company's histories. American Motors Corp. produced a number of models that used such "recycled" names.

In September 1969, AMC introduced a new compact car called the Hornet, a highly respected name found in the Hudson branch of AMC's ancestry. The new Hornet was offered only in two- or four-door sedans. But, for its second year, there was a four-door station wagon added to the Hornet line. In 1971, a Hornet station wagon was new to any line of cars, past or present, bearing that name. To easily identify the Hornet wagon from its sedan siblings, it was called the Sportabout.

It is interesting that when the Hudson Hornet made its debut, the name Sportabout was being used by Dodge. When Dodge brought out its first postwar cars in 1949, there were three series. The lowest priced Wayfarer line was smaller than the other Dodges, and consisted of three body types unobtainable elsewhere in its family. There was a three-window business coupe, a sloped-back two-door sedan and a roadster. The latter was an interesting revival of the once popular softtop car.

The 1949 Dodge roadster seated only three. But, its uniqueness lay in the hand- operated folding top. Like roadsters of the past, early 1949 models did not have roll-up side glass. Instead, they had detachable plexiglas windows that fitted in or detached from the doors. Plexiglass was a state of the art feature not found on prewar roadsters. This was probably an industry first.

Two different sources say the detachable plexiglas windows disappeared in either mid-1949 or 1950. "*The Dodge Story*," by Thomas A. McPherson (Crestline) says "ventipanes and roll-up windows replaced detachable windows during the year (1949)" and backs this with a photo. However, an article in *Special Interest Autos* (May/June 1976) claims roll-up windows became optional in 1950.

In any case, the roadster returned for 1950 with grille and detail changes. Something was added. It had a Sportabout nameplate that was not used on other Wayfarers. This name continued in 1951.

Mechanically, it was no different from the contemporary Meadowbrook/Coronet. The engine was a 230 cid/103 hp flathead six. It featured Chrysler's smooth, but sluggish, Fluid Drive transmission. Gyro-matic transmission was available in 1951. At $90, it was claimed to be "America's lowest-priced automatic transmission." It was not fully automatic; a four-speed affair, it would shift on its own from first to second and third to fourth, but the driver had to shift from second to third.

Wayfarers rode a 115-inch wheelbase, 8-1/2 inches shorter than other Dodges. Overall length was around 196.3 inches for the three years. That made this Dodge very close to the size of a Chevrolet.

Price of the Wayfarer ragtop was $1,727 for its first two years, then $1,884 in its final year. That made it the lowest-priced soft-top car (except Crosley). It was priced a little less than the Nash Rambler. In top condition today, it's worth twice as much to collectors.

Wayfarer roadsters were not produced in great numbers, and they were not the rarest Dodge model, except in 1951.

Production totals were 5,420 for 1949, down to 2,903 for 1950, and finally down to 1,002 for 1951.

Although the Wayfarer continued unchanged for 1952, there was no roadster and no Dodge with the Sportabout name.

At the same time that AMC introduced the Hornet for 1971, they also introduced the Matador. It was a replacement for their Rebel. But, the Matador name was also taken from an earlier Dodge (though a later model than the Sportabout).

Matador was a model Dodge offered only for 1960. Again Dodge went to two sizes of cars, and the Matador was the lower-priced of the larger models. It was around $200 less than comparable models in the top Polara series. Prices ranged from $2,930 for the four-door sedan to $3,354 for the nine-passenger station wagon. Two- and four-door hardtops, and a six-passenger wagon rounded out the Dodge Matador series.

Standard engine was the 361 cid/295 hp V-8, but the 383 cid/325 hp Polara engine could be ordered. TorqueFlite automatic transmission was offered at an additional $211. Wheelbase was 122 inches and overall length was 214.8 inches, which was identical to the Polara's dimensions.

Exact production figures are hard to come by. The best information seems to indicate that approximately 23,600 Matador sedans and hardtops were built.

In addition, there were an undetermined number of station wagons produced.

One of the most individualistic cars American Motors produced was the Pacer. It was a sub-compact car, at least in length. But, it was wider than any other AMC car at the time. With its rounded shape and large glass area, it has been described as a drop of water wearing overalls.

AMC used the Pacer name, which had been one of the main series of none other than the Edsel. When introduced with unprecedented fanfare for 1958, Edsel came in two sizes, and each size was offered in two versions. The more deluxe of the smaller series was called the Pacer. In fact, Pacer was one of the names seriously considered for the car itself instead of Edsel.

Pacer was offered in a four-door sedan, two- and four-door hardtops and in a convertible. Like its cheaper running mate, the Ranger, overall length was 213.17 inches. The wheelbase of 118 inches was also identical to the Ranger. And power was also shared with the Ranger. It came from a 361 cid/303 hp V8 engine, which had a 10.5:1 compression ratio. There were no higher compression ratios that year.

Base prices for the Pacer ranged from $2,700 for the sedan to $3,766 for the convertible. That was about $200 to $300 more than comparable Ranger prices. For 1959, things changed quite a bit. Because of its far-less-than-expected sales total, the variety of Edsel models was drastically cut. The entire Pacer series was dropped at the end of 1958.

A total of 19,204 Edsel Pacers were produced. That sum consisted of 6,534 two-door hardtops, 6,730 sedans, 5,026 four-door hardtops, and 914 convertibles. (Production estimates from *The Standard Catalog of American Cars 1946-1975.*)

It seems a little strange that AMC, a company that strived so hard to be unique and independent, did not select more distinctive names for the Hornet, Hornet Sportabout, Matador, and Pacer. Of course, this also serves to underline the fact that AMC was a firm with a long history of "marching to its own drum." That's certainly a big reason that marque collectors love AMC products.

The term "Sportabout" was used to promote the early postwar Dodge Wayfarer roadster with snap-in side glass, rather than roll-up windows.

AMC's 1973 Hornet hatchback had hot looks, although it was not a true big-block high-performance machine like its namesake.

AMC snapped up the Sportabout name to use on the station wagon version of its Hornet. Seen here is the 1973 model with D/L trim.

A name likely to be revised a third time is Pacer. Its first postwar use was as a series designation for the 1958 Edsel, a marketplace failure.

AMC latched onto the Matador nameplate years after Dodge had discarded it. Shown here is the 1975 Matador four-door sedan.

AMC used the AMX as a springboard to the youth market. The concept worked, but the car was built only in limited numbers. The 1968 coupe had a production run of 6,725, the two-seat AMX's second highest. It's a collector item now.

AMX led AMC into youth market: Now it's a hot `street classic'

By Larry G. Mitchell

The field of "late-model" collector cars has expanded to the point of having a name of its own. They are also called Special Interest cars. These terms cover a broad range of cars not recognized as "Classics" or antiques.

Someone coined the term "street classic" to expand the classification. It was meant to cover cars from the 1955 Chevy to Bricklins to the latest Corvette. What makes a street classic is a fairly recent car that is rare and popular but still driven on the street in some capacity. Some may be used even as daily transportation. Road Runners, early Mustangs, GTOs, Marlins, and Ford retractables are a few examples. Most any sporty, relatively low-production hardtop or convertible with options can qualify. It should be in good original condition, too.

A fine example of one of the best street classics to come along was also a genuine high-performance sports car. It came to us from American Motors Corp. (AMC) a few years ago. It is the original two-passenger AMX.

Most everyone can name a couple of America's recent two-seat sports cars if you ask them. They will tell you Corvette and the early T-bird rather quickly, but are at a loss for anything else. Some bring up Kurtis or Kaiser-Darrin but can't recall any more. These people are also quick to point out that hybrids, such as the last two, are extremely low in production and were highly uncommon, even when new. They are surprised when you inform them that AMC actually built a small, high-performance, two-seat sports car.

The original two-passenger AMX was developed from a group of AMC experimental designs and prototypes back in 1965. AMC was searching for cars with flair and performance to ease them out of the four-door economy-sedan image they had for many years. Designs developed from this era also gave birth to the Javelin, Hornet, Gremlin and Sportabout models that temporarily pulled AMC's business up from the depths. Credit for this was due to many people, from key stockholders to the crack designers in the styling studios.

In order to really hit the dealers and the new car buyers

hard, AMC had to do something bold and daring. Its choice was the AMX two-seater. It was picked to spearhead the new generation of cars. The concept was to create a car with clean and simple lines that would be unmistakable in appearance. It was not meant to be confused with cars offered from other manufacturers.

A coupe with a long hood and rakish fastback roofline was chosen as the body type to go with. The really distinguishing characteristic that set it apart from Mustangs, Camaros, Barracudas and others was a very short 97 inch wheelbase. The other sporty cars usually had 105 inch or greater wheel spans.

A vehicle this short had no room for rear passengers behind its two front bucket seats; it was an honest-to-gosh, small, two-place sports car. AMC's next step was to find a power plant to back up the gutsy AMX looks.

The company's engineering department had a new, thin wall, high-output V-8 under development. It was a smallblock of 290 cid in 1966 and went up to 343 cid/280 hp in 1967. The boys at the motor division did their homework. This light, small engine block was developed into a whopping 390 cid monster with 315 hp and big-block torque of 425 lbs.-ft.

Officially crowned the "AMX 390" engine, it came with a chrome air cleaner lid, chrome valve covers and chrome oil filler cap to let you know it was special. The smaller 290 and 343 cid versions were available in the AMX, too, but the 390 option was relied upon as "leader of the pack." It showed the way to other automaker's muscle cars.

Featuring construction techniques and materials only found in the hottest versions of GM, Ford and Chrysler muscle cars, the 390 featured: forged steel crankshaft and rods, hi-dome aluminum pistons and screw-in rocker arm studs. Its distributor was up front for easy access. The whole engine compartment was laid out for simplicity.

As for the AMX interior, a cozy cockpit was needed. It had to offer the choice of a no-frills, all-out performance look or, on the opposite end, plenty of creature comforts.

The option list also included intake and exhaust manifolding, a variety of transmissions and axle ratios down to a 5.55. For the AMX consumer who did not want a spartan hot rod, there was a Rallye gauge pack that offered power steering, power disc brakes, air conditioning, leather interior, automatic transmission with full console, tinted glass all around, AM/FM stereo radio (with or without eight-track tape player), tilt steering wheel, Rallye rims, and more.

AMX buyers found even the standard equipment very impressive. It included such items as reclining buckets, armrests, carpets (even in the large cargo area behind the seats), tachometer, heavy-duty suspension, four-barrel carburetion on all V-8s, dual exhausts, heavy-duty Borg Warner T-10 four-speed, Hurst shifter, fat front anti-roll bar and rear axle 'torque links' to stop axle wind-up and wheel hop and promote flat handling.

The AMX was a car to be reckoned with. AMC was ready for the muscle car/pony car wars.

In February 1968, the AMX was officially offered to the car buying public at a base price of $3,297. That was pretty reasonable for what they buyer got. It was not quite a Corvette, but it sure wasn't a Firebird, Camaro or Mustang either. It was definitely something different...more like the two-seat Thunderbird than anything else in recent history. The big difference between the two was the AMX was a closed coupe and the T-bird was a roadster.

AMC still clung to the idea of making family cars that seated four or more people. Certain key management personnel had hedged on the two-passenger AMX concept way back in 1966. They pushed for a new four-passenger sporty car to directly compete with the Mustang-type cars.

Unfortunately, AMC was too small. It couldn't really afford another totally new car for 1968. Management was forced to meet the sporty car challenge and to try to woo buyers hooked on early Mustangs. The answer was to expand on the original AMX by making it a longer car and providing a back seat. To land economy-minded buyers, this sister car was to have a basic six-cylinder engine instead of the AMX's standard V-8. To accomodate four-place seating major tooling costs, the little two-passenger AMX was stretched 12 inches in the wheelbase and roofline. The rear pillars, back glass, side glass and rear quarter panels were altered, too.

With cosmetic changes in the grille and hood area, the four-passenger Javelin was born in time for the 1968 model year. It actually appeared six months ahead of the AMX that sired it.

The Javelin was very similar, in many respects, to the AMX. But, it was still quite a different car with a distinct concept behind it. Even today, some collectors confuse the Javelin with the AMX two-seater, a problem caused by later badge engineering practices that created the AMX-Javelin.

The AMX came on the automotive scene rather quickly, but it still racked up an impressive record in motor sports. Craig and Lee Breedlove took the new AMX racing. In a matter of months, using hand-built speed parts, they achieved numerous wins.

One car visited a large, oval test track in Texas for a workout. It performed flawlessly. Racing against the clock for 24 straight hours, the AMX proceeded to shatter 106 national and international speed and endurance records. Breedlove circled the large oval track for 24 continuous hours at an average speed of 140.7 mph without any special aerodynamics or spoilers. The old record for a factory-stock small-block (under 400 cid) had been broken by more than 20 mph.

The Breedloves then went on to Bonneville for some

The rear view of the 1968 AMX coupe was set off by a buttressed sail panel form and dual exhausts. Rear fender badge indicates this example features the 390 cid V-8 below its hood.

From the start, the two-seat AMX was used in racing. The cars were very competitive in both SCCA Trans-Am events and quarter-mile competition. Note the exhaust exiting below the door. (Photo courtesy Vince Ruffolo Collection)

The 1968 Javelin (top) hit the market six months before the AMX (bottom), though both sprang from the same design project. People still confuse early models of both marques. (Photo courtesy Vince Ruffolo Collection)

straight line runs. Running in dangerously wet weather, the same AMX still managed an official USAC timed run of 189 mph. At that point, the car was still accelerating! Unofficially the AMX went over the 200 mph mark on one run.

In 1969 SCCA competition, an AMX crossed the finish line seconds behind a Corvette for national class honors. It was a full season of amateur racing, too. On autocross and slalom circuits across the country, the AMX took trophy after trophy for Sunday racers. Today, one of the early SCCA cars that was originally backed by the St. Louis Dealers Association, is being campaigned in vintage racing by George Doughtie, of Georgia.

In drag racing, modified AMXs blistered dragstrips regularly, shutting down the competition. Professional drivers like Shirley Shahan and amateurs like Lou Downy set class drag records year after year regionally and nationally. Even today, the AMX remains a formidable straight line opponent.

Records were not the new car's only accomplishments. In 1969, the AMX won *Car & Driver's* "Reader's Poll" as the best sports/GT car over such challengers as the Triumph TR-6 and Lotus Elan.

The little AMX from AMC has made its mark across the face of automotive history. It fulfilled the mission of showing the world that AMC had a change of heart and wanted to enter the youth market. It set the stage for other new, youth-oriented models such as Javelins, Hornets, Gremlins and Pacers. Unfortunately, with insurance companies cracking down on muscle cars, and the threat of government restrictions looming, high-performance cars were doomed.

The two-seat AMX suffered, saleswise, from such influences. Only 6,725 were made in 1968, followed by 8,293 in 1969 and 4,116 in 1970. Then in 1971, the car's image was compromised slightly when AMC hung the AMX badge on a jazzed-up Javelin.

The new Javelin-AMX was a great car in its own right. However, it could never be what the original two-seat sports car represented. The years have gone by now and the original AMX has been recognized by more and more automotive historians for what it was. Collectors are actively seeking well-preserved examples of this graceful, clean-lined sports car. Unfortunately, many of the two-seaters fell victim to misuse. They have been beat, butchered and abused. With only 19,134 made in three short years, possibly 10,000 AMX two-seaters still exist in serviceable condition. Only a few hundred of those were estimated to be in very good to excellent condition, back in 1976.

The totals must be much lower today. With the Corvette's 40-year production at over a million units and the two-seat T-bird's three-year production near 55,000, the two-passenger AMX is, by far, the rarest of the true American sports cars. It is now a real street classic.

On Feb. 13, 1969, AMC released this photo of its new high-performance model, advising that it would have its first public showing at the Chicago Auto Show, March 8-16, 1969. (Photo courtesy Vince Ruffolo Collection)

SC/Rambler Hurst was hot "tot"

By Robert C. Ackerson

When Pontiac launched the modern performance car era with the 1964 GTO, AMC had just introduced a new version of the Rambler American. This little "tot" of an automobile had a heritage dating back to 1950. When the first (postwar) Rambler appeared, Pontiac Motor Div.'s only "performance" engine was the optional "high-head" straight eight developing 113 hp. It had a bit steeper compression that the standard straight eight.

Curiously, both Pontiac and Rambler enjoyed considerable sales success during the late '50s and early '60s, but for totally dissimilar reasons. Pontiac, as has been well documented, shook off its "old maid's car" image of conservative styling and engineering to capitalize on the youth market. Rambler, in contrast, found success in marketing conservative "family" cars.

By 1957, PMD was offering V-8s of 347 cid with 317 hp. Until General Motor's January, 1963 performance ban, it was a force to be reckoned with by would-be NASCAR drivers and drag racing competitors. This flow of creativity wasn't easily dissuaded from following its natural course. Thus, the first of the great GTOs arrived in 1964.

Meanwhile, back in Detroit (where AMC had its corporate headquarters) and Kenosha (where the AMC assembly plant was located), American Motors had been making a bundle of money with automobiles that featued compact dimensions, good fuel economy and styling that, if not exciting or original,
at least wasn't terribly objectionable. In 1963, when the larger Rambler and Rambler Ambassadors were given new "uniside" bodies, *Motor Trend* presented AMC with its "Car of the Year" award. The magazine cited Ramblers as "This year's best examples of outstanding design achievement and engineering leadership."

In the tracks of the attractive new Ramblers came a drastically restyled 1967 AMC American compact. Many of its body components were from AMC's larger models, but they came together on the Rambler's 100 inch wheelbase chassis in a very pleasing manner.

These changes, which AMC management had hoped would maintain the company's sales momentum, arrived during one of those transitional periods in automobile history. Customer preferences and interests were shifting fairly rapidly. This left some automobiles out of the mainstream of sales activity.

AMC did not wake up to the new interest in high-performance as quickly as other automakers. Thus, its profits spiraled downward. In 1964, they totalled $26.2 million. The following year, they slipped to $5.2 million.

AMC didn't have feet of clay. In 1965, some preliminary steps were taken that eventually led to Javelins and AMXs. New models took time to develop, though. In the interim, AMC's "Sensible Spectaculars" had to fill the void.

These Ramblers were decidedly more sensible than spectacular, but as part of AMC's rebuilding campaign, a new family of thin wall V-8s was developed to replace the old 287 and 327 cid engines that dated back to 1956. Although AMC's new V-8 was never known for its high rpm capability, it possessed the virtue of light weight and respectable torque and horsepower ratings.

During the 1966 model run, AMC built a small number of special Rambler Rogue two-door hardtop models. They were powered by 290 cid/225 hp versions of this V-8. Although some critics took exception to AMC's audacity in uttering "Rambler" and "Rogue" in the same breath, others looked beyond names and labels to recognize the Rogue V-8's respectable performance.

Back in 1957, AMC had produced a few Rambler Rebels with 255 hp V-8s, but its new, mid-`60s performance program was to become, out of necessity, considerably more ambitious than the 1957 effort. By 1967, it lead to Rambler Americans with 343 cid/280 hp "Typhoon" V-8s. AMC still lacked a lot of basic performance know-how, but no one could question its determination to learn and, if necessary, to learn the hard way.

With the introduction of the Javelin as a 1968 model, AMC jumped into Trans-Am competition in a big way. Before long, the company was offering its customers a fairly comprehensive list of power and handling options.

This new mood at AMC, plus the entrepreneurial spirit of George Hurst, lead to the development of the ultimate Rambler. It was called the 1969 SC/Rambler Hurst.

When this "custom-built Rambler Rogue Hardtop" (as AMC described it) was seen on Feb. 13, 1969 it was jointly announced by AMC and Hurst Performance, Inc. as a Hurst-modified limited quantity offering. This was a beneficial tie-in for AMC, since the Hurst name carried considerable weight in drag racing circles. However, this relationship also had a sound foundation. Hurst had indicated a keen interest in turning out a small batch of Hurst-AMXs. AMC counter-proposed converting the "new" Rambler (the American name had been dropped and Rambler was used on 103 inch wheelbase models) into a hot mini-car.

The initial reaction from Hurst was reported in CAR LIFE (May 1969) as less than enthusiastic. "It sounded no better...at first," said Car Life "to Hurst, than it did to us. The very idea of a 390 cid engine over the front wheels of the little Rambler..."

But, in reality, there were no sound reasons not to use the 390 V-8 if AMC was intent in making a super car out of the Rambler. Compared to the 290 cid V-8, the larger engine weighed only 35 pounds more. lThis small extra burden upon the Rambler's front wheels had a negligible affect upon its handling.

AMC was anything but reluctant to tout the SC/Rambler Hurst as a muscle car. It was, said AMC, "designed for stock drag racing." With a suggested retail price of $2,998, it was one of America's all-time great performance bargains. Its outrageous red, white and blue exterior, huge cold-air scoop, hood pins and mag-type wheels made it as discreet as an X-rated movie.

We're not suggesting however, that AMC was delving into automotive pornography with the SC/Rambler. On the contrary it had all the virtues of a home-grown, honest-to-gosh American street racer. With 315 hp, four-speed Borg-Warner T-10 gearbox, Hurst shifter, power front disc brakes, 3.54:1 twin-grip differential, stiffer suspension and "special-tone" (glass packed) mufflers, it had plenty of the "right stuff."

The SC/Ramblert wasn't a particularly great handler, but then it's hard to recall any Rambler that was really a standout in this area. You could also find fault with its austere interior. On the other hand, AMC was slapping stickers with that under $3,000 price on the side window. Just about nobody could find fault with that.

With the standard 3.54 rear axle, the SC/Rambler, AMC claimed, could turn the quarter-mile in 14.3 seconds. CAR LIFE reported its test car's quarter-mile time as 14.2 seconds. That translated into a 100.8 mph speed. The SC/Rambler's zero to 30 mph and zero to 60 mph times were equally impressive at 2.4 and 6.3 seconds, respectively.

There was additional performance on tap for SC/Rambler owners who installed the optional 302-degree cam and stronger valve springs that AMC offered for the 390 cid V-8, since those changes boosted its horsepower to 347. For those who carefully perused AMC's Group 19 performance catalog, there were such items as a high-rise aluminum manifold and dual four-barrel carburetors. They could provide upwards of 400 hp to propel the 3,100 pound SC/Rambler.

AMC and Hurst originally planned to produce just one thousand SC/Ramblers, but during the 1969 model run, a total of 1,512 were assembled. Since the Rambler was replaced by the Hornet, in 1970, the SC/Rambler's life span was a short one-year. This makes the model among the rarest of "regular production" super cars today.

An earlier performance car built under the Rebel name was the 1957 Rambler Rebel. This four-door hardtop was promoted with a stasndard 255 hp four-barrel V-8 and optional 288 hp fuel-injected V-8. However, the Bendix electronic fuel-injection system was either never issued or subject to an early recall. The carburated Rebel was still pretty hot. (Courtesy Vince Ruffolo)

The AMC Concept II featured a wedge-shaped design, soft integrated bumpers and headlamps that were concealed by flush, sliding panels. (Photo courtesy Vince Ruffolo Collection).

AMC sought success with innovative designs

By Robert C. Ackerson

"The only certain thing you can say about American Motors Corp." a Wall St. analyst remarked during the company's twilight years, "is that it'll be a subject for study in graduate business schools for the next two generations".

The above quote appeared in the April 8, 1967 issue of *Business Week*. In it lies the seeds of an interesting story. Since its creation on May 1, 1954, AMC had more "burial notices" and premature obituaries written about it than perhaps any other major American corporation. Yet, it managed to survive for many years.

As a case study of how the smallest competitor in an industry dominated by giants managed to stay out from under the giant's foot, AMC's modern history is well worthy of examination. In addition, the role that its predecessor companies played in automotive history hold great interest to hobbyists.

In 1979, during AMC's 25th anniversary year, I spent some time discussing AMC's past/present/future directions with the late Dick Teague. The company was still alive and kicking at the time and Teague was then vice-president in charge of AMC styling.

My recollections of talking with Teague bring back memories of his friendliness and frankness. In fact, his openess about future plans was somewhat surprising to me, given the obvious need to keep plans for future products under wraps at that late stage of corporate history.

According to Teague, between 1954 and 1979, AMC had been extremely fortunate to have "heads up" management that forged ahead with a perspective that turned economic challenges into opportunities.

AMC's first president was George M. Mason, who had served as president and chairman of the board of Nash-Kelvinator since 1937. One of Mason's first actions as chief executive was to have work commence, in conjunction, with the Edward G. Budd Co., on a new economical and lightweight automobile. It premiered, in the fall of 1940, as the Nash 600. This new automobile was the first mass produced car in America to feature what is called today "unit-body" construction.

After the introduction of the new postwar Nash in 1949, Mason promoted the "Nash Experimental International" or NXI compact. He was assisted by George Romney, who joined Nash-Kelvinator in 1948 as assistant to the president.

129

The NXI was the 84 inch wheelbase forerunner of the Nash Metropolitan, which was actually produced in England by Austin. It was marketed from 1954 through 1962.

The Metropolitan might not have been an "international automobile" in the same sense as the Ford Fiesta, but Mason's venture into this area has to be recognized for its historical significance and as an early example of Nash-Kelvinator's ability to move into a territory still virgin to the intrusion of the "Big 3."

In March, 1950, came another historic event in Nash history; the introduction of the Rambler. Offered for sale in the same month as the first Rambler was some 50 years earlier, the 1950 model was not merely the first mass-produced postwar compact, but luxurious, too. It avoided the Spartan nature of Kaiser's Henry J and provided an attractive exterior appearance, quality interior appointments and a peppy 82 hp six-cylinder engine. It was a small car with the style, luxury and performance that traditionally appealed to Americans.

Sales of the Rambler, started at 20,872 in 1950 and increased to 57,555 the following year. In 1953, a slight decline took place to 53,055. In 1954, the year of the merger of Nash and Hudson into American Motors Corp., 36,231 Ramblers were built.

The union of Nash and Hudson was barely five months old when George Mason died on Oct. 8, 1954. He was 63. Four days later, George Romney was elected president and chairman of the board of American Motors. Romney was faced with a crisis in terms of sales and mounting losses. However, he believed in the public's confidence in American Motors and its ability to survive.

Responding boldly on both fronts, Romney quickly assembled a young, highly motivated management team that operated in a streamlined infrastructure. His confidence and faith in both himself and AMC acted as a corporate catalyst, though sales continued to lag at first.

In 1953, Nash and Hudson had 3.56 percent of the market. This had declined to 2.14 percent in 1954 and 1.91 percent in 1955, when AMC reported a loss of $69 million. The following year, while the Big Three were having a grand time, AMC suffered a loss of $19,746,243.

Yet, Romney had not been sitting still. He was not wringing his hands in quiet desperation. Instead, he had closely studied the composition of the market and concluded that beneath the glitter and glamour surrounding the success of the Big 3's automobiles, there was a small and growing tide of buyer discontent. The country was not experiencing a gas shortage or a recession, but a recession would come in the late '50s.

The nation was not unreceptive to the idea of an economy car. This was the niche that Romney discovered and he made the most of it. In 1968, AMC president Roy Chapin described Romney's efforts of a decade earlier as nothing short of "rebuilding the company...under conditions of extreme difficulty, but at the same time, under conditions of great opportunity".

While keeping the big Hudsons and Nashs in production through 1957, Romney felt that AMC's future rested with the all-new Ramblers introduced in 1956. On a short 108 inch wheelbase, they had styling that was a bit gaudy, but strongly original. They had personality. In face of the Big 3's obsession with larger and more outlandish cars, Romney initiated what soon became a highly effective counterattack.

Rambler ads assailed competitors' big, gas guzzling cars. They lauded Ramblers as cars that gave the customer big car room, ride and comfort combined with European small car economy and handling ease. In spite of sales of 114,084 vehicles, AMC lost nearly $12 million in 1957. However, a new momentum had been established. Romney had taken the initiative and had no intention of losing it.

Sales nearly doubled in 1958. They reached 217,332 cars. That translated into a tidy profit of $26 million. AMC made money for the first time since its creation. In addition to its 108 inch wheelbase model, Romney resurrected the original 100 inch wheelbase Rambler, which had been dropped after 1955. Very modestly restyled to distinguish it from the earlier models and renamed the American, it sold 42,196 units in 1958. Sales exceeded 90,000 in 1959 and 120,000 in 1960.

Total AMC sales were equally impressive. They reached 368,464 in 1959. For the fiscal year ending Sept. 30, 1959, the company pointed with justifiable pride to a profit of $60,341,823 on net sales of $869,849,704.

A lot of new ideas and thinking underlined the turnaround. For example, it's interesting to note the owners manual for the 1960 Rambler American pictured and described how to make minor repairs and adjustments. This predated Ford's similar, and highly publicized Maverick owners manual by 10 years.

After this peak level of profits, AMC did not maintain its momentum. Automobile historians, Wall Street analysts and corporate specialists do not fully agree on what happened. One thesis is that the company began to suffer due to the Big 3's entrance into the compact car field. This could be called the "Big Foot" theory and includes the postulate that whenever a small producer carves out and expands a previously untapped market niche GM, Ford and Chrysler are sure to follow suit. A second explanation centers around the belief that the nation plays a cat and mouse, hot and cold romance with the small economy car. As the economy improves, its appeal falls off in inverse relation to the nation's economic prosperity.

Another explanation for AMC's decline was given by an industry analyst who wrote, in 1967, "to succeed in the auto business you need an unerring judgement about future tastes...that is, a good crystal ball and a lot of money." AMC guessed wrong. Its decline was as rapid as its ascent. In

Designer Dick Teague envisioned this Concept AM/Van "dream truck" in the late 1970s. The proposed 1980s four-wheel drive RV was two inches shorter than a VW Beetle. (Photo courtesy Vince Ruffolo Collection)

Designed as a four-passenger sports hatchback for the '80s, the Grand Touring stressed a sports-luxury image with features such as leather upholstery and real wire wheels. (Photo courtesy Vince Ruffolo Collection.)

1965, profits totalled just over $52 million. They disappeared altogether in 1966, when a loss of $12,648,170 for the year was reported. AMC also lacked the financial resources to catch up.

A fourth, and possibly most accurate, frank analysis of AMC's ills of the mid-'60s came from Roy D. Chapin, Jr. in 1968. Speaking about the 1963-1966 era at AMC, Chapin said, "For several years, we pursued a policy of trying to expand our lines, quite properly I think, but we got to the point where people seemed to have the idea that as a company, there was nothing very new or original in what we were doing...that we were trying to do exactly like the other fellow. And the problem was that they felt that we might not do it quite as well, and probably not quite as soon as the others."

In 1962, George Romney left AMC to campaign for the governor's office in Michigan. Roy Abernathy replaced him as president. Four years later, Richard Cross resigned as the company's board chairman. Replacing Cross in this capacity was Robert B. Evans. The introduction of new Rambler Rebel and Ambassador models in 1967 failed to stem the tide. While not unattractive, the press tended to view them as "me too" cars. From a 6.4 percent share of the market in 1960, AMC found itself with only a 2.8 percent share of the pie by early '67.

Once again a time of reckoning was at hand. On Jan. 9, 1967 Roy Abernathy retired and was succeeded as president and chief operating officer by William V. Luneburg. The company's new board chairman and chief executive officer was now Roy D. Chapin, Jr. In 1909, Mr. Chapin's father had helped launch the Hudson Motor Car Company. From 1910 to 1923 and again from 1933 until his death in 1936, the elder Chapin served as Hudson's president.

In a sense, Roy Chapin, Jr. inherited a situation similar to that which Romney had encountered back in 1954. The company was building a lineup of cars that apparently found little favor with the public. Its work force contained individuals who began to question AMC's ability to survive. As if that wasn't enough, staggering losses were literally bleeding the company to death. For fiscal year 1967, AMC lost $75.8 million.

The single most important item Chapin had to contend with was the need to revive public confidence in AMC. The company still enjoyed an abundance of public goodwill. He viewed this as one of AMC's greatest assets. What had to be done was to find some way to tap that reservoir and capitalize upon it. However, time was of the essence.

Speaking about those days, Chapin recalled, in 1968, "We had to move quickly. We had to show people that we were alive. We had to make people sit up and take notice of us because we were experiencing the worst kind of lack of interest. The worst thing that can happen to you in business is that people stop paying attention to you, and that was the position AMC was in."

Chapin's strategy to turn the ship around involved directing the leadership's energy toward an attitude change among three groups. First, those who were leaders in the financial world were targeted. Heads of banks that AMC was in debt to and leaders of other financial organizations whose attitude toward American Motors would have a bearing on the company's future, had to be convinced that the firm had a future. So did the company's stockholders. Chapin went to great lengths to demonstrate that the leaders at AMC knew what they were doing and that they were following a sound plan-of-action. He showed that they possessed the confidence and discipline to follow the plan.

Chapin made tremendous progress in reducing AMC operating expenses. This reduced the break-even point to below 275,000 cars. Chapin recognized that AMC had to be capable of operating profitably at a low production level. Then, it could get on with the business of expanding sales volume.

To the financial world, this marked Chapin and Luneberg as men who knew what they were doing. To subordinates at AMC, it was a strong indication that they meant business. Both points were underscored by their promotion of young men into key positions. They were given the opportunity to suggest changes and provided with the opportunity to implement them and the authority to be responsible for the consequences.

In order to regain public confidence and establish an image of AMC as a company with a future, Chapin, Luneberg and their lieutenants talked to people. In the first 100 days of their administration, they traveled across the country twice and addressed over 50 gatherings of employees, stock holders, dealers, bankers and media. In Chapin's words, "we had to plan, to persuade, and to act, and to do all three simultaneously".

The most effective way to get the public to notice AMC was to take steps that would attract favorable mass media attention. If such a step had an immediate and favorable impression upon the sales of cars, so much the better.

Six weeks after taking over, Chapin and Luneberg announced major price reductions ranging from $154-$234 on the then slow-selling Rambler American models. The results were immediate and favorable. Sales of Americans increased at a dramatic pace. This, coupled with highly effective ads that stressed the Rambler's competitive price with a certain German import and its performance and passenger accommodations advantages, gave AMC a much needed shot in the arm.

For the 1968 model year, AMC contracted with Wells, Rich, Greene for its ads. This relative newcomer to the advertising world had quickly gained a reputation for imaginative and effective campaigns. Particularly in the case of the Javelin, it produced some of the best automotive commercials ever seen on television.

As a result of these and other moves by Chapin and Lune-

Another dream car that could have seen the assembly line in the 1980s was the Dick Teague-designed Concept I, which highlighted functionality, compact size and energy efficient operation. (Photo courtesy Vince Ruffolo Collection)

Envisioned as a scaled-down version of the CJ-5, the Concept Jeep II was two feet shorter and nine inches lower than the production Jeep and had a diminutive 76 inch wheelbase. (Photo courtesy Vince Ruffolo Collection)

berg, AMC's 1968 fiscal year saw a profit of $11.8 million. This represented a financial turnaround of $87 million from 1967. In the years that followed, AMC made steady and substantial progress. There was a loss of $58.2 million in 1970, because the Hornet was introduced that year. However, this investment gave AMC a highly competitive compact car. The end of Rambler production at Kenosha was marked on June 30, 1969. It was a historic and sentimental day at AMC. In total, from 1950-1969, Rambler production was 4,204,925 cars.

In April, 1970, AMC scored a major coup over its competitors by introducing the Gremlin as the first U.S. sub-compact car. Its total production exceeded 700,000, making it the most popular single model ever built by AMC.

The year 1971 saw corporate momentum sustained by the introduction of the Hornet Sportabout and documented by both record sales of $1.2 billion and a profit of $10.2 million. The following year, AMC did even better, earning a profit of more than $30 million dollars. The introduction of the Hornet Hatchback, in 1973, helped push Hornet sales beyond the 100,000 mark for the first time.

AMC seemed unstoppable. Profits of $44.5 million were marked 1973. It was one of the best years ever. Yet, within five years, the tables turned again. By 1978, AMC appeared to be in imminent danger of going out of business or leaving the automobile market entirely to the Big 3.

However, after talking to Dick Teague in 1978, one couldn't help but feel that another turnaround was possible. Taking a close look at where AMC had been and where it was headed (at that time), Teague projected the belief that the company planned to be in the automobile business for a long time. He made it clear that AMC was going to be the source of a number of rather interesting and exciting new models if the company survived.

Starting at the top, it was not at all inappropriate to compare AMC president Gerald Meyers' outlook and business philosophy with those of George Mason, George Romney and Roy Chapin, Jr. Like these men, Meyers seemed to have a strong handle on the problems facing his company 15 years ago. "He knows what has to be done and he knows how to do it," I thought.

Teague pointed out that if AMC was to survive and prosper, it had to offer good new products that were also unique. I asked him if he felt that the market was too confined by government regulations to allow AMC to use some of its successful formulas from the past...especially discovering a niche in the marketplace and making the most of it. His answer was an emphatic "yes."

As Teague saw the market from his AMC styling boss's role, there were still pockets of consumer interest with needs for different types of automobiles. He felt AMC could take advantage of them. In fact, Teague stressed that AMC was looking at a number of areas where other companies underresponded or avoided entering.

Since AMC's track record of innovation was extremely strong, it had to be counted as a plus. Yet, as automotive history so painfully proves, once a successful formula is achieved, the competition will soon jump in. As Teague put it, "they either try to bully you or pummel you with their vehicles. In effect, this is what has happened to AMC over the past few years."

In face of this, Teague stressed that AMC had no intention of giving up and running out the door. Instead it wanted to improve its cars and to "keep a leg up on the other guys." As a concrete example of this, Teague pointed to the 1978 AMC Concord.

On the surface, the Concord appeared to be little more than a restyled Hornet. Yet, its sales climbed at a brisk pace that made some of AMC's strongest critics concede they were a bit hasty in forecasting the car's demise. According to Teague, the source of the Concord's success was that, as more and more buyers moved down the size scale, replacing their older, larger automobiles with smaller ones, they still retained a desire for luxury. The Concord, which was billed as a luxury compact (like the original Rambler) was the right car for its time. Much of this could be dismissed as mere press copy if not for one important point; namely that "conquest" sales (sales made to car buyers who were switching brands) of the Concord were running very high.

It was difficult for AMC to compete with the Big Three on a dollar-for-dollar, line-for-line, product-for-product basis. However, Teague felt that AMC could have produced many unique automobiles that fit new niches.

Teague saw the line dividing a commonplace automobile from a distinctive product as a very narrow one. A slightly different twist, he contended, was all that was needed to achieve innovation. He felt a corporation that could offer a little more than its competitors could enjoy considerable dividends from small, but attention-getting and worthwhile innovations.

The design head pointed to the AMC "Buyer Protection Plan," as one example. Since its introduction in the 1972 model year, this warranty program brought AMC a degree of customer confidence that could not have been achieved at a cost 10 times as great.

Of course, Teague knew that the competitions would fight back with their own warranty upgrades, in effect "stealing the bone" away from AMC. "(Then) you have to find another bone," he advised. He was expressing his view that AMC had to keep coming up with innovations and and intended to do so. "In this business, you have to run like hell to keep even", said Teague. "And we plan on running. We have our track shoes on."

Looking toward the future in a design sense, Teague also commented about AMC's "Concept 80" car-of-the-future proposals. He suggested that a close look at the concept cars offered a clue to what could be coming off the assembly lines at Kenosha in the not too distant future. Teague said that the ideas reflected by the cars were very much alive and represented serious production possibilities, regardless of AMC's dire financial situation in 1978.

Teague told *Old Cars* that no major future product plans had been cancelled. The major development in this area was to make some rearrangements in the introduction dates of new models. In some cases they were advanced and in others moved back in time.

Teague knew that past success didn't guarantee future success, but felt that it certainly weighted the odds in AMC's favor. "Consumers have been quick to recognize and purchase a quality product that offers a little more than its competition," he noted. "AMC's past achievements in this type of merchandising strongly suggest that, very shortly, it will be offering automobiles which fit that definition."

The Electron was another of AMC's "Concept '80" proposals. This 85 inch long, 46 inch high three-passenger commuter vehicle would use a lightweight lithium battery system for power. (Photo courtesy Vince Ruffolo Collection)

Renegade Relatives: AMC Jeep

The ruggedness and dependability of the original Jeep helped the allies win World War II. These qualities were carried over into the civilian marketplace after the war ended.

Jeep celebrated 50 years of leadership in 1991

From the halls of Montezuma to the shores of Malibu, Chrysler Corp.'s legendary Jeep brand celebrated its 50th year when it entered the 1991 model year.

"The engineers and designers that turned the Jeep vehicle from a military weapon into the world's first civilian utility vehicle would probably be amazed at today's phenomenal sport utility market," said Chairman Lee A. Iacocca.

"Over the past 50 years," he added, "Jeep has been transformed from a vehicle of military necessity to the ultimate fun machine." Jeep history officially began in 1941 when Willys-Overland, of Toledo, Ohio, began production of the world-famous Jeep after winning the bidding competition to produce military vehicles for the U.S. Army.

As early as 1938, the U.S. Army was searching for a vehicle to replace the motorized tricycle/side-car used for advance reconnaissance duties. After an official army request in 1940, three companies responded with prototype vehicles...Willys-Overland's Quad, Bantam's Blitz Buggy and Ford Motor Co.'s GP.

In June 1941, the Quad was selected as the standard, and after incorporation of several features of the other models, became the Jeep MB. From 1941-1946, Willys-Overland built more than 368,000 units.

The Jeep name distinguished itself in every World War II theater. So distinguished, in fact, that Willys-Overland registered the Jeep name as a trademark internationally.

The civilian evolution of the Jeep vehicle and four-wheel drive recreational lifestyles actually began before the end of World War II. Already in 1944, Willys-Overland was preparing to build Jeep vehicles for agricultural purposes.

Willys-Overland's civilian plans led to the prototype CJ1A and then production of the first Jeep Universal, the 1945 CJ2A. The CJ2A's civilian features included a tailgate and automatic windshield wipers.

In 1946, Willys-Overland introduced the first all-steel station wagon. A six-cylinder engine and four-wheel drive were added in 1949, making it the forerunner of today's Jeep Cherokee.

Willys-Overland then introduced the 1947 Jeep two-wheel drive and four-wheel drive pickup trucks, the 1948 two-door Jeepster phaeton, and expanded the Universal line with the 1949 CJ3A. Willys-Overland ushered in the 1950s with a new half-ton front-wheel drive pickup truck and a new one-ton four-wheel-drive pickup...an exclusive in its field.

A combination of a one-ton pickup truck and four-wheel drive made this 1950 Jeep model a leader in its field. It was one of the civilian applications that Willys-Overland derived for its Jeep.

The name Willys-Overland was gone by 1953 when the Henry J. Kaiser Co. acquired the assets of Willys-Overland for $60 million, renaming it Willys Motors, Inc.

In 1954, the CJ5 was introduced and became such a popular four-wheel drive vehicle that it endured, albeit with improvements, all the way to 1983. While similar to the CJ2A, the CJ5 featured softer styling lines, including rounded body contours.

Kaiser's major influence on the Jeep lineup began in 1955 with a desire to broaden the utility market. Seven years later, Kaiser introduced the J-series of Jeep Wagoneer two-wheel drive and four-wheel drive station wagons and four-wheel drive Jeep Gladiator pickup trucks.

The Wagoneer represented the first from-the-ground-up civilian sport utility wagon.

In March 1963, Willys Motors became known as Kaiser Jeep Corp.

In the fall of 1966, the Jeepster line, originally built from 1948-50, was redesigned as the Jeepster Commando. The station wagon, convertible, pickup truck and roadster models were the first small four-wheel drive vehicles with automatic transmissions.

On Feb. 5, 1970, American Motors Corp. (AMC) acquired Kaiser Jeep for $70 million and changed the company's name to Jeep Corp. In 1971, a portion of Jeep Corp.'s business was transferred to a newly created subsidiary, AM General Corp., which specialized in military vehicles, buses and postal vehicles.

In 1972, all Jeep vehicles, from the Universals to the Wagoneers, came equipped with AMC-built engines, including 304 or 360 cid V-8 engines.

The 1973 Wagoneer introduced Quadra-Trac to the market. With its unique controlled-slip differential, it was a "first" for a Jeep vehicle...a completely automatic full-time four-wheel drive system.

The Jeepster Commando models, without the Jeepster name as of 1972, were eliminated in 1974, while the Cherokee name joined the J-series as a sporty two-door model. A four-door Cherokee was available by 1977.

Also in 1974, Jeep Renegade, previously built in limited-edition performance models, became a regular CJ5 production model.

In 1976, the CJ7 was introduced, offering the first molded plastic top and steel doors. Both CJ7 and CJ5 were built until 1983, when the CJ5 was discontinued.

In late 1983, Jeep introduced the XJ platform in Cherokee two- and four-door models and four-door Wagoneer sportwagons. The senior J-series Wagoneer became the Grand Wagoneer.

The XJ models offered the only four-door compact sport utility and two sophisticated four-wheel drive systems...SelecTrac and shift-on-the-fly CommandTrac.

Off the XJ platform came the Jeep Comanche pickup, in fall 1985. A short-wheelbase version of the Comanche was added one year later.

In January 1986, the CJ7 was discontinued as its consumers sought more passenger car features. Jeep responded with Wrangler, which shared the utility, durability and open-body profile of CJ7, but had more in common, mechanically, with Jeep Cherokee.

Chrysler acquired AMC on Aug. 5, 1987, creating the Jeep/Eagle Div. On March 22, 1990, the one-millionth Jeep XJ, a 1990 Cherokee Limited, was built at Jeep's Toledo Assembly Plant. In 10 years of production, the Cherokee has become the preeminent vehicle of its kind.

As Jeep marked its 50th year, the Cherokee strengthened its leadership role with the addition of a new 4.0-liter/190 hp High-Output PowerTech Six. The Jeep Renegade name also reemerged in 1991. This ultimate Wrangler, powered by a 180 hp inline six, represented the most powerful vehicle in its class.

In January 1992, production of a new Jeep sport utility vehicle, code named ZJ, began at Chrysler Corp.'s new Jefferson Assembly Plant in Detroit.

The fifth generation of the Jeep Universal was launched in late 1954 when the 1955 CJ5 appeared. This example is equipped with the optional hardtop enclosure.

Jeep became part of American Motors Corp. in 1970 and updated the styling, technology and model lineup. Representing one of the new AMC variants is the 1977 CJ7 Renegade shown here.

The 1987 Jeep Wrangler did share the familiar open-body profile of the CJ7, although it had more in common, technically, with the Jeep Cherokee.

In 1966, Kaiser Jeep launched the Jeepster Commando in station wagon, convertible, pickup truck and roadster models. The station wagon is pictured here.

Ten years after the name was introduced, the Cherokee nameplate launched the first sportwagon off the new Jeep XJ platform. This AMC creation became the most popular and profitable vehicle in history.

SIGNIFICANT DATES IN JEEP® HISTORY

THE WILLYS-OVERLAND YEARS: 1940-53

Oct. 1940	Willys-Overland delivers Quad prototypes to U.S. Army.
June 1941	Army selects MA model over Bantam and Ford vehicles.
July 1941	Willys wins bid to produce Jeep vehicles for U.S. Army.
1944	Willys builds 22 prototypes of civilian Jeep vehicles, code named CJ-LA.
Aug. 1945	First civilian Jeep vehicle, CJ2A, introduced.
1946	First all-steel station wagon/sedan delivery vehicles introduced.
1947	Jeep two-wheel and four-wheel drive pickup trucks introduced.
1948	Two-door Jeepster introduced. Discontinued after 1950.
1949	Four-wheel drive and six-cylinder engine added to all-steel wagons.
1949	CJ3A introduced.
1950	New 1/2-ton FVM and one-ton four-wheel drive pickup trucks and two engines, the four cylinder Hurricane and six-cylinder Lightning, introduced.
June 1950	Willys registers Jeep name as trademark in U.S. and internationally.
1951	Jeep MB military vehicle replaced by M38AI model.

THE KAISER YEARS 1953-1970

Apr. 1953	H. J. Kaiser Co. acquires Willys assets, renaming it Willys Motors, Inc.
1954	Jeep CJ5 introduced.
1955	Two-wheel drive Dispatcher Surrey introduced, based on CJ5.
1956	FC (Forward Cab) truck series introduced.
1959	Longer wheelbase CJ6 introduced.
Fall 1962	J-Series Wagoneer, Gladiator pickup and panel delivery introduced.
Mar. 1963	Willys Motors, Inc. renamed Kaiser Jeep Corp. 1965 "Dauntless" V-6 engine introduced on CJ models.
Dec. 1965	Super Wagoneer model introduced.
Fall 1966	Jeepster Commando line introduced.

THE AMERICAN MOTORS Corp. YEARS: 1970-1987

Feb. 1970	American Motors Corp. (AMC) acquires Kaiser Jeep Corp. Jeep Corp. becomes a subsidiary of AMC.
1971	AMC creates a second subsidiary, AM General Corp.
1972	AMC-built V-8 engines offered on all Jeep vehicles.
1973	Quadra-Trac, first automatic four-wheel drive system, introduced on Wagoneer.
1974	Cherokee name added to J Series, Jeep Renegade (CJ5) becomes regular production unit and Commando discontinued.
1976	Jeep CJ7 introduced joining CJ5 production.
1978	Wagoneer Limited model of J-series utility wagon introduced.
1982	Scrambler, the first small four-wheel drive pickup introduced (is CJ8 internationally).
Fall 1983	Jeep XJ, Cherokee and Wagoneer, introduced; J-series Wagoneer renamed Grand Wagoneer; CJ5 discontinued.
Fall 1985	Comanche pickup truck introduced.
Jan. 1986	Jeep CJ7 discontinued.
Spring 1986	Jeep Wrangler introduced.

THE CHRYSLER Corp. YEARS: 1987-PRESENT

Aug. 1987	Chrysler Corp. acquires AMC.
1988	J-series pickup truck discontinued.
Mar. 1990	One-millionth Jeep XJ produced at Toledo (Ohio) Assembly.
July 1990	1992 consolidation of all current Jeep manufacturing at Toledo (Ohio) Assembly.
Fall 1990	Jeep Renegade introduction.
Jan. 1992	Jeep "ZJ" production to begin at new Jefferson Assembly Plant (Detroit).

HOW TO USE THIS CATALOG

APPEARANCE AND EQUIPMENT: Word descriptions identify cars by styling features, trim and (to a lesser extent) interior appointments. Most standard equipment lists begin with the lowest-priced model, then enumerate items added by upgrade models and option packages. Most lists reflect equipment available at model introductions.

I.D. DATA: Information is given about the Vehicle Identification Number (VIN) found on the dashboard. VIN codes show model or series, body style, engine size, model year and place built. Beginning in 1981, a standardized 17 symbol VIN is used. Earlier VINs are shorter. Locations of other coded information on the body and/or engine block may be supplied. Deciphering those codes is beyond the scope of this catalog.

SPECIFICATIONS CHART: The first column gives series or model numbers. The second gives body style numbers revealing body type and trim. Not all cars use two separate numbers. Some sources combine the two. Column three tells number of doors, body style and passenger capacity ('4-dr Sed-6P' means four-door sedan, six-passenger). Passenger capacity is normally the maximum. Cars with bucket seats hold fewer. Column four gives suggested retail price of the car when new, on or near its introduction date, not including freight or other charges. Column five gives the original shipping weight. The sixth column provides model year production totals or refers to notes below the chart. In cases where the same car came with different engines, a slash is used to separate factory prices and shipping weights for each version. Unless noted, the amount on the left of the slash is for the smallest, least expensive engine. The amount on the right is for the least costly engine with additional cylinders. 'N/A' means data not available.

ENGINE DATA: Engines are normally listed in size order with smallest displacement first. A 'base' engine is the basic one offered in each model at the lowest price. 'Optional' describes all alternate engines, including those that have a price listed in the specifications chart. (Cars that came with either a six of V-8, for instance, list the six as 'base' and V-8 'optional'). Introductory specifications are used, where possible.

CHASSIS DATA: Major dimensions (wheelbase, overall length, height, width and front/rear tread) are given for each model, along with standard tire size. Dimensions sometimes varied and could change during a model year.

TECHNICAL DATA: This section indicates transmissions standard on each model, usually including gear ratios; the standard final drive axle ratio (which may differ by engine or transmission); steering and brake system type; front and rear suspension description; body construction; and fuel tank capacity.

OPTIONAL EQUIPMENT LISTS: Most listings begin with drivetrain options (engines, transmissions, steering/suspension and mechanical components) applying to all models. Convenience/appearance items are listed separately for each model, except where several related models are combined into a single listing. Option packages are listed first, followed by individual items in categories: comfort/convenience; lighting; mirrors; entertainment; exterior; interior; then wheels/tires. Contents of some options packages are listed prior to the price; others are described in the Appearance/Equipment text. Prices are suggested retail, usually effective early in the model year. ('N/A' indicates prices are unavailable). Most items are Regular Production Options (RPO), rather than limited-production (LPO), special-order or dealer-installed equipment. Many options were available only on certain series or body types or in conjunction with other items. Space does not permit including every detail.

HISTORY: This block lists introduction dates, total sales and production amounts for the model year and calendar year. Production totals supplied by auto-makers do not always coincide with those from other sources. Some reflect shipments from factories rather than from actual production or define the model year a different way.

HISTORICAL FOOTNOTES: In addition to notes on the rise and fall of sales and production, this block includes significant statistics, performance milestones, major personnel changes, important dates and places and facts that add flavor to this segment of America's automotive heritage.

JEEP — 1986 SERIES — (ALL ENGINES): An historic milestone in light-duty truck history was the end of production of the CJ Universal Jeep during model year 1986. At the start of the season, the 1986 sales brochure gave little hint that the CJ series would be phased-out. "CJ has always been special," said the promotional copy. "And for 1986 it makes more sense when you consider its long list of standard features, its exceptional fuel economy and its affordable price." The end came in Jan. 1986, closing 40 years of civilian production during which some 1.6 million units were built. Naturally, there were no significant changes in the 1986 CJ-7 model.

I.D. DATA: VIN located on the top left surface of the instrument panel. There are 17 symbols. The first three symbols indicate the manufacturer, make and vehicle type. The engine type is indicated by the fourth symbol: B=126 cid (2.1 liter) OHC-4 turbo diesel; H=150 cid (2.5-liter) four-cylinder TBI engine; W=173 cid (2.8 liter) V-6; C=258 cid (4.2 liter) six-cylinder; N=360 cid (5.9 liter) V-8. The next letter identifies the transmission type. The series and type of vehicle are identified by the sixth and seventh symbols, which are the same as the model number. The eighth character identifies the GVW rating with the next serving as the check digit. Then follows the model year (tenth symbol G) and assembly point codes. The last six digits are sequential production numbers.

Model	Body Type	Price	Weight	Prod. Total
Wagoneer Sport Wagon — (L4/TBI) — (1/2-Ton) — (4x4)				
75	4d Sta Wag	13,360	3,039	—
75	4d Ltd Sta Wag	18,600	3,234	—
Wagoneer Sport Wagon — (V-6) — (1/2-Ton) — (4x4)				
75	4dr Sta Wag	14,607	3,104	—
75	4dr Ltd Sta Wag	19,037	3,299	—
Grand Wagoneer — (L-6) — (1/2-Ton) — (4x4)				
15	4dr Sta Wag	21,350	4,252	16,252

ENGINE (Turbo Diesel; Optional: 65/66): Inline. OHV. Four-cylinder. Cast iron block. Bore & stroke: 2.99 x 3.5 in. Displacement: 126 cid/2.1L. Taxable horsepower: 18.34. Hydraulic valve lifters. Turbo charged.
ENGINE (Standard: Wrangler/66; 65; 87): Inline. OHV. Four-cylinder. Cast iron block. Bore & stroke: 3.88 x 3.19 in. Displacement: 150 cid/2.5L. Compression ratio: 9.2:1. Brake horsepower: 117 at 5000 rpm. Taxable horsepower: 24.04. Hydraulic valve lifters. Throttle Body Injection (EFI). Torque (Compression): 135 lbs.-ft. at 3500 rpm.

CHASSIS (CJ-7): Wheelbase: 93.5 in. Overall length: 153.2 in. Height: (hardtop) 71.0 in. Front tread: 55.8 in. Rear tread: 55.1 in. Tires: P205/75R-15 in.
CHASSIS (Scrambler): Wheelbase: 103.4 in. Width: 65.3 in. Overall length: 166.2 in. Height: 70.8 in. Front tread: 55.8 in. Rear tread: 55.1 in. Tires: P205/75R15 in.
CHASSIS (Wrangler): Wheelbase: 93.4 in. Overall length: 152 in. Height: (hardtop) 69.3 in. Front tread: 58 in. Rear tread: 58 in. Tires: P215/75R-15 in.

TECHNICAL (CJ-7/Scrambler): Selective synchromesh transmission. Speeds: 4F/1R. Floor-shift. Single dry disc clutch. Shaft drive. Semi-floating rear axle. Overall ratio: 3.54:1. Manual front disc/rear drum brakes. 15 x 6 in., five-bolt pressed steel wheels.
TECHNICAL (J-10): Selective synchromesh transmission. Speeds: 4F/1R. Floor-mounted gearshift. Single dry disc clutch. Shaft drive. Rear axle: semi-floating. Overall ratio: 2.73:1. Power front disc/rear drum brakes. 15 x 6 in. pressed steel wheels.
TECHNICAL (J-20): Automatic transmission. Speeds: 3F/1R. Column-mounted gearshift. Shaft drive. Rear axle: full-floating. Overall ratio: 3.73:1. Power front disc/rear drum brakes. 16.5 x 6 in. pressed steel wheels.

OPTIONS (Wrangler): Carbureted 4.2L six-cylinder engine. Three-speed automatic transmission. Hardtop (standard with Laredo). Tilt steering. (Sport Decor Group): includes, most standard features, plus AM/FM monaural radio; black side cowl carpet; special Wrangler hood decals; special Wrangler lower body side stripes; P215/75R15 Goodyear all-terrain Wrangler tires; conventional size spare with lock and convenience group. (Laredo Hardtop Group): includes richer interior trim; AM/FM monaural radio; Buffalo-grain vinyl upholstery; front and rear carpeting; center console; extra-quiet insulation; leather-wrapped Sport steering wheel; special door trim panels and map pockets; chrome front bumper; rear bumperettes; grille panel; headlamp bezels; tow hooks; color-keyed wheel flares; full-length mud guards; integrated steps; deep tinted glass; OSRV door mirrors; bumper accessory package; special hood and body side stripes; convenience group; 15 x 7 in. aluminum wheels; P215/75R-15 Goodyear Wrangler RWL radial tires, spare tire and matching aluminum spare wheel. Rear trac-lok differential. Air conditioning. Extra-quiet insulation. Full-carpets. Halogen fog lamps. Power steering. Cruise Control (6-cyl. only). Leather-wrapped Sport steering wheel. Electric rear window defogger (hardtop). Heavy-duty suspension. Heavy-duty cooling. Aluminum wheels. Off-Road equipment package. Conventional spare tire. Metallic exterior paints.

HISTORICAL: Introduced: Fall, 1985; (Wrangler) May 1986. Innovations: New Comanche mini-pickup series. New Wrangler replaces CJ-Series in Jan. 1986 (considered a 1987 model). Model year production (excluding AM General products) totaled 243,406. This included: 25,929 CJ-7s; 122,968 Cherokee XJs; 13,716 Wagoneer XJs; 45,219 Comanche pickups; 1,657 J-10/J-20 pickups; 17,665 Grand Wagoneers and 16,252 of the 1987 Wranglers. A California Jeep dealer launched an unsuccessful nationwide publicity campaign to "Save the Jeep CJ." Joseph E. Cappy became president of AMC during 1986. Cappy referred to the new Comanche as "AMC's first state-of-the-art-product in the two-wheel-drive market, which accounts for 75 percent of the light truck market." Other names considered for the new compact pickup were Renegade, Commando, Wrangler and Honcho. AMC also announced that a total of five years had been spent in development of the 1987 Wrangler as an up-to-date replacement for the famous Jeep CJ.

137

RAMBLER
1897-1913

RAMBLER — Chicago, Illinois — (1897-1900) / Kenosha, Wisconsin — (1902-1913) — Rambler was the name of the bicycle produced in Chicago, prior to the turn of the century, by Thomas B. Jeffery and R. Philip Gormully. They operated the second largest bicycle factory in the U.S. (Col. Albert Pope on the East Coast being first.) The partners manufactured tires as well, as the G & J Tire Co., which ultimately became part of U.S. Rubber.

Of the two men, it was Thomas Jeffery who was the most ardent about entering the automotive age. His young son Charles T. Jeffery enthusiastically urged him on. In 1897, Tom Jeffery built his first single-cylinder gasoline car; in 1898 Charles Jeffery built two considerably more sophisticated machines. Aside from brief mentions in the press that Jeffery, along with E.C. Stearns and George N. Pierce were "among the bicycle manufacturers who are experimenting with the motor vehicle," little attention was initially paid to these vehicles.

In 1900, the cars were displayed at the automobile shows in Chicago, in September, and in New York City, in November. Reporters recognized a good story when they saw it. The Jeffery-designed car was alternately referred to in the press as the G & J or Rambler. It carried no plaque, so the confusion was understandable, though both Thomas and Charles Jeffery preferred the latter designation. However called, its features of a front-mounted engine and left-hand drive were very advanced for an American car of the period.

Press reaction was enthusiastic, though *The Motor Age* wondered about the marketability of steering from the left side of a car: "Whether this will become popular remains to be seen," the magazine commented. "It has many points in its favor, however."

By now Thomas and Charles Jeffery had made two important decisions. The first was to sell out their bicycle business — following the sudden death of Philip Gormully — to the American Bicycle Co. (the conglomerate engineered by Col. Pope, which was an attempt to monopolize bicycle manufacture in this country.) The second was to buy a huge factory in Kenosha, Wis., from which to launch themselves wholeheartedly into the automobile business. They retained rights to the Rambler name.

The car left behind in their Chicago factory was produced, for a while in 1901, by the American Bicycle people, as the Hydro-Car. No doubt, it was a variation of the earliest Thomas Jeffery car. The car taken to Kenosha was the son's more advanced design, which Charles advanced further by replacing the tiller with a steering wheel.

However, Thomas Jeffery had sudden second thoughts about the public acceptance of an automobile with front-mounted engine, left-hand drive and wheel steering. This delayed the onset of production. When the new Rambler was introduced in, February 1902, it had its engine mounted under the seat and was steered by tiller from the right side. Still, it was a honey of a car.

"Its low price, $750, almost warrants someone expecting something infinitely inferior...," *The Motor World* reported. "The vehicle is plainly a high class one...rare value for the money." Plainly, too, Tom Jeffery's conservatism paid off; a total of 1,500 cars were produced in 1902, a figure exceeded only by Ransom Olds with his Curved Dash runabout.

Unlike Ransom Olds, Thomas Jeffery did not long remain content with a one-lunger. In 1904, Jeffery built 2,342 cars, some of them high-powered two-cylinder versions with front-mounted engines. All Ramblers had steering wheels now. The year following, the company made an even more drastic change. It discontinued the single-cylinder car at midyear and focused all efforts on three larger two-cylinder cars, priced from $1,200 to $3,000.

Sales in 1905 increased to 3,807 cars. A Rambler four was introduced in 1906, and so was a certain savoir faire to Rambler advertising. From "The Right Car at the Right Price," Rambler burst forth with "June Time is Rambler Time" and other evocative phrases. These were courtesy of Edward S. Jordon, an employee who would rise to become general manager and secretary of Jeffrey, before leaving to give the world the Jordon Playboy and "Somewhere West of Laramie."

By now, the Jeffery company was an industry leader and the Kenosha factory, in which it turned out its Rambler, was both the biggest in the country and the best equipped. Thomas Jeffery was sitting on top of the world. Mass production, however, never interested him. However, the fortune he made in mass producing high-quality, medium-priced cars was a splendid one anyway.

On April 2, 1910, while on vacation in Italy, Thomas Jeffery died of a heart attack. His will stipulated that his business, which had previously traded under the name Thomas B. Jeffery & Co., would now be incorporated as the Thomas B. Jeffery Co., although ownership remained entirely with the Jeffery family. His son, Charles, took over as president and some changes followed.

Production capacity was raised by about 500 cars per month (Thomas had preferred 3,000 per year, more or less). Fancy new model designations, such as Country Club, Knickerbocker and Valkyrie (Ned Jordon's idea, naturally) were applied to various Rambler models. Not changed, however, was the use of right-hand steering. The biggest change came in 1914. It was a new car altogether. The decision was a gutsy one. Rambler was among the oldest and most respected nameplates in the industry, but now it was no more. The new car from Kenosha would be called the Jeffery.

1901 RAMBLER

1901 Rambler, Runabout (WLB)

RAMBLER MODEL A/B- ONE-CYLINDER: Models A and B, built in 1901, were experimental and were not offered for sale.
I.D. NUMBERS: [Both models] None.

Model No.	Body Type & Seating	Price	Weight	Prod. Total
A/B	Runabout-2P	750	1200	2

ENGINE: Data unavailable.
TECHNICAL: Data unavailable.
CHASSIS: Data unavailable.
HISTORY: Experimental car designed by Charles Jeffery, son of Thomas B. Jeffery. Manufactured by Thomas B. Jeffery & Co., Kenosha, Wis.

1902 RAMBLER

1902 Rambler, Model C Runabout (OCW)

RAMBLER — MODEL C/D — ONE-CYLINDER: The Model D was identical to the Model C, except that it had a leather top, rubber side curtains and a storm apron.
I.D. NUMBERS: [Both models] 1-1500.

Model No.	Body Type & Seating	Price	Weight	Prod. Total
C	Runabout-2P	750	1200	NA
D	Stanhope-2P	825	1250	NA

ENGINE: One-cylinder. B x S.: 4.5 x 6. NACC HP.: 8.1.
TECHNICAL: Data unavailable.
CHASSIS: Data unavailable.
HISTORY: First year of mass production for Thomas B. Jeffery & Co. Model year production (approx.): 1500 cars.

1903 RAMBLER

1903 Rambler, Runabout (Chester L. Krause)

RAMBLER — MODEL E/F — ONE-CYLINDER: Basically a continuation of the 1902 Rambler line. The F had a leather top, rubber side curtains and a storm apron.
I.D. NUMBERS: [Model E] 1501-2850; [Model F] 2850-5192.

Model No.	Body Type & Seating	Price	Weight	Prod. Total
E	Runabout-2P	750	1400	1350
F	Stanhope-2P	800	1400	2343

ENGINE: One-cylinder. B x S.: 5 x 6. NACC HP.: 8.1.
TECHNICAL: Data unavailable.
CHASSIS: Wheelbase: [E] 78 inches; [F] 78 inches.
HISTORY: Second year of mass production for Thomas B. Jeffery & Co. Model year production (approx.): 3,693 cars.

1904 RAMBLER

1904 Rambler, Rear Entrance Tonneau (OCW)

RAMBLER — MODEL G/H — ONE-CYLINDER: Basically a continuation of the 1903 Rambler line. The Stanhope became a touring. A panel delivery truck was added.

RAMBLER — MODELS J/K/L — TWO-CYLINDER: The two-cylinder Rambler was over twice as powerful as the one-cylinder, for less than a 50 percent higher price. Typical standard equipment (Model L) included 30 inch wood artillery wheels; four full elliptic springs; 3-1/2 inch tires; two powerful brakes; tools and equipment (oilers, tire pump, tire repair kit); Solar triple-top brass oil sidelamps, No. 1 Solar brass headlamp; Solar brass taillight; brass tube horn; two willow baskets; canopy top with beveled plate glass swinging front; and waterproof side curtains.

I.D. NUMBERS: [All models] 2850-5192.

Model No.	Body Type & Seating	Price	Weight	Prod. Total
[ONE-CYL.]				
G	Roadster-2P	750	1500	NA
H	Roadster Touring-5P	850	1500	NA
D-1	Panel Delivery-2P	850	1500	NA
[TWO-CYL.]				
J	Roadster-2P	1100	1800	NA
J-2	Roadster	1100	1800	NA
K	Touring-5P	1200	1800	NA
K-2	Touring-5P	1200	1800	NA
L	Touring-5P	1350	1725	NA
L-2	Touring-5P	1350	1725	NA

ENGINE: [One-cylinder]: B x S.: 5 x 6. NACC HP.: 7. [Two-cylinder] B x S.: 5 x 6. NACC HP.: 16.

TECHNICAL: Data unavailable.

CHASSIS: Wheelbase: [G/H] 81 inches; [J/K/L] 84 inches.

HISTORY: Two-cylinder series added. Calendar year production (approximate): 2,343. Prices given were for factory pickup sales. "An automobile that stands wear without constant repair," said a 1904 advertisement. Speed range of the Model L was six to 40 mph.

1905 RAMBLER

1905 Rambler, Type I "Surrey" Tonneau Touring (JAC)

RAMBLER — Type I/II — [18/20 HORSEPOWER] — TWO-CYLINDER: A larger new touring with a 90 inch wheelbase was designated the Surrey I. The Surrey II was the same basic car with a 10 inch longer wheelbase. A limousine was also marketed. Apparently, this was the first closed body Rambler.

I.D. NUMBERS: [All models] 5193-9000.

Model No.	Body Type & Seating	Price	Weight	Prod. Total
[18/20 HP]				
Surrey I	Touring-5P	1200	2000	NA
Surrey II	Touring-5P	1650	2000	NA
Lim.	Limousine-5P	3000	3000	NA

ENGINE: [Two-cylinder] B x S.: 5 x 6. NACC HP.: 18/20.

TECHNICAL: Data unavailable.

CHASSIS: Wheelbase: [Type I] 90 inches; [Type II] 100 inches; [Limo] 112 inches.

HISTORY: Calendar year production (approximate): 3,807. First closed Rambler introduced. Rambler used the term "Surrey" to describe what was actually a Tonneau Touring car with cape cart top.

1906 RAMBLER

1906 Rambler, Type III "Surrey" Tonneau Touring (OCW)

RAMBLER — Type III/IV — 18/20 HORSEPOWER — TWO-CYLINDER: The 90 inch wheelbase and 100 inch wheelbase Tonneau Touring cars with cape cart tops comprised this carryover series.

RAMBLER — MODEL 14 — 20 HP — FOUR-CYLINDER: A five-passenger "Light Four" on a 106 inch wheelbase. Has a new four-cylinder engine.

RAMBLER — MODEL 15/16 — 40 HP — FOUR-CYLINDER: Rambler's first four-cylinder line was on a 112 inch wheelbase. A closed limousine was included, as well as a pair of touring cars. The 1906 Rambler Model 15 was advertised as a high-power car with a four-cylinder motor, 35/40 horsepower, sliding gear transmission and all modern features, but simplified to the practical service of non-professional operators. "Our catalog will interest you and a personal examination will convince you that it is the car of the year," said the ad copy. The Model 15 was of open front entrance design, with doors fitted on its rear tonneau.

RAMBLER — MODEL 17 — 10/12 HP — TWO-CYLINDER: A short wheelbase runabout with light horsepower rating. Not shown in AMC specifications book.

RAMBLER — MODEL 19 — 18 HP — TWO-CYLINDER: Probably a renumbered version of the Type I or Type II.

RAMBLER — MODEL 147 — 20 HP — FOUR-CYLINDER: An improved version of the Model 14 with the new four-cylinder engine.

I.D. NUMBERS: [All 18 HP models and Model 15] 5193-9000; [Models 14/16/147] 9001-15,300.

Model No.	Body Type & Seating	Price	Weight	Prod. Total
[10/12 HP]				
17	Runabout-2P	800	NA	NA
[18/20 HP]				
Surrey III	Touring-5P	1350	2200	NA
Surrey IV	Touring-5P	1350	2200	NA
Model 19	Touring-2P	1250	2200	NA
[20/25 HP LIGHT FOURS]				
14	Touring-5P	1750	2300	NA
147	Touring-5P	1750	2400	NA
[35/40 HP FOURs]				
15	Touring-5P	2500	2600	NA
16	Limousine-5P	3000	3000	NA

ENGINES:

[Model 17] Two-cylinder. B x S.: 4-1/2 x 5 inches. NACC HP.: 10/12

[Type III/IV Model 19] Two-cylinder. B x S.: 5 x 6 inches. NACC HP.: 18/20.

[Model 14] Four-cylinder. B x S.: 4-1/2 x 4-1/2 inches. NACC HP.: 20/25.

[Model 147] Four-cylinder. B x S.: 4 x 4-1/2 inches. NACC HP.: 20.

[Models 15/16] Four-cylinder. B x S.: 5 x 5-1/2 inches. NACC HP.: 35/40 HP.

TECHNICAL: Data unavailable.

CHASSIS:

[Type III/IV] W.B.: 96 inches.

[Model 17] W.B.: 88 inches.

[Model 19] W.B.: 96 inches.

[Model 14/147] W.B.: 106 inches.

[Model 15/16] W.B.: 112 inches.

HISTORY: Calendar year production (approximate): 6,299. Offered by Thomas B. Jeffery & Co. with main office and factory in Kenosha, Wis. and sales branches in Boston, Mass.; Philadelphia, Pa.; Milwaukee, Wis.; Chicago, Ill.; San Francisco, Calif.; and New York City. The company's new Model 15 was also sold in 1907.

1907 RAMBLER

1907 Rambler, Model 21 Roadster (JAC)

RAMBLER — MODEL 27 — 16.2 HORSEPOWER — TWO-CYLINDER: A 90 inch wheelbase light two-cylinder roadster with an exclusive engine.

RAMBLER — MODEL 21/22 — 20 HP — TWO-CYLINDER: Technically, a continuation of the "Surrey" line of 1905-1906 using just the larger 100 inch wheelbase.

RAMBLER — MODEL 24/248 — LIGHT FOUR — 25/30 HP — FOUR-CYLINDER: Technically, a continuation of the Model 14 with two inch longer wheelbase.

RAMBLER — MODEL 25/245 — 35/40 HP — FOUR-CYLINDER: Technically a continuation of the Models 15/16 with largest Rambler engine and wheelbase.

I.D. NUMBERS: [All models 15] numbers 9001-15,300 continued from 1906.

Model No.	Body Type & Seating	Price	Weight	Prod. Total
[16.2 HP]				
27	Roadster-2P	950	2000	NA
[18/20 HP]				
21	Touring-5P	1350	2000	NA
22	Roadster-2P	1250	2000	NA
[25/30 HP LIGHT FOURS]				
24	Touring-5P	2000	2600	NA
248	Touring-5P	1900	2600	NA
[35/40 HP FOURS]				
25	Touring-5P	2500	2600	NA
245	Touring-5P	2500	2600	NA

ENGINES:

[Model 27] Two-cylinder. B x S.: 4-1/2 x 5 inches. NACC HP.: 16.2

[Model 21/22] Two-cylinder. B x S.: 5 x 6 inches. NACC HP.: 18/20.

[Model 24/248] Four-cylinder. B x S.: 4-1/2 x 4-1/2 inches. NACC HP.: 25/30.

[Model 25/245] Four-cylinder. B x S.: 5 x 5-1/2 inches. NACC HP.: 40.

[Models 15/16] Four-cylinder. B x S.: 5 x 5-1/2 inches. NACC HP.: 35/40 HP.

TECHNICAL: Selective sliding gear transmission.

CHASSIS:

[Model 27] W.B.: 90 inches.

[Model 21/22] W.B.: 100 inches.

[Model 24/248] W.B.: 108 inches.

[Model 254/245] W.B.: 112 inches.

HISTORY: Model 15 also built in 1907.

1908 RAMBLER

1908 Rambler, Model 36 Limousine (OCW)

RAMBLER — MODEL 31/37 — 22 HORSEPOWER — TWO-CYLINDER: The old 5 x 6 two-cylinder Rambler engine uprated to 22 NACC hp and installed in a larger 106 inch wheelbase. Quite large for a two-cylinder car.

RAMBLER — MODEL 34/34A/36 — LIGHT FOUR — 32 HP — FOUR-CYLINDER: Technically, a continuation of the Models 24/248 with four inch longer wheelbase and slight horsepower rating boost. Heftier prices and weights apparent.

RAMBLER — MODEL 345 — 40 HP — FOUR-CYLINDER: A virtually unchanged carryover of the Model 245 with largest Rambler engine and wheelbase. Even has same price and weight.

I.D. NUMBERS: [Model 345] numbers 9001-15,300 continued from 1907; [All other models] 15603 to 17200.

Model No.	Body Type & Seating	Price	Weight	Prod. Total
[TWO-CYLINDER]				
31	Touring-5P	1400	2300	NA
37	Roadster-2P	1200	2300	NA
[LIGHT FOUR]				
34	Touring-5P	2250	2800	NA
34A	Roadster-3P	2250	2700	NA
36	Limousine-5P	3250	3000	NA
[40 HP FOUR]				
345	Touring-5P	2500	2600	NA

ENGINES:

[Model 31/37] Two-cylinder. B x S.: 5 x 6 inches. NACC HP.: 22.

[Model 34/34A/36] Four-cylinder. B x S.: 4-1/2 x 4-1/2 inches. NACC HP.: 32.4.

[Model 345] Four-cylinder. B x S.: 5 x 5-1/2 inches. NACC HP.: 40.

TECHNICAL: Selective sliding gear transmission.

CHASSIS:

[Model 31/37] W.B.: 106 inches.

[Model 34/34A/36/] W.B.: 112 inches.

[Model 345] W.B.: 112 inches.

HISTORY: New series production: 1,597 cars.

1909 RAMBLER

1909 Rambler, Model 45 Tonneau Touring (JAC)

RAMBLER — MODEL 41/47 — 22 HORSEPOWER — TWO-CYLINDER: The old 5 x 6 two-cylinder Rambler engine again in the large 106 inch wheelbase. A continuation of the 1908 two-cylinder with only refinements.
RAMBLER — MODEL 44/44A/44CC — LIGHT FOUR — 32 HP — FOUR-CYLINDER: Technically, a continuation of previous Light Four with one new Close-Coupled sedan model. Slightly heavier, but prices basically unchanged.
RAMBLER — MODEL 45 — 40 HP — FOUR-CYLINDER: The 245/345 drive train on a massive 123 inch wheelbase chassis. Much heavier, but no price hike. Touring becomes a seven-passenger model.
I.D. NUMBERS: [All models] 20001-21963.

Model No.	Body Type & Seating	Price	Weight	Prod. Total
[TWO-CYLINDER]				
41	Touring-5P	1350	2300	NA
47	Roadster-2P	1200	2300	NA
[LIGHT FOUR]				
44	Touring-5P	2250	2800	NA
44-A	Roadster-3P	2250	2800	NA
44CC	C.C. Sedan-5P	2250	2800	NA
[40 HP FOUR]				
45	Touring-7P	2500	3400	NA
45CC	C.C. Sedan-4P	2500	3400	NA

ENGINES:
[Model 41/47] Two-cylinder. B x S.: 5 x 6 inches. NACC HP.: 22.
[Model 44] Four-cylinder. B x S.: 4-1/2 x 4-1/2 inches. NACC HP.: 32.4.
[Model 45] Four-cylinder. B x S.: 5 x 5-1/2 inches. NACC HP.: 40.
TECHNICAL: Selective sliding gear transmission.
CHASSIS:
[Model 41/47] W.B.: 106 inches.
[Model 44] W.B.: 112 inches.
[Model 45] W.B.: 123 inches.
HISTORY: Calendar year production: 1,963 cars.

1910 RAMBLER

1910 Rambler, Model 53 Roadster (OCW)

RAMBLER — MODEL 53 — LIGHT FOUR — 32 HP — FOUR-CYLINDER: Technically, a continuation of previous Light Four on a three inch larger wheelbase chassis. This is now the base series Rambler, with two-cylinder cars dropped.
RAMBLER — MODEL 54 — 40 HP — FOUR-CYLINDER: This car used the big Rambler engine in a 117 inch wheelbase chassis.
RAMBLER — MODEL 55 — 40 HP — FOUR-CYLINDER: This line used the big Rambler four in the biggest chassis and featured a new five-passenger limousine.
I.D. NUMBERS: [All models] 22501-24744.

Model No.	Body Type & Seating	Price	Weight	Prod. Total
[LIGHT FOUR]				
53	Touring-5P	1800	3200	NA
53-Rd	Roadster-2P	1800	3200	NA
[40 HP FOUR/117 in. w.b.]				
54	Touring-5P	2250	3400	NA
54CC	C.C. Sedan-4P	2250	3400	NA
[40 HP FOUR/123 in. w.b.]				
55	Touring-7P	2500	3400	NA
55CC	C.C. Sedan-4P	2350	3300	NA
55L	Limousine-5P	3750	3500	NA

ENGINES:
[Model 53] Four-cylinder. B x S.: 4-1/2 x 4-1/2 inches. NACC HP.: 32.4.
[Model 54/55] Four-cylinder. B x S.: 5 x 5-1/2 inches. NACC HP.: 40.
TECHNICAL: Selective sliding gear transmission.
CHASSIS:
[Model 53] W.B.: 109 inches.
[Model 54] W.B.: 117 inches.
[Model 55] W.B.: 123 inches.
HISTORY: Calendar year production: 2,243 cars.

1911 RAMBLER

1911 Rambler, touring (JAC)

RAMBLER — MODEL 63 — LIGHT FOUR — 32 HP — FOUR-CYLINDER: Technically, a continuation of previous Light Four with another three inch wheelbase stretch. New body styles included a coupe and a town car.
RAMBLER — MODEL 64 — 40 HP — FOUR-CYLINDER: This car, with the big Rambler engine, was the 54 with a three inch longer wheelbase and new Toy Tonneau and Landaulet styles.
RAMBLER — MODEL 65 — 40 HP — FOUR-CYLINDER: The "Big Four" line got a bigger-by-five-inches stance for 1911 and a new Toy Tonneau model. In addition to the touring, the limousine was now a seven-passenger job.
I.D. NUMBERS: [All models] 25000-28000.

Model No.	Body Type & Seating	Price	Weight	Prod. Total
[LIGHT FOUR]				
63	Touring-5P	2175	3200	NA
63-Rdr	Roadster-2P	2105	3200	NA
63C	Coupe-4P	2605	3500	NA
63TC	Town Car-7P	2880	3500	NA
[40 HP FOUR/120 in. w.b.]				
64	Touring-5P	2775	3400	NA
64TT	Toy Tonneau-4P	2775	3400	NA
64Land	Landaulet-7P	3650	3500	NA
[40 HP FOUR/128 in. w.b.]				
65	Touring-7P	3050	3400	NA
65TT	Toy Tonneau-4P	3050	3400	NA
65Lim	Limousine-7P	4150	3500	NA

ENGINES:
[Model 63] Four-cylinder. B x S.: 4-1/2 x 4-1/2 inches. NACC HP.: 32.4.
[Model 64/55] Four-cylinder. B x S.: 5 x 5-1/2 inches. NACC HP.: 40.
TECHNICAL: Selective sliding gear transmission.
CHASSIS:
[Model 63] W.B.: 112 inches.
[Model 64] W.B.: 120 inches.
[Model 65] W.B.: 128 inches.
HISTORY: Calendar year production: 3,000 cars.

1912 RAMBLER

1912 Rambler, Cross Country Touring (JAC)

RAMBLER — MODEL 73 — LIGHT FOUR — 32 HP — FOUR-CYLINDER: Technically, a continuation of previous Light Four a massive eight inch wheelbase stretch. New body styles included a limousine and close-coupled.

RAMBLER — MODEL 74 — 40/50 HP — FOUR-CYLINDER: This car, with the big Rambler engine, was based on the 64 with no change in wheelbase. New models included a close-coupled and the "Valkrie" four-passenger Torpedo Touring. Standard equipment included a gas tank; lamps; horn; tools; and jack.

RAMBLER — MODEL 75/76 — 40/50 HP — FOUR-CYLINDER: The "Big Four" was technically similar to last year's Model 65. The models now had names like Morain, Metropolitan and Cross-Country. Standard equipment included a gas tank; lamps; horn; and tools.

I.D. NUMBERS: [74/75] Numbers 25000-28000 continued from 1910; [Other Models] 28001-31551.

Model No.	Body Type & Seating	Price	Weight	Prod. Total
[LIGHT FOUR]				
73 4CC	Country Club Touring-5P	1650	3600	NA
73-Rd4	Roadster-2P	1650	3600	NA
73 S	Sub CC Torpedo Touring-4P	1650	3600	NA
73 G	Gotham Limousine-7P	2750	3800	NA
73	Sedan-6P	2500	3800	NA
[40 HP FOUR/120 in. w.b.]				
74 CC	Country Club Touring-5P	2250	3400	NA
74	Valkyrie Touring-4P	2250	3400	NA
[40 HP FOUR/128 in. w.b.]				
75 7M	Moraine Touring-7P	2500	3400	NA
76 Tor	CC Torpedo Touring-6P	2850	3800	NA
76 7 7M	Metropolitan Touring-7P	2850	3800	NA
76 K	Limousine-7P	4200	4000	NA

ENGINES:
[Model 73] Four-cylinder. Cast singly. Offset. B x S.: 4-1/2 x 4-1/2 inches. NACC HP.: 32.4 (SAE). Holley carb.
[Model 74/75/76] Four-cylinder. Offset. Cast singly. B x S.: 5 x 5-1/2 inches. NACC HP.: 40 (SAE). Four-cycle. Water-cooled. Holley carburetor. Force feed and splash lubrication system. Vertical tube radiator. Bosch ignition.
TECHNICAL: Selective sliding gear transmission. Speeds: 3F/1R. Band clutch. Shaft drive. Semi-floating rear axle. Right-hand steering and right-hand shifting and braking controls.

CHASSIS:
[Model 73] W.B.: 120 inches. 36 x 4 pneumatic tires front and rear. 16-gallon gas tank on touring and torpedo touring; 28-gallon on roadster.
[Model 74] W.B.: 120 inches. 36 x 4 pneumatic tires front and rear. 16-gallon gas tank on touring and torpedo touring.
[Model 75/76] W.B.: 128 inches. 40 x 4-1/2 pneumatic tires (front/rear) on 18 inch wheels. Semi-elliptic front springs. 3/4-elliptic rear springs. 26-gallon gas tank. Pressed steel frame.
HISTORY: Calendar year production: 3,550 cars.

1913 RAMBLER

1913 Rambler, Cross Country Touring (OCW)

RAMBLER — MODEL 83 — LIGHT FOUR — 32 HP — FOUR-CYLINDER: The only Rambler left was the Model 83 with a Cross-Country Roadster; Cross-Country Touring (in four- or five-passenger configurations); Cross-Country sedan; and Gotham Limousine. All used the 32.4 (NACC) horsepower four and a 120 inch wheelbase.

I.D. NUMBERS: [83] Numbers 31552-35987.

Model No.	Body Type & Seating	Price	Weight	Prod. Total
[LIGHT FOUR]				
83 4CC	Cross Country Touring-5P	1875	3600	NA
83-Cross	Cross Country Touring-4P	1875	3600	NA
83 Rd	Cross Country Roadster-2P	1815	3600	NA
83 Sedan	Cross Country Sedan-6P	2575	3800	NA
83 G	Gotham Limousine-7P	2750	3800	NA

ENGINE: [Model 83] Four-cylinder. Cast singly. Offset. B x S.: 4-1/2 x 4-1/2 inches. NACC HP.: 32.4 (SAE). Holley carb.
TECHNICAL: Selective sliding gear transmission. Speeds: 3F/1R. Band clutch. Shaft drive. Semi-floating rear axle. Right-hand steering and right-hand shifting and braking controls.
CHASSIS: [Model 83] W.B.: 120 inches. 36 x 4 pneumatic tires front and rear. 16-gallon gas tank on touring and torpedo touring; 28-gallon on roadster.
HISTORY: Calendar year production: 4,435 cars. Last year for Rambler name until revived by Nash for 1950 compact car. In 1914, the trade name Jeffery was adopted by Thomas B. Jeffery & Co. to identify the cars it manufactured.

JEFFERY
1913-1918

JEFFERY — Kenosha, Wisconsin — (1913-1917) — "To the end that his name may remain in the memories of men, we have named our new car the Jeffery." Charles Jeffery was speaking of his father, Thomas Jeffery, the man who had brought the Rambler to Wisconsin and who, as a local historian put it, changed Kenosha "from a prairie to a city."

In the years since 1902, Thomas Jeffery had also constructed the largest automobile factory in the United States and had built the Rambler into one of the most respected and successful American automobiles. In 1910, he had died suddenly of a heart attack. Most likely, the principal motivating factor behind Charles Jeffery's decision to rename the car was sentiment; certainly discarding a name as revered as Rambler was a gutsy decision. However, it appears that sometime during 1913, Jeffery badges began to show up on what had previously been the Model 83 Rambler.

Officially the Jeffery marque started in 1914, when three lines of cars were offered. One, the C-4, was simply the 1913 Rambler Model 83 Cross Country five-passenger touring with a Jeffery nameplate. (Even the range of serial numbers for the two cars is the same.) The others were the four-cylinder Model 93 and the six-cylinder Model 96.

The Model 93 and Model 96 Jefferys were new cars. The first to appear was the 93, a 40 hp (32.4 NACC horsepower) monobloc four available on a 116 inch wheelbase. It was soon joined by the 96, a 48 hp six on a 128 inch chassis. Both featured left-hand drive. This was a feature of the first experimental cars built by Charles Jeffery at the turn of the century, but Thomas Jeffery had decided against using it on production Ramblers.

New for the Thomas B. Jeffery Co., too, was a truck called the Quad. Serial numbers indicate that 4,435 Model 83s were made in 1913. After the changes in name and product that came in 1914, production leaped to 10,417 Jeffery cars and 3,096 Jeffery Quads.

In 1915, the four was refined. Worm drive was a new feature. Two Model 96 sixes were carried over. They were joined by the Model 106 (using the same engine), plus the Model 104 line of four cars with a larger six that was based on the four-cylinder engine's dimensions. However, with the war in Europe, and the likelihood that America would be drawn into it, Charles Jeffery chose to concentrate his company's efforts on trucks. Some 7,600 of them were produced, with only half that number of the Jeffery cars.

In May of that year, Charles Jeffery embarked upon a trip which would forever change his life. The ship on which he set sail for another of his routine fact-finding treks to the "Continent" was the Lusitania. He was one of the 761 who survived the torpedoes. Yet, the memory of the four harrowing hours he spent in the icy waters, before being picked up by a trawler, remained horrifically in his mind.

Conceivably, too, the memory of Thomas B. Jeffery's sudden death brought to Charles a vivid realization of his own mortality. During the summer of 1916, at the age of 40, he decided to retire and spend the rest of his life in personal pursuits. Charles Jeffery sold his company to another Charles . . . whose last name was Nash.

1913 JEFFERY

1913 Jeffery, Model 83 five-passenger touring (Dr. Art Burrichter)

1913 JEFFERY — MODEL 83 — FOUR: The 1913 Jeffery (a.k.a. Rambler 83) was a large car with a squarish cowl, large sidelights in the cowl and huge 9-1/4 in. drum style headlamps trimmed in black and nickel. It had a new radiator of distinctive curved-top design with seven curved horizontal moldings decorating it. The radiator had 12,000 sq. in. of cooling surface and was topped by a radiator cap of exclusive design. The touring model had what was called a "straight line torpedo" body. It included a roomy tonneau seat four feet wide with 31 inches of legroom. It was 27 inches from the front seat to the dash. The name Cross-Country (which would appear on AMC/Rambler station wagons years later) was used to identify four open models and one sedan. Also available was the seven-passenger "Gotham" limousine. Ad copywriters described the vehicle as a big car of exceeding beauty with fenders of sweeping grace and a radiator of new, distinctive design. It had doors that were 20 inches wide, which opened fully with no outside latches. The standard finish was English Purple Lake, referred to as "a rare shade of deep maroon." It was trimmed in nickel with the hood, fenders and filler panels in black enamel and nickel. Soft upholstery covered seats with eight inch deep cushions made of the finest selected long hair. Rear cushions had 45 double-acting steel coil springs. Traditional factory and aftermarket reference sources date the Jeffery marque from 1914 and list the Model 83 as a Rambler only. However, some period sources do exist that show this car as a Jeffery, separate from the Rambler. One is the Wisconsin "Motor Vehicle Weights Guidebook" (1924). Another is "Motor Age Passenger Car Serial Numbers" (1920). In addition, classic car dealer Dr. Art Burrichter, had a 1913 Jeffery for sale several years ago. It is also a fact that the 1914 Jeffery C-4 was actually the Model 83, as both cars had identical specifications and used the same range of serial numbers. Since it is possible that some Model 83s got Jeffery nameplates, we are listing the car here. Standard equipment on the Model 83 included Bosch Duplex ignition; fine large black and nickel headlamps; gas tank; black and nickel side and tail oil lamps; large tool box with complete tool outfit; jack; tire pump; robe rail; and footrest.

I.D. NUMBERS: Serial numbers to left on front frame cross-member. Serial numbers for 1913 Rambler Model 83 to 1914 Jeffery C-4 were 31552 to 35987.

Model No.	Body Type & Seating	Price	Weight	Prod. Total
CROSS-COUNTRY				
83	4dr Tr-5P	1875	3600	NA
83	4dr Tr-4P	1875	3600	NA
83	2dr Roadster-2P	1815	3600	NA
83	4dr Sedan-6P	2575	3800	NA
GOTHAM				
83	4dr Limousine-7P	2825	3800	NA

NOTE 1: Apparently, this line was produced as a Rambler built by the Thomas B. Jeffery Co., with the five-passenger touring carried over as the 1914 Jeffery C-4 and serial numbers continued.
NOTE 2: The serial numbers suggest production of 4,435 units under both the Rambler and Jeffery names during the 1913-1914 period.
ENGINE: L-head. Four cylinder. Cast en bloc. Cast iron. B x S.: 4-1/2 x 4-1/2 in. Disp.: (approximately 350 cid). NACC HP.: 32.4. Max. HP.: 42.
TECHNICAL: Electric starting. Cone clutch. Four-speed selective sliding gear transmission. Bevel gear final drive.
CHASSIS: I-beam front axle, forward-set. Drop frame and spring suspension. Front springs: 39 in. Rear springs: 52 in. Wheelbase (W.B.): 120 in. Wheels: 36 in. 34 x 4 in. tires on demountable rims. Spare tire on demountable rim carried in rear. Braking surface: 400 sq. in.
HISTORICAL: An advertisement of 1912 promoted the Rambler Cross Country as "The most comfortable car in America selling below $2,500." According to the maker, the car could creep along at four miles-per-hour in New York City's Fifth Ave. traffic or hit 50 mph on the open road. It took Abbey Hill, in New York, on high gear with five passengers, starting at 22 mph and going 30 mph at top. It took Viaduct Hill on high, starting at 25 mph, dropping to 12 mph at the crest, and going 18 mph at the top (passing two higher-priced cars while climbing.) In Philadelphia, the Model 83 challenged City Line Hill in high gear. At Kingston, N.Y., it climbed State St. Hill with six passengers aboard.

1914 JEFFERY

1914 Jeffery, two-passenger roadster (WLB)

JEFFERY — MODEL C-4 — FOUR: The 1914 Jeffery was the same large car that was the Rambler Model 83 a year earlier. Seen again were a squarish cowl, large sidelights in the cowl, huge 9-1/4 in. drum style headlamps trimmed in black and nickel and a radiator of curved-top design with seven curved horizontal moldings decorating it. The "straight line torpedo" touring was the only model offered. It included a roomy tonneau seat four feet wide with 31 inches of legroom. Twenty inch wide doors opened fully on concealed latches. Finish was English Purple Lake (deep maroon) trimmed in nickel and black enamel. Soft leather upholstery covered seats with eight inch deep cushions made of the finest selected long hair. Rear cushions had 45 double-acting steel coil springs. Standard equipment on the Model 83 included Bosch Duplex ignition; fine large black and nickel headlamps; gas tank; black and nickel side and tail oil lamps; large tool box with complete tool outfit; jack; tire pump; robe rail; and footrest.
JEFFERY — MODEL 93 — FOUR: This was a line of five smaller wheel-based four-cylinder cars. They were characterized by a high and short hood, torpedo headlamps and beltline that met the windshield at a right angle. The Model 93 engine was smaller. It had 10 less NACC (National Automobile Chamber of Commerce) horsepower than the Model 83/C-4 engine.
JEFFERY — MODEL 96 — SIX: A larger car available to Jeffery buyers in 1914 was the Model 96. Judging from the engine bore and stroke specifications, the massive six-cylinder motor was basically the same as the Model 93's four-cylinder power plant, with an extra "jug" of two cylinders. There were six body styles, including a large seven-passenger limousine.

I.D. NUMBERS: Serial numbers to left on front frame cross-member. [Model C-4]: 31552 to 35987; [Model 93]: 40000 to 46200; [Model 96]: 38000 to 40000.

Model No.	Body Type & Seating	Price	Weight	Prod. Total
JEFFERY C-4 [CROSS-COUNTRY FOUR]				
C-4	4dr Tr-5P	1875	3600	Note 1
JEFFERY MODEL 93 [FOUR]				
93	4dr Tr-5P	1550	2900	Note 2
93	2dr Roadster-2P	1550	3000	Note 2
93	All-Weather-4P	1950	2900	Note 2
93	4dr Sedan-4P	2350	3000	Note 2
(93 K)				
93	4dr Limousine-7P	3000	3000	Note 2
JEFFERY MODEL 96 [SIX]				
96	2dr Roadster-2P	2250	3750	Note 3
96	4dr Tr-5P	2250	3765	Note 3
96	4dr Tr-6P	2300	3765	Note 3
96	4dr Tr-7P	2350	3765	Note 3
96	4dr Sedan-5P	3250	3900	Note 3
96	4dr Limousine-7P	3700	4000	Note 3

NOTE 1: The Rambler Cross-Country Touring became the 1914 Jeffery C-4 and serial numbers suggest production of 4,435 units under both the Rambler and Jeffery names during the 1913-1914 period.
NOTE 2: Serial numbers indicate approx. 2,000 built.
NOTE 3: Serial numbers indicate approx. 6,200 built.
ENGINES
[C-4] L-head. Four-cylinder. Cast en bloc. Cast iron. B x S.: 4-1/2 x 4-1/2 in. Disp.: (approximately 350 cid). NACC HP.: 32.4. Max. HP.: 42.
[93] L-head. Four-cylinder. Cast en bloc. Cast iron. B x S.: 3-3/4 x 5-1/4 in. Disp.: 231.9 cid. NACC HP.: 22.5. Max. HP.: 38. Carb.: Rayfield. Pump cooling. Bosch ignition. U.S. L. electric starting/lighting.
[96] L-head. Six-cylinder. Cast in pairs. Cast iron block. B x S.: 3-3/4 x 5-1/4 in. Disp.: 347.9 cid. NACC HP.: 33.7. Max. HP.: 48. Carb.: Rayfield. Pump cooling. Bosch ignition. U.S.L. electric starting/lighting.
TECHNICAL
[C-4] Electric starting. Cone clutch. Four-speed selective sliding gear transmission. Bevel gear final drive.
[93] Electric starting. Cone clutch. Four-speed selective sliding transmission. Bevel gear final drive. Left-hand steering. Central controls. Bosch Duplex type ignition.
[96] Electric starting. Dry disc clutch. Four-speed selective sliding gear transmission. Spiral bevel gear final drive. Left-hand steering. Central controls for shifting and braking. Bosch ignition.
CHASSIS
[C-4] Wheelbase: 120 in. Tires: 34 x 4.
[93] Wheelbase: 116 in. Tires: 34 x 4 (front and rear). Rear springs: 3/4- elliptic.
[96] Wheelbase: 128 in. Tires: 37 x 4.5 (front and rear). Rear springs: 3/4-elliptic.
HISTORY: The Official Specifications 1902-1963 Rambler (and its predecessors) published by American Motors Corp. listed the 1914 model as the first Jeffery. In other sources indicated in the 1913 section above, 1914 is shown as the second year of the marque. Annual production was recorded as 10,417 automobiles and 3,096 Jeffery Quad trucks.

145

1915 JEFFERY

1915 Jeffery, Chesterfield Six Roadster (JAC)

JEFFERY — MODEL 93-2 — FOUR: This was a line of five smaller wheel-based four-cylinder cars. They were characterized by a high and short hood, torpedo headlamps and beltline that met the windshield post at a right angle. The backs of the touring car's front seat jutted up high above the beltline. This was the only four-cylinder Jeffery line this year. The factory specifications booklet cited above identifies the 93-2 touring as a "Chesterfield." However, sources, such as MoToR, indicate that the Chesterfield was the touring car in the Model 104 six-cylinder line.
JEFFERY — MODEL 96-2 — SIX: The larger car available to Jeffery buyers in 1914, the Model 96-2, was carried over in just two body styles for 1915. They were the roadster and touring. These models were not shown in MoToR's annual show issue and were most likely discontinued after cars left in inventory were sold out.
JEFFERY — MODEL 106 — SIX: The 106 used the same engine as the Model 96 (which was the six-cylinder version of the Model 93 four). In fact, common serial numbers indicate that it was really just an extended version of the Model 96 chassis. It was not related to the new Model 104 six, which had an entirely different motor. One feature of this model that set it apart was a long, long wheelbase.
JEFFERY — MODEL 104 — SIX: This six-cylinder Jeffery series was the only really new product line for 1915. Some sources refer to it simply as the Jeffery Six. The touring car was called the "Chesterfield Six." It was characterized by a high hood, curved cowl top, low one-piece windshield and curved tonneau style front seat with lower seatbacks than those of the Model 93. The hoodsides were smooth on the Chesterfield. The line also included three other body types.
I.D. NUMBERS: Serial numbers to left on front frame cross-member. [93-2] 53000 to 60500; [96-2 and 106]: 52000 to 53000; [104]: 47000 to 48500.

Model No.	Body Type & Seating	Price	Weight	Prod. Total
JEFFERY MODEL 93-2 [FOUR]				
93-2	4dr Tr-5P	1500	2900	Note 1
93-2	2dr Roadster-2P	1500	2900	Note 1
JEFFERY MODEL 96-2 [SIX]				
96-2	4dr Tr-5P	2300	3765	Note 2
96-2	2dr Roadster-2P	2350	3765	Note 2
JEFFERY MODEL 106 [SIX]				
106	4dr Tr-7P	2400	3850	Note 2
JEFFERY MODEL 104 [SIX]				
104	4dr Chesterfield Tr-5P	1650	2900	Note 3
104	2dr Roadster-2P	1650	2900	Note 3
104	All Weather-2P	1950	2900	Note 3
104	4dr Sedan-5P	2450	3000	Note 3

NOTE 1: Production: Approximately 7,500.
NOTE 2: Production: Approximately 1,000 (Model 96-2 and Model 106).
NOTE 3: Production: Approximately 1,500.
ENGINES
[93-2] L-head. Four-cylinder. Cast en bloc. Cast iron. B x S.: 3-3/4 x 5-1/4 in. Disp.: 231.9 cid. NACC HP.: 22.5. Max. HP.: 38. Carb.: Rayfield. Pump cooling. Bosch ignition. U.S.L. electric starting/lighting.
[96-2/106] L-head. Six-cylinder. Cast in pairs. Cast iron block. B x S.: 3-3/4 x 5-1/4 in. Disp.: 347.9 cid. NACC HP.: 33.7. Max. HP.: 48. Carb.: Rayfield. Pump cooling. Bosch ignition. U.S.L. electric starting/lighting.
[104] L-head. Six-cylinder. Cast en bloc. Cast iron block. B x S.: 3 x 5 in. Disp.: 212.6 cid. NACC HP.: 21.6. Max. HP: NA. Carb.: NA. Pump cooling. Bosch single ignition. Bijur electric starting/lighting.
TECHNICAL
[93-2] Electric starting. Cone clutch. Four-speed selective sliding transmission. Bevel gear final drive. Left-hand steering. Central controls. Bosch Duplex type ignition.
[96-2/106] Electric starting. Dry disc clutch. Four-speed selective sliding gear transmission. Spiral bevel gear final drive. Left-hand steering. Central controls for shifting and braking. Bosch ignition.
[104] Electric starting. Dry disc clutch. Four-speed selective sliding gear transmission. Worm gear final drive. Left-hand steering. Central controls for shifting and braking. Bosch ignition.
CHASSIS
[93-2] Wheelbase: 116 in. Tires: 34 x 4 (front and rear). Rear springs: 3/4- elliptic.
[96-2] Wheelbase: 128 in. Tires: 37 x 4.5 (front and rear). Rear springs: 3/4-elliptic.
[106] Wheelbase: 133-1/2 in. Tires: 34 x 4-1/2 in. (front and rear). Rear springs: 3/4-elliptic.
[104] Wheelbase: 122 in. Tires: 34 x 4 (front and rear). Rear springs: Cantilever.
HISTORY: Jeffery began to concentrate on the manufacture of four-wheel drive Quad trucks for military use in the Mexican War.

1916 JEFFERY

1916 Jeffery, Model 462 four-door touring (JAC)

JEFFERY — MODEL 462 — FOUR: At the beginning of 1916, Jeffery introduced the four-cylinder Model 462 for the New York Auto Show. It was obviously based on the 93/93-2 with identical bore and stroke, weight and wheelbase. The hood had vertical louvers on its lower side panels and retained a high feature line. The headlamps were again torpedo style. The cowl was curved. The front seatbacks stuck up high over the beltline.
JEFFERY — MODEL 661 — SIX: The 661 was an all-new six-cylinder model, apparently introduced after the start of the model year. A dry disc clutch replaced the cone type. The rear axle gear ratio was changed from 4.15:1 to 4.50:1.
I.D. NUMBERS: Serial number located to left of front frame cross-member. [462] 57,000 to 60,500. [661] 68,000 to 69,108

Model No.	Body Type & Seating	Price	Weight	Prod. Total
JEFFERY MODEL 462 [FOUR]				
462 Rds	2dr Roadster-2P	1000	2800	Note 1
462-5	4dr Tr-5P	1000	2800	Note 1
462-7	2dr Tr-7P	1035	2825	Note 1
462 Sed	4dr Sedan-5P	1165	2850	Note 1
462 S-7	4dr Sedan-7P	1200	2875	Note 1
462	4dr Sedan-5P	2450	3000	Note 1
JEFFERY MODEL 661 [SIX]				
661	4dr Tr-5P	1435	3250	Note 2
661	2dr Roadster-2P	1465	3250	Note 2

NOTE 1: Production: Approximately 3,500.
NOTE 2: Production: Approximately 1,108.
ENGINES
[462] L-head. Four-cylinder. Cast en bloc. Cast iron. B x S.: 3-3/4 x 5-1/4 in. Disp.: 231.9 cid. NACC HP.: 22.5. Max. HP.: 38. Carb.: Stromberg (gravity fed). Helical gear camshaft drive. Pump cooling. Bosch ignition. Bijur electric starting/lighting.
[661] L-head. Six-cylinder. Cast in pairs. Cast iron block. B x S.: 3-1/2 x 5-1/4 in. Disp.: 303 cid. NACC HP.: 29.4. Max. HP.: 40. Carb.: Rayfield (vacuum fed). Pump cooling. Chain camshaft drive.
TECHNICAL
[462] Bijur electric starting. Cone clutch. Three-speed selective sliding transmission. Semi-floating rear axle with 4.15:1 ratio on high. Semi-elliptic front springs. Three-quarter elliptic rear springs. Force feed and splash lubrication. Pump cooling. Bosch single type ignition.
[661] Bijur electric starting. Dry disc clutch. Three-speed selective sliding transmission. Semi-floating rear axle with 4.50:1 ratio on high. Hotchkiss drive. Semi-elliptic rear springs. Force feed and splash engine lubrication. Pump cooling.
CHASSIS
[462] Wheelbase: 116 in. Tires: 34 x 4 (front and rear). Rear springs: 3/4- elliptic.
[661] Wheelbase: 125 in. Tires: 34 x 4 (front and rear). Rear springs: Semi-elliptic.
HISTORY: Jeffery continued to concentrate on the manufacture of four-wheel drive Quad trucks for military use. A primary customer for these units was the British government, as rumblings of war began to be heard in Europe.

1917 JEFFERY

JEFFERY — MODEL 472 — FOUR: The Model 462 became the 472 for 1917. It now had a vacuum fuel system. A short, high hood with a slight slant to the front remained a design characteristic, but the beltline no longer met the windshield at a right angle. It now curved smoothly up to meet the base of the windshield stanchion. The front seatbacks no longer jutted high above the body; their tops could just be seen. Vacuum fed fuel delivery was a new four-cylinder feature.
JEFFERY — MODEL 671 — SIX: The 671 had a high hood that ran straight back to the cowl without a curve. The beltline met the windshield at about a 60-degree angle. It had torpedo headlights. The hood had vertical louvers. The engine had the same bore, but a smaller stroke, than the 661.

1917 Jeffery, Model 671 four-door seven-passenger touring (JAC)

I.D. NUMBERS: Serial number located to left of front frame cross-member. [472] 61000 to 62027; [472-2] 76000 to 80000; [671 Touring] 71800 to 86000; [671 Roadster] 71796 to 71800; and [671 Sedan] 86000 to 92999.

Model No.	Body Type & Seating	Price	Weight	Prod. Total
JEFFERY MODEL 472 [1ST Series]				
472 Rdr	2dr Roadster-2P	1065	2800	Note 1
472	4dr Tr-7P	1095	2800	Note 1
JEFFERY MODEL 472 [2nd Series]				
472-2	2dr Tr-7P	1095	2800	Note 2
472-2 Sedan	4dr Sedan-7P	1260	2875	Note 2
JEFFERY MODEL 671 [SIX]				
671 Rdr	2dr Roadster-2P	1335	3050	Note 3
671	4dr Tr-7P	1365	3080	Note 3
671 Sedan	4dr Sed-7P	1530	3080	Note 3

NOTE 1: Production: Approximately 1,027.
NOTE 2: Production: Approximately 4,000.
NOTE 3: Production: Approximately 14,200 combined.

ENGINES

[472/472-2] L-head. Four-cylinder. Cast en bloc. Cast iron. B x S.: 3-3/4 x 5-1/4 in. Disp.: 231.9 cid. NACC HP.: 22.5. Max. HP.: 38. Carb.: Stromberg V (vacuum fed). Helical gear camshaft drive.

[671] L-head. Six-cylinder. Cast in pairs. Cast iron block. B x S.: 3-1/2 x 4-5/8 in. Disp.: 267 cid. NACC HP.: 29.4. Max. HP.: 53. Carb.: Rayfield (vacuum fed). Pump cooling. Chain camshaft drive.

TECHNICAL

[472/472-2] Dry disc clutch. Three-speed selective sliding transmission. Bijur electrics. Semi-floating rear axle with 4.50:1 ratio on high. Semi-elliptic front springs. Three-quarter elliptic rear springs. Hotchkiss drive. Force feed and splash lubrication. Pump cooling.

[671] Bijur electric starting. Dry disc clutch. Three-speed selective sliding gear transmission. Semi-floating rear axle with 4.50:1 ratio on high. Hotchkiss drive. Semi-elliptic rear springs. Force feed and splash engine lubrication. Pump cooling.

CHASSIS

[462] Wheelbase: 116 in. Tires: 34 x 4 (front and rear). Rear springs: 3/4- elliptic.

[661] Wheelbase: 125 in. Tires: 34 x 4 (front and rear). Rear springs: Semi-elliptic.

HISTORY: *MoToR* magazine published a list of 149 pleasure car makers in 1918. Nash Motors Co., of Kenosha, Wis., was now listed as manufacturer of the Jeffery. The company was one of four Wisconsin automakers listed that year. There were others, but the list was of American Licensed Automobile Manufacturers (ALAM) member companies.

1918 JEFFERY

JEFFERY — MODEL 671 — SIX: The 671 was carried over into early 1918, with no indications that any changes were made. The 1902-1963 specifications book published by AMC in 1962 listed serial numbers for 1918 Jefferys. However, the annual show number of *MoToR* did not include Jeffery prices or specifications.

I.D. NUMBERS: Serial number located to left of front frame cross-member. [671] 86000 to 92999.

Model No.	Body Type & Seating	Price	Weight	Prod. Total
JEFFERY MODEL 671 [SIX]				
671 Rdr	2dr Roadster-2P	1435	3050	Note 1
671	4dr Tr-7P	1465	3080	Note 1
671 Sedan	4dr Sedan-7P	1630	3080	Note 1

NOTE 1: Production: Approximately 6,999.

ENGINES

[671] L-head. Six-cylinder. Cast in pairs. Cast iron bloc. B x S.: 3-1/2 x 4-5/8 in. Disp.: 267 cid. NACC HP.: 29.4. Max. HP.: 53. Carb.: Rayfield (vacuum fed). Pump cooling. Chain camshaft drive.

TECHNICAL

[671] Bijur electric starting. Dry disc clutch. Three-speed selective sliding gear transmission. Semi-floating rear axle with 4.50:1 ratio on high. Hotchkiss drive. Semi-elliptic rear springs. Force feed and splash engine lubrication. Pump cooling.

CHASSIS

[661] Wheelbase: 125 in. Tires: 34 x 4 (front and rear). Rear springs: Semi-elliptic.

HISTORY: The Jeffery 671 became the Nash 671 in 1918.

LAFAYETTE
1920-1924

LaFAYETTE — Indianapolis, Ind./Milwaukee, Wis. — (1920-1924) — At first, no one mentioned who the president of the new company was. There were a good many Cadillac people involved, including chief engineer D. McCall White and his assistant J.W. Applin, sales manager Earl C. Howard and advertising manager Leo Burnett. Jim Storrow and Lee Higginson, stalwart financial men of General Motors, were putting up much of the money. But who was behind it all?

The Lafayette Motors Company of Mars Hill (Indianapolis), Ind., was founded in October of 1919. In January of 1920, when the new Lafayette was introduced, the man whose idea it was came out of the closet. It was Charles W. Nash, the automotive King of Kenosha, Wis., who had purloined the Cadillac people in order to produce a luxury car of his own.

Nash's Lafayette venture was entirely separate from Nash Motors. D. McCall White told the press, with a blatant slap at his former employer, that the Lafayette was "far ahead of any previous automobile built under (my) supervision."

The car was powered by a 90 hp V-8, and was offered in a variety of open and closed body styles on a 132 inch wheelbase, with price tags beginning in the $5,000 range. Production began in August of 1920, with the Lafayette introduced as a 1921 model. It met the postwar depression head-on. "There have been far better times to introduce a motorcar," Charlie Nash rued in the fall of 1921, as he announced the sale of just 700 Lafayettes had been realized.

In April of 1922, there were rumors that Lafayette and Pierce-Arrow would merge. Nash was to become chairman of the board of the combined companies, but neither Lafayette nor Pierce-Arrow could come to terms. In July, Lafayette announced its forthcoming move from the Mars Hill factory (which had been used by the government to produce hand grenades during World War I) to Milwaukee, Wis. The declared reasons for the move were a "desire for a closer geographical coordination of Nash activities and the need for greater factory space." Only the former reason was really valid.

By now, the venture had been reorganized into Lafayette Motors Corp, with Nash Motors as the largest stockholder. Only about 1,000 cars had thus far been sold. By now, too, D. McCall White had left. So had Leo Burnett, who remained in Indianapolis to work for an advertising agency and to begin thinking about starting his own such firm.

Former Packard engineer Earl G. Gunn became Lafayette's chief engineer by January 1923, when the company was installed in its new Milwaukee factory. But the end was near. In August, a letter to Lafayette stockholders, bearing the signatures of Jim Storrow and Charlie Nash, announced that "the company has not been able to overcome the difficulty of unprofitable operation" and asked that they agree to the sale of Lafayette to Ajax Motors Co. for $225,000. The stockholders did.

Ajax was a new organization owned entirely by Nash Motors. Ajax would market a new car at the precise opposite end of the automotive scale from the Lafayette, but no one knew that in 1923. Charlie Nash was simply playing cat-and-mouse again. After pouring $2 million of Nash Motors money into the Lafayette, he gave up on it.

Never again would Nash produce an all-out luxury car, although there would be a later model of the Nash called the Lafayette. The last of 2,267 Lafayettes were built in early 1924. Soon thereafter, the machinery from the Lafayette plant was shipped to the old Mitchell plant in Racine, Wis. Charlie Nash had just bought the building to house production of the new Ajax. Nash resurrect the Lafayette name, in 1934, but for a lower-priced Nash derivative.

1920 LAFAYETTE

LA FAYETTE — MODEL 134 — V-8: One very early advertisement showed a large Torpedo Touring with beveled front fenders, torpedo headlamps and high, thin, vertical hood louvers. The radiator grille featured vertical shutters. On the radiator badge was a cameo bearing the image of the French General Lafayette. The ad spoke of competent engineering and announced that the Lafayette was "available for ownership." It said that automobiles were in the hands of company distributors and going forth to private ownership. Unfortunately, not many did go forth. Standard equipment included drum headlamps (on most models other than the Torpedo); Waltham speedometer; Kellogg power tire pump; Waltham clock; leather upholstery; Firestone tires; and Gabriel shock absorbers. Drednaut top material was used.

I.D. NUMBERS: Branham indicated that serial numbers were stamped on the patent plate located on the left side of the dash. Other sources list the front floorboard as the location. Numbers for 1920 models were approximately 1001 to 1250; approximate, since Lafayette did not designate cars by year model. The motor number was stamped on a patent plate on the left front side of the dash.

Model No.	Body Type & Seating	Price	Weight	Prod. Total
134	4dr Tr-7P	5000	4340	NA
134	4dr Torpedo Tr-4P	5000	4230	NA
134	4dr Cpe-4P	6300	4538	NA
134	4dr Vest Sedan-7P	6500	4535	NA
134	4dr Limousine-7P	6500	4700	NA

NOTE 1: Prices according to MoTor Age (1920); MoTor (1921-1924); weights according to Wisconsin Motor Vehicle Weights guidebook (1924).
NOTE 2: Production was approximately 250 cars.
NOTE 3: Passenger capacities based on Official Specifications 1902-1963 Rambler and its predecessors, published by AMC in 1963.
ENGINE: L-head. 108-degree vee-block. Cylinders cast in banks of four. Cast aluminum block. Five main bearings. B x S.: 3-1/4 x 5-1/4. Displ.: 348.4 cid. NACC HP.: 33.80. Brake HP.: 90 at 2700 rpm. Carburetor: (Own) two-inch. Pressure and splash lubrication. Pressure fuel feed. Delco ignition.
TECHNICAL: Transmission of unit type. 3F/1R. Disc type clutch. Torque tube drive. Floating rear axle. Gear ratios: (First) 3.04:1; (Second) 1.71:1; (Final) 4.50:1.
CHASSIS: Parish & Bingham frame with 6-1/2 inch side rails of 3/16-inch steel and five cross-members. W.B.: 132 in. Wheels: 36 in. wood artillery spoke. Front springs: 42 inches long semi-elliptic. Rear springs: 60 inches long semi-elliptic. Rear wheel brakes. Worm-and-sector steering
OPTIONS: Front bumper. Rear bumper. Heater. Traffic light.
HISTORY: About 250 cars were built in 1920. Other key employees at Lafayette included chief draftsman L.H. Menges and works engineer J.P. Robertson.

1921 LAFAYETTE

LA FAYETTE — MODEL 134 — V-8: The Lafayette was carried over from 1920 with no changes to speak of. However, drum headlamps may have been used on all models this year. (Actually, the only Lafayette picture we have seen with torpedo headlamps was a drawing of the car in the early ad). Also, this is the first year that most sources show a roadster. For 1921, closed Lafayettes remained tall cars that stood high off the ground. All models had generally uninspired performance and styling, compared to other cars in their price class, such as Cadillacs and Pierce-Arrows.

1921 Lafayette, Model 134 roadster (JAC)

I.D. NUMBERS: Branham indicated that serial numbers were stamped on the patent plate located on the left side of the dash. Other sources list the front floorboard as the location. Numbers for 1921 models were approximately 1251 to 1905; approximate, since Lafayette did not designate cars by year model. The motor number was stamped on a patent plate on the left front side of the dash.

Model No.	Body Type & Seating	Price	Weight	Prod. Total
134	2dr Roadster-3P	5625	4170	NA
134	4dr Tr-7P	5625	4340	NA
134	4dr Torpedo Tr-4P	5625	4230	NA
134	4dr Cpe-4P	7200	4638	NA
134	4dr Vest Sedan-7P	7400	4535	NA
134	4dr Limousine-5P	7500	4700	NA

NOTE 1: Prices according to MoToR (1921-1924); weights according to "Wisconsin Motor Vehicle Weights" guidebook (1924).
NOTE 2: Production was approximately 250 cars.
NOTE 3: Passenger capacities based on Official Specifications 1902-1963 Rambler and its predecessors, published by AMC in 1963. Other sources indicate different passenger capacities (i.e. Two-passenger roadster).
ENGINE: L-head. 108-degree vee-block. Cylinders cast in banks of four. Cast aluminum block. Five main bearings. B x S.: 3-1/4 x 5-1/4. Displ.: 348.4 cid. NACC HP.: 33.80. Brake HP.: 90 at 2700 rpm. Carburetor: (Own) two-inch. Pressure and splash lubrication. Pressure fuel feed. Delco ignition.
TECHNICAL: Transmission of unit type. 3F/1R. Disc type clutch. Torque tube drive. Floating rear axle. Gear ratios: (First) 3.04:1; (Second) 1.71:1; (Final) 4.50:1.
CHASSIS: Parish & Bingham frame with 6-1/2 inch side rails of 3/16-inch steel and five crossmembers. W.B.: 132 in. Wheels: 36 in. wood artillery spoke. Front springs: 42 inches long semi-elliptic. Rear springs: 60 inches long semi-elliptic. Rear wheel brakes. Worm-and-sector steering.
OPTIONS: Front bumper. Rear bumper. Heater. Traffic light.
HISTORY: Calendar year production: 655 cars. The Lafayette engine was tested to 4,000 rpm out of the car. It was good for 3,000 rpm installed in the chassis.

1922 LAFAYETTE

1922 Lafayette, Model 134 Vestibule Sedan (JAC)

LA FAYETTE — MODEL 134 — V-8: The Lafayette was carried over from 1921 with no appearance changes to speak of. A carburetor of a new brand was used and the horsepower rating increased to 100. Standard equipment included Delco lighting; the company's own headlamps with plain glass front lenses; ammeter; Klaxon horn; Exide battery (six-volt); Alemite lubrication system; Waltham speedometer; Kellogg tire pump; Waltham clock; Drednaut top material; Gabriel shock absorbers; Goodyear tires; Firestone rims; and Boyce motometer.

I.D. NUMBERS: Branham indicated that serial numbers were stamped on the patent plate located on the left side of the dash. Other sources list the front floorboard as the location. Numbers for 1922 models were approximately 1906 to 2468; approximate, since Lafayette did not designate cars by year model. The motor number was stamped on a patent plate on the left front side of the dash.

Model No.	Body Type & Seating	Price	Weight	Prod. Total
134	2dr Roadster-3P	4850	3950	NA
134	4dr Tr-7P	4850	4150	NA
134	4dr Torpedo Tr-4P	4850	4050	NA
134	4dr Cpe-4P	6250	4350	NA
134	4dr Vest Sedan-7P	6500	4620	NA
134	4dr Limousine-5P	6750	4700	NA
134	4dr Vest Limousine-7P	NA	4800	NA

NOTE 1: Prices according to; MoToR (1921-1924); weights according to Wisconsin Motor Vehicle Weights guidebook (1924).
NOTE 2: Production was approximately 563 cars.
NOTE 3: Passenger capacities based on Official Specifications 1902-1963 Rambler and its predecessors, published by AMC in 1963. Other sources indicate different passenger capacities (i.e. Two-passenger roadster).
ENGINE: L-head. 108-degree vee-block. Cylinders cast in banks of four. Cast aluminum block. Five main bearings. B x S.: 3-1/4 x 5-1/4. Displ.: 348.4 cid. C.R.: 4.80:1. NACC HP.: 33.80. Brake HP.: 100 at 2750 rpm. Carburetor: (Johnson) two-inch. Pressure and splash lubrication. Pressure fuel feed. Delco ignition.
TECHNICAL: Transmission of unit type. 3F/1R. 7-3/4 inch multi-disc type clutch. Torque tube drive. Floating rear axle. Gear ratios: (First) 3.04:1; (Second) 1.71:1; (Final) 4.50:1.
CHASSIS: Parish & Bingham frame with 6-1/2 inch side rails of 3/16-inch steel and five cross-members. W.B.: 132 in. Wheels: 36 in. wood artillery spoke. Front springs: 42 inches long semi-elliptic. Rear springs: 60 inches long semi-elliptic. Rear wheel brakes. Worm-and-sector steering.
OPTIONS: Front bumper. Rear bumper. Heater. Traffic light.
HISTORY: Calendar year production: 563 cars. Some sources credit the 1921 Lafayette

with being the first American car rated at 100 hp. This is not true, however. The following 1920 cars had over the magic number: Singer Model 20 (102 hp at 1950 rpm); Revere Model A (105 hp at 2600 rpm); Porter (125 hp at 2600 rpm); and McFarlan Model 127 (112 hp at 2400 rpm).

1923 LAFAYETTE

1923 Lafayette, Model 134 four-passenger coupe (JAC)

LA FAYETTE — MODEL 134 — V-8: The early 1923 Lafayette was carried over from 1922 with no appearance changes to speak of. Cars built later in the year had pull-type hood latches and cowl belts. The Vestibule Sedan was now called the sedan and a new Imperial Vestibule Limousine was added to the line. Standard equipment included: Six-volt battery; Waltham speedometer; Kellogg tire pump; Waltham clock; leather upholstery; Drednaut top material; Gabriel shock absorbers; various brands of tires on Firestone rims; Boyce motometer; Gas gage; Johnson rearview mirror; Trico windshield wipers; cigar lighter; and trunk (on Sport models).

I.D. NUMBERS: Branham indicated that serial numbers were stamped on the patent plate located on the left side of the dash. Other sources list the front floorboard as the location. Numbers for 1923 models were approximately 2469 to 2624; approximate, since Lafayette did not designate cars by year model. The motor number was stamped on a patent plate on the left front side of the dash.

Model No.	Body Type & Seating	Price	Weight	Prod. Total
134	2dr Roadster-3P	3985	3995	NA
134	4dr Tr-7P	4400	4160	NA
134	4dr Torpedo Tr-4P	4300	4060	NA
134	4dr Cpe-4P	5500	4360	NA
134	4dr Sedan-7P	5500	4445	NA
134	4dr Limousine-5P	5750	4525	NA
134	4dr Imp Vest Limousine-7P	6250	4632	NA

NOTE 1: Prices according to *MoToR* (1921-1924); weights according to "Wisconsin Motor Vehicle Weights" guidebook (1924).
NOTE 2: Production was approximately 156 cars.
NOTE 3: Passenger capacities based on Official Specifications 1902-1963 Rambler and its predecessors, published by AMC in 1963. Other sources indicate different passenger capacities (i.e. Two-passenger roadster).
ENGINE: L-head. 108-degree vee-block. Cylinders cast in banks of four. Cast aluminum block. Five main bearings. B x S.: 3-1/4 x 5-1/4. Displ.: 348.4 cid. C.R.: 4.80:1. NACC HP.: 33.80. Brake HP.: 100 at 2750 rpm. Carburetor: (Johnson) two-inch. Drilled crankshaft. Pressure and splash lubrication. Pressure fuel feed. Delco ignition.
TECHNICAL: Transmission of unit type. 3F/1R. 7-3/4 inch multi-disc type clutch. Torque tube drive. Floating rear axle. Gear ratios: (First) 3.04:1; (Second) 1.71:1; (Final) 4.58:1.
CHASSIS: Parish & Bingham frame with 6-1/2 inch side rails of 3/16-inch steel and five cross-members. W.B.: 132 in. Wheels: 36 in. wood artillery spoke. Front springs: 42 inches long semi-elliptic. Rear springs: 60 inches long semi-elliptic. Rear wheel brakes. Worm-and-sector steering.
OPTIONS: Front bumper. Rear bumper. Heater. Traffic light. Jack.

HISTORY: Calendar year production: 156 cars. This was the worst year for Lafayette sales.

1924 LAFAYETTE

LA FAYETTE — MODEL 134 — V-8: Production was moved to Milwaukee and a minor change was the use of Hall headlamps with Bausch & Lomb lens. Final gear ratio was lowered to 4.58:1. The Vestibule Sedan name returned on the five-passenger closed car, while the term vestibule was dropped from the Imperial Limousine's description. Most likely, these were name changes only. Standard equipment included: Six-volt battery; Waltham speedometer; Kellogg tire pump; Waltham clock; leather upholstery; Drednaut top material; Gabriel shock absorbers; various brands of tires on Firestone rims; Boyce motometer; Gas gage; Johnson rearview mirror; Trico windshield wipers; cigar lighter; and trunk (on Sport models).

1924 Lafayette, Model 134 four-door touring (HAC)

I.D. NUMBERS: Branham indicated that serial numbers were stamped on the patent plate located on the left side of the dash. Other sources list the front floorboard as the location. Numbers for 1924 models were approximately 2625 to 3267; approximate, since Lafayette did not designate cars by year model. The motor number was stamped on a patent plate on the left front side of the dash.

Model No.	Body Type & Seating	Price	Weight	Prod. Total
134	2dr Roadster-3P	5000	3995	NA
134	4dr Tr-7P	5000	4268	NA
134	4dr Torpedo Tr-4P	5000	4232	NA
134	4dr Cpe-4P	6300	4438	NA
134	4dr Vest Sedan-7P	6500	4450	NA
134	4dr Limousine-5P	6500	4525	NA
134	4dr Imp Limousine-7P	6750	4632	NA

NOTE 1: Prices according to *MoToR* (1921-1924); weights according to "Wisconsin Motor Vehicle Weights" guidebook (1924).
NOTE 2: Production was approximately 643 cars.
NOTE 3: Passenger capacities based on Official Specifications 1902-1963 Rambler and its predecessors, published by AMC in 1963. Other sources indicate different passenger capacities (i.e. Two-passenger roadster).
ENGINE: L-head. 108-degree vee-block. Cylinders cast in banks of four. Cast aluminum block. Five main bearings. B x S.: 3-1/4 x 5-1/4. Displ.: 348.4 cid. C.R.: 4.80:1. NACC HP.: 33.80. Brake HP.: 100 at 2750 rpm. Carburetor: (Johnson) two-inch. Drilled crankshaft. Pressure and splash lubrication. Pressure fuel feed. Delco ignition.
TECHNICAL: Transmission of unit type. 3F/1R. 7-3/4 inch multi-disc type clutch. Torque tube drive. Floating rear axle. Gear ratios: (First) 3.04:1; (Second) 1.71:1; (Final) 4.58:1.
CHASSIS: Parish & Bingham frame with 6-1/2 inch side rails of 3/16-inch steel and five cross-members. W.B.: 132 in. Wheels: 36 in. wood artillery spoke. Front springs: 42 inches long semi-elliptic. Rear springs: 60 inches long semi-elliptic. Rear wheel brakes. Worm-and-sector steering.
OPTIONS: Front bumper. Rear bumper. Heater. Traffic light. Jack.
HISTORY: Calendar year production: 643 cars. Average gas mileage of the Lafayette was 10 mpg, according to *MoToR*. In 1924, Charlie Nash closed the Indianapolis plant. Production was moved to Milwaukee, but 1924 was the last year. Total output for 1920 to 1924 was 2,267 Lafayettes. J.W. Applin became a Durant dealer in Franklin, Ind. D. McCall White moved to Landis Engineering & Mfg. Co., Waynesboro, PA.

NASH
1917-1957

Born in 1864 in Illinois and abandoned by his parents at age six, Charles W. Nash was "bound out" by a district court to work for a Michigan farmer. He was to receive room, board and three months of schooling a year, until age 21. Then, $100, a new suit of clothes and freedom would be his. But, Charlie Nash ran away, at age 12, to get a paying job on another Michigan farm.

Nash learned the carpenter's trade and clerked in a grocery store in Flint, Mich. In the early 1890s, he was the fastest cushion stuffer at the Flint Road Cart Co., owned by William C. Durant and J. Dallas Dort. By 1895, he was managing the Durant-Dort Carriage Co. By 1910, he was heading the Buick Motor Car Co. By 1912, he was the president of General Motors.

Rags to riches was a popular theme in novels of this period. Charlie Nash had managed to out do Alger Horatio. His career at General Motors ended in June 1916, like many did, when he resigned following a policy dispute with Billy Durant.

Nash's next step was a logical one. He traveled to Kenosha, Wis. and, with former GM man James Storrow, bought the Thomas B. Jeffery Co. Formerly the producer of the Rambler, and then the Jeffery, it was one of the oldest, best-known and largest automobile companies in the industry. The purchase price, purportedly, was $9 million.

On July 29, 1916, Nash Motors Co. was born. The Jeffery was continued in production for a while, with Nash nameplates appearing on the cars from the summer of 1917. Indeed, the first Nash remained a badge-engineered Jeffery.

The first real Nash arrived on April 18, 1918. It was a six designed by Erik Wahlberg, whom Nash had hired away from Oakland as chief engineer. That the engine was of overhead valve design was no surprise, given Nash's Buick experience. That it and the Nash chassis (which featured Hotchkiss drive and semi-elliptic suspension all around) were formidably clean and tidy was commented upon with considerable favor in the trade press.

Charlie Nash was a stickler for conservative neatness in car design and company management. Sales of 10,000-plus cars, in 1918, more than doubled the following year to 27,000. In its first 15 months of operation, Nash Motors netted over $2 million.

Prior to the World War I Armistice, the four-wheel drive truck, which was begun as a Jeffery product, was continued as the Nash Quad. It was sold alongside a standard Nash truck. But, Charlie Nash began phasing out commercial vehicle manufacture by the 1920s, as he launched a several-pronged attack in the production car field.

Introduced in 1920 was a brand-new car, in the $5,000 price range, called the LaFayette. It was initially built in Indianapolis, Ind. The same November, in Kenosha, the Nash assembly line began humming with a new 35 hp four-cylinder model. In effect, it was the 67 hp Nash Six minus two cylinders and nine inches in wheelbase.

Nash could sell this car for several hundred dollars less than his standard, medium-priced products, and sales climbed. For 1922, the four was provided with rubber engine mounting and a "Carriole" sedan model at $1,350. This was five dollars more expensive than the Essex coach, which had been introduced a few weeks earlier, allowing the Hudson Motor Car Co. to legitimately claim honors in pioneering the closed car in the popular-price field. Still, a net profit of $7.6 million, for 1922, had to make Charlie Nash feel good.

Sales for 1923 passed the 50,000 mark for the first time. That volume of business was good for a net profit of $9.3 million. The LaFayette venture, which had moved into a new factory in Milwaukee, Wis., by January 1923, had proved a commercial disaster, however. After pouring $2 million into it, Charlie Nash abandoned LaFayette early in 1924, moving the machinery from Milwaukee to the old Mitchell plant in Racine.

Nash had just managed to outbid Hupmobile for this factory. There, from 1925 into 1926, he would produce another new car at the opposite end of the scale from the LaFayette. This was the $865-$995 Ajax.

Charles Nash wasn't particularly thrilled with the less than 25,000 Ajaxes sold during its first year. However, since more than 85,000 Nashes were delivered during the same period, the new car's identity crisis problem was easily solved. The L-head Ajax six became the Nash Light Six during 1926.

Joining the Special Six and Advanced Six, with overhead valve engines, the Light Six made Nash products six-cylinders across the board. In June of 1928 Twin-Ignition arrived in Kenosha on the big Nash. It increased horsepower measurably and made the Advanced Six a genuine 80 mph car, while the Special Six was good for 75 mph. The Standard Six (formerly Light Six), with a conventional single-plug-per-cylinder ignition system, shared its bigger brothers' invar-strutted aluminum pistons and 5.0:1 compression ratio (up from 4.5:1), creating a car now capable of an easy 70 mph.

These new Nashes were extremely fine cars, with handsome Seaman bodies and price tags of $885-$2,190. This made for some of the greatest bargains in the industry that year. More than 138,000 Nashes were produced in 1928, and Nash Motors made over $20 million.

Charlie Nash was 64 years old when the stock market crashed. Already, he had introduced, for the 1930 model year, one of the most splendid Nashes of all...the Twin-Ignition Eight. It was an overhead valve inline eight with nine main bearings, 298.6 cid and 100 hp at 2900 rpm. It could do "80 mph in three blocks," the billboards would say, and the price range was just $1,675 to $2,385.

Because Charlie Nash had not run wild during the 1920s, he was better prepared for the 1930s. His assets-to-liabilities ratio was by far the most exemplary in the automobile industry. In 1931, when most companies lost money, Nash Motors turned a profit of $4.8 million. For 1932, Nash and crew came up with even nicer Nashes boasting Ride-Control, freewheeling, five-point rubber-insulated engine suspension and synchromesh gearshifting.

Nineteen thirty-two was an awful year in America, with industry-wide auto production a mere 13 percent of the pre-depression figure, but in Kenosha, Charlie Nash's company made a million dollars. That year, in order to allow himself more time for long-range planning, Nash elevated associate Earl McCarty to company president. Nash retained the chairman of the board role, which he had assumed after James Storrow's death in 1926.

In 1933, Nash marked time with cars not noticeably different from 1932. For the first time in history, Nash Motors lost money. The situation was repeated in 1934, though for a different reason...Charlie Nash had spent a fortune re-tooling for the new 1934 cars.

151

The LaFayette name was revived for a car which now carried Nash into the low-priced ($595-$695) field. It used the L-head engine of the former Big Six. Although promoted under its own name until 1937, it was really a Nash model. It was just as much a Nash as the big Ambassador. It was big, with wheelbases up to 142 inches, but priced at the opposite end of the scale.

In July 1936, Charlie Nash purchased Seaman Body Corp., of Milwaukee, Wis. He had bought a half interest in Seaman in 1919). That August, Nash Motors was 21 years old and Charlie Nash — now 72 — was growing tired. When Earl McCarty retired, Nash invited George W. Mason, the vice-president of Kelvinator Corp., to become Nash president. Mason agreed, as long as Nash bought Kelvinator. The deal was done.

This provided comedians of the day with fodder for bad jokes: ice cube trays would now become standard equipment for Nash cars, Kelvinator refrigerators would get four-wheel brakes. But, when the laughter died down, Nash had sold nearly 86,000 cars for 1937 (the best year of the decade) and raked in a $3.5 million profit.

The recession year of 1938 was bad for both cars and refrigerators, and the Nash-Kelvinator marriage had a $7.5 million loss. Under Mason's direction, Nash continued to build fine cars, however. They had such interesting features as the Weather-Eye controlled ventilation system and overdrive (first offered in 1936).

The year 1940 saw a 21 car production run of a nifty Ambassador Eight Cabriolet styled by Alexis de Sakhnoffsky (who earlier had kibitzed with chief engineer Wahlberg on the 1934 Nash line) and the availability, in England, of a Perkins diesel-engined Nash.

Since 1936, Nash had offered cars with a rear seat that converted into a bed. But, it was 1941 which brought the biggest news from Nash...the Ambassador 600. It replaced the LaFayette and boasted unitized body construction, torque tube drive, 75 bhp, 25-30 mpg and a $750-$850 price tag. The public response was terrific. In 1941, Nash sales topped 80,000 cars.

Following Pearl Harbor, Nash cars ceased to be built in Kenosha. Pratt & Whitney aviation engines were manufactured, instead, for the duration of World War II. With the coming of peace, Nash was in a solid position as an independent automaker and the postwar years would mark the beginning of a new era for the company.

The 1946 Nash 600 carried its prewar sheet metal on a 112 inch wheelbase chassis powered by an L-head six. It was marketed together with the fancier Ambassador, which used an overhead valve engine in a 121 inch wheelbase car. The Ambassador chassis was quite different from that of the 600. The smaller car had unitized construction, while the Ambassador retained the body-on-frame type. Also, the 600 had coil springs all around, while the Ambassador used semi-elliptic springs at the rear.

Strong sales were registered throughout the period, due primarily to America's dire need for cars in a time when manufacturing was artificially restricted. The ability to have some product on hand was helpful and probably played a role in having a unique honor bestowed upon Nash for the first and last time in 1947. The company was asked to supply the Official Pace Car for the Indianapolis 500-Mile Race. It was an Ambassador Sedan.

Nash, unlike other independent automakers, avoided a major restyling in the early post World War II years. The 600 became its bread-and-butter model. The Ambassador had richer appointments inside and out. These handsome looking cars were similar in style to contemporary Cadillacs, but Nash was headed in a new postwar direction and would continue building this type of vehicle only through 1948. On June 6, 1948, Charles Nash died at age 84. Two months earlier, on April 1, George Romney had arrived with a new idea of what a Nash should be.

Dramatically new for 1949, the Nash Airflyte series offered unit-construction in a fastback body that looked like a baby Packard. These cars carried over features popular with buyers, such as an overhead valve six in the Ambassador series, full coil spring suspension, large 15 inch tires and the efficiency of an overdrive transmission. Styling features included fully enclosed front and rear wheelhousings, one-piece windshields, fold-down travel bed seats and the 'Uniscope' dash instrument cluster that put all the gauges in one place.

In his plans for the changing marketplace of the '50s, George Mason saw that the early postwar seller's market would not last and he tried to merge all the independent automakers together. He was unable to do so, but did accomplish the introduction of the first compact car in 1950. Mason's small car interest was shared by George Romney, who joined Nash-Kelvinator in 1948 as Mason's administrative assistant.

Nash re-introduced the Rambler nameplate in 1950, affixing it to the midyear compact model built off a new platform. It first came only as an equipment- loaded convertible-landau. This was basically like a two-door hardtop with a soft-top. When the roof was lowered on rails, the windows and frames remained in a fixed-upright position. The Rambler's engine was the original Nash 600 motor. It could go even further on a gallon of gas in the small Rambler.

A second model was soon added. This two-door station wagon was announced at the same price as the convertible. Mason's idea to make the compact marketable was to introduce fancier models first to gain respectability for the compact. This followed the old Nash advertising motto 'Built up to a standard, not down to a price' and it worked.

The year 1951 brought a two-door hardtop to the Rambler line. It was called the Country Club coupe. This was the first modern compact hardtop. It had loads of trim and a fashionable continental spare tire. The larger models were basically carried over in 1951, with a few detail changes.

Nash's Golden Anniversary was celebrated in 1952. There was new styling for the Ambassador and Statesman. A new notch back body created by Italy's Pininfarina was the fashion hit of the marque's 50th year. The fastback look was gone and eye appeal was greatly enhanced.

Shortly before George Mason's death in 1954, his dream of joining all the independent auto companies into one came partly true. Nash and Hudson merged, in May, 1954, to form American Motors. Then, it fell to George Romney to lead the fledgling enterprise in its formative and perhaps most significant years. Also in 1954, a two-door sedan version of the Rambler American was finally introduced, along with a four-door sedan.

It took George Romney several years to fully integrate Nash and Hudson operations and fully develop AMC. During the interim, some of the autocracy of the two firms was retained. Both Nashes and Hudsons continued to appear in separate showrooms. Each kept some styling distinctions, although the basic Nash body was used for both. Hudson dealers were also supplied with a "badge-engineered" Rambler. It had Hudson nameplates, hubcaps and grille medallions. Both companies also sold badge-engineered versions of the first American-designed sub-compact. However, this pint-sized Metropolitan was built by Austin, in England. Like the hybrid Nash-Healey, also British-made, it falls beyond the scope of this catalog of "American" automobiles.

The Rambler and Ambassador were both restyled for 1955. Full front wheel openings were the major change for the smaller car, along with new exterior trim design. The Ambassador and Statesman received a new slab-cornered front end treatment. It had larger, but not full, front wheel cutouts along with a wraparound windshield and larger oval grille that incorporated the headlights. The Packard Clipper V-8 was made available in the full-sized Nash Ambassador, along with Packard's Ultramatic transmission. Nash sixes continued to use GM's Hydra-Matic.

The big news for 1956 was a completely new type of Ramblers that debuted in the spring. They were based on the 1954 four-door models' 108-inch wheelbase. Cars in this series featured many Rambler firsts, including a new overhead valve six, two-tone exterior sweep panel treatments and optional Hydra-Matic. The fancier and larger economy class cars accounted for 82,000 of the 104,000 units sold during the year.

The grand finale came for both Nash and Hudson in 1957. By this stage of the merger program, Hudson production in Detroit, Mich. had ceased completely and almost all AMC assemblies were done at the Nash facilities in Kenosha, Wis. The Ambassador was face-lifted for the last time, with new emphasis placed on sportiness, luxury and V-8 power. In fact, AMC had its own V-8 since the middle of 1956.

No amount of effort was quite enough to turn the tide. In fact, merely over 10,000 Nashes found buyers during the calendar year. The same basic product in Hudson trim sold just over 4,000.

When 1958 rolled around, both marques were relegated into the history books. The Nash and Hudson names were dropped at the last minute, just before the new models were introduced. Rambler became the name of the marque. The timing for the change was perfect. In 1958, rising auto prices and a general economic recession combined to boost buyer interest in small cars. The only such domestic product around at that time was the Rambler. For awhile, it would be a success.

1918 NASH

NASH — 680 — SIX: The 1918 Nash had a low hood line; short, vertical hood louvers; a painted radiator shell that was somewhat rounded on top; a slanted windshield on open models and shell headlamps.

I.D. DATA: Serial numbers on left front cross-member, just in back of radiator. Starting: 100101. Engine numbers on right front of flywheel housing just behind starting motor.

Model No.	Body Type & Seating	Price	Weight	Start. Ser. No.
681	4dr Tr-5P	1295	2930	100101
682	4dr Tr-7P	1545	3040	111601
683	2dr Roadster-4P	1295	2930	121001
684	4dr Sedan-5P	1985	3455	100108
685	2dr Cpe-4P	2085	3225	94501

1918 Nash Model 681 touring (JAC)

NOTE 1: Models 681 and 683 have a 121 in. wheelbase. Other models have a 127 in. wheelbase.
ENGINE: Inline. Overhead valves. Cast en bloc. Six. Cast iron block. B & S: 3-1/4 x 5 in. Disp.: 248.9 cid Brake HP 55 at 2400 rpm. NACC HP: 25.35. Main bearings: Three. Valve lifters: Solid. Carb.: Marvel.
CHASSIS: W.B.: 121 in; 127 in. Rear/Rear Tread: 56 in./56 in. Tires: 34 x 4.
TECHNICAL: Selective, sliding gear transmission. Speeds: 3F/1R. Floor shift controls. Single dry plate clutch. Spiral bevel drive. Semi-floating rear axle. Overall ratio: 4.50:1. Two-wheel external mechanical brakes. Artillery wheels.
HISTORICAL: Introduced: September 1, 1917. Enclosed overhead valve mechanism. Calendar year production: 10,283. The president of Nash was C.W. Nash. First Nash automobile. In its first full year of production, Nash accounted for 1.1 percent of all new car production in the U.S.

1919 NASH

1919 Nash, Model 681 touring (JAC)

NASH — 680 — SIX: The 1919 Nash was identical to the 1918 model.
I.D. DATA: Serial number on left front cross-member, just behind radiator. Starting: 106430. Engine numbers on right front of flywheel housing behind starting motor.

Model No.	Body Type & Seating	Price	Weight	Start. Ser. No.
681	4dr Tr-5P	1395	2930	106430
682	4dr Tr-7P	1545	3040	111769
683	2dr Tr-4P	1395	2930	121910
684	4dr Sedan-5P	2085	3455	120118
685	2dr Cpe-4P	2085	3225	44035
686	2dr Roadster-2P	1490	2800	131851
687	4dr Sport Tr-4P	1595	2950	133351

NOTE 1: Models 681, 683, 686, and 687 have 121 in. wheelbase. Others have 127 inch wheelbase

ENGINE: Inline. Overhead valves. Cast en bloc. Six. Cast iron block. B & S: 3-1/4 x 5 in. Disp.: 248.9 cid Brake HP 55 at 2400 rpm. NACC HP: 25.35. Main bearings: Three. Valve lifters: Solid. Carb.: Marvel.
CHASSIS: W.B.: 121 in. & 127 in. Rear/Rear Tread: 56 in./56 in. Tires: 33 x 4. (34 x 4-1/2 on 127 w.b. models).
TECHNICAL: Selective, sliding gear transmission. Speeds: 3F/1R. Floor shift controls. Single dry plate clutch. Spiral bevel drive. Semi-floating rear axle. Overall ratio: 4.50:1. Two-wheel external mechanical brakes. Artillery wheels.
HISTORICAL: Introduced: September 1, 1918. Calendar year production: 27,081. The president of Nash was C.W. Nash. Virtually unchanged from the previous year. Nash accounted for 1.6 percent of U.S. new car production for 1919. Nash also built 4,090 trucks during 1919, and during this year purchased a half-interest in the Seaman Body Corp., Milwaukee, Wis.

1920 NASH

1920 Nash, Model 681 touring (JAC)

NASH — 680 — SIX: The 1920 Nash was identical to the 1918-1919 models.
I.D. DATA: Serial numbers on right rear engine girder at transmission base effective July 1, 1920. Earlier cars had serial number on left front cross-member, just in back of radiator. Starting: 139330. Engine numbers on right front of flywheel housing, just behind starting motor.

Model No.	Body Type & Seating	Price	Weight	Start. Ser. No.
681	4dr Tr-5P	1490	2930	139330
682	4dr Tr-7P	1640	3040	113661
684	4dr Sedan-7P	2575	3455	144334
685	2dr Cpe-4P	2350	3225	144985
686	2dr Roadster-2P	1490	2800	132580
687	4dr Sport Tr-4P	1545	2950	134276

NOTE 1: Models 681, 686 and 687 had 121 in. wheelbase, other models have 127 in. wheelbase.
ENGINE: Inline. Overhead valves. Cast en bloc. Six. Cast iron block. B & S: 3-1/4 x 5 in. Disp.: 248.9 cid Brake HP 55 at 2400 rpm. NACC HP: 25.35. Main bearings: Three. Valve lifters: Solid. Carb.: Marvel.
CHASSIS: W.B. 121 in. and 127 in. Rear/Rear Tread: 56 in./56 in. Tires: 33 x 4. (34 x 4-1/2 on 127 inch wheelbase models).
TECHNICAL: Selective, sliding gear transmission. Speeds: 3F/1R. Floor shift controls. Single dry plate clutch. Spiral bevel drive. Semi-floating rear axle. Overall Ratio: 4.50:1. Two-wheel external mechanical brakes. Artillery wheels.
HISTORICAL: Introduced Sept. 1, 1920. Calendar year production was 35,084. The president of Nash was C.W. Nash. Nash Motors earned $7,007,471 during 1920. Nash accounted for 1.9 percent of U.S. auto production, 1920 essentially unchanged from 1918-1919 models. A new assembly plant was opened in Milwaukee, Wis.

1921 NASH

NASH — 680 — SIX: The 1921 Nash 680 was identical to 1918-1920 models.
I.D. DATA: Serial numbers on right rear engine girder at transmission base. Starting: 175874. Engine numbers on right front of flywheel housing just back of starting motor.

Model No.	Body Type & Seating	Price	Weight	Start. Ser. No.
681	4dr Tr-5P	1695	3068	175874
682	4dr Tr-7P	1875	3198	167177
684	4dr Sedan-7P	2895	3708	178543
685	2dr Cpe-4P	2650	3403	179003
686	2dr Roadster-2P	1695	2988	179551
687	4dr Sport Tr-4P	1850	3098	177655

NOTE 1: Models 681, 686 and 687 had 121 in. wheelbase. Other models had a 127 in. wheelbase.

1921 Nash, Model 682 touring (JAC)

ENGINE: Inline. Overhead valves. Cast en bloc. Six. Cast iron block. B & S: 3-1/4 x 5 in. Disp.: 248.9 cid Brake HP 55 at 2400 rpm. NACC HP: 25.35. Main bearings: Three. Valve lifters: Solid. Carb.: Marvel.
CHASSIS: W.B.: 121 in.; 127 in. Rear/Rear Tread: 56 in./56 in. Tires: 33 x 4. (34 x 4-1/2 on 127 in. wheelbase models).
TECHNICAL: Selective sliding gear transmission. Speeds: 3F/1R. Floor shift controls. Single dry plate clutch. Spiral bevel drive. Semi-floating rear axle. Overall ratio: 4.50:1. Two-wheel mechanical brakes. Wood artillery wheels.
NASH — 40 — FOUR: The new four-cylinder Nash models were similar in appearance to 680 series models except for their shorter stubbier hoods.
I.D. DATA: Serial numbers on front frame cross-member just behind radiator on left side. Starting: 1000. Engine numbers on left side of crankcase just behind starting motor.

Model No.	Body Type & Seating	Price	Weight	Start. Ser. No.
41	4dr Tr-5P	1395	2502	1000
42	2dr Roadster-2P	1395	2432	1000
43	2dr Cpe-3P	1985	2732	1000
44	4dr Sedan-4P	2185	2942	1000
45	2dr Cabriolet-2P	1545	2676	1000

ENGINE: Overhead valves. Inline. Four. Cast iron block. B & S: 3-1/4 x 5 in. Disp.: 165.9 cid. Brake HP 35 at 2200 rpm. NACC HP: 16.9. Main bearings: Two. Valve lifters: Solid. Carb.: Schebler.
CHASSIS: W.B.: 112 in. Rear/Rear Tread: 56 in./56 in. Tires: 32 x 4.
TECHNICAL: Selective sliding gear transmission. Speeds: 3F/1R. Floor shift controls. Single dry plate clutch. Spiral bevel drive. Semi-floating rear axle. Overall ratio: 4.50:1. Two-wheel mechanical brakes. Wood artillery wheels.
HISTORICAL: Introduced: Sept. 1, 1920 (Series 680); Sept. 25, 1920 (Series 40). A new, four-cylinder series took Nash into a lower price range. Calendar year production: 20,850. The president of Nash was C.W. Nash. Severe postwar recession led to a sharp drop in production and sales. Nash accounted for 1.4 percent of U.S. auto production in 1921. On Feb. 14, 1921, the cylinder bore of the Series 40 engine was increased to 3-3/8 in, raising the displacement to 178.9 cid and the horsepower to 36.75 at 2800 rpm. Starting serial no.: [178.9 cid] 1782.

1922 NASH

1922 Nash, Model 43 coupe (JAC)

SERIES 690 — SIX: Styling similar to 680 series of 1918-1921, except taller radiator, drum headlamps.
NASH FOUR — SERIES 40: Identical to 1921 series 40.
I.D. DATA: [Series 690] Serial numbers on right rear engine girder at transmission base. Starting: 195754. Engine numbers on right front of flywheel housing just back of starting motor. [Series 40] Serial numbers on front frame cross-member just back of radiator, left side. Starting: 4511. Engine numbers on left side of crankcase just behind starting motor.

Model No.	Body Type & Seating	Price	Weight	Prod. Total
691	4dr Tr-5P	1545	2930	195754
692	4dr Tr-7P	1695	2950	208441
693	4dr Sedan-5P	2040	3430	217938
694	4dr Sedan 7P	2695	3455	198240
695	2dr Victoria-4P	2395	3255	200013
696	2dr Roadster-2P	1525	2805	207559
697	4dr Sport Tr-4P	1695	3037	205729
41	4dr Tr-5P	1045	2502	4511
42	2dr Roadster-2P	1025	2432	4511
43	2dr Cpe-3P	1645	2732	4511
44	4dr Sedan-5P	1835	2942	4511
45	2dr Cabriolet-5P	1245	2676	4511
46	2dr Carriole-5P	1350	2824	4511

ENGINE: [Nash 690] Overhead valves. Inline. Six. Cast iron block. B & S: 3-1/4 in. x 5 in. Disp.: 248.9 cid. C.R.: 4.2:1. Brake HP: 55 at 2400 rpm. Taxable HP: 25.35. Main bearings: Three. Valve lifters: Solid. Carb.: Marvel. [Nash 40] Overhead valves. Inline. Four. Cast iron block. B & S: 3-3/8 in. x 5 in. Disp.: 178.9 cid. C.R.: 3.8:1. Brake HP: 36.75 at 2800 rpm. Taxable HP: 18.23. Main bearings: Two. Valve lifters: Solid. Carb.: Schebler.
CHASSIS: [Series 690] W.B.: 121 in. and 127 in. Rear/Rear Tread: 56 in./56 in. Tires: 33 x 4. (34 x 4-1/2 on 127 in. wheelbase models). [Series 40] W.B.: 112 in. Rear/Rear Tread: 56 in./56 in. Tires: 33 x 4.
TECHNICAL: Selective Sliding gear transmission. Speeds: 3F/1R. Floor shift controls. Single dry plate clutch. Spiral bevel. Semi-floating rear axle. Overall ratio: 4.50:1. Mechanical brakes on two wheels. Wood artillery wheels.
HISTORICAL: Introduced Oct. 5, 1921. Innovations: First use of rubber engine mountings (Series 40). Two-door Carriole (Series 40) brought closed-car comfort to a lower-priced field. Calendar year production: 41,652. The president of Nash was C.W. Nash. Production doubled the 1921 figure; Nash accounted for 1.7 percent of U.S. auto production and placed eigth in the industry. Nash claimed to manufacture a higher percentage of its components than any other automaker.

1923 NASH

NASH SIX — SERIES 690: Identical to 1922 model, except disc wheels now optional, and cowl ventilator used.
NASH FOUR — SERIES 40: Identical to 1921-1922 series 40.
I.D. DATA: Series 690 serial numbers on right rear engine girder at transmission base. Starting: 226409. Engine numbers on right front of flywheel housing just back of starting motor. Series 40 serial numbers on front frame cross-member just back of radiator, left side. Starting: 19436.

1923 Nash, Series 690 four-door sedan (OCW)

Model No.	Body Type & Seating	Price	Weight	Start Ser. No.
691	4dr Tr-5P	1240	3030	226409
692	4dr Tr-7P	1390	3150	226882
693	4dr Sedan-5P	2040	3430	218364
694	4dr Sedan-7P	2190	3580	232021
695	2dr Victoria-4P	1890	3330	232241
696	2dr Roadster-2P	1210	2930	232589
697	4dr Sport Tr-4P	1645	3530	227921
698	4dr Cpe-5P	2090	3550	231401
41	4dr Tr-5P	935	2720	19436
42	2dr Roadster-2P	915	2600	19436
46	2dr Carriole-5P	1275	2910	19436
47	4dr Sedan-5P	1445	3090	19436
48	4dr Sport Tr-5P	1195	2980	19436

ENGINE: See 1922 data.
TECHNICAL: Same as 1922 except for overall drive ratios, [Series 40] 4.89:1; [690] 4.50:1. Wood artillery wheels.
OPTIONS: Steel disc wheels.
HISTORICAL: Introduced: Oct. 17, 1922. Calendar year registrations: 41,838. Calendar year production: 41,652. The president of Nash was C.W. Nash. Nash accounted for 1.6 percent of U.S. auto production during 1923.

1924 NASH

1924 Nash Model 40 Carriole Sedan (OCW)

NASH SIX — SERIES 690: Identical to 1922-1923 Nash 690 Series.
NASH FOUR — SERIES 40: Identical to 1921-1923 Series 40.
I.D. DATA: Series 690 serial numbers on right rear engine girder at transmission base. Starting: 256987. Engine numbers on right front of flywheel housing just back of starting motor. Series 40 serial numbers on front frame cross-member just back of radiator, left side. Starting: 34577.

Model No.	Body Type & Seating	Price	Weight	Start. Ser. No
691	4dr Tr-5P	1240	3120	256987
692	4dr Tr-7P	1390	3230	248120
693	4dr Sedan-5P	2040	3550	248526
694	4dr Sedan-7P	2190	3700	240423
695	2dr Victoria-4P	1990	3440	240778
696	2dr Roadster-2P	1240	3030	254852
697	4dr Sport Tr-4P	1645	3530	251417
698	4dr Cpe-5P	2090	3550	259009
699	4dr Sport Sedan-5P	1640	3400	269265
41	4dr Tr-5P	935	2720	34577
42	2dr Roadster-2P	915	2600	34577
46	2dr Carriole-5P	1275	2910	34577
47	4dr Sedan-5P	1445	3090	34577
48	4dr Sport Tr-5P	1145	2800	34577
49	2dr Cpe-2P	1165	2750	34577

ENGINE: [Series 690] Identical in all these respects to 1922-1923 Series 690. [Series 40] Identical in all these respects to 1922-1923 Series 40.
CHASSIS: Unchanged in these respects from 1922-1923.
TECHNICAL: Unchanged except for another increase in overall drive ratio. Overall ratio: [690] 4.50; [40] 5.50.
HISTORICAL: Introduced: July 20, 1923. Innovations: Industry's first use of electric dashboard clock (optional). Calendar year registrations: 47,571. Calendar year production: 53,626. The president of Nash was C.W. Nash. Nash accounted for 1.7 percent of U.S. auto production in 1924. Last year for the Nash Four. On Feb. 27, 1924 Nash purchased the Racine, Wis. plant of the bankrupt Mitchell Motors Co.

1925 NASH

1925 Nash roadster (OCW)

NASH ADVANCED SIX — SERIES 160: Much updated from previous model with balloon tires; nickeled radiator shell; long, thin hood louvers and long visor. Closed models have boxy configuration, particularly at the rear.
NASH SPECIAL SIX — SERIES 130: Similar in styling to 160 series, but smaller.
AJAX — SERIES 220 — SIX: Vertical styling, very similar to contemporary Nash models, was seen on the Ajax. However, it was a substantially smaller car. Long visors characterized sedans. Drum headlamps and a low, flat hood were other Ajax attributes.
This car was merchandised as a separate marque and was never called a Nash-Ajax when it was new.
I.D. DATA: Series 160 serial numbers on right-side of front motor support. Starting: 288001. Ending: 330125. Engine numbers on engine block near starting motor. Series 130 serial numbers on top right side rear motor support or on left rear spring hanger. Starting: 51001. Ending: 81503. Engine numbers on engine block beside starting motor. Ajax serial numbers on right front spring hanger. Starting: 1001. Ending: 2331. Engine numbers on left-side engine block, upper forward corner.

Model No.	Body Type & Seating	Price	Weight	Start Ser. No.
NASH LINE				
161	4dr Tr-5P	1375	3400	288001
162	4dr Tr-7P	1525	3480	290351
163	2dr Sedan-5P	1485	3550	308325
164	4dr Sedan-7P	2290	3830	290601
165	2dr Victoria-4P	2090	3640	297901
166	2dr Roadster-2P	1375	3320	290791
168	4dr Cpe-4P	2190	3750	291281
169	4dr Sedan-5P	1695	3860	291571
131	4dr Tr-5P	1095	2960	51001
132	2dr Roadster-2P	1095	2870	54573
133	2dr Sedan-5P	1225	3120	51001
134	4dr Sedan-5P	1545	3270	64990
AJAX LINE				
21	4dr Sedan-5P	995	2410	1001
51	4dr Tr-5P	865	2210	1001

ENGINE: Series 160 Engine: Overhead valve. Inline. Six. Cast iron block. B & S: 3-1/4 in. x 5 in. Disp.: 248.9 cid. Brake HP.: 60 at 2400 rpm. Taxable HP.: 25.35. Main bearings: Three. Valve lifters: Solid. Carb.: Marvel U4S. Series 130 Engine: Overhead valve. Inline. Six. Cast iron block. B & S: 3-1/8 in. x 4-1/2 in. Disp.: 207.0 cid. Brake HP.: 46 at 2200 rpm. Taxable HP.: 23.44. Main bearings: 3. Valve lifters: solid. Carb.: Marvel U3S. Ajax Engine: L-head. Inline. Six. Cast iron block. B & S: 3 in. x 4 in. Disp.: 169.6 cid. C.R.: 4.5:1. Brake HP.: 40 at 2400 rpm. Main bearings: 7. Valve lifters: solid. Carb.: Carter.
CHASSIS: [Series 160] W.B.: 121 in. and 127 in. Tires: 33 x 6.00. [Series 130] W.B.: 112-1/2 in. Tires: 31 x 5.25. [Series 220] W.B.: 108 in. Front/rear Tread: 56 in./56 in. Tires: 30 x 4.75.
TECHNICAL: Selective sliding gear transmission. Speeds: 3F/1R. Floor shift controls. Single dry plate clutch. Spiral bevel drive. Semi-floating rear axle. Overall ratio: 4.50:1 (160); 4.88:1 (130). Mechanical brakes on four wheels. Steel disc wheels. Ajax: Selective sliding gear transmission. Speeds: 3F/1R. Floor shift controls. Single dry plate clutch. Spiral bevel drive. Semi-floating rear axle. Overall ratio: 4.6:1. Mechanical brakes on four wheels. Steel disc wheels.
HISTORICAL: Introduced: Aug. 1, 1924. Innovations: Special Six replaced four-cylinder Nash. First use by Nash of four-wheel brakes. Calendar year registrations: 73,384. Calendar year production: 85,428. The president of Nash was C.W. Nash. Sales up 50 percent over 1924. Nash held 2.4 percent of new car registrations for the year. Ajax: Introduced May 1, 1925. Innovations: Nash's first entry in the $1,000 market; first L-head engine; first seven bearing crankshaft. At time of introduction, the Ajax was the only car in its price class with four-wheel brakes. Calendar year production: 10,693 (1925 only). The president of Nash was C.W. Nash. About 27,300 Ajax cars were sold before the model name was changed to Nash Light Six on May 1, 1926.

1925 Nash, Advanced Six touring (AB)

1926 NASH

NASH ADVANCED SIX — SERIES 260: Smoother, more rounded lines than 1925 models, particularly roofline at rear. Shorter visor than before.
NASH SPECIAL SIX — SERIES 230: Similar to 260 Series, but smaller.
NASH LIGHT SIX — SERIES 220: Identical, except in name, to the 1925-1926 Ajax.
I.D. DATA: Series 260 serial numbers on right-side of front motor support. Starting: 330126. Engine numbers on engine block near starting motor. Series 230 serial numbers on top right-side rear motor support or on left rear spring hanger. Starting: 75276. Engine numbers on engine block beside starting motor. Series 220 serial numbers on right cross member, just ahead of rear spring rear bracket. Starting: 2332-R. Ending: 28373. Engine numbers on left-side engine block, upper forward corner.

1926 Nash Light Six four-door sedan (OCW)

Model No.	Body Type & Seating	Price	Weight	Start. Ser. No.
261	4dr Tr-5P	1375	3400	330126
262	4dr Tr-7P	1525	3480	330126
263	2dr Sedan-5P	1485	3550	330126
264	4dr Sedan-7P	2290	3830	336404
265	2dr Victoria-4P	2090	3640	336404
266	2dr Roadster-2P	1375	3320	337932
268	4dr Cpe-4P	2190	3750	337932
269	4dr Sedan-5P	1525	3650	354114
231	4dr Tr-5P	1135	2960	75276
232	2dr Roadster-2P	1115	2870	81509
233	2dr Sedan-5P	1265	3120	75276
234	4dr Spl Sedan-5P	1545	3300	75276
235	2dr Cpe-2P	1165	3030	92070
236	2dr Roadster-2/4P	1225	2980	A20247
239	4dr Sedan-5P	1315	3170	98405
221	4dr Tr-5P	865	2210	2332
224	4dr Sedan-5P	995	2410	2332
225	2dr Cpe-2P	925	2310	2332

ENGINE: [Series 260] Overhead valve. Inline. Six. Cast iron block. B & S: 3-7/16 in. x 5 in. Disp.: 278.4 cid. Brake HP.: 60 at 2400 rpm. Taxable HP.: 28.37. Main bearings: Seven. Valve lifters: Solid. Carb.: Marvel U4S. [Series 230] Overhead valve. Inline. Six. Cast iron block. B & S: 3-1/8 in. x 4-1/2 in. Disp.: 207.4 cid. Brake HP.: 47 at 2200 rpm. Taxable HP.: 23.44. Main bearings: Seven. Valve lifters: Solid. Carb.: Marvel U38. [Series 220] L-head. Inline. Six. Cast iron block. B & S: 3 in. x 4 in. Disp.: 169.6 cid. C.R.: 4.5:1 Brake HP.: 40 at 2400 rpm. Taxable HP.: 21.6. Main bearings: Seven. Valve lifters: Solid. Carb.: Carter.
CHASSIS: [Series 260] W.B.: 121 in., 127 in. Tires: 33 x 6.00. [Series 230] W.B.: 112-1/2 in. Tires: 31 x 5.25. [Series 220] W.B.: 108 in. Front/rear Tread: 56 in./56 in. Tires: 30 x 4.75.
TECHNICAL: Selective sliding gear transmission. Speeds: 3F/1R. Floor shift controls. Single dry plate clutch. Spiral bevel drive. Semi-floating rear axle. Overall ratio: [260] 4.5:1; [230] 4.9:1; [220] 4.6:1. Mechanical brakes on four wheels. Steel disc wheels.
HISTORICAL: Introduced: June 1, 1925 [260], July 1, 1925 [230], May 1, 1926 [220]. Innovations: First use of seven-bearing crankshafts in Nash's overhead valve engines [Series 260/230], as well as L-head [Series 220]. Calendar year registrations: 98,804. Calendar year production: 135,520. The president of Nash was C.W. Nash. Light Six [Series 220] represented a continuation of the Ajax under a new name. Evidently as the result of the name change, sales took a great leap forward. 60 percent increase in Nash production for the year. Nash held 3.6 percent of new car registrations for the year.

1926 Nash, Light Six (formerly Ajax) sedan (AB)

1927 NASH

NASH — ADVANCED SIX — SERIES 260: Little change in styling from 1926 model.
NASH — SPECIAL SIX — SERIES 230: Little changed in styling from 1926 model.

1927 Nash, four-door sedan (OCW)

1927 Nash Mason

NASH — LIGHT SIX — SERIES 220: Less boxy than 1926 model: styling similar to 230 series, but smaller.
I.D. DATA: Series 260 serial numbers on right side of front motor support. Starting: 386972. Engine numbers on engine block near starting motor. Series 230 serial numbers on top right-side rear motor support or on left rear spring hanger. Starting: A26276. Engine numbers on engine block beside starting motor. Series 220 serial numbers on right frame member just ahead of rear spring rear bracket. Starting: R28374. Engine numbers on left-side engine block, upper forward corner.

Model No.	Body Type & Seating	Price	Weight	Start. Ser. No.
260	2dr Cpe.-2/4P	1775	3580	419570
261	4dr Tr-5P	1340	3400	386972
262	4dr Tr-7P	1490	3480	386972
263	2dr Sedan-5P	1425	3550	386972
264	4dr Sedan-7P	2090	3830	386972
265	2dr Victoria-4P	1790	3640	386972
266	2dr Roadster-2/4P	1475	3390	386972
267	4dr Ambassador Sedan-5P	2090	3800	408304
268	4dr Cpe-4P	1990	3750	386972
269	4dr Sedan-5P	1525	3650	386972
270	4dr Spl Sedan-5P	1695	3650	410709
271	4dr Sport Tr-5P	1540	3500	415998
231	4dr Tr-5P	1135	2980	A26276
232	2dr Roadster-2P	1115	2900	A26276
233	2dr Sedan-5P	1215	3150	A26276
235	2dr Bus Cpe-2P	1165	3030	A26276
236	2dr Roadster-2/4P	1225	2980	A26276
237	4dr Cavalier Sedan-5P	1695	3330	A42260
239	4dr Sedan-5P	1315	3170	A26276
240	4dr Special Sedan-5P	1485	3250	A44332
241	2dr Cabriolet-2/4P	1290	3070	A46894
221	4dr Tr-5P	865	2275	R28374
223	2dr Sedan-5P	925	2410	R46146
224	4dr Sedan-5P	995	2475	R28374
225	2dr Cpe-2P	925	2310	R28374
227	4dr Deluxe Sedan-5P	1085	2550	R48852

ENGINE: [Series 260] Overhead valve. Inline. Six-cylinder. Cast iron block. B & S: 3-7/16 in. x 5 in. Disp.: 278.4 cid. C.R.: 4.6:1. Brake HP.: 69 at 2500 rpm. Taxable HP.: 28.4. Main bearings: Seven. Valve lifters: Solid. Carb.: Marvel U4S. [Series 230] Overhead valve. Inline. Six-cylinder. Cast iron block. B & S: 3-1/4 in. x 4-1/2 in. Disp.: 224.0 cid. C.R.: 4.69:1. Brake HP.: 52 at 2600 rpm. Taxable HP.: 25.3. Main bearings: Seven. Valve lifters: Solid. Carb.: Marvel U38. [Series 220] L-head. Inline. Six. Cast iron block. B & S: 3 in. x 4 in. Disp.: 169.6 cid. C.R.: 4.5:1. Brake HP.: 40 at 2400 rpm. Taxable HP.: 21.6. Main bearings: Seven. Valve lifters: Solid. Carb.: one-inch. Carter 82S/89S.
CHASSIS: [Series 260] W.B.: 121 in. and 127 in. Tires: 33 x 6.00. [Series 230] W.B.: 112-1/2 in. Tires: 31 x 5.25. [Series 220] W.B.: 108 in. Tires: 30 x 4.75.
TECHNICAL: Selective sliding gear transmission. Speeds: 3F/1R. Floor shift controls. Single dry plate clutch. Spiral bevel drive. Semi-floating rear axle. Overall ratio: 4.5:1 [260]; 4.67:1 [230]; 4.77:1 [220] Mechanical brakes on four wheels. Steel disc wheels.
HISTORICAL: Introduced: [260] June 1, 1926, [220] July 6, 1926, [230] Aug. 1, 1926. Innovations: Larger bore, more power, series 260 and 230. Calendar year registrations: 109,979. Calendar year production: 122,606. The president of Nash was C.W. Nash. Two new premium-level sedans, the Ambassador [Series 260] and Cavalier [Series 230] featured more rounded rear contours, a styling feature picked up by all Advanced Six and Special Six Sedans for 1929. Nash held 4.2 percent of industry registrations for 1927.

1927 Nash, rumbleseat Cabriolet (OCW)

1928 NASH

NASH — ADVANCED SIX — SERIES 360: Similar in styling to 1927 model, but with taller radiator.
NASH — SPECIAL SIX — SERIES 330: Similar in appearance to advanced (360) series, but smaller.
NASH — STANDARD SIX — SERIES 320: Similar in styling to 360 and 330 series but much smaller.

1928 Nash, Advanced Six Model 370 four-door sedan (JAC)

I.D. DATA: Series 360 serial numbers on right-side of front motor support. Starting: 423612. Ending: 452099. Engine numbers on engine block near starting motor. Series 330 serial numbers on top right-side rear motor support or on left rear spring hanger. Starting: A58246. Ending: A87449. Engine numbers on engine block beside starting motor. Series 320 serial numbers on right frame member just ahead of rear spring rear bracket. Starting: R71557. Ending: R119558. Engine numbers on left-side engine block, upper forward corner.

Model No.	Body Type & Seating	Price	Weight	Start. Ser. No.
360	2dr Cpe-2/4P	1775	3650	423612
361	4dr Tr-5P	1340	3400	423612
362	4dr Tr-7P	1440	3500	423612
363	2dr Sedan-5P	1425	3620	423612
364	4dr Sedan-7P	1990	3830	423612
364I	4dr Imperial Sedan-7P	2165	3900	440770
365	2dr Victoria-4P	1595	3640	423612
366	2dr Roadster-2/4P	1475	3400	423612
367	4dr Ambassador Sedan-5P	1925	3820	422617
370	4dr Sedan-5P	1545	3650	423612
371	4dr Sport Tr-5P	1540	3500	423612
331	4dr Tr-5P	1135	2980	A58246
333	2dr Sedan-5P	1215	3150	A58246
335	2dr Cpe-2P	1165	3030	A58246
335R	2dr Cpe-2/4P	1245	3030	A58246
336	2dr Roadster-2/4P	1225	2980	A58246
338	4dr Landau Sedan-5P	1445	3380	A68271
340	4dr Spl Sedan-5P	1335	3250	A58246
341	2dr Cabriolet-2/4P	1290	3070	A58246
342	2dr Victoria-4P	1295	3170	A70804
320	4dr Sedan-5P	995	2500	R71557
321	4dr Tr-5P	865	2325	R71557
322	2dr Cabriolet-2/4P	995	2505	R91928
323	2dr Sedan-5P	895	2450	R71557
325	2dr Cpe-2P	875	2345	R71557
328	4dr Landau Sedan-5P	1085	2610	R71557

ENGINE: [Series 360] Overhead valve. Inline. Six. Cast iron block. B & S: 3-7/16 in. x 5 in. Disp 279.0 cid. C.R.: 4.6:1. Brake HP.: 70 at 2400 rpm. Taxable HP.: 28.35. Main bearings: Seven. Valve lifters: Solid. Carb.: Marvel U. [Series 330] Overhead valve. Inline. Six. Cast iron block. B & S: 3-1/4 in. x 4-1/2 in. Disp: 224.0 cid. C.R.: 4.69:1. Brake HP.: 52 at 2600 rpm. Taxable HP.: 25.35. Main bearings: Seven. Valve lifters: Solid. Carb.: Marvel. [Series 320] L-head. Inline. Six. Cast iron block. B & S: 3-1/8 in. x 4 in. Disp: 184.1 cid. C.R.: 4.5:1. Brake HP.: 45 at 2600 rpm. Main bearings: Seven. Valve lifters: Solid. Carb.: Carter DRHO.
CHASSIS: [Series 360] W.B.: 121 in. and 127 in. Tires: 32 x 6.00. [Series 330] W.B.: 112-1/2 in. Tires: 30 x 5.25. [Series 320] W.B.: 108 in. Tires: 30 x 5.00.
TECHNICAL: Selective, sliding gear transmission. Speeds: 3F/1R. Floor shift controls. Single dry plate clutch. Spiral bevel drive. Semi-floating rear axle. Overall ratio: 4.50:1 (360); 4.88:1 (330); 4.77:1 (320). Mechanical brakes on four wheels. Steel disc wheels.
OPTIONS: Single sidemount. Dual sidemounts.
HISTORICAL: Introduced June 29, 1927. Calendar year registrations: 115,172. Calendar year production: 138,137. The president of Nash was C.W. Nash. Introduction in June 1928 of the stylish 1929 Nash 400 series led to a sharp increase in sales for the calendar year. Nash's best year to date — not surpassed until 1949. (Nash held 3.67 percent of the 1928 automobile market.)

1929 NASH

NASH — ADVANCED SIX — SERIES 460: Tall, narrow radiator; one-piece fenders; chromed, bowl-shaped headlamps with slight peak at top; tall, single row of vertical hood louvers; new "fish scale" radiator badge; double-bar bumpers.

1929 Nash, Special Six four-door sedan (AB)

NASH — SPECIAL SIX — SERIES 430: Similar to Advanced Six, except smaller and has double row of vertical hood vents.
NASH — STANDARD SIX — SERIES 420: Same general styling theme as Advanced Six and Special Six series, but considerably smaller. No peak on bowl-shaped headlamps; single row of vertical hood louvers.
I.D. DATA: [Series 460] Serial numbers on right-side front motor support. Starting: 452100. Ending: 496399. Engine numbers on engine block near starting motor. [Series 430] Serial numbers on top right-side rear motor support or of left rear spring hanger. Starting: A87450. Ending: B37581. Engine numbers on engine block beside starting motor. [Series 420] Serial numbers on right frame member just ahead of rear spring rear bracket. Starting: R119559. Engine numbers on left-side engine block, upper forward corner.

Model No.	Body Type & Seating	Price	Weight	Start. Ser. No.
460	2dr Cpe-2/4P	1775	3710	452100
461	2dr Cabriolet-2/4P	1660	3675	452100
462	4dr Phaeton-7P	1550	3700	452100
463	2dr Sedan-5P	1480	3760	452100
464	4dr Sedan-7P	1990	3970	452100
465	4dr Limousine-7P	2190	4010	452100
467	4dr Ambassador Sedan-5P	1925	3940	452100
470	4dr Sedan-5P	1550	3700	452100
431	4dr Phaeton-5P	1250	3150	A87450
433	2dr Sedan-5P	1260	3400	A87450
434	2dr Cpe-2/4P	1315	3250	A87450
435	2dr Cpe-2P	1245	3250	A87450
436	2dr Roadster-2/4P	1345	3200	B18953
440	4dr Sedan-5P	1345	3400	A87450
441	2dr Cabriolet-2/4P	1345	3260	A87450
442	2dr Victoria-4P	1345	3300	A87450
444	4dr Sedan-7P	1645	3530	B18339
420	4dr Sedan-5P	955	2725	R119559
421	4dr Phaeton-5P	935	2500	R119559
422	2dr Cabriolet-2/4P	955	2550	R119559
423	2dr Sedan-5P	885	2625	R119559
425	2dr Cpe-2P	885	2500	R119559
428	4dr Landau Sedan-5P	995	2725	R119559

ENGINE: [Series 460] Overhead valve. Inline. Six-cylinder. Cast iron block. B & S: 3-7/16 in. x 5 in. Disp: 278.4 cid. C.R.: 5.1:1. Brake HP.: 78 at 2900 rpm. Taxable HP.: 28.4. Main bearings: Seven. Valve lifters: Solid. Carb.: Marvel U 1-1/4 in. updraft. [Series 430] Overhead valve. Inline. Six. Cast iron block. B & S: 3-1/4 in. x 4-1/2 in. Disp: 224.0 cid. C.R.: 5.15:1. Brake HP.: 65 at 2900 rpm. Taxable HP.: 25.3. Main bearings: Seven. Valve lifters: Solid. Carb.: Marvel U 1-1/4 in. updraft. [Series 420] L-head. Inline. Six. Cast iron block. B & S: 3-1/8 in. x 4 in. Disp: 184.1 cid. C.R.: 5.0:1. Brake HP.: 50 at 2800 rpm. Taxable HP.: 23.4. Main bearings: Seven. Valve lifters: Solid. Carb.: 1-1/8 in. Carter DRJH updraft.
CHASSIS: [Series 460] W.B.: 121 in. and 130 in. Tires: 32 x 6.00. [Series 430] W.B.: 116 in. & 122 in. Tires: 29 x 5.50. [Series 420] W.B.: 112-1/4 in. Tires: 30 x 5.00.
TECHNICAL: Selective sliding gear transmission. Speeds: 3F/1R. Floor shift controls. Single dry plate clutch. Spiral bevel drive. Semi-floating rear axle. Overall ratio: [460] 4.5:1; [430] 4.8:1; [420] 4.7:1. Mechanical brakes on four wheels. Wood artillery wheels. [460/420] 20 in.; [430] 19 in.
OPTIONS: Single sidemount. Dual sidemounts with wire wheels ($125). Leather upholstery ($25). Five wire wheels ($40).
HISTORICAL: Introduced June 1, 1928. Innovations: First year for Twin Ignition (overhead valve models only) with two spark plugs firing each cylinder. First use by Nash of chromed brightwork. Calendar year registration: 105,146. Calendar year production: 116,622. The president of Nash was C.W. Nash. Attractive new Nash styling was evidently inspired by the LaSalle. Increased horsepower, all series. Longer wheelbases, Standard, Special and most

Advanced series models. Rumors were reported of a possible Nash-Packard merger. Nash held 2.7 percent of new car registrations.

1929 Nash, Model 460 coupe (JAC)

1930 NASH

1930 Nash, Model 498 dual cowl phaeton (OCW)

NASH — TWIN-IGNITION EIGHT — SERIES 490: Similar in styling to 1929 Series 460, but larger, narrower chrome band at top of radiator shell as seen from front; figure eight with inverted wings above Nash badge; automatic shutters covering radiator. (Largest, heaviest Nash built to date.)
NASH — TWIN-IGNITION SIX — SERIES 480: Similar to Series 490, but smaller.
NASH — SINGLE SIX — SERIES 450: Similar in styling to 1929 Series 420, except for narrow chrome band at top of radiator shell, as seen from front.
I.D. DATA: [Series 490] Serial numbers on frame near starting motor. Starting: 496400. Ending: 509200. Engine numbers on engine block near starting motor. [Series 480] Serial numbers on top right-side rear motor support or on left rear spring hanger. Starting: B37582. Ending: B54927. Engine numbers on engine block beside starter motor. [Series 450] Serial numbers on right frame member just ahead of rear spring rear bracket. Starting: R216590. Ending: R249707. Engine numbers on left-side engine block, upper forward corner.

Model No.	Body Type & Seating	Price	Weight	Start. Ser. No.
490	4dr Sedan-5P	1695	4000	496400
491	2dr Cabriolet-2/4P	1775	3840	496400
492	2dr Cpe-2P	1775	3900	496400
492R	2dr Cpe-2/4P	1845	3950	496400
493	2dr Sedan-5P	1625	3950	496400
494	4dr Sedan-7P	2085	4170	496400
495	4dr Limousine-7P	2260	4210	496400
497	4dr Ambassador Sedan-5P	1995	4050	496400
498	4dr Phaeton-7P	1845	3770	505422
498S	4dr Sport Phaeton-5P	1975	3840	505965
499	2dr Victoria-5P	1945	3950	496400
480	4dr Sedan-5P	1385	3535	B37582
481	2dr Cabriolet-2/4P	1355	3350	B37582
482	2dr Cpe-2P	1295	3400	B37582
482R	2dr Cpe-2/4P	1345	3450	B37582
483	2dr Sedan-5P	1295	3535	B37582
484	4dr Sedan-7P	1695	3750	B37582
485	4dr Limousine-7P	1920	3760	B47085
486	2dr Roadster-2/4P	1415	3250	B37582
488	4dr Phaeton-7P	1425	3450	B37582
488S	4dr Sport Phaeton-5P	1545	3720	B37582
489	2dr Victoria-4P	1385	3400	B37582
450	4dr Sedan-5P	985	2850	R216590
451	2dr Cabriolet-2/4P	985	2600	R216590
452	2dr Cpe-2P	915	2650	R216590
452R	2dr Cpe-2/4P	955	2700	R216590
453	2dr Sedan-5P	915	2750	R216590
455	4dr Landau Sedan-5P	1125	2900	R216590
456	2dr Roadster-2/4P	945	2550	R216590
457	4dr Deluxe Sedan-5P	1075	2900	R216590
458	4dr Phaeton-5P	975	2650	R216590

ENGINE: [Series 490] Overhead valve. Straight eight. Cast iron block. B & S: 3-1/4 in. x 4-1/2 in. Disp. 298.6 cid. C.R.: 5.25:1. Brake HP.: 100 at 3200 rpm. Taxable HP.: 33.8. Main bearings: Nine. Valve lifters: Solid. Carb.: Marvel two-inch. [Series 480] Overhead valve. Inline. Six. Cast iron block. B & S: 3-3/8 in. x 4-1/2 in. Disp.: 242.0 cid. C.R.: 5.0:1. Brake HP.: 74.5 at 2800 rpm. Taxable HP.: 27.3. Main bearings: Seven. Valve lifters: Solid. Carb.: Marvel 1-1/4 in. [Series 450] -head. Inline. Six. Cast iron block. B & S: 3-1/8 in. x 4-3/8 in. Disp. 201.3 cid. C.R.: 5.0:1. Brake HP.: 60 at 2800 rpm. Taxable HP.: 23.4. Main bearings: Seven. Valve lifters: Solid. Carb.: Carter 1-5/16 in.
CHASSIS: [Series 490] W.B.: 124 in. and 133 in. Front/rear Tread: 56 in./58 in. Tires: 6.50 x 19. [Series 480] W.B.: 118 in. and 128-1/4 in. Front/rear Tread: 56-3/4 in./58-1/4 in. Tires: 5.50 x 19. [Series 450] W.B.: 114-1/4 in. Front/rear Tread: 56 in./57-1/4 in. Tires: 5.00 x 19.
TECHNICAL: Selective sliding gear transmission. Speeds: 3F/1R. Floor shift controls. Single dry plate clutch. Spiral bevel drive. Semi-floating rear axle. Overall ratio: 4.5:1 (490, 480); 4.7:1 (450). Mechanical brakes on four wheels. Wood artillery wheels. Wheel size: 19 in.
OPTIONS: Single sidemount. Dual sidemounts with wire wheels ($87). Five wire wheels ($40).
HISTORICAL: Introduced Oct. 1, 1929. Innovations: [Series 490] was first Nash eight and largest Nash built to date. Longer wheelbases than corresponding 1929 models. First use by Nash of radiator shutters. Calendar year registrations: 51,086. Calendar year production: 54,605. The president of Nash was C.W. Nash. Nash held 1.95 percent of new car registrations for 1930.

1931 NASH

NASH — SERIES 890 — EIGHT: Nearly identical in styling to 480 series of 1930, but parking lamps mounted on fenders.
NASH — SERIES 880 — EIGHT: Similar in styling to six-cylinder 480 series of 1930, but with longer hood to accommodate straight eight.

1931 Nash, Series 880 Convertible Sedan (HC)

NASH — SERIES 870 — EIGHT: Similar in styling to 1930 "single six," series 450, but with longer hood to accommodate straight-8 engine. Figure "8" flanked by inverted wings above Nash radiator badge.
NASH — SERIES 660 — SIX: Identical in appearance except shorter hood, no "winged 8" radiator emblem.
I.D. DATA: [Series 890] Serial numbers on frame, adjacent to starting motor. Starting: 509201. Ending: 515399. Engine numbers on engine block right side, near starting motor. Starting: B70124. Ending: B74370. [Series 880] Serial numbers on frame adjacent to starting motor. Starting: B54928. Ending: B61757. Engine numbers on right-side engine block near starting motor. Starting: B70124. Ending: B74370. [Series 870] Serial numbers on frame, adjacent to starting motor. Starting: X1001. Ending: X13116. Engine numbers on left-side engine block next to generator. Starting: XE1000. Ending: XE13184. [Series 660] Serial numbers on frame, adjacent to starting motor. Starting: R249708. Ending: R261948. Engine numbers on left-side engine block next to generator. Starting: E1000. Ending: E13290.

Model No.	Body Type & Seating	Price	Weight	Start. Ser. No.
890	4dr Sedan-5P	1565	4000	509201
891	2dr Cabriolet-2/4P	1695	3840	509201
892	2dr Cpe-2P	1695	3900	509201
892R	2dr Cpe-2/4P	1745	3950	509201
894	4dr Sedan-7P	1925	4170	509201
895	4dr Limousine-7P	2025	4210	509201
897	4dr Ambassador Sedan-7P	1825	4050	509201
898	4dr Phaeton-7P	1595	3880	513589
899	2dr Victoria-5P	1765	3950	509201
880	4dr Sedan-5P	1295	3360	B54928
881	2dr Convertible Sedan-5P	1325	3275	B58597
882	2dr Cpe-2P	1245	3200	B54928
882R	2dr Cpe-2/4P	1285	3250	B54928
887	4dr Town Sedan-5P	1375	3400	B54928
870	4dr Sedan-5P	995	3000	X1001
871	2dr Convertible Sedan-5P	1075	2950	X8171
872	2dr Cpe-2P	945	2870	X1001
872R	2dr Cpe-2/4P	975	2920	X1001
877	4dr Spl Sedan-5P	955	3000	X1001
660	4dr Sedan-5P	845	2800	R249708
662	2dr Cpe-2P	795	2600	R249708
662R	2dr Cpe-2/4P	825	2650	R249708
663	2dr Sedan-5P	795	2740	R249708
668	4dr Sport Phaeton-5P	895	2640	R249708

ENGINE: [Series 890] Overhead valve. Inline. Eight. Cast iron block. B & S: 3-1/4 in. x 4-1/2 in. Disp.: 298.6 cid. C.R.: 5.25:1. Brake HP.: 115 at 3600 rpm. Taxable HP.: 33.8. Main

bearings: Nine. Valve lifters: Solid. Carb.: Stromberg UUR-2. [Series 880] Overhead valve. Inline. Eight-cylinder. Cast iron block. B & S: 3 in. x 4-1/4 in. Disp.: 240.0 cid. C.R.: 5.25:1. Brake HP: 88.5 at 3400 rpm. Taxable HP: 28.8. Main bearings: Nine. Valve lifters: Solid. Carb.: Marvel. Torque: 100 lbs.-ft. at 200 rpm. [Series 870] L-head. Eight-cylinder. Cast iron block. B & S: 2-7/8 in. x 4-3/8 in. Disp.: 227.2 cid. C.R.: 5.0:1. Brake HP: 78 at 3300 rpm. Taxable HP: 26.4. Main bearings: Nine. Valve lifters: Solid. Carb.: Stromberg E-2. Torque: 92 lbs.-ft. at 1600 rpm. [Series 660] L-head. Inline. Six-cylinder. Cast iron block. B & S: 3-1/8 in. x 4-3/8 in. Disp.: 201.3 cid. C.R.: 5.0:1. Brake HP: 65 at 3200 rpm. Taxable HP: 23.4. Main bearings: Seven. Valve lifters: Solid. Carb.: Carter DRT-08. Torque: 92 lbs.-ft. at 1600 rpm.
CHASSIS: [Series 890] W.B.: 124 in. and 133 in. Front/rear Tread: 56-1/4 in./58 in. Tires: 6.50 x 19. [Series 880] W.B.: 121 in. Front/rear Tread: 55-15/16 in./58-1/8 in. Tires: 5.50 x 18. [Series 870] W.B.: 116-1/4 in. Front/rear Tread: 56 in./58-3/8 in. Tires: 5.25 x 19. [Series 660] W.B.: 114-1/4 in. Front/rear Tread: 56 in./58-3/8 in. Tires: 5.00 x 19.
TECHNICAL: Selective sliding gear transmission. Speeds: 3F/1R. Floor shift controls. Single dry plate clutch. Spiral bevel drive. Semi-floating rear axle. Overall ratio: 4.5:1 (890); 4.72:1 (880); 5.1:1 (870, 660). Mechanical brakes on four wheels. Artillery wheels (wire opt., extra cost). Rim size: [880] 18 in.; [All other models] 19 in.
OPTIONS: Single sidemount. Dual Sidemounts.
HISTORICAL: Introduced Oct. 1, 1930. Innovations: First Nash "flathead eight." Calendar year registrations: 39,366. Calendar year production: 38,616. The president of Nash was E.H. McCarty. Despite falling sales, Nash earned $4,808,000 during fiscal 1931. Nash held 2.06 percent of new car registrations for the year.

1932 NASH

1932 Nash, Model 1094 seven-passenger sedan (OCW)

NASH — SERIES 990 — EIGHT: Basically a carryover of the 1931 Nash 890 series, but with a veed grille instead of the shuttered flat radiator. Headlamps bullet-shaped. Single-bar bumpers.
NASH — SERIES 980 — EIGHT: Similar in appearance to 990 series, but smaller.
NASH — SERIES 970 — EIGHT: Similar in styling to 870 series of 1931, but with slightly veed grille, instead of flat radiator front.
NASH — SERIES 960 — SIX: Virtually identical in appearance to 970 series, but has two inch shorter hooder.
NASH — AMBASSADOR EIGHT/ADVANCED EIGHT — SERIES 1090: Second series 1932 car. Slanted windshield with visor eliminated; more sweeping lines (especially fenders); semi-beavertail rear quarter. Ambassador Eight models on 142 inch wheelbase (longest ever offered by Nash). Advanced Eight models similar, but shorter bodies on 133 inch wheelbase. Vertical veed grille. Bullet-shaped, chrome-plated headlamps.
NASH — SPECIAL EIGHT — SERIES 1080: Similar in appearance to 1090 series, but smaller; less brightwork on radiator shell.
NASH — STANDARD EIGHT — SERIES 1070: Slanted windshield; veed, vertical grille; sweeping fender lines; semi-beavertail rear section.
NASH — BIG SIX — SERIES 1060: Similar in appearance to 1070 series, but shorter hood and front fenders.
I.D. DATA: [Series 990] Serial numbers on frame, adjacent to starting motor. Starting: 515400. Ending: 519299. Engine numbers on right-side engine block near starting motor. Starting: 398700. Ending: 402599. [Series 980] Serial numbers on frame, adjacent to starting motor. Starting: B61758. Ending: B66800. Engine numbers on right-side engine block near starting motor. Starting: B74371. Ending: B79449. [Series 970] Serial numbers on frame, adjacent to starting motor. Starting: X13117. Ending: X21317. Engine numbers on left-side engine block next to generator. Starting: XE13185. Ending: XE21416. [Series 960] Serial numbers on frame, adjacent to starting motor. Starting: R261949. Ending: R267735. Engine numbers on left-side engine block next to generator. Starting: R261949. Ending: R267735. [Series 1090] Serial numbers on frame, adjacent to starting motor. Starting: 519300. Ending: 521190. Engine numbers on right-side engine block near starting motor. [Series 1080] Serial numbers on frame, adjacent to starting motor. Starting: B66800. Ending: B70020. Engine numbers on right-side engine block near starting motor. [Series 1070] Serial numbers on frame, adjacent to starting motor. Starting: X21318. Ending: X25386. Engine numbers on left-side engine block next to generator. [Series 1060] Serial numbers on frame, adjacent to starting motor. Starting: R267736. Ending: R274299. Engine numbers on left-side engine block next to generator.

Model No.	Body Type & Seating	Price	Weight	Start. Ser. No.
990	4dr Sedan-5P	1565	4000	515400
991	2dr Cabriolet-2/4P	1695	3840	515400
992	2dr Cpe-2P	1695	3900	515400
992R	2dr Cpe-2/4P	1745	3950	515400
994	4dr Sedan-7P	1925	4170	515400
995	4dr Limousine-7P	2025	4210	515400
996	4dr LWB Sedan-5P	1825	4100	515400
997	4dr Ambassador Sedan-5P	1825	4050	515400
998	2dr Phaeton-7P	1595	3880	515400
999	2dr Victoria-5P	1765	3950	515400
980	4dr Sedan-5P	1295	3360	B61758
981	2dr Convertible Sedan-5P	1325	3275	B61758
982	2dr Cpe-2P	1245	3200	B61758
982R	2dr Cpe-2/4P	1285	3250	B61758
987	4dr Town Sedan-5P	1375	3400	B61758
970	4dr Sedan-5P	995	3000	X13117
971	2dr Convertible Sedan-5P	1075	2950	X13117
972	2dr Cpe-2P	945	2870	X13117
972R	2dr Cpe-2/4P	975	2920	X13117
977	4dr Spl Sedan-5P	955	3000	X13117
960	4dr Sedan-5P	845	2800	R261949
962	2dr Cpe-2P	795	2600	R261949
962R	2dr Cpe-2/4P	825	2650	R261949
963	2dr Sedan-5P	795	2740	R261949
968	4dr Sport Phaeton-5P	895	2640	R261949
1090	4dr Sedan-5P	1595	4350	519300
1091	2dr Convertible Roadster-2/4P	1795	4270	519300
1092	2dr Cpe-2P	1695	4210	519300
1092R	2dr Cpe2/4P	1695	4300	519300
1093	4dr Convertible Sedan-5P	1875	4470	519300
1094	4dr Sedan-7P	1955	4600	519300
1095	4dr Limousine-7P	2055	4650	519300
1096	4dr Sedan-5P	1855	4510	519300
1097	4dr Brougham-5P	1855	4470	519300
1099	2dr Victoria-5P	1785	4300	519300
1080	4dr Sedan-5P	1320	3870	B66800
1081	2dr Convertible Roadster-2/4P	1395	3750	B66800
1082	2dr Cpe-2P	1270	3710	B66800
1082R	2dr Cpe-2/4P	1320	3800	B66800
1083	4dr Convertible Sedan-5P	1475	4000	B66800
1089	2dr Victoria-5P	1395	3840	B66800
1070	4dr Sedan-5P	1015	3400	X21318
1071	2dr Convertible Roadster-2/4P	1055	3270	X21318
1072	2dr Cpe-2P	965	3250	X21318
1072R	2dr Cpe-2/4P	1015	3300	X21318
1073	4dr Convertible Sedan-5P	1095	3275	X21318
1077	4dr Town Sedan-5P	975	3400	X21318
1060	4dr Sedan-5P	840	3200	R267736
1061	2dr Convertible Roadster-2/4P	895	3120	R267736
1062	2dr Cpe-2P	777	3050	R267736
1062R	2dr Cpe-2/4P	825	3100	R267736
1063	2dr Convertible Sedan-5P	935	3125	R267736
1067	4dr Town Sedan-5P	825	3150	R267736

1932 Nash Ambassador Eight, four-door sedan (JAC)

ENGINE: [Series 990] Overhead valve. Inline. Eight-cylinder. Cast iron block. B & S: 3-1/4 in. x 4-1/2 in. Disp.: 298.6 cid. C.R.: 5.25:1. Brake HP: 115 at 3600 rpm. Taxable HP: 33.8. Main bearings: Nine. Valve lifters: Solid. Carb.: Stromberg UUR2. Torque: 100 lbs.-ft. at 1200 rpm. [Series 980] Overhead valve. Inline. Eight-cylinder. Cast iron block. B & S: 3 in. x 4-1/4 in. Disp.: 240.0 cid. C.R.: 5.25:1. Brake HP: 94 at 3400 rpm. Taxable HP: 28.8. Main bearings: Nine. Valve lifters: Solid. Carb.: Stromberg UUR2. Torque: 100 lbs.-ft. at 1200 rpm. [Series 970] L-head. Inline. Eight-cylinder. Cast iron block. B & S: 2-7/8 in. x 4-3/8 in. Disp.: 227.2 cid. C.R.: 5.00:1. Brake HP: 78 at 3200 rpm. Taxable HP: 26.4. Main bearings: Nine. Valve lifters: Solid. Carb.: Stromberg EE-2. Torque: 92 lbs.-ft. at 1600 rpm. [Series 960] L-head. Inline. Six. Cast iron block. B & S: 3-1/8 in. x 4-3/8 in. Disp.: 201.3 cid. C.R.: 5.00:1. Brake HP: 65 at 3200 rpm. Taxable HP: 23.4. Main bearings: Seven. Valve lifters: Solid. Carb.: Carter 1-5/16 in. Torque: 92 lbs.-ft. at 1600 rpm. [Series 1090] Overhead valve. Inline. Eight-cylinder. Cast iron block. B & S: 3-3/8 in. x 4-1/2 in. Disp.: 322.0 cid. C.R.: 5.25:1. Brake HP: 125 at 3600 rpm. Taxable HP: 36.4. Main bearings: Nine. Valve lifters: Solid. Carb.: Stromberg UUR2. [Series 1080] Overhead valve. Inline. Eight-cylinder. Cast iron block. B & S: 3-1/8 in. x 4-1/4 in. Disp.: 260.8 cid. C.R.: 5.25:1. Brake HP: 100 at 3400 rpm. Taxable HP: 31.2. Main bearings: Nine. Valve lifters: Solid. Carb.: Stromberg UUR2. [Series 1070] L-head. Inline. Eight-cylinder. Cast iron block. B & S: 3 in. x 4-3/8 in. Disp.: 247.4 cid. C.R.: 5.1:1. Brake HP: 85 3200 rpm. Taxable HP: 28.8. Main bearings: Nine. Valve lifters: Solid. Carb.: Stromberg EE-22. [Series 1060] L-head. Inline. Six. Cast iron block. B & S: 3-1/8 in. x 4-3/8 in. Disp.: 201.3 cid. C.R.: 5.1:1. Brake HP: 70 at 3000 rpm. Taxable HP: 23.4. Main bearings: Seven. Valve lifters: Solid. Carb.: Stromberg E2.
CHASSIS: [Series 990] W.B.: 124 in. and 133 in. Front tread/rear tread: 56-1/4 in./58 in. Tires: 6.50 x 19. [Series 980] W.B.: 121 in. Front tread/rear tread: 55-15/16 in./58-1/8 in. Tires: 6.00 x 18. [Series 970] W.B.: 116-1/4 in. Front tread/rear tread: 56 in./58-3/8 in. Tires: 5.25 x 19. [Series 960] W.B.: 114-1/4 in. Front tread/rear tread: 56 in./58-3/8 in. Tires: 5.00 x 19. [Series 1090] W.B.: 133 in. & 142 in. Front tread/rear tread: 57-1/4 in./58 in. Tires: 7.00 x 18. [Series 1080] W.B.: 128 in. Front tread/rear tread: 58-3/4 in./60-3/4 in. Tires: 6.50 x 17. [Series 1070] W.B.: 121 in. Front tread/rear tread: 56-1/2 in./59-3/8 in. Tires: 5.50 x 18. [Series 1060] W.B.: 116 in. Front tread/rear tread: 56-1/2 in./60 in. Tires: 5.25. x 18.
TECHNICAL: Selective sliding gear transmission. Speeds: 3F/1R. Floor shift controls. Single dry plate clutch. [1080/1090] Worm drive. [All others] Spiral bevel drive. Semi-floating rear axle. Overall ratio: [990/1909] 4.50:1; [980] 4.46:1; [1080] 4.43:1; [970/960] 4.73:1; [1070] 4.44:1; [1060] 4.70:1. Mechanical brakes on four wheels. Artillery or wire wheels. Wheel size: [990/ 970/960] 19 in.; [980/1090/1070/1060] 18 in.; [1080] 17 in. Drivetrain Options: Freewheeling.
OPTIONS: Single sidemount. Dual sidemount. Sidemount cover(s).
HISTORICAL: Introduced: [First Series] June 1, 1931; [Second series] March 1, 1932. Innovations: Larger engines in all second series cars. Nash's first four-door convertible sedans [1090 and 1080 series]. Worm drive [1080/1090 series]. Calendar year registrations: 20,233 [both series]. Calendar year production: 17,696 [both series]. The president of Nash was E.H. McCarty. First time Nash offered two series in one model year. A dismal year for sales, due to the depression, but Nash made money in 1932. It was the only auto

manufacturer, apart from General Motors, to do so. Profit for the fiscal year came to $1,029,552. Nash held 1.85 percent of industry registrations for the year.

1933 NASH

NASH — AMBASSADOR EIGHT — SERIES 1190: Identical in appearance to the 1090 series of 1932, except for a medallion with the number "8" mounted on headlamp tie bar.
NASH — ADVANCED EIGHT — SERIES 1180: Similar in appearance to 1190 series, but smaller; less brightwork on radiator shell than 1190.
NASH — SPECIAL EIGHT — SERIES 1170: Identical in appearance to series 1070 standard eight of 1932.
NASH — STANDARD EIGHT — SERIES 1130: Identical in appearance to 1932 series 1060 Big Six.
NASH — BIG SIX — SERIES 1120: Identical in appearance to 1932 series 1060 Big Six.

1933 Nash, convertible sedan (OCW)

I.D. DATA: [Series 1190] Serial numbers on frame, opposite starting motor. Starting: 521191. Ending: 521800. Engine numbers on crankcase bell housing. Starting: 404491. Ending: 404992. [Series 1180] Serial numbers on frame, opposite starting motor. Starting: B70021. Ending: B70800. Engine numbers on crankcase bell housing. Starting: B82671. Ending: B83420. [Series 1170] Serial numbers on frame, right-side. Starting: X25387. Ending: X26099. Engine numbers on left-side near generator. Starting: XE25413. Ending: XE28189. [Series 1130] Serial numbers on frame, right-side. Starting: X26100. Ending: X28100. Engine numbers on right-side below valve cover. Starting: XE26000. Ending: XE28001. [Series 1120] Serial numbers on frame, right-side. Starting: R274300. Ending: R278900. Engine numbers on right-side below valve cover. Starting: E25700. Ending: E30211.

Model No.	Body Type & Seating	Price	Weight	Start. Ser. No.
1190	4dr Sedan-5P	1575	4350	521191
1191	2dr Convertible Roadster-2/4P	1645	4270	521191
1192R	2dr Cpe-2/4P	1545	4300	521191
1193	4dr Convertible Sedan-5P	1875	4470	521191
1194	4dr Sedan-7P	1955	4600	521191
1195	4dr Limousine-7P	2055	4650	521191
1196	4dr LWB Sedan-5P	1855	4510	521191
1197	4dr Brougham-5P	1820	4470	521191
1199	2dr Victoria-5P	1785	4300	521191
1180	4dr Sedan-5P	1320	4000	B70021
1181	2dr Convertible Roadster-2/4P	1395	3750	B70021
1182	2dr Cpe-2P	1255	3710	B70021
1182R	2dr Cpe-2/4P	1275	3800	B70021
1183	4dr Convertible Sedan-5P	1575	3870	B70021
1189	2dr Victoria-5P	1395	3840	B70021
1170	4dr Sedan-5P	1015	3400	X25387
1171	2dr Convertible Roadster-2/4P	1055	3270	X25387
1172	2dr Cpe-2P	965	3250	X25387
1172R	2dr Cpe-2/4P	1015	3300	X25387
1173	2dr Convertible Sedan-5P	1095	3275	X25387
1177	4dr Town Sedan-5P	975	3400	X25387
1130	4dr Sedan-5P	845	3200	X26100
1131	2dr Convertible Roadster-2/4P	900	3050	X26100
1132	2dr Cpe-2P	830	3050	X26100
1132R	2dr Cpe-2/4P	845	3100	X26100
1133	2dr Convertible Sedan-5P	945	3150	X26100
1137	4dr Town Sedan-5P	830	3175	X26100
1120	4dr Sedan-5P	745	3125	R274300
1121	2dr Convertible Roadster-2/4P	810	3000	R274300
1122	2dr Cpe-2P	725	3000	R274300
1122R	2dr Cpe-2/4P	745	3050	R274300
1123	2dr Convertible Sedan-5P	845	3100	R274300
1127	4dr Town Sedan-5P	695	3125	R274300

ENGINE: [Series 1190] Overhead valve. Inline. Eight. Cast iron block. B & S: 3-3/8 in. x 4-1/2 in. Disp.: 322.0 cid. C.R.: 5.2:1. Brake HP.: 125 at 3600 rpm. Taxable HP.: 36.4. Main bearings: Nine. Valve lifters: Solid. Carb.: Stromberg UUR-2. [Series 1180] Overhead valve. Inline. Eight. Cast iron block. B & S: 3-1/8 in. x 4-1/4 in. Disp.: 260.8 cid. C.R.: 5.2:1. Brake HP.: 100 at 3200 rpm. Taxable HP.: 31.2. Main bearings: Nine. Valve lifters: Solid. Carb.: Stromberg UUR-2. [Series 1170] L-head Inline. Eight. Cast iron block. B & S: 3 in. x 4-3/8 in. Disp.: 247.4 cid. C.R.: 5.1:1. Brake HP.: 85 at 3200 rpm. Taxable HP.: 28.8. Main bearings: Seven. Valve lifters: Solid. Carb.: 1-7/16 in. Stromberg. [Series 1130] L-head Inline. Eight. Cast iron block. B & S: 3 in. x 4-3/8 in. Disp.: 247.4 cid. C.R.: 5.1:1. Brake HP.: 80 at 3200 rpm. Taxable HP.: 28.8. Main bearings: Nine. Valve lifters: Solid. Carb.: Stromberg EX-2. Torque: 92 lbs.-ft. at 1600 rpm. [Series 1120] L-head. Inline. Six. Cast iron block. B & S: 3-1/4 in. x 4-3/8 in. Disp.: 217.8 cid. C.R.: 5.3:1. Brake HP.: 75 at 3200 rpm. Taxable HP.: 25.3. Main bearings: Seven. Valve lifters: Solid. Carb.: Stromberg EX-2.

CHASSIS: [Series 1190] W.B.: 133 in. and 142 in. Tires: 7.00 x 18. [Series 1180] W.B.: 128 in. Tires: 6.50 x 17. [Series 1170] W.B.: 121 in. Tires: 5.50 x 18. [Series 1130] W.B.: 116 in. Tires: 5.50 x 17. [Series 1120] W.B.: 116 in. Tires: 5.50 x 17.

TECHNICAL: Selective sliding gear transmission. Speeds: 3F/1R. Floor shift controls. Single dry plate clutch. [1190/1180] Worm drive; [all others] Spiral bevel. Semi-floating rear axle. Overall ratio: [1190] 4.5:1; [1180] 4.71:1; [1170/1130] 4.44:1; [1120] 4.70:1. Mechanical brakes on four wheels. Wire wheels. Wheel size: [1190/1170] 18 in.; [all others] 17 in.

OPTIONS: Single sidemount. Dual sidemounts. Sidemount cover(s).

HISTORICAL: Introduced Dec. 1, 1932. Innovations: "Big Six" engine bored, horsepower raised from 70 to 75. Calendar year registrations: 11,353. Calendar year production: 14,973. The president of Nash was E.H. McCarty. Some badge engineering and a new eight on the "Big Six" chassis. Otherwise the 1933 line was carried over from the 1932 second series. Nash production lowest since 1918, and for the first time in its history, the company lost money. (The loss came to $1,188,863.) Share of new car registrations: 0.76 percent.

1933 Nash, Model 1181 convertible roadster (AB)

1934 NASH

1934 Nash Ambassador four-door sedan with C.W. Nash and E.H. McCarty (AB)

NASH — AMBASSADOR EIGHT — SERIES 1290: High-styled body, designed by Count Alexis De Sakhnoffsky, featuring deep-skirted fenders; ribs running length of hood; horizontal door ventilators on hood sides; chromed, bullet-shaped headlamps; full beavertail rear section.

NASH — ADVANCED EIGHT — SERIES 1280: Identical in appearance to Ambassador models (1290 series), but shorter.

NASH — BIG SIX — SERIES 1220: Identical in appearance to 1280 series, but shorter hood and front fenders.

LA FAYETTE — SERIES 110: Similar in design to Nash models, but smoother, less elaborate. (No embossing on hood and fenders, for instance.) Painted headlamp shells on standard models, chromed on others.

I.D. DATA: [Series 1290] Serial numbers on right-side frame, under hood. Starting: 521801. Ending: 523253. Engine numbers on right-side engine block. Starting: 405819. Ending: 406555. [Series 1280] Serial numbers on right-side frame, under hood. Starting: B70801. Ending: B75001. Engine numbers on right-side engine block. Starting: B85513. Ending: B87709. [Series 1220] Serial numbers on right-side frame, under hood. Starting: R278901. Ending: R294724. Engine numbers on right-side engine block. Starting: E35651. Ending: E46124. LaFayette serial numbers on right frame, under hood. Starting: L1001. Ending: L13700. Engine numbers on right front of engine block, just below valve cover. Starting: LE501. Ending: LE13200.

Model No.	Body Type and Seating	Price	Weight	Start. Ser. No.
1290	4dr Sedan-5P	1575	4330	521801
1293	4dr Brougham-5P	1625	4360	521801
1294	4dr Sedan-7P	1955	4590	521801
1295	4dr Limousine-7P	2055	4640	521801
1296	4dr LWB Sedan-5P	1955	4500	521801
1297	4dr LWB Brougham-5P	1820	4460	521801
1280	4dr Sedan-5P	1065	3540	B70801
1282	2dr Cpe-2P	1045	3460	B70801
1282R	2dr Cpe-2/4P	1065	3510	B70801
1283	4dr Brougham-5P	1085	3570	B70801
1287	4dr Town Sedan-5P	1035	3540	B70801
1288	4dr Brougham Sedan-5P	1145	3570	B70801
1220	4dr Sedan-5P	785	3370	R278901
1222	2dr Cpe-2P	765	3290	R278901
1222R	2dr Cpe-2/4P	785	3340	R278901
1223	4dr Brougham-5P	795	3400	R278901
1227	4dr Town Sedan-5P	745	3370	R278901
1228	4dr Brougham Sedan-5P	865	3400	R278901
110	4dr Sedan-5P	695	3030	L1001
112	2dr Cpe-2P	635	2925	L1001
112R	2dr Cpe-2/4P	675	2970	L1001
113	4dr Brougham-5P	745	3050	L10131
115	2dr Tr Sedan-5P	685	3030	L1001
116	2dr Std Sedan-5P	595	2970	L1001
117	4dr Std Sedan-5P	645	3000	L4026
118	4dr Std Brougham-5P	695	3050	L10133

ENGINE: [Series 1290] Overhead valve. Inline. Eight-cylinder. Cast iron block. B & S: 3-3/8 in. x 4-1/2 in. Disp.: 322.0 cid. C.R.: 5.25:1. Brake H.P.: 125 at 3600 rpm. Taxable H.P.: 36.4. Main bearings: Nine. Valve lifters: Solid. Carb.: Stromberg UUR-2. Torque: 115 lbs.-ft. at 400 rpm. [Series 1280] Overhead valve. Inline. Eight-cylinder. Cast iron block. B & S: 3-1/8 in. x 4-1/4 in. Disp.: 260.8 cid. C.R.: 5.25:1. Brake HP: 100 at 3400 rpm. Taxable HP: 31.2. Main bearings: Nine. Valve lifters: Solid. Carb.: Stromberg EE-22. Torque: 110 lbs.-ft. at 350 rpm. [Series 1220] Overhead valve. Inline. Six-cylinder. Cast iron block. B & S: 3-3/8 in. x 4-3/8 in. Disp.: 234.8 cid. C.R.: 5.25:1. Brake HP: 88 at 3200 rpm. Taxable HP: 27.3. Main bearings: Seven. Valve lifters: Solid. Carb.: Stromberg EX-32. Torque: 100 lbs.-ft. at 350 rpm. [LaFayette] L-head. Inline. Six-cylinder. B & S: 3-1/4 in. x 4-3/8 in. Disp.: 217.8 cid. C.R.: 5.3:1. Brake HP: 75 at 3200 rpm. Taxable HP: 25.3. Main bearings: Seven. Valve lifters: Solid. Carb.: Marvel B.
CHASSIS: [Series 1290] W.B. 133 in. and 142 in. O.L.: 214-1/4 in. Front/rear tread: 57-1/4 in./60-1/4 in. Tires: 7.00 x 17. [Series 1280] W.B.: 121 in. O.L.: 198-3/16 in. Front/rear tread: 57-7/8 in./60 in. Tires: 6.50 x 16. [Series 1220] W.B.: 116 in. O.L.: 194-1/2 in. Front/rear tread: 57-7/8 in./60 in. Tires: 5.50 x 17 (6.25 x 16 opt.). [Series LaFayette (110)] W.B.: 113 in. Front/rear tread: 56-1/2 in./59-11/16 in. Tires: 5.50 x 17 (6.25 x 16 optional).
TECHNICAL: Selective sliding gear transmission. Speeds: 3F/1R. Floor shift controls. Single dry plate clutch. [Series 1290] Worm drive; [All others] Spiral bevel drive. Semi-floating rear axle. Overall ratio: [1290] 4.72:1; [1280] 4.10:1; [1220] 4.44:1; [LaFayette] 4.70:1. Mechanical brakes on four wheels. Steel artillery wheels. Wheel size: [Series 1280] 16 in.; [Others] 17 in. Baker axleflex independent front suspension.
OPTIONS: Single sidemount. Dual sidemounts. Sidemount cover(s). Fender skirts. Radio. Heater. Clock. Cigar lighter. Radio antenna. Trunk. Trunk rack.
HISTORICAL: Introduced: Oct. 1, 1933 (Nash), Jan. 10, 1934 (LaFayette). Innovations: First year for low-priced LaFayette. All Nash cars now using Overhead valve engines. L-head engine of previous Big Six now used in LaFayette. Draft-free ventilation system all models. Calendar year registrations: 14,315 (Nash), 9,301 (LaFayette). Calendar year production: 28,664 (Nash and LaFayette). The president of Nash was E.H. McCarty. The one millionth Nash — a 1227 Town Sedan — was produced on April 27, 1934. A contest was held to find the oldest Nash automobile still in use by its original owner. Nash no. 1,000,000 was the prize. It went to Dr. E. O. Nash (no relation to Charlie Nash) of Pueblo, Col. His car, the 517th Nash built, had traveled 215,580 miles. Losses for the fiscal year came to $1,625,078. Share of industry registrations: 1.25 percent (Nash and LaFayette combined). Styled by Count Alexis De Sakhnoffsky.

1935 NASH

1935 Nash, four-door sedan (JAC)

NASH — AMBASSADOR and ADVANCED EIGHTS — SERIES 3580: Fastback "Aeroform" styling; streamlined fenders; sharply sloping pressed steel grille; recessed spare tire behind flush door; teardrop headlamps. Ambassador name applied to top trim line.
NASH — ADVANCED SIX — SERIES 3520: Identical in appearance to series 3580, except for shorter hood and front fenders.

LAFAYETTE — SERIES 3510: Styling similar to 1934 LaFayette, with the following changes: horizontal louvers replaced vent doors in hood sides; extremely convex headlamp lenses.
I.D. DATA: [Series 3580] Serial numbers on right-side frame, under hood. Starting: B75010. Ending: B77324. Engine numbers on right-side engine block. Starting: B87710. Ending: B90024. [Series 3520] Serial numbers on right-side frame, under hood. Starting: R294725. Ending: R303300. Engine numbers on right-side engine block. Starting: E46125. Ending: E54700. LaFayette serial numbers on right frame, under hood. Starting: L13701. Ending: L23100. Engine numbers on right front of engine block, just below valve cover. Starting: LE13201. Ending: LE22600.

Model No.	Body Type and Seating	Price	Weight	Start. Ser. No.
3580	4dr Sedan-6P	1165	3750	B75010
3585	2dr Victoria-6P	1115	3660	B75010
3588	4dr Ambassador Sedan-6P	1290	3750	B75010
3589	2dr Ambassador Victoria-6P	1240	3660	B75010
3520	4dr Sedan-6P	945	3630	R294725
3525	2dr Victoria-6P	895	3540	R294725
3510	4dr Std Sedan-5P	670	3000	L13701
3512	2dr Std Cpe-2P	585	2925	L13701
3512R	2dr Spt Cpe-2/4P	700	2970	L13701
3513	4dr Std Brougham-5P	700	3050	L13701
3515	2dr Std Tr Sedan-5P	650	3030	L13701
3516	4dr Std Sedan-5P	620	2970	L13701
3517	4dr Spl Sedan-5P	720	3030	L13701
3518	4dr Spl Brougham-5P	750	3050	L13701

ENGINE: [Series 3580] Overhead valve. Inline. Eight-cylinder. Cast iron block. B & S: 3-1/8 in. x 4-1/4 in. Disp.: 260.8 cid. C.R.: 5.25:1. Brake HP: 100 at 3400 rpm. Taxable HP: 31.25. Main bearings: Nine. Valve lifters: Solid. Carb.: Stromberg EE-22. Torque: 110 lbs.-ft. at 350 rpm. [Series 3520] Overhead valve. Inline. Six-cylinder Cast iron block. B & S: 3-3/8 in. x 4-3/8 in. Disp.: 234.8 cid. C.R.: 5.25:1. Brake HP: 88 at 3200 rpm. Taxable HP: 27.3. Main bearings: Seven. Valve lifters: Solid. Carb.: Stromberg EX-32. Torque: 100 lbs.-ft. at 350 rpm. [LaFayette] L-head. Inline. Six-cylinder Cast iron block. B & S: 3-1/4 in. x 4-3/8 in. Disp.: 217.7 cid. C.R.: 5.54:1. Brake HP: 80 at 3200 rpm. Taxable HP.: 25.3. Main bearings: Seven. Valve lifters: Solid. Carb.: Marvel B.
CHASSIS: [Series 3580] W.B. 125 in. O.L.: 207 in. Front/rear tread: 57-7/8 in./60 in. Tires: 6.50 x 16. [Series 3520] W.B.: 120 in. O.L.: 202 in. Front/rear tread: 57-7/8 in./60 in. Tires: 6.25 x 16. [Series 3510 (LaFayette)] W.B.: 113 in. O.L.: 189-1/2 in. Front/rear tread: 56-1/2 / 59-11/16. Tires: (5.50 x 17 (Standard). (6.00 x 16 (Special).
TECHNICAL: Selective sliding gear transmission. Speeds: 3F/1R. Floor shift controls. Single dry plate clutch. Spiral bevel drive. Semi-floating rear axle. Overall ratio: [3500] 4.1:1; [3520] 4.4:1 (3520); [LaFayette] 4.7:1. Hydraulic brakes (Nash), Mechanical brakes (LaFayette) on four wheels. Steel artillery wheels. Wheel size: 16 in. (LaFayette standard model — 17 in.) Overdrive (Nash models only).
OPTIONS: Single sidemount. Dual Sidemounts (LaFayette only). Sidemount cover(s). Fender skirts. Radio. Heater. Clock. Cigar Lighter. Radio Antenna.
HISTORICAL: Introduced Jan. 1, 1935. Innovations: Big, 322 cid "90" series gone from lineup. Number of body styles sharply reduced. First use of hydraulic brakes (Nash only; LaFayette still used mechanicals). First year for all-steel body (Nash only). Calendar year registrations: 17,739 Nash, 17,445 LaFayette. Calendar year production: 44,637 (Nash and LaFayette). The president of Nash was E.H. McCarty. Loss for the year came to $610,227. Share of industry registrations: 1.29 percent.

1935 Nash LaFayette, two-door coupe (OCW)

1936 NASH

NASH — AMBASSADOR SUPER EIGHT — SERIES 3680: Similar to 1935 model, but with steel top; die-cast chrome waterfall grille extending to top of hood; chromed, zeppelin-shaped die-cast vent grilles on hood sides; extruded trunk lid.
NASH — AMBASSADOR SIX — SERIES 3620: Identical in appearance to series 3680 Ambassador Super Eight.
NASH 400 — SERIES 3640: Sharply sloping "alligator" hood with pressed steel grille; embossed steel disc wheels; fastback styling with no outside opening to luggage/spare tire compartment (sedan, Victoria); extruded trunk lid (touring sedan and Victoria); seamless steel top.
NASH 400 DELUXE — SERIES 3640A: Similar to series 3640 with the following exceptions: chromed, die-cast waterfall grille extending to top of hood. Side-opening hood. Chromed, zeppelin-shaped die-cast vent grilles on hood sides. Steel artillery wheels.
LAFAYETTE — SERIES 3610: First series: pressed steel grille, embossed disc wheels. Second series: die-cast, chromed waterfall grille, steel artillery wheels. Both series: styling similar to 3640 and 3640A series, except for different front end treatment, side-opening hood.

1936 Nash, 400 four-door sedan (OCW)

I.D. DATA: [Series 3680] Serial numbers on right-side frame, under hood. Starting: B77325. Ending: B80026. Engine numbers on right-side engine block. Starting: B90025. Ending: B92726. [Series 3620] Serial numbers on right-side frame, under hood. Starting: R303301. Ending: R309300. Engine numbers on right-side engine block. Starting: E54701. Ending: E60700. [Series 3640] Serial numbers on right-side frame, under hood. Starting: C1001. Ending: C9500. Engine numbers on right-side engine block. Starting: CE501. Ending: CE9000. [Series 3640A] Serial numbers on right-side frame, under hood. Starting: C9501. Ending: C23000. Engine numbers on right-side engine block. Starting: CE9001. Ending: CE22500. [LaFayette] Serial numbers on right frame, under hood. Starting: L23101. Ending: L50780. Engine numbers on right front of engine block, just below valve cover. Starting: LE22601. Ending: LE50277.

Model No.	Body Type and Seating	Price	Weight	Start. Ser. No.
3680	4dr Sedan-6P	995	3820	B177325
3685	2dr Victoria-6P	945	3730	B177325
3620	4dr Sedan-6P	885	3710	R303301
3625	2dr Victoria-6P	835	3629	R303301
3640	4dr Sedan-6P	765	2970	C1001
3642	2dr Cpe-3P	675	2900	C1001
3642R	2dr Cpe-3/5P	725	2960	C1001
3643	2dr Tr Victoria-6P	745	2970	C1001
3645	2dr Victoria-6P	715	2950	C1001
3648	4dr Tr Sedan-6P	790	3000	C1001
3640A	4dr Sedan-6P	765	3020	C9501
3641A	2dr Cabr-3/5P	800	3000	C9501
3642A	2dr Cpe-3P	675	2950	C9501
3642AR	2dr Cpe-3/5P	725	3010	C9501
3643A	2dr Tr Victoria-6P	745	3020	C9501
3645A	2dr Victoria-6P	715	3000	C9501
3648A	4dr Tr Sedan-6P	790	3050	C9501
3610	4dr Sedan-6P	675	2950	L23101
3611	2dr Cabr-3/5P	740	2930	L23101
3612	2dr 3W Cpe-3P	595	2880	L23101
3612R	2dr 3W Cpe-3/5P	650	2930	L23101
W3612	2dr 5W Cpe-3P	610	2880	L23101
W3612R	2dr 5W Cpe-3/5P	665	2930	L23101
3613	2dr Tr Victoria-6P	655	2950	L23101
3615	2dr Victoria-6P	625	2930	L23101
3618	4dr Tr Sedan-6P	700	2980	L23101

NOTE 1: Questionable whether Model 3685 was actually produced.

1936 Nash, Lafayette Sport Cabriolet (HAC)

ENGINE: [Series 3680] Overhead valve. Inline. Eight-cylinder. Cast iron block. B & S: 3-1/8 in. x 4-1/4 in. Disp.: 260.8 cid. C.R.: 5.25:1. Brake HP.: 102 at 3400 rpm. Taxable HP.: 31.25. Main bearings: Nine. Valve lifters: Solid. Carb.: Stromberg EE-1. Torque: 110 lbs.-ft. at 350 rpm. [Series 3620] Overhead valve. Inline. Six-cylinder Cast iron block. B & S: 3-3/8 in. x 4-3/8 in. Disp.: 234.8 cid. C.R.: 5.70:1. Brake HP.: 93 at 3400 rpm. Taxable HP.: 27.3. Main bearings: Seven. Valve lifters: Solid. Carb.: Stromberg Ex-2 or AX-2. Torque: 125 lbs.-ft. at 350 rpm. [Series 3640] L-head. Inline. Six-cylinder Cast iron block. B & S: 3-3/8 in. x 4-3/8 in. Disp.: 234.8 cid. C.R.: 5.61:1. Brake HP.: 90 at 3400 rpm. Taxable HP.: 27.3. Main bearings: Seven. Valve lifters: Solid. Carb.: Stromberg EX-22. Torque: 125 lbs.-ft at 350 rpm. [Series 3640A] (Mechanically identical to Series 3640.) [LaFayette] L-head. Inline. Six-cylinder Cast iron block. B & S: 3-1/4 in. x 4-3/8 in. Disp.: 217.7 cid. C.R.: 5.61:1. Brake HP.: 83 at 3200 rpm. Taxable HP.: 25.3. Main bearings: Seven. Valve lifters: Solid. Carb.: Marvel B2 (first series), Stromberg AX2 (second series).

CHASSIS: [Series 3680] W.B.: 125 in. O.L.: 207-1/8 in. Front/rear tread: 58 in./60 in. Tires: 6.50 x 16. [Series 3620] W.B.: 125 in. O.L.: 207-1/8 in. Front/rear tread: 58 in./60 in. Tires: 6.25 x 16. [Series 3640, 3640A] W.B.: 117 in. O.L.: 191-1/8 in. Front/rear tread: 58 in./60-1/4 in. Tires: 6.00 x 16. [Series 3610] W.B.: 113 in. Front/rear tread: 58 in./60-1/4 in. Tires: 6.00 x 16.

TECHNICAL: Selective sliding gear transmission. Speeds: 3F/1R. Floor shift controls. Single dry plate clutch. Spiral bevel drive. Semi-floating rear axle. Overall ratio: 4.44 (Ambassador Super 8), 4.11:1 all others. Hydraulic brakes on four wheels. Steel artillery wheels (steel disc, series 3640 and first series LaFayette). Wheel size: 16 in.

OPTIONS: Fender skirts. Radio. Heater. Clock. Cigar lighter. Radio antenna.

HISTORICAL: Introduced: Nov. 15, 1935 [Ambassadors, 3680 and 3610]; May 20, 1935 [(3640]; Juno 15, 1935 [(LaFayette]; Oct. 15, 1935 [3640A and second series LaFayette] Innovations: Nash "400" series first to feature cast intake manifolds in engine block. "Double Bed" conversion offered in 400 and LaFayette sedans and Victorias. Calendar year registrations: 43,070. Calendar year production: 53,038. The president of Nash was E.H. McCarty. Nash bought the remaining half interest in the Seaman Body Corp. during 1936. A profit of $1,020,708 was posted for the year. Share of industry registrations: 1.27 percent.

1937 NASH

1937 Nash, Model 3788 sedan and Babe Ruth (JAC)

NASH — AMBASSADOR EIGHT — SERIES 3780: Veed, die-cast grille. Vertical grille bars. Highly ornate radiator ornament. Chrome spear on hood sides. Integral trunk on sedans and Victorias. Split windshield.

NASH — AMBASSADOR SIX — SERIES 3720: Identical to series 3780 except with less elaborate radiator ornament consisting of stylized wing; shorter hood and front fenders.

LAFAYETTE — SERIES 3710 — SIX: Identical in styling to senior Nashes with the following exceptions: radiator ornament with circular theme; horizontal bars in grille; chevrons on hood sides; shorter hood and front fenders than Ambassador series.

I.D. DATA: [Series 3780] Serial numbers on right-side frame, under hood. Starting: B80031. Ending: B86030. Engine numbers on right-side engine block. Starting: B92731. Ending: B98730. [Series 3720] Serial numbers on right-side frame, under hood. Starting: R309311. Ending: R324310. Engine numbers on right-side engine block. Starting: E60711. Ending: E75710. [LaFayette] Serial numbers on right-side frame under hood. Starting: L50781 or H1001. Ending: L106280 or H10500. NOTE: Prefix "H" indicates car was assembled in Kenosha. "L" indicates assembled in Racine. Engine numbers right-side engine block. Starting: LE50281 or H1001 or HE501. Ending: LE105780 or H10500 or HE10000.

Model No.	Body Type & Seating	Price	Weight	Start. Ser. No.
3781	2dr Cabriolet-3/5P	960	3640	B80031
3782	2dr Cpe-3P	855	3590	B80031
3782R	2dr Cpe3/5P	895	3640	B80031
3782A	2dr All-Purpose Cpe3/5P	910	3610	B80031
3783	2dr Victoria-6P	895	3690	B80031
3788	4dr Sedan-6P	945	3720	B80031
3721	2dr Cabriolet-3/5P	860	3320	R309311
3722	2dr Cpe-3P	755	3290	R309311
3722R	2dr Cpe-3/5P	795	3320	R309311
3722A	2dr All-Purpose Cpe-3/5P	810	3310	R309311
3723	2dr Victoria-6P	795	3380	R309311
3728	4dr Sedan-6P	845	3400	R309311
3711	2dr Cabriolet-3/5P	740	3180	L50781
3712	2dr Cpe-3P	595	3140	L50781
3712R	2dr Cpe-3/5P	650	3190	L50781
3712A	2dr All-Purpose Cpe-3/5P	660	3160	L50781
3713	2dr Victoria-6P	650	3200	Note 1
3718	4dr Sedan-6P	700	3240	Note 1

NOTE 1: L50781 or H1001.
NOTE 2: All-Purpose Coupes replaced Rumbleseat Coupes.

ENGINE: [Series 3780] Overhead valve. Inline. Eight-cylinder. Cast iron block. B & S: 3-1/8 in. x 4-1/4 in. Disp.: 260.8 cid. C.R.: 5.64:1. Brake HP.: 105 at 3400 rpm. Taxable HP.: 31.25. Main bearings: Nine. Valve lifters: Solid. Carb.: Stromberg EE-1. Torque: 196 lbs.-ft. at 1800 rpm. [Series 3720] Overhead valve. Inline. Six. Cast iron block. B & S: 3-3/8 in. x 4-3/8 in. Disp.: 234.8 cid. C.R.: 5.67:1. Brake HP.: 93 at 3400 rpm. Taxable HP.: 27.3. Main bearings: Seven. Valve lifters: Solid. Carb.: Stromberg EX-32. Torque: 174 lbs.-ft. 1600 rpm. [LaFayette] L-head. Inline. Six. Cast iron block. B & S: 3-3/8 in. x 4-3/8 in. Disp.: 234.8 cid. C.R.: 5.6:1. Brake HP.: 90 at 3400 rpm. Taxable HP.: 27.3. Main bearings: Seven. Valve lifters: Solid. Carb.: Stromberg AX-2. Torque: 171 lbs.-ft. at 1200 rpm.

CHASSIS: [Series 3780] W.B.: 125 in. O.L.: 204-7/16 in. Front/rear tread: 58 in./60 in. Tires: 7.00 x 16. [Series 3720] W.B.: 121 in. O.L.: 200-7/16 in. Front/rear tread: 58 in./60-1/4 in. Tires: 6.25 x 16. [Series 3710] W.B.: 117 in. O.L.: 196-7/16 in. Front/rear tread: 58 in./60-1/4 in. Tires: 6.00 x 16.

1937 Nash, Ambassador Six Cabriolet (AB)

TECHNICAL: Selective sliding gear transmission. Speeds: 3F/1R. Floor shift controls. Single dry plate clutch. Spiral bevel drive. Semi-floating rear axle. Overall ratio: 4.11:1. Hydraulic brakes on four wheels. Steel artillery wheels. Wheel size: 16 in. Overdrive.
OPTIONS: Fender skirts. Radio. Heater. Clock. Cigar Lighter. Radio Antenna. Bed Conversion.
HISTORICAL: Introduced: Oct. 1, 1936. Calendar year registrations: 70,571. Calendar year production: 85,949. The president of Nash was George Mason. Merger of Nash Motors with Kelvinator Corp. to form Nash-Kelvinator Corp. effected this year. Nash "400" and LaFayette combined into a single series known (for 1937 only) as the "Nash-LaFayette 400." Profit for the year came to $3,640,747. Share of industry registrations: 2.03 percent.

1938 NASH

1938 Nash, four-door sedan (JAC)

NASH — AMBASSADOR EIGHT — SERIES 3880 — EIGHT: Body styling similar to 1937 model, but with painted, bright-trimmed grille with horizontal bars. Series name on side of hood, wedge-shaped headlamps mounted on radiator sides. "Nash" in vertical letters at top of radiator grille.
NASH — AMBASSADOR SIX — SERIES 3820: Identical in appearance to series 3880, except for shorter hood and front fenders.
NASH — LAFAYETTE — SERIES 3810 — SIX: Identical in appearance to senior Nashes, except for bullet-shaped headlamps and shorter hood/front fenders.
I.D. DATA: [Series 3880] Serial numbers on right front frame member. Starting: B86031. Ending: B88999. Engine numbers on right front side, engine block. Starting: B98731. Ending: B101699. [Series 3820] Serial numbers on right front frame member. Starting: R324311. Ending: R331399. Engine numbers on engine block, right front. Starting: E75711. Ending: E82799. [LaFayette] Serial numbers on right front frame member. Starting: L106281 or H10501. Ending: L128294 or H19449. Note: "L" indicates car assembled in Racine; "H" indicates assembly in Kenosha. Engine numbers on engine block, right front. Starting: LE105781 or HE10001. Ending: LE128424 or HE18949.

Model No.	Body Type & Seating	Price	Weight	Start. Ser. No.
3881	2dr Cabriolet-3/5P	1240	3620	B86031
3882	2dr All-Purpose Cpe-3/5P	1165	3640	B86031
3883	2dr Victoria-6P	1150	3780	B86031
3885	2dr Cpe-3P	1120	3580	B86031
3888	4dr Sedan-6P	1200	3790	B86031
3821	2dr Cabriolet-3/5P	1090	3340	R324311
3822	2dr All-Purpose Cpe-3/5P	1015	3360	R324311
3823	2dr Victoria-6P	1000	3450	R324311
3825	2dr Cpe-3P	970	3300	R324311
3828	4dr Sedan-6P	1050	3460	R324311
3811	2dr Del Cabriolet-3/5P	940	3240	Note 1
3812	2dr Del All-Purpose Cpe-3/5P	860	3230	Note 1
3813	2dr Del Victoria-6P	855	3290	Note 1
3814	2dr Del Cpe-3P	820	3160	Note 1
3815	2dr Cpe-3P	770	3120	Note 1
3816	2dr Sedan-6P	805	3190	Note 1
3817	4dr Sedan-6P	850	3200	Note 1
3818	4dr Del Sedan-6P	900	3300	Note 1

NOTE 1: L106281 or H10501.

1938 Nash, LaFayette coupe (OCW)

ENGINE: [Series 3880] Overhead valve. Inline. Eight-cylinder. Cast iron block. B & S: 3-1/8 in. x 4-1/4 in. Disp.: 260.8 cid. C.R.: 6.00:1. Brake HP.: 115 at 3400 rpm. Taxable HP.: 31.2. Main bearings: Nine. Valve lifters: Solid. Carb.: Stromberg EE7. Torque: 200 lbs.-ft. at 1200 rpm. [Series 3820] Overhead valve. Inline. Six-cylinder. Cast iron block. B & S: 3-3/8 in. x 4-3/8 in. Disp.: 234.8 cid. C.R.: 6.00:1. Brake HP.: 105 at 3400 rpm. Taxable HP.: 27.3. Main bearings: Seven. Valve lifters: Solid. Carb.: Stromberg EX32. Torque: 190 lbs.-ft at 1050 rpm. [LaFayette] L-head. Inline. Six-cylinder. Cast iron block. B & S: 3-3/8 in. x 4-3/8 in. Disp.: 234.8 cid. C.R.: 5.83:1. Brake HP.: 95 at 3400 rpm. Taxable HP.: 27.3. Main bearings: Seven. Valve lifters: Solid. Carb.: Stromberg EX22 or AX2. Torque: 175 lbs.-ft. at 1000 rpm.
CHASSIS: [Series 3880] W.B.: 125 in. O.L.: 204-11/16 in. Front/rear tread: 58 in./61-3/8 in. Tires: 7.00 x 16. [Series 3820] W.B.: 121 in. O.L.: 200-11/16 in. Front/rear tread: 58 in./60-1/4 in. Tires: 6.25 x 16. [Series 3810] W.B.: 117 in. O.L.: 196-11/16 in. Front/rear tread: 58 in./60-1/4 in. Tires: 6.00 x 16.
TECHNICAL: Selective sliding gear transmission. Speeds: 3F/1R. Floor shift controls. Single dry plate clutch. Spiral bevel drive. Semi-floating rear axle. Overall ratio: 4.11:1. Hydraulic brakes on four wheels. Steel disc wheels. Wheel size: 16 in. Drivetrain Options: Hill-holder ($10). Overdrive ($50). Dash-mounted, vacuum-operated shift ($30).
OPTIONS: Fender skirts ($13). Radio ($49). Heater ($30). Clock. Cigar lighter. Radio antenna. White sidewall tires, Model 3810 ($20). White sidewall tires, Model 3820 ($22.50). White sidewall tires, Model 3880 ($27.50). Banjo steering wheel, w/horn ring ($11). Bed conversion.
HISTORICAL: Introduced Oct. 15, 1937. Innovations: Optional dash-mounted, vacuum-controlled gearshift. Calendar year registrations: 31,814. Calendar year production: 32,017. The president of Nash was George Mason. Share of industry registrations: 1.68 percent. "400" designation dropped from LaFayette name. Essentially, the LaFayette had become the base Nash series; the radiator badge read simply "Nash."

1939 NASH

1939 Nash, Ambassador Six four-door sedan (JAC)

NASH — AMBASSADOR EIGHT — SERIES 3980: Tall, narrow grille with wide-spaced horizontal bars, suggesting that Nash styling this year may have been inspired by the LaSalle. Rectangular headlamps inset in fenders. Vertical Nash emblem at center of trunk lid. Nash script on hubcaps.
NASH — AMBASSADOR SIX — SERIES 3920: Identical in appearance to Series 3980, except for shorter hood and front fenders.
NASH — LAFAYETTE — SERIES 3910 — SIX: Identical in appearance to senior Nashes, except for shorter hood and front fenders. Small LaFayette body plate behind front fender, just above running board.
I.D. DATA: [Series 3980] Serial numbers on right front frame member. Starting: R89000. Ending: R106299. Engine numbers on engine block, right front. Starting: B101700. Ending: B105551. [Series 3920] Serial numbers on right front frame member. Starting: R331400. Ending: R339999. Engine numbers on engine block, right front. Starting: E82800. Ending: E339999. [Series 3910 LaFayette] Serial numbers on right front frame member. Starting: H19450. Ending: H56999. Engine numbers on upper left front, engine block. Starting: HE18950. Ending: HE56499.

Model No.	Body Type & Seating	Price	Weight	Start. Ser. No.
3980	4dr Tr Sedan-6P	1235	3800	B89000
3981	2dr Cabriolet-3/5P	1295	3740	B89000
3982	2dr All-Purpose Cpe-3/5P	1210	3710	B89000
3983	2dr Victoria-6P	1205	3770	B89000
3985	2dr Cpe-3P	1175	3720	B89000
3988	4dr Sedan-6P	1235	3800	B89000
3920	4dr Tr Sedan-6P	985	3470	R331400
3921	2dr Cabriolet-3/5P	1050	3430	R331400
3922	2dr All-Purpose Cpe-3/5P	960	3360	R331400
3923	2dr Victoria-6P	955	3420	R331400
3925	2dr Cpe-3P	925	3370	R331400
3928	4dr Sedan-6P	985	3450	R331400
3910	4dr Del Sedan-6P	885	3350	H19450
3911	2dr Del Cabriolet-3/5P	950	3340	H19450
3912	2dr Del All-Purpose Cpe-3/5P	860	3260	H19450
3913	2dr Del Victoria-6P	855	3320	H19450
3914	2dr Del Cpe-3P	825	3270	H19450
3915	2dr Cpe-3P	770	3200	H19450
3916	2dr Sedan-6P	810	3250	H19450
3917	4dr Sedan-6P	840	3290	H19450
3918	4dr Del Tr Sedan-6P	885	3350	H19450
3919	4dr Tr Sedan-6P	840	3285	H19450

ENGINE: [Series 3980] Overhead valve. Inline. Eight-cylinder. Cast iron block. B & S: 3-1/8 in. x 4-1/4 in. Disp.: 260.8 cid. C.R.: 6.00:1. Brake HP.: 115 at 3400 rpm. Taxable HP.: 31.2. Main bearings: Nine. Valve lifters: Solid. Carb.: Carter 436S. Torque: 200 lbs.-ft. at 1200 rpm. [Series 3920] Overhead valve. Inline. Six-cylinder. Cast iron block. B & S: 3-3/8 in. x 4-3/8 in. Disp.: 234.8 cid. C.R.: 6.00:1. Brake HP.: 105 at 3400 rpm. Taxable HP.: 27.3. Main bearings: Seven. Valve lifters: Solid. Carb.: Carter 435S. Torque: 190 lbs.-ft. at 1050 rpm. [Series LaFayette] L-head. Inline. Six-cylinder. Cast iron block. B & S: 3-3/8 in. x 4-3/8 in. Disp.: 234.8 cid. C.R.: 6.30:1. Brake HP.: 99 at 3400 rpm. Taxable HP.: 27.3. Main bearings: Seven. Valve lifters: Solid. Carb.: Stromberg EE-1. Torque: 179 lbs.-ft. at 1200 rpm.

1939 Nash Ambassador Eight, convertible coupe (HAC)

CHASSIS: [Series 3980] W.B.: 125 in. O.L.: 208-1/4 in. Front/rear tread: 55-3/4 in./61-3/8 in. Tires: 7.00 x 16. [Series 3920] W.B.: 121 in. O.L.: 204-1/4 in. Front/rear tread: 58 in./60-1/4 in. Tires: 6.25 x 16. [Series 3910] W.B.: 117 in. O.L.: 200-1/4 in. Front/rear tread: 58 in./60-1/4 in. Tires: 6.00 x 16.
TECHNICAL: Selective sliding gear transmission. 3F/1R. Steering column controls. Single dry plate clutch. Hypoid drive. Semi-floating rear axle. Overall Ratio: 4.10:1. Hydraulic brakes on four wheels. Steel disc wheels. Wheel size: 16 in. Drivetrain Options: Hill-Holder. Overdrive.
OPTIONS: Fender skirts. Radio. Heater. Clock. Cigar lighter. Radio Antenna. Fog lamps. Deluxe steering wheel. Bed conversion.
HISTORICAL: Introduced Oct. 15, 1938. Calendar year registrations: 54,050. Calendar year production: 65,662. The president of Nash was George Mason. Share of industry registrations: 2.86 percent. Styled by George W. Walker.

1940 NASH

1940 Nash, Convertible Cabriolet (AB)

NASH — AMBASSADOR EIGHT — SERIES 4080: Tall, narrow grille, similar to 1939 Model except with thin, closely-spaced horizontal bars. Sealed beam headlamps. Nash script, lower right corner of trunk lid, and on hood sides just behind grille.

NASH — AMBASSADOR SIX — SERIES 4020: Identical in appearance to 4080 Series except for shorter hood and front fenders.
NASH — LAFAYETTE — SERIES 4010 — SIX: Identical in appearance to Senior Nash Series except for shorter hood and front fenders. LaFayette body plate just above running board, behind front fender.
I.D. DATA: [Series 4080] Serial numbers on right front frame member. Starting: B106300. Ending: B110000. Engine numbers on engine block, right front. Starting: B105800. Ending: B109049. [Series 4020] Serial numbers on right front frame member. Starting: R340000. Ending: R353000. Engine numbers on engine block, right front. Starting: E339500. Ending: E352017. [Series LaFayette] Serial numbers on right front frame member. Starting: H57000. Ending: H76055. Engine numbers on upper left front, engine block. Starting: HE56500. Ending: H102862.

Model No.	Body Type & Seating	Price	Weight	Start. Ser. No.
4080	4dr Tr Sedan-6P	1195	3710	B106300
4081	2dr Cabriolet-3/5P	1295	3640	B106300
4082	2dr All-Purpose Cpe-3/5P	1170	3575	B106300
4083	2dr Sedan-6P	1165	3620	B106300
4085	2dr Cpe-3P	1135	3555	B106300
4088	4dr Sedan-6P	1195	3705	B106300
4020	4dr Tr Sedan-6P	985	3385	R340000
4021	2dr Cabriolet-3/5P	1085	3410	R340000
4022	2dr All-Purpose Cpe-3/5P	960	3295	R340000
4023	2dr Sedan-6P	955	3350	R340000
4025	2dr Cpe-3P	925	3290	R340000
4028	4dr Sedan-6P	985	3380	R340000
4010	4dr Tr Sedan-6P	875	3280	H57000
4011	2dr Cabriolet-3/5P	975	3310	H57000
4012	2dr All-Purpose Cpe-3/5P	850	3190	H57000
4013	2dr Sedan-6P	845	3235	H57000
4015	2dr Cpe-3P	795	3190	H57000
4018	4dr Sedan-6P	875	3275	H57000

1940 Nash, Ambassador Eight Cabriolet by Alexis de Sakhnoffsky (AB)

ENGINE: [Series 4080] Overhead valve. Inline. Eight-cylinder. Cast iron block. B & S: 3-1/8 in. x 4-1/4 in. Disp.: 260.8 cid. C.R.: 6.0:1. Brake HP.: 115 at 3400 rpm. Taxable HP.: 31.2. Main bearings: Nine. Valve lifters: Solid. Carb.: Carter WDO-4655. Torque: 200 lbs.-ft. at 1200 rpm. [Series 4020] Overhead valve. Inline. Six-cylinder. Cast iron block. B & S: 3-3/8 in. x 4-3/8 in. Disp.: 234.8 cid. C.R.: 6.00:1. Brake HP.: 105 at 3400 rpm. Taxable HP.: 27.3. Main bearings: Seven. Valve lifters: Solid. Carb.: Carter 435S. Torque: 190 lbs.-ft. at 1050 rpm. [Series LaFayette] L-head. Inline. Six-cylinder. Cast iron block. B & S: 3-3/8 in. x 4-3/8 in. Disp.: 234.8 cid. C.R.: 6.30:1. Brake HP.: 99 at 3400 rpm. Taxable HP.: 27.3. Main bearings: Seven. Valve lifters: Solid. Carb.: Carter 458S. Torque: 179 lbs.-ft. at 1200 rpm.
CHASSIS: [Series 4080] W.B.: 125 in. O.L.: 207-3/16 in. Front/rear tread: 57-11/16 in./61-3/8 in. Tires: 7.00 x 15. [Series 4020] W.B.: 121 in. O.L.: 203-3/16 in. Front/rear tread: 56-7/8 in./60-1/4 in. Tires: 6.25 x 16. [Series 4010] W.B.: 117 in. O.L.: 199-3/16 in. Front/rear tread: 56-7/8 in./60-1/4 in. Tires: 6.00 x 16.
TECHNICAL: Selective sliding gear transmission. Speeds: 3F/1R. Steering column controls. Single dry plate clutch. Hypoid drive. Semi-floating rear axle. Overall Ratio: 4.10:1. Hydraulic brakes on four wheels. Steel disc wheels. Wheel size: 4080: 15 in., 4020 and 4010: 16 in. Drivetrain options: Hill-Holder. Overdrive ($55).
OPTIONS: Fender skirts. Radio. Heater. Clock. Cigar lighter. Radio antenna. Bed conversion.
HISTORICAL: Introduced Sept. 15, 1939. Calendar year registrations: 52,853. Calendar year production: 63,617. The president of Nash was George Mason. Share of industry registrations: 1.5 percent. A profitable year, but a disappointing one in terms of sales.

1940 Nash, LaFayette coupe (AB)

1941 NASH

1941 Nash, four-door sedan (OCW)

NASH — AMBASSADOR EIGHT — SERIES 4180: Pointed prow; split die-cast grilles on either side, featuring thin vertical ribs. Script on trunk reading "Nash 8."
NASH — AMBASSADOR SIX — SERIES 4160: Identical to series 4180 except script on trunk reads "Nash 6."
NASH — AMBASSADOR "600" — SERIES 4140 — SIX: Identical to series 4180, except hood is substantially shorter; script on trunk reads "Nash 600;" shrouded rear fenders.
I.D. DATA: [Series 4180] Serial numbers on right front frame member. Starting: B110001. Ending: B113500. Engine numbers on engine block, right front. Starting: B110001. Ending: B113500. [Series 4160] Serial numbers on right front frame member. Starting: R353001. Ending: R383400. Engine numbers on engine block, right front. Starting: R353001. Ending: R383400. [Series 4140] Serial numbers on right frame member just ahead of dash. Starting: K5001. Ending: K55100. Engine numbers on right side of crankcase toward front. Starting: K5001. Ending: K55100.

Model No.	Body Type & Seating	Price	Weight	Start. Ser. No.
4180	4dr Tr Sedan-6P	1151	3475	B110001
4181	2dr Cabriolet-3/5P	1215	3580	B110001
4183	2dr Brougham-5P	1081	3400	B110001
4187	4dr Spl Sedan-6P	1051	3450	B110001
4188	4dr Sedan-6P	1101	3455	B110001
4160	4dr Tr Sedan-6P	1030	3300	R353001
4161	2dr Cabriolet-3/5P	1095	3430	R353001
4162	2dr Cpe-3P	905	3310	R353001
4163	2dr Brougham-5P	974	3235	R353001
4165	2dr Spl Cpe-3P	855	3180	R353001
4167	4dr Spl Sedan-6P	930	3300	R353001
4168	4dr Sedan-6P	985	3300	R353001
4169	2dr Spl Sedan-6P	898	3320	R353001
4140	4dr Tr Sedan-6P	860	2655	K5001
4142	2dr Cpe-3P	783	2500	K5001
4143	2dr Brougham-5P	810	2575	K5001
4145	2dr Spl Cpe-3P	731	2490	K5001
4146	2dr Spl Sedan-6P	745	2630	K5001
4147	4dr Spl Sedan-6P	780	2615	K5001
4148	4dr Sedan-6P	810	2630	K5001
4149	2dr Sedan-6P	777	2640	K5001

1941 Nash, Ambassador Eight Convertible Coupe (HAC)

ENGINE: [Series 4180] Overhead valve. Inline. Eight-cylinder. Cast iron block. B & S: 3-1/8 in. x 4-1/4 in. Disp.: 260.8 cid. C.R.: 6.30:1. Brake HP: 115 at 3400 rpm. Taxable HP: 31.2. Main bearings: Nine. Valve lifters: Solid. Carb.: Carter 511S. Torque: 200 lbs.-ft. at 1600 rpm. [Series 4160] Overhead valve. Inline. Six-cylinder. Cast iron block. B & S: 3-3/8 in. x 4-3/8 in. Disp.: 234.8 cid. C.R.: 6.30:1. Brake HP: 105 at 3400 rpm. Taxable HP: 27.3. Main bearings: Seven. Valve lifters: Solid. Carb.: Carter 435S. Torque: 195 lbs.-ft. at 1600 rpm. [Series 4140] L-head. Inline. Six-cylinder. Cast iron block. B & S: 3-1/8 in. x 3-3/4 in. Disp.: 172.6 cid. C.R.: 6.87:1. Brake HP: 75 at 3600 rpm. Taxable HP: 23.4. Main bearings: Four. Valve lifters: Solid. Carb.: Carter 513S. Torque: 136 lbs.-ft. at 1200 rpm.
CHASSIS: [Series 4180] W.B.: 121 in. O.L.: 200-3/4 in. Front/rear tread: 57 in./61-1/4 in. Tires: 6.50 x 16. [Series 4160] W.B.: 121 in. O.L.: 200-3/4 in. Front/rear tread: 57-1/2 in./60-1/2 in. Tires: 6.25 x 16. [Series 4140] W.B.: 112 in. O.L.: 194 in. Front/rear tread: 56 in./59-3/4 in. Tires: 5.50 x 16.
TECHNICAL: Selective sliding gear transmission. Speeds: 3F/1R. Steering column controls. Single dry plate clutch. Hypoid drive. Semi-floating rear axle. Overall ratio: 4.10:1.

Hydraulic brakes on four wheels. Steel disc wheels. Wheel size: 16 in. Drivetrain Options: Hill-holder ($13). Overdrive ($50 — 4140) ($55 — 4160/4180).
OPTIONS: Fender skirts (except 4140). Deluxe Radio ($45). Custom radio ($65). Heater (Weather-Eye) ($35). Clock. Cigar lighter. Radio antenna. Bed equipment, standard ($17.50). Bed equipment, deluxe ($24.50). Two-tone paint ($10.50). Deluxe steering wheel ($15).
HISTORICAL: Introduced Oct. 1, 1940. Innovations: Ambassador "600" (4140) series pioneered unitized body/frame construction; first low-priced car with coil springs all around; sliding pillar type independent front suspension similar to Lancia design; up to 30 mpg claimed (the 600 name came from 20 gallon fuel tank times 30 mpg or 600 miles per tankful of gas). Conventional construction and suspension used on 4160, 4180 series. Last year for "Twin Ignition." Calendar year registrations: 77,824. Calendar year production: 80,428. The president of Nash was George Mason. Share of industry registrations: 2.09 percent. The Ambassador 600 marked Nash's re-entry into the low-priced field. Sales sharply up.

1942 NASH

1942 Nash, Ambassador four-door sedan (OCW)

NASH — AMBASSADOR EIGHT — SERIES 4280: Similar to 1941 model, but with stainless steel grille at center featuring short, horizontal blades. Parking lamps in fenders, above headlamps. Chromed fender crowns and fender trim to match grille on deluxe equipped cars only. Script on trunk reads "Nash 8."
NASH — AMBASSADOR SIX — SERIES 4260: Identical in appearance to series 4280, except script on trunk reads "Nash 6."
NASH — AMBASSADOR 600 — SERIES 4240 — SIX: Identical in appearance to series 4280, except hood and front fenders are several inches shorter; script on trunk reads "Nash 600;" rear fenders shrouded.
I.D. DATA: [Series 4280] Serial numbers on right front frame member. Starting: B114001. Ending: B115000. Engine numbers on engine block, right front. Starting: B114001. Ending: B115000. [Series 4260] Serial numbers on right front frame member. Starting: R384001. Ending: R393090. Engine numbers on engine block, right front. Starting: R384001. Ending: R393090. [Series 4240] Serial numbers on right front frame member just ahead of dash. Starting: K56001. Ending: K77660. Engine numbers on right side of crankcase toward front. Starting: K56001. Ending: K77660.

Model No.	Body Type & Seating	Price	Weight	Start. Ser. No.
4280	4dr Tr Sedan-6P	1209	3465	B114001
4282	2dr Cpe-3P	1134	3350	B114001
4283	2dr Brougham-5P	1174	3385	B114001
4288	4dr Sedan-6P	1184	3465	B114001
4289	2dr Sedan-6P	1164	3485	B114001
4260	4dr Tr Sedan-6P	1159	3335	R384001
4262	2dr Cpe-3P	1084	3200	R384001
4263	2dr Brougham-5P	1124	3230	R384001
4268	4dr Sedan-6P	1134	3335	R384001
4269	2dr Sedan-6P	1114	3265	R384001
4240	4dr Tr Sedan-6P	993	2655	K56001
4242	2dr Cpe-3P	918	2540	K56001
4243	2dr Brougham-5P	958	2580	K56001
4248	4dr Sedan-6P	968	2655	K56001
4249	2dr Sedan-6P	948	2605	K56001

ENGINE: [Series 4280] Overhead valve. Inline. Eight-cylinder. Cast iron block. B & S: 3-1/8 in. x 4-1/4 in. Disp.: 260.8 cid. Compression ratio: 6.60:1. Brake HP: 115 at 3400 rpm. Taxable HP: 31.2. Main bearings: Nine. Valve lifters: Solid. Carb.: Carter WDO-538S. Torque: 200 lbs.-ft. at 1600 rpm. [Series 4260] Overhead valve. Inline. Six-cylinder. Cast iron block. B & S: 3-3/8 in. x 4-3/8 in. Disp.: 234.8 cid. C.R.: 6.50:1. Brake HP: 105 at 3400 rpm. Taxable HP: 27.3. Main bearings: Seven. Valve lifters: Solid. Carb.: Carter WA1-462S. Torque: 203 lbs.-ft. at 1600 rpm. [Series 4240] L-head. Inline. Six-cylinder. Cast iron block. B & S: 3-1/8 in. x 3-3/4 in. Disp.: 172.6 cid. C.R.: 6.87:1. Brake HP: 75 at 3600 rpm. Taxable HP: 23.4. Main bearings: Four. Valve lifters: Solid. Carb.: Carter WDO-513S. Torque: 138 lbs.-ft. at 1200 rpm.
CHASSIS: [Series 4280] W.B.: 121 in. O.L.: 205-1/2 in. Front/rear tread: 57 in./60-1/4 in. Tires: 6.50 x 16. [Series 4260] W.B.: 121 in. O.L.: 205-1/2 in. Front/rear tread: 57-1/2 in./60-1/2 in. Tires: 6.25 x 16. [Series 4240] W.B.: 112 in. O.L.: 196-1/2 in. Front/rear tread: 56 in./59-3/4 in. Tires: 5.50 x 16.
TECHNICAL: Selective sliding gear transmission. Speeds: 3F/1R. Steering column controls. Single dry plate clutch. Hypoid drive. Semi-floating rear axle. Overall ratio: 4.11:1. Hydraulic brakes on four wheels. Steel disc wheels. Wheel size: 16 in. Hill-Holder. Overdrive.
OPTIONS: Bumper Front (standard). Rear Bumper (standard). Single sidemount (N/A). Dual Sidemount (N/A). Sidemount covers (N/A). Fender skirts (except 4240). Bumper Guards (standard). Radio ($65). Heater ($35). Clock ($10.50). Cigar Lighter ($2.10). Radio

Antenna (included with radio). Seat Covers. Spotlight ($17.75). Cowl lamps (N/A). Bulb Horn (N/A). Bed Equipment ($21). Oil filter ($9). Outside mirror, right or left ($2.45). Fog lights ($12).

HISTORICAL: Introduced Oct. 1, 1941. Calendar year production: 5,428. The president of Nash was George Mason. Very short production year due to U.S. entry into World War II. (Production ceased Feb. 1, 1942).

1946 NASH

1946 Nash '600' Coupe, 6-cyl

NASH 600 — SERIES 40 — (6-CYL): Nash, like most manufacturers after World War II, brought out cars based on prewar models with minor modifications. A significant change in the 600 was the use of a conventional wishbone front suspension. It replaced the Lancia-inspired 1941-1942 "sliding-pillar" type. Other changes included a wider front grille, the movement of turn signals from atop the front fenders (in 1942) to a position inboard of the front headlights, and a new hood crest showing the Nash coat of arms. This emblem was less stylized and ornate than in 1942. Like the 1941 original, the new '600' had unitized construction. The model name reflected the fact that Nash claimed the car would go 600 miles on a full 20-gallon tank of gas. Technical features included sealed-in manifolding; full-pressure lubrication; full-length water jackets; steel-strut aluminum alloy pistons; extra-hard cylinder blocks; double-automatic spark control; air-cooled voltage generator; radial balanced crankshaft; three-point rubber-insulated engine mounting; crankshaft vibration dampener; and 20-gallon fuel tank. Standard equipment varied by model as follows: [Deluxe] spare wheel and tire; no-draft ventilation; hi-test safety glass; Deluxe bumpers; twin bumper guards front and rear; dual horns, sun visors and windshield wipers; front door armrests; dome light; cigar lighter; instrument panel ashtray; and center-rear compartment ashtray. [Four-door sedan] rear quarter compartment ashtray and robe cord. [Two- and four-door sedan] assist cords. [Custom] Wind-up type clock; metal rear fender gravel pads; door locks; glove compartment locks; Deluxe steering wheel; rear quarter vent windows; combination plastic and lacquered radio grille; rotary door locks; sealed beam headlights; stainless steel running board moldings; carpet insert in front floor mat and voltage control generator.

NASH I.D. NUMBERS: Vehicle Identification Plate located on the right-hand side of the cowl below the hood contains motor-serial number and number codes for model, paint and trim. Motor serial number matches the serial number. Model number has four symbols. The first two symbols indicate model year, 46=1946. The third symbol indicates series: 4=600; 6=Ambassador. The fourth symbol indicates body style. (All four numbers appear in Body/Style Number column of charts below). Motor numbers located on right-hand side of crankcase towards front and on front upper left-hand side of block. Motor-serial numbers for 1946 were: [600] K-77701 to K-135801; [Ambassador] R-393101 to R-429001.

NASH 600 SERIES 40

Model Number	Body/Style Number	Body Type & Seating	Factory Price	Shipping Weight	Production Total
40	4640	4-dr Trk Sed-6P	1342	2740	7,300
40	4643	2-dr Brgm Sed-6P	1293	2685	8,500
40	4648	4-dr FsBk Sed-6P	1298	2780	42,300

1946 Nash, Ambassador four-door sedan, 6-cyl (TVB)

NASH AMBASSADOR — SERIES 60 — (6-CYL): The Ambassador shared the looks and detail changes seen on the 600, but had a nine inch longer wheelbase. Instead of unitized construction, it featured separate chassis/frame engineering. However, the sheet metal and body construction were the same as for 600s from the cowl back. The car's extra nine inches of sheet metal was originally planned for a Nash overhead valve straight eight. The inline eight engine was dropped from production after World War II. The Ambassador now used a slightly updated overhead valve six with 112 hp. Technical features included sealed-in manifolding; full-pressure lubrication; full-length water jackets; steel-strut aluminum alloy pistons with four rings; extra-hard cylinder block; double automatic spark control; air-cooled voltage generator; radial-balanced crankshaft with vibration dampener; four-point rubber insulated engine mounting; oil filter; six quart crankcase; and 20-gallon fuel tank.

AMBASSADOR 60 SERIES

Model Number	Body/Style Number	Body Type & Seating	Factory Price	Shipping Weight	Production Total
60	4660	4-dr Trnk Sed-6P	1511	3335	3,875
60	4663	2-dr Brgm Sed-6P	1453	3260	4,825
60	4664	4-dr Suburb-6P	1929	3470	275
60	4668	4-dr FsBk Sed-6P	1469	3360	26,925

ENGINES

(NASH 600 SIX) Inline. L-head six: Cast iron block. Displacement: 172.6 cid. Bore and stroke: 3-1/8 x 3-3/4 inches. Compression ratio: 6.8:1. Brake hp: 82 at 3800 rpm. Four main bearings. Solid valve lifters. Carburetion: Carter one-barrel WA1-611S.

(AMBASSADOR SIX) Inline six: Overhead valves. Cast iron block. Displacement: 234.8 inches. Bore and stroke: 3-3/8 x 4-3/8 inches. Compression ratio: 6.8:1. Brake hp: 112 at 3400 rpm. Seven main bearings. Solid valve lifters. Carburetion: Carter one-barrel Model YF.

CHASSIS FEATURES: Wheelbase: (600) 112 inches; (Ambassador) 121 inches. Overall length: (600) 199-9/16 inches; (Ambassador) 208-9/16 inches. Front tread: (600) 56-13/16 inches; (Ambassador) 57-1/2 inches. Rear tread: (600) 59-3/4 inches; (Ambassador) 60-1/2 inches. Tires: (600) 6.00 x 16; (Ambassador) 6.50 x 15.

OPTIONS: Foam rubber cushions. Conditioned air system. Vacuum booster pumper. Radio and antenna. Directional signals. Oil bath air cleaner. Oil filter. There were no available engine options. The standard gearbox was a three-speed manual type with overdrive offered as an option on the Ambassador only.

HISTORICAL FOOTNOTES: From 1946-1948 Nash produced a wood-bodied four-door sedan in the style of the Chrysler Town & Country. A total of 1,000 of these wood covered 'Suburbans' were built in this three-year period. The fastback sedan was called the Slip Stream. Nash built 98,769 cars in calendar 1946 for eigth place in auto industry sales rankings.

1947 NASH

1947 Nash, '600' four-door sedan, 6-cyl

NASH 600 — SERIES 40 — (6-CYL): The Nash 600 for 1947 received few changes from 1946. The front grilles were widened again and new raised center hubcaps were used.

NASH I.D. NUMBERS: Vehicle Identification Plate located on the right-hand side of the cowl below the hood contains motor-serial number and number codes for model, paint and trim. Motor serial number matches the serial number. Model number has four symbols. The first two symbols indicate model year, 47=1947. The third symbol indicates series: 4=600; 6=Ambassador. The fourth symbol indicates body style. (All four numbers appear in Body/Style Number column of charts below). Motor numbers located on right-hand side of crankcase towards front and on front upper left-hand side of block. Motor-serial numbers for 1946 were: [600] K-136001 to K-153244; [Ambassador] R-429201 to R-440922.

Motor-serial numbers matched serial numbers.

NASH 600 SERIES 40

Model Number	Body/Style Number	Body Type & Seating	Factory Price	Shipping Weight	Production Total
40	4740	4-dr Trk Sed-6P	1464	2786	21,500
40	4743	2-dr Brgm-6P	1415	2731	12,100
40	4748	4-dr FsBk Sed-6P	1420	2826	27,700

1947 Nash, Ambassador, four-door sedan, 6-cyl (IMS)

NASH AMBASSADOR — SERIES 60 — (6-CYL): The only changes for the 1947 Ambassador was the addition of the same front grille as used on Nash 600s, plus the same raised center hubcaps. The four-door Nash Surburban was distinguished by wooden side panels. The fastback sedan was called the Slip Stream.

AMBASSADOR 60 SERIES

Model Number	Body/Style Number	Body Type & Seating	Factory Price	Shipping Weight	Production Total
60	4760	4-dr Trk Sed-6P	1809	3387	15,927
60	4763	2-dr Brgm-6P	1751	3312	8,673
60	4764	4-dr Sub-6P	2227	3522	595
60	4768	4-dr FsBk Sed-6P	1767	3412	14,505

ENGINES
(NASH 600 SIX) Inline. L-head six: Cast iron block. Displacement: 172.6 cid. Bore and stroke: 3-1/8 x 3-3/4 inches. Compression ratio: 6.8:1. Brake hp: 82 at 3800 rpm. Four main bearings. Solid valve lifters. Carburetion: Carter one-barrel WA1-611S.
(AMBASSADOR SIX) Inline six: Overhead valves. Cast iron block. Displacement: 234.8 inches. Bore and stroke: 3-3/8 x 4-3/8 inches. Compression ratio: 6.8:1. Brake hp: 112 at 3400 rpm. Seven main bearings. Solid valve lifters. Carburetion: Carter one-barrel Model YF.

CHASSIS FEATURES: Wheelbases: (600) 112 inches; (Ambassador) 121 inches. Overall length: (600) 199-9/16 inches; (Ambassador) 208-9/16 inches. Front tread: (600) 56-13/16 inches; (Ambassador) 57-1/2 inches. Rear tread: (600) 59-3/4 inches; (Ambassador) 60-1/2 inches. Tires: (600) 6.00 x 16; (Ambassador) 6.50 x 15.

OPTIONS: Foam rubber cushions. Cruising gear (Ambassador only). Conditioned air system. Vacuum booster pump. Radio and antenna. Directional signals. Oil bath air cleaner. Oil filter (600 only). There were no available optional engines. The standard gearbox was a three-speed manual type with Warner Gear overdrive available at extra cost.

HISTORICAL FOOTNOTES: Production for calendar year 1947 increased to 113,315 cars and Nash came in 10th in sales. A one-of-a-kind 12-passenger Nash limousine was built to carry executives and VIPs around the plant. It had four doors on each side. A Nash Ambassador was Official Pace Car for the Indianapolis 500-Mile Race. The five percent of total U.S. auto sales earned by Nash this season was a strong showing for an independent manufacturer. New assembly sites in El Segundo, Calif. and Toronto, Ontario, Canada were acquired by Nash-Kelvinator Corp. this year. George W. Mason was Chairman of the Board and President of the company. Production of the 1947 line commenced in December, 1946.

1948 NASH

1948 Nash, '600' four-door sedan, 6-cyl

NASH 600 SERIES 40 — (6-CYL): Changes to the 1948 Nash consisted of the removal of a chrome molding just below the beltline, giving the cars a clean sided look. Hoodside moldings did not run as far forward and the hood badge design was changed. In addition, the model line was expanded to meet an anticipated upsurge in buyer demand. Included, for the first time since before World War II, was a three-passenger business coupe and upgraded Custom versions of the two-door Brougham and the four-door Slip Stream and Trunk sedan.

NASH I.D. NUMBERS: Vehicle Identification Plate located on the right-hand side of the cowl below the hood contains motor-serial number and number codes for model, paint and trim. Motor serial number matches the serial number. Model number has four symbols. The first two symbols indicate model year, 48=1948. The third symbol indicates series: 4=600 Super/Deluxe; 5=600 Custom; 6=Ambassador Super; 7=Ambassador Custom. The fourth symbol indicates body style. (All four numbers appear in Body/Style Number column of charts below). Motor numbers located on right-hand side of crankcase towards front and on front upper left-hand side of block. [600 SERIES] Serial numbers ran from K196901 to K259792. Motor numbers ran from KE55001 to KE120132 and no longer matched serial numbers. [AMBASSADOR SERIES] Serial numbers ran from R468501 to R514594. Motor serial numbers ran from RE40001 to RE82095.

NASH 600 SERIES 40

Model Number	Body/Style Number	Body Type & Seating	Factory Price	Shipping Weight	Production Total
SUPER LINE/DELUXE COUPE					
40	4840	4-dr Trk Sed-6P	1587	2786	25,103
40	4843	2-dr Brgm-6P	1538	2731	11,530
40	4848	4-dr FsBk Sed-6P	1534	2826	25,044
40	4842	2-dr Bus Cpe-3P	1478	2635	925
CUSTOM LINE					
40	4850	4-dr Trk Sed-6P	1776	2786	346
40	4853	2-dr Brgm-6P	1727	2731	170
40	4858	4-dr FsBk Sed-6P	1732	2826	332

1948 Nash, Ambassador two-door convertible, 6-cyl.

NASH AMBASSADOR — SERIES 60 — (6-CYL): The 1948 Ambassador also had the chrome molding just below the beltline removed for a cleaner side appearance. In addition, the model line was also expanded to meet an anticipated surge in sales caused by the return to full postwar production after the settling of labor disputes and the relaxation of materials restrictions. Added were Custom versions of the two-door Brougham and the four-door Slip Stream (fastback) and Trunk sedan. Also, for the first time since 1941 a convertible was added to the line, but only 1,000 were made. This was to be the last full-sized Nash convertible ever made, although a 1950 Nash Rambler compact size convertible would be produced. Not until 1965 would Nash Motor's successor, AMC, build a full-sized convertible again.

AMBASSADOR 60 SERIES

Model Number	Body/Style Number	Body Type & Seating	Factory Price	Shipping Weight	Production Total
SUPER LINE					
60	4860	4-dr Trk Sed-6P	1916	3387	14,248
60	4863	2-dr Brgm-6P	1858	3312	7,221
60	4864	4-dr Sub-6P	2239	3522	130
60	4868	4-dr Fst 8k Sed-6P	1874	3412	14,777
CUSTOM LINE					
60	4870	4-dr Trk Sed-6P	2105	3387	4,102
60	4873	2-dr Brgm Sed-6P	2047	3312	929
60	4878	4-dr FsBk Sed-6P	2063	3412	4,143
60	4871	2-dr Conv-6P	2355	3465	1,000

ENGINES
(NASH 600 SIX) Inline. L-head six: Cast iron block. Displacement: 172.6 cid. Bore and stroke: 3-1/8 x 3-3/4 inches. Compression ratio: 6.8:1. Brake hp: 82 at 3800 rpm. Four main bearings. Solid valve lifters. Carburetion: Carter one-barrel WA1-6625.
(AMBASSADOR SIX) Inline six: Overhead valves. Cast iron block. Displacement: 234.8 inches. Bore and stroke: 3-3/8 x 4-3/8 inches. Compression ratio: 6.8:1. Brake hp: 112 at 3400 rpm. Seven main bearings. Solid valve lifters. Carburetion: Carter one-barrel model YF.

CHASSIS FEATURES: Wheelbase: (600) 112 inches; (Ambassador) 121 inches. Overall length: (600) 200 inches; (Ambassador) 209-3/16 inches. Front tread: (all) 57-1/2 inches. Rear tread (600) 59-11/16 inches; (Ambassador) 60-1/2 inches. Tires: (600) 6.40 x 15 Super Cushion; (Ambassador) 6.50 x 15; 7.10 x 15 Super Cushion optional.

OPTIONS: Foam rubber cushions. Cruising gear (Ambassador only). Conditioned air system. Vacuum booster pump. Radio and antenna. Directional signals. Oil bath air cleaner. Oil filter (600 only). There were no available optional engines. The standard gearbox was a three-speed manual type with Warner Gear overdrive available at extra cost.

HISTORICAL FOOTNOTES: Charles Nash died on June 6, 1948 at the age of 84. The 1948 Ambassadors were the last Nashes to use separate frames. Production began at the El Segundo, Calif. factory this year. The starting month for production of Nash products built to 1948 specifications was November 1947. This would be the last year that Nash used the term "Brougham" to describe its club coupe. (Starting with the 1949 Nash Airflyte, the Brougham model became a two-door sedan featured two individual "lounge-chair" style seats, angled towards the center of the car, with a permanent center armrest between them.) Calendar year production peaked at 118,621 units and 3.04 percent of total domestic sales for the entire industry. George W. Mason retained the top corporate posts and George Romney served as Nash-Kelvinator Vice-President. The fastback sedan was again called a Slip Stream.

1949 NASH

1949 Nash 600 four-door sedan, 6-cyl

NASH 600 — SERIES 40 — (6-CYL): Nash introduced its first totally redesigned postwar car line in 1949. The Nash Airflyte series, as it was called, featured single-unit construction, one-piece curved windshield, 'Uniscope' gauge cluster (a pod atop the steering column containing all instruments) and fully reclining front seatbacks. In 1949, Nash was the first U.S. manufacturer of mass produced autos to totally commit to unitized, single-unit construction, and one of the first in the world to do so. These 'bathtub' Nashes (as they were known) were styled with an eye toward aerodynamics. At 60 mph in wind tunnel tests, the Airflyte had only 113 pounds of drag. (In comparison a similar looking 1949 Packard had around 171 pounds of drag.)

NASH I.D. NUMBERS: Vehicle Identification Plate located on the right-hand side of the cowl below the hood contains motor-serial number and number codes for model, paint and trim. Motor serial number matches the serial number. Model number has four symbols. The first two symbols indicate model year, 49=1949. The third symbol indicates series: 2=600 Special; 4=600 Super; 5=600 Custom; 6=Ambassador Super; 7=Ambassador Custom; 9=Ambassador Super Special. The fourth symbol indicates body style. (All four numbers appear in Body/Style Number column of charts below). Motor numbers located on right-hand side of crankcase towards front and on front upper left-hand side of block. [600 SERIES] The starting serial number for the Nash 600 of 1949 was K-260501 for cars assembled in Kenosha, Wis.. Unassembled export serial numbers began 4KD-1401. Starting serial numbers for El Segundo, Calif. was KC-1001. Starting engine serial number for all 1949 Nash 600s was S-1001. [AMBASSADOR SERIES] The starting serial number for the Ambassador six of 1949 was: R-515501 for Kenosha, Wis.; 6KD-1501 for unassembled export and RC-1001 for El Segundo. Starting engine serial numbers for all 1949 Ambassadors was A-1001.

NASH 600 SERIES 40

Model Number	Body/Style Number	Body Type & Seating	Factory Price	Shipping Weight	Production Total
SUPER SPECIAL LINE					
40	4923	2-dr Brgm-5P	1846	2960	2,564
40	4928	4-dr Sed-6P	1849	2960	23,606
40	4929	2-dr Sed-6P	1824	2935	9,605
SUPER LINE					
40	4943	2-dr Brgm-5P	1808	2960	2,954
40	4948	4-dr Sed-6P	1811	2950	31,194
40	4949	2-dr Sed-6P	1786	2935	17,006
CUSTOM LINE					
40	4953	2-dr Brgm-5P	1997	2970	17
40	4958	4-dr Sed-6P	2000	2985	199

1949 Nash, Ambassador Custom four-door sedan, 6-cyl (TVB)

NASH AMBASSADOR — SERIES 60 — (6-CYL): The Nash Ambassador for 1949 shared all the styling changes of the 1949 Nash 600. The difference was largely that the Ambassador had a nine inch longer wheelbase due to its longer front end. Coil spring suspension and torque tube drive were featured on both the Ambassador and the Nash 600. The same three series or trim levels were available in the Ambassador: Super, Super Special and Custom.

NASH AMBASSADOR SERIES 60

Model Number	Body/Style Number	Body Type & Seating	Factory Price	Shipping Weight	Production Total
SUPER LINE					
60	4963	2-dr Brgm-5P	2191	3390	1,541
60	4968	4-dr Sed-6P	2195	3385	17,960
60	4969	2-dr Sed-6P	2170	3365	4,602
CUSTOM LINE					
60	4973	2-dr Brgm-5P	2359	3415	1,837
60	4978	4-dr Sed-6P	2363	3415	6,539
60	4979	2-dr Sed-6P	2338	3400	691

Model Number	Body/Style Number	Body Type & Seating	Factory Price	Shipping Weight	Production Total
SUPER SPECIAL LINE					
60	4993	2-dr Brgm-5P	2239	3390	807
60	4998	4-dr Sed-6P	2243	3385	6,777
60	4999	2-dr Sed-6P	2218	3365	2,072

ENGINES
(NASH 600 SIX) Inline. L-head six: Cast iron block. Displacement: 172.6 cid. Bore and stroke: 3-1/8 x 3-3/4 inches. Compression ratio: 6.8:1. Brake hp: 82 at 3800 rpm. Four main bearings. Solid valve lifters. Carburetion: Carter one-barrel WA1-6945.
(AMBASSADOR SIX) Inline six: Overhead valves. Cast iron block. Displacement: 234.8 inches. Bore and stroke: 3-3/8 x 4-3/8 inches. Compression ratio: 6.8:1. Brake hp: 112 at 3400 rpm. Seven main bearings. Solid valve lifters. Carburetion: Carter one-barrel WA1-683S.

CHASSIS FEATURES: Wheelbase: (600) 112 inches; (Ambassador) 121 inches. Overall length: (600) 201 inches; (Ambassador) 210 inches. Front tread: (all) 5411/16 inches. Rear tread: (600) 59-11/16 inches; (Ambassador) 60-1/2 inches. Tires: (600) 6.40 x 15; (Ambassador) 7.10 x 15.

OPTIONS: Bed, single ($19); double ($39). Exhaust pipe extensions ($2). Fog lights, pair ($13). Spotlight, door mounted ($20). Back-O-Matic Lights, pair ($9). Fuel purifier ($2). License plate frames ($2). Electric snap up gas cap ($5). Magnalite trouble light ($4). Non-glare mirror ($4). Rearview mirror, right and left ($5). Visor vanity mirror ($1). Deluxe fiber seat covers (interwoven pattern, blue color) in two-door sedan and four-door sedan ($25); in Broughams ($31). Deluxe fiber Sportster type seat covers in all two-door and four-door sedans ($28); in Broughams ($35). Rayon twill seat covers with Maroon, Blue or Brown color, in all sedan ($48); in Broughams ($51). Trim rings ($12). Wheel discs ($17). Tissue Dispensers ($3). Hood ornaments ($9). Windshield washer ($7). Rear window wiper ($14). Front grille guard ($25). Rear grille guard ($20). Fender guard, pair, ($10). Nash Karvisor ($26). Directional signals ($16). Radio ($82). Manual antenna ($7). Vacuum antenna ($14). Warner Gear overdrive.

HISTORICAL FOOTNOTES: The 1949 Nash line began production in October, 1948. A total of 130,000 units was produced for the model year. Calendar year production was counted at 142,592 cars or 2.78 percent of the total domestic auto business. Nine Ambassadors were built by the engineering department (under the direction of M.F. Moore, vice-president of Nash Research) with automatic transmissions. Three-passenger models and the wood-veneered Suburban sedan were dropped this year. Nash was America's 10th ranked maker of the season.

1950 NASH

1950 Nash, Statesman, four-door sedan, 6-cyl

NASH STATESMAN — SERIES 40 — (6-CYL): The Nash 600 was renamed Statesman for 1950 and its L-head six-cylinder engine had a one-quarter inch larger stroke. This increased displacement from 172.6 cid to 184 cid and increased horsepower from 82 to 85. One significant styling change made this year was a much larger rear window, which was very helpful to the driver considering that the fastback design made seeing the traffic behind very difficult. The bumper guards grew slightly thicker and the cars had a Statesman script on their front fenders. Seat belts were available with the Statesman and the Ambassador in their first use on a U.S. built car. Both the Statesman and the Ambassador came in two basic trim levels, Super and Custom for 1950. The Statesman also had a low-priced 'line leader' Deluxe Business Coupe. Custom models had rear seat armrests; carpets; courtesy lights; a Custom steering wheel; and full wheel discs. Some 1950 models were built in the new Canadian factory. Production figures for Canada are unavailable.

NASH I.D. NUMBERS: Vehicle Identification Plate located on the right-hand side of the cowl below the hood contains motor-serial number and number codes for model, paint and trim. Motor serial number matches the serial number. Model number has four symbols. The first two symbols indicate model year, 50=1950. The third symbol indicates series: 2=Rambler; 3=Statesman; 4=Super Statesman; 5=Custom Statesman; 6=Ambassador Super; 7=Ambassador Custom. The fourth symbol indicates body style. (All four numbers appear in Body/Style Number column of charts below). Motor numbers located on right-hand side of crankcase towards front and on front upper left-hand side of block. [STATESMAN]: Serial numbering for Statesman sixes built in Kenosha started at K-340001 to K-436892; in El Segundo, KC-9501 to KC-23007; in Canada, KT-1001 and up; and for unassembled export units 4KD-2301 and up. Engine numbers for 1950 Statesman sixes began at S-92001 to S-205947. [AMBASSADOR]: Serial numbering for Ambassador sixes built in Kenosha started at R-556001 to R-599704; in El Segundo, RC-3501 to RC-8488; and for unassembled export units RD-2101 and up. (No Ambassadors assembled in Canada). Engine numbers for 1950 Ambassadors began at A-46001 to A-96151. [RAMBLER] The starting serial number was D-1001 to D-12263. Engine numbers were F-1001 to F-12574.

NASH STATESMAN SERIES 40

Model Number	Body/Style Number	Body Type & Seating	Factory Price	Shipping Weight	Production Total
NASH STATESMAN SERIES 40					
40	5032	2-dr Bus Cpe-3P	1633	2830	1,198
SUPER LINE					
40	5043	2-dr Brgm-5P	1735	2940	1,489
40	5048	4-dr Sed-6P	1738	2965	60,090
40	5049	2-dr Sed-6P	1713	2930	34,196
CUSTOM LINE					
40	5053	2-dr Brgm-5P	1894	2965	132
40	5058	4-dr Sed-6P	1897	2990	11,500
40	5059	2-dr Sed-6P	1872	2950	2,693

1950 Nash, Ambassador Custom four-door sedan, 6-cyl

NASH AMBASSADOR — SERIES 60 — (6-CYL): The 1950 Nash Ambassador had a longer hood than the 1949 model, as well as the enlarged back window. Otherwise, there were not many significant changes. An Ambassador script appeared on the fenders for identification. The year's major innovation was the introduction of a GM built Hydra-Matic transmission, available only on the 1950 Nash Ambassador. A new cylinder head design was also introduced for the 234.8 cid overhead valve Ambassador engine and raised the output to 115 hp. Custom models differed from the Supers by featuring a rear seat with a folding center armrest; front floor carpeting; courtesy lights; a Custom steering wheel; and large wheel discs.

NASH AMBASSADOR SERIES 60

Model Number	Body/Style Number	Body Type & Seating	Factory Price	Shipping Weight	Production Total
SUPER LINE					
60	5063	2-dr Brgm-5P	2060	3335	716
60	5068	4-dr Sed-6P	2064	3350	27,523
60	5069	2-dr Sed-6P	2039	3325	7,237
CUSTOM LINE					
60	5073	2-dr Brgm-5P	2219	3385	108
60	5078	4-dr Sed-6P	2223	3390	12,427
60	5079	2-dr Sed-6P	2198	3365	1,045

1950 Nash, Rambler two-door convertible, 6-cyl

NASH RAMBLER — SERIES 10 — (6-CYL): Nash introduced the compact Nash Rambler Convertible Landau in March, 1950. The first cars built by Nash's predecessor, the Thomas B. Jeffery Co. of Kenosha, Wis., also used this name. The compact car had a 100 inch wheelbase and used the 82 hp six-cylinder engine from the Nash 600. The first model introduced was the two-door Convertible Landau. A two-door station wagon was introduced two months later on June 23. Both models came loaded with options such as radio and antenna; Custom steering wheel; turn signals; wheel discs; electric clock; courtesy lights; Custom upholstery; and foam seat cushions. On the convertible only, a sliding top (in black or tan fabric) could be raised over 'bridge beam' side rails above the doors. The Rambler used Hotchkiss drive, unlike the torque tube drive of the conventional Nash.

NASH RAMBLER SERIES 10

Model Number	Body/Style Number	Body Type & Seating	Factory Price	Shipping Weight	Production Total
10	5021	2-dr Cus Conv-5P	1808	2430	9,330
10	5024	2-dr Cus Sta Wag-5P	1808	2515	1,712
10	5026	2-dr Cus Trk Sed	—	—	6
10	5114	2-dr Sup Sta Wag	—	—	1
10	5124	2-dr Cus Sta Wag	—	—	1
10	5121	2-dr Cus Conv-5P	1808	2430	378

NOTE 1: 378 convertibles built in Calif. July/Aug. 1950 possibly prototypes.

ENGINES
(STATESMAN SIX) Inline. L-head six. Cast iron block. Displacement: 184.0 cid. Bore and stroke: 3-1/8 x 3-3/4 inches. Compression ratio: 7.0:1. Brake hp: 85 at 3800 rpm. Four main bearings. Solid valve lifters. Carburetor: Carter one-barrel Model WAI-694S.
(AMBASSADOR SIX) Inline. Overhead valve six. Displacement: 234.8 cid. Bore and stroke: 3-3/8 x 4-3/8 inches. Compression ratio: 7.3:1. Brake hp: 115 at 3400 rpm. Seven main bearings. Solid valve lifters. Carburetor: Carter one-barrel type WA-1 Model 746S.
(RAMBLER SIX) Inline. L-head six. Cast iron block. Displacement: 172.6 cid. Bore and stroke: 3-1/8 x 3-3/4 inches. Compression ratio: 7.25:1. Brake hp: 82 at 3800 rpm. Four main bearings. Solid valve lifters. Carburetor: Carter one barrel type YF model 757S.

CHASSIS FEATURES: Wheelbase: (Statesman) 112 inches; (Ambassador) 121 inches; (Rambler) 100 inches. Overall length: (Statesman) 201 inches; (Ambassador) 210 inches; (Rambler) 176 inches. Front tread: (Statesman and Ambassador) 54-11/16 inches; (Rambler) 53-1/4 inches. Rear tread: (Statesman) 59-11/16 inches; (Ambassador) 60-1/2 inches; (Rambler) 53 inches. Tires: (Statesman) 6.40 x 15; (Ambassador) 7.10 x 15; (Rambler) 5.90 x 15.

OPTIONS: Electric clock (standard on Custom models). Mechanical clock. Overdrive. Two-tone colors. Directional signals (standard on Custom models). Emergency brake alarm. Fender signal. Floor mat pads. Foam rubber seat cushion (standard on custom). Fuel purifier. Locking gas cap. Electric locking gas cap. Fender guards. Grille guards. Trunk guards. Hood ornament. Hydra-Matic transmission (Ambassador). License frames. Back-O-Matic lights. Fog lights, (pair). Spotlight, (front door). Spotlight (with rearview mirror). Trouble light. Mattress. Zipper case for bed. Rearview mirror. Deluxe rearview outside mirror. Visor vanity mirror. Oil filter (Statesman and Ambassador). Heavy-duty oil bath air cleaner (Statesman and Ambassador). Opto shade inside sun shield. Custom radio with manual antenna. Custom radio with vacuum antenna. Deluxe radio with vacuum antenna. Rear speaker for Deluxe radio. Rear door safety locks. Front divided seatback for four-door sedan w/bed (standard on Custom models). Reclining front seat. Fiber Regal seat covers. Fiber Majestic seat covers. Top-Flyte Custom rayon seat covers in maroon, blue and green. Custom steering wheel (standard equipment in Custom models). White sidewall tires, size 6.40 x 14 four-ply. White sidewall tires, size 6.40 x 15 six-ply. White sidewall tires, size 7.10 x 15 four-ply. White sidewall tires, size 7.10 x 15 six-ply. Tissue dispenser packet. Tool pouch. Set of five stainless steel chrome trim rings. Leather upholstery. Outside window visors for two-door; four-door. Visor shade. Weather Eye conditioned air system. Five chrome stainless steel wheel discs (standard on Custom models). Window screen for two-door; four-door. Windshield washer. Rear window wiper.

HISTORICAL FOOTNOTES: Production of 1950 Nash products began in September 1949. A total of 145,782 Statesman and Ambassador models, along with 26,000 Ramblers, were produced for the model year. Sales were counted at 191,865 cars, putting Nash tenth in the auto sales race. This broke the all-time record for Nash production. The official model introduction date for the new Rambler convertible was April 14, 1950. Together, the two Rambler body styles helped Nash achieve the assembly of 7.1 percent of all convertibles and 3.6 percent of all station wagons built in the U.S. in calendar 1950. The compact line was an immediate success. Nash calendar year production leaped to 189,543 cars or 2.84 percent of total auto industry sales. The three-passenger Nash was reintroduced this year, as a back seat-less business car intended strictly for commercial use. Several Nash models competed successfully in stock car races during 1950. Their greater fuel economy meant that less pit stops were required. Many Nash owners reported 25-30 mpg in normal operation, according to the company's advertisements.

1951 NASH

NASH STATESMAN — SERIES 40 — (6-CYL): The 1951 Nash Statesman featured a new "electric shaver" type grille and side marker lights along with new rear fenders and fender lights. Statesman scripts were on the front fender for identification. Super models had basic features. Customs also had foam seat cushions; front floor carpets; courtesy lights; a rear seat center armrest; Custom steering wheel; and full wheel discs. Hydra-Matic automatic transmission made its debut on the Statesman in 1951.

NASH I.D. NUMBERS: Vehicle Identification Plate located on the right-hand side of the cowl below the hood contains motor-serial number and number codes for model, paint and trim. Motor serial number matches the serial number. Model number has four symbols. The first two symbols indicate model year, 51=1951. The third symbol indicates series: 0=Rambler Utility; 1=Rambler Super 2=Rambler Custom; 3=Statesman; 4=Super/Deluxe Statesman; 5=Custom Statesman; 6=Ambassador Super; 7=Ambassador Custom. The fourth symbol indicates body style. (All four numbers appear in Body/Style Number column of charts below). Motor numbers located on right-hand side of crankcase towards front and on front upper left-hand side of block. [NASH STATESMAN SERIES 40 I.D. NUMBERS]: Starting serial number for 1951 Nash Statesman models built at Kenosha, Wis. was K-438001 to K518763; 4KD-3201 and up for unassembled export; KC-23501 and up for El Segundo, Calif. and KT-22501 and up for cars made in the Toronto, Ontario, Canada assembly plant. Starting engine numbers for the Statesman six were S-207001 to S-306795for all assembly points. [NASH AMBASSADOR SERIES 60 I.D. NUMBERS]: Starting serial number for the 1951 Nash Ambassador models built at Kenosha, Wis. was R-600501 to R-655753; 6KD-2071 and up for unassembled exports and RC-8701 and up for El Segundo, Calif. Ambassadors with serial numbers RC were not manufactured at the Canadian plant. Starting engine numbers for all Ambassador six engines were A-97001 to A-160453. [NASH RAMBLER SERIES I.D. NUMBERS]: The starting serial number for 1951 Ramblers built in Kenosha, Wis. was D-12501 to D-78917; DKD-1201 for unassembled export and DC1501 for El Segundo, Calif.

Starting engine numbers for 1951 Ramblers were F-1001 to F-83778. The Rambler Deliveryman utility wagon line had serial numbers D66495 and up/motor numbers F-69802 and up.

1951 Nash, Statesman four-door sedan, 6-cyl.

NASH STATESMAN SERIES 40

Model Number	Body/Style Number	Body Type & Seating	Factory Price	Shipping Weight	Production Total
SUPER/DELUXE LINES					
40	5132	2-dr Del Bus Cpe-3P	1710	2835	52
40	5143	2-dr Brgm-5P	1812	2935	152
40	5148	4-dr Sed-6P	1815	2970	52,325
40	5149	2-dr Sed-6P	1790	2930	22,261
CUSTOM LINE					
40	5153	2-dr Brgm-5P	1971	2950	38
40	5158	4-dr Sed-6P	1974	2990	14,846
40	5159	2-dr Sed-6P	1949	2940	2,141

1951 Nash, Ambassador Custom four-door sedan, 6-cyl

NASH AMBASSADOR SERIES 60 — (6-CYL): The 1951 Nash Ambassador received the same revised front grille and side marker lights and new rear fenders as the 1951 Statesman. The major difference from the Statesman was the Ambassador's nine inch longer front end. Supers had basic features. Customs had a rear seat with a folding center armrest; courtesy lamps; front carpeting; Custom steering wheel; and large wheel discs.

NASH AMBASSADOR SERIES 60

Model Number	Body/Style Number	Body Type & Seating	Factory Price	Shipping Weight	Production Total
SUPER LINE					
60	5163	2-dr Brgm-5P	2158	3370	40
60	5168	4-dr Sed-6P	2162	3410	34,935
60	5169	2-dr Sed-6P	2137	3370	4,382
CUSTOM LINE					
60	5173	2-dr Brgm-5P	2317	3395	37
60	5178	4-dr Sed-6P	2321	3445	21,071
60	5179	2-dr Sed-6P	2296	3380	1,118

1951 Nash, Rambler Country Club two-door hardtop coupe, 6-cyl

NASH RAMBLER — SERIES 10 — (6-CYL): The major change for 1951 was the introduction of a new Rambler model called the Country Club hardtop. This was the first two-door compact hardtop to be introduced in the U.S. This body style had been popularized by the introduction of two-door hardtops in several GM car-lines in 1949.

NASH RAMBLER SERIES 10

Model Number	Body/Style Number	Body Type & Seating	Factory Price	Shipping Weight	Production Total
10	5104	2-dr Utl Wag-3P	1673	2415	1,569
10	5114	2-dr Sta Wag-5P	1723	2515	5,568
10	5117	2-dr HT Cpe-5P	—	—	1
10	5121	2-dr Conv-5P	1837	2430	14,881
10	5124	2-dr Cus Sta wag-5P	1837	2515	28,617
10	5126	2-dr Clb Sed-5P	—	—	50
10	5127	2-dr Cty Clb-5P	1968	2420	19,317

ENGINES
(STATESMAN SIX) Inline. L-head six. Cast iron block. Displacement: 184.0 cid. Bore and stroke: 3-1/8 x 3-3/4 inches. Compression ratio: 7.0:1. Brake hp: 85 at 3800 rpm. Four main bearings. Solid valve lifters. Carburetor: Carter one-barrel Model YF-824S.
(AMBASSADOR SIX) Inline. Overhead valve six. Displacement: 234.8 cid. Bore and stroke: 3-3/8 x 4-3/8 inches. Compression ratio: 7.3:1. Brake hp: 115 at 3400 rpm. Seven main bearings. Solid valve lifters. Carburetor: Carter one-barrel type WA-1 Model 746S.
(RAMBLER SIX) Inline. L-head six. Cast iron block. Displacement: 172.6 cid. Bore and stroke: 3-1/8 x 3-3/4 inches. Compression ratio: 7.25:1. Brake hp: 82 at 3800 rpm. Four main bearings. Solid valve lifters. Carburetor: Carter one barrel type YF model 757S.

CHASSIS FEATURES: Wheelbase: (Statesman) 112 inches; (Ambassador) 121 inches; (Rambler) 100 inches. Overall length: (Statesman) 201 inches; (Ambassador) 211 inches; (Rambler) 176 inches. Front tread: (Statesman and Ambassador) 54-11/16 inches; (Rambler) 53-1/4 inches. Rear tread: (Statesman) 59-11/16 inches; (Ambassador) 60-1/2 inches; (Rambler) 53 inches. Tires: (Statesman) 6.40 x 15; (Ambassador) 7.10 x 15; (Rambler) 5.90 x 15.

OPTIONS: Radio. Vacuum control antenna. Spotlight with mirror. Door-top outside mirror. Door mount outside mirror. Back-up lights. Fog lights. Non-glare rearview mirror. Visor vanity mirror. Curb-L-Arm. Custom seat covers. License plate frame. Exhaust extension. Protect-O-Mat for floor. Automatic windshield washer. Rear window wiper. Opto-shade for windshield. Vent shades. Outside windshield visor. Magnalite trouble light. Gas cap lock. Tissue dispenser. Bed mattress. Plastic screens.

HISTORICAL FOOTNOTES: In 1951, Nash introduced Rambler hardtop, suburban and Deliveryman (utility wagon) models; suspended operations in Canada and received a defense contract to built Pratt & Whitney aero engines. Korean War allocations prevented introduction of the Rambler four-door sedan this season. Annual model introductions were held on Sept. 22, 1950. The Nash Rambler Country Club hardtop was introduced, as an addition to the line, on June 28, 1951. In November 1951, the company received permission from the Economic Stability Agency to increase prices. The subsequent jump was $64 in Rambler; $48-55 in Statesman and $61-66 in Ambassador retail prices. Model year production counted 125,203 standard-sized models and 80,000 compact Ramblers, with the production run beginning in September 1950. Calendar year production hit 161,209 units or 3.02 percent of total American industry output. Over 82,731 Nash products were assembled with optional Overdrive transmission, while another 64,775 cars had automatic trans. (GM Hydra-Matic). The new Deliveryman was a wagon type vehicle, with only one seat, intended strictly for commercial package carrying work. In most reference sources it was listed as a truck and described as a utility wagon, which is how it appears in the specifications charts above. During 1951, Nash made 6.9 percent of all convertibles, 3.9 percent of all hardtops and 15.2 percent of all station wagons produced in the U.S. During the calendar year, 25,962 automatic transmission attachments were sold, while 82,731 cars had the optional Warner Gear overdrive.

1952 NASH

NASH STATESMAN — SERIES 40 — (6-CYL): The year 1952 marked the 50th anniversary of Nash Motor Co. and it's predecessor Thomas B. Jeffery Co. Nash used the occasion to introduce a totally redesigned line of big cars called Nash Golden Airflytes. These were partly styled by Italian designer Pinin Farina and had more conventional lines than the 1949-1951 models. The Statesman and the Ambassador again shared sheet metal from the cowl back. The 1952 Statesman had its wheelbase increased to 114-1/4 inches and the engine was stroked 1/4 inch to 195.6 cid. This increase in displacement boosted the Statesman engine's output to 88 hp at 3800 rpm.

NASH I.D. NUMBERS: Vehicle Identification Plate located on the right-hand side of the cowl below the hood contains motor-serial number and number codes for model, paint and trim. Motor serial number matches the serial number. Model number has four symbols. The first two symbols indicate model year, 52=1952. The third symbol indicates series: 0=Rambler Utility; 1=Rambler Super/Custom; 4=Super Statesman; 5=Custom Statesman; 6=Ambassador Super; 7=Ambassador Custom. The fourth symbol indicates body style. (All four numbers appear in Body/Style Number column of charts below). Motor numbers located on right-hand side of crankcase towards front and on front upper left-hand side of block. [NASH STATESMAN SERIES 40 I.D. NUMBERS]: Starting serial number for 1952 Statesman models built at Kenosha was K-519001 to K-562291; for unassembled export KD-4301 and up; for El Segundo KC-37001 to KC-42976; and for Toronto, Canada, KT-6101 and up. Engine numbers for the Statesman sixes were S-308001 to S-361836 for all assembly points. [NASH AMBASSADOR SERIES 60 I.D. NUMBERS]: Starting serial number for 1952 Ambassador models built at Kenosha was R-656001 to R-691337; for unassembled export 6KD3501 and up; for El Segundo RC-14501 to RC-18798. There was no Canadian production for this series. Engine numbers for the Ambassador sixes were A-165001 to A-205789 for all assembly points. [NASH RAMBLER SERIES 10 I.D. NUMBERS]: Staring serial number for 1952 Rambler models built at Kenosha was D-79501 to D-127045; for unassembled export DKD-2301 and up; and for El Segundo DC-4101 to DC-8914. Starting engine number for all 1952 Nash Ramblers was F-85001 to F-139394. [NASH-HEALEY I.D. NUMBERS]: Serial numbers N-2001 to N-2109 were used on cars built at Warwick, England as 1951 models. Motor Numbers were NHA-1001 and up. The beginning number for cars built at Warwick as 1952 models was N-2086. However, the beginning 1952 number was then followed by N-2103, N-2104, N-2106 and N-2200 with numbers following in sequence after N-2200. Motor Numbers for 1952 were NHA-1088 and up, but did not necessarily follow each other in sequence. The model number for the Nash-Healey was 25162 (1951) and 25262 (1952). The first two symbols indicate the series. The third symbol indicates model year. The fourth and fifth symbols are the equivalent of a body style number.

171

NASH STATESMAN SERIES 40

Model Number	Body/Style Number	Body Type & Seating	Factory Price	Shipping Weight	Production Total
SUPER LINE					
40	5245	4-dr Sed-6P	2178	3045	27,304
40	5246	2-dr Sed-6P	2144	3025	6,795
CUSTOM LINE					
40	5255	4-dr Sed-6P	2332	3070	13,660
40	5256	2-dr Sed-6P	2310	3050	1,872
40	5257	2-dr Cty Clb HT-6P	2433	3095	869

1952 Nash, Ambassador Custom two-door sedan, 6-cyl

AMBASSADOR — SERIES 60 (6-CYL): The 1952 Nash Ambassador shared the same styling and design changes as the 1952 Nash Statesman, the major difference being the seven inch longer front end. The 1952 Nash Ambassador came in two series: Super and Custom. Supers had basic features. Customs added foam seat cushions; two-tone upholstery; an electric clock; directional signals; chrome wheel discs; and front and rear courtesy lights. For 1952, the Ambassador engine was bored 1/8 inch, yielding 252.6 cid.

NASH AMBASSADOR SERIES 60

Model Number	Body/Style Number	Body Type & Seating	Factory Price	Shipping Weight	Production Total
SUPER LINE					
60	5265	4-dr Sed-6P	2557	3430	16,838
60	5266	2-dr Sed-6P	2521	3410	1,871
CUSTOM LINE					
60	5275	4-dr Sed-6P	2716	3480	19,585
60	5276	2-dr Sed-6P	2695	3450	1,178
60	5277	2-dr HT Cpe-6P	2829	3550	1,228

1952 Nash, Rambler Greenbriar two-door station wagon, 6-cyl (TVB)

NASH RAMBLER — SERIES 10 — (6-CYL): The 1952 Nash Rambler line received no major changes from 1951. Custom models came with the Nash "Weather-Eye" conditioned air system and a radio as standard equipment. The Greenbrier station wagon was an upgraded model with two-tone paint and richer trim.

NASH RAMBLER SERIES 10

Model Number	Body/Style Number	Body Type & Seating	Factory Price	Shipping Weight	Production Total
10	2204	2-dr Utl Wag-5P	1842	2415	1,248
10	5214	2-dr Sta Wag-5P	2003	2515	2,970
10	5221	2-dr Conv-5P	2119	2430	3,108
10	5224	2-dr Sta Wag-5P	2119	2515	19,889
10	5227	2-dr Cty Clb-5P	2094	2420	25,785

NOTE 1: Station wagon production included 4,425 Greenbriers.

NASH-HEALEY — SERIES 25 — SIX: The Nash-Healey sports car had a special two-passenger open body made of aluminum, an adjustable steering wheel and leather upholstery. The first Nash-Healey, model 25162, used the 234.8 cid Nash six and is sometimes called a "1951" model. The second version, model number 25262, switched to the 252.6 cid "LeMans" engine with dual carburetion. The English-built sports car's styling traits included a grille of outward curved vertical chrome bars entirely circled by a heavy chrome molding. There were model designations on the front fenders and in back of the wheel opening. The full hood had a unique hatch cover (air scoop) at the center with a vertical grille at the opening. These cars were built at Warwick, England and sold by Nash dealers. Styling by Pinin Farina was seen on 1952 models and late in the year, the more powerful LeMans engine was released. According to *Ward's Automotive Yearbook 1953* the official introduction date of the Nash-Healey, in the U.S., was Feb. 16, 1951. Since the Nash-Healey was of foreign manufacture, it will not be covered fully in this catalog, beyond the data given below.

NASH-HEALEY SERIES N-25

Model Number	Body/Style Number	Body Type & Seating	Factory Price	Shipping Weight	Production Total
N-25	25162	2-dr Spts Conv-2P	4063	2690	104
N-25	25262	2-dr Spts Conv-2P	5909	2750	150

NOTE 1: Style 25162 usually considered a 1951 model; 25262 the 1952 model.

ENGINES
(STATESMAN SIX): Inline. L-head six. Cast iron block. Displacement: 195.6 cid. Bore and stroke: 3-1.8 x 4-1/4 inches. Compression ratio: 7.10:1. Brake hp 88 at 3800 rpm. Four main bearings. Solid valve lifters. Carburetor: Carter one-barrel Type YF model 824S.
(AMBASSADOR SIX) Inline six. Overhead valves. Displacement: 252.6 cid. Bore and stroke: 3.50 x 4.375 inches. Compression ratio: 7.3:1. Brake hp: 120 at 3700 rpm. Seven main bearings. Solid valve lifters. Carburetor: Carter one-barrel Type WA (side-inlet).
(RAMBLER SIX) Inline. L-head six. Cast iron block. Displacement: 172.6 cid. Bore and stroke: 3-1/8 x 3-3/4 inches. Compression ratio: 7.25:1. Brake hp: 82 at 3800 rpm. Four main bearings. Solid valve lifters. Carburetor: Carter one-barrel type YF model 757S.
(NASH-HEALEY SIX) Inline six. Overhead valves. Displacement: 234.8 cid. Bore and stroke: 3.375 x 4.375. Compression ratio: 8.0:1. Brake hp: 125 at 4000 rpm.
(LE MANS SIX) Inline. LeMans six. Overhead valves. Displacement: 252.6 cid. Bore and stroke: 3.50 x 4.375. Compression ratio: 8.0: 1. Brake hp: 140 at 4000 rpm.
CHASSIS FEATURES: Wheelbase: (Statesman) 114-1/4 inches; (Ambassador) 121-1/4 inches; (Rambler) 100 inches. Overall length: (Statesman) 202-1/4 inches; (Ambassador) 209-1/4 inches; (Rambler) 176 inches. Front tread: (Statesman) 55-1/2 inches; (Ambassador) 55-5/8 inches; (Rambler) 55-3/8 inches. Rear tread: (Statesman) 59-11/16 inches; (Ambassador) 60-1/2 inches; (Rambler) 53 inches. Tires: (Statesman) 6.70 x 15; (Ambassador) 7.10 x 15; (Rambler Super) 5.90 x 15; (Rambler Custom) 6.40 x 15.
OPTIONS: Radio. Vacuum control antenna. Spotlight with mirror. Door-top outside mirror. Door mount outside mirror. Back-up lights. Fog lights. Non-glare rearview mirror. Visor vanity mirror. Curb-L-Arm. Custom seat covers. License plate frame. Exhaust extension. Protect-O-Mat for floor. Automatic windshield washer. Rear window wiper. Opto-shade for windshield. Vent shades. Outside windshield visor. Magnalite trouble light. Gas cap lock. Tissue dispenser. Bed mattress. Plastic screens. Rambler grille guard. Windshield washer. Rambler trunk guard. Trunk light. Rooftop carrier. Overdrive indicator. Visor pouch. Continental tire mount. Statesman 7.35:1 high-compression cylinder head. Ambassador 8.25:1 high-compression cylinder head. Rambler 7.6:1 high-compression cylinder head.
HISTORICAL FOOTNOTES: The 1952 Nash Ambassador and Statesman models were introduced March 14, 1952. The updated Rambler appeared on April 1, 1952. The Nash-Healey (sometimes considered a 1951 model) made its American debut as a 1952 Nash offering. Production hit a peak of 152,141 units or 3.51 percent of American auto sales. Model year production included 99,086 Statesman/Ambassador models and 55,055 Ramblers. Over 20 percent of all Nash products, or 28,950 cars, had Hydra-Matic Drive this year. The optional Warner Gear overdrive was installed in 74,535 units. The Nash-Healey took first place in its class in the French Grand Prix, at LeMans, plus a third place overall. These racing models used the LeMans "Dual-Jetfire" Ambassador engine, later released as a production car power plant.

1953 NASH

1953 Nash, Custom Statesman two-door sedan, 6-cyl

NASH STATESMAN — SERIES 40 — (6-CYL): The 1952 Golden anniversary styling for the Statesman went almost unchanged for 1953. The only outward change was the addition of vertical chrome stripes in the fresh air intake just below the front windshield. Supers had basic features. Customs added foam seat cushions; two-tone upholstery; an electric clock; directional signals; chrome wheel discs; and front and rear courtesy lights.
NASH I.D. NUMBERS: Vehicle Identification Plate located on the right-hand side of the cowl below the hood contains motor-serial number and number codes for model, paint and trim. Motor serial number matches the serial number. Model number has four symbols. The first two symbols indicate model year, 53=1953. The third symbol indicates series: 0=Rambler Utility; 1=Rambler Super 2=Rambler Custom; 3=Statesman; 4=Super Statesman; 5=Custom Statesman; 6=Ambassador Super; 7=Ambassador Custom. The fourth symbol indicates body style. (All four numbers appear in Body/Style Number column of charts below). Motor numbers located on right-hand side of crankcase towards front and on front upper left-hand side of block. [NASH STATESMAN SERIES 40 I.D. NUMBERS]: Starting serial numbers for the Statesman six were K-563501 to K-615291 for Kenosha, Wis.; 4KD-4801 and up for unassembled export; KC-43001 to KC-47173 for El Segundo, Calif. and KT-6901 and up for Toronto, Canada. Engine serial numbers were S-365001 to S-425053. [NASH AMBASSADOR SERIES 60 I.D. NUMBERS]: Starting serial numbers for the Ambassador six were R-692101 to R-721686 for Kenosha, Wis.; 6KD-3901 and up for unassembled export and RC-19001 to RC-21684 for El Segundo, Calif. Starting engine serial numbers were A-210001 to A-243959. (LeMans Dual-Jetfire engine numbers pre-

fixed by LMA. [NASH RAMBLER SERIES 10 I.D. NUMBERS]: Starting serial numbers for 1953 were D-127501 to D-155727 for Kenosha, Wis. built cars; DKD-2701 and up for unassembled export and DC-9001 to DC-12299 for El Segundo, Calif. Starting engine serial numbers were F-140001 to F-166688 for cars without Hydra-Matic and H-1001 for cars with Hydra-Matic.

Model Number	Body Style Number	Body Type & Seating	Factory Price	Shipping Weight	Production Total
SUPER LINE					
40	5345	4-dr Sed-6P	2178	3045	28,445
40	5346	2-dr Sed-6P	2144	3025	7,999
CUSTOM LINE					
40	5355	4-dr Sed-6P	2332	3070	11,476
40	5356	2-dr Sed-6P	2309	3050	1,305
40	5357	2-dr HT Cpe-6P	2433	3095	7,025

NOTE 1: Beginning 1953, Nash often publicized FAP (factory as-delivered prices) rather than ADP (advertised delivered prices) retail costs. FAP did not include federal tax and some preparation and handling charges, while ADP did. Earlier editions of this catalog reflected FAPs for 1953 and up models. This edition shows the ADPs based on March issues of industry trade journals through 1956.

1953 Nash, Ambassador Country Club two-door hardtop coupe, 6-cyl

AMBASSADOR — SERIES 60 — (6-CYL): The 1952 Golden anniversary styling from 1952 for the 1953 Ambassador went almost unchanged for 1953. The only outward change was the addition of vertical chrome strips in the fresh air intake just below the front windshield. The 1953 Nash Ambassador again came in two series: Super and Custom. Supers had basic features. Customs added foam seat cushions; two-tone upholstery; an electric clock; directional signals; chrome wheel discs; and front and rear courtesy lights. In addition, a new dual-carburetor 'Le Mans Dual Jetfire' engine (similar to the motor used in the Nash-Healeys raced at LeMans and named after them) was made optional on 1953 Ambassadors. This engine produced 140 hp at 4000 rpm.

NASH AMBASSADOR SERIES 60

Model Number	Body/Style Number	Body Type & Seating	Factory Price	Shipping Weight	Production Total
SUPER LINE					
60	5365	4-dr Sed-6P	2557	3430	12,489
60	5366	2-dr Sed-6P	2521	3410	1,273
CUSTOM LINE					
60	5375	4-dr Sed-6P	2716	3480	12,222
60	5376	2-dr Sed-6P	2695	3450	428
60	5377	2-dr HT Cpe-6P	2829	3550	6,438

1953 Nash, Rambler Custom two-door hardtop, 6-cyl

NASH RAMBLER — SERIES 10 — (6-CYL): The 1953 Nash Rambler was completely restyled for 1953 and distinguished by a lowered hood, single bar front grille and enclosed front and rear fenders. The new styling was credited to Pinin Farina and had many of the styling features of the 1952 and 1953 Golden Anniversary Ambassador and Statesman. Custom models came with the Nash "Weather-Eye" conditioned air system and a radio as standard equipment. Custom convertible and Country Club hardtop included continental spare tire. Dual-Range Hydra-Matic became an option available for the first time on the 1953 Nash Rambler with Hydra-Matic equipped cars receiving an engine with five more brake hp than manual transmission cars.

NASH RAMBLER SERIES 10

Model Number	Body/Style Number	Body Type & Seating	Factory Price	Shipping Weight	Production Total
10	2304	2-dr Deliveryman	—	—	9
10	5314	2-dr Sta Wag-5P	2003	2555	1,114
10	5321	2-dr Conv-5P	2150	2590	3,284
10	5324	2-dr Sta Wag-5P	2119	2570	10,598
10	5327	2-dr CtyClb-5P	2125	2550	15,255

NOTE: Includes three model 5306 — two-door sedan Deluxe.
NOTE: Includes 3536 Greenbriar Station Wagon and 7035 DiNoc Station Wagons.

ENGINES
(STATESMAN SIX): Inline. L-head six. Cast iron block. Displacement: 195.6 cid. Bore and stroke: 3-1/8 x 4-1/4 inches. Compression ratio: 7.45:1. Brake hp: 100 at 3800 rpm. Four main bearings. Solid valve lifters. Carburetor: Carter two-barrel WCD-2034S.
(AMBASSADOR SIX): Inline six. Overhead valves. Displacement: 252.6 cid. Bore and stroke: 3.50 x 4.375 inches. Compression ratio: 7.3:1. Brake hp: 120 at 3700 rpm. Seven main bearings. Solid valve lifters. Carburetor: Carter one-barrel Type YH-895-S or YH-895-SA).
(AMBASSADOR "LE MANS" SIX): Inline. LeMans six. Overhead valves. Displacement: 252.6 cid. Bore and stroke: 3.50 x 4.375 inches. Compression ratio: 8.0:1. Brake hp: 140 at 4000 rpm. Carburetors: (Front) Carter YH 973S; (Rear) Carter YH 974S.
(RAMBLER SIX): Inline. L-head six. Cast iron block. Displacement: 184.1 cid. Bore and stroke: 3-1/8 x 4 inches. Compression ratio: 7.25:1. Brake hp: 85 at 3800 rpm. Four main bearings. Solid valve lifters. Carburetor: Carter one-barrel type YF model 2014S.
(RAMBLER HYDRA-MATIC SIX): Inline. Six-cylinder. L-head. Cast iron block. Displacement: 195.6 inches. Compression ratio: 7.3:1. Four main bearings. Solid lifters. Brake hp: 90 at 3800 rpm. Carburetion: Carter one-barrel Model YF.
CHASSIS FEATURES: Wheelbase: (Statesman) 114-1/4 inches; (Ambassador) 121-1/4 inches; (Rambler) 100 inches. Overall length: (Statesman) 202-1/4 inches; (Ambassador) 209-1/4 inches; (Rambler Super) 176 inches; (Rambler Custom with continental tire extension) 185-3/8 inches. Front tread: (Statesman) 55-1/2 inches; (Ambassador) 55-5/8 inches; (Rambler) 55-3/8 inches. Rear tread: (Statesman) 59-11/16 inches; (Ambassador) 60-1/2 inches; (Rambler) 53 inches. Tires: (Statesman) 6.70 x 15; (Ambassador) 7.10 x 15; (Rambler Super) 5.90 x 15; (Rambler Custom) 6.40 x 15.
OPTIONS: Weather-Eye conditioned air system. Reclining seat and twin bed. Hydra-Matic automatic transmission. Automatic overdrive. Radio with twin speakers. Solex glass. White sidewall tires. Two-tone paint. Power steering (Ambassador only). Oil bath air cleaner. Rambler higher-compression engine with Hydra-Matic transmission.
HISTORICAL FOOTNOTES: Calendar year production amounted to 93,504 Nash models and 41,825 Ramblers. In calendar year 1953, Nash built 3,501 convertibles; 13,533 station wagons; and 34,356 hardtops.

1954 NASH
METRO

NASH STATESMAN — SERIES 40 — (6-CYL): The 1954 Nash Statesman carried over the 1952-1953 Golden Anniversary styling with a minor facelift of the body. Changes included a new front concave grille and new chrome head light bezels. Continental rear tire carriers were added as standard on all Custom models. New interiors and instrument panels on both the Statesman and Ambassador appeared for 1954 also.
NASH I.D. NUMBERS: Vehicle Identification Plate located on the right-hand side of the cowl below the hood contains motor-serial number and number codes for model, paint and trim. Motor serial number matches the serial number. Model number has four symbols. The first two symbols indicate model year, 54=1954. The third symbol indicates series: 0=Rambler Deluxe; 1=Rambler Super; 2=Rambler Custom; 4=Statesman Super; 5=Statesman Custom; 6=Ambassador Super; 7=Ambassador Custom. The fourth symbol indicates body style. (All four numbers appear in Body/Style Number column of charts below). Motor numbers located on right-hand side of crankcase towards front. [NASH STATESMAN SERIES 40 I.D. NUMBERS]: Starting serial numbers for the Nash Statesman were K-615501 for Kenosha, Wis.; 4KD-5101 for unassembled export, KC-47201 for El Segundo, Calif. and KT-9101 for Toronto, Canada. Starting engine serial numbers were J-1001 for 1954. [NASH AMBASSADOR SERIES 60 I.D. NUMBERS]: Starting serial numbers for the Nash Ambassador were R-722501 for Kenosha, Wis.; 6KD-4201 for unassembled export and RC-22001 for El Segundo, Calif. Starting engine serial numbers were A246001 for 1954. [NASH RAMBLER SERIES 10 I.D. NUMBERS]: Starting serial numbers for the 1954 Nash Rambler were D-171501 for Kenosha, Wis.; DKD-3001 for unassembled export, and DC-12301 for El Segundo, Calif. Starting engine serial numbers for the 1954 Nash Rambler were F-170001.

NASH STATESMAN SERIES 40

Model Number	Body/Style Number	Body Type & Seating	Factory Price	Shipping Weight	Production Total
SUPER LINE					
40	5445	4-dr sedan-6P	2178	3045	11,401
40	5446	2-dr Sed-6P	2130	3025	1,855
CUSTOM LINE					
40	5455	4-dr Sed-6P	2362	3095	4,219
40	5456	2-dr Sed-6P	2340	3050	24
40	5457	2-dr HT Cpe-6P	2468	3120	2,726

1954 Nash, Ambassador Custom four-door sedan, 6-cyl

NASH AMBASSADOR — SERIES 60 — (6-CYL): The 1954 Nash Ambassador, with a seven inch longer wheelbase than the 1954 Nash Statesman, shared the same changes as the Statesman for 1954. They included a new front concave grille and new chrome head light bezels, new instrument panel and new interior. Continental rear tire carriers were standard on all Custom models.

NASH AMBASSADOR SERIES 60

Model Number	Body/Style Number	Body Type & Seating	Factory Price	Shipping Weight	Production Total
SUPER LINE					
60	5465	4-dr Sed-6P	2412	3430	7,433
60	5466	2-dr Sed-6P	2360	3025	283
CUSTOM LINE					
60	5475	4-dr Sed-6P	2595	3095	10,131
60	5477	2-dr HT Cpe-6P	2730	3120	3,581

1954 Nash, Rambler Country Club two-door hardtop coupe, 6-cyl

NASH RAMBLER — SERIES 10 — (6-CYL) — The 1954 Nash Rambler received no major appearance changes for 1954. However several new models were added to the lineup. New for 1954 were a four-door sedan and a four-door station wagon on a new longer 108 inch wheelbase, also added were a Deluxe and Super two-door sedan on the 100 inch wheelbase as low-line price leaders. To cut costs, radios and heaters were changed from standard equipment to options.

NASH RAMBLER SERIES 10

Model Number	Body/Style Number	Body Type & Seating	Factory Price	Shipping Weight	Production Total
10	2404	2-dr Utl Wag-5P	1444	2425	56
10	5406	2-dr Del Sed-5P	1550	2425	7,273
10	5414	2-dr Suburb-5P	1945	2520	504
10	5415	4-dr Super Sed-5P	1995	2570	4,313
10	5416	2-dr Super Sed-5P	1865	2425	300
10	5417	2-dr CtyClb-5P	1945	2465	1,071
10	5421	2-dr Conv-5P	2125	2555	221
10	5424	2-dr Sta Wag-5P	2095	2535	2,202
10	5425	4-dr Cus Sed-5P	2175	2630	7,640
10	5427	2-dr Cus HT Cpe-5P	2095	2515	3,612
10	5428	4-dr Cus Sta Wag-5P	2200	2715	9,039

ENGINES
(STATESMAN SIX): Inline. L-head six. Cast iron block. Displacement: 195.6 cid. Bore and stroke: 3-1/8 x 4-1/4 inches. Compression ratio: 8.5:1. Brake hp: 110 at 4000 rpm. Four main bearings. Solid valve lifters. Carburetor: Carter two-barrel WCD-2098S; or Carter YF-2137-S; or Carter YF-2137-SA.
(AMBASSADOR SIX) Inline six. Overhead valves. Displacement: 252.6 cid. Bore and stroke: 3.50 x 4.375 inches. Compression ratio: 7.6:1. Brake hp: 130 at 3700 rpm. Seven main bearings. Solid valve lifters. Carburetor: Carter one-barrel Type YH-895-S or YH-895-SA).
(LEMANS "DUAL-JETFIRE" SIX) Inline. LeMans six. Overhead valves. Displacement: 252.6 cid. Bore and stroke: 3.50 x 4.375. Compression ratio: 8.0:1. Brake hp: 140 at 4000 rpm. Carburetors: (Front) Carter YH 973S; (Rear) Carter YH 974S.
(RAMBLER SIX) Inline. L-head six. Cast iron block. Displacement: 184.1 cid. Bore and stroke: 3-1/8 x 4 inches. Compression ratio: 7.25:1. Brake hp: 85 at 3800 rpm. Four main bearings. Solid valve lifters. Carburetor: Carter one- barrel type YF model 2014S.
(RAMBLER HYDRA-MATIC SIX) Inline. Six-cylinder. L-head. Cast iron block. Displacement: 195.6 inches. Compression ratio: 7.3:1. Four main bearings. Solid lifters. Brake hp: 90 at 3800 rpm. Carburetion: Carter one-barrel Model YF.
CHASSIS FEATURES: Wheelbase: (Statesman) 114.3 inches; (Ambassador) 121.3 inches; (Rambler 10) 100 inches; (Rambler four-door) 108 inches. Overall length: (Statesman) 202.3 inches; (Statesman with continental kit) 212.3 inches; (Ambassador) 209.3 inches; (Ambassador with continental kit) 219.3 inches; (Rambler 10 Super) 178.3 inches; (Rambler 10 Custom with continental tire extension) 185.4 inches; (Rambler four-door) 186.3 inches; (Rambler four-door with continental kit) 193.4 inches. Front tread: (Statesman) 55.5 inches; (Ambassador) 59.7 inches; (Rambler) 53.4 inches. Rear tread: (Statesman) 55.6 inches; (Ambassador) 60.5 inches; (Rambler) 53 inches. Tires: (Statesman) 6.70 x 15; (Ambassador) 7.10 x 15; (Rambler Super) 5.90 x 15; (Rambler Custom) 6.40 x 15.
OPTIONS: Rambler (general prices). Oil bath air cleaner ($7.50). Airliner reclining seats only ($111.45); and twin beds ($18.00). Air mattresses for twin bed ($15.00). Electric clock ($17.00). Two-tone color ($16.00). Directional signals ($16.00). Le Mans Dual-Jetfire engine for Ambassador (price n.a.). Foam cushions front or rear ($10.00); both front and rear ($20.00). Solex glass ($19.00). Hydra-Matic transmission ($179.00). Overdrive ($104.00). Power brakes (price n.a.). Power steering (price n.a.). Radio and antenna standard Rambler. Custom steering wheel ($14.00); with overdrive power pass ($18.95). Heavy springs and shocks ($8.00). White sidewall tires, size 6.40 x 15 ($25.00). White sidewall tires, size 5.90 x 15 ($22.00). Upholstery options ($80.00). Electric window lifts (Ambassador only price n.a.). Wheel discs four ($21.00). Weather-Eye (standard Rambler) optional Statesman and Ambassador.
HISTORICAL FOOTNOTES: Historical note: On April 22, 1954 Nash and Hudson merged to form American Motors Corp. The official date for the beginning of AMC was May 1. George Mason hoped to bring Studebaker and Packard into the new corporation as well, but he passed away before this could be accomplished. Nash, in 1954, was the first to introduce low price air-conditioning systems to mass market autos. The system Nash invented for use then is the basis for all modern auto air-conditioning systems. Previously, air-conditioning was very expensive and available only on very expensive autos in limited quantities and filled half the trunk area. Nash's system was much more compact and easily serviced and integrated into the underhood area of the car. Calendar year production included 37,779 Ramblers and 29, 371 Nash. The total of 67,150 cars made Nash America's 13th largest automaker. Model year production was 62,911 units made to 1954 specifications. The LeMans "Dual-Jetfire" six was advertised as an engine "that has won many Grand Prix d'Endurance awards." In calendar year 1954, Nash was credited with building 6,065 hardtops, 11,800 station wagons; and four convertibles.

1955 NASH
HUDSON

NASH STATESMAN — SERIES 40 — (6-CYL): The 1955 Nash Statesman had a completely revised version of 1952's Golden Anniversary styling. A "Scena-Ramic" wraparound windshield appeared. Another new feature was a long character molding from front to rear fenders. The headlights were enclosed in a redesigned oval grille. The new concave grille had multiple chrome dividers. Custom models had the continental spare tire mount.
NASH I.D. NUMBERS: Vehicle Identification Plate located on the right-hand side of the cowl below the hood contains motor-serial number and number codes for model, paint and trim. Motor serial number matches the serial number. Model number has four symbols. The first two symbols indicate model year, 55=1955. The third symbol indicates series: 0=Rambler Deluxe; 1=Rambler Super; 2=Rambler Custom; 4=Statesman Super; 5=Statesman Custom; 6=Ambassador Super; 7=Ambassador Custom. The fourth symbol indicates body style. (All four numbers appear in Body/Style Number column of charts below). Motor numbers located on right-hand side of crankcase towards front. [NASH STATESMAN SERIES 40 I.D. NUMBERS]: Starting serial numbers for the Statesman six were K-635001 to K-649123 for Kenosha, Wis.; 4KD-5401 and up for unassembled export; KC-48101 and up for El Segundo, Calif. and KT-10501 and up for Toronto, Canada. Starting engine serial numbers were S-440001 to S-453100 for single-carburetor jobs and J-30001 to J-33070 for dual-carburetor jobs. [NASH AMBASSADOR SERIES 60 I.D. NUMBERS]: Starting serial numbers for the Ambassador six were R-742901 to R-757866 for Kenosha, Wis.; 6KD-4401 and up for unassembled export and RC-23001 and up for El Segundo, Calif., with no Canadian production of this series. Starting engine serial numbers were A-270001 to A-278691 for single carburetor Super Six and LMA-270001 to LMA-277002 for dual-carburetor jobs. [NASH AMBASSADOR SERIES 80 I.D. NUMBERS]: Starting serial numbers for the Ambassador V-8s were V-1001 to V-11444 for Kenosha, Wis.; 8KD-1001 and up for unassembled export and VC-1001 and up for El Segundo, Calif. Starting engine serial number for all Ambassador Eights was P-1001 to P-11444. [NASH RAMBLER SERIES 10 I.D. NUMBERS]: Starting serial numbers for Ramblers were D-205001 to D-276099 for Kenosha, Wis.; DKD-3701 and up for unassembled export and DC-15001 to (Nash) DC-23326/(Hudson) DC-23325 for El Segundo, Calif. Starting engine serial number for all Ramblers was H45001 to H-131414.

NASH STATESMAN SERIES 40

Model Number	Body/Style Number	Body Type & Seating	Factory Price	Shipping Weight	Production Total
SUPER LINE					
40	5545-1	4-dr Sup Sed-6P	2215	3134	12,877
CUSTOM LINE					
40	5547-2	2-dr HT Cpe-6P	2495	3220	1,395
40	5545-2	4-dr Cus Sed-6P	2385	3204	Note 1

NOTE 1: Custom four-door production included with Super four-door total.

1955 Nash, Ambassador Country Club two-door hardtop, V-8

AMBASSADOR — SERIES 60 — (6-CYL): The 1955 Nash Ambassador six received the same type of appearance changes as the 1955 Nash Statesman including: new wrap-around windshield, new long character moldings from front to rear fenders; and headlights enclosed in a redesigned oval-concave grille.

NASH AMBASSADOR SERIES 60

Model Number	Body/Style Number	Body Type & Seating	Factory Price	Shipping Weight	Production Total
SUPER					
60	5565-1	4-dr Sup Sed-6P	2480	3538	13,809
CUSTOM LINE					
60	5567-2	2-dr HT Cpe-6P	2795	3593	1,395
60	5565-2	4-dr Cus Sed-6P	2675	3576	Note 1

NOTE 1: Custom four-door production included with Super four-door total.

NASH AMBASSADOR SERIES 80 — (8-CYL): The year 1955 marked the introduction of an overhead valve V-8 engine in the Ambassador line. The new power plant was a 320 cid engine purchased from Packard. It was only available with Twin Ultramatic transmission. The 1955 Nash Ambassador V-8 was distinguished by V-8 emblems on its rear fenders and Ambassador (Custom or Super) V-8 emblems on front fenders. Styling was otherwise the same as on the Ambassador six.

NASH AMBASSADOR SERIES 80

Model Number	Body/Style Number	Body Type & Seating	Factory Price	Shipping Weight	Production Total
SUPER					
80	5585-1	4-dr Sup Sed-6P	2775	3795	8,805
CUSTOM LINE					
80	5587-2	2-dr HT Cpe-6P	3095	3839	1,775
80	5585-2	4-dr Cus Sed-6P	2965	3827	Note 1

NOTE 1: Custom four-door production included with Super four-door total.

1955 Nash, Rambler Country Club two-door hardtop, 6-cyl (TVB)

NASH RAMBLER — SERIES 10 — (6-CYL): The 1955 Nash Rambler received a minor facelift over 1954. New features included the addition of a new cellular grille and full wheel cutouts in the front fenders. Both Nash and Hudson marketed versions of the Rambler in 1955.

NASH RAMBLER SERIES 10

Model Number	Body/Style Number	Body Type & Seating	Factory Price	Shipping Weight	Production Total
FLEET (100" WHEELBASE)					
10	2504	2-dr Utl Wag-5P	—	2500	14
DELUXE (100" W.B. TWO-DOOR/108" W.B. FOUR-DOOR)					
10	5512	2-dr Del Bus Sed-3P	1328	2400	43
10	5516	2-dr Del Clb Sed-5P	1585	2432	Note 1
10	5515	4-dr Del Sed-6P	1695	2567	Note 1
SUPER (100" W.B. TWO-DOOR/108" W.B. FOUR-DOOR)					
10	5516-1	2d Sup Clb Sed-5P	1683	2450	8979
10	5514-1	2-dr Sup Sub-5P	1869	2532	2379
10	5515-1	4-dr Sup Sed-6P	1798	2570	15,998
10	5518-1	4-dr Sup Cr Cty-6P	1807	2675	Note 1
CUSTOM (108" W.B. TWO-DOOR/108" W.B. FOUR-DOOR)					
10	5517-2	2-dr Cus Cty Clb HT-5P	1995	2518	2,993
10	5515-2	4-dr Cus Sed-6P	1989	2606	Note 1
10	5518	4-dr Cus Sta Wag-5	2098	2685	25,617

NOTE 1: Production included with same Style Number in other trim lines.

ENGINES
(STATESMAN SIX): Inline. L-head six. Cast iron block. Displacement: 195.6 cid. Bore and stroke: 3-1/8 x 4-1/4 inches. Compression ratio: 7.45:1. Brake hp: 100 at 3800 rpm. Four main bearings. Solid valve lifters. Carburetor: Carter one-barrel YF-22585.
(STATESMAN SIX DUAL CARB): Inline. Overhead six. Cast iron block. Displacement: 195.6 cid. Bore and stroke: 3-1/8 x 4-1/4 inches. Compression ratio: 8.0:1. Brake hp: 110 at 4000 rpm. Four main bearings. Solid valve lifters. Carburetor: Carter one-barrel (front) YH-973S; (rear) YH-974S.
(AMBASSADOR SIX) Inline six. Overhead valves. Displacement: 252.6 cid. Bore and stroke: 3.50 x 4.375 inches. Compression ratio: 7.6:1. Brake hp: 130 at 3700 rpm. Seven main bearings. Solid valve lifters. Carburetor: Carter one-barrel Type YH-895-S).
(LEMANS "DUAL-JETFIRE" SIX) Inline. LeMans six. Overhead valves. Displacement: 252.6 cid. Bore and stroke: 3.50 x 4.375. Compression ratio: 7.6:1. Brake hp: 140 at 4000 rpm. Carburetors: (Front) Carter YH 973S; (Rear) Carter YH 974S.
(RAMBLER SIX) Inline. Six-cylinder. L-head. Cast iron block. Displacement: 195.6 inches. Compression ratio: 7.3:1. Four main bearings. Solid lifters. Brake hp: 90 at 3800 rpm. Carburetion: Carter one-barrel Model YF.
(AMBASSADOR V-8) V-8. Overhead valves. Cast iron block. Displacement: 320 inches. Bore and stroke: 313/16 x 3-1/2 inches. Compression ratio: 7.8:1. Brake hp: 208 at 4200 rpm. Non-adjustable hydraulic valve lifters. Five main bearings. Carburetor: Carter two-barrel Model WGD.

CHASSIS FEATURES: Wheelbase: (Statesman) 114.3 inches; (Ambassador) 121.3 inches; (Rambler 10) 100 inches; (Rambler four-door) 108 inches. Overall length: (Statesman) 202.3 inches; (Statesman with continental kit) 212.3 inches; (Ambassador) 209.3 inches; (Ambassador with continental kit) 219.3 inches; (Rambler 10 Super) 178.3 inches; (Rambler 10 Custom with continental tire extension) 185.4 inches; (Rambler four-door) 186.3 inches; (Rambler four-door with continental kit) 193.4 inches.. Front tread: (Statesman) 55.5 inches; (Ambassador) 59.7 inches; (Rambler) 53.4 inches. Rear tread: (Statesman) 55.6 inches; (Ambassador) 60.5 inches; (Rambler) 53 inches. Tires: (Statesman) 6.70 x 15; (Ambassador) 7.10 x 15; (Rambler Super) 5.90 x 15; (Rambler Custom) 6.40 x 15.

OPTIONS: [Rambler]: Rambler Oil bath air cleaner ($8). Reclining seats ($11); with twin bed ($18). Air mattress for twin bed ($15). All season air conditioning ($345). Electric clock ($17). Two-tone color ($16). Directional signals ($16). Foam seat cushions front or rear ($10). Foam seat cushions front and rear ($20). Solex glass ($19). Hood ornament ($11). Hydra-Matic transmission ($179). Overdrive ($104). Radio and antenna ($76). Heavy-duty springs ($8). White sidewall tires, size 6.40 x 15 ($27). Black sidewall tires, size 6.40 x 15 ($37). White sidewall tires, size 6.40 x 15, six-ply tubeless construction($70). Upholstery ($55). Vacuum booster pump ($3). Weather-Eye heater ($74). [Nash]: Radio. Electric antenna. Visor vanity mirror. Non-glare rearview mirror. Outside mirror. Spotlight and mirror. Wire wheelcovers. Back-up lights. Windshield washer. Fog lights. Rear window wiper. Trunk light. Electric clock. Air mat. Hand spotlight. Plastic screens. Luggage carrier. Door top shades. Curb-L-arms. Sola-cell cooling system. Dyna-Flyte dual plate distributor. Oil filter. Fuel filter. Gas filler guard. Hood ornament. Door edge guards. Exhaust extension. License plate frame.

HISTORICAL FOOTNOTES: Nash had a slow start in 1955, but, once started, moved along at a warm pace. The company wound-up the year with model year sales of 109,102 units. Calendar year output was 83,852 Ramblers and 51,315 Nashes for 10th place in the industry. Dealer contests and sales promotions were instrumental in stimulating deliveries. A total of 81,237 Nash/Hudson units were built, highest run for the line in history. An added feather in Rambler's cap was its consistent holding of number one spot in used car value, as reflected in NADA reports. Nash's more powerful, 'Speedline' styled 1956 models were unveiled Nov. 17, 1955. Dealer introductions took place Nov. 23, 1954. Two new engines were introduced: the top-powered 220 hp Ambassador Jetfire V-8 with Twin Ultramatic and a Statesman six overhead valve power plant. Model year production began October 1954 and included 40,133 Statesman/Ambassadors and 56,023 Ramblers. Calendar year production included 7,442 two-door hardtops and 28,163 station wagons.

1956 NASH

NASH STATESMAN — SERIES 40 — (6-CYL): The 1956 Nash Statesman received a major facelift. The front and rear fenders were restyled. There were larger, more visible front running lights and, also, new taillights. A revised hood ornament and one-piece rear windows were seen. Chrome side stripping revisions included a shallow 'Z' shape on the side of the car and outline moldings on the hood and rear fender sides. The only Statesman model available for 1956 was the four-door Super sedan. Its engine was redesigned to an overhead valve configuration.

NASH I.D. NUMBERS: Vehicle Identification Plate located on the right-hand side of the cowl below the hood contains motor-serial number and number codes for model, paint and trim. Motor serial number matches the serial number. Model number has four symbols. The first two symbols indicate model year, 56=1956. The third symbol indicates series: 0=Rambler Deluxe; 1=Rambler Super; 2=Rambler Custom; 4=Statesman Super; 5=Statesman Custom; 6=Ambassador Super; 7=Ambassador Custom. The fourth symbol indicates body style. (All four numbers appear in Body/Style Number column of charts below). Motor numbers located on right-hand side of crankcase towards front. [NASH STATESMAN SERIES 40 I.D. NUMBERS]: Starting serial numbers for the Statesman six were K-649201 for Kenosha, Wis.; 4KD-5701 for unassembled export and KT-11101 for Toronto, Canada. Starting engine serial number was DB-1001. [NASH AMBASSADOR SERIES 60 I.D. NUMBERS]: Starting serial numbers for the Ambassador six were R-757901 for Kenosha, Wis. and 6KD-4601 for unassembled export. Starting engine serial numbers were A-279001 for single-carburetor jobs and LMA-277001 for the dual-carburetor jobs. [NASH AMBASSADOR SERIES 80 I.D. NUMBERS]: Starting serial numbers for Ambassador Eights were V-11501 for Kenosha, Wis., and 8KD-1101 for unassembled export. Starting engine serial numbers were P-21001 for Ambassador Eights. [NASH AMBASSADOR SPECIAL SERIES 50 I.D. NUMBERS]: Starting serial numbers were U-1001 for Kenosha, Wis., UKD-1001 for unassembled export and UT-1001 for cars built in Toronto, Canada. Starting engine serial number was G-1001 for all series 50 Ambassador V-8s. [NASH RAMBLER SERIES 10 I.D. NUMBERS]: Starting serial numbers for 1956 Ramblers were D-276101 for Kenosha, Wis.; DKD-5601 for unassembled export and KT-5401 for Toronto, Canada. Starting engine serial number for all Ramblers was B1001.

NASH STATESMAN SERIES 40

Model Number	Body/Style Number	Body Type & Seating	Factory Price	Shipping Weight	Production Total
40	5645-1	4-dr Sup Sed-6P	2385	3199	7,438

NASH AMBASSADOR — SERIES 60 — (6-CYL): The 1956 Nash Ambassador six was available in only one body style, the four-door Super sedan. It shared all the styling changes of 1956 Statesman models on the seven inch longer wheelbase of the Ambassador platform. An Ambassador Super script appeared on front fenders.

NASH AMBASSADOR SERIES 60

Model Number	Body/Style Number	Body Type & Seating	Factory Price	Shipping Weight	Production Total
60	5665-1	4-dr Sup Sed-6P	2689	3555	5,999

1956 Nash, Ambassador Custom four-door sedan, V-8

NASH AMBASSADOR — SERIES 80 — (8-CYL): The 1956 Nash Ambassador Eight shared the same styling changes as the Ambassador six. The Super V-8 sedan was equipped and trimmed similar to the Super Six sedan. Customs had the name "Ambassador Custom" on the front fenders and "Ambassador Country Club" on the hardtops. On later production models, vertical chrome moldings are added to the front fenders of Customs. The V-8 was, again, a Packard built engine, but a larger 352 cubic inch displacement block was used. This motor was only available with Packard's Ultramatic transmission attached.

175

NASH AMBASSADOR SERIES 80

Model Number	Body/Style Number	Body Type & Seating	Factory Price	Shipping Weight	Production Total
80	5685-1	4-dr Sup Sed-6P	2956	3748	3,885
80	5685-2	4-dr Cus Sed-6P	3195	3846	Note 1
80	5687-2	2-dr Cus HT Cpe-6P	3338	3854	796

NOTE 1: Production of Super and Custom four-door is counted as a single total.

NASH AMBASSADOR SPECIAL — SERIES 50 — (8-CYL): The Nash Ambassador Special V-8 was introduced as a midyear 1956 model. It came out in April with a V-8 engine of AMC's own design and manufacture. The Ambassador Special was available in three models, a Super four-door sedan, a Custom four-door sedan and a two-door hardtop coupe. Supers had single side rub-rail moldings and chrome moldings across the front of the hood and fenders. Power brakes, an "Airliner" reclining seat and continental tire mounting were standard on Custom models. They also had double chrome side rub-rail moldings enclosing a separate color area and a chrome band across the front of the hood and fenders, down to the bumper wings. On later production models, vertical chrome moldings are added to the front fenders of Customs.

NASH AMBASSADOR SERIES 50

Model Number	Body/Style Number	Body Type & Seating	Factory Price	Shipping Weight	Production Total
50	5655-1	4-dr Sup Sed-6P	2355	3397	4,145
50	5655-2	4-dr Cus Sed-6P	2462	3418	Note 1
50	5657-2	2-dr Cus Sed-6P	2541	3567	706

NOTE 1: Production of Super/Custom four-doors is counted as a single total.

1956 Nash, Rambler Custom Cross Country four-door wagon, 6-cyl

NASH RAMBLER — SERIES 10 — (6-CYL): The 1956 Nash Rambler received a major redesign of the long-wheelbase four-door sedan and station wagon. The short-wheelbase cars were dropped. (They would reappear, with a few minor changes, as the 1958 American.) The 1956 models were totally redesigned on the outside, with a new oval-shaped grille housing the headlights located inside the grille. Running lights (front parking lights that stayed on even when the headlights were turned on) were set high in each front fender. They complemented the new rear fenders and revisions to the rear deck. Chrome trim and color treatments with three-tone color combinations were available and a wraparound rear window appeared. Also introduced was the first four-door hardtop station wagon. Nash production was discontinued at the El Segundo, Calif. plant.

NASH RAMBLER SERIES 10

Model Number	Body/Style Number	Body Type & Seating	Factory Price	Shipping Weight	Production Total
DELUXE (108" WHEELBASE)					
10	5615	4-dr Del Sed-6P	1829	2891	21,966
SUPER (108" WHEELBASE)					
10	5615-1	4-dr Sup Sed-6P	1939	2906	Note 1
10	5618-1	4-dr Sup Sta Wag-6P	2233	2992	21,554
CUSTOM (108" WHEELBASE)					
10	5613-2	4-dr Cty Clb Wag-6P	2491	3095	402
10	5615-2	4-dr Cus Sed-6P	2056	2990	Note 1
10	5618-2	4-dr Cus Crs Cty Wag-6P	2326	3110	Note 1
10	5619-2	4-dr Cus HT Sed-6P	2224	2990	2,155

NOTE 1: If two models have same first four symbols in Body Style Numbers the production of cars with different levels of trim is not separately broken out.

BASE ENGINES

(STATESMAN SIX) Six-cylinder. Overhead valves. Cast iron block. Displacement: 195.6 cid. Bore and stroke: 3-1/8 x 4-1/4 inches. Compression ratio: 7.47:1. Brake hp: 130 at 4500 rpm. Four main bearings. Solid valve lifters. Carburetor: Carter two barrel WCD 2350S or Stromberg BVX-25 Model 380288.. A 12-volt electrical system was used for the first time.
(AMBASSADOR SIX) Inline six. Overhead valves. Displacement: 252.6 cid. Bore and stroke: 3.50 x 4.375 inches. Compression ratio: 7.6:1. Brake hp: 135 at 3700 rpm. Seven main bearings. Solid valve lifters. Carburetor: Carter one-barrel Type YH 2368S).
(LEMANS "DUAL-JETFIRE" SIX) Inline. LeMans six. Overhead valves. Displacement: 252.6 cid. Bore and stroke: 3.50 x 4.375. Compression ratio: 7.6:1. Brake hp: 145 at 4000 rpm. Carburetors: (Front/Rear) Carter YH 2369S.
(AMBASSADOR/PACKARD V-8). V-8. Overhead valves. Displacement: 352 cid. Bore and stroke: 4 x 3-1/2 inches. Compression ratio: 9.55:1. Brake hp: 220 at 4600 rpm. Five main bearings. Hydraulic lifters. Carburetion: Carter two-barrel WGD 2231S.
(AMBASSADOR/AMC V-8). V-8. Overhead valve. Cast iron block. Displacement: 250 cid. Bore and stroke: 3-1/2 x 3-1/4 inches. Compression ratio: 8.0:1. Brake hp: 190 at 4900 rpm. Five main bearings. Non-adjustable hydraulic lifters. Carburetion: Carter two-barrel WGD.
(RAMBLER SIX) Six-cylinder. Overhead valve. Cast Iron block. Displacement: 195.6 cid. Bore and stroke: 3-1/8 x 4-1/4 inches. Compression ratio: 7.47:1. Brake hp: 120 at 4200 rpm. Four main bearings. Solid valve lifters. Carburetion: Carter one-barrel Model AS-2349S.

CHASSIS FEATURES: Wheelbase: (Rambler) 108 inches; (Statesman) 114.3 inches; (Ambassador 50/60) 121.3 inches. Overall length: (Rambler) 191.1 inches; (Rambler Custom with continental kit) 198.9 inches; (Statesman) 202.3 inches; (Ambassador 50/60) 209.3 inches. Front tread: (Rambler) 57.8 inches; (Statesman/Ambassador 50 and 60) 56.6 inches. Rear tread: (Rambler) 58 inches; (Statesman) 59.7 inches; (Ambassador 50/60) 60.5 inches. Tires: (Rambler) Four-ply tubeless, size 6.40 x 15; (Statesman) 6.70 x 15; (Ambassador 50) 7.60 x 15; (Ambassador 60) 7.10 x 15.

OPTIONS: (All models) Radio. Hood ornament. Outside mirror. Non-glare rearview mirror. Spotlight mirror. Backup lights. Windshield washer. Air screens. Center pillar overlay (four-door sedan). Locking gas cap. Filler neck guard. Cross country cargo straps. Power brakes. Exhaust extension. Curb indicators. Trunk light. Door top shades. Door edge guards. Rear window wiper (Statesman and Ambassador). Oil filter. Solo-cell. Seat belts. Rear door safety locks.

HISTORICAL FOOTNOTES: Dealer introductions of 1956 models took place on Nov. 22, 1955. Cars were sold by both Hudson and Nash dealers and the Series 10 models were known as American Motors Ramblers. The Hudson and Nash products were comparable, except for hood medallions and their "N" or "H" wheelcover insignias. The company's automotive division sustained a sizable loss while its appliance division enjoyed its most profitable year since 1950. During the year the corporation completed the sale of idle plants and equipment (El Segundo, Calif. and the Hudson-Gratiot plant in Detroit) for a net amount of $5.3 million. Year 1956 saw production by American Motors of 104,190 cars (79,166 Ramblers, 17,842 Nashes, 7,182 Hudsons). American Motors Corp. produced its 2,000,000th single-unit construction car March 27, 1956. Genuine leather trims were available in the following 1956 models: Ambassador Six and Ambassador V-8 sedans; Ambassador Country Club hardtop; Rambler sedan; Rambler Country Club station wagon; Rambler four-door hardtop and Rambler Country Club station wagon.

1957 NASH

1957 Nash, Ambassador four-door sedan, V-8

NASH AMBASSADOR SERIES 80 — (V-8): The Nash Ambassador for 1957 was available only with a 327 cubic inch AMC V-8 engine in two-door hardtops and four-door sedan. Super and Custom trim levels were provided. The Nash Ambassador six and Statesman six were discontinued. The new Ambassador received a major facelift incorporating the first four-beam headlight system used on any U.S. car. Also seen was completely new front end styling, including a new cellular grille, front parking lights on top of the front fenders and new lightning streak side trim. The Ambassador Super had its name on the front fenders in script, small hubcaps and single lightning streak side trim with no upper beltline molding. Ambassador Customs had scripts with that name on the fenders, dual molding lightning streak trim and full wheelcovers. This was the last year for Nash production.

NASH I.D. NUMBERS: Vehicle Identification Plate located on the right-hand side of the cowl below the hood contains motor-serial number and number codes for model, paint and trim. Motor serial number matches the serial number. Model number has four symbols. The first two symbols indicate model year, 57=1957. The third symbol indicates series: 1=Rambler; 8=Ambassador Super; 7=Ambassador Custom. The fourth symbol indicates body style. (All four numbers appear in Body/Style Number column of charts below). Suffixes were added behind the four numbers on some models, as follows: 1=Super; 2=Custom. No suffix means Deluxe. Motor numbers located on right-hand side of crankcase towards front.

[NASH AMBASSADOR SERIES 80 I.D. NUMBERS]: Starting serial number for the Nash Ambassador was V-16501, with all Nash Ambassadors built in Kenosha, Wis.. Starting engine serial number is N1001. [RAMBLER SIX SERIES 10 I.D. NUMBERS]: Starting serial number for the 1957 Rambler six was D-341001. Starting engine serial numbers were B-73001 for the standard engine and CB-2001 for the optional six-cylinder engine. [RAMBLER V-8 SERIES 20 I.D. NUMBERS]: Starting serial numbers for the 1957 Rambler V-8 was A-1001. Starting engine number was G7501. [RAMBLER REBEL SERIES 30 I.D. NUMBERS]: Starting serial numbers for the Rambler Rebel was F-1001. Starting engine serial number was CN1001.

NASH AMBASSADOR SERIES 80

Model Number	Body/Style Number	Body Type & Seating	Factory Price	Shipping Weight	Production Total
80	5785-1	4-dr Sup Sed-6P	2821	3639	3,098
80	5785-2	4-dr Cus Sed-6P	3011	3701	5,627
80	5787-1	2-dr Sup HT-6P	2910	3655	608
80	5787-2	2-dr Cus HT-6P	3101	3722	997

NOTE 1: Prices given for 1957 models are ADP figures from *Official Automobile Guide* (87th edition) effective Jan. 1, 1957.

1957 Nash, Super Cross Country four-door station wagon, 6-cyl

1957 Nash, Rebel four-door hardtop, V-8

RAMBLER SIX SERIES 10 — (6-CYL): The Rambler six for 1957 continued the 108 inch wheelbase with a few minor changes. Included were new vertical, front running lights, with horizontal bright metal dividers positioned below the headlights; a new wing-shaped ornament on top of the rectangular grille section and the elimination of side color accent trim running over the roof. Three series were again available. Deluxe models had the lowest level of trim and equipment and were essentially built for fleet customers. Super series models carried a single, full-length side molding with the word Super, in script, on the rear fenders. Deluxe models came with no series name or side moldings. The Custom series models came with Rambler Custom scripts on the front fenders and dual side moldings, with a round medallion at the forward end.

RAMBLER SIX SERIES 10

Model Number	Body/Style Number	Body Type & Seating	Factory Price	Shipping Weight	Production Total
DELUXE LINE					
10	5715	4-dr DeL Sed-6P	1961	1962	9,402
10	5718	4-dr DeL Sta Wag	2291	3034	75
SUPER LINE					
10	5715-1	4-dr Sup Sed-6P	2123	2914	16,320
10	5718-1	4-dr Sup Sta Wag	2410	3042	14,083
10	5719-1	4-dr Sup HT-6P	2208	2936	612
CUSTOM LINE					
10	5715-2	4-dr Cus Sed-6P	2213	2938	10,520
10	5718-2	4-dr Cus Sta Wag	2500	3076	17,745

NOTE 1: The Deluxe station wagon, Style 5718, was for fleet use only.

1957 Nash, Custom four-door hardtop sedan, V-8

RAMBLER V-8 SERIES 20 — (V-8): The Rambler, for 1957, was also available with a V-8 engine of 250 cid. The same four-door station wagon and sedan styles were offered with this brand new Rambler power plant. Super and Custom trim levels were provided. Super series models carried a single, full-length side molding with the word Super, in script, on the rear fenders. Deluxe models came with no series name or side moldings. The Custom series models came with Rambler Custom scripts on the front fenders and dual side moldings, with a round medallion at the forward end.

RAMBLER V-8 SERIES 20

Model Number	Body/Style Number	Body Type & Seating	Factory Price	Shipping Weight	Production Total
20	5723-2	4-dr Cus HT Sta Wag	2715	3409	182
20	5725-1	4-dr Sup Sed-6P	2253	3223	3,555
20	5725-2	4-dr Cus Sed-6P	2343	3259	3,199
20	5728-1	4-dr Sup Sta Wag	2540	3359	2,461
20	5728-2	4-dr Cus Sta Wag-6P	2630	3392	4,560
20	5729-2	4-dr Cus HT Sed-6P	2428	3269	485

RAMBLER REBEL — 30 SERIES — (V-8): The 1957 Rambler Rebel used the Ambassador 327 cid engine in a Rambler V-8 body. This limited-production car was available exclusively in light silver-gray metallic finish. It had black nylon and silver gray vinyl upholstery. However, many of the cars were later repainted by dealers, due to excessive fading of the silver gray paint. The 1957 Rebel featured a side molding of bronze/gold anodized aluminum, which ran the full length of the car. The four-door hardtop body style was the only one available. The Rebel was the first attempt by American Motors to build a high performance car. In fact, this was the first time a large engine had been placed in a true intermediate-sized chassis (an idea Pontiac would find great success with in the GTO) by any automaker. In an April, 1957, *Motor Trend* test it was found that the only car capable of a faster 0-to-60 time than the Rebel was the fuel-injected Corvette. Fuel-injection had actually been planned for the 1957 Rebel with 288 hp possible. However, problems with the electric control unit prevented its production. Thus, a mildly reworked 327 cid Ambassador engine was used.

RAMBLER REBEL SERIES 30

Model Number	Body/Style Number	Body Type & Seating	Factory Price	Shipping Weight	Production Total
30	5739-2	4-dr Cus HT Sed-6P	2786	3353	1,500

BASE ENGINES

[AMBASSADOR EIGHT] V-8: Overhead valves. Displacement: 327 cid. Bore and stroke: 4 x 3-1/4 inches. Compression ratio: 9.0:1. Brake hp: 255 at 4700 rpm. Five main bearings. Hydraulic valve lifters. Carburetor: Carter four-barrel model WCFB-2593SA. Dual exhaust. 12-volt electrical system.

(RAMBLER SIX) Six-cylinder. Overhead valve. Cast Iron block. Displacement: 195.6 cid. Bore and stroke: 3-1/8 x 4-1/4 inches. Compression ratio: 8.25:1. Brake hp: 125 at 4200 rpm. Four main bearings. Solid valve lifters. Carburetion: Carter one-barrel model AS-2580S.

(RAMBLER EIGHT) V-8. Overhead valve. Cast iron block. Displacement: 250.1 cid. Bore and stroke: 3-1/2 x 3-1/4 inches. Compression ratio: 8.0:1. Brake hp: 190 at 4900 rpm. Five main bearings. Non-adjustable hydraulic lifters. Carburetion: Carter two-barrel model WGD-2352-SA.

[RAMBLER REBEL EIGHT] V-8: Overhead valves. Cast iron block. Displacement: 326.7 cid. Bore and stroke: 4 x 3-1/4 inches. Compression ratio: 9.0:1. Brake hp: 255 at 4700 rpm. Five bearings. Solid valve lifters. Carburetion: Carter four-barrel model WCBF-2593-SA.

CHASSIS FEATURES: [NASH AMBASSADOR] Wheelbase: 121.3 inches. Overall length: 209.3 inches (219.3 inches with continental tire mount). Front tread: 59.1 inches. Rear tread: 60.5 inches. Tires: 8.00 x 14. [RAMBLER]: Wheelbase: 108 inches. Overall length: (Station wagon) 193.61 inches; (four-door sedan) 191.14 Inches; (four-door sedan with continental tire mount 198.89 inches). Front tread: 57-3/4 inches. Rear tread: 58 inches. Tires: (six) 6.40 x 15; (V-8) 6.70 x 15.

OPTIONS: [AMBASSADOR] Power steering. Power brakes (standard on Custom). Power-lift windows. Weather-Eye heating and ventilating system. All-season air conditioning. Airliner reclining seats (standard on Custom). Electric clock (standard on Customs). Oil filter. Oil bath air cleaner. Hydra-Matic. Automatic overdrive. Continental tire mount. Twin speaker radio. Whitewall tubeless tires. Solex glass. Back-O-Matic lights. Special leather seat trim (for Custom only). Two-tone paint. Three-tone paint (Custom only). Heavy-duty springs and shocks. Factory applied undercoating. Padded sun visors (standard on Custom). Dealer-installed seat belts. [RAMBLER] Six-cylinder engine with two-barrel carburetor and Power Pack (135 hp). Weather-Eye heating and ventilating system. All-Season air conditioning. Radio. Airliner reclining seats. Rear foam cushions (standard on all except Deluxe). Padded instrument panel and padded sun visors. Directional signals. Electric clock. Cigarette lighter (standard on all except Deluxe). Continental tire mount (not available on station wagon). Chrome wheelcovers. Power steering. Power brakes. Automatic or overdrive transmission. Oil filter. Oil bath air cleaner on six-cylinder. Solex glass. Hood ornament. Special leather seat trim (for Customs only). Back-O-Matic lights. Windshield washer. Size 6.70 x 15 tires. Heavy-duty springs and shocks. Seat belts. Travel rack leather straps. Child guard door locks.

HISTORICAL FOOTNOTES: Dealer introductions for the 1957 models were held Oct. 25, 1956. Calendar year production of 3,561 Nash automobiles gave the marque a .06 percent market share. The Rambler nameplate did somewhat better with calendar year production of 109,178 cars for a 1.78 percent slice of the pie.

NASH-HEALEY
1951-1954

While most new car models are developed after lengthy deliberations, the impetus for the Nash-Healey stemmed from a chance encounter. Late in 1949, George Mason, the president of Nash-Kelvinator, was sailing home from Europe on the liner Queen Elizabeth. Also on board was Donald Healey, of British sports car fame.

At the time, Mason was seeking an image boost for his Nash models, which were considered a bit stodgy. Healey was in the market for a source of engines for his sports car, preferring the new Cadillac V-8. No V-8s could come from the Nash organization then, but the two agreed on the value of a British-American hybrid sports car using Nash mechanical components.

A prototype two-seater roadster appeared at 1950 auto shows, in both London and Paris. Panelcraft produced the car's aluminum body, which sat atop an ordinary Healey chassis with trailing link front suspension. Under the hood was a modified Nash Ambassador six-cylinder engine with a hotter camshaft, aluminum cylinder head, higher (8.1:1) compression, and dual SU carburetors. Nash torque tube drive fed the power to the back wheels.

The slab-sided body wore a grille, headlamps and other body items that originated in the Ambassador Airflyte sedan. Production began before the end of 1950. Early examples were sold in Britain, continental Europe and Canada, but before long, the U.S. became the sole outlet.

In February 1951, the new sports car debuted at the Chicago Auto Show, wearing a price tag of $4,063. Nash-Healeys quickly took up racing, powered by the LeMans Dual-Jetfire Ambassador six that would later go into domestically built sedans.

The very first Nash-Healey, wearing a special monoposto body, took ninth place in the 1,950 Mille Miglia and a fourth at the LeMans 24 Hours event. That experimental model averaged 87.6 mph in the 2,100 mile race. The same car, with a new hardtop body, wound up third in its class (sixth overall) at LeMans the next year. At the 1952 LeMans, an open version ranked third overall, taking second in the Index of Performance. Nash-Healey was also fourth in its class (seventh overall) at the 1952 Mile Miglia. Then, in 1953, a Nash-Healey convertible finished 11th overall at LeMans.

For its second (1952) season, Pinin Farina did a restyle of the roadster. He was also creating the design for 1952 Nash sedans. The new Nash-Healey had a lower, one-piece windshield and rear fender bulges, minimizing the slab-sided appearance. By this time, a steel body replaced the aluminum one. A larger engine, with Carter carburetors (and 15 more horsepower), came during the year.

Nash sent drivetrains and other parts to Healey, whose company did the rolling chassis. Then, in Italy, Pinin Farina installed the body and finished assembly. In 1953, a LeMans hardtop on a longer wheelbase (with rear quarter windows) joined the open model, which was now called a convertible. For its final year, only the LeMans version remained, wearing a three-piece wraparound rear window and with its price cut sharply. All told, over its four-year life, 506 Nash-Healeys were produced.

1951 NASH-HEALEY

1951 Nash-Healey roadster

SERIES 25 — SIX: Debuting in the U.S. on Feb. 16, 1951, the Nash-Healey roadster wore a two-seat body with body panels and other structural parts made of aluminum. At the center of the broad, low hood was a small air scoop with a vertical chromed grille. The front end appearance was similar to the familiar Nash Airflyte. Its grille consisted of outward curved vertical chrome bars, within a heavy chrome molding. Model designations were evident on the front fenders, to the rear of the wheel opening. The roadster measured only 38 inches to the top of its hood. Inside was a leather-finished instrument panel. The adjustable single seat was upholstered in English leather over latex foam cushions, while the folding fabric top contained a soft plastic rear window. Side windows made of hard plastic lowered into the door panels. The spare tire and luggage compartment were reached through a nearly horizontal deck. Standard equipment included an adjustable steering wheel, turn signals and chrome wheelcovers. Under the hood rested a 125 hp version of the 234.8 cid (3,847 cc.) Nash Dual-Jetfire Ambassador six-cylinder engine with 8.1:1 compression, an aluminum racing head (with sealed-in intake manifold) and twin SU horizontal carburetors. Premium fuel was required. Coil springs were installed at all four wheels, with the Healey trailing link suspension up front. Each front wheel was mounted on a swinging arm pivoted ahead of the wheel centerline. At the rear, coil springs worked with direct-acting shock absorbers, similar to those in the Nash Ambassador. Duo servo brakes were installed, along with a 3.54:1 axle and 6.40 x 15 whitewall tires. The three-speed transmission came with standard overdrive.

I.D. DATA: Serial number is located on right-hand top fender panel, under the hood. Engine number is located on right top front of block. Serial number range (1951 models): N-2001 to N-2109. Starting engine number: NHA-1001.

Body Type	Model & Seating	POE Price	Weight (lbs.)	Prod. Total
25162	2dr Spt Rds-2P	4063	2600	104

NOTE 1: Weight shown was announced as "curb weight" by the manufacturer. Other sources gave a figure of 2,690 pounds.

ENGINE DATA: [BASE SIX] Inline. Overhead valve. Six-cylinder. (Nash Dual-Jetfire). Cast iron block and aluminum head. Displacement: 234.8 cid (3,847 cc.). Bore & Stroke: 3.375 x 4.375 in. (86 x 111 mm.). Compression ratio: 8.1:1. Brake Horsepower: 125 at 4000 rpm. Torque: 210 lbs.-ft. at 1600 rpm. Seven main bearings. Solid valve lifters. Two SU horizontal carburetors.

CHASSIS DATA: Wheelbase: 102 in. Overall Length: 170 in. Height: 55.5 in. Width: 66 in. Front Tread: 53 in. Rear Tread: 53 in. Standard Tires: 6.40 x 15 whitewall.

TECHNICAL: Layout: front-engine, rear-drive. Transmission: three-speed manual (with overdrive). Standard final drive ratio: 3.54:1 (overall net, 2.48:1). Suspension (front): Healey trailing link with coil springs and anti-roll bar. Suspension (rear): rigid axle with coil springs and track bar. Brakes: hydraulic, front/rear drum (Bendix duo servo). Body construction: aluminum body on steel frame. Fuel Tank: 16 gallon.

PERFORMANCE: Top Speed: 102-104 mph (factory initially claimed an estimated 125 mph). Acceleration (0-60 mph): about 12 seconds. Manufacturer: Nash Motors, Kenosha, Wis. and Donald Healey Motor Co., Ltd., Warwick, England. Distributor: Nash Motors Div. of Nash-Kelvinator Corp., Detroit, Mich.

HISTORY: An early press release called the Nash-Healey the "first American sports car introduced by an established automobile manufacturer since the mid-'20s," adding that prices would be "substantially higher" than other Nash models. Nash estimated top speed, rather optimistically, at 125 mph. The car was scheduled to undergo exhaustive engineering tests at Daytona, Salt Lake Flats and Indianapolis Speedway. Production/sales were supposed to be limited in 1951, until after the American market was thoroughly explored. Nash-Healeys were assembled by Healey at Warwick, England and sold by Nash dealers in the U.S. Tom McCahill, of Mechanix Illustrated, wrote that he'd never driven a sports car that handled better or gave the driver so much control. Motor Trend advised, "the Nash-Healey rides far better then the average sports car, without any apparent ill effect upon handling qualities."

1952 NASH-HEALEY

1952 Nash-Healey roadster with George Mason and Pinin Farina

SERIES 25 — SIX: Since Pinin Farina was performing a restyling, in Italy, for the 1952 Nash sedan line, doing a revision of the Nash-Healey roadster seemed a reasonable extra task. This second version looked less box-like than the first, with its lower windshield. The bulged back fenders had minimal fins. Inboard headlamps sat within a thick rounded rectangular front opening that held a grille made of just two horizontal bars with a round insignia in the center. Small round parking lights stood at the fender tips. Atop the hood was a tiny center scoop. This time, a steel body was used, rather than aluminum. Power initially came from the same 234.8 cid (3,847 cc.) engine as the original, but a larger 252.6 cid engine, with Carter carburetors, became available during the year. As before, the engine and mechanical components were supplied to Healey by Nash. After completing the chassis, Healey sent the result to Farina in Italy for body installation.

I.D. DATA: Serial number is located on right fender panel, under the hood. Engine number is located on right front of block. Serial number range (1952 models): N-2200 up (a few with lower numbers N-2086, 2103, 2104 and 2106 were produced:). Starting serial number with 253 cid engine was N-2250. Starting engine number: NHA-1088 (numbers did not necessarily follow in sequence).

Body Type	Model & Seating	POE Price	Weight (lbs.)	Prod. Total
25262	2dr Spt Rds-2P	5868	2750	150

ENGINE DATA:
[EARLY 1952 BASE SIX]: Inline. Overhead valve. Six-cylinder. (Nash Dual-Jetfire). Cast iron block and aluminum head. Displacement: 234.8 cid (3,847 cc.). Bore & Stroke: 3.375 x 4.375 in. (86 x 111 mm.). Compression ratio: 8.1:1. Brake Horsepower: 125 at 4000 rpm. Torque: 210 lbs.-ft. at 1600 rpm. Seven main bearings. Solid valve lifters. Two SU horizontal carburetors.
[LATE 1952 BASE SIX]: Inline. Overhead valve. Six-cylinder. (LeMans Dual-Jetfire). Cast iron block and aluminum head. Displacement: 252.6 cid (4,140 cc.). Bore & Stroke: 3.50 x 4.375 in. (88.9 x 111 mm.). Compression ratio: 8.0:1. Brake Horsepower: 140 at 4000 rpm. Torque: 230 lbs.-ft. at 2000 rpm. Seven main bearings. Solid valve lifters. Two Carter carburetors.

CHASSIS DATA: Wheelbase: 102 in. Overall Length: 170.75 in. Height: 48.65 in. Width: 64 in. Front Tread: 53 in. Rear Tread: 54.9 in. Standard Tires: 6.40 x 15 whitewall.

TECHNICAL: Layout: front-engine, rear-drive. Transmission: three-speed manual (with overdrive). Steering: walking beam type. Suspension (front): Healey trailing link with coil springs and anti-roll bar. Suspension (rear): rigid axle with coil springs and track bar. Brakes: hydraulic, front/rear drum. Body construction: steel body on steel frame.

PERFORMANCE: Top speed: about 104 mph. Acceleration (0-60 mph): 11.5-14.5 sec. Manufacturer: Nash Motors, U.S.A.; Donald Healey Motor Co. Ltd., Warwick, England; and Pinin Farina, Turin, Italy. Distributor: Nash Motors Div. of Nash-Kelvinator Corp., Detroit, Mich.

1953-1954 NASH-HEALEY

SERIES 25 — SIX: Production of the two-seat convertible continued with its Pinin Farina body and 252.6 cid (4,140 cc.) engine, which produced 140 hp. It was joined, in 1953, by a LeMans hardtop with a steel top and rear quarter windows, which rode a longer 108 inch wheelbase. That hardtop, too, had been styled by Pinin Farina and was shown for the first time at the Chicago Auto Show in March 1953. Its low hood/high fender contour extended over the cowl, through the windshield and rear window, onto the rear deck, without interruption. Trailing rear fender fins accented the flowing front-to-rear lines. Rear fenders rose slightly above the rear deck, forming a molded part of the body. Headlamps were mounted within a racing air scoop grille. Smoothly rounded front fenders extended forward of the grille line. Standard equipment included leather upholstery, whitewall tires, tachometer, wheelcovers, light and ashtray. For 1954, only the hardtop remained available. A three-piece wraparound rear window was added. Enthusiasts and collectors call these Pinin Farina hardtops, rather than LeMans hardtops.

1953 Nash-Healey hardtop coupe

I.D. DATA: Serial number is located on right fender panel, under the hood. Engine number is located on right front of block. Starting serial number: (1953 convertible): N-2290; (1953 hardtop) N-3000. Serial numbers for 1954 models included N-3027, 3036, 3037, 3039-3041, 3043, 3046, 3067-3069 and 3071 up. Starting engine number: (1953 convertible) NHA-1203; (1953 hardtop) NHA-1223.

Body Type	Model & Seating	POE Price	Weight (lbs.)	Prod. Total
25362	2dr Spt Conv-2P	5909	2750	Note 1
25367	2dr HT Cpe-2P	6399	2970	Note 1

NOTE 1: A total of 162 Nash-Healeys were produced in 1953; 90 in 1954.
NOTE 2: List price of convertible dropped to $5,555 in 1954, while the hardtop sold for only $5,128. Prices as low as $4,721 were published in U.S. directories for final models.
ENGINE DATA: [BASE SIX] Inline. Overhead valve. Six-cylinder. (LeMans Dual-Jetfire). Cast iron block and aluminum head. Displacement: 252.6 cid (4,140 cc.). Bore & Stroke: 3.50 x 4.375 in. (88.9 x 111 mm.). Compression ratio: 8.0:1. Brake Horsepower: 140 at 4000 rpm. Torque: 230 lbs.-ft. at 2000 rpm. Seven main bearings. Solid valve lifters. Two Carter carburetors.

1954 Nash-Healey hardtop

CHASSIS DATA: Wheelbase: (convertible) 102 in.; (hardtop) 108 in. Overall Length: (convertible) 170.75 in.; (hardtop) 180.5 in. Height: (convertible) 48.65 in.; (hardtop) 55 in. Width: (convertible) 64 in.; (hardtop) 65.9 in. Front Tread: 53 in. Rear Tread: 54.9 in. Standard Tires: 6.40 x 15 whitewall.
TECHNICAL: Layout: front-engine, rear-drive. Transmission: three-speed manual (with overdrive). Standard final drive ratio: 4.1:1. Steering: walking beam type. Suspension (front): Healey trailing link with coil springs and anti-roll bar. Suspension (rear): rigid axle with coil springs and track bar. Brakes: hydraulic, front/rear drum. Body construction: aluminum body on steel frame. Fuel tank: 20 gallon.
MAJOR OPTIONS: Weather-Eye heater.
PERFORMANCE: Similar to 1952. Manufacturer: Nash Motors; Donald Healey Motor Co. Ltd., Warwick, England; and Pinin Farina, Turin, Italy. Distributor: Nash Motors Div., Nash-Kelvinator Corp., Detroit, Mich.
HISTORY: The last Nash-Healey was produced in August 1954.

METROPOLITAN
1954-1962

Hardly the fastest car of its day and not the most agile either, the Nash/Hudson Metropolitan qualifies as one of the cutest. Its curious appeal continues today. In addition to the Metro's aesthetic attributes, it served as an early example international joint ventures, through which a car mixed mechanical components and styling from two countries for marketing in a third nation, which supplied the original concept. In that respect, it was a forerunner to the modern Cadillac Allante with a low-budget pricetag, even for the mid-1950s. The Metropolitan can also be considered the first U.S. subcompact. It was made available just after the formation of American Motors (merging Nash and Hudson) in May 1954.

The Metropolitan's drivetrain came from England, in the form of the Austin A40 overhead-valve four. In fact, the entire car was assembled by Austin in Britain, from a design by Italian Pinin Farina. Development costs were about $2 million (considered an amazingly modest sum at that time). The production Metropolitan evolved from the Nash Experimental International (NXI) prototype. This, in turn, became the NKI (Nash-Kelvinator International). The idea reached back to the period just after World War II, when Nash president George W. Mason ordered research to begin on an economy car. He looked over a design submitted by independent stylist Bill Flajole. It was based on the chassis and running gear of the tiny Flat 500. Early rumors suggested it would use a Flat engine.

George Romney (later to become head of American Motors) displayed the prototype NXI to dealers and the press, at a series of private functions starting in January 1950, to gauge public reaction. Among other features, the hood and fenders of the NXI lifted, as a unit, to give easy engine access. The grille and nose stayed in place. *Motor Trend* reported, in 1950, that the car was expected to sell for $1,000. Late in 1953 arrangements were made with Austin to begin volume production. Bodies would be built in Birmingham, England by Fisher & Ludlow, then shipped to the Austin plant at Longbridge. Metropolitans went on sale early in 1954. A second series, the 1500, debuted in April 1956. It had a bigger engine (Austin A50) and more power, as well as bright two-tone color schemes.

Metropolitans were marketed under both Nash and Hudson badges, until Hudson's demise. After 1957, the American Motors Corp. (AMC) name was used. Then, Metropolitan became a separate make, gaining such extras as a trunk lid and vent windows in 1959. Metropolitans remained in full production into 1960, with a few leftovers available in the U.S. for two more years. A total of 94,986 were produced in all.

1954-55 METROPOLITAN

SERIES A (1954)/B (1955) — FOUR: The two-seat Metropolitan came in hardtop coupe and convertible body styles, the former with two-tone paint. One distinctive feature was the cutout at the top of each door. A low, one-piece rear-hinged hood held a decorative air scoop. The grille consisted of a single die-cast horizontal bar with central medallion. Fenders stood taller than the hood. Above the hood, at the cowl, was a fresh-air intake for the optional Weather-Eye system. Round park/signal lights stood directly below the headlamps. Both front and rear wheels were partly enclosed (a Nash trademark at this time). Wraparound front and rear bumpers included standard guards. A continental (external) spare tire mounting was standard equipment. It came with a vinyl cover. The rear license plate bracket and light were mounted on the spare wheel's hubcap. Square-cut doors had push-button handles. Chrome window frames could be lowered completely. Inside, upholstery was done in leather and nylon cord. A single cluster, ahead of the driver, held the speedometer, fuel gauge, and warning lights. A manual choke was standard and the glove compartment was open. The tubular steel seat frame adjusted four inches, the rear cushion could be removed and the seatback swung forward to give access to the luggage area (which offered more than six cubic feet of storage space). The 12-volt battery rested beneath a mat covering the barely existent back seat, described by *Motor Trend* as "actually little more than a padded shelf." Standard equipment included dual Lucas horns, twin sun visors, ashtray, lighter, 17 inch steering wheel, glove box, map light, and rearview mirror. The 1200 cc. (73 cid) Austin A40 engine produced 42 hp. The sole transmission was a three-speed manual unit with column-mounted gearshift lever (which actually emerged from the instrument panel). A single-plate Borg & Beck clutch was used. Front suspension was Nash's "Air-flex" setup, with a coil spring between the wheelhousing and upper control arm. Metropolitans also adopted the Nash "Airflyte" unit body construction. Standard body colors for 1954 included Spruce green, Canyon red, Caribbean blue, and Croton green. The hardtop came only in Mist gray on the upper body. The vinyl convertible top came in tan or black, depending on body color. Upholstery was Old Ivory leather with beige nylon-faced Bedford cord (black/yellow striped). Options included a Weather-Eye Conditioned Air System, radio with manual antenna and whitewall tires. A flat one-piece windshield and three-piece curved rear window were standard with both body styles.

1954 Metropolitan

I.D. DATA: Chassis serial number is located on the right side of the cowl, under the hood. Engine number is on right center of block. Starting chassis number: (Series A) E-1001; (Series B) E-11001. Starting engine number: (Series A) 1G-881459; (Series B) GS-510003.

Body Type	Model & Seating	POE Price	Weight (lbs.)	Prod. Total
METROPOLITAN NASH/HUDSON				
541	2dr Conv-2/3P	1469	1785	Note 1
542	2dr HT Cpe-2/3P	1445	1825	Note 1

NOTE 1: A total of 13,095 Metropolitans were produced during 1954 (including 743 early examples built in 1953), followed by 6,096 in 1955.
NOTE 2: American Motors claimed three-passenger capacity for the Metropolitan's front seat, which is a tight squeeze; the miniature utility seat in back was barely suitable for a child.
NOTE 3: Prices shown were valid during 1954 and 1955.
NOTE 4: Shipping weights shown are for initial models.
ENGINE: [BASE FOUR] Inline. Overhead valve. Four-cylinder (Austin A40). Cast iron block and head. Displacement: 73.2 cid (1200 cc.). Bore & Stroke: 2.58 x 3.50 in. (65.5 x 88.9 mm.). Compression Ratio: 7.2: 1. Brake hp: 42 at 4500 rpm. Torque: 62 lbs.-ft. at 2400 rpm. Three main bearings. Solid valve lifters. One Zenith downdraft carburetor.
CHASSIS DATA: Wheelbase: 85 in. Overall length: 149.5 in. Height: 54.5 in. Width: 61.5 in. Front tread: 45.3 in. Rear tread: 44.8 in. Wheel type: pressed steel disc. Standard tires: 5.20 x 13.
TECHNICAL: Layout: front-engine, rear-drive. Transmission: three-speed manual (column shift). Gear ratios: (1st) 2.436: 1; (2nd) 1.535: 1; (3rd) 1.00: 1; (Rev.) 3.489: 1. Standard final drive ratio: 4.625:1. String: cam and lever. Suspension (front): Airflex with direct-acting, extra-long coil springs. Suspension (rear): rigid axle with semi-elliptic leaf springs. Brakes: Girling hydraulic, front/rear drum. Body construction: steel unibody (Airflyte). Fuel tank: 10.5 gallon.
MAJOR OPTIONS: Radio ($60). Weather-Eye heater ($69). Whitewall tires.
PERFORMANCE: Top speed: 70 mph claimed: about 77 mph in early test. Acceleration (0-60 mph): as much as 23-27.5 seconds and as little as 19.2 seconds reported. Fuel mileage: about 30-32 mpg city, 37-40 mpg highway.
PRODUCTION/SALES: Approximately 6,617 Metropolitans (401 Hudson versions) were sold in the U.S. during 1954. Manufacturer: The Austin Motor Co. Ltd., Longbridge, Birmingham, England (bodies from Fisher & Ludlow Ltd., Birmingham, England). Distributor: American Motors Corp., Detroit, Mich.
HISTORY: In a fuel-economy test prior to the cars debut, a Metropolitan achieved 41.57 mpg at an average speed of 34.83 mph, going 24 hours non-stop. In an endurance test from the same period, the Metropolitan averaged 61.24 mph, covering 1469.7 miles in 24 hours. Observed by NASCAR, these tests were held at the one-mile asphalt track in Raleigh, N.C. *Motor Trend* called the new Metropolitan a scaled-down version of everything good in a Nash, which is saying plenty. Initial dealer stocks, according to the company, were sold out the day after its introduction.

1956-62 METROPOLITAN

SERIES B — FOUR: Availability of the original Metropolitan with 42 hp engine continued into 1957, but it was eclipsed by the higher-powered 1500. See previous listing for engine data.
SERIES 1500 — FOUR: A larger and more powerful Austin A50 engine went into the second-generation Metropolitan, which otherwise remained largely unchanged. The 1489 cc. (91 cid) overhead valve four produced 52 hp (24 percent more than the initial A40 engine). That gave a top speed a boost to about 78 mph. A larger clutch was installed (eight inch rather than the former 7.25 inch). Metropolitans displayed a new cellular style oval grille with heavily chromed surround molding. It had a round Nash or Hudson medallion in the center. Hoods no longer had an air scoop. Bodyside trim was modified to give the car a fresh look. The bodyside molding began at the headlamps, dipped to the lower portion of the car (just behind the door) and continued horizontally to the rear. That molding served as a separation line for two-tone paint. Standard upper body colors were Caribbean green, Sunburst yellow, and Coral red (above the molding and on the hood, deck and windshield posts). The second color (Snowberry white) went below the molding. The standard continental spare tire was covered in black vinyl, trimmed in white vinyl. Inside, gray/black upholstery had off-white vinyl trim. Window vent wings and a trunk lid (for outside access) were added during 1959, along with tubeless tires. An Austin A55 engine went into final Metropolitans. They were rated at 55 bhp.
I.D. DATA: Chassis serial number is located on the right side of the cowl, under the hood. Engine number is on right center of block. Starting serial number (Series 1500): E-21008. Starting engine number: 1H-14004.

Body Type	Model & Seating	POE Price	Weight (lbs.)	Prod. Total
METROPOLITAN 1500 (NASH/HUDSON/AMC)				
561	2dr Conv-2/3P	1551	1803	Note 1
562	2dr HT Cpe-2/3P	1527	1843	Note 1

NOTE 1: A total of 9,068 Metropolitans were produced during 1956, followed by 15,317 in 1957; 13,128 in 1958; 22,309 in 1959; and 13,103 in 1960. A final 853 examples became available in 1962 and 412 in 1962.
NOTE 2: AMC claimed three-passenger seating capacity.
NOTE 3: Prices shown were valid in 1956. Prices rose to $1,673 for the hardtop and $1,697 for the convertible by 1960. Later examples, without a trunk lid, sold for $47 less.

1960 Metropolitan convertible

ENGINE DATA: [BASE FOUR] Inline. Overhead valve. Four-cylinder (Austin A50). Cast iron block and head. Displacement: 90.9 cid (1489 cc.). Bore & Stroke: 2.875 x 3.50 in. (73 x 88.9 mm.). Compression Ratio: 7.2:1. Brake hp: 52 at 4500 rpm. Torque: 77 lbs.-ft. at 2500 rpm. Three main bearings. Solid valve lifters. One Zenith downdraft carburetor.
CHASSIS DATA: Wheelbase: 85 in. Overall length: 149.5 in. Height: 54.5 in. Width: 61.5 in. Front tread: 45.3 in. Rear tread: 44.8 in. Wheel type: pressed steel disc. Standard tires: 5.20 x 13 (later, 5.60 x 13).
TECHNICAL: Layout: front-engine, rear-drive. Transmission: three-speed manual (column shift). Standard Final Drive Radio: 4.3:1. Steering: cam and lever. Suspension (front): A-arms with coil springs. Suspension (rear): rigid axle with semi-elliptic leaf springs. Brakes: Girling hydraulic, front/reciprocating drum. Body construction: steel unibody (Airflyte). Fuel tank: 10.5 gallon.
MAJOR OPTIONS: Weather-Eye heater. Radio. Whitewall tires.
PERFORMANCE: Top speed: (1500) 78 mph claimed. Acceleration (0-60 mph): (1500) 19.5-22.0 seconds (0-50 mph in 16.1 seconds). Fuel mileage: 40 mpg claimed; 32-39 mpg reported.
PRODUCTION/SALES: Approximately 7,145 Metropolitans were sold in the U.S. during 1956, followed by 11,791 in 1957, and 8,657 in 1961. Manufacturer: The Austin Motor Co. Ltd., Longbridge, Birmingham, England (bodies from Fisher & Ludlow Ltd., Birmingham, England). Distributor: American Motors Corp. Detroit, Mich.
HISTORY: The Metropolitan 1500 went on display April 9, 1956 at Nash and Hudson dealers. After the Nash and Hudson names faded away in 1957, Metropolitans were marketed as American Motors models, sold by Rambler dealers.

HUDSON
1910-1957

HUDSON — Detroit, Mich. — (1910-1942 et. seq.) — The first advertisement appeared in *The Saturday Evening Post* on June 19, 1909, the first car left the factory on July 8th. By the following July, more than 4,000 Hudsons had been sold for the biggest first-year business yet recorded in the automobile industry. It was an auspicious beginning, but it wasn't really surprising.

The men behind the Hudson were Roy D. Chapin and Howard E. Coffin, veterans of Olds Motor Works who had just recently built the Thomas-Detroit and the Chalmers-Detroit. Joining them in partnership, in 1908, were George W. Dunham and Roscoe B. Jackson, also graduates of Olds. Putting up the money was Joseph L. Hudson, probably the only member of the venture who had any misgivings. His niece was Roscoe Jackson's wife and gentle family pressure, combined with the fact that $90,000 wasn't an outlandish sum for the man who owned Detroit's most successful department store, proved ultimately persuasive.

On. Feb. 24, 1909, the Hudson Motor Car Co. was organized with a capital stock of $100,000. By spring, the firm had bought up the Selden patent license of the defunct Northern and had moved into the factory of the defunct Aerocar. By summer, the company was in business. The car which enjoyed that record-breaking first year success was a 20 hp four??? It was good for 50 mph, priced at $900 and offered only as a racy and brassy little roadster on a 110-inch wheelbase. It was introduced as a 1910 model.

Though it carried the soon-to-be-famous Hudson triangle, the Model 20 Hudson's emblem was brass. It turned white with the introduction of the Model 33 for 1911. The 33 was a larger monobloc four on a 114-3/4 inch wheelbase with prices now edging upwards of $1,000. There was a spartan, but speedy, new two-seater called the "Mile-a-Minute" roadster. Guaranteeing that level of performance was quite a feat in 1912.

All this was a prelude, however. Coffin (the engineering genius of the Hudson group) was already working on the car that would transform Hudson's auspicious beginning into a long-term solid industry success. By now, the company had moved into its brand new Albert-Kahn designed factory at the corner of Jefferson and Conner Avenues in Detroit. It was there that the prototype of the first Hudson six was completed in July of 1912.

The new car was introduced for 1913. It promised 65 mph in touring trim for $2,450. This was followed, in 1914, by a lightweight six offering similar performance for only $1,550. Prior to Hudson, fast-paced sixes had primarily been the preserve of luxury car manufacturers. By 1914, all Hudson were sixes. Over 10,000 were built that year; over 12,000 the year following, as the company began advertising itself as the "world's largest manufacturer of six-cylinder cars."

At the New York Automobile Show, in January 1916, the Super Six arrived. It was truly super, with an entirely new L-head engine and improved cylinder head design for a 5.0:1 compression ratio on standard fuel and a puissant 76 bhp at 3,000 rpm. In April, at Daytona Beach, Fla., Ralph Mulford drove a Super Six to a new one-mile straightaway stock car record of 102.5 mph; in May, at Sheepshead Bay, N.Y., the 24-hour stock speed record was taken at a 75.8 mph average, a mark that would stand for 15 years! In August, Mulford's climb up Pikes Peak in 18 minutes 25 seconds represented a new class record to stand for eight years. In September, Mulford (with co-drivers named Vincent and Patterson) got into a Super Six seven-passenger touring car in San Francisco, Calif. Five days, three hours and 31 minutes later they were in New York City to break the cross-country record previously set in a Marmon automobile. Then, they turned around and drove back to San Francisco, establishing America's first ever double-transcontinental mark.

Hudson sales doubled that year, too. They topped more than 25,000 cars. The Super Six would remain the solid rock around which Hudson fortunes revolved for a decade. It was followed in 1928 by the 91 hp Special Six, an F-head that was an amalgam of the meritorious engineering features of the Essex four- and six-cylinder engines, as well as the venerable Hudson motor.

The Essex, introduced for 1918, was Hudson's low-priced companion car. By the mid-'20s, it had forcefully carried forward the Hudson focus on closed car production, which began in 1916, when Hudson sold more sedans than either Ford or Chevrolet. The offering of a closed coach (two-door sedan) for less than the cost of a touring car was heretofore unheard of in the auto industry.

Be they open or closed, some Hudsons of the later '20s and early '30s carried custom bodies. Originally, they came from Biddle & Smart, of Amesbury, Mass. Later, Murray and Briggs and occasionally LeBaron built them. In 1930, the chassis on which these, and all Hudsons, were built was an eight-cylinder one. The engine was essentially the Essex six with two cylinders added. It was called the Great Eight, though at 80 hp, it was less powerful than the company's big six. However, it became heftier in horses subsequently, and a big six was returned to the line, too.

In 1931, Frank Spring, formerly of Murphy Body Co., joined Hudson as styling director. By now, most of the original team members who had started the company had retired or died, except for Roy CHapin, who had a short sojourn in Washington as President Hoover's Secretary of Commerce. Chapin returned to Hudson in 1933, when the company's hottest car was the Terraplane.

The depression had set in. Like every other business in America, Hudson faced a lukewarm financial outlook, at best. The strain of keeping the company afloat seemed to hasten Chapin's death, at age 56, in February 1936.

A.E. Barit took over as Hudson's president. He had begun with the company, in 1910, as a stenographer. Stuart G. Baits, who started as a draftsman, in 1915, became vice-president. The 1934 Hudsons had introduced Axle-Flex, a semi-independent front suspension. Then, 1935 brought the "Electric Hand," a semi-vacuum-powered automatic gearshift built by Bendix. It was not a particularly good idea.

Though beauty was in the proverbial beholder's eyes, the look of the 1940 models, with their Symphonic Styling, was nice and tidy. Tidy, too, were records set at Bonneville by a stock 1940 Hudson, which now had virtually fully independent front suspension. Driver John Cobb captured virtually every AAA Class C closed car record from one mile to 3,000 kilometers and from one to 12 hours. The half-day mark was completed at a terrific 91.29 mph.

Despite the vagaries of the depression, the Hudson Motor Car Co. had survived the '30s in good order. In 1941, the firm made a profit of $3,756,000. On Feb. 5, 1942, Hudson ceased building cars. For the duration of World War II, its factories turned out machine guns and aircraft components. Hudson's best prewar year had been 1929, with over 300,000 cars produced. It was a figure the company would never approach again. Although there would be some very interesting and spirited Hudsons built in the postwar period, the glory years of the company were already behind it.

Hudson retained both its L-head six and eight engines with the Super Six accounting for two-thirds of Hudson's 1946 production. The wheelbase remained unchanged at 121 inches, but optional transmissions included Drive Master, Vacumotive Drive and overdrive. Except for minor exterior and interior changes, models were unchanged for 1947 and Hudson registered profits both years.

For 1948, Hudson introduced one of the greatest postwar designs with its unit-body Hudson, which was to continue through 1954. Being low and sleek, it had a low center of gravity and handled exceptionally well. Its dropped floorpan earned it the nickname Step-Down. It was offered in four models — Commodore Six and Eight and Super Six and Eight — and sat on a 124 inch wheelbase. This same year, Hudson introduced a new engine, the 262 cid Super Six, which developed 121 hp at 4000 rpm. In 1951 this same engine evolved into the '308' Hornet powerplant which, from 1951 through 1954, was the king of stock car racing.

Although the Step-Down proved to be one of America's most roadable cars from 1948-1954, Hudson lacked sufficient funds to add new models to the series, and combined with a lack of innovation, principally the lack of a V-8 engine in subsequent years, found sales dropping through the early 1950s.

Hudson introduced a Pacemaker model in 1950 which used a de-stroked version of the flathead Super Six and sold over 60,000 units. Mention should be make of Pacemakers' five inch shorter wheelbase and tighter turning radius. The Commodore Six and Eight were continued and all models offered optional Drive Master, Supermatic Drive and overdrive transmissions. Hydra-Matic was an option from 1951 on.

Carrying the same pricetag as the Commodore Eight, the legendary Hudson Hornet was introduced in 1951. Available in four body styles, the Hornet's six-cylinder powerplant produced 145 hp at 3800 rpm in stock form. In the hands of skilled tuners, though, it was capable of considerably more and the most noted of these, Marshall Teague, achieved 112 mph from a NASCAR-certified stock Hornet. In 1953 Hudson also offered factory severe-usage options which were designed for racing application. Racing items were listed as "Export" options! These included 'Twin H-Power' for improved breathing and a '7-X' racing engine which combined Twin H-Power with other high-performance options to produce about 210 hp.

Tim Flock was the 1952 NASCAR champion in a Hornet. In 1953, Marshall Teague's Hornet won 12 of 13 AAA stock car events and drivers Herb Thomas, Dick Rathmann, Frank Mundy and Al Keller drove Hornets to 65 NASCAR victories through 1954. Although Hudson added and subtracted series throughout this period, its inability to add new body styles hurt sales. When the Hornet and the Hollywood hardtop were added in 1951, Hudson dropped the standard Pacemaker and Super Eight. In 1952, the Wasp replaced the Super Six and all the Commodores were dropped the next year. A new Hornet Special of 1954 failed to increase sales.

Hudson's ill-fated compact Jet appeared in 1953 and the luxurious Jet Liner of 1954 sold poorly. However, it inspired a two-passenger Grand Turismo, built on the Jet chassis and called the Italia. It was designed by Hudson's Frank Spring and built by Carrozzeria Touring of Milan, Italy. Powered by a 114 hp Jet engine, the Italia had an aluminum body with a wraparound windshield, doors cut into the roof, fender scoops for brake cooling, flow-through ventilation and a leather interior. In addition to the prototype, 25 production Italias were made, plus a four-door 'X-161' which was built on the Hornet's 124 inch wheelbase chassis.

By late 1953, Hudson sales were slumping and the company merged with Nash and moved production to Kenosha, Wis., after closing its Detroit plant on Oct. 30, 1954. The all new 1955 Hudson was really a restyled Nash using Hudson's 1954 dashboard instruments. Hudson front suspension components and Dual-Safe brakes were retained. Wasps were powered by the former '202' Jet engine and the big six was retained for the Hornet. The Hornet V-8 used a Packard 208 hp engine and a line of Ramblers and Metropolitans was offered.

In 1956, American Motors introduced its own 190 hp V-8 for the Hornet Special and a line of Hudson Ramblers, but modest styling and engineering advances contributed to decreasing sales. The 1957 Hornets were two inches lower and used 14 inch wheels and the new 327 cid AMC V-8 with a four-barrel carb and dual exhaust. Cars so-equipped were excellent performers. Like other Hudson, they also carried price reductions, but it was still to be Hudson's last year. AMC decided to drop Hudson and Nash to concentrate on Rambler.

1910 HUDSON

HUDSON — MODEL 20 — FOUR: The Model Twenty roadster was an extremely handsome automobile that Hudson advertised as "Strong — Speedy — Roomy — Stylish." In the first public announcement, the Twenty was touted as an automobile that was far and away superior to its competitors. Included in its initial price were two headlamps, generator, side oil lamps, a tool set and a horn. Both the Hudson's appointments and design belied its low $900 price. The early models were finished in maroon with black striping and fenders. The interior was blue-black leather. The radiator, steering column, side lamp brackets, hubcaps and side control levers were finished in brass and the entry step plate was of aluminum. Hudson pointed with pride to the features the Twenty shared with far more expensive automobiles. For example, its sliding gear transmission was "such as you find on the Packard, Peerless, Pierce, Lozier and other high grade cars," according to ads. The Hudson engine was described as "the Renault type" and Renault Motors were the pride of France. The larger, 110 inch wheelbase Fore-Door roadster and touring models were less graceful than the roadster, but their body styles enabled Hudson to expand its share of the market. By the end of 1910, it ranked 17th in U.S. auto sales.

I.D. DATA: Serial numbers found in two locations, on plate on front seat riser and on right side of frame. Starting: 1. Ending: 7100. Engine No. Location: NA. Starting Engine No.: NA. Ending Engine No.: NA. Initially the Twenty was powered by an engine supplied by the Atlas Motor Co., of Indianapolis, Ind. As production increased, the Buda Company of Harvey, Ill. also supplied engines.

Model No.	Body Type & Seating	Price	Weight	Prod. Total
20	Open Rds-3P	900	1800	4000
20	2dr Rds-3P	1200	1800	1000
20	2dr Tr-5P	1150	2000	2099

NOTE 1: Beginning in January 1910, the price of the Model 20 Open Roadster was increased to $1,000.
NOTE 2: The Model 20 two-door roadster differed from the open roadster by its having two small doors. However, the door on the right side was not operational due to the position of the outside controls.
NOTE 3: Production total of the Model 20 Open Roadster included approximately 1,100 cars built in 1909 (but announced as 1910 models) by Hudson.
NOTE 4: Price on the Model 20 two-door touring included three oil lamps, two gas lamps, generator, horn, tire repair kit, tools and jack.

ENGINE: Inline Four. Cast iron block. B & S: 3-3/4 x 4-1/2 in. Disp. 198.8 cid. Brake HP: 20. ALAM H.P.: 22.5. Main bearings: 2. Valve lifters: mechanical. Carb. Holley, Mayer and Stromberg Model B carburetors were used.

CHASSIS: [Model 20 Roadster] W.B.: 100 in. Tires: 32 x 3 front 32 x 3-1/2 rear. [Model 20 Fore-Door Roadster] W.B.: 110 in. Tires: 32 x 35. [Model 20 Touring] W.B.: 110 in. Tires: 32 x 35.

*Note — There was only one series — Model 20 — but the three different body types warrant identification. After January 1910, the 20 designation was abandoned and the three Hudsons were advertised as either the roadster, touring car or Fore-Door roadster.

TECHNICAL: Sliding gear transmission. Speeds: 3F/1R. Floor shift controls. Leather-faced cone clutch. Shaft drive. Semi-floating rear axle. Overall Ratio: NA. Mechanical brakes on two wheels. Wooden wheels. Rim size: 32 in.

OPTIONS: Bosch magneto, top, Presto-O-Lite tank, rumbleseat available as a group option for the roadster and priced at $150. "Zig-Zag" windshield ($40). Rumbleseat or 25 gallon fuel tank (Fore-Door roadster).

HISTORICAL: Introduced June 1909. In October 1909, a preliminary (10 lap Massapequa Cup) Vanderbuilt Cup race was won by Hudson. The Hudson set the fastest lap and finished fourth. Hudson offered a competition model, the Express, which won a 24 hour race held at Seattle, Wash. From July 1909 thru December 1909 a total of 1,108 cars were shipped. Calendar year production was 7,100 cars. Hudson made 4,556 shipments to dealers during the 1910 calendar year. Model year production was approximately 8,200 units. The president of Hudson was Roy D. Chapin.

1911 HUDSON

1911 Hudson, Model 33 phaeton (OCW)

HUDSON — MODEL 33 — FOUR: The Model 33 was an entirely new Hudson designed by Howard E. Coffin. As such it is usually accepted as the first of the true Hudson automobiles. Its monobloc engine had its intake and exhaust valves on opposite sides of the cylinder head and they were completely enclosed. The Model 33's clutch, with its cork facings, was enclosed in an oil filled unit, along with its disc. This was expensive to manufacture, but it soon gave the Model 33 a deserved reputation for smooth operation. The Model 33's styling was conventional, with high crowned fenders, exposed controls and suspension components. However, the distinctive shape of its radiator with the already well-known Hudson triangle (with large letters spelling HUDSON across its surface) made it easy to pick out a Model 33 from the competition.

HUDSON — MODEL 20 — FOUR: No changes were made in the Model 20 for 1911.

I.D. DATA: Car number on front seat riser plate and on right side of frame: Starting: [Model 20] 7101; [Model 33] 7501. Ending: [Model 20] 9000; [Model 33] 15000. Engine numbers on right front of cylinder block. Starting Engine No.: NA. Ending Engine No.: NA.

Model No.	Body Type & Seating	Price	Weight	Prod. Total
20	2dr Roadster-2P	1000	1800	400
20	Fore Door Roadster-3P	1200	1800	NA
33	2dr Tr-5P	1250	2250	1500
33	3dr Fore Door Tr-5P	1600	2250	500
—	2dr Pony Tonneau	1300	2360	3500
—	3dr Torpedo Tr-5P	1350	2460	2000

ENGINE: [Model 20] Inline Four. Cast iron block. B & S: 3-3/4 x 4-1/2 in. Disp. 198.8 cid. Brake HP: 26. NACC HP: 22.5. Main bearings: 2. Valve lifters: mechanical. Carb.: Stromberg Model B. [Model 33] Inline Four. Cast iron block. B & S: 4 x 4-1/2 in. Disp. 226 cid. Brake HP: 33. NACC HP: 26.6. Main bearings: 2. Valve lifters: mechanical. Carb.: Stromberg.

CHASSIS: [Model 20] W.B.: 110 in. Tires: 32 x 3-1/2. [Model 33] W.B.: 114 in. Tires: 34 x 3-1/2.

TECHNICAL: Sliding gear transmission. Speeds: 3F/1R. Floor shift controls. [Model 20]: leather-faced clutch; [Model 33]: cork insert, wet clutch. Shaft drive. Semi-floating rear axle. Overall ratio: NA. Mechanical brakes on two wheels. Wood wheels. Rim size: [Model 20] 32 in. [Model 33] 34 in.

OPTIONS: Front bumper. Rear bumper. Double rumbleseat [Model 20]. 25 gallon fuel tank.

HISTORICAL: Introduced October 1910. Hudson made 6,486 shipments to dealers during the 1911 calendar year. Model year production was approximately 7,900. The president of Hudson was Roy D. Chapin. The Model 20 engine was built by Buda Company. The Model 33 engine was manufactured by the Continental Motor Manufacturing Co. of Muskegan, Mich. During 1911, a new plant was built by Continental in Detroit. It then became Hudson's only source of engines.

1912 HUDSON

1912 Hudson, Model 33 touring (FSA)

HUDSON — MODEL 33 — FOUR: The Model 20 was dropped from the 1912 line and all Hudsons were Model 33s. There were no major mechanical or styling changes, but a total of seven body types were available, including the 60 mph "Mile-A-Minute" roadster. Unlike the standard roadster, this two-seater was fitted with 32 inch wheels and was not equipped with doors or a windshield. An interesting piece of standard equipment was a 100 mph speedometer. Hudson continued to tout the engineering virtues of the Model 33 in 1912. Potential customers were warned to "beware of unsafe motor car purchases" that were out of date, due to rapid engineering advances. In contrast, the Model 33 was depicted as a car that possessed more advanced features than any other automobile. In mid-model year, the Disco self starter (built by the Disco Company of Grand Rapids, Mich.) became standard equipment for the Model 33. With a weight of just 4-1/2 pounds and only 12 moving parts, it was an appropriate addition to the Model 33, which Hudson claimed had "approximately 1,000 fewer parts" than the average car.

I.D. DATA: Car number on front seat riser plate and right side of frame. Starting: 15001. Ending: 27200 upward. Engine number on left side of front motor mount. Hudson continued to use Continental built engines.

185

Model No.	Body Type & Seating	Price	Weight	Prod. Total
NA	3dr Tr-5P	1600	2757	NA
NA	3dr Torpedo-4P	1600	2737	NA
NA	Mile-A-Minute Rds-2P	1600	NA	NA
NA	2dr Commercial Rds-2P	1600	2631	NA
NA	4dr Limousine-7P	2750	NA	NA
NA	2dr Cpe-2P	2250	NA	NA
NA	3dr Torpedo-5P	1600	2737	NA

ENGINE: Inline. Four-cylinder. Cast iron block. B & S: 4 x 4-1/2 in. Disp. 226 cid. Brake HP: 33. NACC HP: 26.6. Main bearings: 2. Valve lifters: mechanical. Carb. Stromberg.
CHASSIS: [Model 33] W.B.: 114-1/2 in. Front/rear Tread: 56 in. Tires: 34 x 4.
TECHNICAL: Sliding gear transmission. Speeds: 3F/1R. Floor shift controls. Cork inserts, wet clutch. Shaft drive. Semi-floating rear axle. Mechanical brakes on two wheels. Wood wheels. Rim size: 34 in.
OPTIONS: Front bumper. Rear bumper. All Model 33s, except the Mile-A-Minute Roadster, were delivered with a top, windshield, Bosch dual ignition Prest-O-Lite gas tank or generator as standard equipment.

1912 Hudson, Model 33 limousine (WRG)

HISTORICAL: Introduced July 1911. Hudson made 5,708 shipments to dealers during the 1912 calendar year. The president of Hudson was Roy D. Chapin.

1913 HUDSON

1913 Hudson, Model 54 five-passenger torpedo (OCW)

HUDSON - MODEL 37 - FOUR: 1913 Hudsons were easily identified by their lack of an externally mounted crank. However, a crank was packed in with the Hudson's standard equipment tool kit. In addition, the longer wheelbase of the Model 37 set it apart from the older Model 33. The most stylish Hudson in either the Model 37 or Model 54 lines was the Torpedo. Compared to the touring car body, its cowl was extended slightly and a shorter windshield was installed. The front seat and steering wheel were also repositioned slightly further back.
HUDSON - MODEL 54 - SIX: The Model 54 introduced Hudson's new six-cylinder engine. Its success enabled Hudson to both proclaim itself "the world's largest producer of six-cylinder automobiles" and declare, in August, 1913, a stock dividend of 100 percent. The six-cylinder Hudson's top speed was approximately 65 mph and it could reach 58 mph in 30 seconds. With its 127 inch wheelbase, the Model 54 was easily set apart from the Model 37 line. Both Model 37 and Model 54 were equipped with a Delco starting system in place of the older and less-efficient Disco acetylene-gas unit.
I.D. DATA: Car number on front seat riser plate and on right side of frame. Starting: [Model 37] 30001; [Model 54] 45001. Ending: [Model 37] 39200; [Model 54] 56000. Engine number on left side of cylinder block [Model 54], right front of cylinder block [Model 37].
The Model 37 engine was supplied by Continental, which gave it a Model C designation.

1913 Hudson, Model 37 coupe (WRG)

Model No.	Body Type & Seating	Price	Weight	Prod. Total
37	2dr Rds-2P	1875	3173	NA
37	3dr Torpedo-5P	1875	3350	NA
37	3dr Tr-5P	1875	3390	NA
37	2dr Cpe-3P	2350	3408	85
37	3dr Limousine-7P	3250	3680	41
54	2dr Roadster-2P	2450	3588	NA
54	3dr Torpdeo-5P*	2450	3748	NA
54	3dr Tr-7P	2600	3870	NA
54	3dr Tr-5P	2450	3823	NA
54	3dr Limo-7P	3750	4110	59
54	2dr Cpe-3P	2950	3933	15

NOTE 1: The three-door Torpedo was re-designated the phaeton in February 1913.

ENGINE: [Model 54] Inline. Six-cylinder. Cast iron block. B & S: 4-1/8 x 5-1/4 in. Disp.: 421 cid. Brake HP: 54 at 1500 rpm. NACC HP: 40.84. Main bearings: 3. Valve lifters: mechanical. [Model 37] Inline. Four-cylinder. Cast iron block. B & S: 4-1/8 x 5-1/4 in. Disp. 280.6 cid. Brake HP: 37 1500 rpm. NACC HP: 27.23. Main bearings: 3. Valve lifters: mechanical.
CHASSIS: [Model 37] W.B.: 118 in. Front/Rear Tread: 56 in. Tires: 34 x 4. [Model 54] W.B.: 127 in. Front/Rear Tread: 56 in. Tires: 36 x 4-1/2.
TECHNICAL: Sliding gear transmission. Speeds: 3F/1R. Floor shift controls. Cork insert, wet clutch. Shaft drive. Semi-floating rear axle. Mechanical brakes on two wheels. Wooden wheels. Rim size: 34 in. [Model 37] 36 in. [Model 54].
OPTIONS: Front bumper. Rear bumper. Standard equipment both of the Model 37 and Model 54 was very extensive, including electric starting and lights, illuminated dash, mohair top, side curtains, "rain vision" windshield, speedometer, clock and demountable rims. Jump seats for touring cars were $40.
HISTORICAL: Model 37 introduced in July, 1912; Model 54 in August, 1912. Inno

vations: Delco starting system. Hudson made 6,404 shipments to dealers during the 1913 calendar year. The president of Hudson was Roy D. Chapin.

1914 HUDSON

HUDSON - MODEL SIX-40 - SIX: Hudson was strictly a manufacturer of six-cylinder automobiles in 1914 and, while the new 6-40 Series was larger than the four-cylinder models they replaced, their prices were reduced. The new engine's dimensions of 3-1/2 inch bore and 5-inch stroke pointed towards future American practice and offered both better fuel economy and more power than the old four. Appearance changes included enclosed hinges on all models and a reshaped radia

tor with smooth curves.

HUDSON - MODEL SIX-54 - SIX: Hudson's 6-54 senior six now had a long 135 inch wheelbase and featured new styling, along with the 6-40, new styling. The Hudson grille still carried the familiar triangular logo, but its rounded form, more gently curved fenders and the smooth lines of the cowl were clear indicators that Hudson was abandoning the angular appearance of earlier mod

els. Also adding to the Hudson's visual appeal was its windshield, which, for the first time, was designed as a fully integrated part of its body.
I.D. DATA: Car number on front seat riser plate and on right side of frame. Starting: [Six-40] 63001; [Six-54] 565001. Ending: [Six-40] 77201; [Six-54] 62500. Engine number on left side of cylinder block.

Note: The Six-54 engine was a Continental 6C model. The Six-40 engine was designated 7N by Continental.

1914 Hudson, Model 40 landau (WRG)

1914 Hudson, Model 46 touring (WRG)

Model No.	Body Type & Seating	Price	Weight	Prod. Total
6-40	2dr Roadster-2P	1750	2822	NA
6-40	4dr Phaeton-6P	1750	2977	NA
6-40	2dr Cabriolet-2P	1950	2976	NA
6-40	4dr Tr-5P	1750	2968	NA
6-40	4dr Tr-5P (RHD)	1750	2974	NA
6-40	4dr Phaeton-7P	2250	3939	NA
6-54	2dr Sedan-5P	3100	4100	NA

ENGINE: [Model 6-40] Inline. Six cylinder. Cast iron block. B & S: 3-1/2 x 5 in. Disp. 288.5 cid. Brake hp.: 40. SAE. HP: 29.4. Main bearings: 3. Valve lifters: mechanical. [Model 6-54] Inline. Six cylinder. Cast iron block. B & S: 4-1/8 x 5-1/4 in. Disp.: 421 cid. Brake hp.: 54 at 1500 rpm. NACC HP: 40.84. Main bearings: 3. Valve lifters: mechanical.

CHASSIS: [Model 6-40] W.B.: 123 in. Rear/Rear Tread: 56 in. Tires: 34 x 4. [Model 6-54] W.B.: 135 in. Rear/Rear Tread: 56 in. Tires: 36 x 4-1/2.

1914 Hudson, Model 54 sedan (JAC)

TECHNICAL: Sliding gear transmission. Speeds: 4F/1R, 3F/1R (Model 6-40). Floor shift controls. Cork insert, wet clutch. Shaft drive. Semi-floating rear axle. Mechanical brakes on two (rear) wheels. Artillery type wood wheels. Rim size: [6-54] 36 in.; [6-40] 34 in.

OPTIONS: Front bumper. Rear bumper. Wire wheels, radiator cap with Hudson triangle I.D.

HISTORICAL: Introduced Aug. 1913 — 6-54, Nov. 1913 — 6-40. Hudson made 10,261 shipments to dealers during the 1914 calendar year. The president of Hudson was Roy D. Chapin.

1915 HUDSON

1915 Hudson, Model 40 seven-passenger touring (OCW)

HUDSON — MODEL SIX-40 — SIX: The new Hudson models were externally identified by their honeycomb type radiators (previous models were fitted with horizontal-finned versions) and higher-mounted headlamp tie-bar. All models also had a smoother radiator/hood line. Open cars (roadster and phaeton) had a new, non-folding two-piece windshield design whose upper portion pivoted from the top of the side brackets. On the Six-40 models, a revamped interior arrangement placed the gas pedal between the brake and clutch pedals. A new electric horn was activated by the steering wheel center-mounted button. Mechanically, the Six-40 for 1915 had a new tubular driveshaft and a tapered frame that allowed for a slightly shorter turning radius. Other improvements to the Six-40 included a horsepower increase to 42, a cast en bloc manifold, more efficient preheating of carburetor air and a hollow, rather than solid, driveshaft.

HUDSON — MODEL SIX-54 — SIX: Changes in the Six-54 were limited to the use of the honeycomb radiator and smoother radiator/hoodline as adopted by the Six-40.

I.D. DATA: Car number on front seat riser plate and on right side of frame. Starting: Six-40 73501; Six-54 59001. Ending: Six-40 90000; Six-54 62000. Engine number on left side of cylinder block.

Model No.	Body Type & Seating	Price	Weight	Prod. Total
Six-40	2dr Roadster-3P	1550	2772	NA
Six-40	4dr Phaeton-7P	1550	2922	NA
Six-40	2dr Cabriolet-3P	1750	2946	NA
Six-40	2dr Cpe-4P	2150	3162	NA
Six-40	4dr Limousine Landaulet-7P	2700	3432	NA
Six-40	4dr Limousine-7P	2550	3362	NA
Six-40	4dr RHD Phaeton-7P	1550	2922	NA
Six-54	4dr Phaeton-7P	2350	3965	NA
Six-54	2dr Sedan-5P	3100	NA	NA
Six-54	4dr Limousine-7P	3500	4226	NA

ENGINE: [Model Six-40] Inline. Six-cylinder. Cast iron block. B & S: 3-1/2 x 5 in. Disp.: 288.5 cid. Brake HP: 42. NACC HP: 29.4. Main bearings: 4. Valve lifters: mechanical. [Model Six-54] Inline. Six-cylinder. Cast iron block. B & S: 4-1/2 x 5-1/4 in. Disp.: 421 cid. Brake HP: 55 at 1500 rpm. NACC. HP: 40.84. Main bearings: 3. Valve lifters: mechanical.

CHASSIS: [Model Six-40] W.B.: 123 in. Frt/Rear Tread: 56 in. Tires: 34 x 4. [Model Six-54] W.B. 135 in. Frt/Rear Tread: 56 in. Tires: 36 x 4-1/2.

TECHNICAL: Sliding gear transmission. Speeds: 4F/1R, 3F/1R [Model Six-40]. Floor shift controls. Cork insert, wet clutch. Shaft drive. Semi-floating rear axle. Mechanical brakes on two rear wheels. Artillery-type wood wheels. Rim size: [6-54] 36 in.; [6-40] 34 in.

OPTIONS: Front bumper. Rear bumper. Wire wheels.

HISTORICAL: Introduced June 1914. Hudson made 12,864 shipments to dealers during the 1915 calendar year. The president of Hudson was Roy D. Chapin.

1916 HUDSON

HUDSON — MODEL SIX-40 — SERIES G-SIX: The Six-40's model run was a short one in 1916. It lasted from June 1915 to January 1916. Then it, as well as the unchanged Six-54, was replaced by the Super Six, which Hudson regarded as a 1917 model. The Six-40 featured both styling and engineering changes. The beltline was given a gentle curve to further enhance the "yacht-line" styling that had been an advertised feature since 1913. Both entry into and riding in a Hudson became a bit more enjoyable due to wider doors and a roomier interior. On open models, the upper portion of the beltline was leather-covered. Hudson also touted its new "Ever-Lustre" finish. In a veiled reference to rust and corrision, it noted that the new finish "combats, as never before, the main cause of depreciation." In preparation for the new Super Six and a change of philosophy regarding annual model changes, Hudson identified the last of the Six-40s as Series G.

1916 Hudson, Super Six Touring Victoria (OCW)

I.D. DATA: Car number on front seat riser plate and on right side of frame. Starting: G10001. Ending: G40000. Engine number on left side of cylinder block.

Model No.	Body Type & Seating	Price	Weight	Prod. Total
Series G	2dr Roadster-3P	1350	2900	NA
Series G	4dr Phaeton-7P	1350	3033	NA
Series G	2dr Cabriolet-3P	1650	3009	NA
Series G	2dr Cpe-4P	2000	3240	NA
Series G	2dr Tr Sedan-7P	1875	3330	NA
Series G	4dr Limousine-7P	2450	3535	NA
Series G	4dr Town Car-7P	—	3370	NA

ENGINE: Inline. Six-cylinder. Cast iron block. B & S: 3-1/2 x 5 in. Disp.: 288.5 cid. C.R.: 5.0:1. Brake HP: 42. NACC. HP.: 29.4. Main bearings: 4. Valve lifters: mechanical.
CHASSIS: [Model Six-40] W.B.: 123 in. Front/rear Tread: 56 in. Tires: 34 x 4.
TECHNICAL: Sliding gear transmission. Speeds: 3F/1R. Floor shift controls. Cork insert, wet clutch. Shaft drive. Semi-floating rear axle. Mechanical brakes on two (rear) wheels. Artillery-type wood wheels. Rim size: 34 in.
OPTIONS: Front bumper. Rear bumper. Wire wheels.
HISTORICAL: Introduced June 1916. Hudson made 25,772 shipments to dealers during the 1916 calendar year. The president of Hudson was Roy D. Chapin.

1916 Hudson, Super Six touring (WGR)

HUDSON - SUPER SIX - SERIES H: Hudson made motoring history with introduction of the Super Six. It was powered by the first Hudson-built engine and established a new benchmark by which to measure the performance of production cars. Hudson patented the four-bearing crankshaft fitted with eight counter-weights. Other notable advances included larger valves, a high 5.0:1 compression ratio and excellent valve porting. Hudson called it the "greatest motor ever built." Produced through 1926, it established Hudson as a manufacturer whose six-cylinder engines made a mockery of the performance claims made by producers of far more expensive automobiles.

I.D. DATA: Car numbers on front seat riser and on right side of frame. Starting: Series HH-1. Ending: Series HH-99999. Engine number on left side of cylinder block.

Model No.	Body Type & Seating	Price	Weight	Prod. Total
H	2dr Roadster-2P	1375	3170	NA
H	4dr Phaeton-7P	1375	3385	NA
H	2dr Cabriolet-3P	1675	3310	NA
H	dr Tr Sedan-7P	1900	3600	NA
H	4dr Limousine-7P	2500	3750	NA
H	4dr Landaulet Limo-7P	2750	—	NA
H	4dr Town Car-7P	2500	3660	NA
H	4dr Town Car Landaulet	2750	—	NA

ENGINE: Inline. Six-cylinder. Cast iron block. B & S: 3-1/2 x 5 in. Disp.: 289 cid. C.R.: 5.0:1. Brake HP: 76 at 2450 rpm. NACC HP: 29.4. Main bearings: 4. Type of valve lifters: mechanical. Carb.: Hudson sidedraft.
CHASSIS: [Series H] W.B.: 125-1/2 in. Front/rear tread: 56 in. Tires: 35 x 4-1/2.
TECHNICAL: Sliding gear transmission. 3F/1R. Floor shift. Cork inserts, wet clutch. Shaft drive. Semi-floating rear axle. Mechanical brakes on two rear wheels. Artillery type wood wheels. Rim size: 34 in.
OPTIONS: Front bumper. Rear bumper. Wire wheels. Spare tire. The Series H Phaeton had disc wheels and a motometer as standard equipment.
HISTORICAL: Introduced Jan. 16, 1916, following a series of Super Six speed runs made on a Long Island (New York) racetrack in December, 1915. Series H shipments to dealers were included in the 25,772 shipments made during calendar 1916. A Super Six set a new transcontinental record between San Francisco and New York, then turned around and made it back to San Francisco in comparable time.

1917 HUDSON

1917 Hudson Super Six touring (JAC)

1917 Hudson Super Six coupe (WRG)

HUDSON - SUPER SIX - SERIES J: Beginning Dec. 1, 1916, the Series J and Series 4J replaced the Series H. They had built-in radiator shutters and a Boyce motometer. In addition, a different upholstery pattern was used on open phaetons and the limousine had a straight front door line and squared-off window edges.

I.D. DATA: Car number on front seat riser plate and on right side of frame. Starting: [J] 1; [4J] 75000. Ending: [J] 96499; [4J] 97999. Engine number on left side of cylinder block/ motor mount.

Model No.	Body Type & Seating	Price	Weight	Prod. Total
NA	4dr Phaeton-4P	1750	3180	NA
NA	4dr Phaeton-7P	1650	3220	NA
NA	1r Tr Sedan	2175	3450	NA
NA	2dr Cabr-3P	1950	3195	NA
NA	4dr Limousine-7P	2925	3715	NA
NA	4dr Landau Limousine-7P	3025	3760	NA
NA	4dr Town Car-7P	3400	2925	3530
NA	4dr Landau Town Car-7P	3025	3585	NA
NA	4dr Tr Sedan-5P	NA	NA	NA
NA	dr Landau Runabout P	2350	3250	NA
NA	4dr Sedan-7P	NA	3700	NA
NA	4dr Tr Limousine-4P	3150	3655	NA
NA	4dr Full-Folding Landau-4P	NA	NA	NA

ENGINE: Inline. Six-cylinder. Cast iron block. B & S: 3-1/2 x 5 in. Disp.: 289 cid. C.R.: 5.0:1. Brake HP: 76 at 2450 rpm. NACC HP: 29.4. Main bearings: 4. Valve lifters: mechanical. Carb.: Hudson sidedraft.
CHASSIS: Wheelbase: 125-1/2 in. Frt/Rear Tread: 56 in. Tires: 34 x 4-1/2.
TECHNICAL: Sliding gear transmission. Speeds: 3F/1R. Floor shift controls. Multiple disc, cork inserts, running in oil. Shaft drive. Semi-floating rear axle. Mechanical brakes on two (rear) wheels. Wooden spoke wheels with detachable rims. Rim size 34 inches.
OPTIONS: Front bumper. Rear bumper. Wire wheels. Spare tire.
HISTORICAL: Introduced December 1, 1917. The president of Hudson was Roy D. Chapin. The Hudson Super Six continued its winning ways in racing with a four-car team enjoying numerous 1917 victories.

1918 HUDSON

1918 Hudson Super Six touring limousine (JAC)

HUDSON — SUPER SIX — SERIES M: Changes in the Super Six format were very limited. The rear doors on the phaeton models now were rear-hinged and closed models were not equipped with external windshield visors.

I.D. DATA: Car number on front seat riser plate and on right side of frame. Starting: M5000. Ending: M97499. Engine number on left side of cylinder block/motor mount.

Model No.	Body Type & Seating	Price	Weight	Prod. Total
NA	4dr Phaeton-4P	2050	3180	NA
NA	4dr Phaeton-7P	1950	3400	NA
NA	2dr Landau Runabout-2P	2350	3250	NA
NA	2dr Cabriolet-3P COUPE	2650	3500	NA
NA	2dr Cpe-4P	2850	3450	NA
NA	4dr Sedan-7P	2750	3700	NA
NA	4dr Limousine-7P	3400	3715	NA
NA	4dr Landau Limousine-7P	3500	3760	NA
NA	4dr Town Car-7P	3400	3605	NA
NA	4dr Full-Folding Landau- P	4250	3765	NA
NA	4dr Tr Limousine- P	3150	NA	NA
NA	4dr Landau Town Car-7P	3500	NA	NA

ENGINE: Inline. Six-cylinder. Cast iron block. B & S: 3-1/2 x 5 in. Disp.: 289 cid. C.R.: 5.0:1. Brake HP: 76 at 2450 rpm. NACC HP: 29.4. Main bearings: 4. Valve lifters: mechanical. Carb.: Stewart Warner.

CHASSIS: Wheelbase: 125-1/2 in. Frt/Rear Tread: 56 in. Tires: 34 x 4-1/2.

The four-passenger Phaeton and Landau Runabout used 32 x 4-1/2 tires. The Landau Town Car, Town Car and Limousine were equipped with 33 x 5 tires.

TECHNICAL: Sliding gear transmission. Speeds: 3F/1R. Floor shift controls. Multiple disc, cork inserts, running in oil. Shaft drive. Semi-floating rear axle. Mechanical brakes on two (rear) wheels. Wooden spoke wheels with detachable rims. Rim size 34 inches.

OPTIONS: Front bumper. Rear bumper. Wire wheels. Runningboard mats. Windshield mounted spotlight. Leather top for Landau Runabout.

1918 Hudson Super Six full-folding landau (WRG)

HISTORICAL: Introduced December 1917. Hudson made 12,526 shipments to dealers during the 1918 calendar year. The president of Hudson was Roy D. Chapin. The Hudson racing team was disbanded in August 1917.

1919 HUDSON

HUDSON — SUPER SIX — SERIES O: All closed body Series O Hudsons had large external sun visors and their front doors were rear-hinged. Common to all Hudsons were 12-spoke front wheels. Major chassis revisions included seven-inch side frame rails, a larger and sturdier rear axle, plus brakes with measurements of 2-1/2 x 15 inches. The Hudson's single taillight was moved from its 1918 left rear fender location to the rear cross-member on 1919 models. Also identifying the Series O Hudsons were their higher gearshift and four hinge doors (replacing doors with three hinges).

I.D. DATA: Car number on front seat riser plate and on right side of frame. Starting: 5000. Ending: 90999. Engine number on left side of cylinder block/motor mount.

Model No.	Body Type & Seating	Price	Weight	Prod. Total
NA	4dr Phaeton-4P	2075	3320	NA
NA	4dr Phaeton-7P	1975	3475	NA
NA	2dr Cabriolet- P COUPE	2450	3500	NA
NA	2dr Cpe- P	2950	3530	NA
NA	4dr Sedan-7P	2775	3775	NA
NA	4dr Tr Limousine- P	3300	3730	NA
NA	4dr Limousine-7P	3650	3800	NA
NA	4dr Landau Limousine	NA	NA	NA
NA	4dr Town Car-7P	NA	NA	NA
NA	4dr Landau Town Car-7P	NA	NA	NA

1919 Hudson Super Six Special sedan (WRG)

ENGINE: Inline. Six-cylinder. Cast iron block. B & S: 3-1/2 x 5 in. Disp.: 289 cid. C.R.: 5.0:1. Brake HP: 76 at 2450 rpm. NACC HP: 29.4. Main bearings: 4. Valve lifters: mechanical. Carb.: Stewart Warner.

CHASSIS: Wheelbase: 125-1/2 in. Frt/Rear Tread: 56 in. Tires: 34 x 4-1/2.

TECHNICAL: Sliding gear transmission. Speeds: 3F/1R. Floor shift controls. Multiple disc, cork inserts, running in oil. Shaft drive. Semi-floating rear axle. Mechanical brakes on two (rear) wheels. Wooden spoke wheels with detachable rims.

OPTIONS: Front bumper. Rear bumper. Wire wheels. Runningboard mats. Windshield mounted spotlight. Leather top for Landau Runabout.

HISTORICAL: Introduced May 1919. Hudson made 18,175 shipments to dealers during the 1919 calendar year. The president of Hudson was Roy D. Chapin. Six Hudson racers attempted to qualify for the 1919 Indy 500. Ira Vail's Hudson qualified at 94.1 mph and finished eighth. Just behind was car no. 21, a Hudson driven by Dennis Hickey, who averaged 80.22 mph for the 500 miles. Another Hudson, driven by Ora Haibe, started in 26th position and was credited with 14th place at the end of the race. Car no. 5, which was powered by a modified Hudson engine, had a qualifying average of 99.80 mph and started in 14th position. However, it left the race on the 14th lap with a broken connecting rod.

1920 HUDSON

HUDSON — SUPER SIX — 10-O/11-O/12-O: Hudson produced the Super Six in three series, yet they were little changed from 1919. Most models were slightly heavier than their year-old counterparts. Although not used throughout the entire model run, a new front fender tie-bar was introduced that positioned the headlights noticeably higher.

1920 Hudson Super Six coupe (OCW)

HUDSON — SUPER SIX — 10-0/11-0/12-0: Hudson produced the Super Six in three series, yet they were little changed from 1919. Most models were slightly heavier than their year-old counterparts. Although not used throughout the entire model run, a new front fender tie-bar was introduced that positioned the headlights noticeably higher.

I.D. DATA: Car number on front seat riser plate and on right side of frame. Starting: 5000. Ending: 91999. Engine number on left side of cylinder block/motor mount.

Model No.	Body Type & Seating	Price	Weight	Prod. Total
NA	4dr Phaeton-4P	2600	3405	NA
NA	4dr Phaeton-7P	2600	3575	NA
NA	2dr Cabriolet-2P COUPE	NA	3550	NA
NA	2dr Cpe-P	3575	3620	NA
NA	4dr Sedan-7P	3400	3815	NA
NA	4dr Tr Limousine-5P	3925	3840	NA
NA	4dr Limousine-7P	4275	3860	NA

NOTE 1: Hudson made substantial price cuts during 1920.
ENGINE: Inline. Six-cylinder. Cast iron block. B & S: 3-1/2 x 5 in. Disp.: 289 cid. C.R.: 5.0:1. Brake HP: 76 at 2450 rpm. NACC HP: 29.4. Main bearings: 4. Valve lifters: mechanical. Carb.: Stewart Warner.
CHASSIS: Wheelbase: [All] 125-1/2 in. Frt/Rear Tread: 56 in. Tires: 34 x 4-1/2.
TECHNICAL: Sliding gear transmission. Speeds: 3F/1R. Floor shift controls. Multiple disc, cork inserts, running in oil. Shaft drive. Semi-floating rear axle. Mechanical brakes on two (rear) wheels. Wooden spoke wheels with detachable rims.
OPTIONS: Front bumper. Rear bumper. Wire wheels. Runningboard mats. Spotlight.
HISTORICAL: Introduced December 1919. Hudson made 22,268 shipments to dealers during the 1920 calendar year. The president of Hudson was Roy D. Chapin.

1921 HUDSON

1921 Hudson Super Six four-passenger coupe (OCW)

HUDSON — SUPER SIX: From 1921 through 1923, Hudson annually produced, in effect, two separate lines of automobiles. The initial 1921 models were essentially carried over from 1920. Then, in September, revamped 1921 models were introduced. In turn, these became 1922 models, until May of 1922, when new models (also considered 1922s) were brought out. The second run of 1921 models were identified by a revamped interior featuring a new steering wheel and instrument panel which placed all the instruments in a panel that was center-mounted on the dash. In addition, the classic H-shaped shifting gate was replaced with a rotating ball arrangement. Hudson also rearranged its foot controls by moving the accelerator from its position between the clutch and brake to the more logical location adjacent to the right side of the brake. External styling revisions of the second series Hudsons included heavier fenders with a more pronounced overlap and the installation of splash shields beneath the radiator.

I.D. DATA: Car number on front frame channel, dash and frame side. Starting: 100000. Ending: 499999 (serial number range 1921-23). Engine number on left side of cylinder block.

Model No.	Body Type & Seating	Price	Weight	Prod. Total
NA	4dr Phaeton-4P	NA	3405	NA
NA	4dr Phaeton-7P	NA	3575	NA
NA	2dr Cabriolet-2P	NA	3550	NA
NA	2dr Cpe-3P	3275	3620	NA
NA	4dr Sedan-7P	3400	3815	NA
NA	4dr Tr Limousine-5P	3625	3840	NA
NA	4dr Limousine-7P	4000	3860	NA

NOTE 1: Hudson reduced its 1921 prices in June and August.
ENGINE: Inline. Six-cylinder. Cast iron block. B & S: 3-1/2 x 5 in. Disp.: 289 cid. C.R.: 5.0:1. Brake HP: 76 at 2450 rpm. NACC HP: 29.4. Main bearings: 4. Valve lifters: mechanical. Carb.: Stewart Warner.
CHASSIS: Wheelbase: 125-1/2 in. Frt/Rear Tread: 56 in. Tires: 34 x 4-1/2.
TECHNICAL: Sliding gear transmission. Speeds: 3F/1R. Floor shift controls. Multiple disc, cork inserts, running in oil. Shaft drive. Semi-floating rear axle. Mechanical brakes on two (rear) wheels. Wooden spoke wheels with detachable rims.
OPTIONS: Front bumper. Rear bumper. Wire wheels. Runningboard mats. Radiator shutters. Spotlight.
HISTORICAL: Introduced December 1, 1920. Revamped models introduced September 1921. Hudson made 13,721 shipments to dealers during the 1921 calendar year. The president of Hudson was Roy D. Chapin. A competition prepared Hudson won the Penrose trophy at Pikes Peak Hill Climb with a time of 19 minutes and 16.1 seconds.

1922 HUDSON

HUDSON — SUPER SIX — SERIES 9: Hudson moved out of the postwar sales depression in a strong fashion. In July, price reductions randing from $50 to $100 were announced and company president Roy D. Chapin reported that "the volume of shipment is now so great that certain savings have been effected in costs and the public will be given the benefit." Stockholders also received a share of Hudson's renewed prosperity since, in September, a dividend of 50 cents per share on non-par and $2.50 per share on par capital stock was declared. Although the new Essex coach captured the bulk of the public's attention, Hudson also introduced this new body style. During the first portion of the year, new drum-shaped headlamps were adopted. Among the second line of Super Sixes, debuting in May 1922, was a sedan with a Biddle & Smart body. Its styling, far more elegant than the Fisher-built body it replaced, pointed to the Classic Hudsons yet to come. Improvements to Hudson's long-lived Super Six included the adoption of aluminum pistons and a Morse timing chain in place of the older helical drive. Second series Hudsons had their battery placed under the front seat.

1922 Hudson Super Six phaeton (HAC)

I.D. DATA: Car number on front frame channel, dash and frame side. Starting: 100000. Ending: 499999 (serial number range 1921-23). Engine number on left side of cylinder block.

Model No.	Body Type & Seating	Price	Weight	Prod. Total
NA	4dr Phaeton-4P	1695	3395	NA
NA	4dr Phaeton-7P	1745	3445	NA
NA	2dr Cabriolet-2P COUPE	2295	3550	NA
NA	2dr Coach-5P	1625	3435	NA
NA	2dr Cpe-P	2570	3620	NA
NA	4dr Sedan-7P	2650	3785	NA
NA	4dr Tr Limousine-5P	2920	3870	NA
NA	4dr Limousine-7P	3495	3860	NA
NA	4dr Spds-4P	1525	3310	NA
NA	4dr Phaeton-7P	1575	3455	NA
NA	4dr Sedan-5P	2295	3720	NA

NOTE 1: Five-passenger four-door Sedan has Biddle & Smart body.
ENGINE: Inline. Six-cylinder. Cast iron block. B & S: 3-1/2 x 5 in. Disp.: 289 cid. C.R.: 5.0:1. Brake HP: 76 at 2450 rpm. NACC HP: 29.4. Main bearings: 4. Valve lifters: mechanical. Carb.: Stewart Warner.
CHASSIS: Wheelbase: 125-1/2 in. Frt/Rear Tread: 56 in. Tires: 34 x 4-1/2.
TECHNICAL: Sliding gear transmission. Speeds: 3F/1R. Floor shift controls. Multiple disc, cork inserts, running in oil. Shaft drive. Semi-floating rear axle. Mechanical brakes on two (rear) wheels. Wooden spoke wheels with detachable rims.
OPTIONS: Front bumper. Rear bumper. Wire wheels. Runningboard mats. Radiator shutters.
HISTORICAL: Introduced December, 1921. Second version May 1922. Hudson made 28,242 shipments to dealers during the 1922 calendar year. The president of Hudson was Roy D. Chapin. Hudsons finished first at the Pikes Peak Hill Climb with a time of 20 minutes and five seconds.

1923 HUDSON

1923 Hudson Super Six coach (JAC)

HUDSON — SUPER SIX — SIX: Except for minor detail modifications, which were incorporated as running changes during the year, the 1923 Hudsons were virtually identical to the 1922 models. Among the revisions were the use of McKee "Spreadlight" headlight lenses, plus an extended length, 28 inch gearshift. This was an extremely profitable year for Hudson. Its fiscal year profits totalled nearly $14.5 million, up from the $12.6 million level of the previous year. Twice during the year, the company declared an extra 25-cents dividend in addition to the regular 50-cents dividend. The Hudson Motor Car Company also made some significant changes in its top level management structure. Roy D. Chapin, after serving as president for 13 years, became chairman of the board. His successor was Roscoe B. Jackson. Replacing Jackson as vice-president and treasurer was William J. McAneeny.

I.D. DATA: Car number on front frame channel, dash and frame side. Starting: 100000. Ending: 499999 (serial number range 1921-23). Engine number on left side of cylinder block.

Model No.	Body Type & Seating	Price	Weight	Prod. Total
NA	4dr Speedster-4P	1295	3395	NA
NA	4dr Phaeton-7P	1350	3445	NA
NA	2dr Coach-5P	1375	3433	NA
NA	2dr Cpe-_P	2570	3620	NA
NA	4dr Sedan-5P	1895	3620	NA
NA	4dr Sedan-7P	2095	3720	NA

NOTE 1: Five-passenger four-door Sedan has Biddle & Smart body.
NOTE 2: The two-door coupe and seven-passenger sedan bodies were phased out of production during 1922.

ENGINE: Inline. Six-cylinder. Cast iron block. B & S: 3-1/2 x 5 in. Disp.: 289 cid. C.R.: 5.0:1. Brake HP: 76 at 2450 rpm. NACC HP: 29.4. Main bearings: 4. Valve lifters: mechanical. Carb.: Stewart Warner.

CHASSIS: Wheelbase: 125-1/2 in. Frt/Rear Tread: 56 in. Tires: 34 x 4-1/2.

TECHNICAL: Sliding gear transmission. Speeds: 3F/1R. Floor shift controls. Multiple disc, cork inserts, running in oil. Shaft drive. Semi-floating rear axle. Mechanical brakes on two (rear) wheels. Wooden spoke wheels with detachable rims.

1923 Hudson Super Six limousine (WRG)

OPTIONS: Front bumper. Rear bumper. Spotlight. Steel disc wheels (25.00) Spare wheel. Under seat heater. Radiator shutters.

HISTORICAL: Introduced December 1922. Hudson made 46,337 shipments to dealers during the 1923 calendar year. The president of Hudson was Roscoe B. Jackson.

1924 HUDSON

1924 Hudson Super Six five-passenger sedan (JAC)

HUDSON — SUPER SIX — SIX: The initial line of Hudson Super-Six automobiles were virtually unchanged from those offered in the latter part of 1923. In mid-June they were replaced by significantly altered models. Styling changes included windshields with a curved lower edge and a raised hood line from radiator to cowl. A new fender crease line began to take Hudson away from the more rigid look of the early twenties. A longer, 127.5 inch wheelbase also contributed to this effect. Mechanical changes were also extensive. New, smaller 33 inch wheels were fitted with balloon type, 33 x 6.20 tires. To accommodate this change, the Hudson's steering and suspension were modified. Although its horsepower rating of 76 remained unchanged, the Super Six engine now had a separate intake manifold mounted on the right side. Previously the manifold had been on the opposite side of the engine and cast integrally with the block. The Hudson-made side draft carburetor was also replaced by a Detroit Lubricator model.

I.D. DATA: Car number on front frame channel, dash and frame side. Starting: 500001. Ending: 562016. Engine number on left side of cylinder block.

Model No.	Body Type & Seating	Price	Weight	Prod. Total
NA	4dr Sedan-5P	1895	3590	NA
NA	4dr Sedan-7P	2145	3675	NA
NA	4dr Sedan-5P DELUXE	2145	3605	NA
NA	4dr Speedster-4P	1400	3275	NA
NA	4dr Phaeton-7P	1500	3400	NA
NA	4dr Sedan-5P DELUXE	2150	3585	NA
NA	2dr Coach-5P	1395	3385	NA
NA	4dr Sedan-7P	2250	3640	NA

NOTE 1: Five- and seven-passenger sedans have Biddle & Smart body.

1924 Hudson Super Six seven-passenger phaeton (WRG)

ENGINE: Inline. Six-cylinder Cast iron block. B & S: 3-1/2 x 5 in. Disp.: 289 cid. C.R.: 5.0:1. Brake HP: 76 at 2450 rpm. NACC HP: 29.4. Main bearings: 4. Valve lifters: mechanical. Carb.: Detroit Lubricator.

CHASSIS: Wheelbase 127-1/2 in. Tires: 33 x 6.20. 125" 1st SERIES

TECHNICAL: Sliding gear transmission. Speeds: 3F/1R. Floor shift controls. Multiple disc, cork inserts, running in oil. Shaft drive. Semi-floating rear axle. Mechanical brakes on two (rear) wheels. Wooden spoke wheels with detachable rims.

OPTIONS: Front bumper. Rear bumper. Spotlight. Auxiliary seats & carpeting in sedan (115.00). Steel disc wheels. Wire wheels. Radiator shutters.

HISTORICAL: Introduced December 1923. New summer models introduced June 1924. Hudson made 59,427 shipments to dealers during the 1924 calendar year. The president of Hudson was Roscoe B. Jackson.

191

1925 HUDSON

1925 Hudson Super Six Brougham (OCW)

HUDSON — SUPER SIX — SIX: Hudson did not make any significant changes in the appearance of its automobiles for 1925. However, beginning in June 1925 a very handsome Biddle & Smart bodied Brougham model became available and proved to be the most popular Biddle & Smart Hudson ever offered. Hudson also revised its Coach model during 1925 by reshaping its body to accept thinner side pillars and a windshield with a curved lower edge. In January a shift from 33 x 6.20 to 33 x 6.00 tires was announced. A total output of 269,474 Hudson and Essex automobiles put the Hudson Motor Car Co. in third position behind Chevrolet and Ford for the 1925 calendar year.

I.D. DATA: Car number on front frame channel, dash and frame side. Starting: 562017. Ending: 672227. Engine number on left side of -cylinder block/motor mount.

Model No.	Body Type & Seating	Price	Weight	Prod. Total
NA	2dr Coach-4P	1345	3385	NA
(1)	2dr Coach-4P	1165	3385	NA
(2)	4dr Sedan-5P 3W	1695	3585	NA
(2)	4dr Sedan-7P	1650	3640	NA
(2)	4dr Brougham-4P 2W	1450	3425	NA
NA	4dr Phaeton-7P	1200	3400	NA

NOTE (1): This Coach replaced the older Coach in March 1925. Its Biddle & Smart body was constructed of aluminum.
NOTE (2): Biddle & Smart body.

1925 Hudson Super Six seven-passenger sedan (WRG)

ENGINE: Inline. Six-cylinder. Cast iron block. B & S: 3-1/2 x 5 in. Disp.: 289 cid. C.R.: 5.0:1. Brake HP: 76 at 2450 rpm. NACC HP: 29.4. Main bearings: 4. Valve lifters: mechanical. Carb.: Detroit Lubricator.
CHASSIS: No series designations assigned. Wheelbase 127-1/2 in. Tires: 33 x 6.20 — replaced by 33 x 6.00 starting January 1925.
TECHNICAL: Sliding gear transmission. Speeds: 3F/1R. Floor shift controls. Multiple disc, cork inserts, running in oil. Shaft drive. Semi-floating rear axle. Mechanical brakes on two (rear) wheels. Wooden spoke wheels with detachable rims.
OPTIONS: Front bumper. Rear bumper. Auxiliary seats and carpeting (Sedan). Steel Tru-Arc disc wheels. Wire wheels. Radiator shutters.
HISTORICAL: Introduced December 1924. Hudson made 109,840 shipments to dealers during the 1925 calendar year. The president of Hudson was Roscoe B. Jackson.

1926 HUDSON

1926 Hudson Super Six touring (WRG)

1926 HUDSON — SUPER SIX: Although Hudson carried the style and design of its 1925 models in 1926, it was nonetheless an important year for Hudson. Recognition of the role Hudson had played in automotive history in developing the Essex Coach came from *The New York Times* (Jan. 10, 1926), which noted, "The flood of new, small closed sixes is one of the outstanding features of the year...that the light, economical six makes a definite and potent appeal cannot be doubted. Indeed the Hudson and Essex organization has one of the most remarkable achievements of 1925 to its credit in the production of 250,000 cars of both makes in almost entirely closed models." Hudson also moved boldly to expand its corporate base in 1926. It opened a new $3 million body plant and the first product was a revised coach for both Hudson and Essex. Mechanical changes consisted of revamped carburetor and intake manifold that Hudson said would improve fuel consumption by two miles per gallon.

I.D. DATA: Car number on front frame channel, dash and frame side. Starting: 672228. Ending: 713809*. Engine number on left side of -cylinder block/motor mount.
* In a letter to its dealers dated Sept. 10, 1926, Hudson explained that it was instituting a yearly model classification starting with definite serial numbers. Thus, the following cars were considered 1927 models: Hudson coaches beginning with serial number 713810, sedans starting with serial number 714674 and Broughams starting with serial number 716440.

Model No.	Body Type & Seating	Price	Weight	Prod. Total
NA	4dr Brougham-4P	1450	3425	—
NA	4dr Brougham-5P	1395	3495	—
NA	4dr Sedan-7P	1650	3640	—
NA	4dr Touring-7P	1300	3395	—
NA	2dr Coach-4P	1165	3470	—
NA	2dr Spl Coach	1150	3440	—

NOTE 1: A five-passenger Brougham with a Biddle & Smart body replaced the older Brougham midway through the model year.
NOTE 2: The four-door touring had the Biddle & Smart body.
NOTE 3: The two-door Special Coach was fitted with a Hudson-built body approximately two inches lower than the older version.
ENGINE: L-head. Inline. Six-cylinder. Cast iron block. B & S: 3-1/2 x 5 in. Disp.: 289 cid. C.R.: 5.0:1. Brake HP: 76 at 2450 rpm. NACC HP: 29.4. Main bearings: 4. Valve lifters: mechanical. Carb.: Detroit Lubricator.
CHASSIS: Wheelbase: 127-1/2 in. Tires: 33 x 6.00.
TECHNICAL: Sliding gear transmission. Speeds: 3F/1R. Floor shift controls. Multiple disc, cork inserts, running in oil. Shaft drive. Semi-floating rear axle. Mechanical brakes on two (rear) wheels. Wooden spoke wheels with detachable rims.
OPTIONS: Front bumper. Rear bumper. Spotlight. Radiator shutters.
HISTORICAL: Introduced December, 1925. Hudson made 70,261 shipments to dealers during the 1926 calendar year. The president of Hudson was Roscoe B. Jackson.

1927 HUDSON

HUDSON — MODEL O/MODEL S — SIX: Hudsons for 1927 were radically changed automobiles. They were fitted with new 18 inch wheels, four wheel brakes, new rear suspension and styling that featured a higher radiator-hood line and fenders of full crown design. Headlamps were bullet shaped and contributed, along with a four inch reduction in height, to Hudson's very attractive styling. Hudson also offered new exterior color choices to take full advantage of its new look. On models offered after late June, a full-length beltline molding, painted in a contrasting body color, dramatically improved Hudson's appearance. The Model S Hudsons were mounted on a 118 inch wheelbase, while the Model Os continued the 127-3/8 inch wheelbase. A startling development was the replacement of the veteran Super Six engine with a six-cylinder engine of F-head design. Also breaking with previous practice was the adoption of a single-plate clutch in place of the older multi-disc unit. Hudson had spent some $7 million to expand output of its factory to 1,800 cars per nine-hour day and Hudson chairman Roy D. Chapin understandably looked to the future with optimism. He noted however that, "Buyers are now insistent that cars shall excel in appearance and convenience, as well as in the fundamental qualities. The demand for improved performance is widespread and is being met by better design, material and workmanship."

1927 Hudson Model O Custom roadster (WRG)

I.D. DATA: Car numbers on front frame channel, dash and frame side. Starting: [Model O] 750000, [Model S] 1001. Ending: [Model O] 803568, [Model S] 12269. Starting: 713810 (carryover 1926 cars sold as 1927 models) Engine numbers on left side of cylinder block/motor mount. Starting: 438230. Ending: NA.

Model No.	Body Type & Seating	Price	Weight	Prod. Total
STANDARD LINE 127				
O	2dr Coach-5P	1285	3505	NA
O	4dr Sedan-5P	1385	3620	NA
CUSTOM LINE 127				
O	2dr Roadster-2P	1500	3480	NA
O	4dr Sedan-5P	1750	3755	NA
O	4dr Phaeton-7P	1600	3565	NA
O	4dr Sedan-7P	1850	3870	NA
O	4dr Brougham-4P	1575	3660	NA
STANDARD LINE 118"				
S	2dr Coach-5P	1175	3510	NA
S	4dr Sedan-5P	1285	3590	NA

NOTE 1: All Model O Custom Hudsons had Biddle & Smart bodies.

1927 Hudson Model O Custom Murphy Victoria (WRG)

ENGINE: F-head. Inline. Six-cylinder. Cast iron block. B & S: 3-1/2 x 5 in. Disp.: 288.5 cid. C.R.: 5.0:1. Brake HP: 92 at 3200 rpm. NACC HP: 29.4. Main bearings: 4. Valve lifters: mechanical. Carb.: Marvel 1-1/4 in.

CHASSIS: [Model O] Wheelbase 127-3/8 in. O.L.: 188 in. Tires: 31 x 6.00. [Model S] Wheelbase 118 in. O.L.: 178-1/2 in. Tires: 31 x 6.00.

TECHNICAL: Sliding gear transmission. Speeds: 3F/1R. Floor shift controls. Single disc, cork inserts, running in oil. Shaft drive. Semi-floating rear axle. Overall Ratio: 4.09:1 (Model S) and 4.45:1 (Model O). Bendix mechanical brakes on four wheels. Wooden spoke wheels with detachable rims. Rim size: 19 x 4-1/2 in.

OPTIONS: Trunk. Bumpers, front and rear. Radiator shutters.

HISTORICAL: Introduced January 1927. Innovations: new F-head engine, four wheel brakes. Hudson made 66,034 shipments to dealers during the 1927 calendar year. The president of Hudson was Roscoe B. Jackson. The Hudson F-head engine's official rating of 92 hp is regarded as very conservative by Hudson historians. This is especially true of a revised version which became available in July 1927. It had new manifolding, an altered head design and relocated spark plugs and intake valves. Claims of a 100 mph top speed for the 1927 Hudson were not uncommon and Barney Oldfield drove a 1927 coach for 1,000 miles, at a Culver City, Calif. racetrack, at a speed in excess of 76 mph.

1928 HUDSON

1928 Hudson Model S Standard roadster (JAC)

HUDSON — MODEL O/MODEL S — SIX: The 1928 Hudsons were handsome automobiles. The use of a higher and more slender radiator, vertical engine louvers and larger, parabolically shaped headlights gave them a stately, almost aristocratic appearance. With the motometer moved to the dash, a sculptured hood ornament took its place and small saddle lamps were now mounted on the cowl. An industry first was Hudson's new steering wheel, constructed of a hard rubber shell and a solid steel core. Formed with finger scallops, it was colored ebony black to match the finish of the instrument panel. The spark, throttle, light and horn controls were placed at the steering wheel's center. In place of the transmission lock previously used, the Hudson was now equipped with an Electrolock ignition system. Although the Hudson's basic chassis structure was unchanged, two tubular cross-members were added. During 1928, a number of Hudsons were fitted with custom-built bodies by Murphy Body Co. of Pasadena, Calif. They were extremely handsome. Murphy was also responsible for the basic design of two of the most attractive Hudson production models, the Victoria and Landau Sedan. The Model S Hudsons were mounted on a 118-1/2 inch wheelbase and were available in standard and custom form. The Model O versions also had standard and custom styles and a 127-3/8 wheelbase.

1928 Hudson Model O convertible landau sedan (WRG)

I.D. DATA: Car numbers on front frame channel, dash and frame side. Starting: [Model S] 12270, [Model O] 803569. Ending: [Model S] NA, [Model O] 825406. Engine numbers on left motor mount, side of cylinder block.

Model No.	Body Type & Seating	Price	Weight	Prod. Total
STANDARD LINE 2/4P				
S	2dr Cpe-2P	1295	3525	NA
S	2dr Coach-5P	1250	3575	NA
S	4dr Sedan-5P -2W	1325	3645	NA
CUSTOM LINE				
S	2dr Roadster-3P	1295	3355	NA
O	4dr Sedan-5P -2W	1450	3720	NA
O	4dr Phaeton-7P	1650	3630	NA
O	2dr Victoria-4P	1650	3710	NA
O	4dr Landau Sedan-5P	1650	3780	NA
O	4dr Sedan-7P	1950	3945	NA

ENGINE: F-head. Inline. Six-cylinder. Cast iron block. B & S: 3-1/2 x 5 in. Disp.: 288.5 cid. C.R.: 5.0:1. Brake HP: 92 at 3200 rpm. NACC HP: 29.4. Main bearings: 4. Valve lifters: mechanical. Carb.: Marvel, 1-1/4 inch.

CHASSIS: [Model O] W.B.: 127 3/8 in. O.L.: 188 in. Tires: 31 x 6.00. [Model S] W.B.: 118 1/2 in. O.L.: 178 1/2 in. Tires: 31 x 6.00.

TECHNICAL: Sliding gear transmission. Speeds: 3F/1R. Floor shift controls. Single disc, cork inserts, running in oil. Shaft drive. Semi-floating rear axle. Overall Ratio: 4.09:1 (Model S); 4.45:1 (Model O). Bendix Mechanical brakes on four wheels. Wooden spoke wheels with detachable rims. Rim size: 18 in.

OPTIONS: Front bumper. Rear bumper. Leather upholstery. Triplex shatterproof windshield glass. Trunk. Radiator shutters.

HISTORICAL: Introduced Jan. 1928. Hudson made 52,316 shipments to dealers during the 1927 calendar year. The president of Hudson was Roscoe B. Jackson.

193

1928 Hudson Model O Murphy Coupe (WRG)

1929 HUDSON

1929 Hudson, Model R coupe (WRG)

HUDSON — MODEL R — SIX: In a move which undoubtedly upset old line Hudson fans, but apparently did not harm sales, Hudson abandoned the time-honored Super Six label and identified its 1929 offering as The Greater Hudson. All Hudsons had bodies approximately four inches longer, with the Model R versions having a 122.5 wheelbase. Among Hudson's 1929 styling highlights were larger windshields of shatterproof glass and narrower corner posts. The Landau Sedan and Victoria Model R Hudsons had Biddle & Smart bodies.

1929 Hudson Model L Biddle & Smart Club sedan (WRG)

HUDSON — MODEL L — SIX: The 139 inch wheelbase Model L chassis was used exclusively for custom-built bodies supplied by Biddle & Smart. All Model L Hudsons were fitted with five wire wheels as standard equipment. All Hudsons had a higher radiator, cowl and hood line, plus larger diameter headlights. Included in the list of 64 improvements Hudson claimed for 1929 were hydraulic, double-acting shock absorbers, self-energizing brakes and silenced roof construction. Hudsons were delivered with a long list of standard equipment, such as an electric gas and oil gauge, windshield wiper, rearview mirror and the electrolock anti-theft device.

I.D. DATA: Car numbers on front frame channel, dash and frame side. Starting: [Model R] 825407, [Model L] 41384. Ending: [Model R] 893401, [Model L] 46598. Engine numbers on left side of cylinder block/motor mount.

1929 Hudson Model L Murphy limousine (WRG)

Model No.	Body Type & Seating	Price	Weight	Prod. Total
R	2dr Phaeton-5P	1350	3495	NA
R	2dr Conv Cpe- 2/4P	1450	3580	NA
R	4dr Std Sedan-5P -3W	1175	3785	NA
R	2dr Cpe- 2/4P -3W	1195	3610	NA
R	2dr Coach-5P	1095	3680	NA
R	4dr Town Sedan-5P -3W	1375	3795	NA
R	2dr Victoria-5P	1500	3795	NA
R	4dr Landau Sedan-5P -2W	1500	3825	NA
R	2dr Roadster- 2/4P	NA	NA	NA
L	4dr Club Sedan-5P -2W	1850	4140	NA
L	4dr Limousine-7P -3W	2100	4290	NA
L	4dr Sedan-7P -3W	2000	4260	NA
L	4dr Sport Phaeton-4P D/C	2200	3795	NA
L	4dr Phaeton-7P	1600	3760	NA

ENGINE: F-head. Inline. Six-cylinder. Cast iron block. B & S: 3-1/2 x 5 in. Disp.: 288.5 cid. Brake HP: 92 at 3200 rpm. Taxable/ALAM/NACC HP: 29.4. Main bearings: 4. Valve lifters: mechanical. Carb.: Marvel 1-1/4 inch.

CHASSIS: [Model R] W.B.: 122.5 in. Front/Rear Tread: 56/57.5 in. Tires: 31 x 6.50. [Model L] W.B.: 139 in. Front/Rear Tread: 56/57.5 in. Tires: 31 x 6.00.

TECHNICAL: Sliding gear transmission. Speeds: 3F/1R. Floor shift controls. Single disc, cork insert, running in oil. Shaft drive. Semi-floating rear axle. Bendix mechanical brakes on four wheels. Wood spoke wheels with detachable rims. Rim size: 19 in.

OPTIONS: Front bumper. Rear Bumper. Steel sidemount cover(s). Eight-day clock. Cigar Lighter. Spotlight. Five wire wheels. 12 spoke demountable wood wheels (10 spoke standard). Radiator shutters. Trunk. Special trunk with fitted luggage. Ball-jointed tire mount mirrors. Protect-a-hood. Lap robes. Spring covers. Window awnings. Spare tire locks.

1929 Hudson Model L Biddle & Smart Special phaeton (WRG)

HISTORICAL: Introduced January 1929. Hudson made 71,179 shipments to dealers during the 1929 calendar year. 139 inch wheelbase models with Biddle & Smart bodies only. The president of Hudson was William J. McAneeny.

1930 HUDSON

HUDSON — MODEL T — EIGHT: Styling changes for 1930 were limited to details such as the use of hood doors instead of louvers and a thinner radiator shell with a simulated cap. In addition the headlights were now mounted on a curved bar. The Model T Hudsons were mounted on a 119 inch wheelbase and were highlighted by the new Sun Sedan body, which provided the open-air motoring pleasure of a phaeton with the comfort of a two-door, five-passenger closed car. Its interior featured special upholstery without pleats. Common to all 1930 Hudsons were wider fenders, chrome-plated trim, beaded body beltline, a lower overall height and runningboard shields.

1930 Hudson Model T roadster (WRG)

1930 Hudson Model U LeBaron close-coupled sedan (OCW)

HUDSON — MODEL U — EIGHT: The Model U wheelbase at 126 inch was a full 13 inches shorter than that of the 1929 Model L. A controversial move was Hudson's decision to drop its F-head six and replace it with an L-head straight eight. This engine was to be the only eight-cylinder engine ever offered by Hudson and it would remain in production until 1952. Perhaps in anticipation of controversy, Hudson called its 1930 offering the Hudson Great Eight. With 80 horsepower, it was less powerful than the old F-head six, but was installed in an automobile that was significantly lighter. Hudson claimed, "It strikes off the shackles of bulk and useless weight."

I.D. DATA: Car numbers on front frame channel, dash and frame side. Starting: [Model T] 893402; [Model U] 46599. Ending: [Model T] 914292; [Model U] 57114. Engine numbers on left side of cylinder block.

Model No.	Body Type & Seating	Price	Weight	Prod. Total
T	2dr Cpe-2P	885	3010	NA
T	4dr Phaeton-5P	965	2940	NA
T	2dr Roadster-4P	995	2870	NA
T	2dr R/S Cpe-4P	925	3060	NA
T	4dr Std Sedan-5P	1025	3200	NA
T	2dr Sun Sedan-5P	1045	3100	NA
T	2dr Coach-5P	895	3080	NA
U	4dr Phaeton-7P	1160	3080	NA
U	4dr Brougham-5P	1195	3210	NA
U	4dr Touring Sedan-5P	1145	3270	NA
U	4dr Sedan-7P	1295	3385	NA

NOTE 1: During 1930 Hudson made significant price reductions in an effort to spur sales. The prices cited represent the lowest levels reached.
ENGINE: L-head. Inline. Eight-cylinder. Cast iron block. B & S: 2-3/4 x 4-1/2 in. Disp.: 213.8 cid. C.R.: 5.78:1. Brake HP: 80 at 3400 rpm. NACC HP: 24.2. Main bearings: 5. Valve lifters: mechanical. Carb.: Marvel 10-776.
CHASSIS: [Model T] W.B.: 119 in. Tires: 18 x 5.50. [Model U] W.B.: 126 in. Tires: 18 x 5.50.
TECHNICAL: Sliding gear transmission. Speeds: 3F/1R. Floor shift controls. Single disc, cork insert, running in oil. Shaft drive. Semi-floating rear axle. Bendix mechanical brakes on four wheels. Wooden spoke wheels with steel rims.
OPTIONS: Spotlight. Single side mount. Trunk. Spare tire. Radiator shutters.
HISTORICAL: Introduced January 1930. A car known as the Marr Special, consisting of an Essex chassis and a Hudson eight-cylinder engine, qualified for the 1930 Indianapolis 500 at a speed of 106.185 mph. Its driver was Chet Miller. Starting in 15th position, Miller finished a respectable 10th, averaging 89.58 mph. This was the last Hudson-engined car to complete the Indianapolis 500-mile Race. A Hudson also won the 1930 Tour de France. Hudson made 36,674 shipments to dealers during the calenday year. The president of Hudson was William J. McAneeny.

1931 HUDSON

HUDSON — SERIES T — EIGHT: The Series T version of the Hudson Greater Eight, as 1931 Hudsons were advertised, had a 119 inch wheelbase. Although styling changes were not extensive, they were significant enough to represent an evolutionary step away from the severely angular lines characteristic of the 1920s. The grille insert was now a very fine mesh and a new front headlight tie bar gave the headlights a freestanding appearance and a lower position. The fenders were more deeply flanged. At the front, they swept downward to further enclose the wheels. Also redesigned for 1931 were Hudson bumpers, hubcaps, runningboards and exterior hardware. At the rear a new rectangular window was used. Also identifying the 1931 Hudson were belt moldings that extended the full length of the hood, to the radiator shell, and at the rear, were positioned higher than before.

1931 Hudson Series T Sport roadster (WRG)

HUDSON — SERIES U — EIGHT: The Series U Hudson had a 126 inch wheelbase and was easily identified by its standard left front fender spare tire well. The hood louvers on all models were placed higher than in 1930. A sure sign of the times was Hudson's decision to discontinue standard radiator shutters. They were still available as an accessory where there were extremely cold winters. However, Hudson noted that they had "been virtually useless, as evidenced by the fact that few of particularly the newer members of the Hudson family ever made use of them." Interior revisions common to all 1931 Hudsons included a hand-brake lever with a pawl and ratchet design that Hudson promised was rattleproof. Hudson offered new choices of Bedford cord, flat fabrics, mohairs and velour interiors and all models had walnut finishes on the dash and side garnish moldings. Rear seat occupants enjoyed at least two inches of additional legroom in the 1931 Hudsons, with some sedans offering as much as five more inches. In June, freewheeling was introduced as a $35 option. It was operated by a small lever, placed just behind the gearshift.

I.D. DATA: Car numbers on front frame channel, dash and frame side. Starting: Series T 914293; Series U 57115. Ending: Series T 930769; Series U 62883. Engine numbers on left side of cylinder block.

Model No.	Body Type & Seating	Price	Weight	Prod. Total
T	4dr Phaeton-5P	1095	2865	NA
T	2dr Cpe-2P	875	2865	NA
T	2dr R/S Cpe-2/4P	925	2955	NA
T	2dr R/S Spt Cpe-2/4P	1065	3145	NA
T	2dr Coach-5P	895	2975	NA
T	4dr Town Sedan-5P	945	3055	NA
T	4dr Std Sedan-5P	995	3115	NA
815	2dr Spt Roadster-2P	995	2675	NA
815	2dr R/S Roadster-2/4P	NA	NA	NA
U	4dr Phaeton-7P	1295	3055	NA
U	4dr Touring Sedan-5P	1145	3190	NA
U	4dr Family Sedan-7P	1195	3230	NA
U	4dr Spl Sedan-5P	1325	3430	NA
U	4dr Sedan-7P	1450	3375	NA
U	4dr Club Sedan-5P	1445	3235	NA
U	4dr Brougham-5P	1225	3190	NA
U	4dr Del Brougham-5P	1375	3480	NA

ENGINE: L-head. Inline. Eight-cylinder. Cast iron block. B & S: 2-7/8 x 4-1/2 in. Disp.: 233.7 cid. C.R.: 5.8:1. Brake HP: 87 at 3600 rpm. NACC HP: 26.4. Main bearings: 5. Valve lifters: mechanical. Carb.: 1-1/2 inch Marvel 10-951.
CHASSIS: [Series T] W.B.: 119 in. Tires: 18 x 5.50. [Series U] W.B.: 126 in. Tires: 18 x 5.50.

1931 Hudson Series U Brougham (OCW)

TECHNICAL: Sliding gear transmission. Speeds: 3F/1R. Floor shift controls. Single disc, cork insert, running in oil. Shaft drive. Semi-floating rear axle. Bendix mechanical brakes on four wheels. Wooden spoke wheels. Free-Wheeling (35.00). Startix.
OPTIONS: Heater. Chrome windshield frame. Dual windshield wipers. Twin taillights. White sidewall tires. Spare tire cover. Wire wheels.
HISTORICAL: Introduced November 1930. In August 1931, the Hudson and Essex became the first American cars available with Startix. This device automatically started and if needed re-started the car after the driver turned the ignition key to the on position. Clutch plate now constructed from duraluminum. Hudson made 17,487 shipments to dealers during the 1931 calendar year. The president of of Hudson was William J. McAneeny. The Marr Special averaged 89.58 mph to finish 10th, at Indy. The driver was Chet Miller.

1932 HUDSON

1932 Hudson Series T rumbleseat coupe (WRG)

HUDSON — SERIES T (STANDARD) — EIGHT: The 1932 Greater Eight Hudsons were easily identified by their elegant V-shaped grille with prominent vertical bars, single-piece bumper and triangular-shaped head, cowl and taillights. Frank Spring who was to influence the appearance of virtually every future Hudson gave the 1932 models a fresh look by the use of gracefully sweeping fenders and gentler body curves. All new Hudsons had a new instrument panel with larger gauges and a knob allowing the driver to adjust the ride control of the shock absorbers. The Standard models had a 119 inch wheelbase and were equipped with a single windshield wiper and taillight. A choice of either painted wood or wire wheels was offered.

1932 Hudson Series U seven-passenger sedan (OCW)

HUDSON — SERIES U (STERLING) — EIGHT: Sterling Series Hudsons were mounted on a 126 inch wheelbase and were equipped with standard dual wipers, taillights, and white sidewall tires.

1932 Hudson Series L Custom phaeton (WRG)

HUDSON — SERIES L (MAJOR) — EIGHT: The 132 inch wheelbase Hudson had the same standard features of the Series U. Thus, customers had a choice of either natural-finish wood wheels or wire wheels.

I.D. DATA: Car numbers on front frame channel, dash and frame side. Starting: [Series T] 930770, [Series U] 62884, [Series L] 25001. Ending: [Series T] 936702, [Series U] 68332, [Series L] 25116. Engine numbers on left side of cylinder block.

Model No.	Body Type & Seating	Price	Weight	Prod. Total
STANDARD LINE				
T	2dr R/S Spt Cpe-4P	1115	3215	NA
T	4dr Town Sedan-5P -2W	1050	3270	NA
T	2dr Conv-2P/4P	1195	3085	NA
T	2dr Coach-5P	1025	3190	NA
T	4dr Sedan-5P -3w	1095	3285	NA
T	2dr Cpe-2P	995	—	NA
T	2dr R/S Cpe-4P	1045	—	NA
T	2dr R/S Spl Cpe-2/4P	1115	3215	NA
STERLING LINE				
U	2dr Suburban-5P	1275	3350	NA
U	4dr Spl Sedan-5P -3w	1295	3415	NA
MAJOR LINE 7P PHAETON				
L	4dr Touring Sedan-5P -3w	1445	3475	NA
L	4dr Club Sedan-5P -3w	1495	3555	NA
L	4dr Brougham-5P -2w	1495	3560	NA
L	4dr Sedan-7P -3w	1595	3590	NA

ENGINE: Inline. Eight cylinder. Cast iron block. B & S: 3 x 4-1/2 in. Disp.: 254.4 cid. C.R.: 5.8:1. Brake HP: 101 at 3600 rpm. Main bearings: 5. Valve lifters: mechanical. Carb.: 1-1/2 inch Marvel 10-996.
CHASSIS: [Series T] W.B.: 199 in. [Series U] W.B.: 126 in. Tires: 17 x 6.00. [Series L] W.B.: 132 in. Tires: 17 x 6.50.
TECHNICAL: Sliding gear transmission. Speeds: 3F/1R. Floor shift controls. Single disc, cork inserts, running in oil. Shaft drive. Semi-floating rear axle. Overall Ratio: 4.64 (5.0:1 opt.). Bendix mechanical brakes on four wheels. Wooden artillery wheels. Drivetrain Options: Free-wheeling and selective automatic clutch.
OPTIONS: Sidemount cover(s) metal, fabric. Clock. Radio Antenna. Wire wheels. Chrome grille cover. Chrome hood doors. Shatterproof glass. Leather upholstery. Locking glove box. Whitewall tires. Trunk (and trunk rack). Double stoplights and taillights. Side and center rear armrests.
HISTORICAL: Introduced January 1932. Hudson made 7,777 shipments to dealers during the 1932 calendar year. The president of Hudson was William J. McAneeny. Two Hudson Specials were entered at Indy. Chet Miller's car qualified at 111.053 mph, Al Miller's at 110.129. Neither car finished. The Chet Miller car was out on lap 125, that of Al Miller on lap 66.

1933 HUDSON

1933 Hudson Series E four-door sedan (WRG)

HUDSON — SUPER SIX — SERIES E — SIX: The 1933 Hudson Super Six was essentially the 1932 Essex Pacemaker. Distinguishing them from other Hudsons for 1933 were their triangular-shaped headlights and single row of hood louvers. The first six-cylinder Hudson since 1923 featured a two-piece aluminum and iron cylinder head with either a 6.2:1 or 7.1:1 compression ratio.
HUDSON — STANDARD EIGHT — SERIES T — EIGHT: The Standard Eight Hudson was mounted on a 119 inch wheelbase and retained door-type hood ventilators.

1933 Hudson Series L Brougham (OCW)

HUDSON — MAJOR EIGHT — SERIES L — EIGHT: Production of Hudsons reached its nadir in 1933. Only 2,401 were assembled, in spite of more attractive styling and significant engineering improvements. The Series L, on a 132 inch wheelbase, had a more luxurious interior than the Standard Eight. Aside from their greater length, the Major Eight Hudsons were distinguished from the Standard Eights by their wide runningboard trim and dual chromed horns mounted on each side of the grille. After losing $5,429,350 in 1932, Hudson experienced another loss of $4,409,903 (on sales of $23,521,458) in 1933.

I.D. DATA: Car number on dash, right rear frame cross-member. Starting: [Series E] 1300501, [Series T] 936703, [Series L] 251117. Ending: [Series E] 1301462, [Series T] 938029, [Series L] 251679. Engine number on left side of cylinder block opposite number one cylinder.

Model No.	Body Type & Seating	Price	Weight	Prod. Total
HUDSON SUPER SIX				
E	2dr Cpe-4P	735	2845	NA
E	2dr Bus Cpe-2P	695	2780	NA
E	2dr Coach-5P	695	2900	NA
E	4dr Sedan-5P	765	2980	NA
E	2dr R/S Conv-4P	845	2800	NA
E	4dr Phaeton-5P	835	—	NA
HUDSON PACEMAKER (STANDARD EIGHT)				
T	2dr R/S Cpe-4P	995	3190	NA
T	2dr Coach-5P	975	3245	NA
T	4dr Sedan-5P	1045	3345	NA
T	2dr R/S Conv-4P	1145	3145	NA
HUDSON PACEMAKER (MAJOR EIGHT)				
L	4dr Tr Sedan-5P	1250	3485	NA
L	4dr Club Sedan-5P	1350	3630	NA
L	4dr Sedan-7P	1350	3605	NA
L	4dr Brougham-5P	1350	3650	NA

ENGINE: [SERIES T/L] Inline. Eight-cylinder. Cast iron block. B & S: 3 x 4-1/2 in. Disp.: 254.4. cid. C.R.: 5.8:1. Brake HP: 101 at 3600 rpm. Main bearings: 5. Valve lifters: mechanical. Carb.: updraft Marvel 10-1535.
[OPTIONAL ENGINE]: Inline. Eight-cylinder. Cast iron block. B & S: 3 x 4-1/2 in. Disp.: 254.4. cid. C.R.: 7.1:1. Brake HP: 110 at 3600 rpm. Main bearings: 5. Valve lifters: mechanical. Carb.: updraft Marvel 10-1535.
ENGINE: [SUPER SIX] Inline. Six-cylinder. Cast iron block. B & S: 2-15/16 x 4-3/4 in. Disp.: 193.1 cid. C.R.: 6.2:1. Brake HP: 73 at 3200 rpm. Main bearings: 3. Valve lifters: mechanical. Carb.: updraft Marvel 10-1532.
[OPTIONAL ENGINE]: Inline. Six-cylinder. Cast iron block. B & S: 2-15/16 x 4-3/4 in. Disp.: 193.1 cid. C.R.: 7.1:1. Brake HP: 80 at 3200 rpm. Main bearings: 3. Valve lifters: mechanical. Carb.: updraft Marvel 10-1532.
CHASSIS: [Series E] W.B.: 113 in. Tires: 18 x 5.25. (17 x 5.50 opt). [Series L] W.B.: 132 in. Tires: 17 x 6.50. [Series T] W.B.: 119 in. Tires: 17 x 6.00.
TECHNICAL: Sliding gear transmission. Speeds: 3F/1R. Floor mounted shift controls. Single plate, cork inserts, running in oil. Shaft drive. Semi-floating rear axle. Overall ratio: 4.64 (5.1 opt.). Bendix mechanical on four wheels. Wire spoke wheels. Rim size: Super Six, 18 inch; Standard Eight, 17 inch; Major Eight, 17 inch. Super Six drivetrain options were: automatic clutch, 17 x 5.50 tires, 17 inch wheels.
OPTIONS: Wire wheels. Chrome grille cover. Chrome hood doors. Shatterproof glass. Leather upholstery. Locking glove box. White wall tires. Trunk, trunk rack. Double stoplights and taillights. Side and center rear armrests.
HISTORICAL: Introduced January, 1933. Innovations: standard equipment on all Hudsons included freewheeling, Startix, and adjustable steering column. Hudson made 2,401 shipments to dealers during the 1933 calendar year. The president of Hudson was William J. McAneeny. Four cars powered by Hudson engines qualified at Indianapolis. Car no. 29, driven by Gene Haustein, qualified in 28th position. It retired after 197 laps and was credited with a 15th place finish. Car no. 28 was driven by Chet Miller. Qualifying at 112.025 mph, it began the race in 32nd position. Just after Miller was relieved by Shorty Cantlon, it fell by the wayside with a broken connecting rod. Al Miller's no. 19 car started 24th and also retired, with a blown connecting rod, on lap 161. It was placed 20th in the final standings. A fourth Hudson-powered car, no. 59, was qualified by Ray Campbell at 108.65 mph. Starting the race in 37th position, it retired after 24 laps with an oil leak.

1934 HUDSON

1934 Hudson Series LT coupe (WRG)

HUDSON — STANDARD EIGHT — SERIES LL: The 1934 Hudsons had all new styling with wide flowing fenders, longer hoods and, in sedans and coaches, a reverse-curve rear section that allowed the spare tire to be stored within the body. Hudson described this new look as "Streamlined in Wind-Sculptured Steel." For the first time, Hudson offered a factory-installed radio and, as a no-cost option, the Axle-Flex semi-independent front suspension. Other improvements included a new dash panel which placed the instruments closer to the driver, an improved interior ventilation system and an improved synchromesh for the three-speed transmission. An interesting innovation was the use of three-beam headlights. The third beam was intended to serve as a cornering light and was controlled, along with the usual high and low beams, by a toe switch. Although sales rebounded strongly during 1934, Hudson reported a loss of $3,239,201 for the year. Hudson Standard Eight models in the LL series had Bedford cord interiors.

HUDSON — DE LUXE EIGHT — SERIES LLU: The LLU models were distinguished from the LL versions by their broadcloth upholstery and front fenderside lamps. Both types had six hood doors, fender striping, bright-work trim around the windshield and front door vent panes.

HUDSON — EIGHT — SERIES LT: The LT models lacked front fender parking lights. Their windshields were given bright molding trim and front door vent panes were fitted. The Sport Roadster model was equipped with dual horns and taillights in a chrome finish. In addition it had two windshield wipers, while other models carried a single wiper for the driver.

1934 Hudson Series LU deluxe sedan and Lowell Thomas (WRG)

HUDSON — DE LUXE EIGHT — SERIES LU: The Deluxe models were equipped with dual front fenderside lamps, dual chrome-finish taillights, dual windshield wipers and dual horns.

HUDSON — CHALLENGER — SERIES LTS — EIGHT: These ultra-low priced models were introduced in June. They were identified by their three-door hoods and lack of front door vent panes. They were fitted with a manual, rather than automatic clutch, and lacked such features as an interior sun visor.

I.D. DATA: Car numbers on firewall; right rear frame cross-member. Starting: [Series LT/Series LU] 950000; [Series LTS] 964463; [Series LLU/SeriesLL] 252000. Ending: [Series LT/Series LU] 968679; [Series LTS] 968679; [Series LL/Series LLU] 256158. Engine numbers on left side of cylinder block opposite number one cylinder.

Model No.	Body Type & Seating	Price	Weight	Prod. Total
STANDARD LINE				
LT	2dr Cpe-4P	775	2795	NA
LT	2dr Bus Cpe-2P	695	2720	NA
LT	2dr Cpe-2P	725	2750	NA
LT	2dr R/S Conv-4P	835	2815	NA
LT	2dr R/S Spt Rds-4P	—	2845	NA
LT	2dr Coach-5P	745	2855	NA
LT	2dr Victoria-5P	785	2850	NA
LT	4dr Sedan-5P	805	2905	NA
LT	2dr Comp Sedan-5P	845	2930	NA
LT	2dr BIT Coach-5P	745	—	NA
LT	4dr BIT Sedan-5P	805	—	NA
CHALLENGER LINE				
LTS				
LTS	4dr Sedan-5P	765	2910	NA
LTS	2dr Cpe-2P	685	2720	NA
LTS	2dr Cpe-5P	735	2765	NA
LTS	2dr Conv-4P	800	2785	NA
LTS	2dr Coach-5P	705	2800	NA

NOTE 1: After August, five-passenger sedan has built-in trunk at same price.

LL	4dr Tr Sedan-5P	970	2950	NA
LL	4dr Comp Tr Sedan-5P	1000	2975	NA
LLU	4dr Club Sedan-5P	1070	3080	NA
LLU	4dr Comp Club Sedan-5P	1125	3110	NA
LLU	4dr Brougham-5P	1145	3075	NA
LU	2dr R/S Cpe-4P	855	2850	NA
LU	2dr Cpe-2P	815	2805	NA
LU	2dr R/S Conv-4P	900	2835	NA
LU	4dr Sedan-5P	895	2930	NA
LU	4dr Comp Sedan-5P	935	2955	NA
LU	2dr Coach-5P	835	2870	NA
LU	2dr Comp Victoria	875	2895	NA
LU	4dr BIT Sedan	895	—	NA
LU	2dr BIT Coach	835	—	NA

NOTE 1: In June 1935, production ceased on the following models: LTS four-door sedan-5P and two-door Coach-5P; LU four-door Compartment Sedan-5P and two-door Compartment Victoria.
NOTE 2: Series LU four-door sedan with built-in trunk (BIT Sedan) and two-door Coach with built-in trunk (BIT Coach) were introduced in August 1935.

ENGINE: [Series LL/Series LLU] L-head. Straight. Eight-cylinder. Chrome alloy block. B & S: 3 x 4-1/2 in. Disp.: 254.4 cid. C.R.: 6.25:1. Brake HP: 113 at 3800 rpm. Taxable/ALAM/NACC HP: 28.8. Main bearings: 5. Valve lifters: mechanical. Carb.: Carter downdraft one-barrel 2825. Optional Engine: L-head. Straight. Eight-cylinder. Chrome alloy block. B & S: 3 x 4-1/2. Disp.: 254.4 cid. C.R.: 7.0:1. Brake HP 121 3800 rpm. Taxable/ALAM/NACC HP: 28.8. Main bearings: 5. Valve lifters: mechanical. Carb.: Carter downdraft one-barrel 2825.

1934 Hudson Series LT convertible coupe (OCW)

ENGINE: [Series LT/Series LTS/Series U] Straight L-head. Inline. Eight-cylinder. Chrome alloy block. B & S: 3 x 4.5 in. Disp.: 254.4 cid. C.R.: 5.75:1. Brake HP 108 at 3800 rpm. Taxable HP. 28.8. Main bearings: 5. Valve lifters: mechanical. Carb.: Carter downdraft one-barrel 2825. [Optional Engine]: Straight L-head. Inline. Eight-cylinder. Chrome alloy block. B & S: 3 x 4.5 in. Disp.: 254.4 cid. C.R.: 7.0:1. Brake HP 121 at 3800 rpm. Taxable/ALAM/NACC H.P.: 28.8. Main bearings: 5. Valve lifters: mechanical. Carb. Carter downdraft one-barrel 2825.

CHASSIS: [Series LL] W.B.: 123 in. O.L.: 197 in. Front/Rear Tread: 56/57.5 in. Tires: 16 x 6.50. [Series LT] W.B.: 116 in. O.L.: 194 in. Front/Rear Tread 56/56 in. Tires: 16 x 6.25. [Series LU] W.B.: 116 in. O.L.: 194 in. Front/Rear Tread 56/56 in. Tires: 16 x 6.25. [Series LLU] W.B.: 123 in. O.L.: 197 in. Front/Rear Tread 56/57.5 in. Tires: 16 x 6.50. [Series LTS] W.B.: 116 in. O.L.: 194 in. Front/Rear Tread 56/56 in. Tires: 16 x 6.25.

TECHNICAL: Sliding gear transmission. Speeds: 3F/1R. Floor shift controls. Single plate, cork insert, running in oil. Shaft drive. Semi-floating rear axle. Overall Ratio: 4.11:1. Bendix mechanical brakes on four wheels. Steel artillery wheels. Rim size: 16 in. Drivetrain options: Hill-holder. Automatic clutch.

OPTIONS: Dual Sidemount (except for Challenger). Bumper guards front and rear ($11 or $9.50 in primer; $2.15 each set). Radio. Heater (standard $12.50 or Deluxe $15.50). Electric clock ($13.50). Cigar lighter ($1.25 or $2). Stainless steel wheel moldings ($5 to $9). Underhood battery charger ($9.75). Chrome plated exhaust extension ($1.95). Luggage rack ($9.50). Rightside tail & stop light. Twin horns. Dual windshield wipers ($10). Inside right side visor. Trunk light.

HISTORICAL: Introduced January 1934. Innovations: Axle-flex three beam headlights. Hudson made 27,130 shipments to dealers during the 1934 calendar year. The president of Hudson was Roy D. Chapin. Freewheeling was discontinued. The Martz Special, powered by a 257 cid Hudson engine, was qualified for the Indianapolis 500, by Gene Haustein, at 109.426 mph. The car started in 31st position and was credited with finishing in 30th place after being wrecked in an accident on lap 13. Another Hudson-powered car, no. 13, was the first alternate to the qualifiers, but it didn't race.

1935 HUDSON

1935 Hudson Series GH convertible coupe (OCW)

HUDSON CUSTOM EIGHT — HHU SERIES: Although the styling of the 1935 Hudson was very similar to the 1934 model, there were many improvements that have prompted Hudson historians to regard the 1935s as the first modern Hudsons. The most dramatic advance was the industry's first all-steel body, which Hudson introduced prior to General Motor's Turret Top. The most predominant external changes, aside from the steel roof, were the new bullet-shaped headlight shells, larger rear windows and narrow hood louver panels. The use of flatter rear leaf springs enabled overall height to be lowered 1.5 inches. On July 8, 1935 the Hudson Motor Car Co. celebrated its 26th anniversary with the production of its 2,262,810 automobile. Also during 1935, Roy Chapin established the 20-Year Club for Hudson employees with that number of years of service. They received a solid gold pin with Hudson and 20 engraved on one side and their name on the other. Jan. 5, 1935 was the 25th anniversary of Roy D. Chapin's election to the presidency of Hudson. The Custom Eight had the exterior features of the Deluxe Eights, plus chrome trim for the runningboards and three chrome stripes on the trailing edge of the front fenders. Interior trim appointments were also upgraded. Standard equipment included the Electric Hand transmission.

HUDSON SPECIAL EIGHT — HT SERIES: The Special Eight was Hudson's lowest priced line of eight-cylinder models and had a 117 inch wheelbase. These cars had painted headlights shells and single windshield wipers.

HUDSON DE LUXE EIGHT — HU SERIES: The Deluxe Eight models shared the 117 inch chassis with the Special Eight series. However, they were equipped with dual windshield wipers and chrome plated headlight shells.

SPECIAL COUNTRY CLUB EIGHT — HTL SERIES: This series combined the painted headlight shells and single windshield wiper of the Special Eight series with the long, 124 inch wheelbase of the DeLuxe Eight models.

DE LUXE COUNTRY CLUB EIGHT — HUL SERIEAS: The HUL series cars shared their chromed headlight shells with the HU models, but they were given a higher quality interior.

HUDSON BIG SIX — GH SERIES: The Hudson Big Six was easily identified by its grille mesh with horizontal inserts.

I.D. DATA: Car numbers on dash, right rear frame cross-member. Starting: [GH] 53101; [HHU] 56101; [HT] 54101; [HU] 55101; [HTL] 57101; [HUL] 58101. Ending: [GH] 537724; [HHU] 561560; [HT] 547250; [HU] 553197; [HTL] 571066; [HUL] 58221. Canadian built cars inserted a 'C' between series and serial number. Engine numbers on left rear side of cylinder block. Starting: 70000. Ending: 78999.

Model No.	Body Type & Seating	Price	Weight	Prod. Total
HHU	4dr Club Sedan-5P	1025	3130	720
HHU	4dr Suburban Sedan-5P	1057	3145	720
HHU	4dr Brougham-6P	1095	3055	720*
HHU	4dr Tr Brougham-6P	1127	3070	720
HU	2dr Coach-5P	875	2880	3096
HU	2dr Tr Brougham	907	2895	3096
HU	4dr Sedan	935	2945	3096
HU	4dr Suburban Sedan-5P	967	2960	3096
HU	2dr Bus Cpe-2P	845	2790	3096*
HU	2dr R/S Coupe-4P	895	2855	3096
HU	2dr R/S Conv-2P	955	2805	3096
HTL	2dr Tr Brougham-5P	812	2855	965
HTL	2dr Coach-5P	780	2840	965
HTL	4dr Sedan-6P	840	2890	965
HTL	4dr Suburban Sedan-6P	872	2905	965*
HUL	4dr Suburban Sedan-5P	1007	3030	9923
HUL	4dr Club Sedan-5P	975	3015	9923
HUL	4dr Brougham-5P	1025	3055	9923
HUL	4dr Tr Brougham-5P	1052	3070	9923*
GH	2dr R/S Cpe-3P	740	2665	7623
GH	2dr Bus Cpe-2P	695	2600	7623
GH	4dr Sedan-5P	770	2780	7623
GH	2dr Coach-5P	710	2720	7623
GH	4dr Suburban Sedan-5P	802	2795	7623
GH	2dr Conv	790	2640	7623*
GH	2dr Tr Brougham	742	2735	7623
HT	2dr Coach-5P	780	2840	7149
HT	2dr Tr Brougham	812	2855	7149
HT	4dr Sedan	840	2890	7149
HT	4dr Suburban Sedan-5P	872	2905	7149
HT	2dr Bus Cpe-2P	760	2740	7149
HT	2dr R/S Cpe-4P	810	2810	7149
HT	2dr R/S Conv-2P	860	2765	7149*

(*) Total applies to series production, not model production.

1935 Hudson Series HUL four-door Custom Brougham (WRG)

ENGINE: [Hudson Eight] L-head. Inline. Eight-cylinder. Chrome alloy block. B & S: 3 x 4.5. Disp.: 254 cid. C.R.: 6.0:1. Brake HP: 113 at 3800 rpm. Taxable H.P.: 28.8. Main bearings: 5. Valve lifters: mechanical. Carb.: Carter 330S. [Optional Engine]: L-head. Inline. Eight-cylinder. Chrome alloy block. B & S: 3 x 4.5 in. Disp.: 254 cid. C.R.: 7.0:1. Brake HP: 124 at 4000 rpm. Main bearings: 5. Valve lifters: mechanical. Carb.: Carter 330S.

ENGINE: [Hudson Six] L-head. Inline. Six-cylinder. Chrome alloy block. B & S: 3 x 5 in. Disp.: 212 cid. C.R.: 6.25:1. Brake HP: 93 at 3800 rpm. Taxable HP.: 21.6. Main bearings: 3. Valve lifters: mechanical. Carb.: Carter 329S. [Optional Engine]: L-head. Inline. Six-cylinder. Chrome alloy block. B & S: 3 x 5 in. Disp.: 212 cid. C.R.: 7.0:1. Brake HP: 100 at 3800 rpm. Main bearings: 3. Valve lifters: mechanical. Carb.: Carter 329S.

CHASSIS: [Series HHU] W.B.: 124 in. Tires: 16 x 6.50. [Series HU] W.B.: 117 in. Tires: 16 x 6.25. [Series HTL] W.B.: 124 in. Tires: 16 x 6.50. [Series HUL] W.B.: 124 in. Tires: 16 x 6.50. [Series HT] W.B.: 117 in. Tires: 16 x 6.25. [Series GH] W.B.: 116 in. Tires: 6 x 16.

TECHNICAL: Manual, sliding gear transmission. Speeds: 3F/1R. Floor shift controls. Single disc, cork insert, running in oil. Shaft drive. Semi-floating rear axle. Overall Ratio: 4.19:1. Mechanical: Rotary Equalizer brakes on four wheels. Steel wheels. (Wooden spokes on 124 inch wheelbase models). Drivetrain Options: Electric Hand, standard on Custom Eights, optional on all other Hudsons and Terraplanes. ($20.42). Axelflex independent front suspension. Adjustable steering column. Automatic clutch ($10.21). Automatic clutch and Electric Hand ($28.08).

OPTIONS: Radio ($51.81). High-compression head, six-cylinder ($18.50). High- compression head, eight-cylinder ($22). Startix ($8.50). Luggage carrier ($10). Tune up kit ($4.50). Zenith radio ($44). Twin air horns ($11.50). Front and rear seat covers ($7.50). Leather upholstery ($18.81). Wire wheels.

HISTORICAL: Introduced December 1934. Rotary-Equalized brakes. Optional Electric Hand pre-selector shifting mechanism. Hudson made 29,476 shipments to dealers during the 1935 calendar year. The president of Hudson was Roy D. Chapin. Two Hudson powered cars were entered in the Indy, but neither qualified. In February 1935, Sir Malcolm Campbell set seven new records at Daytona Beach with a Hudson sedan. These included the flying mile at 88.207 mph, flying kilometer at 00.207 mph, flying five miles at 88.051 mph, flying five kilometers at 88.105 mph and the standing start one mile at 68.18 mph. In April, 36 new AAA records were captured by a Hudson 8 sedan at Muroc Dry Lake in California. Among these records set by Wilbur Shaw, Babe Stapp and Al Gordon was a 93.03 mph speed for five miles and an average of 85.8 mph for 1,000 miles. In addition the Hudson captured every record in its engine displacement class (up to 3000 kilometers), as well as four unlimited class records for closed cars.

Model No.	Body Type & Seating	Price	Weight	Prod. Total
63	2dr R/S Cpe-3/5P	755	7810	NA
63	2dr Bus Cpe-3P	710	2730	NA
63	2dr Conv Cpe-3P	810	2870	NA
63	2dr Brougham-6P	730	2830	NA
63	2dr Tr Brougham-6P	755	2830	NA
63	4dr Sedan-6P	785	2880	NA
63	4dr Tr Sedan-6P	810	2880	NA
64	2dr Bus Cpe-3P	760	2865	NA
64	2dr R/S Cpe-3-5P	810	2965	NA
64	2dr Conv Cpe-3P	875	3000	NA
64	2dr Brougham-6P	790	2985	NA
64	2dr Tr Brougham-6P	815	2985	NA
64	4dr Sedan-6P	830	3045	NA
64	4dr Tr Sedan-6P	855	3045	NA
66	4dr Sedan-6P	855	3110	NA
66	4dr Tr Sedan-6P	880	3110	NA
65	2dr Bus Cpe-3P	845	2915	NA
65	2dr R/S Cpe-3-5P	895	3000	NA
65	2dr Conv Cpe-3-5P	970	3045	NA
65	2dr Brougham-6P	885	3030	NA
65	2dr Tr Brougham-6P	910	3030	NA
65	2dr Sedan-6P	925	3075	NA
65	4dr Tr Sedan-6P	950	3075	NA
67	4dr Sedan-6P	950	3140	NA
67	4dr Tr Sedan-6P	975	3140	NA

1936 HUDSON

1936 Hudson Series 63 coupe (OCW)

HUDSON — CUSTOM SIX — SERIES 63: A new styling motif for Hudson, with rounded surfaces, highly domed fenders and an ornate front grille design, was introduced in 1936. All series had longer wheelbases and body widths. An important technical advance was Hudson's adoption of a hydraulic braking system. It incorporated a mechanical unit operating on the rear wheels. If the front pedal traveled beyond the 3/4 point of its maximum distance, this system would function. Replacing the Axleflex system as Hudson's answer to General Motors' independent front suspension system was Radial Safety Control. A solid front axle was continued, but two radius arms attached to the axle. They pivoted to the frame side members and controlled its movement. This allowed softer leaf springs to be used. The result was steadier steering and a smoother ride. Net profits for the year totalled $3,305,616. All Custom models (six- or eight-cylinder) had chrome hubcaps and front fender medallions. Standard upholstery texture was worsted boucle cloth with a green-gray color.

HUDSON — CUSTOM EIGHT — SERIES 65/SERIES 67: With the exception of their wheelbases, cars in these two series of Custom Eight Hudson were identical. Full-sized hubcaps with Hudson Eight lettering were standard. Standard equipment on the Custom Eight included a radio, with its antenna positioned under the car.

1936 Hudson Series 64 rumbleseat convertible coupe (WRG)

HUDSON — DE LUXE EIGHT — SERIES 64/SERIES 66: The Deluxe Eight shared the 120 and 127 inch wheelbase chassis with the Custom Eight Series. These cars had small standard hubcaps that made them easy to identify.

I.D. DATA: Car number on dash and right rear frame cross-member. Starting: [63] 63101; [64] 64101; [66] 66101. Ending: [63] 639820; [64] 645456; [66] 663543. Engine number on left side of cylinder block. Car number on plate mounted on firewall. Starting: [65] 65101; [67] 67101. Ending: [65] 652514; [67] 675004. Engine numbers on left side of cylinder block. Starting: [Six-cylinder] 79000; [Eight-cylinder] 1008. Ending: [Six-cylinder] 89999; [Eight-cylinder] 17634. Canadian-built cars included letter C.

1936 Hudson Series 65 touring sedan and Sir Malcom Campbell (WRG)

ENGINE: [All Hudson Eights] L-head. Inline. Eight-cylinder. Chrome alloy block. B & S: 3 x 4.5 in. Disp.: 254 cid. C.R.: 6.0:1. Brake HP: 113 at 3800 rpm. Taxable HP: 28.8. Main bearings: 5. Valve lifters: mechanical. Carb.: Carter 330S. [Optional Engine]: L-head. Straight. Eight-cylinder. Chrome alloy block. B & S: 3 x 4.5 in. Disp.: 254 cid. C.R.: 7.0:1. Brake HP: 124 4000 rpm. Main bearings: 5. Valve lifters: mechanical. Carb.: Carter 330S. [Custom Six] L-head. Inline. Six-cylinder. Chrome alloy block. B & S: 3 x 5 in. Disp.: 212 cid. C.R.: 6.25:1. Brake HP 93 at 3800 rpm. Taxable HP: 21.6. Main bearings: 3. Valve lifters: mechanical. Carb.: Carter 329S. [Optional Engine]: L-head. Inline. Six-cylinder Chrome alloy block. B & S: 3 x 5 in. Disp.: 212 cid. C.R.: 7.0:1. Brake HP 100 at 3800 rpm. Main bearings: 3. Valve lifters: mechanical. Carb.: Carter 329S.

CHASSIS: [63] W.B.: 120 in. Tires: 16 x 6.00. [64] W.B.: 120 in. Tires: 16 x 6.25 in. [65] W.B.: 120 in. Tires: 16 x 6.25. [66] W.B.: 127 in. Tires: 16 x 6.25. [67] W.B.: 127 in. Tires: 16 x 6.25.

TECHNICAL: Sliding gear transmission. Speeds: 3F/1R. Floor shift controls. Single disc, cork insert, running in oil. Shaft drive. Semi-floating rear axle. Overall ratio: 4.19:1. Hydraulic brakes on four wheels. Pressed steel wheels. Rim size: 16 in. Drivetrain Options: Electric hand preselector transmission.

OPTIONS: Bumper Guards. Radio. Heater. Mohair upholstery. Fender skirts. Full wheel covers. High compression cylinder heads. Startix. Luggage carrier. Tune up kit. Zenith radio. Twin air horns. Front and rear seat covers. Leather upholstery. Wire wheels.

HISTORICAL: Introduced November 1935. Innovations: radial safety control front suspension, hydraulic brakes. Hudson made 25,409 shipments to dealers during the 1936 calendar year. The president of Hudson was A. Edward Barit (elected president in February 1936 following death of Roy D. Chapin).

1937 HUDSON

HUDSON CUSTOM SIX — SERIES 73 — SIX: The Custom Six wheelbase was extended to 122 inches for 1937. Appearance of the Custom Six was identical to that of the Deluxe and Custom Eight models. The only exception was the Hudson 6 identification on the Custom Six grille. All the restyled Hudsons were described by their manufacturer as possessing "useful beauty" with interiors of "drawing room luxury." Overall body width was increased by five inches and wheelbases for all series were lengthened by two inches. Overall height was reduced by the same amount. These new dimensions were accompanied by the use of a stronger 7-1/4 inch deep double drop frame. In addition to their greater size, the 1937 Hudsons were set apart from previous models by their front hinged front doors, reshaped hood louvers and simpler grille design. In place of the rumbleseat, in coupes was a transversely-mounted jump seat. Detail changes included placement of the battery within the engine compartment and improvements in Hudson's Radial Safety front suspension and Electric Hand gear shift mechanism. The downturn in the nation's economy had a negative impact upon Hudson sales. At year's end, Hudson's profit was a slim $670,716.

1937 Hudson Series 74 convertible coupe (WRG)

HUDSON DE LUXE EIGHT — SERIES 74/SERIES 76: Deluxe Eight models carried front fender medallions, running board trim strips and Hudson 8 identification on the front grille. The Series 74 Hudsons had a 122 inch wheelbase, while that of the Series 76 models extended to 129 inches.

HUDSON CUSTOM EIGHT — SERIES 75/SERIES 77: Custom Eight models were externally identical to Deluxe Eights. However, their interiors differed significantly. That of the Custom Eight had a knobby twist upholstery (leather was used for all Hudson convertible interiors), standard equipment radio, cigarette lighter and electric clock. With the exception of its 129 inch wheelbase, the Series 77 Hudson was identical to the Series 75.

1937 Hudson Series 77 four-door sedan (WRG)

I.D. DATA: Car numbers on plate mounted on firewall. Starting: [73] 73101; [74] 74101; [75] 75101; [76] 76101; [77] 77101. Ending: [73] 766913; [74] 745728; [75] 753374; [76] 761197; [77] 773752. Canadian-built chassis serial numbers have a C after the first two digits, which identify the series. Engine numbers on left side of cylinder block. Starting: [Six-cylinder] 90000; [Eight-cylinder] 18000. Ending: [Six-cylinder] 97082; [Eight-cylinder] 31693. Canadian-built cars included letter C.

Model No.	Body Type & Seating	Price	Weight	Prod. Total
73	2dr Victoria Cpe-4P	765	2865	—
73	2dr Bus Cpe-3P	695	2760	—
73	2dr Cpe-3P	720	2805	—
73	2dr Brougham-5P	740	2925	—
73	2dr Tr Brougham-5P	765	2925	—
73	2dr Conv Cpe-4P	820	2870	—
73	4dr Sedan-6P	790	2990	—
73	4dr Tr Sedan-6P	815	2990	—
73	2dr Conv Brougham-6P	900	2945	—
74	2dr Conv Brougham-6P	965	3125	—
74	2dr Conv Cpe-4P	885	3020	—
74	2dr Brougham-5P	800	3105	—
74	2dr Tr Brougham	825	3105	—
74	4dr Tr Sedan-6P	865	3135	—
74	4dr Sedan-6P	840	3135	—
74	2dr Cpe-3P	770	3010	—
74	2dr Victoria Cpe-4P	820	3055	—
~~74~~	~~4dr Sedan-6P~~	~~865~~	~~3205~~	
~~74~~	~~4dr Tr Sedan~~	~~890~~	~~3205~~	
76	4dr Sedan-6P	865	3205	—
76	4dr Tr Sedan-6P	890	3205	—
77	4dr CC Sedan-6P	965	3260	—
77	4dr CC Tr Sedan-6P	990	3260	—
75	2dr Victoria Cpe-4P	905	3085	—
75	2dr Conv Cpe-4P	980	3070	—
75	2dr Cpe-3P	855	3055	—
75	2dr Conv Brougham-6P	1060	3160	—
75	2dr Brougham-6P	895	3135	—
75	2dr Tr Brougham-6P	920	3135	—
75	4dr Tr Sedan-6P	965	3195	—
75	4dr Sedan	940	3195	—

NOTE 1: Prices increased later in the model year.

ENGINE: [Custom Six] L-head. Inline. Six-cylinder. Chrome alloy block. B & S: 3 x 5 in. Disp.: 212 cid. C.R.: 6.25:1. Brake HP: 101 at 4000 rpm. Taxable HP: 21.6. Main bearings: 3. Valve lifters: mechanical. Carb.: two-barrel Carter. [Optional Engine] L-head. Inline. Six-cylinder. Chrome alloy block. B & S: 3 x 5 in. Disp.: 212 cid. C.R.: 7.0:1. Brake HP: 107 at 4000 rpm. Taxable HP: 21.6. Main bearings: 3. Valve lifters: mechanical. Carb.: two-barrel Carter. [Standard Deluxe/Custom Eight Engine]: L-head. Straight. Eight-cylinder. Chrome alloy block. B & S: 3 x 4-1/2 in. Disp.: 254.47 cid. C.R.: 6.25:1. Brake HP: 122 at 4200 rpm. Taxable HP: 28.8. Main bearings: 5. Valve lifters: mechanical. Carb.: two-barrel Carter.

CHASSIS: [73] W.B.: 122 in. O.L.: 199 in. Tires: 16 x 6.00 (15 x 7.00 optional).[74] W.B.: 122 in. O.L.: 199 in. Tires: 16 x 6.25. (15 x 7.00 optional). [76] W.B.: 129 in. O.L.: 203 in. Tires 16 x 6.25. (15 x 7.00 optional). [75] W.B.: 122 in. O.L.: 199 in. Tires: 16 x 6.25 (15 x 7.00 optional). [77] W.B.: 129 in. O.L.: 203 in. Tires: 16 x 6.25 (15 x 7.00 optional).

TECHNICAL: Manual synchromesh transmission. 3F/1R. Floor shift. Single plate, cork inserts, running in oil. Shaft drive. Semi-floating rear axle. Overall Ratio: 4.11:1. Hydraulic brakes on four wheels. Steel drop center type wheels. Drivetrain Options: "Hydraulic Hill Hold." Electric Hand (pre-selector).

OPTIONS: Single sidemount. Dual sidemounts. Fender skirts. Bumper guards. Radio. Heater. Clock (standard on Custom Eight). Cigar lighter. Seat covers. Spotlight. Twin air horns. Vacuum assist fuel pump.

HISTORICAL: Introduced November 1937. Under AAA supervision, Hudson broke all existing Class C, closed stock car records from 10 to 2,000 miles and from one to 24 hours. In addition, new records were set in the unlimited class for all distances from 500 to 2,000 miles and from six to 24 hours. In the latter run, the Hudson Deluxe Eight Brougham covered 2,104.22 miles in 24 hours at an average speed of 87.67 mph. The Hudson's speed of 93.03 mph across the Muroc Dry Lake was the fastest speed a closed, Class C car had ever attained in a five mile run. In addition to the 38 new stock car records set on this outing, Hudson had earlier claimed seven additional new marks in speed at Daytona Beach. Hudson made 19,848 shipments to dealers during the 1937 calendar year. The president of Hudson was Abraham Edward Barit.

1938 HUDSON

1938 Hudson 112 two-door coach (OCW)

HUDSON 112 — SERIES 89 — SIX: The Hudson 112 was introduced in Jan. 1938. Standard and Deluxe models were available on a 112 inch wheelbase. The more expensive Deluxe version had a walnut grain finish on the dash and window moldings, in place of the painted finish found on the standard models. In addition, stainless steel trim was used for the Deluxe interior. A more attractive upholstery with a pleated surface was also used. Both versions were powered by a 175 cid six-cylinder engine. At Bonneville, the Hudson 112 set numerous Class D records including 80.50 mph for one hour and a 12 day, 20,327 mile run at 70.58 mph. It was also the pace car for the Indianapolis 500.

HUDSON CUSTOM SIX — SERIES 83: All 1938 Hudsons featured a new grille with larger horizontal bars divided by a single vertical bar. This arrangement was less complicated than the form used in 1937. The Custom Six displayed a new front bumper with a center indentation that gave it a two-piece appearance.

HUDSON CUSTOM EIGHT — SERIES 85: The Custom Eight was, with the exception of its engine, identical to the Custom Six. The only difference was represented by the two Custom Eight Series 87 Country Club models that used a 129 wheelbase, rather than 122 inch wheelbase.

1938 Hudson Series 84 coupe (WRG)

HUDSON DE LUXE EIGHT — SERIES 84: The Deluxe Eight shared its 122 inch chassis with the Custom Six and Eight Series. In addition, its interior appointments were those of the Custom Six.

I.D. DATA: [Hudson 112]: Car numbers on plate mounted on firewall. Starting: 8928566. Ending: 8956040. Engine numbers on left side of cylinder block. Starting: same as serial numbers. [Hudson Six]: Car numbers on right front door post. Starting: 83131. Ending: 8356040. Engine numbers on left side of cylinder block. Starting: same as serial numbers. [Hudson Deluxe Eight]: Car numbers on right front door post. Starting: 84101. Ending: 8456040. Engine numbers on front of cylinder block on right side. Starting: same as serial numbers. [Hudson Custom Eight Country Club]: Car numbers on right front door post. Starting: 87161. Ending: 8756040. Engine numbers on front of cylinder block on right side. Starting: same as serial numbers. Canadian-built cars included letter "C."

Model No.	Body Type & Seating	Price	Weight	Prod. Total
SERIES 89				
(STANDARD LINE) 89 112				
89	2dr Cpe-3P	694	2500	NA
89	2dr Victoria Cpe-4P	740	2540	NA
89	2dr Conv Brougham-6P	886	2610	NA
89	4dr Sedan-6P	755	2620	NA
89	4dr Tr Sedan-6P	775	2625	NA
89	2dr Conv Cpe-3P	835	2545	NA
89	2dr Brougham-6P	724	2595	NA
89	2dr Tr Brougham-6P	743	2600	NA
(DE LUXE LINE) 112				
89	2dr Cpe-3P	704	2500	NA
89	2dr Victoria Cpe-4P	750	2540	NA
89	2dr Conv Brougham-6P	891	2610	NA
89	4dr Sedan-6P	765	2620	NA
89	4dr Tr Sedan-6P	785	2625	NA
89	2dr Conv Cpe-3P	840	2545	NA
89	2dr Brougham-6P	734	2595	NA
89	2dr Tr Brougham-6P	753	2600	NA

NOTE 1: The Standard models comprised the initial 112 offerings. After May 1, 1938, they were designated Standard models and Deluxes with walnut grain finished instrument panels and interior trim were introduced.

Model No.	Body Type & Seating	Price	Weight	Prod. Total
CUSTOM SIX — SERIES 83 122				
83	2dr Cpe-3P	909	2825	NA
83	2dr Conv Cpe-3P	1041	2895	NA
83	2dr Victoria Cpe-5P	995	2880	NA
83	2dr Conv Brougham-6P	1104	2975	NA
83	2dr Brougham-6P	948	2935	NA
83	2dr Tr Brougham-6P	968	2940	NA
83	4dr Sedan-6P	984	3005	NA
83	4dr Tr Sedan-6P	1005	3010	NA
DE LUXE EIGHT — SERIES 84 122				
84	2dr Victoria Cpe-5P	1031	3060	NA
84	2dr Cpe-3P	990	3010	NA
84	2dr Conv Cpe-3P	1121	3060	NA
84	2dr Conv Brougham-6P	1185	3140	NA
84	4dr Sedan-6P	1060	3155	NA
84	4dr Tr Sedan-6P	1080	NA	NA
84	2dr Brougham-6P	1028	3115	NA
84	2dr Tr Brougham-6P	1049	3120	NA
CUSTOM EIGHT — SERIES 85 122				
85	2dr Cpe-3P	1080	3020	NA
85	2dr Victoria Cpe-5P	1131	3080	NA
85	2dr Brougham-6P	1134	3140	NA
85	2dr Tr Brougham-6P	1155	3145	NA
85	4dr Sedan-6P	1171	3190	NA
85	4dr Tr Sedan-6P	1191	3195	NA
CUSTOM EIGHT COUNTRY CLUB — SERIES 87 129				
87	4dr Sedan-6P	1199	3270	NA
87	4dr Tr Sedan	1219	3275	NA

1938 Hudson Series 87 touring sedan (WRG)

ENGINE: [Eight] L-head. Straight. Eight cylinder. Chrome alloy block. B & S: 3 x 4.5 in. Disp.: 254 c.i.d. C.R.: 6.25:1. Brake HP: 122 at 4200 rpm. Taxable HP: 28.8. Main bearings: 5. Valve lifters: mechanical. Carb.: two-barrel, downdraft Carter WDO 402S. [Hudson 112] L-head. Inline. Six cylinder. Chrome alloy block. B & S: 3 x 4-1/8 in. Disp.: 175 cid C.R.: 6.50:1. Brake HP: 83 at 4000 rpm. Taxable HP: 21.6. Main bearings: 3. Valve lifters: mechanical. Carb.: Carter 411S (early, 417S (late). [Hudson Six] L-head. Six cylinder. Chrome alloy block. B & S: 3 x 5 in. Disp.: 212 cid Comp.: 6.25:1. Brake HP: 101 at 4000 rpm. Taxable HP: 21.6. Main bearings: 3. Valve lifters: mechanical. Carb.: two-barrel Carter WDO downdraft. [Optional]: L-head. Six cyl. Chrome alloy block. B & S: 3 x 5 in. Disp.: 212 cid C.R.: 7.0:1. Brake HP: 107. Main bearings: 3. Valve lifters: mechanical. Carb.: Dual Downdraft Carter WDO.
CHASSIS: [112] W.B.: 112 in. O.L.: 186 in. Height: 70 in. Frt/Rear Tread: 56/59 in. Tires: 16 x 5.50. [Hudson Custom Six]: W.B.: 122 in. O.L.: 197-3/4 in. Frt/Rear Tread: 56/59 in. Tires: 16 x 6.00. (15 x 7.00 opt). [Hudson Custom Six]: W.B.: 122 in. O.L.: 197-3/4 in. Frt/Rear Tread: 56/59 in. Tires: 16 x 6.00 (15 x 7.00 opt). [Hudson Deluxe Eight]: W.B.: 122 in. O.L.: 197-3/4 in. Frt/Rear Tread: 56/59 in. Tires: 16 x 6.00 (15 x 7.00 opt). [Hudson Custom Eight Country Club]: W.B.: 129 in. Frt/Rear Tread: 56/59 in. Tires: 16 x 6.00 (15 x 7.00 opt).
TECHNICAL: Sliding gear, synchromesh transmission. 3F/1R. Floor shift. Single disc cork insert, running in oil. Shaft drive. Semi-floating rear axle. Overall Ratio: 4.11:1. Bendix Hydraulic brakes on four wheels. Steel disc wheels. Drivetrain Options: Selective Automatic shift. Automatic clutch.

OPTIONS: Deluxe heater. Custom seven-tube Hudson-RCA Victor DB-38 radio. Deluxe six-tube Hudson-RCA Victor SA-38 radio. Custom hot water heater. Hydraulic hill-hold.
HISTORICAL: Introduced January 1938. Hudson made 51,078 shipments to dealers during the 1938 calendar year. The president of Hudson was A.E. Barit.

1939 HUDSON

1939 Hudson 112 Convertible Brougham (OCW)

HUDSON 112 — SERIES 90 — SIX: The Hudson 112 was the only 1939 Hudson that retained the side-hood mounted headlights of earlier years. Artificial catwalk grille panels were new, as was the steering column mounted shift lever.

HUDSON PACEMAKER — SERIES 91 — SIX: The Pacemaker Series was introduced in March 1939 and is often regarded as the successor to the Terraplane. As were other Hudsons, the Pacemaker was equipped with an Auto-Poise Control suspension. This consisted of a bar attached to the frame, across the front of the chassis. Its ends were angled backward to form arms that attached to the wheel spindles. The result was a torsional effect that pulled the wheels back to a center location whenever they moved away from a straight-ahead position.

HUDSON SIX — SERIES 92: Interior and dash appointments of the Series 92 were of a higher quality than those of the Series 91. Both series shared painted catwalk grilles, narrow body beltline trim and a relatively short hood ornament. In common with all Hudsons, the Six was equipped with the forward hinged hood introduced on the 1938 Hudson 112.

HUDSON COUNTRY CLUB SIX — SERIES 93: Standard equipment on all Country Club and convertible models were Airfoam latex rubber cushions. Exterior appointments included chrome catwalk grilles, front fender chrome spears placed above the headlights, wider beltline molding and small arrowhead shaped lights in the leading edge of the side hood trim.

1939 Hudson Country Club Eight convertible (WRG)

HUDSON COUNTRY CLUB EIGHT — SERIES 95: These Hudsons were identical to the Series 93 versions with the exception of their front bumper center section which carried a plate reading Hudson Eight. Those on the Series 93 read Hudson.

HUDSON COUNTRY CLUB EIGHT — SERIES 97: The two models in this series were mounted on 109 inch wheelbase chassis and were distinguished by their taupe cashmere cloth interiors.

I.D. DATA: Car number on plate mounted on firewall. Starting: [Series 90] 90101, [Series 91] 9132576, [Series 92] 92101. Ending: [Series 90] 9054902, [Series 91] 9154902, [Series 92] 9254902. Engine numbers on right side of cylinder block. Starting: same as serial numbers. Car numbers on plate mounted on firewall. Starting: [Series 93] 93101, [Series 95] 95101, [Series 97] 97101. Ending: [Series 93] 9354902, [Series 95] 9554902, [Series 97] 9754902. Engine numbers on right side of cylinder block. Starting: same as serial numbers. Canadian-built cars included letter C.

Model No.	Body Type & Seating	Price	Weight	Prod. Total
90	2dr Cpe-3P	745	2587	NA
90	2dr Conv Cpe-3P	886	2627	NA
90	2dr Conv Brougham-6P	936	2732	NA
90	2dr Tr Brougham-6P	775	2682	NA
90	4dr Tr Sedan-6P	806	2712	NA
90	2dr Victoria Cpe-4P	791	2622	NA
90	2dr Utility Cpe-3P	750	2714	NA
90	2dr Traveler Cpe-3P	695	2544	NA
90	2dr Utility Coach-6P	725	2634	NA
90	4dr Station Wagon-6P	931	2880	NA

112"

NOTE 1: The two-door Utility Coupe, Traveler Coupe and Utility Coach, as well as the four-door Station Wagon were classified as Series 90 "business" cars.

91	4dr Tr Sedan-6P	854	2867	NA
91	2dr Tr Brougham-6P	823	2832	NA
91	2dr Victoria Cpe-4P	844	2752	NA
91	2dr Cpe-3P	973	2717	NA
92	2dr Cpe-3P	823	2757	NA
92	2dr Victoria Cpe-5P	869	2787	NA
92	2dr Conv Cpe-3P	972	2782	NA
92	2dr Conv Brougham-6P	1032	2892	NA
92	2dr Tr Brougham-6P	856	2847	NA
92	4dr Tr Sedan-6P	898	2897	NA
93	2dr Cpe-3P	919	2848	NA
93	2dr Victoria Cpe-5P	967	2893	NA
93	2dr Conv Cpe-3P	1052	2898	NA
93	4dr Tr Sedan-6P	995	3023	NA
93	2dr Conv Brougham-6P	1115	2983	NA
93	2dr Tr Brougham-6P	960	2968	NA
95	2dr Victoria Cpe-5P	1051	3053	NA
95	2dr Conv Cpe-3P	1138	3033	NA
95	2dr Conv Brougham-6P	1201	3123	NA
95	2dr Tr Brougham-6P	1049	3138	NA
95	4dr Tr Sedan-6P	1079	3193	NA
97	4dr Cus Sedan-6P	1175	3268	NA
97	4dr Cus Sedan-7P	1430	3378	NA

118 (Series 91,92) · 122 (Series 93,95) · 129 (Series 97)

1939 Hudson 112 four-door touring sedan (WRG)

ENGINE: [Series 95/97] L-head. Inline. Eight-cylinder. Chrome alloy block. B & S: 3 x 4-1/2 in. Disp.: 254 cid. C.R.: 6.25:1. Brake hp.: 122 at 4200 rpm. Taxable hp.: 28.8. Main bearings: 5. Valve lifters: mechanical. Carb.: Carter two-barrel, downdraft WDO 402S. [Series 93] L-head. Inline. Six-cylinder. Chrome alloy block. B & S: 3 x 5 in. Disp.: 212 cid. Comp.: 6.25:1. Brake hp. 96 at 3600 rpm. Main bearings: 3. Valve lifters: mechanical. Carb.: two-barrel Carter downdraft WDO. [Series 91] L-head. Inline. Six-cylinder. Chrome alloy block. B & S: 3 x 5 in. Disp. 212 cid. C.R.: 6.25:1. Brake hp.: 96 at 3900 rpm. Main bearings: 3. Valve lifters: mechanical. Carb.: one-barrel Carter. [Series 92] L-head. Inline. Six-cylinder. Chrome alloy block. B & S: 3 x 5 in. Disp.: 212 cid. C.R.: 6.25:1. Brake hp.: 96 3900 rpm. Main bearings: 3. Valve lifters: mechanical. Carb.: one-barrel Carter. [Optional Engine]: L-head. Inline. Six-cylinder. Chrome alloy block. B & S: 3 x 5 in. Disp.: 212 cid. C.R.: 6.25:1. Brake hp 101. Main bearings: 3. Valve lifters: mechanical. Carb.: two-barrel Carter WDO downdraft. [Series 90] L-head. Inline. Six-cylinder. Chrome alloy block. B & S: 3 x 4-1/2 in. Disp.: 175 cid. C.R.: 6.5:1. Brake hp.: 86 at 4000 rpm. Taxable hp.: 21.6. Main bearings: 3. Valve lifters: mechanical. Carb.: one-barrel Carter WDO.

CHASSIS: [Series 90] W.B.: 112 in. O.L.: 187-7/8 in. Front/rear Tread: 50/59-1/2 in. Tires: 6.00 x 16. [Series 91] W.B.: 118 in. O.L.: 193-7/16 in. Front/rear Tread: 56/59-1/2 in. [Series 93] W.B.: 122 in. O.L.: 199 in. Front/rear Tread: 56/59-1/2 in. Tires: 6.25 x 16. [Series 92] W.B.: 118 in. Front/rear Tread: 56/59-1/2 in. [Series 95] W.B.: 122 in. O.L.: 199 in. Front/rear Tread: 56/59-1/2 in. Tires: 6.50 x 16. [Series 97] W.B.: 129 in. O.L.: 206 in. Front/rear Tread: 56/59-1/2 in. Tires: 6.50 x 16.

TECHNICAL: Sliding gear transmission. Speeds: 3F/1R. Steering column controls. Single disc, cork inserts, running in oil. Shaft drive. Semi-floating rear axle. Overall ratio: 4.11:1. Bendix hydraulic brakes on four wheels. Steel, drop center type wheels. Rim size: 16 in. Selective automatic shift. Automatic clutch.

OPTIONS: Single sidemount. Heater (custom and deluxe). Custom radio. Deluxe radio. Foglights. Air electric horns. Side mirrors. Seat covers.

HISTORICAL: Introduced November 1938. Innovations: Airform seat cushions, Auto-Poise Control, interior hood release. Hudson made 82,161 shipments to dealers during the 1939 calendar year. The president of Hudson was A.E. Barit.

1940 HUDSON

1940 Hudson Super Six Victoria coupe (OCW)

HUDSON — GENERAL DESCRIPTION: Hudson's styling was extensively revised for 1940. The almond-shaped headlights of 1939 gave way to circular sealed beam units. The shape of the headlight frame and the positioning of the parking lights directly below the main lamps, gave both arrangements a similar appearance. The Hudson's grille form had been steadily evolving towards a strictly horizontal design for several years. Now, it took another strong step in that direction, as a two section arrangement with horizontal bars was adopted. Additional horizontal bars ran the length of the body adding an impression of lowness and fleetness. At the rear, a larger, single-piece window was adopted. All models were offered with or without runningboards at no extra cost. Leading the list of Hudson mechanical and technical improvements was a full independent front suspension consisting of coil springs, unequal length A arms and hydraulic shock absorbers. Also new, were five-inch longer rear semi-elliptic springs with a total length of 60 inches. Optional was a semi-electric Warner overdrive, replacing the Selective Automatic Shift. Hudson's Fluid-Cushioned automatic clutch was available with or without overdrive. The Hudson 112 series was dropped for 1940, with a new line of four Hudson Six Travelers taking its place. Hudson produced 86,865 autos, with losses totalled $1,507,780.

HUDSON SUPER SIX — SERIES 41: Super-Six models carried Hudson lettering on the sides of their hood, along with the small triangular lights of the Series 43 and 44 models. In addition, their greater length made it easy to separate them from the Series 40T and 40P.

HUDSON TRAVELER SIX — SERIES 40T: The Series 40T models, like all 1940 Hudsons, were equipped with sealed beam headlamps. Since this line was Hudson's least expensive, it was fitted with sliding glass. Its interior featured taupe worsted boucle. The Traveler was easily set apart from other Hudsons by its lack of front vent panes.

HUDSON DELUXE SIX — SERIES 40P: The higher price of the Deluxe Six, compared to the Traveler Six, was justified by its higher quality brown taupe stripe broadcloth upholstery and roll-down windows.

HUDSON EIGHT — SERIES 44: The Series 44 Hudsons were identical in appearance to the Series 41 Super Six models.

HUDSON DELUXE EIGHT — SERIES 45: The two models in this series had a higher quality interior than the one used in Hudson Eights.

HUDSON COUNTRY CLUB EIGHT — SERIES 47: These Hudsons were identified by their added length, wider door (front) and distinctive front grille with a rectangular opening in its center section. The front fenders were crowned with chrome strips and the taillights were given added chrome trim. Included as standard equipment were directional signals.

HUDSON COUNTRY CLUB SIX — SERIES 43: These three Hudsons were, with the exception of the engine and running gear, identical to the Country Club Eight. Interior features included a standard two-tone brown and tan Hockanum woolen upholstery. Door panels were fitted with chrome scruff plates and the pleated sections of the seats and door panels had horizontal dividers.

I.D. DATA: Car numbers on plate mounted on firewall. Starting: [Series 40] 40101, [41] 41250, [43] 43370, [44] 44294, [45] 451752, [47] 47167. Ending: [40] 4089192, [41] 4189192, [43] 4389192, [44] 4489192, [45] 4589192, [47] 4789192. * Includes 40T, 40P and 40C (commercial cars). Engine numbers on right side of cylinder block. Starting: Series 40 same as serial numbers, [41/43] 43101, [44/45] 45101. Ending: Series 40 same as serial numbers, [41/43] 4389192, [44/45] 4589192. Canadian-built cars included letter C.

Model No.	Body Type & Seating	Price	Weight	Prod. Total
40T	2dr Cpe-3P	670	2800	NA
40T	2dr Victoria Cpe-4P	750	2830	NA
40T	2dr Tr Sedan-6P	735	2895	NA
40T	4dr Sedan-6P	763	2940	NA
40P	2dr Cpe-3P	745	2840	NA
40P	2dr Victoria Cpe-4P	791	2865	NA
40P	2dr Conv Cpe-5P	930	2860	NA
40P	2dr Tr Sedan-6P	775	2930	NA
40P	2dr Conv Sedan-6P	955	2920	NA
40P	4dr Tr Sedan-6P	806	2965	NA
41	2dr Cpe-3P	809	2950	NA
41	2dr Victoria Cpe-5P	860	2980	NA
41	2dr Conv Cpe-5P	1087	2980	NA
41	2dr Conv Sedan-6P	1030	3020	NA
41	2dr Tr Sedan-6P	839	3020	NA
41	4dr Tr Sedan-6P	870	3050	NA
43	4dr Tr Sedan-6P	1018	3240	NA
43	4dr Sp. Tr Sedan-6P	1044	3240	NA
43	4dr Sedan-8P	1230	3355	NA
44	2dr Cpe-3P	860	3040	NA

44	2dr Victoria Cpe-4P	942	3075	NA
44	2dr Conv Sedan-6P	1122	3130	NA
44	2dr Conv Cpe-5P	1087	3065	NA
44	2dr Tr Sedan-6P	942	3185	NA
44	4dr Tr Sedan-6P	952	3185	NA
45	2dr Tr Sedan-6P	942	3185	NA
45	4dr Tr Sedan-6P	976	3215	NA
47	4dr Tr Sedan-6P	1118	3285	NA
47	4dr Sp. Tr Sedan-6P	1144	3285	NA
47	4dr Sedan-8P	1330	3400	NA

1940 Hudson Country Club Eight four-door touring sedan (WRG)

ENGINE: [Series 40T/40P] L-head. Inline. Six-cylinder. Chrome alloy block. B & S: 3 x 4-1/8 in. Disp.: 175 cid. C.R.: 6.5:1. Brake hp.: 92 at 4000 rpm. Main bearings: 3. Valve lifters: mechanical. Carb.: Carter 430 SV (early) 461S (late). [Optional Engine] L-head. Inline. Six-cylinder. Chrome alloy block. B & S: 3 x 5 in. Disp.: 212 cid. C.R.: 6.5:1. Brake hp: 102 at 4000 rpm. Main bearings: 3. Valve lifters: mechanical. Carb.: Carter 430SV (early) 461S (late). [Series 41/43] L-head. Inline. Six-cylinder. Chrome alloy block. B & S: 3 x 5 in. Disp.: 212 cid. C.R.: 6.5:1. Brake hp: 102 at 4000 rpm. Main bearings: 3. Valve lifters: mechanical. Carb.: two-barrel Carter 454S. [Series 44/45/47] L-head. Inline. Eight-cylinder. Chrome alloy block. B & S: 3 x 4-1/2 in. Disp.: 254.4 cid. C.R.: 6.5:1. Brake hp.: 128 4200 rpm. N.A.C.C. hp.: 28.8. Main bearings: 5. Valve lifters: mechanical. Carb.: two-barrel Carter 455S.

CHASSIS: [Series 40T/40P] W.B.: 113 in. O.L.: 190-3/8 in. Height: 70.5 in. Front/rear Tread: 56-1/4/59-1/2 in. Tires: 16 x 5.50 — 40T, 16 x 6.00 — 40P, 15 x 7.00 optional. [Series 41] W.B.: 118 in. O.L.: 195-3/8 in. Height: 70.5 in. Front/rear Tread: 56-1/4 / 59-1/2 in. Tires: 16 x 6.00, 15 x 7.00 optional. [Series 43] W.B.: 125 in. O.L.: 202-3/8 in. Height: 70.5 in. Front/rear Tread: 56-1/4 / 59-1/2 in. Tires: 16 x 6.25, 15 x 7.00 optional. [Series 44] W.B.: 118 in. O.L.: 195-3/8 in. Height: 70.5 in. Front/rear Tread: 56-1/4 / 59-1/2 in. Tires: 16 x 6.00, 15 x 7.00 optional. [Series 45] W.B.: 118 in. O.L.: 195-3/8 in. Height: 70.5 in. Front/rear Tread: 56-1/4 / 59-1/2 in. Tires: 16 x 6.00, 15 x 7.00 optional. [Series 47] W.B.: 125 in. O.L.: 202-3/8 in. Height: 70.5 in. Front/rear Tread: 56-1/4 / 59-1/2 in. Tires: 16 x 6.50, 15 x 7.00 optional.

TECHNICAL: Sliding gear transmission. 3F/1R. Column controls. Single disc, cork inserts, running in oil. Shaft drive. Semi-floating rear axle. Overall Ratio: Series 40T/40P — 4.55 all others 4.11. Bendix hydraulic brakes on four wheels. Steel, drop center type wheels. Drivetrain Options: Overdrive. Automatic clutch.

OPTIONS: Turn signals (standard only on Country Club models). Airfoam cushions (optional on Hudson Six models — standard in all others). Weather-Master Fresh Air and Heat Control.

HISTORICAL: Introduced September 1939. 1940 was the first year for Hudson's fully independent suspension. Hudson made 87,900 shipments to dealers during the 1940 calendar year. The president of Hudson was A.E. Barit. From August 23 to August 27, just prior to the introduction date of the 1940 Hudsons, both the Hudson Six and Hudson Eight set numerous new AAA speed records. The Hudson Six established 58 new Class D endurance marks, plus 23 unlimited records for stock cars regardless of size or price. Equipped with overdrive and the optional 3.88:1 rear axle, the Hudson averaged 70.5 mph for 10,000 miles. The Hudson Eight, with the optional 7.0:1 cylinder head, overdrive and 3.88:1 rear axle, was used by John Cobb to set additional Class D records. They included a 10 mile run at 92.89 and a flying mile record of 93.89. Both cars also participated in economy runs. The Hudson Six finished a 1,000 mile run at an average of 29.88 mph with fuel consumption of 32.66 mpg. The Hudson Eight averaged 27.12 mpg for one thousand miles at a constant speed of 29.31 mph.

1941 HUDSON

HUDSON — GENERAL DESCRIPTION: The 1941 Hudsons were substantially redesigned automobiles, sharing only front fender sheet metal and drive line components with 1940 models. All wheelbases were increased three inches and overall height was reduced by two inches, thanks to a flatter roofline. Overall body length was extended by 5-3/4 inches. The handsome, horizontal grille format was continued from 1940, but nine, rather than seven, grille bars were used. Their greater length reduced the size of the center grille divider. Rear deck changes included moving the taillights from the fenders to the quarter panels and adding externally mounted chrome trunk hinges. Convertibles were now equipped with a power top. A new three-speed synchromesh transmission with helical-cut gears was also introduced. In a break with tradition, Hudson introduced Symphonic Styling for 1941, offering the customer a wide selection of interior color combinations that harmonized with the car's exterior color. Although shipments to dealers dropped to 79,529, Hudson's profit was a respectable $3,756,418.

HUDSON TRAVELER SIX DELUXE — SERIES 10T/10P: The 10T and 10P shared the same chassis, wheelbase and engines. Both series were equipped with a single taillight and sun visor, but the 10P was fitted with many other features. These included larger 16 x 6.00 standard tires, gray or tan colored broadcloth upholstery (in place of the 10T's taupe worsted boucle), woodgrained dash and garnish moldings, front window vent panes, rear anti-sway bar, spring covers, bumper guards, rear seat ashtrays and assist straps, plus front armrests.

1941 Hudson Super Six station wagon (OCW)

HUDSON SUPER SIX — SERIES 11: The Super Six was powered by the same engine as the Commodore Six and shared the 121 inch wheelbase chassis with both the Commodore Sixes and Eights. Externally it was identified by Super Six nameplates positioned on each side of the hood, near the windshield base. Its standard equipment was identical to that of the Deluxe Six, but the Super Six interior featured Hockanum Tweed in either tan, gray, or green and the instrument panel and garnish moldings were painted to match one of those colors.

HUDSON COMMODORE SIX (SERIES 12)/COMMODORE EIGHT (SERIES 14): The Commodore Six and Commodore Eight were, except for their engine and accompanying mechanical modifications, identical automobiles. The only exception was the hood identification, which read either Commodore Six or Commodore Eight. Both cars had a 121 inch wheelbase. Crowning their front fenders were long, chrome bars that trailed away from the parking lights. The front bumpers had large guard wings at either end, with similar gravel deflectors mounted on the rear bumpers. The Commodore interior featured Hockanum Twill Cord upholstery in gray, tan, or green and the finish of the instrument panel matched that of the upholstery. The front and rear seatbacks and armrests were leather-finished and a deluxe, 18 inch steering wheel with a chrome horn ring was fitted. Among the Commodore's standard equipment were airfoam seats, twin horns and taillights, large hubcaps and, in sedans, a rear dome light. In addition, the Commodores were offered in special two-tone exterior color combinations.

HUDSON COMMODORE CUSTOM EIGHT — SERIES 15/17: The Commodore Custom Eights were offered in either coupe (121 inch wheelbase) or sedan (128 inch wheelbase) models. They were identified by such external features as the center bar placed between the front bumper guards, wheel trim rings, hood identification script, inward-pointed triangular taillights, and a similarly shaped rear deck emblem. In addition to the equipment supplied to the Commodores, the Custom Eight's standard appointments included two cigarette lighters, a radio and front and rear center armrests. Fancy Hockanum Bedford Cord upholstery (in green, gray or tan) covered double thickness airfoam seat cushions. Additional leather trim was also installed on the interior door edges and front seat corners.

I.D. DATA: Car numbers on plate positioned on right front door hinge pillar post. Starting: [Series 10T] T10101; [Series 10P/10C] P10101; [Series 11/12] 11101; [Series 14/15] 14101; [Series 17] 17101. Ending: [Series 10T] T1092988; [Series 10P/10C] P1092988; [Series 11/12] 1192988; [Series 14/15] 1492988; [Series 17] 1792988. Engine number on stamping on top of the cylinder block between numbers one and two exhaust manifold flanges. Starting: [Series 10T] T10101; [Series 10P/10C] C10101; [Series 11/12] 12101; [Series 14/15] 15101; [Series 17] NA. Ending: [Series 10T] T1092988; [Series 10P/10C] C1092988; [Series 11/12] 1292988; [Series 14/15] 1592988; [Series 17] NA. C included in Canadian cars.

Model No.	Body Type & Seating	Price	Weight	Prod. Total
10T	2dr Club Cpe-6P	788	2840	NA
10T	2dr Cpe-3P	695	2790	NA
10T	2dr Tr Sedan-6P	765	2850	NA
10T	4dr Tr Sedan-6P	793	2900	NA
10P	2dr Club Cpe-6P	848	2895	NA
10P	2dr Cpe-3P	801	2840	NA
10P	2dr Tr Sedan-6P	822	2900	NA
10P	4dr Tr Sedan-6P	856	2950	NA
10P	2dr Conv Sedan-6P	1063	2980	NA
11	2dr Club Cpe-6P	936	2980	NA
11	2dr Cpe-3P	881	2935	NA
11	2dr Tr Sedan-6P	901	3000	NA
11	4dr Tr Sedan-6P	932	3050	NA
11	2dr Conv Sedan-6P	1156	3125	NA
11	4dr Sta Wag-8P	1383	3400	NA
12	2dr Club Cpe-6P	997	3045	NA
12	2dr Cpe-3P	935	3000	NA
12	2dr Conv Sedan-6P	1204	3160	NA
12	2dr Tr Sedan-6P	966	3050	NA
12	4dr Tr Sedan-6P	994	3100	NA
14	2dr Club Cpe-6P	1040	3210	NA
14	2dr Cpe-3P	978	3135	NA
14	2dr Conv Sedan-6P	1254	3350	NA
14	2dr Tr Sedan-6P	1003	3210	NA
14	4dr Tr Sedan-6P	1035	3260	NA
14	4dr Sta Wag-8P	1383	3400	NA
15	2dr Club Cpe-6P	1127	3235	NA
15	2dr Cpe-3P	1064	NA	NA
17	4dr Tr Sedan-6P	1232	3400	NA
17	4dr Sedan-8P	1438	3440	NA

ENGINE: [Series 11/12] L-head. Inline. Six-cylinder. Chrome alloy block. B & S: 3 x 5 in. Disp.: 212 cid. C.R.: 6.25:1. Brake hp.: 102 at 4000 rpm. Taxable hp.: 21.6. Main bearings: 3. Valve lifters: mechanical. Carb.: Carter duplex downdraft 501S. [Series 10T and 10P] L-head. Inline. Six-cylinder. Chrome alloy block. B & S: 3 x 4-1/2 in. Disp.: 175 cid. C.R.: 7.25:1. Brake hp.: 92 at 4000 rpm. Taxable hp.: 21.6. Main bearings: 3. Valve lifters: mechanical. Carb.: Carter single downdraft. [Optional Engine]: L-head. Inline. Six-cylinder. Chrome alloy block. B & S: 3 x 5 in. Disp.: 212 cid. C.R.: 6.5:1. Brake hp.: 102 at 4000 rpm. Taxable/ALAM/NACC hp.: 21.6. Main bearings: 3. Valve lifters: mechanical. Carb.: Carter duplex downdraft. [Series 14 and 15 and 17] L-head. Inline. Eight-cylinder. Chrome alloy block. B & S: 3 x 4-1/2 in. Disp. 254.4 cid C.R.: 6.5:1. Brake hp.: 128 at 4200 rpm. Taxable hp.: 28.8. Main bearings: 5. Valve lifters: mechanical. Carb.: two-barrel Carter 502S.

1941 Hudson Commodore Eight convertible (WRG)

CHASSIS: [Series 10T] W.B.: 116 in. O.L.: 195-1/4 in. Height: 68 in. Frt/Rear Tread: 56-1/4 / 59-1/2 in. Tires: 16 x 5.50 (16 x 6.00 optional). [Series 10P] W.B.: 116 in. O.L.: 195-1/4 in. Height: 68 in. Frt/Rear Tread: 56-1/4 / 59-1/2 in. Tires: 16 x 6.00. [Series 11] W.B.: 121 in. O.L.: 200-1/4 in. Height: 68-3/4 in. Frt/Rear Tread: 56-1/4 / 59-1/2 in. Tires: 16 x 6.00 (16 x 6.50 or 15 x 7.00 optional). [Series 12] W.B.: 121 in. O.L.: 203-1/4 in. Height: 68-3/4 in. Frt/Rear Tread: 56-1/4 / 59-1/2 in. Tires: 16 x 6.25 (16 x 6.50 or 15 x 7.00 optional). [Series 14] W.B.: 121 in. O.L.: 203-1/4 in. Height: 68-3/4 in. Frt/Rear Tread: 56-1/4 / 59-1/2 in. Tires: 16 x 6.25 (16 x 6.50 or 15 x 7.00 optional). [Series 15] W.B.: 121 in. O.L.: 203-1/4 in. Height: 68-3/4 in. Frt/Rear Trend: 56-1/4 / 59-1/2 in. Tires: 16 x 6.25 (16 x 6.50 or 15 x 7.00 optional). [Series 17] W.B.: 128 in. O.L.: 210-1/4 in. Height: 68-3/4 in. Frt/Rear Tread: 56-1/4 / 59-1/2 in. Tires: 16 x 6.50 (15 x 7.00 optional).
TECHNICAL: Sliding gear transmission. Speeds: 3F/1R. Column controls. Single disc, cork inserts, running in oil. Shaft drive. Semi-floating rear axle. Overall ratio: 4.11, 4.55 optional — except series 10T and 10P. Standard axle with overdrive 4.55, 4.11 optional. Bendix hydraulic brakes on four wheels. Steel wheels. Rim size: 16 in. (15 in. optional). Vacumotive Drive automatic clutch ($27.50). Overdrive ($62.50).
OPTIONS: Deluxe radio ($49.75). Custom heater, includes defroster ($26). Clock ($13.50). Spotlight ($17). Junior radio ($29.50). Custom radio ($67.50). Power radio antenna ($6.75). Weathermaster heater and defroster ($36). Directionals ($17.50, $19.50). Seat covers ($7.25). Chrome outside window mouldings ($7.25, 13.00). Deluxe 18" steering wheel ($13.95). Special running board mouldings ($2.25). Air foam seats. Large hubcaps ($6.75). Twin horn.
HISTORICAL: Hudson made 79,529 shipments to dealers during the 1941 calendar year. The president of Hudson was A.E. Barit.
Hudson emphasized its economy of operation, rather than performance, in 1941. In the Gilmore Oil Grand Canyon Run a Hudson Deluxe Six was a class winner with a 24.6 mpg mark. Also, a class champion was the Commodore Eight at 20.18 mpg. Safety Engineering Magazine persented Hudson with its Safety Engineering Trophy for the 1941 model's safety and engineering excellence. In each of 14 categories the Hudson received a perfect score.

1941 Hudson Commodore four-door touring sedan (OCW)

1942 HUDSON

HUDSON — ALL SERIES OVERVIEW: The last of the prewar Hudsons received a fairly substantial facelift. The lower body section now flared out to conceal the remnants of running boards. Pointing the way to the full-width grille arrangements common to most postwar American cars were chrome strips running across the body and extending around the first few inches of the front fender. Also given greater width was the Hudson grille. To accommodate these changes, new front and rear fenders were used. Overall height was reduced by 1.5 inches, due to an altered rear spring and frame design. The Hudson profile also carried new trim consisting of a single, long, bright strip of chrome with shorter strips at either end. However, Dec. 31, 1941 was the end of chrome plated trim for U.S. cars. The only external components exempt were bumpers. After that date, Hudson trim consisted of metal pressings covered with plastic. Hudson's production of 1942 models began on July 21, 1941. By Feb. 5, 1942, when it came to an end, a total of 40,661 cars were assembled.

1942 Hudson Super Six club coupe (OCW)

I.D. DATA: Car numbers on plate mounted on firewall. Starting: [Series 20T] T-20101; [Series 20P] P-20101; [Series 21] 21101; [Series 22] 22101. Ending: [Series 20T] T2041232; [Series 20P] P2041232; [Series 21] 2141232; [Series 22] 2241232. Engine numbers stamped on top of the cylinder block, between numbers one and two exhaust manifold flanges. Starting: same as serial numbers. Car numbers on plate mounted on firewall. Starting: [Series 24] 24101; [Series 25] 25101; [Series 27] 27101. Ending: [Series 24] 2441232; [Series 25] 2541232; [Series 27] 2741232. Engine number stamped on top of the-cylinder block between numbers one and two exhaust manifold flanges. Starting: same as serial numbers. Canadian-built cars included letter C.

1942 Hudson Commodore Eight four-door sedan (OCW)

Model No.	Body Type & Seating	Price	Weight	Prod. Total
22	2dr Club Cpe-6P	1239	3090	NA
22	2dr Cpe-3P	1176	2995	NA
22	2dr Club Sedan-6P	1216	3090	NA
22	4dr Sedan-6P	1246	3145	NA
22	2dr Conv Sedan-6P	1481	3280	NA
24	2dr Club Cpe-6P	1282	3205	NA
24	2dr Cpe-3P	1220	3130	NA
24	2dr Club Sedan-6P	1252	3230	NA
24	4dr Sedan-6P	1291	3280	NA
24	2dr Conv Sedan-6P	1533	3400	NA
25	2dr Club Cpe-6P	1380	3235	NA
25	2dr Cpe-3P	1318	3160	NA
27	4dr Sedan-8P	1510	3395	NA
20T	2dr Club Cpe-6P	965	2845	NA
20T	2dr Cpe-3P	893	2795	NA
20T	2dr Club Sedan-6P	945	2895	NA
20T	4dr Sedan-6P	973	2940	NA
20P	2dr Club Cpe-6P	1034	2900	NA
20P	2dr Cpe-3P	981	2845	NA
20P	2dr Club Sedan-6P	1012	2935	NA
20P	4dr Sedan-6P	1045	2975	NA
20P	2dr Conv-6P	1292	3140	NA
21	2dr Club Cpe-6P	1159	3010	NA
21	2dr Cpe-3P	1102	2950	NA
21	2dr Club Sedan-6P	1132	3035	NA
21	4dr Sedan-6P	1162	3080	NA
21	2dr Conv Sedan-6P	1414	3200	NA
21	2dr Sta Wag-6P	1486	3315	NA

ENGINE: [Series 24/25/27] L-head. Inline. Eight-cylinder. Chrome alloy block. B & S: 3 x 4-1/2 in. Disp.: 254.4 cid. C.R.: 6.5:1. Brake hp.: 128 at 4200 rpm. Taxable hp.: 28.8. Main bearings: 5. Valve lifters: mechanical. Carb.: two-barrel Carter 502S downdraft. [Series 21/22] L-head. Inline. Six-cylinder. Chrome alloy block. B & S: 3 x 5 in. Disp.: 212 cid. C.R.: 6.25:1. Brake hp.: 102 at 4000 rpm. Taxable hp.: 21.6. Main bearings: 3. Valve lifters: mechanical. Carb.: Carter duplex downdraft 501S. [Series 20T/20P] L-head. Inline. Six-cylinder. Chrome alloy block. B & S: 3 x 4-1/2 in. Disp.: 175 cid. C.R.: 7.25:1. Brake hp.: 92 at 4000 rpm. Taxable hp.: 21.6. Main bearings: 3. Valve lifters: mechanical. Carb.: Carter single downdraft 454S. Optional Engine: L-head. Inline. Six-cylinder. Chrome alloy block. B & S: 3 x 5 in. Disp.: 212 cid. C.R.: 6.5:1. Brake hp.: 102 at 4000 rpm. Taxable hp.: 21.6. Main bearings: 3. Valve lifters: mechanical. Carb.: Carter duplex downdraft 501S.
CHASSIS: [Series 20T] W.B.: 116 in. O.L.: 198-1/4 in. Front/Rear Tread: 56-5/16/59-1/2 in. Tires: 16 x 5.50 (16 x 6.00 optional). [Series 20P] W.B.: 116 in. O.L.: 198-1/4 in. Front/Rear Tread: 56-5/16/59-1/2 in. Tires: 16 x 6.00 (16 x 6.50 optional). [Series 21] W.B.: 121 in. O.L.: 207-3/8 in. Front/Rear Tread: 56-5/16/59-1/2 in. Tires: 16 x 6.00 (16 x 6.50, 15 x 7.00 optional). [Series 22] W.B.: 121 in. O.L.: 207-3/8 in. Front/Rear Tread: 56-5/16/59-1/2 in. Tires: 16 x 6.00 (16 x 6.50, 15 x 7.00 optional). [Series 24] W.B.: 121 in. O.L.: 207-3/8 in.

Front/Rear Tread: 56-5/16 / 59-1/2 in. Tires: 16 x 6.25 (16 x 6.50, 15 x 7.00 optional). [Series 25] W.B.: 121 in. O.L.: 207-3/8 in. Front/Rear Tread: 56-5/16 / 59-1/2 in. Tires: 16 x 6.25 (16 x 6.50, 15 x 7.00 optional). [Series 27] W.B.: 128 in. O.L.: 214-3/8 in. Front/Rear Tread: 56-5/16 / 59-1/2 in. Tires: 15 x 6.50 (15 x 7.00 optional).

TECHNICAL: Sliding gear transmission. Speeds: 3F/1R. Steering column controls. Single disc, cork inserts, running in oil. Shaft drive. Semi-floating rear axle. Overall Ratio: Series 20T/20P. 4.55, all others 4.11. With overdrive: Series 20T/20P — 4.87, all others 4.55. Bendix hydraulic brakes on four wheels. Steel drop center type wheels. Drivetrain Options: Drive-Master transmission. Overdrive.

OPTIONS: White sidewall tires. Full chrome hubcaps. Bumper wing guards (standard on Commodore). Sleeper kit.

HISTORICAL: Introduced August 1941. Innovations: Drive-Master was a combination of vacumotive gearshifting, plus an additional power unit which shifted gears by intake manifold vacuum. Cars with this feature had an instrument panel switch with three push-buttons. The off button allowed the car to function in the conventional manner. The VAC button put Vacumotive Drive into action. For full Drive-Master operation, the HDM button was pushed. Model Year production: 40,661 cars. The president of Hudson was A.E. Barit.

1942 Hudson Super Six four-door sedan (WRG)

1946 HUDSON

HUDSON SUPER — SERIES 51 SIX — SERIES 53 EIGHT: Hudson's big 1946 change was a new grille in the old postwar body. It now had a massive upper bar housing a Hudson badge. There were wide indentations at the center of each of the horizontal blades. The nose was smooth and no longer had a molding between the mascot and the grille, though a strip of chrome did run from the windshield base to the mascot. Nameplates alongside the rear of the hood revealed the model identity. The standard list was long, including dual brake system; dual carburetion; Auto Poise control; chrome alloy motor block; oil-cushioned clutch; center point steering; rear lateral stabilizer; teleflash signals; push-button starting; safety locking hood; hand-rubbed lacquer finish; Duo-Flo oiling; high-compression head; large trunk; and Long Life spark plugs. The basic equipment assortment for all models comprised left-hand front door armrests; twin air horns; dashboard and rear ashtrays; woodgrained instrument panel; spring covers; front and rear stabilizers; twin wipers; stoplights; locking glove box; windshield and rear window reveal moldings; new oval headlamp rims with sealed beam bulbs; front door locks; pile carpet; and, for sedans, envelope type front seatback pockets. Supers were trimmed in blue and gray boucle waffle weave cloth.

1946 Hudson, Super Six two-door Convertible Brougham, 6-cyl

VEHICLE I.D. NUMBERS: Serial numbers were on the right door post. Engine numbers were the same and were found on a boss near the top left side of the cylinder block and also between the first two manifold flanges. The first symbol was a '3' in 1946. The second symbol corresponded to the second number in the series/model code. The following group of symbols designated production sequence. Super Sixes were numbered 31101 to 3195099; Commodore Sixes 32101 to 3295062; Super Eights 33101 to 3395085 and Commodore Eights 34101 to 3495100. Body Style Numbers not used through 1955. First columns in charts below show six-cylinder model numbers; second columns show eight-cylinder model numbers.

HUDSON SUPER SIX (SERIES 51)/SUPER EIGHT (SERIES 52)

Model Number	Model Number	Body Type & Seating	Factory Price	Shipping Weight	Production Total
51	53	4-dr Sed-6P	1555/1668	3085/3235	Note 1
51	NA	2-dr Brghm-6P	1511	3030	Note 1
51	53	2-dr Clb Cpe-6P	1553/1664	3015/3185	Note 1
51	NA	2-dr Cpe-3P	1481	2950	Note 1
51	NA	2-dr Conv Brghm-6P	1879	3195	Note 1

NOTE 1: Series production was 61,787 Super Six/3,961 Super Eight with no breakouts.
NOTE 2: Approximately 1,037 Super Six convertible broughams built.
NOTE 3: Sedan and club coupe came only as Super Eight.
NOTE 4: Data above slash for six/below slash for eight.

HUDSON COMMODORE — SERIES 52 SIX — SERIES 54 EIGHT: All basic features and Super Series equipment were standard on Commodores, plus blue gray plain cloth upholstery; air foam seat cushions; carpet insert front floor covering; crank type ventipanes; door step courtesy lights; black filled etched aluminum scuff plates; luggage compartment rubber floor mats; and vertical rear window bars. In addition, Deluxe level appointments included interior hardware; taillamps; passenger assist straps and door handles. Gold finish was used on Commodore instrument dial letters and dash panel finish plates. Chrome-nickel plating brightened the steering column, shift lever, brake hand grip and glove locker box. An 18-inch custom steering wheel with horn ring was used and so were right-hand sun visors and electric clocks. A quick look at the Commodore exterior would reveal front and rear bumper bar extensions; front fender lamps with chrome extension moldings; oversize tires with large hubcaps (wheelcovers); wide body moldings with wheel color stripes; and an extra chrome strip above the belt moldings. Sedans had rear seat center armrests, leather robe hangers and rear dome lamps.

HUDSON COMMODORE SIX (SERIES 52)/COMMODORE EIGHT (SERIES 54)

Model Number	Model Number	Body Type & Seating	Factory Price	Shipping Weight	Production Total
52	54	4-dr Sed-6P	1699/1774	3150/3305	Note 1
52	54	2-dr Clb Cpe-5/6P	1693/1760	3065/3235	Note 1
NA	54	2-dr Conv Brghm-P	2050	3410	Note 1

NOTE 1: Series production was 17,685 Commodore Sixes and 8,193 Commodore Eights with no breakouts.
NOTE 2: Approximately 140 Commodore Eight convertibles brougham were built.
NOTE 3: Convertible brougham (convertible) came only as Commodore Eight.
NOTE 4: Data above slash for six/below slash for eight.

ENGINES

(SIX-CYLINDER) Inline. L-head six-cylinder. Chrome alloy block. Displacement: 212 cid. Bore and stroke: 3 x 5 inches. Compression ratio: 6.5:1. Brake hp: 103 at 4000 rpm. Three main bearings. Solid valve lifters. Carburetor: Carter two-barrel WDO type Model 501S.

(EIGHT-CYLINDER) Inline. L-head eight-cylinder. Chrome alloy block. Displacement: 254 cid. Bore and stroke: 3 x 4-1/2 inches. Compression ratio: 6.5:1. Brake hp: 128 at 4200 rpm. Five main bearings. Solid valve lifters. Carburetor: Carter two-barrel WDO type Model 502S.

1946 Hudson, Commodore Eight four-door sedan, 8-cyl

CHASSIS FEATURES: Wheelbase (all): 121 inches. Overall length: (all) 207-3/8 inches. Front tread (all): 56-5/16 inches. Rear tread (all) 59-1/2 inches. Tires: (Supers) 6.00 x 16; (Commodores) 6.50 x 15.

OPTIONS: Commodore front and rear seat cushions ($17); front only ($9). Commodore front and rear bumper bar extensions ($20). Commodore electric clock ($14). Commodore fender lamps ($16). Commodore horn ring with standard steering wheel ($6). Large hubcaps ($9). Custom 18 inch steering wheel with horn ring ($19). Right side visor ($3). Oversize 6.50 x 15 tires with four large hubcaps ($28). Right front door armrest ($4). Direction indicator for Supers ($26). Red or cream wheel color (no cost). Police car and taxi equipment ($11). Radio ($77). Heavy scale front and rear springs for sixes (no cost). Heavy scale rear springs (no cost). Weather Master heater ($50). Chrome wheel trim rings ($13). Three-quarter leather trim ($32-$53 depending on body style). Three-quarter leather grain trim ($25-$41 depending on body style). Vacumotive Drive ($40). Drive-Master, including Vacumotive Drive ($98). Oil bath air cleaner ($3). Power Dome with eight-cylinder engine ($10). Overdrive transmission ($88). Combination fuel and vacuum pump ($7). Rear axle with 4-1/9:1 gear ratio (no cost). Rear axle with 4-5/9:1 gear ratio (no cost). Factory delivered prices included leather-grain trims. Police/Taxi package included: larger clutch; heavy rear springs; 11 inch brakes; and heavy construction seats. Convertible Broughams included full-leather trim; air foam seats, oversized tires; and extra sun visors. Blue-gray Shadow Weave Cloth (Supers) or blue-gray Bedford Cloth (Commodores) was used in combination with special three-quarter trim options. Note: First 10 equipment items listed above standard on Commodore, optional on Super.

HISTORICAL FOOTNOTES: Production of 1946 Hudsons began on Aug. 30, 1945. Dealer introductions were held Oct. 1, 1945. The company displayed a 1909 Hudson at the Automotive Golden Jubilee in Detroit, Mich. this year. Model year sales hit 95,000 units. Calendar year deliveries peaked at 93,870 cars. Hudson held ninth place on the industry's sales charts. The trunk emblem used in 1946 was made of plastic. Nine standard colors, two extra cost hues and four two-tone combinations were provided for 1946 Hudson. Royal red finish was $23 extra and Nepal ivory was $60 extra. Two-tone selections were all priced at

$18 extra. Eight tire size and construction options were offered at exchange prices between $15 and $72, but whitewalls were not available. A unique cab pickup, with passenger car type sheet metal, was marketed in 1946 and 1947.

1947 HUDSON

HUDSON SUPER — SERIES 171 SIX — SERIES 173 EIGHT: Minor styling changes and equipment revisions were seen on 1947 Hudsons. Plastic trunk emblems were replaced with bright metal types and the corporate logo badge, centered in the upper grille bar, was ever so sightly modified to a larger size. Standard features of Supers included diagonal check Boucle upholstery; single adjustable hinged sun visor; 30-hour wind-up clock; woodgrained window finish moldings; black rubber front floor covering; carryall luggage compartment with vertically housed spare tire; felt trunk mats; cord robe hangers in sedans; new hoodside ornaments; 17 inch steering wheel; latch type front ventipanes; twin standard taillamps; sliding pane rear quarter glass in sedans; and stationary rear quarter glass in club coupes.

VEHICLE I.D. NUMBERS: Serial numbers were on the right door post. Engine numbers were the same and were found on a boss near the top left side of the cylinder block and also between the first two manifold flanges. The numbering system was basically the same as in 1946, with numbers running in single consecutive order, regardless of series. The first three symbols were comprised of the new series/model codes, followed by a group of numbers beginning at 101. Super Sixes were numbered 171101 to 17195100. Commodore Sixes 172101 to 17295099; Super Eights 173101 to 17394992; and Commodore Eights 174101 to 17495088. Body Style Numbers not used through 1955. First columns in charts below show six-cylinder model numbers; second columns show eight-cylinder model numbers.

HUDSON SUPER SIX (SERIES 171)/SUPER EIGHT (SERIES 173)

Model Number	Model Number	Body Type & Seating	Factory Price	Shipping Weight	Production Total
171	173	4-dr Sed-6P	1749/1862	3110/3260	Note 1
171	NA	2-dr Brghm-6P	1704	3055	Note 1
171	173	2-dr Clb Cpe-6P	1744/1855	3040/3210	Note 1
171	NA	2-dr Cpe-3P	1628	2975	Note 1
171	NA	2-dr Conv Brghm-6P	2021	3220	Note 1

NOTE 1: Series production was 49,276 Super Sixes and 5,076 Super Eights with no breakouts.
NOTE 2: Approximately 1,462 Super Six convertible broughams were built.
NOTE 3: Sedan and club coupe were Super Eights only
NOTE 4: Data above slash for six/below slash for eight.

1947 Hudson Super Six two-door sedan

HUDSON COMMODORE — SERIES 172 SIX — SERIES 174 EIGHT: Standard equipment for Commodores included herringbone weave upholstery; electric clock; air foam seat cushions; rear seat center armrest in sedans; cigarette lighter; chrome window finish moldings; instrument dial dimmer (also used 1946); carpet insert rubber front floor mats; rubber trunk mat; leather robe hangers in sedans; side window reveal moldings; rear window bars; auxiliary belt moldings; new hood top ornament with plastic crest; hoodside ornaments; bumper bar wing extensions for front and rear; 18 inch Deluxe steering wheel with horn ring; crank type ventipanes; carryover window glass construction; Deluxe type twin taillamps; and front fender lamps. Commodores also used the new, chrome trunk medallion and heavier molding around the grille top medallion. They again came standard with 6.50 x 15 tires and large hubcaps, which were optional equipment on Supers.

COMMODORE SIX (SERIES 172)/COMMODORE EIGHT (SERIES 174)

Model Number	Model Number	Body Type & Seating	Factory Price	Shipping Weight	Production Total
172	174	4-dr Sed-6P	1896/1972	3175/3330	Note 1
172	174	2-dr Clb Cpe-6P	1887/1955	3090/3260	Note 1
NA	174	2-dr Conv Brghm-6P	2196	3435	Note 1

NOTE 1: Series production was 25,138 Commodore Sixes and 12,593 Commodore Eights with no breakouts were provided.
NOTE 2: Approximately 361 Commodore Eight convertible broughams were built.
NOTE 3: Convertible brougham came only as an eight.
NOTE 4: Data above slash for six/below slash for eight.

ENGINES
(SIX-CYLINDER) Inline. L-head six-cylinder. Chrome alloy block. Displacement: 212 cid. Bore and stroke: 3 x 5 inches. Compression ratio: 6.5:1. Brake hp: 103 at 4000 rpm. Three main bearings. Solid valve lifters. Carburetor: Carter two-barrel WDO type Model 501S.

(EIGHT-CYLINDER) Inline. L-head eight-cylinder. Chrome alloy block. Displacement: 254 cid. Bore and stroke: 3 x 4-1/2 inches. Compression ratio: 6.5:1. Brake hp: 128 at 4200 rpm. Five main bearings. Solid valve lifters. Carburetor: Carter two-barrel WDO type Model 502S.

CHASSIS FEATURES: Wheelbase (all): 121 inches. Overall length (all): 207-3/8 inches. Front tread (all): 56-5/16 inches. Rear tread (all): 59-1/2 inches. Tires: (Supers) 6.00 x 16; (Commodores) 6.50 x 15.

OPTIONS: Air foam seat cushions, front and rear ($17); front only ($9). Front and rear bumper bar extensions ($20). Electric clock ($14). Fendertop lamps ($16). Horn ring with standard 17 inch steering wheel ($6). Large hubcaps ($9). Custom 18 inch steering wheel with horn ring ($19). Right side inside sun visor ($3). Oversize 6.50 x 15 tires with four large hubcaps ($28). Right fender door armrest ($4). Directional indicators for Super ($26). Red or cream wheel colors (no cost). Police and taxi equipment including large clutch; heavy rear springs; 11 inch brakes and heavy construction seats ($11). Radio ($77). Heavy scale front and rear springs for sixes (no cost). Heavy scale rear springs (no cost). Weather Master heater ($50). Chrome wheel trim rings ($13). Three-quarter leather trim ($32-53 extra per body style). Three-quarter leather grain trim ($25-41 per body style). [Note: Above trim options include special blue-gray Shadow Weave Cloth on Supers or special blue-gray Bedford Cloth on Commodores.] Full-leather upholstery with Air Foam seat cushions and an extra sun visor was standard in convertibles. Factory Town delivered prices included three-quarter leather grain trim. Royal red finish ($23). Nepal Ivory finish ($60). Two-tone finish ($18). Eight tire options were available at exchange prices from $14.60 to $72.80. Available sizes included 6.00 x 16 and 7.00 x 15 and all could be had in four or six-ply construction on specific body styles and models. Many tire options included four large hubcaps. Whitewall tires reappeared late in the year. Vacumotive Drive ($40). Drive-Master, including Vacumotive Drive ($98). Oil bath air cleaner ($3). Eight-cylinder Power Dome ($10). Overdrive manual transmission ($88). Combination fuel and vacuum pump ($7). Rear axle with 4-1/9:1 gear ratio (no cost). Rear axle with 4-5/9:1 gear ratio (no cost). Note: First 10 equipment items listed above standard on Commodore, optional on Super.

HISTORICAL FOOTNOTES: Sales of the 1947 Hudson line began in December 1946. Model year totals peaked at approximately 95,000 units. Calendar year sales were 100,393 cars. Hudson was rated the 13th largest producer in the United States. The company built its 3,000,000 car this season, which was proudly displayed and photographed besides a 1909 Hudson which was part of the factory's antique car collection. This latest Hudson milestone was a fancy Commodore Eight convertible photographed wearing white sidewall tires. Six prototype wooden bodied station wagons were built for special use on the factory grounds.

1948 HUDSON

HUDSON SUPER — SERIES 481 SIX — SERIES 483 EIGHT: In November 1947, Hudson introduced a completely new line of slab-sided cars with the famous 'Step Down' design. They were sleek, low and aerodynamic with unit-body construction and many advanced body engineering features. Standard equipment for Super series models included striped Bedford Cord upholstery; gray salt and pepper carpet-like rubber front floor mats; rear carpeting; cord robe hangers in sedans; dark mahogany woodgrained dash on 1948 Super models; window garnish moldings; 30-hour wind-up clocks; 17 inch steering wheel; adjustable sun visors; armrests at ends of all seats; latch type ventipanes; wing type rear quarter windows in sedans; side window reveal moldings; full opening rear quarter windows in club coupes; front parking lamps; twin standard taillamps; carryall trunk with horizontal mount spare tire; luggage compartment floor mat; and hubcaps on Supers.

1948 Hudson Commodore four-door sedan, 8-cyl

VEHICLE I.D. NUMBERS: Serial numbers were on the right front door post. Engine numbers were the same and they were found on the upper front right-hand side of six-cylinder blocks or between the first and second exhaust manifold flanges on eight-cylinder blocks. The first three symbols corresponded to the series/model code, followed by a group of numbers beginning at 101. Super Sixes were numbered 481101 to 481117300; Commodore Sixes 482101 to 482117301; Super Eights 483101 to 48311786 and Commodore Eights 484101 to 484117256. Body Style Numbers not used through 1955. First columns in charts below show six-cylinder model numbers; second columns show eight-cylinder model numbers.

HUDSON SUPER SIX (SERIES 481)/SUPER EIGHTS (SERIES 483)

Model Number	Model Number	Body Type & Seating	Factory Price	Shipping Weight	Production Total
481	483	4-dr Sed-6P	2222/2343	3500/3535	Note 1
481	NA	2-dr Brghm-6P	2172	3470	Note 1
481	483	2-dr Clb Cpe-6P	2219/2340	3480/3495	Note 1
481	NA	2-dr Clb Cpe-3P	2069	3460	Note 1
481	NA	2-dr Conv Brghm-6P	2836	3750	Note 1

NOTE 1: Series production was 49,388 Super Sixes and 5,338 Super Eights with no breakouts.
NOTE 2: Approximately 86 Super Six convertible broughams were built.
NOTE 3: Only the sedan and club coupe came as Super Eights.
NOTE 4: Data above slash for six/below slash for eight.

HUDSON COMMODORE — SERIES 482 SIX — SERIES 484 EIGHT: Standard specifications for the new 'Step Down' Commodores included Broadcloth upholstery (tan with green stripes or gray with blue stripes); air foam seat cushions; taupe colored carpet-like rubber front floor mats; rear compartment carpeting; cloth covered sedan robe hangers; 16 inch rear seat center armrest in club coupe; cigarette lighter; dark walnut and blonde grained instrument panel (two-tone); instrument panel dial dimmer; walnut grain window garnish moldings; twin adjustable swiveling sun visors; Plastic rimmed 18 inch Deluxe steering wheel with horn ring; electric clock; side window reveal moldings; crank type front door ventilating wings; wing type rear quarter window ventilation in sedans; full-opening rear quarter windows in club coupe; rubber trunk mat and Deluxe twin taillamps; and front parking lamps.

HUDSON COMMODORE SIX (SERIES 482)/COMMODORE EIGHT (SERIES 484)

Model Number	Model Number	Body Type & Seating	Factory Price	Shipping Weight	Production Total
482	484	4-dr Sed-6P	2399/2514	3540/3600	Note 1
482	484	2-dr Clb Cpe-6P	2374/2490	3550/3570	Note 1
482	484	2-dr Brghm Cnv-6P	3057/3138	3780/3800	Note 1

NOTE 1: Series production was 27,159 Commodore Sixes and 35,315 Commodore Eights with no breakouts.
NOTE 2: Approximately 49 Commodore Six convertibles and 65 Commodore Eight convertibles were built.
NOTE 3: Data above slash for six/below slash for eight.

ENGINES

(SIX-CYLINDER) Inline. L-head six-cylinder. Chrome alloy block. Displacement: 262 cid. Bore and stroke: 3-9/16 x 4-3/8 inches. Compression ratio: 6.5:1. Brake hp: 121 at 4000 rpm. Four main bearings. Solid valve lifters. Carburetor: Carter two-barrel WDO type Model 647S.

(EIGHT-CYLINDER) Inline. L-head eight-cylinder. Chrome alloy block. Displacement: 254 cid. Bore and stroke: 3 x 4-1/2 inches. Compression ratio: 6.5:1. Brake hp: 128 at 4200 rpm. Five main bearings. Solid valve lifters. Carburetor: Carter two-barrel WDO type Model 647S.

CHASSIS FEATURES: Wheelbase (all) 124 inches. Overall length: (all) 207-1/2 inches. Front tread (all) 58-1/2 inches. Rear tread (all); 55-1/2 inches. Tires: (all) 7.10 x 15.

OPTIONS: Red or cream wheelcovers (no cost). Front fender ornaments on Supers ($6). Front bumper guard on Supers ($13). Convertible Brougham top rear window glass ($21). Rear window reveal moldings on Supers ($4). Weather Control heater ($64). Radio ($84). Foam rubber front seatback ($16). Directional indicators ($20). Commodore steering wheel on Supers ($20). Foam rubber seat cushions on Supers ($28). Wheel trim rings ($13). Hydraulic window regulators on Super Six convertibles ($63)) Large hubcaps ($10). Foam rubber front seat cushions on Supers ($14). Electric clock on Supers ($17). Leather trim options for closed cars were available at prices between $83 and $145, with the cost depending on body style. Leather trims came in Russet, gray or dark red colors. Convertible Broughams came standard with 7.60 x 15 tires, antique grain Maroon leather trim and hydraulic window regulators (except Super Six). Brown cloth and maroon leather or gray cloth and maroon leather trims were no cost convertible options. Convertible tops came in black, gray or maroon. Specific upholstery trim and top colors were recommended with certain exterior body colors, although variations were possible. Note: Equipment choices specifically listed above as Super Series options were standard on Commodores. Tire options included white sidewalls, oversized 7.60 x 15 tires, and extra-ply construction. Six-cylinder aluminum cylinder head ($11). Eight-cylinder aluminum cylinder head ($13). Vacumotive Drive ($44). Drive-Master, including Vacumotive Drive ($112). Oil bath air cleaner, six ($6); Eight ($8). Overdrive manual transmission ($101). Vacuum booster pump ($9). Standard rear axle gear ratio was 4.1:1 on all models. Optional 4-5/9:1 or 4.3:1 gear ratios were available at no extra cost. With Overdrive transmission, the 4-5/9:1 rear axle was standard and 4.1:1 or 4.3:1 axles were no cost options. The 4.1:1 gear ratio rear axle was used in conjunction with DriveMaster.

HISTORICAL FOOTNOTES: The new styling introduced on 1948 models was created by a group of Hudson designers under the direction of Frank Spring. The 1948 line was offered for sale in December 1947. Model year sales hit the 117,200 unit level, while calendar year sales peaked at 142,454 cars. Hudson was ranked as America's 11th largest maker this season. The Vacumotive system automatically controlled the operation of the clutch. The Drive-Master system automatically controlled both clutching and gear shifting operations. Hudson dealers in New York City began one of the first television automobile advertising campaigns in 1947. Some body styles were not available at the beginning of the season. Four-door sedans, brougham sedans and club coupes appeared first in late 1947. The convertible brougham did not show up until August 1948.

1949 HUDSON

HUDSON SUPER — SERIES 491 SIX — SERIES 493 EIGHT: There were very minor annual revisions in the 1949 Hudson models and all were found inside the cars. The Super Series standard equipment list included all previous features, plus non-glare dashboard top finish and leather grained trim on the following items: valance panels under windows; kick pads on all doors; rear quarter panels of broughams and club coupes; recessed panel shelves; and top of armrests. Cloth covered robe hangers were now used in all models, except the three-passenger coupe. A new, ribbed type front rubber floor mat with simulated carpet pattern was seen.

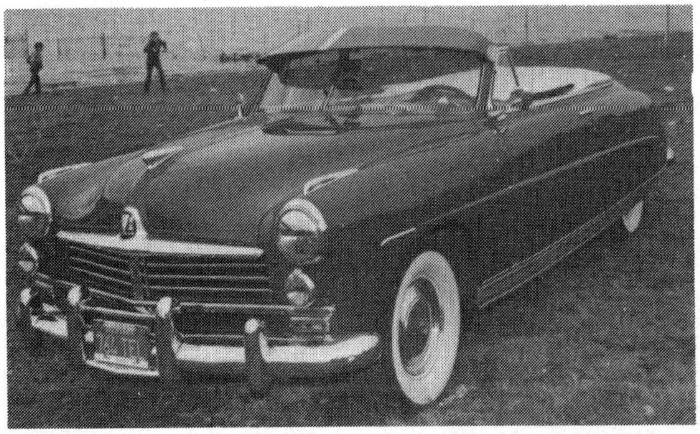

1949 Hudson, Super Six two-door convertible brougham, 6-cyl

VEHICLE I.D. NUMBERS: Serial numbers were on the right front door post. Engine numbers were the same and they were found on the upper front right-hand side of six-cylinder blocks or between the first and second exhaust manifold flanges on eight-cylinder blocks. The first three symbols changed to correspond to new model numbers. Super Sixes were numbered 491-101 to 491159201; Commodore Custom Sixes 492-101 to 492159081; Super Eights 493-101 to 49315919 and Commodore Custom Eights 494-101 to 494159159. Body Style Numbers not used through 1955. First columns in charts below show six-cylinder model numbers; second columns show eight-cylinder model numbers.

HUDSON SUPER SIX (SERIES 491)/SUPER EIGHT (SERIES 493)

Model Number	Model Number	Body Type & Seating	Factory Price	Shipping Weight	Production Total
491	493	4-dr Sed-6P	2207/2296	3555/3565	Note 1
491	493	2-dr Brghm-6P	2156/2245	3515/3545	Note 1
491	493	2-dr Clb Cpe-6P	2203/2292	3480/3550	Note 1
491	NA	2-dr Cpe-3P	2053	3485	Note 1
491	493	2-dr Conv Brghm-6P	2799	3750	Note 1

NOTE 1: Production was 91,333 Super Sixes and 6,365 Super Eights with no breakouts.
NOTE 2: Approximately 1,868 Super Six convertible broughams were built.
NOTE 3: Sedan, brougham sedan and club coupe only came as Super Eights.
NOTE 4: Data above slash for six/below slash for eight.

HUDSON COMMODORE CUSTOM — SERIES 492 SIX — SERIES 494 EIGHT: Upper series models were now called Commodore Customs and looked identical to 1948 Commodores on the outside. The standard equipment list was essentially the same, with the following minor changes: a new brown front floor mat was made of rubber and had a ribbed, simulated carpet pattern. Non-glare finish was used on top of the instrument panel. Envelope type pockets were now used on the front seatbacks of convertibles, instead of only on sedans. Leather graining was seen on the door kick pads and rear quarter panels of broughams and club coupes. Two large parcel compartments, with locks, were now incorporated at each side of the Commodore Custom dashboard, the left-hand locker being new. As usual, Commodore Customs had such distinguishing features as bumper guards; metal hand rails on back of front seat; rear window reveal moldings; and 18 inch Deluxe steering wheel with full circle horn ring. Much of this could be ordered, as optional equipment, on Supers.

COMMODORE SIX (SERIES 492)/COMMODORE CUSTOM EIGHT (SERIES 494)

Model Number	Model Number	Body Type & Seating	Factory Price	Shipping Weight	Production Total
492	494	4-dr Sed-6P	2383/2472	3625/3650	Note 1
492	494	2-dr Clb-Cpe-6P	2359/2448	3585/3600	Note 1
492	494	2-dr Conv Brghm-6P	2952/3041	3780/3800	Note 1

NOTE 1: Series production was 32,715 Commodore Custom Sixes and 28,687 Commodore Custom Eights with no breakouts.
NOTE 2: Approximately 656 Commodore Custom Six convertibles and 596 Commodore Custom Eight convertibles were built.
NOTE 3: Data above slash for six/below slash for eight.

ENGINES

(SIX-CYLINDER) Inline. L-head six-cylinder. Chrome alloy block. Displacement: 262 cid. Bore and stroke: 3-9/16 x 4-3/8 inches. Compression ratio: 6.5:1. Brake hp: 121 at 4000 rpm. Four main bearings. Solid valve lifters. Carburetor: Carter two-barrel WDO type Model 647S.

(EIGHT-CYLINDER) Inline. L-head eight-cylinder. Chrome alloy block. Displacement: 254 cid. Bore and stroke: 3 x 4-1/2 inches. Compression ratio: 6.5:1. Brake hp: 128 at 4200 rpm. Five main bearings. Solid valve lifters. Carburetor: Carter two-barrel WDO type Model 647S.

CHASSIS FEATURES: Wheelbase (all) 124 inches. Overall length: (all) 207-1/2 inches. Front tread (all) 58-1/2 inches. Rear tread (all); 55-1/2 inches. Tires: (all) 7.10 x 15.

OPTIONS: Red or cream wheelcovers (no cost). Front fender ornaments on Supers ($6). Front bumper guard on Supers ($13). Convertible Brougham top rear window glass ($21). Rear window reveal moldings on Supers ($4). Weather Control heater ($64). Radio ($84). Foam rubber front seatback ($16). Directional indicators ($20). Commodore steering wheel on Supers ($20). Foam rubber seat cushions on Supers ($28). Wheel trim rings ($13). Hydraulic window regulators on Super Six convertibles ($63)) Large hubcaps ($10). Foam rubber front seat cushions on Supers ($14). Electric clock on Supers ($17). Leather trim options for closed cars were available at prices between $83 and $145, with the cost depending on body style. Leather trims came in Russet, gray or dark red colors. Convertible Broughams came standard with 7.60 x 15 tires, antique grain Maroon leather trim and hydraulic window regulators (except Super Six). Brown cloth and maroon leather or gray cloth and maroon leather trims were no cost convertible options. Convertible tops came in black, gray or maroon. Specific upholstery trim and top colors were recommended with certain exterior body colors, although variations were possible. Note: Equipment choices specifically listed above as Super Series options were standard on Commodores. Tire options included white sidewalls, oversized 7.60 x 15 tires, and extra-ply construction. Six-cylinder aluminum cylinder head ($11). Eight-cylinder aluminum cylinder head ($13). Vacumotive Drive ($44). Drive-Master, including Vacumotive Drive ($112). Oil bath air cleaner, six ($6); Eight ($8). Overdrive manual transmission ($101). Vacuum booster pump ($9). Standard rear axle gear ratio was 4.1:1 on all models. Optional 4-5/9:1 or 4.3:1 gear ratios were available at no extra cost. With Overdrive transmission, the 4-5/9:1 rear axle was standard and 4.1:1 or 4.3:1 axles were no cost options. The 4.1:1 gear ratio rear axle was used in conjunction with DriveMaster.

HISTORICAL FOOTNOTES: The 1949 Hudson line was introduced to the public in November 1948. Hudson retained its 11th rank in industry with model year sales of 159,100 cars and calendar year sales of 142,462 units. The firm celebrated its 40th anniversary this season.

1950 HUDSON

1950 Hudson, Pacemaker four-door sedan, 6-cyl

HUDSON PACEMAKER — SERIES 500 — SERIES 50A DELUXE: A shorter Hudson was Hudson's new 'baby' in 1950. This Pacemaker was not much smaller than conventional styles, but seemed to be back then. It had the season's new look, which included a grille with four horizontal blades widening as they neared the bumper and twin struts forming a triangle with a company medallion at the top. Though only the grille was drastically changed, all the cars seemed lower. Basic equipment on all models included Durafab plastic interior trims; 18 inch steering wheel; twin, adjustable visors; full-opening rear quarter windows for club coupes; new, two-piece curved Full-View windshield; front dome lamp; lockable parcel compartment; large trunk with mat and horizontal spare and; a rearview mirror. Several Pacemaker features, including lighted grille/hood medallions; standard twin taillamps; latch type front ventipanes; and a new streamlined hood ornament, were shared with Supers. Distinctive Pacemaker equipment included striped, Bedford Cord upholstery; front and rear rubber floor mats; Blue Spruce two-spoke steering wheel; fabric finish dash; seatback pockets in sedans only; ashtrays in front seatback and dash; trumpet horn; and parking lamps under the lower grille bar. Deluxe Pacemakers had a bit of extra trim and slightly richer appointments.

VEHICLE I.D. NUMBERS: Serial numbers were on the right front door post. Engine numbers were the same and they were found on the upper front right-hand side of six-cylinder blocks or between the first and second exhaust manifold flanges on eight-cylinder blocks. Serial numbers followed the same system. The first three symbols correspond with new model numbers. Pacemaker 500s were numbered 500-101 to 500121481; Pacemaker Deluxes 50A-101 to 50A121505; Super Sixes 501-101 to 501121508; Commodore Sixes 502-101 to 502121504; Super Eights 503-101 to 503121491 and Commodore Eights 504-101 to 504121500. Body Style Numbers not used through 1955. First columns in charts below show Pacemaker 500 or six-cylinder model numbers; second columns show Pacemaker Deluxe or eight-cylinder model numbers.

HUDSON PACEMAKER 500 (SERIES 500)/PACEMAKER DELUXE (SERIES 50A)

Model Number	Model Number	Body Type & Seating	Factory Price	Shipping Weight	Production Total
500	50A	4-dr Sed-6P	1933/1959	3510/3520	Note 1
500	50A	2-dr Brghm-6P	1912/1928	3475/3485	Note 1
500	50A	2-dr Clb Cpe-6P	1933/1959	3460/3470	Note 1
500	50A	2-dr Cpe-3P	1807	3445	Note 1
500	50A	2-dr Conv Brghm-6P	2428/2444	3655/3665	Note 1

NOTE 1: Series production was 39,455 Pacemaker 500s and 22,297 Pacemaker Deluxes with no breakouts available.
NOTE 2: Approximately 1,865 Pacemaker 500 convertibles were built.
NOTE 3: Approximately 660 Pacemaker Deluxe convertibles were built.
NOTE 4: The three-passenger coupe came only as a Pacemaker 500.
NOTE 5: Data slash for Pacemaker 500/below slash for Pacemaker Deluxe.

HUDSON SUPER — SERIES 501 SIX — SERIES 503 EIGHT: Supers were basically a 1949 carryover with the newly designed 1950 grille appearance. A small 'spear tip' ornament at the front of the body contour line, just above the wheel opening, served as a Super identifier. A broad sill panel molding was used as the only major bodyside trim. No fendertop ornaments were seen and the new, streamlined hood ornament matched that on Pacemakers. The Super equipment list comprised all basic features and items shared with Pacemakers, plus striped Bedford cloth upholstery; two-tone woodgrained dash; wind-up clock; Cord robe hangers; light tan steering wheel; and door pillar assist straps. In addition, the following items were shared with Custom Commodores: ribbed carpet-like front mats; rear carpets; armrests at seat ends; sedan rear ventipanes; bright metal windows and windshield reveal moldings; larger sedan and brougham rear window; parking lamps in lower grille bar; license lamps in center rear bumper guards; fender skirts; twin air horns, ashtrays in seat ends; and dashboard and envelope style seatback pockets in all models.

HUDSON SUPER SIX (SERIES 501)/SUPER EIGHT (SERIES 503)

Model Number	Model Number	Body Type & Seating	Factory Price	Shipping Weight	Production Total
501	503	4-dr Sed-6P	2105/2189	3590/3605	Note 1
501	503	2-dr Brghm-6P	2068/2152	3565/3575	Note 1
501	503	2-dr Clb Cpe-6P	2102/2186	3555/3560	Note 1
501	NA	2-dr Conv Brghm-6P	2629	3750	Note 1

NOTE 1: Series production included 17,246 Super Sixes and 1,074 Super Eights. No body style breakouts were provided.
NOTE 2: Approximately 464 Super Six Convertible Broughams were built.
NOTE 3: Prices and shipping weights above slashes are for sixes, below slash for eights.

HUDSON CUSTOM COMMODORE — SERIES 502 SIX — SERIES 504 EIGHT: Commodores had upper level trim and enriched interiors. Four bumpers guards were seen, as were front fendertop ornaments. Side trim consisted of the broad body sill panel and a strip of molding that followed the body contour line several inches below it. At the front of this molding were model nameplates, while the rear portion widened and curved into the sill panel behind the enclosed rear wheelhousing. The Custom Commodore equipment list included all basic features and the additional items shared with Supers, plus nylon Bedford Cord upholstery (in tan with brown stripes or blue-gray with blue stripes); foam rubber seat cushions; Durafab covered robe hangers; bright metal seatback hand grips; 16 inch rear seat center armrest; pop-out cigarette lighter; dash dimmer switch; leather grain dash and window garnish molding finish; three-spoke steering wheel; electric clock; crank type ventipanes; two rear dome lights (in sedans and club coupes); and inner and outer bumper guards, front and rear. Basic items on Custom Commodores were slightly upgraded. For example, the rearview mirror was an extra large Deluxe type.

CUSTOM COMMODORE 6 (SERIES 502)/CUSTOM COMMODORE 8 (SERIES 504)

Model Number	Model Number	Body Type & Seating	Factory Price	Shipping Weight	Production Total
502	504	4-dr Sed-6P	2282/2366	3655/3675	Note 1
502	504	2-dr Clb Cpe-6P	2257/2341	3640/3575	Note 1
502	504	2-dr Conv Brghm-6P	2809/2893	3840/3865	Note 1

NOTE 1: Series production included 24,605 Custom Commodore Sixes and 16,731 Custom Commodore Eights with no body style breakouts provided.
NOTE 2: Approximately 700 Custom Commodore Six convertibles and 426 Custom Commodore Eight convertibles were built.
NOTE 3: Factory prices and shipping weights above slashes are for sixes, below slash for eights.

ENGINES

(PACEMAKER SIX) Inline. Six-cylinder. Flathead. Chrome alloy block. Displacement: 232 cid. Bore and stroke: 3-9/16 x 3-7/8 inches. Compression ratio: 6.7:1. Brake hp: 112 at 4000 rpm. Four main bearings. Solid valve lifters. Carburetor: Carter one-barrel WA-1 type Model 749S.

(SUPER/COMMODORE SIX) Inline. Six-cylinder. Flathead. Chrome alloy block. Displacement: 262 cid. Bore and stroke: 3-9/16 x 4-3/8 inches. Compression ratio: 6.7:1. Brake hp: 123 at 4000 rpm. Four main bearings. Solid valve lifters. Carburetor: Type WGD Model 776S with L-shaped air horns.

(SUPER/COMMODORE EIGHT) Inline. eight-cylinder. Flathead. Chrome alloy block. Displacement: 254 cid. Bore and stroke: 3 x 3-1/2 inches. Compression ratio: 6.7:1. Brake hp: 128 at 4200 rpm. Five main bearings. Solid valve lifters. Carburetor: Carter Type WGD Model 773S with L-shaped air horns.

1950 Hudson, Commodore Eight four-door sedan, 8-cyl

CHASSIS FEATURES: Wheelbase: (Pacemaker) 119 inches; (other models) 124 inches. Overall length: (Pacemaker) 201-1/2 inches; (other models) 208-3/32 inches. Front tread: (all) 58-1/2 inches. Rear tread: (all) 55-1/2 inches. Tires: (convertibles) 7.60 x 15; (all others) 7.10 x 15.

OPTIONS: Foam rubber Pacemaker seat cushions. Rear wheelcovers (fender skirts) for Pacemaker. Mechanical or electric clock in Pacemaker. Pacemaker front bumper outer guards. White sidewall tires. Super front bumper outer guards. Foam rubber cushions in Supers. Super side ornamentation. Hydraulic window lifts for Super convertible. Radio. Heater. Overdrive manual transmission ($95). Drive Master semi-automatic transmission ($105). Supermatic ($199). Aluminum cylinder head with 7.2:1 compression ratio. Rear axle ratios including 4.10:1 (standard), 4.55:1 (standard Supermatic) or 3.82:1 gears at no extra cost. Oil bath air cleaner. Vacuum booster pump.

HISTORICAL FOOTNOTES: The new Hudsons were introduced on Nov. 18, 1949. The firm slid to 13th rank in the American industry, with model year sales of 121,408 cars; calendar year output of 143,586 units. Prices were slightly reduced from the previous year and Hudson reported a $12 million profit on sales of $267 million. Canadian production, suspended during war, was resumed at the Hudson factory in Tilbury, Ontario, April, 1950. The convertible brougham, in both Commodore and Hornet series, came with hydraulic windows and leather trim. The same upholstery was featured on Super Series convertibles, but hydraulic window lifts were optional. Convertible top colors were tan, black or maroon. A 'Fold Away' rear window was optional with all convertibles.

1951 HUDSON

1951 Hudson, Pacemaker four-door sedan, 6-cyl

PACEMAKER CUSTOM — SIX — SERIES 4A: The Hudson grille was changed again this year. It now had three horizontal blades. The top two were bowed to meet the bottom bar. A twin-strut triangle was seen near the center. Rectangular parking lights were housed outboard of the main grille bars on either side. The lamp housings were slightly rounded where they wrapped around the Pacemaker body corners, but looked squarer on other models. The sides of Pacemakers were trimmed only by 'spear tip' ornaments (without spears) and broad lower sill panels that stretched, from behind the front wheel opening, to the extreme rear of the car. Standard on Pacemakers were Hudson basics like Twin-Contour wipers; gas gage; Teleflash 'idiot lights;' water temperature gage; windshield/defroster vents; Cushion-Action door latches; theft-proof locks; push-button door handles; windshield and side window reveals; dash ashtray; rearview mirror; twin sun visors; full opening crank-out rear quarter windows in club coupes and broughams; twin stop and taillamps; front dome light; lockable parcel compartment; twin air horns; and illuminated grille medallion. Pacemakers had the rear ashtray in the front seatback and door pillar assist straps in brougham sedans. Upholstery was gray special-weave cord with red and brown stripes and Dura-fab plastic trim.

VEHICLE I.D. NUMBERS: Serial numbers were on the right front door post. Engine numbers were the same and they were found on the upper front right-hand side of six-cylinder blocks or between the first and second exhaust manifold flanges on eight-cylinder blocks. Serial numbers followed the same system. The first two symbols correspond with new model designations and were a number and letter. Pacemaker Customs were numbered 4A-1001 to 132072; Super Customs 5A-1001 to 132246; Commodore Customs (Six) 6A-1001 to 132586; Hornets 7A-1001 to 132915 and Commodore Customs (Eight) 8A-1001 to 132028. Body Style Numbers not used through 1955. First columns in charts below show Pacemaker 500 or six-cylinder model numbers; second columns show Pacemaker Deluxe or eight-cylinder model numbers.

PACEMAKER CUSTOM SIX (SERIES 4A)

Model Number	Model Number	Body Type & Seating	Factory Price	Shipping Weight	Production Total
4A	NA	4-dr Sed-6P	2145	3460	Note 1
4A	NA	2-dr Brghm-6P	2102	3430	Note 1
4A	NA	2-dr Clb Cpe-6P	2145	3410	Note 1
4A	NA	2-dr Cpe-3P	1964	3380	Note 1
4A	NA	2-dr Conv Brghm-6P	2642	3600	Note 1

NOTE 1: Series production included 34,495 Pacemaker Customs. No body style breakouts were provided.
NOTE 2: Approximately 425 Pacemaker convertibles were built.

SUPER CUSTOM — SIX — SERIES 5A: For 1951, Supers were given the new frontal treatment and basically the type of side trim used on 1950 Commodores, but without standard outer grille guards. Small hubcaps were seen. The regular assortment of equipment was identical to that listed for Pacemakers, with only a few exceptions. Variations included rear ashtrays housed in recess panels on the doors and inner rear quarter panels (instead of front seatback) and wing-type ventipanes for sedan rear quarter windows. Upholstery was in tan Bedford cloth with brown and maroon stripes. A new Hollywood model, with two-door pillarless hardtop styling, was introduced in September of 1951 as a late-year addition to the line. A new, rounded corner traphazoid shaped front center grille guard was seen on all 'big' Hudsons, including the Supers.

SUPER CUSTOM SIX (SERIES 5A)

Model Number	Model Number	Body Type & Seating	Factory Price	Shipping Weight	Production Total
5A	NA	4-dr Sed-6P	2287	3565	Note 1
5A	6A	2-dr Brghm-6P	2238	3535	Note 1
5A	NA	2-dr Clb Cpe-6P	2287	3525	Note 1
5A	NA	2-dr Holly HT-6P	2605	3590	Note 1
5A	NA	2-dr Conv Brghm-6P	2827	3720	Note 1

NOTE 1: Series production included 22,532 Super Custom Sixes. No body style breakouts were provided.
NOTE 2: Approximately 1,100 Hollywood two-door hardtops and 282 convertibles were built in this series.

COMMODORE CUSTOM — SERIES 6A SIX — SERIES 8A EIGHT: No longer the flagship of the Hudson fleet, the Commodore Six was priced under a new Hudson Hornet line, while Commodore Eights were marketed at equal-to-Hornet prices. Distinguishing Commodores from lower series were larger front fender nameplates; outer grille guards front and rear; metal hand grips on front seatbacks; rear window reveal moldings and three dimensional weave upholstery with stripes and Antique Crush Dura-fab trim. These features were also used in and on Hornets. The balance of equipment was the same found on Super Customs, plus a 16 inch rear center armrest to provide a two-person seating arrangement. The Commodore convertible came in nine standard or four extra cost colors with dark red or blue genuine top grain leather upholstery and harmonizing leather grain trim. It again had hydraulic window lifts and a hydraulic roof with tan, black or maroon top material. A large, plastic rear window was optional.

CUSTOM COMMODORE SIX (SERIES 6A)/EIGHT (SERIES 8A)

Model Number	Model Number	Body Type & Seating	Factory Price	Shipping Weight	Production Total
6A	8A	4-dr Sed-6P	2480/2568	3600/3620	Note 1
6A	8A	2-dr Clb Cpe-6P	2455/2543	3585/3600	Note 1
6A	8A	2-dr Holly HT-6P	2780/2869	3640/3650	Note 1
6A	8A	2-dr Conv Brghm-6P	3011/3099	3785/3800	Note 1

NOTE 1: Series production included 16,979 Custom Commodore Sixes and 14,243 Custom Commodore Eights. No body style breakouts were provided.
NOTE 2: In the Custom Commodore Six series, approximately 819 Hollywoods and 211 convertibles were built.
NOTE 3: In the Custom Commodore Eight series, approximately 669 Hollywoods and 181 convertibles were built.
NOTE 4: Factory prices and shipping weights above slashes are for sixes, below slash for eights.

HUDSON HORNET — SIX — SERIES 7A: The first of the famed Hudson Hornets was really a Commodore with a special high-performance six and a few distinctive identification and appointment details. Special features included a gold and chrome plated 'Skyliner Styling' hood mascot; pillar assist straps in coupes and sedans; Deluxe robe hanger, hand grips and tailored pockets on back of the lounge-wide front seat; Hornet H-145 type medallions in each front door valance panel; indirectly lighted precision instruments set into a polished chrome dash housing on a leather grained panel with non-glare Dura-fab top; and gleaming, rocketship-shaped "Badges of Power" in front of the bodyside rub moldings and on the trunk. These badges showed a rocket piercing two vertically angled bars, with Hornet lettering turning them into a letter 'H'. Upholstery was the Commodore type and came in tan-brown with gold stripes or blue-gray with blue stripes. Antique Crush type leather grained Dura-fab trim combinations were used. The high-compression, aluminum 'Power-Dome' cylinder head was standard on the Hornet engine, but the regular iron alloy head was a no-cost option.

HUDSON HORNET H-145 SIX (SERIES 7A)

Model Number	Model Number	Body Type & Seating	Factory Price	Shipping Weight	Production Total
7A	NA	4-dr Sed-6P	2568	3600	Note 1
7A	NA	2-dr Clb Cpe-6P	2543	3580	Note 1
7A	NA	2-dr Holly HT-6P	2869	3630	Note 1
7A	NA	2-dr Conv Brghm-6P	3099	3780	Note 1

NOTE 1: Series production included 43,666 Hudson Hornets. No body style breakouts were provided.
NOTE 2: Approximately 2,101 Hollywood two-door hardtops and 551 convertibles were built in this series.

ENGINES

(PACEMAKER SIX) Inline. Six-cylinder. Flathead. Chrome alloy block. Displacement: 232 cid. Bore and stroke: 3-9/16 x 3-7/8 inches. Compression ratio: 6.7:1. Brake hp: 112 at 4000 rpm. Four main bearings. Solid valve lifters. Carburetor: Carter one-barrel WA-1 type Model 749S.

(SUPER/COMMODORE SIX) Inline. Six-cylinder. Flathead. Chrome alloy block. Displacement: 262 cid. Bore and stroke: 3-9/16 x 4-3/8 inches. Compression ratio: 6.7:1. Brake hp: 123 at 4000 rpm. Four main bearings. Solid valve lifters. Carburetor: Type WGD Model 776S with L-shaped air horns.

(COMMODORE EIGHT) Inline. eight-cylinder. Flathead. Chrome alloy block. Displacement: 254 cid. Bore and stroke: 3 x 3-1/2 inches. Compression ratio: 6.7:1. Brake hp: 128 at 4200 rpm. Five main bearings. Solid valve lifters. Carburetor: Carter Type WGD Model 773S with L-shaped air horns.

(HORNET SIX) Inline. L-head six. Chrome alloy block. Displacement: 308 cid. Bore and stroke: 3-13/16 x 4-1/2 inches. Compression ratio: 7.2:1. Brake hp: 145 at 3800 rpm. Four main bearings. Solid valve lifters. Carburetor: Carter two-barrel type WGD model 776S.

1951 Hudson, Hornet four-door sedan, 6-cyl.

CHASSIS FEATURES: Wheelbase: (4A) 119-7/8 inches; (5A, 6A, 7A and 8A) 123-7/8 inches. Overall length: (4A) 201-1/2 inches; (5A, 6A, 7A and 8A) 208-1/2 inches. Front tread: (all) 58-1/2 inches. Rear tread: (all) 55-1/2 inches. Overall width: (4A and 5A) 77-1/16 inches; (6A, 7A and 8A) 77-21/32 inches. Tires: (5A, 6A 7A and 8A convertibles) 7.60 x 15; (4A convertible and all other models) 7.10 x 15.

OPTIONS: Foam rubber Pacemaker seat cushions. Rear wheelcovers (fender skirts) for Pacemaker. Mechanical or electric clock in Pacemaker. Pacemaker front bumper outer guards. White sidewall tires. Super front bumper outer guards. Foam rubber cushions in Supers. Super side ornamentation. Hydraulic window lifts for Super convertible. Radio. Heater. Overdrive manual transmission ($100). Drive Master semi-automatic transmission for Pacemaker/Super only ($99). Supermatic, all except Hornet ($158). Hydra-Matic for Commodore or Hornet ($158). Available rear axles included units with 4.55:1, 4.10:1 and 3.58:1 gear ratios. The Power-Dome cylinder head was optional on lower series at extra cost. The standard cylinder head was optional on Hornets at no extra cost.

HISTORICAL FOOTNOTES: The new Hudson line was introduced in September 1950 and continued in extended production through January 1952. Hollywood hardtops were a late edition to the 1951 line. Model year deliveries hit 131,915 units. Calendar year sales dropped to 92,859 cars. Hudson was ranked 15th in the American industry. A loss of

$1,125,210 was reported on sales volume of $186,050,832. Labor unrest and delays in getting government authorization to raise prices, during the Korean conflict, was responsible for the poor business year. The Hudson factory help support stock car racing efforts with the new Hornets by providing special 'export' and 'severe usage' parts suitable for high-performance applications. Hudsons were able to win 12 of the 41 NASCAR Grand National contests held in 1951. Top Hudson Hornet drivers included Marshall Teague, Herb Thomas, Tim Flock and Dick Rathmann. A special dual carburetion package helped Herb Thomas take a checkered flag in the second Southern 500 Race, at Darlington, S.C., with an 86.21 mph average speed. *Motor Trend* and *Mechanix Illustrated* determined the top speed of the stock 1951 Hudson Hornet at 97 mph. Herb Thomas captured Top Driver honors on the NASCAR circuit this season.

1952 HUDSON

PACEMAKER/WASP — SIX — SERIES 4B/5B: The Pacemaker was slightly downgraded for 1952 and had a plainer look. The twin-strut grille arrangement was deleted and fender skirts were optional. The 'spear tips' had a staggered look and a boomerang shaped fin became the hood mascot. The rear end was spartan, having small oval taillamp lenses and only outer bumper guards. Standard specifications included fancy gray special-weave cord upholstery with red and brown stripes; ribbed rubber floor mats; dark brown painted dash; two-spoke light tan steering wheel; friction type front ventipanes; two assist straps in sedan, one in club coupe; pop-out cigarette lighter; dash and seatback ashtrays; windshield and window reveal moldings; twin air horns; armrests at front seat end (plus rear seat end on sedan and club coupe); seatback pockets; woven trunk mat; and 232 cid engine. Prices were up about $165 over 1951. The new Wasp Six was built off the Pacemaker platform. Wasps were an inch longer, since they came with center rear bumper guards which protruded out that much further. In terms of price, the Wasp replaced the Super. In terms of character, it was to the Pacemaker what the Hornet was to the Commodore Eight: a slightly fancier and more powerful version of the same car. (This is borne out by name changes of 1953, when the Pacemaker was renamed Wasp and the term Super Wasp was applied if the bigger motor was used.) Standard 1952 Wasp specifications included tan special-weave cord upholstery with red and brown stripes; rear compartment carpeting; dark brown leather grain dash; door courtesy lamps; wind-up clock; three-spoke steering wheel with half-circle horn ring; armrests at seat ends, except convertible and brougham rear seat; robe hanger and hand grips on front seatback; friction front ventipanes; windshield and window reveal moldings; rear center guard with license lamp; woven fabric trunk mat; fender skirts; pop-out lighter; dash ashtray (and front seatback type in sedans); distinctive Hudson triangle grille ornament; front fendertop ornaments; seatback pockets; side body rub rail moldings; and twin-strut front grille guard.

VEHICLE I.D. NUMBERS: Serial numbers were on the right front door post. Engine numbers were the same and they were found on the upper front right-hand side of six-cylinder blocks or between the first and second exhaust manifold flanges on eight-cylinder blocks. Serial numbers followed the same system. The first two symbols correspond with new model designations, the second symbol of the prefix changing to a letter 'B'. Pacemakers were numbered 132916 to 202512; Wasps 132916 to 202715; Commodore Sixes 132916 to 198220; Hornets 132916 to 202916 and Commodore Eights 132916 to 200201. Body Style Numbers not used through 1955. First columns in charts below show Pacemaker or six-cylinder model numbers; second columns show Wasp or eight-cylinder model numbers.

1952 Hudson, Wasp two-door Hollywood hardtop, 6-cyl

PACEMAKER SIX (SERIES 4B)/WASP SIX (SERIES 5B)

Model Number	Model Number	Body Type & Seating	Factory Price	Shipping Weight	Production Total
4B	5B	4-dr Sed-6P	2311/2466	3390/3485	Note 1
4B	5B	2-dr Brghm-6P	2264/2413	3355/3470	Note 1
4B	5B	2-dr Clb Cpe-6P	2311/2466	3335/3435	Note 1
4B	NA	2-dr Cpe-3P	2116	3305	Note 1
NA	5B	2-dr Holly HT-6P	2812	3525	Note 1
NA	5B	2-dr Conv Brghm-6P	3048	3635	Note 1

NOTE 1: Series 4B production included 7,486 Pacemakers; no breakouts.
NOTE 2: Series 5B production included 21,876 Wasps; no breakouts.
NOTE 3: Approximately 1,320 Hollywood two-door hardtops and 220 convertibles in Wasp series.
NOTE 4: Data above slash for 4B/below slash for 5B.
NOTE 5: The three-passenger coupe came only as a Pacemaker; the Hollywood and convertible brougham came only as a Wasp.

COMMODORE — SIX — SERIES 6B: The 1952 Commodore line featured new Hudson-Aire identification and appointment items. They included double rub rail moldings that ran along the body contour line, from the front fenders to the rear fenders, with a downward sweep towards the back bumper; twin-strut grille and arrangement; front and rear center bumper guards; front fender nameplates; rocker sill beauty panels; large Deluxe hubcaps and taillights styled to form a continuous horizontal trim line. Standard equipment on sixes was markedly different than on eights. It included six-tone Bedford cord upholstery with tan and brown stripes; leather grain Dura-Fab trim; rear compartment carpets; dark brown leather grain dash; wind-up clock; three-spoke, half-ring steering wheel; armrests at ends of all seats (except convertible and Brougham); center rear seat armrest in sedan, club coupe and Hollywood; pop-out lighter; ashtrays at seat ends; dash ashtray; friction front ventipanes; leather grain window garnish moldings; reveal moldings; woven fabric trunk mat; fender skirts; Hudson triangle hood ornament; seatback pockets; front dome lamp (two side lamps in Hollywood); rear quarter dome lamps in sedan, club coupe and Hollywood; robe hanger and hand grips on seatback and front fendertop ornaments.

COMMODORE — EIGHT — SERIES 8B: The Commodore Eight had the following differences from the Commodore Six: nylon three-dimensional weave upholstery in tan-Brown with gold stripes or blue-Gray with blue stripes; foam rubber seat cushions; front and rear carpets; cord type, Dura-Fab covered robe hangers in all models; instrument lighting dimmer switch; Deluxe steering wheel; electric clock; crank type front ventipanes; printed jute trunk mat and Inline. eight-cylinder motor. The front fender 'spear tips' on sixes were decorated with a number '6'; on eights with a number '8'. Front parking lenses for all Commodores were of the wraparound style seen on Wasps, but not on Pacemakers.

COMMODORE SIX (SERIES 6B)/COMMODORE EIGHT (SERIES 8B)

Model Number	Model Number	Body Type & Seating	Factory Price	Shipping Weight	Production Total
6B	8B	4-dr Sed-6P	2674/2769	3595/3630	Note 1
6B	8B	2-dr Clb Cpe-6P	2647/2742	3550/3580	Note 1
6B	8B	2-dr Holly HT-6P	3000/3095	3625/3660	Note 1
6B	8B	2-dr Conv Brghm-6P	3247/3342	3750/3770	Note 1

NOTE 1: Series 6B production included 1,592 Commodore Sixes; no breakouts.
NOTE 2: Series 8B production included 3,125 Commodore Eights; no breakouts.
NOTE 3: Approximately 100 Hollywood two-door hardtops and 20 convertibles were built in the Commodore Six series.
NOTE 4: Approximately 190 Hollywood two-door hardtops and 30 convertibles were built in the Commodore Eight series.
NOTE 5: Factory prices and shipping weights above slashes are for sixes, below slash for eights.

HUDSON HORNET — SIX — SERIES 7B: The Hornet for 1952 was based on the Commodore Eight. Special features seen on the Hornet included dark blue or brown leather grain window garnish moldings; Hornet 'Flying-H' identification on the side of front fenders and rear deck; gold and chrome hood mascot; Hornet medallions on front door valance panels; and the high-compression H-145 six-cylinder engine. All other specifications matched those of the Commodore Eight. On a model for model basis, the two series were again priced identically, with the 8A models weighing 30 pounds more than Hornets.

HUDSON HORNET SIX (SERIES 7B)

Model Number	Model Number	Body Type & Seating	Factory Price	Shipping Weight	Production Total
7B	NA	4-dr Sed-6P	2749	3600	Note 1
7B	NA	2-dr Clb Cpe-6P	2722	3550	Note 1
7B	NA	2-dr Holly HT-6P	3071	3630	Note 1
7B	NA	2-dr Conv Brghm-6P	3318	3750	Note 1

NOTE 1: Series production included 35,921 Hornets; no breakouts.
NOTE 2: Approximately 2,160 Hollywood two-door hardtops and 360 convertibles were built in the Hornet Series.

ENGINES

(PACEMAKER SIX) L-head six. Chrome alloy block. Displacement: 232 cid. Bore and stroke: 39/16 x 3-7/8 inches. Compression ratio: 6.7:1. Brake hp: 112 at 4000 rpm. Four main bearings. Solid valve lifters. Carburetor: Carter one-barrel type WA-1 model 749S.

(WASP/COMMODORE SIX) Inline. L-head six. Chrome alloy block. Displacement: 262 cid. Bore and stroke: 3-9/16 x 4-3/8 inches. Compression ratio: 6.7:1. Brake hp: 127 at 4000 rpm. Four main bearings. Solid valve lifters. Carburetor: Carter one-barrel type WA-1 model 776S.

(COMMODORE EIGHT) Inline. eight-cylinder. Flathead. Chrome alloy block. Displacement: 254 cid. Bore and stroke: 3 x 3-1/2 inches. Compression ratio: 6.7:1. Brake hp: 128 at 4200 rpm. Five main bearings. Solid valve lifters. Carburetor: Carter Type WGD Model 773S with L-shaped air horns.

1952 Hudson, Hornet four-door sedan, 6-cyl

(HORNET SIX) Inline. L-head six. Chrome alloy block. Displacement: 308 cid. Bore and stroke: 3-13/16 x 4-1/2 inches. Compression ratio: 7.2:1. Brake hp: 145 at 3800 rpm. Four main bearings. Solid valve lifters. Carburetor: Carter two-barrel type WGD model 776S.

CHASSIS FEATURES: Wheelbase: (4B and 5B) 119 inches; (all other models) 124 inches. Overall length: (4B) 201-1/2 inches; (5B) 202-1/2 inches; (all other models) 208-1/2 inches. Front tread: (all) 58-1/2 inches. Rear tread: (all) 55-1/2 inches. Tires: Low-pressure, high-volume Super-Cushion 7.10 x 15 tires were standard on all models except Hornet and Commodore convertibles. External whitewall tires. Size 7.60 x 15 tires. Size 7.60 x 15 was optional on all other models at extra cost. White sidewall tires were optional at extra cost, with availability limited by Korean War production restrictions.

OPTIONS: Fender skirts for Pacemaker. Center bumper guards for Pacemaker. Large hubcaps for Pacemaker and Wasp. Front fender ornaments for Pacemaker. Radio. Radio antenna, roof mount type. Heater. wind-up clock in Pacemaker. Electric clock in Wasp or Commodore Six. External windshield sun visor. Side window sun shields. Wheel trim rings. Oversize tires. Hydraulic window lifts in Wasp convertible (standard in other convertibles). Plastic rear window for convertible. Other standard factory and dealer installed accessories. Pacemaker models were available in five solid colors with six special hues optional at extra cost. All other Hudsons were available in 11 solid colors plus black, 19 two-tones. Convertibles came only in solid colors. Hudson Motor Car Co. released a special 'High-Output' options catalog this year. It included the high compression Pacemaker Six with optional aluminum 7.2:1 compression cylinder head; the super high-output H-127 Wasp/Commodore Six with the same optional head and the high-output Super-Eight engine with this 'Power Dome' cylinder head. Also, the 'Miracle H-Power' Hornet engine was available with both 7.2:1 aluminum head and 6.7:1 iron alloy head, plus a 7.2:1 iron alloy head. A dual-carburetor induction system with dual intake manifolds was released for the Hornet engine as the 'Twin-H' power package. Overdrive transmission was available as an option for $111 extra. Hydra-Matic Drive could be ordered for $175.71 extra. Numerous types of extra-special performance components were offered by the factory to professional stock car racers this year. Iron alloy 'Power Dome' cylinder heads were also available for Super Six and Super Eight motors.

HISTORICAL FOOTNOTES: The 1952 Hudson line was introduced January, 1952. Hudson Hornet stock cars won 27 NASCAR races out of a total 34 Grand Nationals held this year. Hudson drivers in NASCAR included Herb Thomas and Tim Flock. Marshall Teague began driving Hudsons in AAA competition, after taking the 1952 NASCAR Daytona stock car race in a Hornet. The car was torn down after the race and proved to be 100 percent stock. In AAA racing, Teague took 14 checkered flags for Hudson, while other drivers captured a total of five. For the year, the Hornets had captured 40 wins in 48 major stock car races. It was quite a feat! With model year sales of 70,000 cars and calendar year deliveries of 79,117 units, the company's sales rank moved up one notch to 14th position. In May 1952, Hudson announced that it was starting to tool-up for production of a new compact-sized line of 1953 models. This car became the Hudson Jet.

1953 HUDSON

1953 Hudson, Super Jet four-door sedan, 6-cyl

SERIES 1C JET — SERIES 2C SUPER JET — SIX: A down-sized Hudson flew on the scene in 1953. It was called the Jet. Marked by slab sided styling with conventional notch back lines, the Jet looked different from other Hudsons. A fake air scoop decorated the front of the hood. The grille had a flat oval appearance with a chrome molding highlighting the upper opening. Super Jet scripts appeared on the fenders of the more highly trimmed line, which also had an air scoop ornament. Fender skirts were optional on both lines. Standard 'custom car' equipment on the base model included Teleflash 'idiot lights;' water temperature and gasoline gages; Twin-Contour vacuum wipers; defroster vents; rotary door latches; theft-proof locks; push-button door handles; lock buttons; dash ashtray; wing type front ventipanes; twin stop and taillamps; front parking lamps; manual dome light; lockable parcel compartment; twin horns and visors; and lighted ignition switch keyway. Upholstery was done in gray weave worsted striped red and brown. Super Jet extras included oversize tires; wing type rear ventipanes; automatic dome lamps; and a host of features attached to the front seatback, such as pockets, coat hooks, cigarette lighter, robe hanger and ash receiver. Super Jets also had two-tone blue or green woven wool upholstery with Dura-fab leather grain trim.

VEHICLE I.D. NUMBERS: Serial numbers were on the right front door post. Engine numbers were the same and they were found on the upper front right-hand side of six-cylinder blocks or between the first and second exhaust manifold flanges on eight-cylinder blocks. Serial numbers followed the same system. The first two symbols were a prefix corresponding with model designations, with the second symbol changing to a 'C' in 1953. Jets were numbered 202917 to 268963; Super Jets 202917 to 269059; Wasps 202917 to 267518; Super Wasps 202917 to 267451 and Hornets 202917 to 267453. Body Style Numbers not used through 1955. First columns in charts below show lower series; second columns show higher series; Hornet offered only in one series coded in column one.

JET SIX (SERIES 1C)/SUPER JET SIX (SERIES 2C)

Model Number	Model Number	Body Type & Seating	Factory Price	Shipping Weight	Production Total
1C	2C	4-dr Sed-6P	1858/1954	2650/2700	Note 1
NA	2C	2-dr Cpe Sed-6P	1933	2695	Note 1

NOTE 1: Series production included 21,143 Jets and Super Jets combined. No body style breakouts.
NOTE 2: The two-door coupe sedan came only as a Super Jet.
NOTE 3: Factory prices and shipping weights above slash are for Jets, below slash for Super Jets.

WASP SERIES 4C — SUPER WASP SERIES 5C — SIX: The Wasp now became a mid-size Hudson offering with traditional Step Down styling on the 119 inch wheelbase. Appearance changes included deletion of the twin-strut grille guard and the addition of an air scoop hood. Upholstery was in tan weave cord with red and brown stripes and Dura-fab trim. Power came from the former Pacemaker Six. A standard steering wheel, plain-top fenders and small hubcaps were identification features. Super Wasp models were comparable to the 1952 Wasp. They were upholstered in new nylon combinations with special check weave and Dura-fab trim. Two-tone green was standard with six solid exterior colors and twelve two-tone combinations. Two-tone blue was standard with four different solids

and nine two-tones. However, both upholstery choices were optional with opposite colors at no extra cost. Leather upholstery was also a no-cost option on the Super Wasp convertible. Standard equipment on Super Wasps also included a special 127 hp six-cylinder engine; large hubcaps; front fendertop ornaments; combination fuel and vacuum pump; foam rubber front seat cushions; and Deluxe steering wheel.

WASP SIX (SERIES 4C)/SUPER WASP SIX (SERIES 5C)

Model Number	Model Number	Body Type & Seating	Factory Price	Shipping Weight	Production Total
4C	5C	4-dr Sed-6P	2311/2466	3380/3480	Note 1
4C	5C	2-dr Sed-6P	2264/2413	3350/3460	Note 1
4C	5C	2-dr Clb Cpe-6P	2311/2466	3340/3455	Note 1
NA	5C	2-dr Holly HT-6P	2812	3525	Note 1
NA	5C	2-dr Conv Brghm-6P	3048	3655	Note 1

NOTE 1: Series production included 17,792 Wasps and Super Wasps combined. No body style breakouts were provided.
NOTE 2: Approximately 590 two-door Hollywood hardtops and 50 convertible broughams were built in the Super Wasp Series.
NOTE 3: The Hollywood and the convertible brougham came only as Super Wasps.
NOTE 4: Factory prices and shipping weights above slashes are for Wasps, below slash for Super Wasps.

HUDSON HORNET — SIX — SERIES 7C: The 1953 Hudson Hornet was very similar to the previous model bearing the same name, except that the strut bar look was eliminated from the grille and the air scoop hood look was used in its place. Hornets had most equipment used on Wasps, plus front rectangular bumper guards; front outer bumper guards; electric clock; large hubcaps; front and rear foam seat cushions; and hydraulic window regulators for convertibles. The rocketship-shaped Hornet front fender and trunk ornaments were seen again. Special decorator check weave nylon upholstery was featured (same colors as Wasp interior) and a slim three-spoke steering wheel, with specially positioned horn button, was seen. Below the hood was the H-145 six-cylinder engine, with 'Power Dome' aluminum cylinder head standard.

HUDSON HORNET SIX (SERIES 7C)

Model Number	Model Number	Body Type & Seating	Factory Price	Shipping Weight	Production Total
7C	NA	4-dr Sed-6P	2769	3570	Note 1
7C	NA	2-dr Clb Cpe-6P	2742	3530	Note 1
7C	NA	2-dr Holly HT-6P	3095	3610	Note 1
7C	NA	2-dr Conv Brghm-6P	3342	3760	Note 1

NOTE 1: Series production included 27,208 Hornets. No body style breakouts were provided.
NOTE 2: Approximately 910 Hornet Hollywood two-door hardtops were built.

1953 Hudson, Super Wasp four-door sedan, 6-cyl

ENGINES

(JET/SUPER JET SIX) Inline. L-head six. Chrome alloy block. Displacement: 202 cid. Bore and stroke: 3 x 4-3/4 inches. Compression ratio: 7.5:1. Brake hp: 104 at 4000 rpm. Four main bearings. Solid valve lifters. Carburetor: Carter one-barrel type WA-1 Models 2009S or 2009SA.

(WASP SIX) L-head six. Chrome alloy block. Displacement: 232 cid. Bore and stroke: 39/16 x 3-7/8 inches. Compression ratio: 6.7:1. Brake hp: 112 at 4000 rpm. Four main bearings. Solid valve lifters. Carburetor: Carter one-barrel type WA-1 model 749S.

(SUPER WASP SIX) Inline. L-head six. Chrome alloy block. Displacement: 262 cid. Bore and stroke: 3-9/16 x 4-3/8 inches. Compression ratio: 6.7:1. Brake hp: 127 at 4000 rpm. Four main bearings. Solid valve lifters. Carburetor: Carter one-barrel type WA-1 model 774S.

(HORNET SIX) Inline. L-head six. Chrome alloy block. Displacement: 308 cid. Bore and stroke: 3-13/16 x 4-1/2 inches. Compression ratio: 7.2:1. Brake hp: 145 at 3800 rpm. Four main bearings. Solid valve lifters. Carburetor: Carter two-barrel type WGD model 776S.

1953 Hudson, Wasp four-door sedan, 6-cyl

CHASSIS FEATURES: Wheelbase: (Jets) 105 inches; (Wasps) 119 inches; (Hornet) 124 inches. Overall length: (Jets) 180-11/16 inches; (Wasps) 201-1/2 inches; (Super Wasps) 202-1/2 inches; (Hornet) 208-1/2 inches. Front tread: (Jets) 54 inches; (all others) 58-1/2 inches. Rear tread: (Jets) 52 inches; (all others) 55-1/2 inches. Tires: (Jets) 5.90 x 15 (Super Jets) 6.40 x 15; (Hornet convertible) 7.60 x 15; (all others) 7.10 x 15.

OPTIONS: Two-door sedan rear seat armrests ($4). Front rectangular bumper guard ($24). Front outer bumper guards ($15). Electric clock, Jets ($22); Wasps ($19). Exhaust deflector ($2). Direction indicator, Jets ($21); Wasps ($24); Hornet ($24). Large hubcaps ($11). Cigar lighter ($4). Back-up lights, Super Jet ($18); Wasps and Hornet ($24). Glareproof mirror ($5). Outside rearview mirror, Jets ($5); others ($6). Front fendertop ornaments ($7). Large plastic rear window for convertible ($10). Eight-tube push-button radio ($100). Six-tube manual radio ($82). Front foam seat cushions, Jets ($13); Wasp ($14). Rear foam seat cushions (same price per model). Heavy-duty shock absorbers ($14). Solex glass with sunshade windshield ($42). Deluxe steering wheel ($20). Wheel trim rings ($15). Orion convertible top ($134). Window and wing vent shades, except convertible and Hollywood hardtop ($18). Outside sun visor with traffic light viewer ($33). Remote control Weather Control heater, Jets ($73); others ($74). Rear fender skirts [Hudson called them "wheel-covers."] ($15). Custom wheel discs Jets and Wasp ($20); Super Wasp and Hornet ($18). Windshield washer ($11). Hydraulic window regulators for Super Wasp convertible ($67). Hand-buffed genuine leather trim ($132-146 per body style). Note: leather trim not available on base Jet or blue combinations not available on base Wasps. Dura-fab trim ($53). Special solid paint colors on Jets ($27); other models ($28). Two-tone paint combinations on Super Jet only ($27); on Super Wasp or, Hornet only ($31). Note: Special and two-tone colors were not available on base Jet or base Wasp. Tire options for Jets and Super Jets included white-walls, six-ply and Super Jet size on Jet at exchange prices from $6 to $50. Tire options on the Hornet convertible included whitewalls at $41 and six-ply blackwalls at $54 (exchange prices). Tire options on other Hornets, Wasps and Super Wasps included 7.60 x 15 size, whitewalls and special order six-ply choices at exchange prices from $22 to $72. Oil bath air cleaner ($8). Two oil bath air cleaners with 'Twin-H' power package ($16). Aluminum cylinder head, Wasps ($14); Jets ($12). Special 127 hp Super Wasp Six for Wasp ($37). Hydra-Matic Drive ($176). Oil filter ($14). Overdrive for Jets ($102); for other models ($111). Combination fuel and vacuum pump for base Wasp ($12); for Jet/Super Jet ($11). 'Twin-H' Power for all models, twin oil bath air cleaners mandatory on Wasp 4C ($85.60). Available rear axles included units with 4.09:1, 4.10:1 4.55:1, 4.27:1, 3.54:1, 3.31:1 and 3.07:1 gear ratios. Specific applications of axle ratios varied with models and transmissions, but options available were no extra cost. Cost of the high-performance 7-X engine from Hudson Dealer Parts Departments was $385.

HISTORICAL FOOTNOTES: The 1953 Hudson line was introduced in November, 1952. Model year sales were 66,143 units. Calendar year production peaked at 67,089 cars. Hudson was the 15th ranked producer. A special 7-X engine package was released for "severe usage," such as stock car racing. An official power rating was not published, but 200 hp was estimated for cars with this option. Hudsons captured 22 (out of 37) major NASCAR races, with driver Herb Thomas winning championship honors for the season. In AAA competition 13 (out of 16) races went to Hudson. The Hornets took checkered flags in 35 of 53 contests. Rumors of an impending merger between Hudson and another independent manufacturer began circulating in Detroit.

1954 HUDSON

JET SIX — SERIES 1D: The 1954 Jet grille had four ribs on each side of the main blade and a center embossment. Standard equipment was the same as the previous year. Tan worsted weave upholstery with brown and red stripes was featured, in combination with brown Plasti-hide trim. The base model lacked robe cords; courtesy lamps; front seatback pockets; wing type rear ventipanes; coat hooks; and a rear ashtray. Even a cigar lighter was extra. Two-tone paint was not available and the sole upholstery option was gray Plasti-hide with leather trim. On April 12, 1954, a new Family Club sedan was added to the line as a stripped economy model. Priced $216 under the base sedan, this two-door had a non-scoop hood, plainer grille, black rubber windshield surround and even more spartan appointments.

VEHICLE I.D. NUMBERS: Serial numbers were on the right front door post. Engine numbers were the same and they were found on the upper front right-hand side of six-cylinder blocks or between the first and second exhaust manifold flanges on eight-cylinder blocks. The system of numbering Hudsons was changed. Each serial number had seven symbols. The first designated model, the others were the consecutive unit number. The starting number for all series was 269060 and they ran in mixed production. Engine numbers were the same. The serial number was on the right door post and, in Jet models, on the top right frame rail flange near the dash panel. Engine numbers were at the right front corner of the block near the top. Charts below list the lower series Model Number in first column and upper series Model Number in second column when two series are offered.

JETS SIX (SERIES 1D)

Model Number	Model Number	Body Type & Seating	Factory Price	Shipping Weight	Production Total
1D	NA	4-dr Sed-6P	1858	2675	Note 1
1D	NA	2-dr Fam Clb Sed	1621	2635	Note 1
1D	NA	2-dr Utl Sed-6P	1837	2715	Note 1

NOTE 1: Mixed production with Super Jet/Jetliner series below.

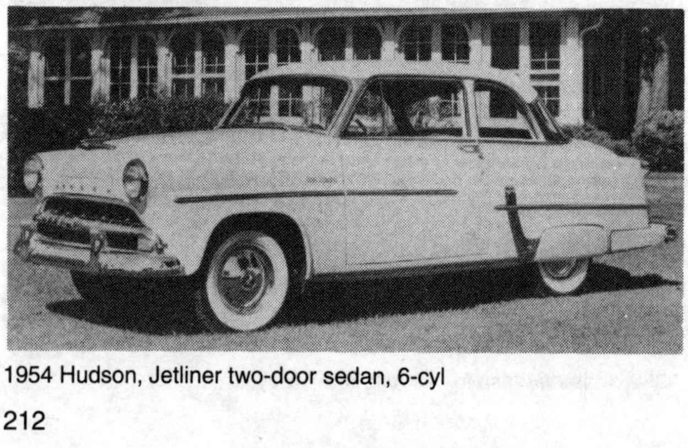

1954 Hudson, Jetliner two-door sedan, 6-cyl

SERIES 2D SUPER JET — SERIES 3D JETLINER — SIX: Block letters spelled out Super Jet on the front fenders of this one-step-up model. Features included hood air scoop ornamentation; horizontal front fender and door moldings; robe cords; wing type rear door ventipanes; front seatback pockets; rear ashtray; courtesy door lights; coat hooks; and cigar lighter. Two-tone green or blue decorator selected worsted upholstery fabrics in a handsome check pattern, with solid Plasti-hide trim, were used. The Jetliner was a new, top-level offering characterized by Jetliner fender block lettering; rear wheelcovers (skirts); Custom wheel discs; rear fender horizontal rub moldings; body sill highlights; bright rear gravel shields; and chrome rear taillight trim. Most Super Jet appointments were included, plus front and rear foam seat cushions and smartly pleated antique white Plastic-hide upholstery with headliner and bolsters of the same material in blue, green or red. Worsted cloth Super Jet combinations were optional at no extra cost.

SUPER JET SIX (SERIES 2D)/JETLINER SIX (SERIES 3D)

Model Number	Model Number	Body Type & Seating	Factory Price	Shipping Weight	Production Total
2D	3D	4-dr Sed-6P	1954/2057	2725/2760	Note 1
2D	3D	2-dr Clb Sed-6P	1933/2046	2710/2740	Note 1

NOTE 1: Series production included 14,224 Jets, Super Jets and Jetliners combined. No body style breakouts available.
NOTE 2: Data above slash for Super Jet/below slash for Jetliner.

WASP SERIES 4D — SUPER WASP SERIES 5D — SIX: New styling, resembling that of the Jets, was applied to regular Hudsons this year. The grille had a heavy, bowed molding tracing the upper radiator opening. There was a full-width, flat horizontal loop surrounding the wedge shaped parking lights at each end. The main bar (top of the loop) was ribbed towards the middle and held a triangular Hudson medallion in a finned housing at its center. Behind this bar was an angled plate with four additional, wide-spaced ribs. Block letters spelled out Hudson below the scoop on the nose of the hood. Wasp or Super Wasp signature scripts were placed on the front fender tips above a full-length horizontal rub molding. Two-door hardtops had Hollywood scripts at the upper rear edge of front fenders. A panoramic one-piece windshield and protruding tip taillamps were new. Bright metal gravel shields with windsplit vents decorated the rear fendersides. Standard equipment on Wasps included fender skirts; cigar lighter; robe cord; front seatback pockets; rear ashtray; and special pattern cloth upholstery with blue or green Plasti-hide trim. Super Wasps had the same features, plus the following additions or changes: large hubcaps; front foam seat cushions; Custom steering wheel; passenger assist handles; crank type front ventipanes on Hollywoods and convertibles; courtesy door lights; combination fuel and vacuum pump; and two-tone blue or green check pattern tweed cloth upholstery with worsted bolster material and Plasti-hide trim. In Super Wasp convertibles blue, maroon or green leather cushions with Plasti-hide side trim was standard. The Super Wasp Hollywood hardtop had brown, blue or green nylon cord seats with snowflake design cloth upholstery and harmonizing Plasti-hide bolsters.

WASP SIX (SERIES 4D)/SUPER WASP SIX (SERIES 5D)

Model Number	Model Number	Body Type & Seating	Factory Price	Shipping Weight	Production Total
4D	5D	4-dr Sed-6P	2256/2466	3440/3525	Note 1
4D	5D	2-dr Clb Sed-6P	2209/2413	3375/3490	Note 1
4D	5D	2-dr Clb Cpe-6P	2256/2466	3360/3475	Note 1
NA	5D	2-dr Holly HT-6P	2704	3570	Note 1
NA	5D	2-dr Conv Brghm-6P	3004	3680	Note 1

NOTE 1: Series production included 11,603 Wasps and Super Wasps combined. No body style breakouts available.
NOTE 2: Industry trade magazines indicate that, on a calendar year basis, Hudson built 2,654 Hollywoods and 222 convertibles (note change from earlier editions of *SCAC*) with no series breakouts.
NOTE 3: Data above slash for Wasp/below slash for Super Wasps.

HORNET SPECIAL SERIES 6D — HORNET SERIES 7D — SIX: Hornets seemed a little more like Super Wasps this year, although the longer 124 inch wheelbase was still used. The appearance change to the Jet-like look brought an end to the front fender rocketship ornaments. Hudson signature scripts were seen on the fenders, but only the trunk lid had a special badge. As on Super Wasps, two-door hardtops also had Hollywood scripts at the high trailing edge of front fenders, above the full-length horizontal body rub moldings. Hornets had most Super Wasp equipment, plus the following additions or changes: crank type front ventipanes on all models; cast aluminum 'high-compression' head; electric clock; foam rubber rear seat cushions; Custom wheel discs; hydraulic window lifts (in convertibles); and special trims. Sedans and club coupes were upholstered in 15 percent nylon worsted Bedford cloth with broadcloth bolsters and Plastihide trim in different shades of the same colors; brown, blue or green. The Hornet Hollywood had similarly toned, snowflake design nylon cord seats with Plasti-hide bolsters. The convertible was done in blue, maroon or green genuine leather (with Plasti-hide side trim). Convertible tops were available in maroon, black or tan. Specific combinations of top colors with car finishes were recommended, but not considered mandatory. Also, the Hollywood hardtop could be had with tri-colored seat and headlining combinations of antique white Plasti-hide and red, blue and green bolsters, at no extra cost. The last models introduced by Hudson, in Detroit, were the Hornet Specials. They appeared March 19, 1954 at prices $115 to $140 lower than comparable Hornets. They had Hornet Special front fender scripts; Hornet engine and subdued level of exterior brightwork, but Super Wasp interior trim.

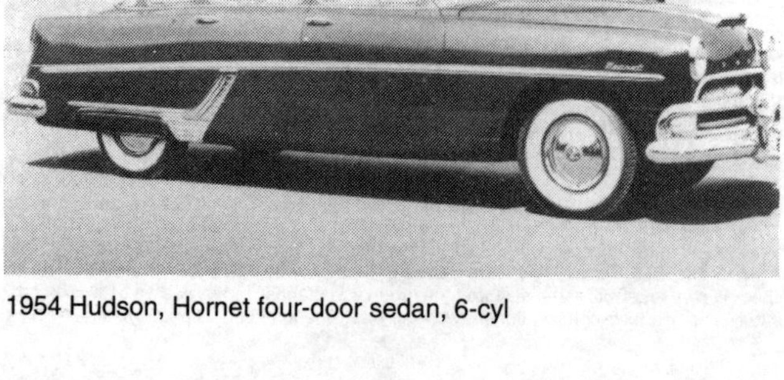

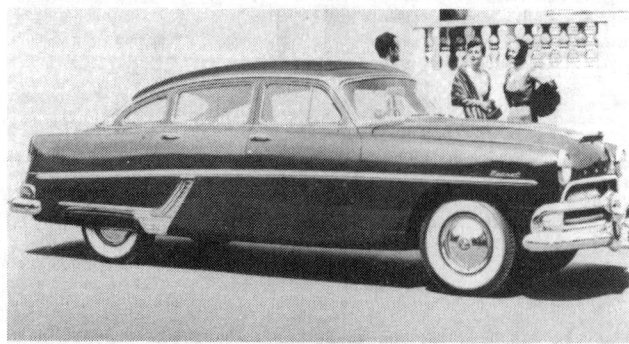

1954 Hudson, Hornet four-door sedan, 6-cyl

HORNET SPECIAL SIX (SERIES 6D)/HORNET SIX (SERIES 7D)

Model Number	Model Number	Body Type & Seating	Factory Price	Shipping Weight	Production Total
6D	7D	4-dr Sed-6P	2619/2769	3560/3620	Note 1
6D	NA	2-dr Clb Sed-6P	2571	3515	Note 1
6D	7D	2-dr Clb Cpe-6P	2619/2742	3505/3570	Note 1
NA	7D	2-dr Holly HT-6P	2988	3655	Note 1
NA	7D	2-dr Conv Brghm-6P	3288	3800	Note 1

NOTE 1: Series production included 24,833 Hornet Specials and Hornets combined. No body style breakouts.
NOTE 2: The club sedan came as a Hornet Special only; the Hollywood hardtop and convertible brougham came as Hornets only.
NOTE 3: Data above slash for Hornet Special/below slash for Hornet.

HUDSON ITALIA — CUSTOM SERIES — SIX: Twenty-six Hudson Italia coupes were built in 1954 on the Jet platform. The Italia body was styled and crafted by Carrozzeria Touring, of Milan, Italy, based on original sketches by Hudson's own Frank Spring. The sporty GT had aluminum coachwork; functional front fender scoops with brake cooling ducts; wraparound windshield; flow through ventilation; contoured leather bucket seats with three different densities of foam for proper support; deep-pile Italian floor carpeting; Borrani wire spoke wheels; white sidewall tires; radio; heater; back-up lights; and turn signals stacked in Jet-tube pipes tunneled into rear fenders. The Italia was announced, as a production model, on Jan. 14, 1954, the same day Hudson's merger with Nash Motors was approved. The 26 cars were actually designed and custom built as four-passenger Grand Touring 'image' cars to steal attention from Chevy Corvettes, Ford T-Birds and Ghia Chrysler show cars. Twenty-five were actually sold as production models, while a coupe prototype and four-door X-161 pilot model were also created. Twenty-one of these cars are known to still exist.

HUDSON ITALIA SIX (CUSTOM SERIES)

Model Number	Model Number	Body Type & Seating	Factory Price	Shipping Weight	Production Total
NA	NA	2-dr GT Spt Cpe-4P	4800	2710	27

HUDSON METROPOLITAN — (SERIES E) — FOUR: On May 1, 1954 Hudson Motor Car Co. became a division of American Motors Corp. Hudson dealers then undertook the sale of four-cylinder Metropolitans, previously marketed as a Nash offering in the United States. Hudson dealers were supplied with replacement grille center inserts having an 'H' instead of an 'M.' These were to be installed in Metropolitans sold as Hudson. The cars were otherwise identical to Nash Metropolitans, which are mentioned elsewhere in this catalog. Even serial and engine numbers were the same. Hudson Metropolitan were marketed through 1956. Since these were early "captive imports," rather than a true domestic automobile, specifications are not listed in this catalog. They can be found in Krause Publication's *Standard Catalog of Imported Cars.*

ENGINES

(JET/SUPER JET SIX) Inline. L-head six. Chrome alloy block. Displacement: 202 cid. Bore and stroke: 3 x 4-3/4 inches. Compression ratio: 7.5:1. Brake hp: 104 at 4000 rpm. Four main bearings. Solid valve lifters. Carburetor: Carter one-barrel type WA-1 Models 2009S or 2009SA.

(WASP SIX) Inline. L-head six. Chrome alloy block. Displacement: 232 cid. Bore and stroke: 3-9/16 x 3-7/8 inches. Compression ratio: 7.0:1. Brake hp: 126 at 4400 rpm. Four main bearings. Solid valve lifters. Carburetor: Carter one-barrel type WA-1 model 749S.

(SUPER WASP SIX) Inline. L-head six. Chrome alloy block. Displacement: 262 cid. Bore and stroke: 3-9/16 x 4-3/8. Compression ratio: 7.0:1. hp: 140 at 4000 rpm. Four main bearings. Solid valves lifters. Carburetor: Carter two-barrel type WGD Model 2115S.

(HORNET 'BIG' SIX) Inline. L-head six. Chrome alloy block. Displacement: 308 cid. Bore and stroke: 3-15/16 x 4-1/2 inches. Compression ratio: 7.5:1. Brake hp: 160 at 3800 rpm. Four main bearings. Solid valve lifters. Carburetor: Carter two-barrel type WGD Model 2115S.

(ITALIA SIX) Inline. L-head six. Chrome alloy block. Displacement: 201.5 cid (202). Bore and stroke: 3.00 x 4.75 inches. Compression ratio: 7.5:1. Brake hp: 114 at 4000 rpm. Four main bearings. Solid valve lifters. Induction system: Twin-H Power package with dual manifolding and carburetion (Carter carburetors).

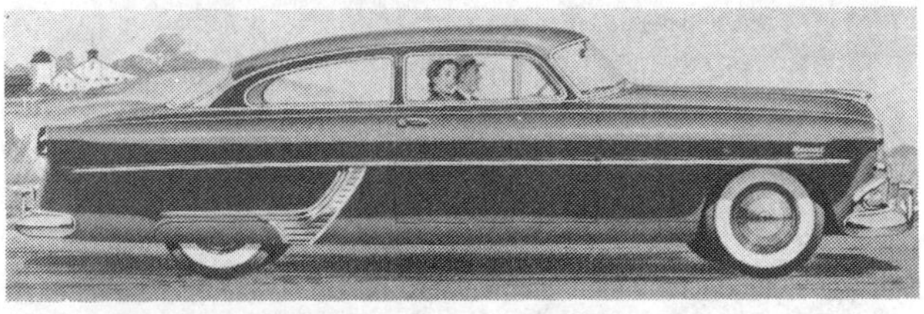

1954 Hudson, Hornet Special two-door sedan, 6-cyl

CHASSIS FEATURES: Wheelbase: (Jets) 105 inches; (Wasps) 119 inches; (Hornets) 124 inches; (Italia) 105 inches. Overall length: (Jets) 180-11/16 inches; (Wasp) 201/1-2 inches; (Super Wasp) 202-15/32 inches; (Hornets) 268-7.8 inches. Front tread: (Jets/Italia) 54 inches; (other models) 58.5 inches. Rear tread: (Jets/Italia) 52 inches; (other models) 55.5 inches. Tires: (Jet) 5.90 x 15; (Super Jet/Jetliner) 6.40 x 15; (Hornet convertible) 7.60 x 15; (other models, except Italia) 7.10 x 15.

OPTIONS: Wasp two-door sedan rear armrest ($4). Power brakes, except Jets ($43). Electric clock, in Jets ($22); in Wasps ($19). Exhaust deflector ($2). Direction indicators, in Jets ($16); in others ($20). Large hubcaps, Wasp only ($11). Jet cigar lighter ($4). Super Jet/Jetliner back-up lights ($18); same on Wasps and Hornets ($24). Glare-proof mirror ($5). Outside rearview mirror (OSRV), in Jets ($5); all others ($6). Plastic convertible rear window ($10). Eight-tube push-button radio ($100). Six-tube manual radio, Jets only ($82). Front foam seats, in Jet/Super Jet ($13); in Wasp ($14). Rear foam seats, Jet/Super Jet ($13); in Wasp/Super Wasp ($14). Extra-heavy-duty shock absorbers, in Jets ($5); other models ($14). Solex glass with sunshade windshield ($33). Heavy scale springs, front and rear or rear only, separate no-cost option in Jets and other models. Power steering, Wasps and Hornets only ($177). Custom steering wheel, in Jet/Wasp ($20); in Super Jet/Jetliner ($19). Wheel trim rings, Jets and Wasps ($15). Orion, convertible top ($134). Window and wing vent shades, except convertible and Hollywood hardtop ($18). Outside visor with traffic light viewer in Wasps and Hornets ($33). Weather Control heater with remote control, in Jets ($73); in others ($74). Rear wheelcovers (skirts), in Jet/Super Jet only ($15). Custom wheel discs, on Jet/Super Jet and base Wasp ($20); on Super Hornet ($18). Wheels painted upper body color, (no cost with two-tone paint). Windshield washer ($11). Hydraulic window regulators in Super Wasp convertible ($67). Safety Group including back-up lights; directionals; glare proof and OSRV mirrors and windshield washer in Super Wasp and Hornet ($66). Safety Group with all above items plus combination fuel/vacuum pump in Wasp ($78); in Super Jet/Jetliner ($66). Safety Group with all above, less back-up lights in Jet ($49). Chrome plated wire wheels, except Jets. Velchrome painted wire wheels, (Special order in all models). The following equipment was available on models indicated, when sold as taxis or Police cars: Extra-wide 2-1/4 inch brake (standard in 5D, 7D; special order in 4D). Heavy-duty clutch (standard in 5D, 7D; special order in 1-2-3-4D). Heavy-duty battery (special order in 1-2-3D). Police/taxi special seat construction (special order in 1-2-3D). Color and Trim Options: Roman bronze or Pasture green special paint (no charge, except Jet/Wasp base models). Algerian blue, Coronation cream, St. Clair gray, or Lipstick red solid colors (extra cost). Model 1D Gray Plasti-hide trim (extra cost). Model 2D blue or green Plasti-hide trim (extra cost). Model 3D worsted upholstery (no cost). Model 4D Pioneer grain leather trim (extra cost). Model 5D maroon Pioneer grain or green Antique grain leather trim (extra cost). Model 5D Hollywood Plasti-hide trim (no cost). Model 7D sedan and club coupe, blue, maroon or green Antique grain leather trim (extra cost). Model 7D Hollywood blue or maroon leather trim (extra cost). Tire options included white sidewalls, oversize or extra-ply construction types at a variety of exchange prices based on series, model and body style. A continental tire extension kit was offered as a dealer installed accessory. Oil bath air cleaner ($8). Twin oil bath air cleaners for 'Twin-H' package mandatory on base Wasp equipped with 'Twin H'($16). Aluminum cylinder head optional on all, except Italia/Hornet, on Jets ($12); on others ($14). Super Wasp Six for base Wasp ($48). Hydra-Matic Drive ($178). Oil filter, no charge on Super Jet/Jetliner with Hydra-Matic; (other models $14). Overdrive transmission, on Jets ($102); on all others ($111). Combination fuel/vacuum pump, on Jets ($11); on all others ($12). 'Twin-H' power with dual carburetion and manifolding, 170 hp, dual air cleaners mandatory on base Wasp ($86).

HISTORICAL FOOTNOTES: Oct. 2, 1953 was the dealer introduction date for 1954 Hudson and Jet Utility sedans. The Jetliner Series was introduced 10 days later. Model year production totaled 51,314 cars. Calendar year production peaked at 32,287 cars, including 4,239 Ramblers. On Jan. 14, 1954, Hudson directors approved a merger with Nash-Kelvinator. On March 24, 1954, Hudson stockholders approved the merger. On April 12, 1954 the Jet Family Club sedan was added to the line. On May 1, 1954. Hudson officially became part of American Motors Corp. Twenty-six days later, Hudson employees were notified that production was being switched to the Nash automobile factory in Kenosha, Wis. On Oct. 30, 1954, the 1954 Hudson model run ended in Detroit. Eleven days later the first Hornet/Rambler departed the Kenosha plant. On Dec. 28, 1954, the first 1955 Hudson Hornet V-8 was built at Kenosha. An era in Hudson's history had ended.

1955 HUDSON

HUDSON RAMBLER — SERIES 5500 — SIX: The year 1955 found 11 Nash Ramblers, thinly disguised with grille badge inserts and hubcaps with a letter 'H' in the center, at Hudson dealerships. Two 'Fleet Specials' were stripped, three-passenger economy jobs with painted lamp rims, rubber windshield surrounds and spartan appointments. Next in price came a Deluxe series with plated headlamp rims, Deluxe front fender scripts, plain air scoop, no hood ornament and slightly up-market interior trim. A hood ornament and air scoop trim band were seen on the Super series, priced in the next higher bracket. At the top of the heap was a Custom series, with Custom front fender scripts, standard continental spare tire, enriched interior and all other Super features. These Hudson Ramblers were identical to the latest editions of the comparable Nash Rambler in features, trim and price.

VEHICLE I.D. NUMBERS: [HUDSON/RAMBLER] Serial numbers were located on a plate attached to the center cowl panel, below the hood. Cars built in Kenosha were numbered D205001 to D-276099. Cars built at El Segundo, California were numbered DC-15001 to DC-23325. Engine numbers were on the upper left corner of the block (with air-conditioning at the left front side of the block) and ran H-45001 to H-131414. Nash and Hudson Ramblers were built in mixed production. Body type identification was possible, with (Body Style) model numbers stamped on a plate located at the right side of dash, below hood. These numbers correspond to the Model Numbers in column two of charts below. [WASP/HORNET SUPER/HORNET CUSTOM] Wasps (built in Kenosha,Wis.) were numbered W1001 to W-8026. Kenosha-built Hornet Sixes were numbered X-1001 to X-7523; V-8s Y-1001 to Y-7170. El Segundo-built Hornet Sixes were numbered XC-1001 to XF-1048; V-8s YC-1001 to YC-1048. Engine number locations were on a machined surface on the left-hand side of block (at side of second cylinder) for Series 40 (Wasp Six) and Series 60 (Hornet Six) and were on a machined surface on right-hand side of block (at side of cylinder number eight below exhaust manifold) for Series 80 (Hornet V-8). Engine numbers were M-1001 to M-8026 (Wasp Six); F-1001 to F-7523 (Hornet Six) and P-1001 to P-7170 (Hornet V-8). Body type identification was now possible with (Body Style) model numbers stamped on a plate at the center of cowl panel, below hood. These numbers correspond to model numbers in column two of charts below.

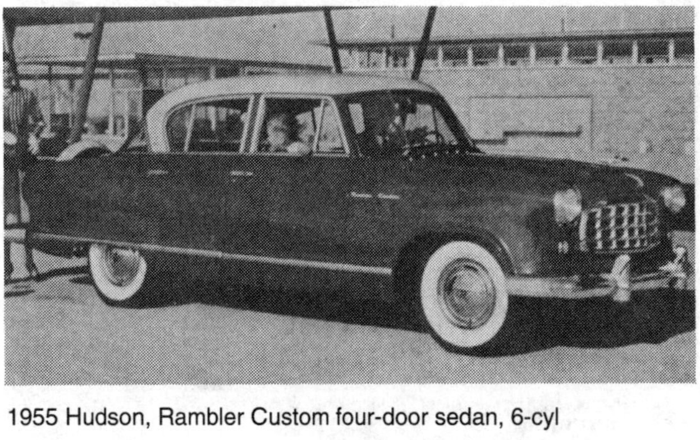

1955 Hudson, Rambler Custom four-door sedan, 6-cyl

HUDSON RAMBLER SIX (SERIES 2500/5500)

Series Number	Body Model Number	Body Type & Seating	Factory Price	Shipping Weight	Production Total
FLEET SPECIAL SERIES					
NA	5504	2-dr Utl Wag-3P	1570	2500	21
NA	5512	2-dr Del Bus Sed	1457	2400	34
DELUXE SERIES					
NA	5515	4-dr Sed-6P	1695	2567	Note 1
NA	5516	2-dr Clb Sed-5P	1585	2432	0
NA	5514	2-dr Sub-5P	1771	2528	0
SUPER SERIES					
NA	5516-1	2-dr Clb Sed-5P	1683	2450	2,970
NA	5514-1	2-dr Sub-5P	1869	2532	1,335
NA	5515-1	4-dr Sed-6P	1798	2570	Note 1
NA	5518-1	4-dr Cty Clb-6P	1975	2675	0
CUSTOM SERIES					
NA	5515-2	4-dr Sed-6P	1989	2606	Note 1
NA	5518-2	4-dr Cr Ctry-6P	2098	2685	12,023
NA	5517-2	dr Cty Clb-5P	1995	2518	601

NOTE 1: Production of 7,210 Hudson Rambler four-door sedans was recorded with only the partial model breakouts shown here available.

WASP SUPER AND CUSTOM — SERIES 40 — SIX: The post-AMC merger Hudsons were introduced at the Chicago Auto Show, Feb. 23, 1955. Styling and engineering, though based on the Nash unit body and platform, were planned to give Hudsons a distinct character. For example, the former Jet engine was under the hood and the Hudson dual braking system was retained. Gone, however, was the traditional Hudson 'crab tread' stance. The Wasp front tread was 3/16 inches less than its rear tread. As far as the sheet metal went, the only panel interchangeable between Nashes and Hudsons was the rear deck lid. Overall styling was pleasant. A massive eggcrate grille filled the area below and between the single headlamps, with an inverted steer horn shaped bar bordering the top. This upper border bar had a Hudson badge set into a housing at its center. Hudson block letters decorated the hood, which no longer had a simulated air scoop. It have a full-width cowl vent near the windshield base. Horizontal moldings stretched across the front fenders and doors. A higher molding swept rearwards from the wraparound windshield post towards the upper back fender region. A stand-up hood ornament and Wasp front fender nameplates were seen on Super models, which also had Super scripts on the sides of the cowl. Customs had a flatter hood ornament and Custom cowl side script and included a continental spare tire as standard. Wrapover rear roof pillars were seen on all models.

WASP SIX AND CUSTOM SIX (SERIES 40)

Series Number	Body Model Number	Body Type & Seating	Factory Price	Shipping Weight	Production Total
STANDARD SERIES					
NA	35545-1	4-dr Sed-6P	2290	3254	Note 1
CUSTOM SERIES 40-2					
NA	35545-2	4-dr Sed-6P	2460	3347	5,551
NA	35547-2	2-dr Holly HT-6P	2570	3362	1,640

NOTE 1: Production of Super Wasp four-door sedans is included in Custom Wasp four-door sedan total.

HORNET SUPER AND CUSTOM — SERIES 60 SIX — SERIES 80 V-8: The Hornet for 1955 had the same styling as the new Wasps on a longer wheelbase platform. Hornet nameplates were seen on the front fenders. Hardtops had Hollywood cowlside scripts as well. Very high quality interiors and standard continental spare tire carriers were regular equipment distinguishing Custom level cars. The Custom interior included 16 inch table-like rear seat armrests; transparent sun visors; roof package net; and padded dashboard. Hornet engine choices included the standard (160 hp) or 'Twin-H' (170 hp) versions of the big 308 cid six or a new V-8 built and supplied by Packard. Packard's new Twin Ultramatic transmission was both standard and mandatory in Hudsons equipped with V-8.

HORNET SUPER AND CUSTOM SIX (SERIES 60)/V-8 (SERIES 80)

Series Number	Body Model Number	Body Type & Seating	Factory Price	Shipping Weight	Production Total
SUPER SIX SERIES 60-1					
NA	35565-1	4-dr Sed-6P	2565	3495	Note 1
CUSTOM SIX SERIES 60-2					
NA	35565-2	4-dr Sed-6P	2760	3562	5,357
NA	35557-2	2-dr Holly HT-6P	2880	3587	1,554

NOTE 1: Production of Hornet Super Six four-door sedans is included in Hornet Custom Six four-door sedan total.

Series Number	Body Model Number	Body Type & Seating	Factory Price	Shipping Weight	Production Total
SUPER V-8 SERIES 80-1					
NA	35585-1	4-dr Sed-6P	2825	3806	Note 1
CUSTOM V-8 SERIES 80-2					
NA	35585-2	4-dr Sed-6P	3015	3846	4,449
NA	35587-2	2-dr Holly HT-6P	3145	3876	1,770

NOTE 1: Production of Hornet Super V-8 four-door sedans is included in Hornet Custom V-8 four-door sedan total.

1955 Hudson, Wasp four-door sedan, 6-cyl

ENGINES

(RAMBLER 'FLYING SCOT' SIX) Inline. L-head six. Cast iron block. Displacement: 195.6 cid. Bore and stroke: 3-1/8 x 4-1/4 inches. Compression ratio: 7.3:1. Brake hp: 90 at 3800 rpm. Four main bearings. Solid valve lifters. Carburetor: Carter one-barrel type YF Model 2014S.

(WASP SIX) Inline. L-head Six. Cast iron block. Displacement: 202 cid. Bore and stroke: 3 x 4-3/4 inches. Compression ratio: 7.5:1. Brake hp: 120 at 4000 rpm. Four main bearings. Solid valve lifters. Carburetor: Carter one-barrel type WA-1 Model 2013B.

(HORNET 'BIG' SIX) Inline. L-head six. Chrome alloy block. Aluminum cylinder head. Displacement: 308 cid. Bore and stroke: 3-13/16 x 4-1/2 inches. Compression ratio: 7.5:1. Brake hp: 160 at 3800 rpm. Four main bearings. Solid valve lifters. Carburetor: Carter one-barrel type WA-1 Model 2113S.

(PACKARD EIGHT) V-8. Overhead valves. Cast iron block. Displacement: 320 cid. Bore and stroke: 3-13/16 x 3-1/2 inches. Compression ratio: 7.8:1. Five main bearings. Brake hp: 208 at 4200 rpm. Five main bearings. Non-adjustable hydraulic valve lifters. Carburetor: Carter two-barrel type WGD Model 2231 SA.

1955 Hudson, Custom Hornet four-door sedan, 6-cyl

1955 Hudson Hornet Hollywood two-door hardtop

CHASSIS FEATURES: Wheelbase: (Rambler four-door) 108 inches; (Rambler two-door) 100 inches; (Wasps) 114-1/4 inches; (Hornets) 121-1/4 inches. Overall length: (Rambler four-door) 186-1/4 inches or 193-3/8 inches with continental spare; (Rambler two-door) 178-1/4 inches or 185-3/8 inches with continental spare; (Wasps) 202-1/4 inches or 212-1/4 inches with continental spare; (Hornets) 209-1/4 inches or 2191/4 inches with continental spare. Front tread: (Rambler) 54-5/8 inches; (all Hudson) 59-1/2 inches. Rear tread: (Rambler) 53 inches; (Wasp) 59-11/16 inches; (Hornet) 60-1/2 inches. Tires: (Rambler) 6.40 x 15; (Wasp) 6.70 x 15; and (Hornet) 7.10 x 15.

OPTIONS: Rambler radio ($76). Rambler heater ($74). Rambler air conditioning ($345). Hudson air conditioning with heater ($395). Hudson twin speaker radio ($98). Hudson Weather Eye heater and defroster ($77). Power steering ($140). Power brakes ($39). Power windows ($128). Continental spare tire, except Customs. Reclining seats. Twin-Travel bed. Air mattress. Detachable window screens. Full wheel discs. Wasp 'Twin-H' six with two (2) Carter one-barrel type WA-1 Model 2013S carburetors, 8.0:1 compression aluminum cylinder head and 130 hp at 4000 rpm. Hornet 'Twin-H' six with two (2) Carter one-barrel type WA-1 carburetors, 7.5:1 compression and 170 hp at 4000 rpm. Hornet Special V-8 with one Carter two-barrel type WGD Model 2352S carburetor, 8.25:1 compression and 215 hp at 4600 rpm. Hydra-Matic transmission ($179). Packard Twin Ultramatic transmission (mandatory on V-8 Hornets and included in factory price of these models). Overdrive manual transmission ($104). Available rear axle gear ratios: (Rambler) 3.77:1 and 4.4:1 (Hudson) 3.15:1, 3.54:1, 3.58:1, and 4.4:1.

HISTORICAL FOOTNOTES: Model year production included 25,214 Ramblers, 7,191 Wasps, 6,911 Hornets, and 6,219 V-8s for a total of 45,535 Hudsons. Sales promotions such as a 'Dealer Volume Investment Fund' and 'Sun Valley Sweepstakes' (for salesmen) helped Hudson move upwards by seven percent in the sales volume ranking charts. A national contest offering new Hudsons and trips to Disneyland as prizes was open to public participation. Although Hudson's headquarter address was still 14250 Plymouth Road, Detroit 32, Mich., all production of 1955 models was quartered at the AMC factory, Kenosha, Wis.

1956 HUDSON

1956 Hudson, Rambler Custom four-door sedan, 6-cyl

HUDSON RAMBLER — (6-CYL) — SERIES 5600: The Hudson Jet was discontinued, but Hudson dealers still had small cars to sell under the AMC 'family plan.' These were totally badge-engineered cars including the imported Metropolitan. "American" cars from American Motors included Hudson Ramblers built at Kenosha, Wis. They were, in fact, identical to 1956 Nash Ramblers, except for having hubcaps with an 'H' in the center and 'H' logo circular grille inserts.

VEHICLE I.D. NUMBERS: The numbering system was the same as for previous models. Letter prefixes indicated model and place of origin. [HUDSON RAMBLER] Serial numbers were D-276101 for Kenosha built cars; DKD-5601 for unassembled export and DT-5401 for Canadian models. [WASP] Serial numbers for Kenosha built Wasps were W-8101 to W-10619. [HORNET] Serial numbers for Kenosha built Hornet Sixes were X-7601 to X-10665. Serial numbers for Kenosha built Hornet V-8s were Y-7201 to Y-10191. [HORNET SPECIAL] Serial numbers for Kenosha built Hornet Special V-8s were Z-1001 to Z-2757. Hudson Rambler engine numbers started at B-1001 and up. Wasp and Hornet engine numbers were as follows: (Wasp Six) M-8701 and up (with M-10001 to M-10100 previously assigned to 1955 models); (Hornet Six) F-8601 and up (Hornet V-8s) P-21001 to P-28804 and (Hornet Special V-8s) G-1001 to G-7288.

HUDSON RAMBLER SERIES

Model Number	Body/Style Number	Body Type & Seating	Factory Price	Shipping Weight	Production Total
DELUXE LINE					
NA	5615	4-dr Sed-6P	1826	2891	Note 1
SUPER LINE					
NA	5615-1	4-dr Sed-6P	1936	2906	Note 1
NA	5618-1	4-dr Sta Wag-6P	2230	2992	Note 1
NA	5613-2	4-dr HT Wag-6P	2491	3095	Note 1
CUSTOM LINE					
NA	5615-2	4-dr Sed-6P	2056	2929	Note 1
NA	5619-2	4-dr HT Sed-6P	2221	2990	Note 1
NA	5618-2	4-dr HT Wag-6P	2326	3110	Note 1

NOTE 1: Production of 1956 Hudson Ramblers was 20,496 with no breakouts.

HUDSON WASP — (SIX) — SERIES 40: A completely new, V-shaped grille with a Hudson medallion set into another V-shaped dip in the center was a new styling trademark for 1956. Other changes included new hood ornaments and new rectangular front parking lamps set into wedge-shaped chrome moldings that accented the V-shape of the grille. There were air-scoop fendertop ornaments and new bodyside rub rail moldings, which also had a V-shaped dip on the rear doors or fenders. In addition, the taillights were redesigned. The Wasp models could be identified by their Wasp nameplates inside the V-shaped dip in the rub rail molding. They could also be spotted by the chrome-enclosed panel on the rear fender sides that was finished in lower body color. Popular features included Deep Coil ride; Triple-Safe hydraulic brakes with a reserve mechanical system); positive action handbrake; Double-Safe single unit construction; tubeless tires; Select-O-Lift starter; drawer-type glove compartment; wraparound windshield and rear window; and double-acting airplane type shock absorbers. The only model available in the 1956 Wasp lineup was the Super trim level four-door sedan. Twin-H power was optionally available.

HUDSON WASP SIX

Model Number	Body Style Number	Body Type & Seating	Factory Price	Shipping Weight	Production Total
SUPER LINE					
NA	35645-1	4-dr Sed-6P	2416	3264	2,519

HUDSON HORNET — (SIX/V-8) — SERIES 50/80: The 1956 Hornet had the same general styling changes as the new Wasp, but came with richer interior appointments and more standard equipment. All Hornet models had identification nameplates in the V-shaped dip in the side rub rail moldings. Custom trim level Hornets had a continental style spare tire and a chrome enclosed, gold-finished panel just to the rear of the V-shaped dip in the side molding. However, Super Hornets did not use either of these features, having the chrome-enclosed rear fender panel painted lower body color. In 1956 Hornet V-8s the Packard-built engine was used. Ultramatic transmission was again a mandatory option with this particular powerplant. Hornet Sixes came with the famous 308 cid Hudson 'Championship' six and could be had with three-speed manual, overdrive, or Hydra-Matic transmission. Hornet Sixes and Series 80 Hornet V-8s were on the longer 121-1/4 inch wheelbase. Super and Custom trim versions were offered as sixes and Customs only could be fitted with the V-8.

HUDSON HORNET SERIES

Model Number	Body/Style Number	Body Type & Seating	Factory Price	Shipping Weight	Production Total
SUPER LINE SIX					
60	35665-1	4-dr Sed-6P	2777	3545	Note 1
CUSTOM LINE SIX					
60	35665-2	4-dr Sed-6P	3019	3636	3,022
60	35667-2	2-dr Holly HT-6P	3136	3646	358
CUSTOM LINE V-8					
80	35685-2	4-dr Sed-6P	3286	3862	1,962
80	35687-2	2-dr Holly HT-6P	3429	3872	1,053

NOTE 1: Production of the Super Six included in total for Custom Six sedan.

HUDSON HORNET SPECIAL — (V-8) — SERIES 50: On March 5, 1956 the Hornet Special returned. It was a different type of car than the Hornet Special of 1954. The early Hornet Specials represented a cheaper version of the standard wheelbase Hornet. The 1956 Hornet Special had a cheaper price, but was actually something of a high-performance car. It came with a new AMC built 250 cid overhead valve V-8 in the 114-1/4 inch wheelbase Wasp chassis. Exterior trim and interior appointments were comparable to Super Hornets. Three-speed manual transmission was standard and both overdrive or Hydra-Matic Drive were available at extra cost.

HUDSON HORNET SPECIAL V-8 SERIES 50

Model Number	Body/Style Number	Body Type & Seating	Factory Price	Shipping Weight	Production Total
SUPER LINE					
50	35655-1	4-dr Sed-6P	2626	3467	1,528
50	35657-1	2-dr Holly HT-6P	2741	3486	229

1956 Hudson, Wasp four-door sedan, 6-cyl

1956 Hudson, Custom Hornet four-door sedan, 6-cyl

ENGINES

(HUDSON-RAMBLER SIX) Inline. L-head six. Cast iron block. Displacement: 195.6 cid. Bore and stroke: 3-1/8 x 4-1/4 inches. Compression ratio: 7.47:1. Brake hp: 120 at 4200 rpm. Four main bearings. Solid valve lifters. Carburetor: Carter type YF one-barrel Model 2014S.

(BASE WASP SIX) Inline. L-head six. Cast iron block. Displacement: 202 cid. Bore and stroke: 3 x 4.75 inches. Compression ratio: 7.5:1. Brake hp: 120 at 4000 rpm. Four main bearings. Solid valve lifters. Carburetor: Carter WA1 one-barrel Model 2009S.

(WASP TWIN-H SIX) Inline. L-head six. Cast iron block. Displacement: 202 cid. Bore and stroke: 3 x 4.75 inches. Compression ratio: 8.0:1. Brake hp: 130 at 4000 rpm. Four main bearings. Solid valve lifters. Carburetor: Two (2) Carter type WA1 one-barrels Model 2013S.

(BASE HORNET SIX) Inline. L-head six. Cast iron block. Displacement: 308 cid. Bore and stroke: 3-13/16 x 4-1/2 inches. Compression ratio: 7.5:1. Brake hp: 165 at 3800 rpm. Four main bearings. Hydraulic valve lifters. Carburetor: Carter Type WGD two-barrel Model 2252S.

215

(HORNET TWIN-H SIX) Inline. L-head six. Cast iron block. Displacement: 308 cid. Bore and stroke: 3-13/16 x 4-1/2 inches. Compression ratio: 7.5:1. Brake hp: 175 at 4000 rpm. Four main bearings. Hydraulic valve lifters. Carburetor: Two (2) Carter Type WA1 one-barrel Model 2113S.

(PACKARD EIGHT) V-8. Overhead valves. Cast iron block. Displacement: 352 cid. Bore and stroke: 4 x 3-1/2 inches. Compression ratio: 9.55:1. Brake hp: 220 at 4600 rpm. Five main bearings. Hydraulic valve lifters. Carburetor: Carter Type WGD two-barrel Model 2231SA.

(HORNET SPECIAL EIGHT) V-8. Overhead valves. Cast iron block. Displacement: 250 cid. Bore and stroke: 3-1/2 x 3-1/4 inches. Compression ratio: 8.0:1. Brake hp: 190 at 4900 rpm. Five main bearings. Hydraulic valve lifters. Carburetor: Carter Type WGD two-barrel Model 2352S.

CHASSIS FEATURES: Wheelbase: (Rambler) 108 inches. (Wasp/Hornet Special) 114.25 inches; (Hornet) 121.25 inches. Overall length: (Rambler) 191.14 inches; (Rambler with Continental tire) 198.89 inches; (Wasp/Hornet Special) 202.25 inches; (Hornet Special with Continental tire) 212.25 inches; (Hornet) 209.25 inches; (Hornet with Continental tire) 219.25 inches; (Rambler wagons) 198.89 inches. Front tread: (Rambler) 57.75 inches; (wasp/Hornet Special) 59.5 inches; (Hornet) 59.5 inches. Rear tread: (Rambler) 58 inches; (Wasp/Hornet Special) 59-11/16 inches; (Hornet) 60.5 inches. Tires: (Rambler) 6.40 x 15; (Wasp/Hornet Special) 6.70 x 15; (Hornet Six) 7.10 x 15; (Hornet V-8) 7.60 x 15.

OPTIONS: Power steering, in Rambler ($80). Air conditioning, in Rambler ($345). Radio, in Rambler ($76). Heater, in Rambler ($74). All-Season air conditioning with heater, in Hudson ($395). Twin speaker radio, in Hudson ($98). Hudson Weather-Eye heater and defroster ($77). Power steering, in Hudson ($140). Power brakes, in Hudson ($39). Power windows, in Hudson ($128). Reclining seats and twin beds. Continental rear mount tire. Directional signals. White sidewall tires. Other standard options and accessories. Note: Power steering, power brakes and power windows were available on Hornets and Wasps only, continental rear-mounted tires and reclining seats with twin beds were standard on all Custom models except station wagons, which were not available with the continental tire. Three-speed manual transmission was standard. Ultramatic automatic transmission was a mandatory option in Hornets with the 352 cid V-8. Overdrive transmission ($107). Automatic transmission ($188). Twin-H carburetion. Heavy-duty air cleaner. Available rear axle gear ratios: 3.07:1; 3.31:1; 3.58:1; 3.54:1; 3.15:1; 4.0:1; 4.09:1; 4.40:1; 4.60:1.

HISTORICAL FOOTNOTES: Introduction date for the 1956 Hudson Wasp, Hornet Six and Hornet V-8 was Nov. 30, 1955. The 1956 Hudson Ramblers were then introduced on Dec. 15, 1955. It wasn't until March 5, 1956 that the Hornet Special V-8 appeared and thereafter began showing up in dealer showrooms. Calendar year output amounted to 10,671 units. Model year production hit its peak at 22,588 assemblies, excluding Hudson Ramblers. Seat belts were an option in 1956 and the 12-volt electrical system was introduced. Some 1956 Hudson factory literature included the weights of various options as follows: Hydra-Matic (105-125 pounds); overdrive (30-50 pounds); radio (13-18 pounds); Weather-Eye (18-22 pounds); air conditioning (120 pounds); power steering (42-50 pounds); power brakes (18 pounds); power windows (12 to 18 pounds) and continental tire carrier (65 pounds).

1957 HUDSON

HUDSON HORNET — (V-8) — SERIES 80: Hudson trimmed its model lineup by an amazing figure of 11 cars in 1957, while its roofline was trimmed two inches. The Hudson Rambler, Wasp and Hornet Special Series were dropped. Other new features included 14 inch wheels; a new Hydra-Matic with parking gear; standard 327 cid V-8 with dual exhausts; ball joint front suspension; interiors restyled along more modern lines (with new materials and colors) and standard padded dashboard in all lines. Styling changes included a new 'V' medallion in the center of the grille; dual-fin front fender ornaments; rear tailfin fenders with vertical lamps and a new side trim treatment with front fender and door accent panels. On Supers, the accent panel was painted. There were Hornet and Hollywood nameplates on the hardtops, as well as rear fender 'H' medallions. The Super sedan had Hornet front door nameplates and no 'H' medallions. Hornet Customs could be identified by textured aluminum insert panels used on the front fenders, between the trim moldings. Nameplate and medallion placements were the same as on comparable Super styles.

1957 Hudson, Hornet four-door sedan, V-8

VEHICLE I.D. NUMBERS: The numbering system was the same as for previous models. Number plates were on a plate under the hood at the top center of dash on Hornets. Body Style Numbers took the form 357( )( ) with the first symbol indicating Hudson; the second and third symbols indicating model year; the fourth symbol indicating the car-line or series and the fifth symbol indicating body type. Serial number Y-10501 to Y-14376 were used. Engine numbers were below the right rear exhaust port on V-8s. Engine numbers were N-1001 and up.

HUDSON HORNET

Model Number	Body/Style Number	Body Type & Seating	Factory Price	Shipping Weight	Production Total
SUPER V-8 LINE					
NA	35785-1	4-dr Sed-6P	2821	3631	1,103
NA	35787-1	2-dr HT Cpe-6P	2911	3655	266
CUSTOM V-8 LINE					
NA	35785-2	4-dr Sed-6P	3011	3678	1,256
NA	35785-2	2-dr HT Cpe-6P	3101	3693	483

NOTE 1: Total 5700 Series production, including exports, was 4,108 units.
NOTE 2: An additional 72 model 35660 Hudson Sixes were shipped overseas in 'knocked-down' form.

ENGINE: V-8. Overhead valves. Cast iron block. Displacement: 327 cid. Bore and stroke: 4 x 3.25 inches. Compression ratio: 9.0:1. Brake hp: 255 at 4700 rpm. Five main bearings. Hydraulic valve lifter. Carburetor: four-barrel.

CHASSIS FEATURES: Wheelbase: (Hornet) 121.25 inches. Overall length: (Hornet) 209.25 inches. Front tread: (Hornet) 59-1/16 inches. Rear tread: (Hornet) 60-1/2 inches. Tires (Hornet) 8.00 x 14.

POWERTRAIN OPTIONS: Three-speed manual transmission was standard. Overdrive transmission ($110). Automatic transmission ($232). Available rear axle gear ratios 3.15:1: 4.10:1.

OPTIONS: Power brakes on Super ($40); on Custom (standard). Power steering ($100). Air conditioning ($415); Power windows ($109.50).

HISTORICAL FOOTNOTES: The last Hudsons were introduced on Oct. 25, 1957.

AMC RAMBLER
1958-1966

In many ways the 1958 and later models of the Rambler marked a second coming of the marque.

The history of the modern Rambler goes back to 1950, when it was offered as the first successful modern compact car. In 1954, American Motors Corp. (AMC) was formed when Nash-Kelvinator and Hudson merged. Just before the start of 1958, AMC decided to discontinue Nash and Hudson. Instead, it would concentrate on selling a revised series of cars, including a new version of the original Rambler compact.

Still affectionately known as the "Nash-Rambler" and heralded as such on the 1958 record "Beep Beep" by the Playmates, the little car enjoyed a fair amount of popularity in its new "American" identity (adopted because "Rambler" had become the name of a larger, fancier model line).

Release of this "old-but-new" American was particularly well-timed. It came out immediately after a 1957 economic recession. As a result, high sales of the model increased AMC's industry market share substantially.

In 1958, AMC offerings included more than the just the American, though. The 1957 Rambler model was facelifted. It had a nine inch stretch in wheelbase and overall length. The Ambassador was AMC's full-sized car, continuing a model begun in 1932 by Nash. The Ambassador used the 327 cid V-8 formerly used in top of the line Nashes, not to be confused with the unrelated Chevrolet 327 cid engine.

AMC was the only major U.S. automaker to increase sales in the recession year of 1958. Every other make was down substantially that year. The consumer trend was to smaller cars and the Rambler was the only U.S. compact car around. AMC was at the right place at the right time.

In 1959, the two-door station wagon was revived in the American line and slight styling changes were made to the Rambler, Rebel and Ambassador models. Rambler registrations doubled for 1959, as the introduction of the Studebaker Lark went virtually unnoticed at the AMC factory in Kenosha, Wis.

Things got more competitive in 1960. New compacts were introduced by Ford, Plymouth and Chevrolet. AMC countered with a four-door American. All companies were emphasizing station wagons for 1960. AMC was right there, too, with 14 of them. They included three-seat models in all, but the American, series. As a result, AMC held its own for 1960 despite threats from the "Big Three."

Even more competition came in 1961 as Pontiac, Oldsmobile, Buick and Dodge introduced new smaller cars. The intermediate-sized models were called Rambler Classics for 1961. But, the most exciting news from AMC was the completely facelifted Rambler American series with a four-door station wagon and a convertible added. The American and Classic series came in three levels of trim: Deluxe, Super, and Custom.

Two things disappeared from 1962 AMCs. One was an optional 250 cid V-8 in Classics. The other was the shrinking of the 117 inch Ambassador wheelbase. It now had the 108 inch span of the Classic, but trim and upholstery differences separated the two car-lines. The Super line was discontinued as a middle-market designation and "400" identification was used for a new-top-of-the line model designation. Two-door sedans were added to the Classic and Ambassador series. It was the first time that Rambler offered the body style in this size car since 1957.

A two-door hardtop was added to the American Series for 1963. The Classic and Ambassador got completely new bodies. A restyling included all-new curved side glass. Wheelbases of the new models increased four inches, but overall length was shorter. Model names were revamped. Deluxe and Custom designations were replaced with numbers: American 220, 330 and 440; Classic 550, 660 and 770; and Ambassador 880 and 990.

In 1964, the Rambler American was completely restyled. It looked somewhat like Chrysler's famous turbine car. The wheelbase was increased to 106 inches. The Classic and Ambassador received a minor facelift, including new front grille and rear end designs. A special model of the Classic called the Typhoon was used to introduce AMC's new Torque-Command six-cylinder engine.

The 1965 American was little changed. Classics and Ambassadors got crisper lines in a major facelift. The Ambassador grew four inches in wheelbase and five overall. Convertibles were added to the Classic 770 and Ambassador 990 lines. Much publicity was devoted to the new Marlin, which was introduced as a midyear 1965 model and competitor to the Mustang and Barracuda. It was essentially the Classic with a fastback roof.

The Marlin and Ambassador continued in 1966, but the Rambler name was not applied to them. They were "AMC" models, now. Ads in magazines had blue and red stripes in the lower right-hand corner signifying the new brand name. The copy advised interested buyers to drive the car of their choice at their "American Motors/Rambler dealer." The rest of the AMC story continues in the next section of this catalog.

1958 RAMBLER

1958 Rambler, American two-door sedan, 6-cyl

RAMBLER AMERICAN — (6-CYL) — SERIES 01: The 1955 Nash Rambler was brought back into production as the 1958 American, joining the model lineup in January. It had minor styling revisions. They included a small grille with a rounded-off rectangular shape and plain bodysides. A two-door sedan was the only model available and came in Super and DeLuxe trim levels. The former featured bright metal windshield and beltline trim to distinguish it from DeLuxes. The 195.6 cid six was used, but had the water pump moved to the front.

RAMBLER I.D. DATA: Serial number under hood on firewall's right dash panel; right-hand wheelhouse panel. Six-cylinder motor number on upper left front corner of engine block. V-8 motor number on top at front of block; or front of block; or center, left side of block, above oil pan. [American] Starting serial number M-1001 for Kenosha, Wis. Starting engine number E-1001. [Rambler Six] Starting serial number D-409001. Starting engine numbers B-145001 (standard engine); or CB9001 (power-pack engine). [Rebel] Starting Serial Number for the Rebel V-8 was A-16001. Starting engine number was G-24001. [Ambassador] Starting Serial Number for the Ambassador V-8 was V-27001. Starting engine number was N-17001.

RAMBLER AMERICAN SERIES 01

Model Number	Body/Style Number	Body Type & Seating	Factory Price	Shipping Weight	Production Total
DELUXE LINE					
01	5802	2-dr Bus Cpe-3P	1775	2439	184
01	5806	2-dr Sed-5P	1789	2463	15,765
SUPER LINE					
01	5806-1	2-dr Sed-5P	1874	2475	14,691

1958 Rambler, Super Cross Country four-door station wagon, 6-cyl

RAMBLER SIX — (6-CYL) — SERIES 10: The Rambler Six received new front and rear fenders. It represented a major restyle of the 108 inch wheelbase Rambler body. The new front fenders featured quad headlights on all, but DeLuxes, which had single headlights standard/dual optional). Several states still had laws against dual headlights. The Rambler's rear fenders featured small, restrained tailfins. This was a move towards conforming with current styling trends. The Rambler line was one of the last to add tailfins to its cars and one of the first to drop them. The Custom and Super models also featured new side trim moldings. All models had Rambler nameplates just above the grille. Custom models had Custom nameplates on the rear deck or tailgate, dual sidespear moldings on front fenders and both doors, with a single molding on the rear fenders. Sedans and hardtops had paint inside dual front moldings and wagons had simulated wood-grain trim. Super models had Super nameplates on front fenders and single full-length bodyside moldings. Deluxe models had no nameplates or side moldings. Deluxes included directional signals, hood or fender ornaments, ashtray, baked enamel colors, fuel filter, bumper jack and wheel lug wrench and hubcaps. In addition, Supers had dual horns, step-on parking brake, cigar lighter and door armrests.

RAMBLER SIX SERIES 10

Model Number	Body/Style Number	Body Type & Seating	Factory Price	Shipping Weight	Production Total
DELUXE LINE					
10	5815	4-dr Sed-6P	2047	2947	12,723
10	5818	4-dr Crs Cty Wag-6P	2376	3056	78
SUPER LINE					
10	5815-1	4-dr Sed-6P	2212	2960	29,699
10	5819-1	4-dr Cty Clb HT-6P	2287	2983	983
10	5818-1	4-dr Crs Cty Wag-6P	2506	3069	26,452
CUSTOM LINE					
10	5815-2	4-dr Sed-6P	2327	2968	16,850
10	5818-2	4-dr Crs Cty Wag-6P	2621	3079	20,131

1958 Rambler, Rebel Custom four-door hardtop sedan, V-8

RAMBLER REBEL — (V-8) — SERIES 20: The Rebel shared the styling of the Rambler six, with the biggest difference being the powertrain. Rebel Customs carried V-8 front fender emblems. New was a deep-dip rustproofing process. A wide side trim panel with a 'half-spear-tip' front shape was used. It had a contrasting beauty insert strip. Split fin hood ornaments were used on both Rebels and Rambler sixes, while side trim of the type described above appeared on Custom models up.

RAMBLER REBEL SERIES 20

Model Number	Body/Style Number	Body Type & Seating	Factory Price	Shipping Weight	Production Total
DELUXE LINE (FLEET)					
20	4825	4-dr Sed-6P	2177	3287	22
SUPER LINE					
20	5825-1	4-dr Sed-6P	2342	3300	2,146
20	5828-1	4-dr Crs Cty Wag-6P	2636	3410	1,782
CUSTOM LINE					
20	5825-2	4-dr Sed-6P	2457	3313	2,595
20	5828-2	4-dr Crs Cty Wag-6P	2751	3418	3,101
20	5829-2	4-dr Cty Clb HT-6P	2532	3328	410

AMBASSADOR — (V-8) — SERIES 80: The 1958 Ambassador was built off of the 108 inch Rambler chassis by adding nine inch longer front end sheet metal. The dropping of the Nash and Hudson names from the Ambassador was a last minute decision made by George Romney and American Motor's upper management. A number of cars were made with Nash and Hudson emblems ahead of the Ambassador nameplates on the sides of the front fenders. Factory photos exist which show the Nash Ambassador nameplates on these cars. Early factory literature also has noticeable airbrushing of the Nash emblems out of the catalogs. The 1958 Ambassador carried model identification just above the grille, on front fenders and on the rear deck lid. Side trim featured dual jet-stream side moldings, which were painted a contrasting color. The tone used for the insert harmonized with body colors used on Super models. Silver aluminum side trim was used inside the moldings on Custom models, which also featured model nameplates on the rear deck lid or tailgate. There were three bright metal windsplits on the rear window pillars of hardtops and sedans and on the wide pillars of station wagons. Super nameplates were on rear fenders of Super models.

AMBASSADOR SERIES 80

Model Number	Body/Style Number	Body Type & Seating	Factory Price	Shipping Weight	Production Total
SUPER LINE					
80	5885-1	4-dr Sed-6P	2587	3456	2,774
80	5888-1	4-dr Sta Wag-6P	2881	3544	1,051
CUSTOM LINE					
80	5885-2	4-dr Sed-6P	2732	3462	6,369
80	5889-2	4-dr HT Sed-6P	2822	3475	1,340
80	5888-2	4-dr Sta Wag-6P	3026	3568	2,742
80	5883-2	4-dr HT Wag-6P	3116	3586	294

RAMBLER BASE ENGINES

(RAMBLER AMERICAN) Inline, L-head Six. Cast iron block. Displacement: 195.6 cid. Bore and stroke: 3-1/8 x 4-1/4 inches. Compression ratio: 8.0:1. Brake hp: 90 at 3800 rpm. Four main bearings. Solid valve lifters. Carburetor: Carter Type YF-2014S one-barrel.

(AMBASSADOR) V-8. Overhead valves. Cast iron block. Displacement: 327 cid. Bore and stroke: 4 x 3-1/4 inches. Compression ratio: 9.7:1. Brake hp: 270 at 4700 rpm. Five main bearings. Hydraulic valve lifters. Carburetor: Holley four-barrel Model 4150C.

(RAMBLER SIX) Inline Six. Overhead valves. Cast iron block. Displacement: 195.6 cid. Bore and stroke: 3-1/8 x 4-1/4 inches. Compression ratio: 8.7:1. Brake hp: 127 at 4200 rpm. Four main bearings. Solid valve lifters. Carburetor: Carter Type YF one-barrel Model 2014S.

(REBEL V-8 ENGINE) V-8. Overhead valves. Cast iron block. Displacement: 250 cid. Bore and stroke: 3-1/2 x 3-1/4 inches. Compression ratio: 8.7:1. Brake hp: 215 at 4900 rpm. Five main bearings. Solid valve lifters. Carburetor: Holley four-barrel Model 4150C.

CHASSIS FEATURES: Wheelbase: (American) 100 inches; (Rambler and Rebel) 108 inches; (Ambassador) 117 inches. Overall length: (American) 178.32 inches; (Rambler and Rebel) 191.14 inches; (Ambassador) 200.14 inches. Add 7.75 inches to overall length for Rambler and Ambassador with continental tire carrier. Front tread: (American) 54.62 inches; (Rambler and Ambassador) 57.75 inches; (Rebel) 58.75 inches. Rear tread: (American) 55 inches; (Rambler) 58 inches; (Rebel and Ambassador) 59.13 inches. Tires: (American) 5.90 x 15; (Rambler six) 6.40 x 15; (Rambler Rebel) 7.50 x 14; (Ambassador) 8.00 x 14.

OPTIONS: Power brakes ($38). Power steering ($85). Air conditioning ($369). Weather-Eye heater ($72). Tinted glass ($27). Wheel discs ($16). Rear view mirror ($4). Two-tone paint ($16). Continental tire ($60). Undercoating ($15). Oil filter ($9). Reclining seat ($15). Whitewall tires ($40). Manual radio ($58). Clock ($18). DeLuxe push-button radio ($90). Rear foam cushion. Left outside rear view mirror. Custom steering wheel. Heavy-duty rear springs and shocks. Windshield washer. Ambassador V-8 Power Saver fan. Power windows. Back-O-Matic lights. Inside anti-glare mirror. Heavy-duty springs and shock absorbers. Padded instrument panel and sun visor. Dealer accessories: Seat belts. Travel rack straps. Child guard door locks. Three-speed manual transmission was standard. Overdrive transmission ($112.50). Flash-O-Matic transmission ($200). Push-button Flash-O-Matic transmission ($219.50). Rambler "Power-Pack" 195.6 cid/138 hp two-barrel engine. Dual exhausts ($18.50). Positive traction rear axle ($39.50). Oil bath air cleaner ($7.15). Heavy-duty radiator ($19.85). Available rear axle gear ratios (standard) 4.10:1; (optional) 4.44:1; 3.55:1 and 3.15:1. Size 6.40 x 15 tires for American.

HISTORICAL FOOTNOTES: The full-sized Ramblers were introduced Oct. 22, 1957 and the American appeared in dealer showrooms the same day. Model year production peaked at 162,182 units. Calendar year sales of 199,236 cars were recorded. George Romney was the chief executive officer of the company this year. R.D. Chapin was executive vice president and general manager of the automotive division. The company made a $26 million profit after two straight years of losses. An expansion program was initiated by year's end.

1959 RAMBLER

1959 Rambler, American two-door sedan, 6-cyl

RAMBLER AMERICAN — (6-CYL) — SERIES 01: The American continued as a compact model with the smooth, rounded styling that dated to 1954. There were front fender namescripts, but no side moldings were seen. The grille had a rectangular shape and was surrounded by a chrome housing with rounded corners. A fine, grid type insert was used, the vertical members looking slightly more prominent. A round medallion was housed in the center. Super models had bright metal trim around the windshield and rear window. A chrome molding decorated the upper beltline. DeLuxes had black rubber windshield and rear window trim and lacked belt moldings. A handful of panel deliveries were built off the station wagon platform. One style had glass rear side windows, the second had metal panels in the same location. The Super two-door sedan was rated for one more passenger than other non-commercial models.

RAMBLER I.D. DATA: Serial number under hood on firewall right dash panel; right-hand wheelhouse panel. Six-cylinder motor number on upper left front corner of engine block. V-8 motor number on top at front of block; or front of block; or center, left side of block, above oil pan. [American] Starting serial number M-32001 for Kenosha, Wis. Starting engine number E-33001. [Rambler Six] Starting serial number D-516001. Starting engine numbers B-227001 (standard engine); or CB36001 (power-pack engine). [Rebel] Starting Serial Number for the Rebel V-8 was A-26001. Starting engine number was G-34501. [Ambassador] Starting Serial Number for the Ambassador V-8 was V-41501. Starting engine number was N-32501.

RAMBLER AMERICAN SERIES 01

Model Number	Body/Style Number	Body Type & Seating	Factory Price	Shipping Weight	Production Total
PANEL DELIVERY LINE					
01	5904-7	2-dr Glass Dly	—	—	3
01	5904-8	2-dr Steel Dly	—	—	3
DELUXE LINE					
01	5902	2-dr Bus Sed-5P	1821	2435	443
01	5904	2-dr Sta Wag-5P	2060	2554	15,256
01	5906	2-dr Sed-5P	1835	2476	29,954
SUPER LINE					
01	5904-1	2-dr Sta Wag-5P	2145	2570	17,383
01	5906-1	2-dr Sed-6P	1920	2492	28,449

RAMBLER SIX — (6-CYL) — SERIES 10: Rambler had new grilles, new arrangements of side trim and re-contoured rear tailfins. The new fender treatment blended the fins into the upper beltline in a smooth, down-curving line. On all models, a horizontal gap ran above the upper grille bar and between the dual headlamps. Stand-up chrome letters spelling the word Rambler appeared in the gap. The manufacturer's name also decorated the deck lid tailgate. Custom models had Custom nameplates at the front of the rear missile-shaped side molding. A narrower, straight molding extended along the upperbody side from just behind the headlamps to the missile-shaped molding. Super models had Super nameplates within the tip of the 'missile.' On these cars the straight front molding stopped just beyond the middle of the front door, without hitting the rear trim. DeLuxes had no series nameplates or side trim and dual headlamps were optional. The lower grille insert was of cellular-grid design.

RAMBLER SIX SERIES 10

Model Number	Body/Style Number	Body Type & Seating	Factory Price	Shipping Weight	Production Total
DELUXE LINE					
10	5915	4-dr Sed-6P	2098	2934	26,157
10	5918	4-dr Crs Cty Wag-6P	2427	3047	422
SUPER LINE					
10	5915-1	4-dr Sed-6P	2268	2951	72,577
10	5919-1	4-dr HT Sed-6P	2343	2961	2,683
10	5918-1	4-dr Crs Cty Wag-6P	2562	3082	66,739
CUSTOM LINE					
10	5915-2	4-dr Sed-6P	2383	2956	35,242
10	5918-2	4-dr Crs Cty Wag-6P	2677	3097	38,761

1959 Rambler, Rebel Custom four-door sedan, V-8

RAMBLER REBEL — (V-8) — SERIES 20: The Rebel shared the same body styling features as the Rambler six. The major external difference was that the Custom models had Rebel V-8 emblems on the front fenders, ahead of the wheel openings. The scripts placed within the missile-shaped rear molding panel carried the name of the trim line, except on the plain-looking DeLuxes, which had no script.

RAMBLER REBEL V-8 SERIES 20

Model Number	Body/Style Number	Body Type & Seating	Factory Price	Shipping Weight	Production Total
DELUXE LINE					
20	5925	4-dr Sed-6P	2228	3283	113
SUPER LINE					
20	5925-1	4-dr Sed-6P	2398	3287	3,488
20	5928-1	4-dr Crs Cty Wag-6P	2692	3398	3,634
CUSTOM LINE					
20	5925-2	4-dr Sed-6P	2513	3295	4,046
20	5929-2	4-dr HT Sed-6P	2588	3338	691
20	5928-2	4-dr Crs Cty Wag-6P	2807	3407	4,427

AMBASSADOR EIGHT (V-8) — SERIES 80: The 1959 Ambassador retained the same basic styling seen in 1958, with changes being comparable to those seen on Rambler and Rambler Rebel. The wide gap above the upper grille bar had Ambassador spelled out in stand-up block letters. Similar lettering appeared on the deck lid or tailgate. Custom models had silver aluminum trim inside dual side moldings. ScotchLite reflecting sections were found on the rear face of the back fenders. The four-door hardtop had Custom Country Club written on the rear fender tips. Other cars, depending on trim line, read Custom or Super in the same spot. The front of the side trim spear had Ambassador lettering at its tip. The side trim, although similar to the Rambler type, had more of a lightning bolt shape than a missile shape. The upper molding 'zigged' (but the lower molding did not 'zag') upwards, just below the rear side window. The Ambassador grille was distinctive in that it had a full-width horizontal central bar, instead of a cellular grid type insert. The bar had a V-shaped dip at its center.

1959 Rambler, Ambassador four-door hardtop sedan, V-8

AMBASSADOR EIGHT SERIES 80

Model Number	Body/Style Number	Body Type & Seating	Factory Price	Shipping Weight	Production Total
DELUXE LINE					
20	5985	4-dr Sed-6P	—	3428	155
SUPER LINE					
20	5985-1	4-dr Sed-6P	2587	3428	4,675
20	5988-1	4-dr Crs Cty Wag-6P	2881	3546	1,782

CUSTOM LINE

20	5985-2	4-dr Sed-6P	2732	3437	10,791
20	5989-2	4-dr Cty Clb HT-6P	2822	3483	1,447
20	5988-2	4-dr Cty Clb HT Wag-6P	3026	3562	578
20	5983-2	4-dr Crs Cty Wag-6P	3116	3591	4,341

RAMBLER BASE ENGINES

(RAMBLER AMERICAN) Inline, L-head Six. Cast iron block. Displacement: 195.6 cid. Bore and stroke: 3-1/8 x 4-1/4 inches. Compression ratio: 8.0:1. Brake hp: 90 at 3800 rpm. Four main bearings. Solid valve lifters. Carburetor: Carter Type YF-2014S one-barrel.

(RAMBLER SIX) Inline Six. Overhead valves. Cast iron block. Displacement: 195.6 cid. Bore and stroke: 3-1/8 x 4-1/4 inches. Compression ratio: 8.7:1. Brake hp: 127 at 4200 rpm. Four main bearings. Solid valve lifters. Carburetor: Carter Type YF one-barrel Model 2014S.

(REBEL V-8 ENGINE) V-8. Overhead valves. Cast iron block. Displacement: 250 cid. Bore and stroke: 3-1/2 x 3-1/4 inches. Compression ratio: 8.7:1. Brake hp: 215 at 4900 rpm. Five main bearings. Solid valve lifters. Carburetor: Holley four-barrel Model 4150C.

(AMBASSADOR) V-8. Overhead valves. Cast iron block. Displacement: 327 cid. Bore and stroke: 4 x 3-1/4 inches. Compression ratio: 9.7:1. Brake hp: 270 at 4700 rpm. Five main bearings. Hydraulic valve lifters. Carburetor: Holley four-barrel Model 4150C.

CHASSIS FEATURES: Wheelbase: (American) 100 inches, (Rambler/Rebel) 108 inches; (Ambassador) 117 inches. Overall length: (American) 178.32 inches; (Rambier/Rebel) 191.15 inches; (Ambassador) 200.15 inches. Front tread: (American) 54.62 inches; (Rambler/Ambassador) 57.75 inches; (Rebel) 58.75 inches. Tires: (American station wagon) 6.40 x 15; (American passenger car) 5.90 x 15; (Rambler six) 6.40 x 15; (Rebel V-8) 7.50 x 14; (Ambassador) 8.00 x 14.

OPTIONS

(AMERICAN) Weather-Eye heater ($72). Oil-bath cleaner ($7.15). Oil filter ($9.30). Solex glass ($26.95). Front or rear foam seat cushions ($9.95). Reclining seat ($14.95). Manual radio and antenna with single speaker ($57.70). Size 5.90 x 15, four-ply Rayon tires on sedans ($32.50); size 6.40 x 15 four-ply Rayon tires on sedans ($52.80); size 6.40 x 15 four-ply Rayon tires on station wagons ($36.05). Undercoating ($14.95). Windshield washer ($11). Wheel discs ($15.25). Electric clock ($15.25). Outside rear view mirror ($3.95). Inside tilt mirror ($4.95). Custom steering wheel ($7.70). Heavy-duty springs and shocks ($4.00). Two-tone paint on sedans ($15.95). Two-tone paint on station wagons ($17.95). Continental tire carrier ($59.50). Heavy-duty cooling system ($12.50). Three-speed manual transmission was standard. Overdrive transmission, in American ($102). Flash-O-Matic automatic transmission, in American ($179)).

(RAMBLER/REBEL) Flash-O-Matic push-button transmission with Rambler six ($199.50); with Rebel V-8 ($219.50). Overdrive ($112.50). Power steering with Rambler six ($69.50); with Rebel V-8 ($79.50). Power windows lifts ($99.50). Power brakes ($37.95). Oil filter ($9.75). Oil bath air cleaner on Rambler six ($7.50). Radio with manual antenna and single speaker ($75.65); with dual speaker ($86.40). Weather-Eye heater and ventilator ($76). Size 6.40 x 15 four-ply tires on Rambler six ($35.90); six 6.70 x 15 four-ply tires on Rambler six ($48.30); size 7.50 x 14 four-ply tires on Rebel ($39.90). Wheel discs ($15.95). Two-tone paint on DeLuxe sedans ($16.95); on Super sedans ($18.95); on Custom sedans ($22.95); on Super wagons ($18.95); on Custom wagons ($29.95) and on station wagon solid with Di-Noc ($49.95). Solex tinted glass ($33). Backup lights ($9.95). Continental tire carrier ($69.50). Front or rear foam cushion ($12.50). Reclining front seat ($25.50). Windshield washer ($11.50). Electric clock ($15.95). Padded instrument panel with visors ($19.95). Rambler six power pack ($19.50). Undercoating ($14.95). Left or right headrests ($12). Air conditioning ($369). Rambler Rebel V-8 Power-Lok differential ($39.50). Inside rearview mirror ($4.95). Outside rear view mirror ($5.25). Rebel dual exhausts ($15.50). DeLuxe six-cylinder dual headlights ($23.50). Rear air coil suspension ($98.50). Self-adjusting brakes ($7.45). Three-speed manual transmission was standard. Overdrive transmission, in American ($102); all other models ($113). Flash-O-Matic automatic transmission, in American ($179); in Rambler six ($200); in Rebel V-8 ($220); in Ambassador ($230). Two-barrel carburetor, in Rambler six ($20). Power-Lok positive traction rear axle, in Rambler ($40); in Ambassador ($43). Heavy-duty air cleaner, in Rambler six ($8); in Ambassador (standard). Optional rear axle gear ratios available at no extra charge.

(AMBASSADOR) Flash-O-Matic push-button transmission ($229.50). Overdrive ($114.50). Power steering ($89.50). Power window lifts ($99.50). Power brakes ($39.95). Radio and antenna with dual speakers ($91.90); on station wagons ($87.10). Weather-Eye heater and ventilator ($82.50). Size 8 x 15 four-ply Rayon tires ($43.55); size 8 x 15 four-ply Nylon tires ($67.30). Wheel discs ($16.95). Two-tone paint on Super and Custom ($22.95). Solid with Di-Noc on station wagons ($59.95). Solex tinted glass ($33). Backup lights ($9.95). Continental tire carrier ($76.50). Rear foam cushion ($14.35). Reclining front seat ($25.50). Windshield washer ($11.50). Electric clock ($17.95). Padded instrument panel and visors ($19.95). Undercoating ($14.95). Air conditioning ($398). Power-Lok differential ($42.50). Inside rear view mirror ($4.95). Outside rear view mirror ($5.25). Power-Saver fan ($19.50). Heavy-duty radiator ($8). Self-adjusting brakes ($7.45). Heavy-duty cooling system ($19.85). Dual exhausts ($18.50). Adjustable front seats ($20). Left and right headrests ($24). Three-speed manual transmission was standard. Power-Lok positive traction rear axle ($43). Heavy-duty air cleaner standard in Ambassador. Optional rear axle gear ratios available at no extra charge.

HISTORICAL FOOTNOTES: The full-sized Ramblers were introduced Oct. 8, 1958 and the Americans appeared in dealer showrooms the same day. Model year production peaked at 374,240 units. Calendar year sales of 401,446 cars were recorded. George Romney was the chief executive officer of the company this year. A record profit of $60,341,823 was earned. Rambler qualified as America's fourth-ranked maker. Resale value on Ramblers was considered high at this time, a point Rambler salesmen often stressed. America was in a recession and a small economy car trend had begun. Rambler benefited greatly, achieving net sales of $869,849,704.

1960 RAMBLER

RAMBLER AMERICAN — (6-CYL) — SERIES 01: Annual styling revisions included removal of upper beltline moldings from below the windows of all Americans. A new four-door sedan was introduced in this series. A rooftop luggage rack was standard on the station wagons. A minor change was an increase in the door opening angle, from 55- to 75-degrees. Supers had a horizontal front spear on the doors and fenders and American front fender scripts. Customs had a similar chrome spear and, like Supers, used chrome windshield and rear window surrounds. DeLuxe models had no side trim or bright metal window surrounds. Custom, Super or DeLuxe scripts appeared at the rear of all cars and all had either American or Rambler American scripts on the front fendersides.

RAMBLER I.D. DATA: VIN on plate on right wheelhouse panel below hood. First symbol identified series: A=Rebel V-8; B=American; C=Rambler; H=Ambassador. Second through fifth symbols were sequential production number. Starting serial numbers were: [American] B-100001; [American for knocked-down export] BK-10001; [Rambler] C-100001; [Rambler for knocked-down export] CK-10001; [Rebel V-8] A-100001 and [Rebel V-8 for knocked-down export] AK-10001; [Ambassador V-8] H-100001; [Ambassador V-8 for knocked-down export]. for Ambassador V-8, HK-10001 for Ambassador V-8 for unassembled export. Body Number plate riveted to left front door hinge pillar has additional data. First line indicates body production sequence number. Second line indicates "Model No." consisting of first two symbols indicating model year and additional symbols indicating body style and series. Complete "Model No." corresponds to column two of charts below. Third line gives trim data. Fourth line gives paint color data. Six-cylinder motor number on upper left front corner of engine block. V-8 motor number on top at front of block; or front of block; or center, left side of block, above oil pan. Starting in late 1959 for the 1960 model year, AMC discontinued engine serial numbers. An "Engine Day Build Code" system gave all engines of a type the same six symbol code. The first symbol of the engine day built code is a single digit number code for the year of manufacture starting with 1=1959; 2=1960; 3=1961; 4=1962; 5=1963; 6=1964; 7=1965; 8=1966; 9=1967; 1-1968; 2-1969; etc. The second and third symbols indicated month of build: 01=Jan.; 02=Feb.; etc. The numbers differed in the fourth symbol, a letter code designating the engine displacement and carburetion. Letter codes were: A=195.6cid/90 hp L-head six; B=195.6 cid/127 hp aluminum OHV six; C=195.6 cid/127 hp cast-iron OHV six; D=195.6 cid/138 hp six 2V; E=327 cid/250 hp V-8 2V; F=327 cid/270 hp V-8 4V; G="287" engine. The fifth and sixth symbols indicated day of manufacture as appropriate. In motor number 201A22: first symbol 2=1960; 01=Jan.; A=195.6 cid/90 hp six and 22=22nd day of month.

RAMBLER AMERICAN SERIES 01

Model Number	Body/Style Number	Body Type & Seating	Factory Price	Shipping Weight	Production Total
DELUXE LINE					
01	6005	4-dr Sed-5P	1844	2474	22,593
01	6006	2-dr Sed-5P	1795	2451	23,960
01	6004	2-dr Sta Wag-5P	2020	2527	12,290
01	6002	2-dr Bus Cpe-3P	1781	2428	630
SUPER LINE					
01	6005-1	4-dr Sed-6P	1929	2490	21,108
01	6006-1	2-dr Sed-5P	1880	2462	17,233
01	6004-1	2-dr Sta Wag-6P	2105	2549	15,093
CUSTOM LINE					
01	6005-2	4-dr Sed-6P	2059	2551	3,272
01	6006-2	2-dr Sed-5P	2010	2523	2,994
01	6004-2	2-dr Sta Wag-5P	2235	2606	1,430

RAMBLER SIX — (6-CYL) — SERIES 10: The Rambler's tailfins were lowered and canted out slightly. A convex lower rear fender contour was new. The Rambler grille was redesigned. It had two rows of 'cells' in its gridwork, instead of three. Stand-up chrome letters still spelled out Rambler in the horizontal space between the head lamps, above the grille. The word Rambler was also on the rear. Customs had a suitable script on the rear also, plus full-length trim moldings that widened into dual moldings between the wheel housings. Supers had different model identifying scripts at the rear and full-length side trim moldings. DeLuxe models had a suitable rear scripts, but no side trim. Dual headlamps were optional on DeLuxes. On all lines, the twin fin hood ornaments of the past were gone and were not replaced.

RAMBLER SIX SERIES 10

Model Number	Body/Style Number	Body Type & Seating	Factory Price	Shipping Weight	Production Total
DELUXE LINE					
10	6015	4-dr Sed-6P	2098	2912	37,666
10	6018	4-dr Sta Wag-6P	2427	3051	24,001
SUPER LINE					
10	6015-1	4-dr Sed-6P	2268	2930	88,004
10	6018-1	4-dr Sta Wag-6P	2562	3054	59,491
10	6018-3	4-dr Sta Wag-8P	2687	3117	8,456
CUSTOM LINE					
10	6015-2	4-dr Sed-6P	2883	2929	38,003
10	6910-2	4-dr HT Sed-6P	2458	2981	3,937
10	6018-2	4-dr Sta Wag-6P	2677	3057	32,092
10	6018-4	4-dr Sta Wag-8P	2802	3137	5,718

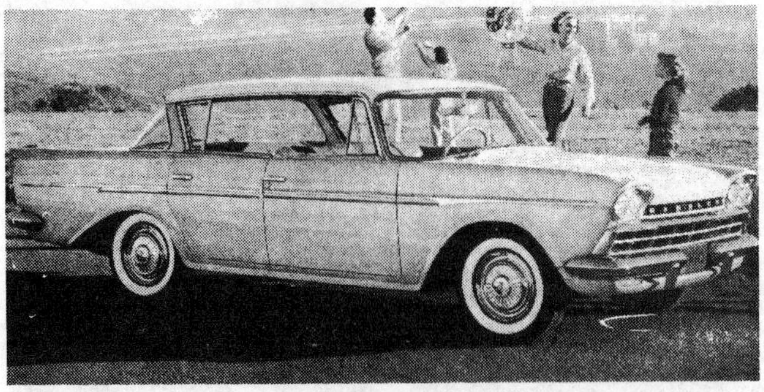

1960 Rambler, Custom Country Club four-door hardtop, 6-cyl

RAMBLER REBEL — (V-8) — SERIES 20: Rebels had the same general styling changes as Rambler sixes. They included lower, canted fins; convex rear fenders; grille with two rows of cellular grids; side trim revisions; and deletion of hood ornaments. Rambler sales features for 1960 included single unit-body construction; deep-dip rustproofing; deep coil spring ride; self-adjusting brakes (optional); push-button automatic gear shift selection (optional); and optional 'Air Liner' reclining seats. Rebel V-8 emblems appeared on front fenders again this year. A new base power plant was used. The former 215 hp four-barrel V-8 (with standard dual exhausts) became optional. A number of Ramblers were built for export this year. These cars were shipped overseas in completely knocked-down (CKD) form, or in other words, unassembled. This meant that the importing country was providing employment for workers who would assemble the cars there. Such cars had a two symbol alphabetical Serial Number prefix, with the second letter 'K' indicating 'CKD' sales.

RAMBLER REBEL V-8 SERIES 20

Model Number	Body/Style Number	Body Type & Seating	Factory Price	Shipping Weight	Production Total
DELUXE LINE					
20	6025	4-dr Flt Sed-6P	2217	3252	143
SUPER LINE					
20	6025-1	4-dr Sed-6P	2387	3270	3,826
20	6028-1	4-dr Sta Wag-6P	2681	3391	3,328
20	6028-3	4-dr Sta Wag-8P	2806	3446	718
CUSTOM LINE					
20	6025-2	4-dr Sed-6P	2502	3278	3,969
20	6029-2	4-dr HT Sed-6P	2577	3319	579
20	6028-2	4-dr Sta Wag-6P	2796	3395	3,613
20	6028-4	4-dr Sta Wag-8P	2921	3447	886

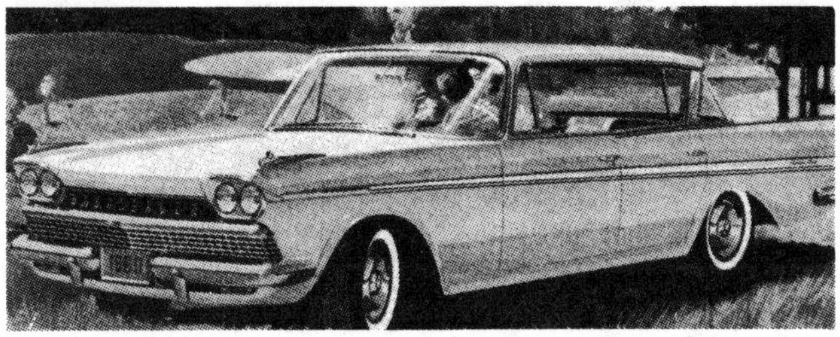

1960 Rambler, Ambassador Country Club four-door HT, V-8

AMBASSADOR SERIES — (V-8) — SERIES 80: On the top line series the letters above the grille spelled Ambassador, a name which appeared on the deck lid or tailgate as well. This series featured the new compound wraparound windshield and a distinctive grille design. Fashioned of aluminum, the lower insert ran fully across the car with a pattern of medium sized square openings stamped out of the metal. Bombsight style front fender ornaments were seen. Side trim consisted of dual moldings running in a tapering line from the middle of the extreme rear body corner and coming to a point just in back of the dual headlamps. On Customs an aluminum beauty panel insert was placed within the moldings and, on all models, the word Ambassador appeared at the tip. In addition, scripts placed on the deck lid of passenger models or tailgate of station wagons identified Supers or Customs. Foam rear seat cushions; full wheel discs; electric clock; padded dashboard and padded sun visors were standard on Customs. A cheap DeLuxe sedan was built in limited numbers for fleet use only and not cataloged with regular cars.

AMBASSADOR EIGHT SERIES 80

Model Number	Body/Style Number	Body Type & Seating	Factory Price	Shipping Weight	Production Total
DELUXE LINE					
80	6085	4-dr Flt Sed-6P	2395	3384	302
SUPER LINE					
80	6085-1	4-dr Sed-6P	2587	3395	3,990
80	6088-1	4-dr Sta Wag-6P	2881	3531	1,342
80	6088-3	4-dr Sta Wag-8P	3006	3581	637
CUSTOM LINE					
80	6085-2	4-dr Sed-6P	2732	3408	10,949
80	6089-2	4-dr Cty Clb HT-6P	2822	3465	1,141
80	6088-2	4-dr Sta Wag-6P	3026	3538	3,849
80	6083-2	4-dr Cty Clb HT Wag-6P	3116	3583	435
80	6088-4	4-dr Sta Wag-8P	3151	3592	1,153

RAMBLER BASE ENGINES

(RAMBLER AMERICAN) Inline, L-head Six. Cast iron block. Displacement: 195.6 cid. Bore and stroke: 3-1/8 x 4-1/4 inches. Compression ratio: 8.0:1. Brake hp: 90 at 3800 rpm. Four main bearings. Solid valve lifters. Carburetor: Carter Type YF-2014S one-barrel.

(RAMBLER SIX) Inline Six. Overhead valves. Cast iron block. Displacement: 195.6 cid. Bore and stroke: 3-1/8 x 4-1/4 inches. Compression ratio: 8.7:1. Brake hp: 127 at 4200 rpm. Four main bearings. Solid valve lifters. Carburetor: Holley model 1904-FC one-barrel.

(REBEL V-8) V-8. Overhead valves. Cast iron block. Displacement: 250 cid. Bore and stroke: 3-1/2 x 3/1-4 inches. Compression ratio: 8.7:1. Brake hp: 200 at 4900 rpm. Five main bearings. Hydraulic valve lifters. Carburetor: Holley model 2040 two-barrel.

(AMBASSADOR EIGHT) V-8. Overhead valves. Cast iron block. Displacment: 327 cid. Bore and stroke: 4 x 3-1/4 inches. Compression ratio: 8.7:1. Brake hp: 250 at 4700 rpm. Five main bearings. Hydraulic valve lifters. Carburetor: Holley model 2040 two-barrel.

CHASSIS FEATURES: Wheelbase: (Rambler/Rebel) 108 inches; (American) 100 inches; (Ambassador) 117 inches. Overall length: (American) 178.32 inches; (Rambler/Rebel) 189.5 inches; (Ambassador) 198.5 inches. Front tread: (Rambler/Rebel) 58.75 inches; (Ambassador) 57.75 inches. Rear tread: (Rambler/Rebel) 59.13 inches; (Ambassador) 59.13 inches. Tires: (American standard passenger car) 5.90 x 15; (American Custom and station wagon) 6.40 x 15; (Rambler six) 6.40 x 15; (Rambler Rebel) 7.50 x 14; (Ambassador) 8.00 x 14.

OPTIONS

(AMERICAN) Anti-freeze ($3.80). Two-tone color on sedans ($15.95); on station wagons ($17.95); special application (except Fleet) ($29.50). Overdrive transmission ($102). Lever control Flash-O-Matic transmission ($178.50). Weather-Eye ($72). Power steering ($69.50). Oil bath air cleaner ($7.15). Oil filter ($9.30). Solex glass ($26.95). Front and rear DeLuxe foam rubber seat cushions ($19.90); front only ($9.95); rear only ($9.95). Reclining seat ($14.65). Manual radio and antenna ($57.70). Size 5.90 x 15 four-ply Rayon white sidewall tires on sedans ($27.40); size 6.40 x 15 four-ply Rayon white sidewall tires on sedans ($44.45); size 6.40 x 15 four-ply nylon white sidewall tires on sedans ($60.80) size 6.40 x 15 four-ply Rayon white sidewall tires on sedans ($30.35); size 6.40 x 15 four-ply nylon white sidewall tires on wagons ($46.70). Undercoating ($14.95). Windshield washer ($11). Backup lights ($9.95); Wheel discs ($15.25). Electric clock ($15.25). Continental tire carrier on sedans ($59.50). Self-adjusting brakes ($7.45). Inside "tilt" rear view mirror ($4.95). Outside left or right rear view mirror ($3.95); left and right ($7.90). Custom steering wheel ($7.70). Twin-Grip differential ($29.50). Pair of license plate frames ($4.05); rear only ($2.05). Heavy-duty front and rear shocks ($2.70). Heavy-duty front and rear shocks and rear springs ($4.00). Heavy-duty radiator ($3.00). Heavy-duty cooling system ($12.50). Three-speed manual transmission was standard. Overdrive transmission ($102). Automatic transmission ($179). Custom American six-cylinder 195.6 cid 125 hp two-barrel engine. Heavy-duty air cleaner. Available rear axle gear ratios.

(RAMBLER/REBEL) Flash-O-Matic push-button transmission on Rambler six ($199.50); on Rebel V-8 ($219.50). Overdrive ($112.50). Power steering on Rambler six ($69.50); on Rebel V-8 ($79.50). Power windows lifts ($99.50). Power brakes ($37.95). Oil filter ($9.75). Oil-bath air cleaner on Rambler six ($7.50). Radio with manual antenna and single speaker ($75.65); with dual speaker ($86.40). Weather-Eye heater and ventilator ($76). Solex tinted glass ($33). Back-up lights ($9.95). Continental tire carrier ($69.50). Front or rear foam cushion ($12.50); reclining front seat ($25.50). Windshield washer ($11.50). Electric clock ($15.95). Padded instrument panel and visors ($19.95). Undercoating ($14.95). Left or right headrests ($12). Air conditioning ($369). Inside rear view mirror ($4.95). Outside rear view mirror ($5.25). DeLuxe six-cylinder dual headlights ($23.50). Rear air coil suspension ($98.50). Self-adjusting: brakes ($7.45). Wheel discs ($15.95). Twin grip differential on six-cylinder ($34.50); on V-8 ($39.50). Parking brake warning light ($3.95). Light package including trunk or cargo, two courtesy, parking and brake warning lights) ($9.95). Two-tone paint on sedans, except Custom ($19.95); two-tone paint on Custom ($24.95); two-tone paint on wagons except Custom ($21.95); two-tone paint on Custom wagons ($29.95); solid with Di-Noc ($59.95). Power Pak on six-cylinder ($19.50); on V-8 ($37.50). Size 6.40 x 15 Rayon white sidewall four-ply tires on six-cylinder except three-seat wagon ($30.30); Size 6.40 x 15 nylon ($46.70); size 6.70 x 15 Rayon ($40.50). Size 7.50 x 14 white sidewall four-ply V-8 except three-seat wagon ($32.10); size 7.50 x 14 nylon ($49.15). Size 6.40 x 15 nylon white sidewall four-ply Captive Air for three-seat wagons on six-cylinder ($28.10); size 7.50 x 14 nylon four-ply on V-8 ($29.50). Three-speed manual transmission was standard. Overdrive transmission ($113). Automatic transmission ($200). Rebel V-8 250 cid 215 hp four-barrel engine ($80). Positive traction rear axle available at extra cost. Heavy-duty air cleaner available on some as standard equipment; others as extra cost. Available rear axle gear ratios.

(AMBASSADOR) Overdrive transmission ($114.50). Push-button Flash-O-Matic transmission ($229.50). Power Pack ($229.50). Weather-Eye ($37.50). Air conditioning and heavy-duty cooling system ($398.00). Power brakes ($39.95). Power steering ($89.50). Power window lifts ($99.50). Solex glass ($99.50). Front foam rubber seat cushions ($14.35). Rear foam rubber seat cushions ($14.35). Reclining seat ($25.50). Left and right headrests ($24). Left or right headrest ($12). Individually adjustable front seats ($20). Radio and antenna with front speakers for wagons ($75.65); for sedans ($91.90). Size 8.00 x 14 white sidewall Rayon four-ply tires for station wagons. (except three-seat wagon) ($35.25); size 8.00 x 14 white sidewall nylon four-ply ($54.15); Captive Air nylon tires, size 8.00 x 14 white sidewall (four tires) ($113.70); Size 8.00 x 14 white sidewall (five tires) ($169.45); size 8.00 x 14 (four) white sidewall for three-seat wagons ($32.40). Undercoating ($14.95). Windshield washer ($11.50). Back-up lights ($9.95). Wheel discs ($16.95). Electric clock ($16.95). Parking brake warning light ($3.95). Padded instrument panel and sun visors ($19.95). Self-adjusting brakes ($7.45). Inside "tilt" rear view mirror ($4.95). Left or right outside rearview mirror ($5.25); left and right outside rearview mirror ($10.50). Power-Saver fan ($19.50). Twin-Grip differential ($42.50). Pair of license plate frames ($4.05); rear only ($2.05). Light package ($9.95). Heavy-duty front and rear shocks ($3.80). Heavy-duty front and rear shocks and front springs with heavy-duty rear springs in sedans ($5.00). Extra heavy-duty springs in sedans ($7.25). Heavy-duty rear springs on wagon ($6.50). Air-Coil ride rear suspension ($98.50). Heavy-duty radiator ($8.00). Heavy-duty cooling system ($19.85). Two-tone, DeLuxe sedans ($22.95); two-tone, Super and Custom sedans ($24.95); two-tone, Super and Custom wagons ($29.95); solid paint with Di-Noc on wagons ($59.95). Three-speed manual transmission was standard. Overdrive transmission ($115). Automatic transmission. Ambassador V-8 327 cid 270 hp four-barrel engine ($90). Positive traction rear axle available at extra cost. Heavy-duty air cleaner available on some as standard equipment; others as extra cost. Available rear axle gear ratios.

HISTORICAL FOOTNOTES: The 1960 Ramblers were introduced Oct. 14, 1959. Model year production peaked at 458,841 units. Calendar year sales of 485,745 cars were recorded. George Romney was the chief executive officer of the company this year. Annual sales topped the billion dollar mark for the first time in the company's six-year history. A new lake front body plant was constructed for 1961 assemblies. Of all 1960 Ramblers built, 49.4 percent had automatic transmission; 8.9 percent had V-8 engines; 9.4 percent had power brakes; 0.4 percent had power windows; 7.5 percent had tinted glass; 3.8 percent had air-conditioning; 1.6 percent had dual exhausts; and 14.6 percent used overdrive transmissions.

1961 RAMBLER

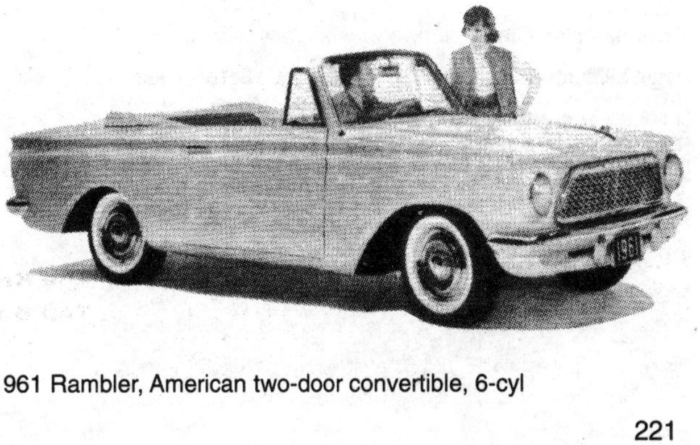

1961 Rambler, American two-door convertible, 6-cyl

RAMBLER AMERICAN — (6-CYL) — SERIES 01: America's recognized economy king was all-new in style and beauty for 1961. The American was even more compact than ever and easier to park. Features included all-welded, single unit construction, deep-dip rustproofing and a ceramic armored muffler and tailpipe guaranteed to the original owner for the life of the car. Styling features included a new trapezoidal grille insert. It was made of perforated aluminum lattice work. The hood carried a stand-up 'R' ornament. The simple rear end design featured circular taillamps and optional round back-up lamps. A new body style was a two-door convertible. Standard equipment on DeLuxe models included air cleaner; front armrests; front ashtray; cigar lighter; dual sun visors; turn signals; black floor mats; black rubber cargo mats on wagons; and black tubeless tires. Super models had all of the above, plus rear door armrests; rear ashtray; colored rubber mats on wagons; colored floor mats; automatic dome switches on front doors; station wagon travel rack; front foam cushion; and chrome horn ring. Custom models had all of the above plus oil-bath air cleaner; colored carpet on station wagons; dual horn; two-tone steering wheel; and wheel discs.

RAMBLER I.D. DATA: VIN on plate on right wheelhouse panel below hood. First symbol identified series: A=Rebel V-8; B=American; C=Rambler; H=Ambassador. Second through last symbols were sequential production number. Starting serial numbers were: [American] B-221001; [American for knocked-down export] CK-10701; [Rambler] C-400001; [Rambler for knocked-down export] CK-10401; [Rebel V-8] A-118001 and [Rebel V-8 for knocked-down export] AK-10001; [Ambassador V-8] H-125001; [Ambassador V-8 for knocked-down export]. for Ambassador V-8, HK-10001 for Ambassador V-8 for unassembled export. Body Number plate riveted to left front door hinge pillar has additional data. First line indicates body production sequence number. Second line indicates "Model No." consisting of two symbols indicating model year and additional symbols indicating body style and series. Complete "Model No." corresponds to column two of charts below. Third line gives trim data. Fourth line gives paint color data. Six-cylinder motor number on upper left front corner of engine block. V-8 motor number on top at front of block; or front of block; or center, left side of block, above oil pan. Continuing in 1961, AMC employed an "Engine Day Build Code" system giving all engines of a type the same six symbol code. The first symbol of the engine day built code is a single digit number code for the year of manufacture starting with 3=1961. The second and third symbols indicated month of build: 01=Jan.; 02=Feb.; etc. The numbers differed in the fourth symbol, a letter code designating the engine displacement and carburetion. Letter codes were: A=195.6 cid/90 hp L-head six; B=195.6 cid/125 hp aluminum OHV six; C=195.6 cid/127 hp cast-iron OHV six; n.a.=195.6 cid/138 hp six 2V; n.a.=250 cid/200 hp V-8; n.a.=250 cid/215 hp V-8; E=327 cid/250 hp V-8 2V; F=327 cid/270 hp V-8 4V; G="287" engine. The fifth and sixth symbols indicated day of manufacture as appropriate. In motor number 301A22: first symbol 3=1961; 01=Jan.; A=195.6 cid/90 hp six and 22=22nd day of month.

RAMBLER AMERICAN SERIES 01

Model Number	Body/Style Number	Body Type & Seating	Factory Price	Shipping Weight	Production Total
DELUXE LINE					
01	6102	2-dr Flt Bus Cpe-3P	1831	2454	355
01	6106	2-dr Sed-6P	1845	2480	28,555
01	6105	4-dr Sed-6P	1894	2513	17,811
01	6108	4-dr Sta Wag-6P	2129	2583	7,260
01	6104	2-dr Sta Wag-6P	2080	2539	5,666
SUPER LINE					
01	6105-1	4-dr Sed-6P	1979	2520	15,741
01	6106-1	2-dr Sed-6P	1930	2489	14,349
01	6108-1	4-dr Sta Wag-6P	2214	2590	10,071
01	6104-1	2-dr Sta Wag-6P	2165	2546	5749
CUSTOM LINE					
01	6105-2	4-dr Sed-6P	2109	2578	5,920
01	6105-2	4-dr Cus Sed-5P	—	—	1,629
01	6106-2	2-dr Sed-6P	2060	2547	4,883
01	6107-2	2-dr Conv-6P	2369	2732	10,855
01	6107-2	2-dr Cus Conv-5P	—	2745	2,063
01	6108-2	4-dr Sta Wag-6P	2344	2648	3,679
01	6104-2	2-dr Sta Wag-6P	2295	2607	1,417

NOTE 1: Customs have five-passenger, bucket seating.

1961 Rambler, Classic four-door station wagon, 6-cyl

RAMBLER CLASSIC six — (6-CYL) — SERIES 10: The Rambler six was renamed for 1961. It received a new front end featuring a one-piece rectangular, extruded aluminum grille with the letters Rambler underneath, just above the bumper. The park/turn lights were located just below the front bumper. New front and rear bumpers were also used along the new side trim. DeLuxe Series standard equipment included: turn signals; twin panel ashtrays; air cleaner; front armrests: cigar lighter; dual headlamps; dual sun visors; station wagon travel rack; and five black tubeless tires. Super Series models were equipped the same as above, plus dual horns; rear door armrests; front foam cushions; and rear ashtrays. Custom Series models were equipped with all above, plus wheel discs; electric clock; glove box light; two-tone steering wheel; carpets; and rear vent window. A major engineering change and a first for the industry was the introduction of a die-cast aluminum six-cylinder engine for Classic Custom models. The special die-cast block was made of an innovative aluminum-silicon alloy with centrifugal cast iron cylinders bonded to the block. This was advertised as "America's first Die-Cast aluminum six" and was optional on DeLuxe and Super models in the Classic Series. Other highlights for this line were freshly sculptured side styling and one-piece bumpers.

RAMBLER CLASSIC SIX SERIES 10

Model Number	Body/Style Number	Body Type & Seating	Factory Price	Shipping Weight	Production Total
DELUXE LINE					
10	6115	4-dr Sed-6P	2098	2905	40,398
10	6118	4-dr Sta Wag-6P	2437	3014	19,848
SUPER LINE					
10	6115-1	4-dr Sed-6P	2268	2923	62,563
10	6118-3	5-dr Sta Wag-8P	2697	3087	4,465
10	6118-1	4-dr Sta Wag-6P	2572	3047	38,370
CUSTOM LINE					
10	6115-2	4-dr Sed-6P	2413	2898	26,497
10	6115-2	4-dr Cus Sed-5P	—	2853	2,901
10	6118-2	4-dr Sta Wag-6P	2717	2984	16,394
10	6118-4	5-dr Sta Wag-8P	2842	3023	2,741

NOTE 1: Customs have five-passenger, bucket seating.

RAMBLER CLASSIC EIGHT (V-8) — SERIES 20: Like the Classic six, the Rambler Classic V-8 gave buyers the best of both worlds, big car room and comfort, compact car economy and handling ease. Styling for both lines was the same, as was the availability of three different levels of trim: DeLuxe, Super and Custom. As on American and Classic six models, a chrome signature placed on the right-hand lower corner of the deck lid identified the line that each car was in. A Classic script was attached to the front fendersides, below the tapering dual side moldings. Cars with the V-8 had large badges attesting to the fact located under the namescript directly behind the front wheel opening. This motor was advertised as a 'high-performance' option.

RAMBLER CLASSIC EIGHT SERIES 20

Model Number	Body/Style Number	Body Type & Seating	Factory Price	Shipping Weight	Production Total
DELUXE LINE					
20	6125	4-dr Sed-6P	2227	3237	121
SUPER LINE					
20	6125-1	4-dr Sed-6P	2397	3255	2,156
20	6128-1	4-dr Sta Wag-6P	—	—	1,964
20	6128-3	5-dr Sta Wag-8P	2826	3408	382
CUSTOM LINE					
20	6125-2	4-dr Sed-6P	2512	3252	2,071
20	6128-4	5-dr Sta Wag-8P	2941	3420	382
20	6128-2	4-dr Sta Wag-6P	2816	3378	1,777
CUSTOM 400 LINE					
20	6125-5	4-dr Sed-6P	2662	3283	109

1961 Rambler, Ambassador four-door sedan, V-8

AMBASSADOR EIGHT — (V-8) — SERIES 80: The 1961 Ambassador was promoted as a high-performance luxury compact. The body styling was highly revised at the front end. Although based on Rambler Classic type sheet metal, the front wheel panels were extended into a highly sculptured, bullet-shaped panel which blended into protruding and flat, bullet-shaped fender ends. The entire front panel of the car was set back several inches from the fender tips and incorporated a new one-piece aluminum grille, one-piece front bumper and bold dual headlights. The grille was 'veed' on its horizontal plane and angled backwards on its vertical axis. This gave the front of the car a 'shovel nose' look. There were seven fine horizontal bars, a trapezoidal outer surround and a gold Ambassador script in the lower left corner. An identification shield also appeared at the center of the hood, above the grille. Side trim included series identification scripts (Super or Custom) ahead of the front wheel openings; a horizontal side spear that branched out into a dual molding on the rear doors and fenders; anodized aluminum insert on Customs; identification shields on the roof pillar; and fin-shaped front fendertop ornaments with bright metal extensions towards the rear. The Ambassador DeLuxe had all equipment that was standard on Classic DeLuxe, plus dual horns. The Ambassador Super series has, in addition to Super equipment listed for Classic, hood insulation, glove box light and rear door vent window. Ambassador Custom models have all Classic Custom features, plus Handi-Pak Carrier, padded dash and visors, rear foam cushion and hood insulation.

AMBASSADOR EIGHT SERIES 80

Model Number	Body/Style Number	Body Type & Seating	Factory Price	Shipping Weight	Production Total
DELUXE LINE					
80	6185	4-dr Flt Sed-6P	2395	3343	273
SUPER LINE					
80	6185-1	4-dr Sed-6P	2537	3361	3,299
80	6188-1	4-dr Sta Wag-6P	2841	3493	1,099
80	6188-3	5-dr Sta Wag-8P	2966	3560	277

CUSTOM LINE

80	6185-2	4-dr Sed-6P	2682	3370	9,269
80	6188-2	4-dr Sta Wag-6P	2986	3495	3,010
80	6188-4	5-dr Sta Wag-8P	3111	3566	784

CUSTOM 400 LINE

80	6185-5	4-dr Sed-5P	2812	3387	831

NOTE 1: Customs have five-passenger, bucket seating.
NOTE 2: AMC called wagon with side-hinged tailgate a five-door station wagon.

RAMBLER BASE ENGINES

(AMERICAN SIX) Inline, L-head Six. Cast iron block. Displacement: 195.6 cid. Bore and stroke: 3-1/8 x 4-1/4 inches. Compression ratio: 8.0:1. Brake hp: 90 at 3800 rpm. Four main bearings. Solid valve lifters. Carburetor: Carter Type YF-2014S one-barrel.

(DELUXE/SUPER SIX) Inline L-head six. Cast iron block. Displacement: 195.6 cid. Bore and stroke: 3-1/8 x 4-1 /4 inches. Compression ratio: 8.0: 1. Brake hp: 90 at 3800 rpm. Four main bearings. Solid valve lifters. Carburetor: Carter Type YF one-barrel Model 2014S.

(CUSTOM SIX) Inline six. Overhead valves. Cast iron block. Displacement: 195.6 cid. Bore and stroke: 3-1/8 x 4-1/4 inches. Compression ratio: 8.7:1. Brake hp: 125 at 4200 rpm. Four main bearings. Solid valve lifters. Carburetor: Holley one-barrel Model 1908-FC.

(CLASSIC SIX) Inline six. Overhead valves. Cast aluminum block. Displacement: 195.6 cid. Bore and stroke: 3.125 x 4.25 inches. Compression ratio: 8.7:1. Brake hp: 127 at 4200 rpm. Four main bearings. Solid valve lifters. Carburetor: Holley one-barrel Model 1908FC.

(REBEL V-8) V-8. Overhead valves. Cast iron block. Displacement: 250 cid. Bore and stroke: 3-1/2 x 3/1-4 inches. Compression ratio: 8.7:1. Brake hp: 200 at 4900 rpm. Five main bearings. Hydraulic valve lifters. Carburetor: Holley model 2040 two-barrel.

(AMBASSADOR EIGHT) V-8. Overhead valves. Cast iron block. Displacement: 326.7 cid. Bore and stroke: 4.00 x 3.25 inches. Compression ratio: 8.7:1. Brake hp: 250 at 4700 rpm. Five main bearings. Hydraulic valve lifters. Carburetor: Holley two-barrel Model H-2040.

CHASSIS FEATURES: Wheelbase: (American) 100 inches, (Classic) 108 inches, (Ambassador) 117 inches. Overall length: (American) 173.1 inches; (Classic) 189.8 inches; (Ambassador) 199 inches. Overall width: (American) 70.0 inches; (Classic) 72.4 inches. (Ambassador) 73.6 inches. Tires: (American) 6.00 x 15 inches: (Classic six) 6.50 x 15; (Classic Eight) 7.50 x 14; (Ambassador) 8.00 x 14).

OPTIONS

(AMERICAN) Anti-freeze ($3.80). Two-tone paint on sedans ($15.95); on station wagons ($17.95); Special application (29.50). Overdrive transmission ($102). Lever control Flash-O-Matic ($164.85). Weather-Eye ($74). Overhead valve engine ($59.50). Air conditioning with heavy-duty cooling system ($359). Power brakes ($37.95). Power steering ($72). Oil-bath air cleaner ($7.15). Oil filter ($9.30). Solex glass ($26.95). Front and rear DeLuxe foam rubber seat cushion ($19.90); front only ($9.95); rear only ($9.95). Reclining seat ($25.50). Undercoating ($14.95). Windshield washer ($11). Back-up lights ($9.95). Wheel discs ($15.25). Electric clock ($15.25). Continental tire carrier on sedans ($59.50). Self-adjusting brakes ($7.45). Inside 'tilt' rearview mirror ($4.95). Left- or right-hand outside rearview mirror ($3.95); Left- and right-hand outside rearview mirror ($7.90). Custom steering wheel ($7.70). Twin-Grip differential ($29.50). Pair of license plate frames ($4.05); rear only ($2.05). Heavy-duty front and rear shocks ($2.70). Heavy-duty front and rear shocks and rear springs ($4). Heavy-duty radiator ($3). Heavy-duty cooling system ($12.50). Push-button radio and antenna ($58.50); Manual radio ($53.95). Individually adjusted front seats ($20). Crankcase ventilation system ($3.25). Padded instrument panel ($12.75). Size 6.00 x 15 four-ply Rayon white sidewall tires ($28.35), size 6.50 x 15 four-ply Rayon black sidewall tires ($14.65); size 6.50 x 15 four-ply Rayon white sidewall tires ($46); size 6.50 x 15 four-ply nylon black sidewall tires ($28.40): size 6.50 x 15 four-ply nylon white sidewall tires ($62.90).

(RAMBLER/REBEL) Flash-O-Matic push-button transmission for six-cylinder ($199.50); for V-8 ($219.50). Overdrive ($112.50). Power steering for six-cylinder ($74.00); for V-8 ($79.50). Power window lifts ($99.50). Power brakes ($39.95). Oil filter on Classic six DeLuxe and Super ($9.75). Oil-bath air cleaner for six-cylinder models ($9.75). Radio with manual antenna and front speaker ($69.95); with front and rear speaker ($80.70). Weather-Eye heater and ventilator ($76). Solex tinted glass ($33). Back-up lights ($9.95). Continental tire carrier ($69.50). Front and rear foam cushions ($12.50). Reclining front seat ($25.50). Windshield washer ($11.50). Electric clock ($15.95); Padded instrument panel and visors ($21.50). Undercoating ($14.95). Left or right headrests ($12). Air conditioning ($369). Inside rearview mirror ($5.25). Self-adjusting brakes ($7.45). Wheel discs ($15.95). Twin grip differential for six-cylinder ($34.50): for V-8 models ($39.50). Parking brake warning light ($3.95). Light package ($3.95). Two-tone paint on sedans ($19.95); on Custom ($24.95); on wagons ($21.95); on Custom wagons ($29.95); on Special (except Fleet) ($29.50). Power Pack on two-barrel six-cylinder models ($12); on four-barrel and dual exhaust V-8 models ($47.50). Individually adjustable front seats ($20). Crankcase ventilation system ($3.25). Side hinged tailgate on two-seat wagons ($39.50). Lock-O-Matic door locks on four-doors ($29.85). Pair of license plate frames ($4.05). Front and rear heavy-duty shocks ($3.80); with heavy-duty front and rear springs on sedans ($5). Heavy-duty rear springs on wagons ($7.25). Heavy-duty radiator on six-cylinder ($5); on V-8 ($9.50). Heavy-duty cooling system on six-cylinder ($15.35); on V-8 ($19.85). Aluminum die-cast six-cylinder engine and oil filter on Classic six, DeLuxe and Super ($30).

(AMBASSADOR) Overdrive transmission ($114.50). Push-button Flash-O-Matic transmission ($229.50). Power Pack ($47.50). Weather-Eye ($82.50). Air conditioning and heavy-duty cooling system ($398). Power brakes ($41.95). Power steering ($89.50). Power window lifts ($99.50). Solex glass ($33). Front foam rubber seat cushions ($14.35). Rear foam rubber seat cushions ($14.35). Reclining seat ($25.50). Left and right headrests ($24); Left or right headrest ($12). Individually adjustable front seats ($20). Radio and antenna with front speaker in station wagons ($69.95). Radio and antenna with front and rear speakers in sedans ($86.20). Crankcase ventilation system ($3.25). Undercoating ($14.95). Windshield washer ($11.50). Back-up lights ($9.95). Wheel discs ($16.95). Electric clock ($17.95). Parking brake warning light ($3.95). Padded instrument panel and sun visors ($21.50). Side hinged tailgate for two-seat wagons ($39.50). Lock-O-Matic door locks in four-doors ($29.85). Self-adjusting brakes ($7.45). Inside 'tilt' rearview mirror ($4.95). Left or right outside rearview mirror ($5.25). Left and right outside rearview mirror ($10.50). Power-saver fan ($19.50). Twin-grip differential ($42.50). Pair of license plate frames ($4.05); rear only ($2.05). Light package ($9.95). Heavy-duty front and rear shocks ($3.80). Heavy-duty front and rear shocks and front springs with heavy-duty rear springs in sedans ($5.00). Extra heavy-duty rear springs on sedans ($7.25). Heavy-duty rear springs on wagons ($6.50). Heavy-duty radiator ($9.50). Heavy-duty cooling system ($19.85). Plus a variety of tire and paint options too numerous to list.

HISTORICAL FOOTNOTES: The full-size Ramblers were introduced Oct. 5, 1961. Model year production peaked at 377,900 units. Calendar year sales of 380,525 cars recorded. R.E. Cross was the chief executive officer of the company this year. American Motors Corp. announced the introduction of the Custom 400 entries (bucket seat four-door sedans) in late April 1961.

1962 RAMBLER

RAMBLER AMERICAN — (6-CYL) — SERIES 6200: The 1962 Rambler American had no major styling changes. The Super models were dropped and replaced by a '400' which had the overhead valve engine standard. Bucket seats were optional only with the '400'. A new Canadian assembly plant was opened in Brampton, Ontario, Canada to build Americans and Classics. Standard equipment for DeLuxe Models included air cleaner, oil filter, front armrests, front ashtray, dual sun visors, turn signals, black rubber floor mats, black rubber cargo mats on wagons and five black tubeless tires, self-adjusting brakes, front foam cushion. Custom Models feature all of the above, plus rear door armrests, rear ashtray, colored carpet on wagons, automatic dome switches on front doors, station wagon travel rack, chrome horn ring and cigar lighter. 400 Models had all above plus, dual horn, two-tone steering wheel, wheel discs, chrome front seat trim, vinyl pleated upholstery, padded instrument panel and visors, metallic door panel insert, door scuff plate, trim and vinyl, glove box lock and overhead valve engine including 45 amp. battery.

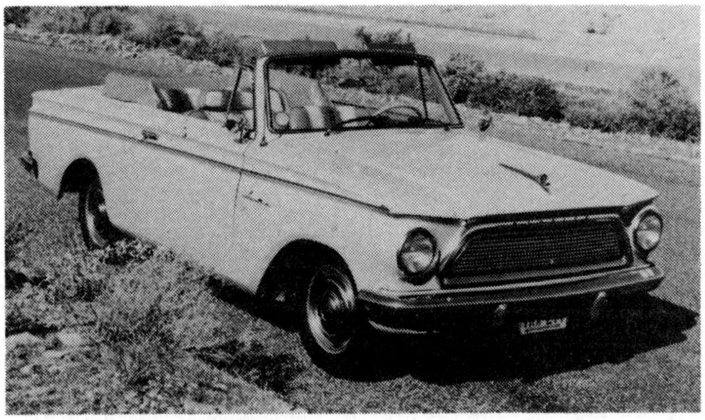

1962 Rambler, American two-door convertible, 6-cyl

RAMBLER I.D. DATA: VIN on plate on right wheelhouse panel below hood. First symbol identified series: B=American; C=Rambler; 2=Super; 5=400; H=Ambassador. Second through last symbols were sequential production number. Starting serial numbers were: [American] B-375001; [American for knocked-down export] BK-13001; [Canadian-built Americans] BT-100201; [Rambler] C-625001; [Rambler for knocked-down export] CK-11501; [Canadian-built Rambler] CT-206001; [Ambassador V-8] H-160001; [Ambassador V-8 for knocked-down export]. for Ambassador V-8, HK-10301 for Ambassador V-8 for unassembled export. Body Number plate riveted to left front door hinge pillar has additional data. First line indicates body production sequence number. Second line indicates "Model No." consisting of first two symbols indicating model year and additional symbols indicating body style and series. Complete "Model No." corresponds to column two of charts below. Third line gives trim data. Fourth line gives paint color data. Six-cylinder motor number on upper left front corner of engine block. V-8 motor number on top at front of block; or front of block; or center, left side of block, above oil pan. Continuing in 1962, AMC employed an "Engine Day Build Code" system giving all engines of a type the same six symbol code. The first symbol of the engine day built code is a single digit number code for the year of manufacture starting with 4=1962. The second and third symbols indicated month of build: 01=Jan.; 02=Feb.; etc. The numbers differed in the fourth symbol, a letter code designating the engine displacement and carburetion. Letter codes were: A=195.6 cid/90 hp L-head six; B=195.6 cid/125 hp aluminum OHV six; C=195.6 cid/127 hp cast-iron OHV six; n.a.=195.6 cid/138 hp six 2V; E=327 cid/250 hp V-8 2V; F=327 cid/270 hp V-8 4V; G=287 engine. The fifth and sixth symbols indicated day of manufacture as appropriate. In motor number 401A22: first symbol 4=1962; 01=Jan.; A=195.6 cid/90 hp six and 22=22nd day of month.

RAMBLER AMERICAN SERIES

Model Number	Body Style Number	Body Type & Seating	Factory Price	Shipping Weight	Production Total
DELUXE LINE					
6200	6202	2-dr Bus Cpe-3P	1832	2454	281
6200	6205	4-dr Sed-6P	1895	2500	17,758
6200	6206	2-dr Sed-6P	1846	2480	29,665
6200	6208	4-dr Sta Wag-6P	2130	2573	6,304
6200	6204	2-dr Sta Wag-6P	2081	2555	4,434
CUSTOM LINE					
6200	6205-2	4-dr Sed-6P	1958	2512	13,884
6200	6206-2	2-dr Sed-6P	1909	2492	12,710
6200	6208-2	4-dr Sta Wag-6P	2190	2600	8,998
6200	6204-2	2-dr Sta Wag-6P	2141	2565	4,398
400 LINE					
6200	6205-5	4-dr Sed-6P	2089	2585	5,773
6200	6206-5	2-dr Sed-6P	2040	2558	4,840
6200	6207-5	2-dr Conv-5P	2344	2735	13,497
6200	6208-5	4-dr Sta Wag-6P	2320	2692	3,134

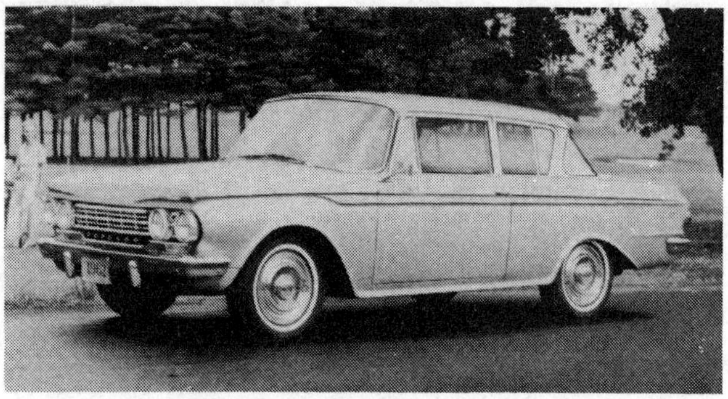

1962 Rambler, Classic two-door sedan, 6-cyl

RAMBLER CLASSIC — (6-CYL) — SERIES 6210: The 1962 Rambler Classic Six had all, but the smallest trace of tailfins removed from the rear end. There were new, round taillights, a new front grille and new side trim and moudings. The die-cast aluminum block was standard on all Classic Sixes and optional on the Custom and DeLuxe lines. Other innovations for the year were a new brake system with tandem master cylinder and an hydraulic tilting front seat. Standard equipment for the Classic Six DeLuxe Series included turn signal, air cleaner, front armrests, cigar lighter, dual headlamps, dual sun visors, station wagon travel rack, front ashtrays, oil filter, front foam cushion, and five black tubeless tires. The Classic Six Custom Series had all above features, plus electric clock, glove box light, carpets, rear armrests, rear ashtrays, dual horns, automatic dome light switch. Three-seat station wagons have four black Captive Air nylon tires. Classic Six 400 Series included all above features, plus padded dash and visors, rear door vent windows, two-tone steering wheel, wheel discs, wagon robe rail and aluminum engine. (Cast iron engine optional at no cost.)

RAMBLER CLASSIC SERIES

Model Number	Body Style Number	Body Type & Seating	Factory Price	Shipping Weight	Production Total
DELUXE LINE					
6210	6215	4-dr Sed-6P	2050	2888	38,082
6210	6216	2-dr Sed-6P	2000	2866	14,811
6210	6218	4-dr Sta Wag-6P	2380	3014	28,203
CUSTOM LINE					
6210	6215-2	4-dr Sed-6P	2200	2898	68,699
6210	6216-2	2-dr Sed-6P	2150	2876	12,652
6210	6218-2	4-dr Sta Wag-6P	2492	3024	53,671
6210	6218-4	5-dr Sta Wag-6P	2614	3094	6,322
400 LINE					
6210	6215-5	4-dr Sed-6P	2349	2853	31,255
6210	6216-5	2-dr Sed-6P	2299	2841	5,521
6210	6218-5	4-dr Sta Wag-6P	2640	2985	21,281

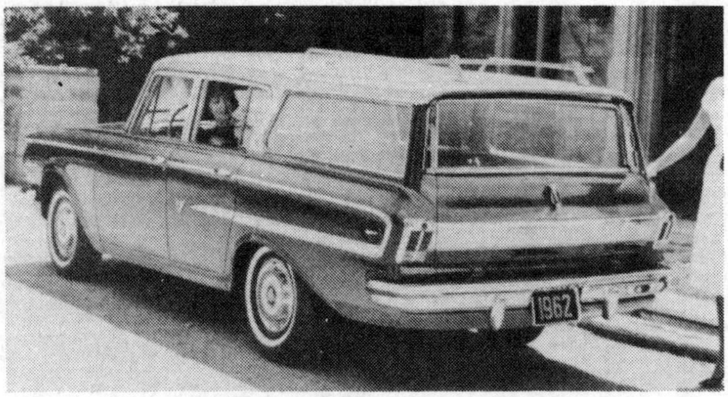

1962 Rambler, Ambassador four-door station wagon, V-8

RAMBLER AMBASSADOR — (V-8) — SERIES 6180: The Rambler Ambassador, for 1962, used the Rambler Classic Six body and running gear with the same minimal styling changes. Both models now shared use of the 108 inch wheelbase. This was basically the 1962 Rambler Rebel V-8 with a new name. At the same time, the 117 inch wheelbase Ambassador was dropped. The Ambassador V-8 DeLuxe came standard with an air cleaner, front armrests. front ashtrays, cigar lighter, dual headlamps, dual horns, oil filter, front foam cushion. The Ambassador V-8 Custom had all features listed for Ambassador DeLuxe, plus rear armrests, rear ashtrays, carpets, electric clock, glove box light, hood insulation, automatic dome light switch, rear door vent windows, station wagon travel rack. The Ambassador V-8 400 had all features listed for Ambassador DeLuxe and Custom, plus Handi-Pak Carrier, padded dash and visors, station wagon robe rail, rear foam cushion, wheel discs and four black nylon Captive Air tires on three seat station wagon.

RAMBLER AMBASSADOR SERIES

Model Number	Body Style Number	Body Type & Seating	Factory Price	Shipping Weight	Production Total
DELUXE LINE (FLEET ONLY)					
6280	6285	4-dr Sed-6P	2336	3249	421
6280	6286	2-dr Sed-6P	2282	3227	45
6280	6288	4-dr Sta Wag-6P	2648	3375	77
CUSTOM LINE					
6280	6285-2	4-dr Sed-6P	2464	3259	7,398
6280	6286-2	2-dr Sed-6P	2410	3237	659
6280	6288-2	4-dr Sta Wag-6P	2760	3385	4,302
400 LINE					
6280	2605-5	3283 Sed-6P	15,120	—	—
6280	6286-5	2-dr Sed-6P	2551	3261	459
6280	6288-5	4-dr Sta Wag-6P	2901	3408	6,401
6280	6288-6	4-dr Sta Wag-8P	3023	3471	1,289

RAMBLER BASE ENGINES

(AMERICAN SIX) Inline, L-head Six. Cast iron block. Displacement: 195.6 cid. Bore and stroke: 3-1/8 x 4-1/4 inches. Compression ratio: 8.0:1. Brake hp: 90 at 3800 rpm. Four main bearings. Solid valve lifters. Carburetor: Carter Type YF-2014S one-barrel.

(DELUXE/SUPER SIX) Inline L-head six. Cast iron block. Displacement: 195.6 cid. Bore and stroke: 3-1/8 x 4-1 /4 inches. Compression ratio: 8.0: 1. Brake hp: 90 at 3800 rpm. Four main bearings. Solid valve lifters. Carburetor: Carter Type YF one-barrel Model 2014S.

(CUSTOM SIX) Inline six. Overhead valves. Cast iron block. Displacement: 195.6 cid. Bore and stroke: 3-1/8 x 4-1/4 inches. Compression ratio: 8.7:1. Brake hp: 125 at 4200 rpm. Four main bearings. Solid valve lifters. Carburetor: Holley one-barrel Model 1908-FC.

(CLASSIC SIX) Inline six. Overhead valves. Cast aluminum block. Displacement: 195.6 cid. Bore and stroke: 3.125 x 4.25 inches. Compression ratio: 8.7:1. Brake hp: 127 at 4200 rpm. Four main bearings. Solid valve lifters. Carburetor: Holley one-barrel Model 1908FC.

(REBEL V-8) Overhead valves. Cast iron block. Displacement: 250 cid. Bore and stroke: 3-1/2 x 3/1-4 inches. Compression ratio: 8.7:1. Brake hp: 200 at 4900 rpm. Five main bearings. Hydraulic valve lifters. Carburetor: Holley model 2040 two-barrel.

(AMBASSADOR EIGHT) V-8. Overhead valves. Cast iron block. Displacement: 326.7 cid. Bore and stroke: 4.00 x 3.25 inches. Compression ratio: 8.7:1. Brake hp: 250 at 4700 rpm. Five main bearings. Hydraulic valve lifters. Carburetor: Holley two-barrel Model H-2040.

CHASSIS FEATURES: Wheelbase: (American) 100 inches; (Classic/Ambassador) 108 inches. Overall length: (American) 173.1 inches; (Classic/Ambassador) 190 inches. Front tread: (American) 54.6 inches; (Classic/Ambassador) 57.8 inches. Rear tread: (American) 55 inches; (Classic/Ambassador) 58 inches. Tires: (American) 6:00 x 15; (Classic) 6.50 x 15; (Ambassador) 7.50 x 14.

OPTIONS

(AMERICAN) Anti-freeze ($4.25). Two-tone paint on sedans ($15.95); on station wagons ($17.95); special application (except fleet) (29.50). Overdrive transmission ($102). Lever control Flash-O-Matic ($164.85). Weather-Eye ($74.20). Overhead valve engine. Oil-bath cleaner and 45 amp. battery ($59.50). Air conditioning with heavy-duty cooling system ($360). Power brakes ($39.95). Power steering ($72.20). Oil-bath air cleaner ($7.15). Solex glass ($26.95); windshield only ($8.95). Seat Cushion, foam rubber, rear ($9.95). Reclining seat ($25.50). Undercoating ($11.95). Windshield washer ($1 1). Back-up lights ($9.95). Wheel discs ($14.95). Electric clock ($15.30). Inside rearview 'tilt' mirror ($4.95). Left or right outside rearview mirror ($3.95); left and right outside rearview mirror ($7.90). Twin-Grip Differential ($29.60). Pair of license plate frames ($3.50); rear only ($1.75). Heavy-duty front and rear shocks ($2.70). Heavy-duty front and rear shocks and rear springs ($4). Heavy-duty radiator ($3.05). Heavy-duty cooling system with radiator fan shroud ($12.50). Push-button radio and antenna ($58.50); manual ($52.50). Individually adjusted front seats ($20). Crankcase ventilation system, required on all California cars ($4.95). Padded instrument panel and visors ($17.60). 6.00 x 15 4-ply Rayon whitewall tires ($28); 6.50 x 15 4-ply Rayon blackwall tires ($8.10); 6.50 x 15 4-ply Rayon whitewall tires ($37.45). Bucket seats ($59.50). Right lounge-tilt seat and headrest ($20.00). Left or right headrests ($12.00); left and right ($24.00). Load levelers only ($25); load levelers with heavy-duty front shocks ($27.10). Two front seat belts ($20.60); four belts, front and rear ($41.20). E-stick transmission (automatic clutch, only available with standard and overdrive transmissions) ($59.50). Light package [trunk, cargo or glove box] with two courtesy lamps, parking brake warning lamp and automatic dome light on DeLuxes ($9.95). Parking brake warning light ($3.95).

(CLASSIC/AMBASSADOR) Air conditioning, All-Season on Classic [tinted glass required] ($370); on Ambassador (tinted glass required) ($399). Back-up lights ($9.95). Bucket seats in 400s ($59.50). Electric clock on Classic DeLuxe ($15.95). Two-tone paint for Sedans ($19.95); for Wagons ($21.95); Special Application ($29.50). Dowguard coolant for Classic ($4.25); for Ambassador ($6.25). Heavy-duty cooling system, on Classic ($15.35); on Ambassador ($19.85). Crankcase ventilation system ($4.95). Aluminum six-cylinder engine on Classic DeLuxe and Custom ($30). Rear foam cushions [standard in Ambassador 400] ($9.95). Individually adjustable front seats ($20); reclining front seat ($25.50). Tinted glass ($37.65); windshield tinted glass ($11.50). Left or right headrest ($12.00). Two license plate frames ($3.50); rear license plate frame ($1.75). Light package ($9.95). Load levelers ($32.15); same with heavy-duty front shocks ($36.00); same with heavy-duty front shocks and springs ($38.70). Inside tilt mirror ($4.95). Left or right outside mirror ($5.30). Padded instrument panel and visors ($21.60). Parking brake warning light ($3.95). Power brakes on Classic ($41.95); on Ambassador ($43.95). Two-door power door locks on Classic ($23.45). Four-door power door locks on Classic and Ambassador ($29.85). Power pack on Classic ($12); on Ambassador ($47.50). Power Saver fan on Classic ($47.50). Power Saver fan on Ambassador ($19.60). Power steering on Classic ($74.20); on Ambassador ($81.20). Power windows ($102.25). Heavy-duty radiator for Classic ($5.05); for Ambassador ($9.55). Single speaker radio and antenna for Classic, Ambassador wagon ($64.95); dual speaker for Classic and Ambassador sedans ($77.25). Two front seat belts ($20.60); four seat belts rear and front ($41.20). Seat lounge tilt with headrest ($20.75). Heavy-duty front and rear shocks ($3.85). Heavy-duty springs and shocks, except wagons ($5.05). Extra-heavy-duty springs and shocks, rear, except wagons ($7.30); wagons ($6.55). Side hinge tailgate on Classic and Ambassador six-passenger wagons ($39.60). Tires: Classic: 6.50 x 15 4-ply whitewall Rayon, except nine-passenger wagons ($29.60); 6.70 x 15 4 -ply black Rayon, except nine-passenger wagons ($8.70); 6.70 x 15 - 4-ply Rayon whitewall, except nine-passenger wagons ($39.90); 6.50 x 15 - 4-ply whitewall nylon Captive Air (four) on nine-passenger wagon ($27.30); 6.70 x 15 4-ply black nylon Captive Air (four) on nine-passenger wagon ($7.50); 6.70 x 15 4-ply whitewall nylon Captive Air (four) on nine-passenger wagon ($36.30). Ambassador: 7.50 x 14 - 4 Ply White SW Rayon exc. 9-pass. wagons ($31.40); 7.50 x 14 - 4 Ply White SW nylon Captive Air (4) on 9-pass. wagons ($28.95); 8.00 x 14 - 4 Ply Black Rayon exc. 9-pass. wagons ($14.20); 8.00 x 14 - 4 Ply White SW Raylon exc. 9-pass. wagons ($48.35); 8.00 x 14 - 4 Ply Black nylon Captive Air (4) on 9-pass. wagons ($12.95); 8.00 x 14 - 4 Ply White SW nylon Captive Air (4) on 9-pass. wagons ($44.50);, plus a variety of tire options too numerous to list here. Flash-O-Matic transmission on Classic ($186.50); on Ambassador ($219.50). Overdrive transmission on Classic ($108.50); on Ambassador ($114.95). Twin Grip differential on Classic ($34.60); on Ambassador ($42.70). Undercoating ($14.95). Weather-Eye heater ($76). Wheel discs (Std. 400) ($14.95). Windshield washer ($11.95).

HISTORICAL FOOTNOTES: The full-sized Ramblers were introduced Oct. 6, 1961 and the compact Americans appeared in dealer showrooms the same day. Model year production peaked at 442,300 units. Calendar years sales of 434,788 cars were recorded. R.E. Cross was the chief executive officer of the company this year. The 2,000,000 Rambler built since 1902 was produced on February 1, 1962 in Kenosha, Wisconsin. In June of the same year a strike by the supplier ended availability of aluminum engines for 1962. The next season, four suppliers and four different versions of such engines were seen. Low installation rate options included V-8 engines (8.2 percent); power brakes (7.9 percent); bucket seats (13.4 percent); air conditioning (6.5 percent) and dual exhausts (1.5 percent).

1963 RAMBLER

1963 Rambler, American two-door convertible, 6-cyl

RAMBLER AMERICAN — (6-CYL) — SERIES 6301: The addition of a pair of two-door hardtops to the 100 inch wheelbase American marked the return of this body style in a Rambler. New grilles and trim were evident. The grille was of the same trapezoidal shape as in 1962, but had an insert with a pattern of closely-spaced vertical bars. At the rear end several changes were noticeable. First, the license plate bracket was moved from the center of the rear deck lid panel to a position below the middle of the back bumper bar. Second, the Rambler lettering across the edge of the 1962 deck lid was replaced with a logo nameplate mounted where the license plate had been. A redesigned power convertible top; transistorized radio; dual-braking system improvements; new wheel discs; a self-adjusting clutch for the semi-automatic 'E-stick' transmission; and better heating and air conditioning units were other notable advances. Model designations were changed to American 220, American 330 and American 440 instead of DeLuxe, Custom and 400 respectively. A new two-door hardtop called the 440-H was introduced. It had a special roof pillar badge; ridged type roof; and contrast-finished side insert panel. Standard equipment on Americans included five 6.00x15 blackwall tires; front foam cushions; front armrests; sun visors; oil filter; and roof luggage rack on station wagons. Custom models also had carpets; a cigarette lighter; and rear armrests; "400" models also had wheelcovers; two-tone steering wheel; glove box lock; padded dash and sun visors; dual horns; and front door dome light switches.

RAMBLER I.D. DATA: VIN on plate on right wheelhouse panel below hood. First symbol identified series: B=American; G=Classic Six; Z=Classic Eight; H=Ambassador Eight. Second through last symbols were sequential production number. Starting serial numbers were: [American] B-515001; [American for knocked-down export] BK-15001; [Canadian-built Americans] BT-110001; [Classic Six] G-100001; [Classic Six for knocked-down export] GK-10001; [Canadian-built Classic Six]; GT-220001 [Classic Eight] Z-100001; [Classic Eight knocked-down for export] ZK-1001; [Ambassador V-8] H-210001; [Ambassador V-8 for knocked-down export] HK-11001 for Ambassador V-8; for Ambassador V-8 for unassembled export. Body Number plate riveted to left front door hinge pillar has additional data. First line indicates body production sequence number. Second line indicates "Model No." consisting of first two symbols indicating model year and additional symbols indicating body style and series. Complete "Model No." corresponds to column two of charts below. Third line gives trim data. Fourth line gives paint color data. Six-cylinder motor number on upper left front corner of engine block. V-8 motor number on top at front of block; or front of block; or center, left side of block, above oil pan. Continuing in 1963, AMC employed an "Engine Day Build Code" system giving all engines of a type the same six symbol code. The first symbol of the engine day built code is a single digit number code for the year of manufacture starting with 5=1963. The second and third symbols indicated month of build: 01=Jan.; 02=Feb.; etc. The numbers differed in the fourth symbol, a letter code designating the engine displacement and carburetion. Letter codes were: A=195.6 cid/90 hp L-head six; B=195.6 cid/125 hp aluminum OHV six; C=195.6 cid/127 hp cast-iron OHV six; n.a.=195.6 cid/138 hp six 2V; E=327 cid/250 hp V-8 2V; F=327 cid/270 hp V-8 4V; G=287 cid/198 hp V-8 engine. The fifth and sixth symbols indicated day of manufacture as appropriate. In motor number 401A22: first symbol 5=1963; 01=Jan.; A=195.6 cid/90 hp six and 22=22nd day of month.

RAMBLER AMERICAN SERIES 6301

Model Number	Body/Style Number	Body Type & Seating	Factory Price	Shipping Weight	Production Total
AMERICAN 220 (DELUXE) LINE					
6301	6305	4-dr Sed-6P	1895	2485	14,419
6301	6306	2-dr Sed-6P	1846	2472	27,780
6301	6302	2-dr Bus Sed-3P	1832	2446	162
6301	6308	4-dr Sta Wag-6P	2130	2549	4,436
6301	6304	2-dr Sta Wag-6P	2081	2528	3,312
AMERICAN 330 (CUSTOM) LINE					
6301	6305-2	4-dr Sed-6P	1958	2500	9,666
6301	6308-2	4-dr Sta Wag-6P	2190	2561	6,848
6301	6304-2	2-dr Sta Wag-6P	2141	2539	3,204
AMERICAN 440 LINE					
6301	6305-5	4-dr Sed-6P	2089	2575	2,937
6301	6306-5	2-dr Sed-6P	2040	2556	1,486
6301	6309-5	2-dr HT-6P	2136	2550	5,101
6301	6309-7	2-dr 440H-4P	2281	2567	9,749
6301	6307-5	2-dr Conv-5P	2344	2743	4,750
6301	6308-5	4-dr Sta Wag-6P	2320	2638	1,874

NOTE 1: The 440H hardtop came with bucket seats and 138 hp 'power pack' six.

RAMBLER CLASSIC SIX — (6-CYL) — SERIES 6310: The Rambler Classic for 1963 was completely redesigned. It had a very clean, lower and narrower box shape on a longer 112 inch wheelbase. The front grille consisted of a double convex surface of fine U-shaped vertical bars with Rambler across the front. It was a one-piece aluminum stamping. New curved glass side windows were used along with new push-button door handles. New series identification numbers were placed in the center of the trunk lid. There was also Rambler identification trim below the trunk lid, A new tri-pose engine mounting system was used. It replaced a four-point system used earlier. At the beginning of the 1962 model year, only a Classic Six was available and it came in Deluxe, Custom and 400 series. After the release of the Classic Eight series, in January, the series names were changed to 550, 660 and 770. The base Classic 550 models compared to Deluxe models and had no side trim. Standard equipment included directional signals; dual headlamps; air cleaner; five 6.50 x 15 blackwall tires; front foam cushions; front armrests; sun visors; and oil filter. The one-step-up Classic 660s compared to Custom models. They had a dual horizontal molding on front fender and single molding from the front door back. In addition to base equipment, "660s" had front door dome light switches; rear armrests; rear ashtrays; glove box lock; electric clock; carpet; dual horns; and luggage rack on station wagons (blackwall Captive Air nylon tires on nine-passenger station wagons). The top-of-the-line Classic 770s compared to the "400" range cars and carried a full-length dual molding with contrasting insert. The Classic 770 had all of this, plus electric clock; padded dash and visors; rear door vent window; wheel discs; two-tone steering wheel; station wagon robe rail; foam cushions; and the die-cast aluminum six (with cast-iron engine as a no-cost option).

1963 Rambler, Classic 770 four-door station wagon, 6-cyl

RAMBLER CLASSIC SIX SERIES 6310

Model Number	Body/Style Number	Body Type & Seating	Factory Price	Shipping Weight	Production Total
CLASSIC SIX 550 LINE					
6310	6315	4-dr Sed-6P	2105	2729	43,315
6310	6316	2-dr Sed-6P	2055	2720	14,417
6310	6318	4-dr Sta Wag-6P	2435	2893	26,261

Series Number	Body/Style Number	Body Type & Seating	Factory Price	Shipping Weight	Production Total
CLASSIC SIX 660 LINE					
6310	6315-2	4-dr Sed-6P	2245	2740	71,646
6310	6316-2	2-dr Sed-6P	2195	2725	11,064
6310	6318-4	5-dr Sta Wag-8P	2609	2885	5,752
6310	6318-2	4-dr Sta Wag-6P	2537	2890	46,282
CLASSIC SIX 770 LINE					
6310	6315-5	4-dr Sed-6P	2349	2686	35,281
6310	6316-5	2-dr Sed-6P	2299	2663	5,496
6310	6318-5	4-dr Sta Wag-6P	2640	2828	19,319

RAMBLER CLASSIC EIGHT — (V-8) — SERIES 6350: The Rambler Classic V-8 shared the same all-new styling as the Rambler Classic Six. The main trim difference was the attachment of V-8 emblems on the front fenders, behind the wheel openings. As on the six-cylinder models, features included redesigned grilles: curved side glass, a sharper silhouette and an advanced type of unitized construction that reduced the number of parts and welds in the assembly by 30 percent. The Classic took on the semi-automatic 'E-stick' transmission and it was in early February that the V-8 (287 cid) was introduced at the Chicago Automobile show as a running addition to the 1963 line that created this new series. Numbering on either the center edge of the deck lid or center of tailgate identified each car as a 550, 660 or 770 Classic. Side molding treatments for each of the lines were the same ones described for the Classic Sixes of a comparable trim level. A 'Twin Stick' five-speed (overdrive) floor-mounted transmission was offered in some models. The right-hand lever was for locking the overdrive function in or out, while the left-hand lever was the floor shifter. Standard equipment for each series was the same as installed in Classic Six series, except for different engine.

RAMBLER CLASSIC V-8 SERIES 6350

Model Number	Body/Style Number	Body Type & Seating	Factory Price	Shipping Weight	Production Total
CLASSIC V-8 550 LINE					
6350	6355	4-dr Sed-6P	2210	3109	3,444
6350	6356	2-dr Sed-6P	2160	3100	992
6350	6358	4-dr Sta Wag-6P	2540	3273	2,318
CLASSIC V-8 660 LINE					
6350	6355-2	4-dr Sed-6P	2345	3120	11,067
6350	6356-2	2-dr Sed-6P	2300	3105	1,369
6350	6358-4	5-dr Sta Wag-9P	2714	3265	1,150
6350	6358-2	4-dr Sta Wag-6P	2642	3340	7,237

CLASSIC V-8 770 LINE

6350	6355-5	4-dr Sed-6P	2454	3130	7,869
6350	6356-5	2-dr Sed-6P	2404	3113	1,341
6350	6358-5	4-dr Sta Wag-6P	2754	3208	4,399

1963 Rambler, Ambassador two-door sedan, V-8

AMBASSADOR — (V-8) — SERIES 6380: The Ambassador was on the same platform as the Rambler Classic and varied mainly in terms of trim. The grille had a horizontal center blade which carried an Ambassador nameplate at its middle. A lower band of horizontal trim decorated the body, between the two wheel housings. Ambassador scripts were carried on the side of the back fenders and the rear treatment included a vertically ribbed, horizontal beauty panel. Rambler Classic type taillamps with chrome division bars were added. There were center-mounted Ambassador nameplates and a logo badge on the top center edge of the trunk. At the beginning of the 1962 model year, Ambassadors were available in Custom and 400 series. After the rearrangement of the Rambler Classic series in January, the Ambassadors also changed to three distinct levels of trim. The Ambassador 800 models were somewhat similar to previous low-rung Customs. Standard equipment was comprised of five 7.50 x 14 blackwall tires; sun visors; air cleaner; front armrests; front ashtray; cigar lighter; dual headlamps; dual horns; oil filter; front foam cushions and station wagon travel rack. Ambassador 880 was a new mid-range series, offering more trim and equipment than the base Ambassador, but not quite as many extras as the former 400. It had all 800 features, plus front and rear armrests; ashtrays; carpets; hood insulation; automatic dome light switch and chrome horn ring. The top of the line Ambassador 990 had everything found on the lower-priced lines, plus, electric clock; padded dash board; padded sun visors; station wagon robe rail; rear front seat cushions; full wheel discs; and Captive Air tires on nine-passenger station wagon.

AMBASSADOR SERIES 6380

Model Number	Body/Style Number	Body Type & Seating	Factory Price	Shipping Weight	Production Total
AMBASSADOR 800 LINE					
6380	6385	4-dr Sed-6P	2391	3140	437
6380	6386	2-dr Sed-6P	2337	3110	41
6380	6388	4-dr Sta Wag-6P	2703	3270	113
AMBASSADOR 880 LINE					
6380	6385-2	4-dr Sed-6P	2519	3145	7,667
6380	6386-2	2-dr Sed-6P	2465	3116	1,042
6380	6388-2	4-dr Sta Wag-6P	2815	3275	4,929
AMBASSADOR 990 LINE					
6380	6385-5	4-dr Sed-6P	2660	3158	14,019
6380	6386-5	2-dr Sed-6P	2606	3132	1,764
6380	6388-6	5-dr Sta Wag-9P	3018	3305	1,687
6380	6388-5	4-dr Sta Wag-6P	2956	3298	6,112

RAMBLER BASE ENGINES

(AMERICAN 220/330) Inline six. Cast iron block. Displacement: 1956 cid. Bore and stroke 3.125 x 4.25 inches. Compression ratio: 8.0:1. Brake hp: 90 at 3800 rpm. Four main bearings. Solid valve lifters. Carburetor: Carter Type RBS one-barrel Model 3487S.

(AMERICAN 440/RAMBLER CLASSIC SIX) Inline six. Overhead valve. Cast iron block. Displacement; 195.6 cid. Bore and stroke: 3.125 x 4.25 inches. Compression ratio: 8.7:1. Brake hp: 126 at 4200 rpm. Four main bearings. Solid valve lifters. Carburetor: Holley one-barrel Model 1909-2555.

(AMERICAN 440H) Inline six. Cast iron or aluminum block. Displacement: 195-6 cid. Bore and stroke: 3.125 x 4.25 inches. Compression ratio: 8.7: 1. Brake hp: 138 at 4500 rpm. Four main bearings. Solid valve lifters. Carburetor: Carter Type WCD two-barrel Model 3434S.

(RAMBLER CLASSIC EIGHT) V-8. Overhead valves. Cast iron block. Displacement: 287 cid. Bore and stroke: 3-3/4 x 3-1/4 inches. Compression ratio: 8.7:1. Brake hp: 198 at 4700 rpm. Five main bearings. Hydraulic valve lifters. Carburetor: Holley two-barrel Model 2209-2699.

(AMBASSADOR EIGHT) V-8. Overhead valves. Cast iron block. Displacement: 326.7 cid. Bore and stroke: 4.00 x 3.25 inches. Compression ratio: 8.7:1. Brake hp: 250 at 4700 rpm. Five main bearings. Hydraulic valve lifters. Carburetor: Holley two-barrel Model H-2040.

CHASSIS FEATURES: Wheelbase: (American) 100 inches; (Classic/Ambassador) 112 inches. Overall length: (American) 173.1 inches; (Classic/Ambassador) 189.3 inches. Front tread: (American) 54.6 inches; (Classic) 58.2 inches; (Ambassador) 58.6 inches. Rear tread: (American) 55 inches; (Classic) 57.4 inches; (Ambassador) 57.5 inches. Tires: (American) 6.00 x 15; (Classic eight-passenger wagon) 7.00 x 14; (other Classics) 6.50 x 15; (Ambassadors) 7.50 x 14.

OPTIONS

(AMERICAN) Antifreeze ($4.25). Two-tone colors on sedans ($15.95); two-tone color on station wagons ($17.95), two-tone color on hardtops except 440H ($37.90); two-tone color on convertible ($21.95), special color application ($29.50). Overdrive transmission ($102). Lever control Flash-O-Matic ($164.85). Weather-Eye ($74.20). Overhead valve engine ($59.50). Air conditioning including heavy-duty cooling system ($360). Power brakes ($39.95). Power steering ($72.20). Oil-bath air cleaner ($7.15). Solex glass ($26.95). Solex windshield only ($8.95). Rear foam rubber seat cushion ($9.95). Reclining seat ($25.50). Undercoating ($14.95). Windshield washer ($11.95). Back-up lights ($10.70). Wheel discs ($14.95). Electric clock ($15.30). Inside 'tilt' rearview mirror ($4.95). Left or right outside rearview mirror ($3.95); left and right outside rearview mirror ($7.90). Twin-grip differential ($29.60). Pair of license frames ($3.50): rear license frame only ($1.75). Front and rear heavy-duty shocks ($2.70). Heavy-duty front and rear shocks and rear springs ($4). Heavy-duty radiator ($3.55). Heavy-duty cooling system ($12.50). Push-button radio and antenna ($58.50); manual radio and antenna ($52.50). Individually adjusted front seats ($20). Padded instrument panel and visors ($17.60). Bucket seats on American 440 ($99.50); left and right bucket seats on American 440 ($24). Rear load levelers ($32.15). Load levelers with heavy-duty front shocks ($34.85). Two front seat belts ($17.85); four front and rear seat belts ($37). E-Stick transmission, automatic clutch available only with standard or overdrive gear boxes ($59.50). Visibility group A ($21.25). Light group ($19.60). Chrome horn ring ($7.70). Heavy-duty battery ($6.50). Air conditioning adaptor group ($20.55). 33 amp. alternator ($12). Vinyl seat upholstery ($15). Left or right-hand lounge-tilt seat and headrest ($21); left- and right-hand lounge-tilt seat and headrest ($42). Power-Pack, twin throat carburetor ($12). Twin-stick floor shift transmission and overdrive ($134.50).

(CLASSIC/AMBASSADOR) All-Season air conditioning ($380-$399). Alternator ($12). Back-up lights ($10.70). Heavy-duty battery ($6.50). Bucket seats ($99.50) Electric clock ($15.95). Two-tone color on sedans ($19.95); two-tone color on wagons ($21.95), special color applications ($29.50). Coolant Dowguard ($4.25-$6.25). Heavy-duty cooling system ($17.20-$19.85). Six-cylinder aluminum engine ($30). Rear foam cushions ($9.95). Adjustable front seats ($20). Reclining front seat ($25.50). Front lounge-tilt seat with headrest ($21). Tinted glass ($39.50-$45.50). Tinted windshield ($15.95). Left or right headrest ($12). Chrome horn ring ($7.70). Light group ($19.60). Load levelers ($32.15). Load levelers with heavy-duty front shocks ($36). Load levels with heavy-duty front shocks and springs ($38.70). Inside 'tilt' mirror ($4.95). Outside left or right mirror ($5.30). Outside rearview remote control mirror ($11.95). Padded instrument panel and visors ($19.95). Power brakes ($42.95-$43.95). Power door locks on two-doors ($23.45); power door locks on four-doors ($29.85). Power Pack ($12.00-$23.75). Power Saver fan ($19.60). Power steering ($19.60-$81.20). Power windows ($102.25). Heavy-duty radiator ($5.55-$9.55). Single speaker radio and antenna ($64.95). Dual speaker radio and antenna ($77.25). Front seat belts ($17.85). Front and rear seat belts ($37). Heavy-duty springs and shocks ($5.05). Extra-heavy-duty springs and rear shocks ($7.30). Heavy-duty springs and shocks ($6.55). Side hinge tailgate ($39.60). Twin stick transmission ($141.00-$147.50). Flash-O-Matic transmission ($186.50-$219.50). Overdrive transmission ($108.50-$114.95). Twin grip differential ($37.50-$42.70). Undercoating ($14.95). Visibility group A ($22.50). Visibility group B ($28.85). Weather-Eye Heater ($78). Wheel discs ($14.95). Windshield washer ($11.95). Numerous tire choices.

HISTORICAL FOOTNOTES: The 1963 Rambler line was introduced Oct. 5, 1962 and the Rambler Classic V-8 appeared in dealer showrooms during February. Model year production peaked at 464,000 units. Calendar year sales of 441,508 cars were recorded in the United States. R.E. Cross was the chief executive officer of the company this year. Options and accessories seeing low installation rates included: power windows (0.5 percent). V-8 engines (1 7 percent); power brakes (8.1 percent): bucket seats (9.6 percent) and air conditioning (8.2 percent). A Rambler took top honors in the Mobilgas Economy Run. The 1963 Rambler line was picked as 'Car of the Year' by *Motor Trend* magazine.

1964 RAMBLER

1964 Rambler, American 440-H two-door hardtop, 6-cyl

RAMBLER AMERICAN — (6-CYL) — SERIES 6401: The Rambler American for 1964 was totally redesigned. Wheelbase was increased from 100 inches to 106 inches. The new styling was cleaner with the corners more rounded. The front fenders were rounded near the front headlights and styled similarly to Chryslers famous experimental turbine car. Front foam seat cushions; a cigarette lighter (except 220); and five 6.45 x 14 blackwall tires were standard in all Americans. 220s and 330s had the 90 horsepower six-cylinder engine. American 440s featured the 125 hp six. In addition, the 440-H had wide reclining bucket seats; rear foam cushions; front seat belts; wheel discs; and, on station wagons, a roof luggage rack.

RAMBLER I.D. DATA: VIN on plate on right wheelhouse panel below hood. First symbol identified series: B=American; G=Classic Six; Z=Classic Eight; H=Ambassador Eight. Second through last symbols were sequential production number. Rambler American Series 6401 starting Serial Numbers were B-650001 for cars built in Kenosha, Wisconsin, BK-16001 for unassembled export and BT-115401 for Brampton, Ontario, Canada. Rambler Classic Series 6410 starting Serial Numbers were G-500001 for cars built in Kenosha, Wisconsin and GK-14001 for unassembled export. Starting Serial Numbers for Rambler Classic V-8s were Z-155001 for cars built in Kenosha, Wisconsin and ZK-11001 for unassembled export. All Classics built in Brampton, Ontario, Canada had a starting Serial Number of GT-239001. Rambler Ambassador Series 6380 starting Serial Numbers were H-255001 for cars built in Kenosha, Wisconsin and HK-12001 for unassembled export. Starting serial numbers for all Rambler Ambassador V-8 built in Brampton, Ontario, Canada were HT-33301.Body Number plate riveted to left front door hinge pillar has additional data. First line indicates body production sequence number. Second line indicates "Model No." consisting of first two symbols indicating model year and additional symbols indicating body style and series. Complete "Model No." corresponds to column two of charts below. Third line gives trim data. Fourth line gives paint color data. Six-cylinder motor number on upper left front corner of engine block. V-8 motor number on top at front of block; or front of block; or center, left side of block, above oil pan. Continuing in 1963, AMC employed an "Engine Day Build Code" system giving all engines of a type the same six symbol code. The first symbol of the engine day built code is a single digit number code for the year of manufacture starting with 6=1964. The second and third symbols indicated month of build: 01=Jan.; 02=Feb.; etc. The numbers differed in the fourth symbol, a letter code designating the engine displacement and carburetion. Letter codes were: A=195.6 cid/90 hp L-head six; B=195.6 cid/125 hp aluminum OHV six; C=195.6 cid/127 hp cast-iron OHV six; n.a.=195.6 cid/138 hp six 2V; E=327 cid/250 hp V-8 2V; F=327 cid/270 hp V-8 4V; G=287 cid/198 hp V-8; L=232 cid/145 hp six. The fifth and sixth symbols indicated day of manufacture as appropriate. In motor number 401A22: first symbol 6=1964; 01=Jan.; A=195.6 cid/90 hp six and 22=22nd day of month.

RAMBLER AMERICAN SERIES 6401

Model Number	Body/Style Number	Body Type & Seating	Factory Price	Shipping Weight	Production Total
AMERICAN 220					
6401	6405	4-dr Sed-6P	1964	2527	18,225
6401	6406	2-dr Sed-6P	1907	2506	32,716
6401	6408	4-dr Sta Wag-6P	2240	2661	8,062
AMERICAN 330					
6401	6405-2	4-dr Sed-6P	2057	2526	19,379
6401	6406-2	2-dr Sed-6P	2000	2504	15,171
6401	6408-2	4-dr Sta Wag-6P	2324	2675	20,587
AMERICAN 440					
6401	6405-5	4-dr Sed-6P	2150	2572	6,590
6401	6409-5	2-dr HT-6P	2133	2596	19,495
6401	6409-7	2-dr 440-H HT-5P	2292	2617	14,527
6401	6407-5	2-dr Conv-6P	2346	2752	8,907

RAMBLER CLASSIC SIX — (6-CYL) — SERIES 6410: The major styling changes for the Classic line was a new grille. It had six stacks of short, bright metal 'dashes' running between dual outboard headlamps. The headlights were horizontally positioned in rounded rectangular housings. The entire ensemble was surrounded by a barbell-shaped chrome grille shell with Rambler lettering stamped into the upper bar. Side trim varied by line. Classic 550 models had a bright rocker panel strip, but no lower beltline molding. Classic 660 models added a horizontal mid-bodyside strip, of constant width, which ran from the headlights to the taillights. The Classic 770 models used the same basic trim, but had a 'butter knife' shaped front tip on the molding. This tip was horizontally ribbed and carried '770' numbering. All three lines had a Classic script plate on the back edge of the rear fender. A 232 cid six-cylinder engine was introduced in a limited number of Classic two-door hardtops beginning in late April 1964. All of these special cars were painted Solar Yellow (with black roofs) and only 2,520 were built. They had distinctive 'Typhoon' rear fender badges, in place of the regular Classic script. However, the Typhoon six-cylinder motor was also provided, as a $59.95 powertrain option, in Classic 770 models. It produced 145 brake hp at 4300 rpm. Standard equipment in all Rambler Classic 550s included front foam cushions; dual headlights; and five 6.95 x 14 blackwall tires. Classic 660s also had automatic dome lights; rear armrests; glove box lock; carpets; and dual horns. Classic 770s had such extras as an electric clock; padded dash and visors; two-tone steering wheel; and full wheel discs.

RAMBLER CLASSIC SIX SERIES 6410

Model Number	Body/Style Number	Body Type & Seating	Factory Price	Shipping Weight	Production Total
CLASSIC 550					
6410	6415	4-dr Sed-6P	2116	2755	21,310
6410	6416	2-dr Sed-6P	2066	2732	6,454
6410	6418	4-dr Sta Wag-6P	2446	2915	13,164
CLASSIC 660					
6410	6415-2	4-dr Sed-6P	2256	2758	37,584*
6410	6416-2	2-dr Sed-6P	2206	2736	3,976
6410	6418-2	4-dr Sta Wag-6P	2548	2916	26,671
CLASSIC 770					
6410	6415-5	4-dr Sed-6P	2360	2763	14,337*
6410	6416-5	2-dr Sed-6P	2310	2740	1,278*
6410	6419-5	2-dr Typ HT-6P	2397	2789	8,996*
6410	6419-7	2-dr HT-5P	2509	2818	2,520
6410	6418-5	4-dr Sta Wag-6P	2651	2921	10,523*

NOTE 1: (*) indicates limited number of cars had optional Typhoon six.

NOTE 2: According to Body Style Number, the following number of Typhoon six attachments were recorded: 6415-2 (6); 6415-5 (2,025); 6416-5 (71); 6418-2 (4); 6418-5 (1,720) and 6419-5 (429).

1964 Rambler, Classic 770 four-door station wagon, V-8

RAMBLER CLASSIC EIGHT — (V-8) — SERIES 6410: The styling and equipment features of the Rambler Classic, when equipped with V-8 power, were the same as for the Classic Six, except for the engine and engine identification badges. The optional V-8 was a 287.2 cid 198 hp. Cars with this engine installed received V-8 fender badges which were placed behind the front wheel housing and under the lower beltline molding.

RAMBLER CLASSIC EIGHT SERIES 6410

Model Number	Body/Style Number	Body Type & Seating	Factory Price	Shipping Weight	Production Total
CLASSIC 550					
6410	6415	4-dr Sed-6P	2221	3115	2,760
6410	6416	2-dr Sed-6P	2171	3092	545
6410	6418	4-dr Sta Wag-6P	2551	3275	2,199
CLASSIC 660					
6410	6415-2	4-dr Sed-6P	2361	3118	11,374
6410	6416-2	2-dr Sed-6P	2311	3096	873
6410	6418-2	4-dr Sta Wag-6P	2653	3276	10,908
CLASSIC 770					
6410	6415-5	4-dr Sed-6P	2465	3123	9,451
6410	6416-5	2-dr Sed-6P	2415	3100	669
6410	6418-5	4-dr Sta Wag-6P	2756	3281	8,835
6410	6419-5	2-dr HT-6P	2502	3149	11,872

RAMBLER AMBASSADOR 990 — (V-8) — SERIES 6480: The 1964 Ambassador 990 looked like a Rambler Classic with some of its teeth knocked out. At the center of each stack of chrome grille dashes, a few dashes were omitted. This left a gap which was filled with Ambassador lettering. Seen on the hood, above the center of the grille, was a winged medallion. Side trim consisted of rocker sill moldings; full-length horizontal lower beltline strips; horizontal chrome slashes on the rear roof pillar; and vertical louvers on the front fenders behind the wheel openings. The deck lid or tailgate was decorated with a horizontal beauty panel that matched the general texture of the front grille. A chrome extension panel appeared between the rear bumper and wheelhousing. Ambassador standard equipment included front foam cushions; dual headlights; five 7.35 x 14 blackwall tires; V-8 engine; electric clock; rear foam cushions; and wheel discs. An optional feature on Ambassador 990 was a bucket seat combination with a special between-the-seat cushion permitting a third person to ride in front. Ambassador scripts were positioned at the lower rear side of the back fenders. The special Ambassador 990-H two-door hardtop came standard with front and rear armrests; front and rear seat cushions; and wheel opening moldings. The 990 convertible had wheel opening moldings and a power top. Ambassador wagons had a rooftop luggage carrier.

1964 Rambler, Ambassador 990-H two-door hardtop, V-8

RAMBLER AMBASSADOR 990 SERIES 6480

Model Number	Body/Style Number	Body Type & Seating	Factory Price	Shipping Weight	Production Total
6480	6485-5	4-dr Sed-6P	2671	3204	9,827
6480	6489-5	2-dr HT-6P	2736	3213	4,407
6480	6489-7	2-dr 990-H HT-6P	2917	3255	1,464
6480	6488-5	4-dr Sta Wag	2985	3350	2,995

RAMBLER BASE ENGINES

(AMERICAN 220/330) Inline six. Cast iron block. Displacement: 195.6 cid. Bore and stroke 3.125 x 4.25 inches. Compression ratio: 8.0:1. Brake hp: 90 at 3800 rpm. Four main bearings. Solid valve lifters. Carburetor: Carter Type RBS one-barrel Model 3708S.

(AMERICAN 440) Inline six. Overhead valve. Cast iron block. Displacement: 195.6 cid. Bore and stroke: 3.125 x 4.25 inches. Compression ratio: 8.7:1. Brake hp: 125 at 4200 rpm. Four main bearings. Solid valve lifters. Carburetor: Holley one-barrel Model 1909-2555-3.

(AMERICAN 440H) Inline six. Cast iron or aluminum block. Displacement: 231.9 cid. Bore and stroke: 3.125 x 4.25 inches. Compression ratio: 8.7:1. Brake hp: 138 at 4600 rpm. Four main bearings. Solid valve lifters. Carburetor: Carter Type WCD two-barrel Model 3706S.

(CLASSIC 550/660/770) Inline six. Cast iron block. Displacement: 195.6 cid. Bore and stroke: 3.125 x 4.25 inches. Compression ratio: 8.7:1. Brake horsepower: 127 at 4200 rpm. Four main bearings. Solid valve lifters. Carburetor: Carter RBS one-barrel Model 2727S.

(RAMBLER CLASSIC EIGHT) V-8. Overhead valves. Cast iron block. Displacement: 287.2 cid. Bore and stroke: 3-3/4 x 3-1/4 inches. Compression ratio: 8.7:1. Brake hp: 198 at 4700 rpm. Five main bearings. Hydraulic valve lifters. Carburetor: Holley 2209-3305 two-barrel.

(AMBASSADOR EIGHT) V-8. Overhead valves. Cast iron block. Displacement: 326.7 cid. Bore and stroke: 3.75 x 3.25 inches. Compression ratio: 8.7:1. Brake hp: 250 at 4700 rpm. Five main bearings. Hydraulic valve lifters. Carburetor: Holley 2300-2442-1 two-barrel.

(AMBASSADOR 990-H) V-8. Overhead valves. Cast iron block. Displacement: 326.7 cid. Bore and stroke: 4.00 x 3.25 inches. Compression ratio: 9.7:1. Brake hp: 270 at 4700 rpm. Five main bearings. Hydraulic valve lifters. Carburetor: Holley four-barrel Model 4150-1957-1.

CHASSIS FEATURES: (American) 106 inches; (Classic/Ambassador) 112 inches. Overall length: (American) 177.3 inches; (Classic station wagon) 190 inches; (Classic passenger cars) 190.5 inches; (Ambassador station wagon) 190 inches; (Ambassador passenger cars)

190.5 inches. Width: (American) 68.6 inches; (all other models) 71.3 inches. Height: (American sedan) 54.5 inches; (Classic sedan) 54.6 inches; (Ambassador sedan) 55.3 inches. Tires: (Ambassador) 7.50 x 14: (American) 6.00 x 14: (Classic) 6.50 x 14.

OPTIONS

(AMERICAN) Antifreeze ($4.25). Two-tone colors on sedans ($15.95); two-tone color on station wagons ($17.95), two-tone color on hardtops except 440H ($37.90); two-tone color on convertible ($21.95), special color application ($29.50). Overdrive transmission ($102). Lever control Flash-O-Matic ($164.85). Weather-Eye ($74.20). Overhead valve engine ($59.50). Air conditioning including heavy-duty cooling system ($360). Power brakes ($39.95). Power steering ($72.20). Oil-bath air cleaner ($7.15). Solex glass ($26.95). Solex windshield only ($8.95). Rear foam rubber seat cushion ($9.95). Reclining seat ($25.50). Undercoating ($14.95). Windshield washer ($11.95). Back-up lights ($10.70). Wheel discs ($14.95). Electric clock ($15.30). Inside 'tilt' rearview mirror ($4.95). Left or right outside rearview mirror ($3.95); left and right outside rearview mirror ($7.90). Twin-grip differential ($29.60). Pair of license frames ($3.50): rear license frame only ($1.75). Front and rear heavy-duty shocks ($2.70). Heavy-duty front and rear shocks and rear springs ($4). Heavy-duty radiator ($3.55). Heavy-duty cooling system ($12.50). Push-button radio and antenna ($58.50); manual radio and antenna ($52.50). Individually adjusted front seats ($20). Padded instrument panel and visors ($17.60). Bucket seats on American 440 ($99.50); left and right bucket seats on American 440 ($24). Rear load levelers ($32.15). Load levelers with heavy-duty front shocks ($34.85). Two front seat belts ($17.85); four front and rear seat belts ($37). E-Stick transmission, automatic clutch available only with standard or overdrive gear boxes ($59.50). Visibility group A ($21.25). Light group ($19.60). Chrome horn ring ($7.70). Heavy-duty batter ($6.50). Air conditioning adaptor group ($20.55). 33 amp. alternator ($12). Vinyl seat upholstery ($15). Left- or right-hand lounge-tilt seat and headrest ($21); left- and right-hand lounge-tilt seat and headrest ($42). Power-Pack, twin throat carburetor ($12). Twin-stick floor shift transmission and overdrive ($134.50).

(CLASSIC/AMBASSADOR) All-Season air conditioning ($380-$399). Alternator ($12). Back-up lights ($10.70). Heavy-duty battery ($6.50). Bucket seats ($99.50) Electric clock ($15.95). Two-tone color on sedans ($19.95); two-tone color on wagons ($21.95), special color applications ($29.50). Coolant Dowguard ($4.25-$6.25). Heavy-duty cooling system ($17.20-$19.85). Six-cylinder aluminum engine ($30). Rear foam cushions ($9.95). Adjustable front seats ($20). Reclining front seat ($25.50). Front lounge-tilt seat with headrest ($21). Tinted glass ($39.50-$45.50). Tinted windshield ($15.95). Left or right headrest ($12). Chrome horn ring ($7.70). Light group ($19.60). Load levelers ($32.15). Load levelers with heavy-duty front shocks ($36). Load levels with heavy-duty front shocks and springs ($38.70). Inside 'tilt' mirror ($4.95). Outside left or right mirror ($5.30). Outside rearview remote control mirror ($11.95). Padded instrument panel and visors ($19.95). Power brakes ($42.95$43.95). Power door locks on two-doors ($23.45); power door locks on four-doors ($29.85). Power Pack ($12.00-$23.75). Power Saver fan ($19.60). Power steering ($19.60-$81.20). Power windows ($102.25). Heavy-duty radiator ($5.55-$9.55). Single speaker radio and antenna ($64.95). Dual speaker radio and antenna ($77.25). Front seat belts ($17.85). Front and rear seat belts ($37). Heavy-duty springs and shocks ($5.05). Extra-heavy-duty springs and rear shocks ($7.30). Heavy-duty springs and shocks ($6.55). Side hinge tailgate ($39.60). Twin stick transmission ($141.00$147.50). Flash-O-Matic transmission ($186.50-$219.50). Overdrive transmission ($108.50-$114.95). Twin grip differential ($37.50-$42.70). Undercoating ($14.95). Visibility group A ($22.50). Visability group B ($28.85). Weather-Eye Heater ($78). Wheel discs ($14.95). Windshield washer ($11.95). Numerous tire choices.

HISTORICAL FOOTNOTES: The 1964 Ramblers were introduced October 1963. Model year registrations peaked at 379,412 units. Calendar year production of 393,863 cars was recorded. R.E. Cross was the chief executive officer of the company this year. A new option was Adjust-O-Tilt steering.

1965 RAMBLER

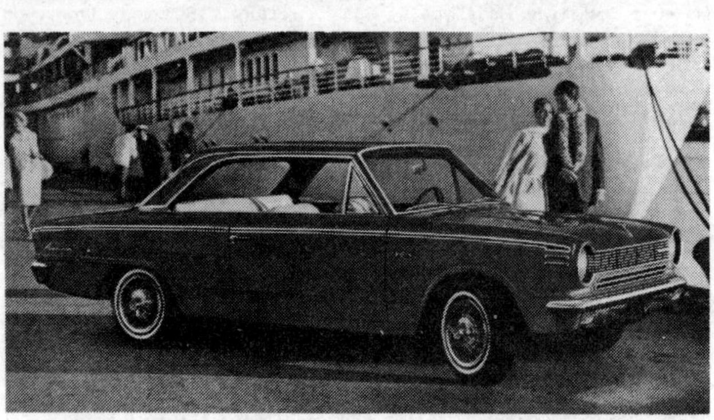

1965 Rambler, American 440 two-door hardtop, 6-cyl

RAMBLER AMERICAN — (SIX) — SERIES 01: The 1965 Rambler American retained the same basic styling as in 1964. The front horizontal bar grille now had indentations made in it. This gave the appearance that it was divided into three parts. Rocker panel moldings were added to all lines, except '220s. New chrome side trim was also used on all, but '220' models, which did not have bodyside moldings. Standard equipment on '220' models included turn signals, heavy-duty lights; front armrests; dual sun visors; rubber floor mats; front foam seat cushions; front seat belts; one front ashtray; dome or pillar lamps; Fresh-Air ventilation; two coat hooks; 60 amp battery and blue/green panel illumination. The '330' models had all of these items, plus rear armrests; cigarette lighter; rear ashtrays; carpets and station wagon luggage rack. The '440' models added richer appointments and trim, plus a lockable glove box. Technical features of all 1965 Ramblers included column-mounted three-speed manual transmission, self-adjusting Double-Safety brakes; oil, gas and fuel pump filters; power booster fuel pump; Anti-Smog system and ceramic armoured exhaust system. A new '232' engine was available and the G-Stick transmission was dropped.

RAMBLER I.D. DATA: VIN on plate on right wheelhouse panel below hood. First symbol identified series: B=American; G=Classic Six; Z=Classic Eight; H=Ambassador Eight. Second through last symbols were sequential production number. [American] Starting Serial Numbers were P-100001 for cars with the '196' six; PK-100001 for export models with the '196' six; W-100001 for cars with the '232' six; WK-100001 for export models with the '232' six. The starting serial numbers for Canadian production were as follows: PT-500001 for '196' six QT-500001 for '199' six, VVT-500001 for '232' six. [Classic] Starting Serial Numbers were J-100001 for cars with the '199' six; JK-100001 for export models with the '199' six; L-1500001 for cars with the '232' six; LK-100001 for export models with the '232' six; Z-275001 for cars with the '287' V-8; ZK-12001 for export models with the '287'; U-100001 for cars with the '327' V-81 UK-100001 for export models with the '327' V-8. The starting serial numbers for Canadian production were as follows: JT-500001 for cars with the '199' six; LT-500001 for cars with '232' six; ZT-500001 for cars with the '287' V-8. [Marlin] Starting serial numbers were 2K-100001 for cars with the '232' six; 4-100001 for cars with the '327' V-8; 4K-10001 for Export models with the '327' V-8. [Ambassador] Starting Serial Numbers were S-100001 for cars with the '232' six and SK-10001 for export models with the same engine; E-100001 for cars with the '287' V-8 and EK-10001 for export models with the same engine; H-100001 for cars with the '327' V-8 and HK-13001 for export cars with the same engine. The starting serial numbers for Canadian built cars were as follows: ST-500001 for cars with the '232': ET-500001 for cars with the '287' V-8; and HT-500001 for cars with the '327' V-8. Body Number plate riveted to left front door hinge pillar has additional data. First line indicates body production sequence number. Second line indicates "Model No." consisting of first two symbols indicating model year and additional symbols indicating body style and series. Complete "Model No." corresponds to column two of charts below. Third line gives trim data. Fourth line gives paint color data. Six-cylinder motor number on upper left front corner of engine block. V-8 motor number on top at front of block; or front of block; or center, left-side of block, above oil pan. Continuing in 1965, AMC employed an "Engine Day Build Code" system giving all engines of a type the same six symbol code. The first symbol of the engine day built code is a single digit number code for the year of manufacture starting with 7=1965. The second and third symbols indicated month of build: 01=Jan.; 02=Feb.; etc. The numbers differed in the fourth symbol, a letter code designating the engine displacement and carburetion. Letter codes were: A=195.6 cid/90 hp L-head six; B=195.6 cid/125 hp six; B=195.6 cid/138 hp OHV six; F=327 cid/270 hp V-8 4V; G=287 cid/198 hp V-8; L=232 cid/145 hp six; L=232 cid/155 hp six; J=199 cid/128 hp six. The fifth and sixth symbols indicated day of manufacture as appropriate. In motor number 401A22: first symbol 7=1965; 01=Jan.; A=195.6 cid/90 hp six and 22=22nd day of month.

RAMBLER AMERICAN SERIES 01

Series Number	Body/Style Number	Body Type & Seating	Factory Price	Shipping Weight	Production Total
220 SERIES					
01	6505	4-dr Sed-6P	2036	2518	13,700
01	6506	2-dr Sed-6P	1979	2495	26,409
01	6508	2-dr Sta Wag-6P	2312	2684	5,224
330 SERIES					
01	6505-2	4-dr Sed-6P	2129	2522	15,148
01	6506-2	2-dr Sed-6P	2072	2490	9,065
01	6508-2	2-dr Sta Wag-6P	2396	2682	12,313
440 SERIES					
01	6505-5	4-dr Sed-6P	2222	2580	5,194
01	6507-5	2-dr Conv-5P	2418	2747	3,882
01	6509-5	2-dr HT Cpe-6P	2205	2596	13,784
01	6509-7	2-dr 440H HT-4P	2327	2622	8,164

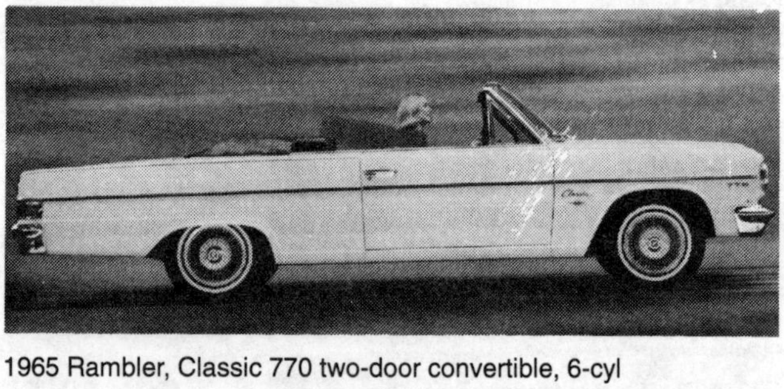

1965 Rambler, Classic 770 two-door convertible, 6-cyl

RAMBLER CLASSIC — (SIX) — SERIES 10: Rambler Classics received a completely restyled front end for 1965,, plus new rear end sheet metal. The front featured new fenders and a grille having 'veed-out' horizontal bars, with three vertical division bars that created four sections. Dual horizontal headlamps flanked the grille on either side. The rear was squared-off and had wraparound, rectangular taillamps. Three trim levels were provided: '550', '660' and '770'. Five body styles were offered, including a new convertible. Standard equipment on '550' models included front armrests; dual visors; cigar lighter; one front ashtray; rubber floor covering (and trunk mat): front foam seat cushions; dome or pillar lights; front seat belts; Fresh Air ventilation; station wagon luggage rack; two coat hooks; 60 amp battery; and blue/green panel lighting. The '660' models had all of the above,, plus rear armrests; two front ashtrays: rear ashtrays; carpets; and locking glove box. The '770' had all above features, except no coat hooks on convertibles. Trim bars were seen on the rear roof pillar and '770' trim and appointments were richer. A 199 cid six was standard in the '550'. A 232 cid six was base powerplant for the other lines. A 287 cid V-8 was extra in '550s,' a 327 cid V-8 in other lines.

RAMBLER CLASSIC SERIES 10

Series Number	Body/Style Number	Body Type & Seating	Factory Price	Shipping Weight	Production Total
550 SERIES					
10	6515	4-dr Sed-6P	2192	2987	30,869
10	6516	2-dr Sed-6P	2142	2963	7,082
10	6518	4-dr Sta Wag-6P	2522	3134	13,759
660 SERIES					
10	6515-2	4-dr Sed-6P	2287	2882	50,638
10	6516-2	2-dr Sed-6P	2282	2991	4,561
10	6518-2	4-dr Sta Wag-6P	2624	3155	32,444

770 SERIES

10	6515-5	4-dr Sed-6P	2436	3029	23,603
10	6519-5	2-dr HT Cpe-6P	2436	3063	14,778
10	6519-7	2-dr 770H HT-5P	2548	3089	5,706
10	6517-5	2-dr Conv-6P	2696	3169	4,953
10	6518-5	2-dr Sta Wag-6P	2727	3180	15,623

MARLIN — (SIX) — SERIES 50: The 1965 Marlin was introduced in February 1965 as a midyear addition to the line. It was basically a Rambler Classic with special fastback roof styling. Different taillights were used, but the grille was of the Classic type with the vertical division bars removed. A special Marlin hood ornament was seen. The Marlin was American Motors Corp.'s answer to the Ford Mustang, but could accommodate six passengers, as opposed to only four in the Mustang.

MARLIN SERIES 50

Series Number	Body/Style Number	Body Type & Seating	Factory Price	Shipping Weight	Production Total
50	6559-7	2-dr HT FsBk-6P	3100	3234	10,327

1965 Rambler, Ambassador 990 four-door sedan, V-8

RAMBLER AMBASSADOR — (SIX) — SERIES 80: The 1965 Ambassador had a totally new front end with a four-inch longer wheelbase than Classics. The front featured vertical, quad headlamps and a grille of numerous horizontal bars which 'veed' slightly outwards along the horizontal plane. Bright metal side trim ran from the rear along the top edge of the body and around the front of the car (crossing the grille). The taillamps wrapped around the body corners and could be seen from both the rear and the side. A convertible, four-door sedans, station wagons, two-door sedans and hardtops were offered in specific lines. A new Ambassador Six marked the first time since 1956 that a six-cylinder motor had been offered in Ambassadors.

RAMBLER AMBASSADOR SERIES 80

Series Number	Body/Style Number	Body Type & Seating	Factory Price	Shipping Weight	Production Total
AMBASSADOR 880 SERIES					
80	6585-2	4-dr Sed-6P	2565	3120	10,564
80	6586-2	2-dr Sed-6P	2512	3087	1,301
80	6588-2	4-dr Sta Wag-6P	2879	3247	3,812
AMBASSADOR 990 SERIES					
80	6585-5	4-dr Sed-6P	2656	3151	24,852
80	6587-5	2-dr Conv-6P	2955	3265	3,499
80	6588-5	4-dr Sta Wag-6P	2970	3268	8,701
80	6589-5	2-dr HT Cpe-6P	2669	3168	5,034
80	6589-7	2-dr 990H HT-5P	2837	3198	6,382

RAMBLER BASE ENGINES

(AMERICAN 220/330) Inline six. Cast iron block. Displacement: 195.6 cid. Bore and stroke: 3.125 x 4.25 inches. Compression ratio: 8.0:1. Brake hp: 90 at 3800 rpm. Four main bearings. Solid valve lifters. Carburetor: Carter Type RBS one-barrel Model 3708S.

(AMERICAN 440/440-H) Inline six. Overhead valve. Cast iron block. Displacement: 195.6 cid. Bore and stroke: 3.125 x 4.25 inches. Compression ratio: 8.7:1. Brake hp: 125 at 4200 rpm. Four main bearings. Solid valve lifters. Carburetor: Holley one-barrel Model 1909-2555-3.

(CLASSIC 550) Inline six. Cast iron block. Displacement: 198.8 cid. Bore and stroke: 3.75 x 3.25 inches. Compression ratio: 8.5:1. Brake horsepower: 128 at 4400 rpm. Four main bearings. Solid valve lifters. Carburetor: Holley 1909-25555-3 one-barrel.

(CLASSIC 660/770/MARLIN SIX) Inline six. Cast iron block. Displacement: 231.9 cid. Bore and stroke: 3.75 x 3.50 inches. Compression ratio: 8.5:1. Brake horsepower: 145 at 4300 rpm. Carburetor: Carter WCD-3882 two-barrel.

(CLASSIC 660/770/MARLIN EIGHT) V-8. Overhead valves. Cast iron block. Displacement: 287 cid. Bore and stroke: 3-3/4 x 3-1/4 inches. Compression ratio: 8.7:1. Brake hp: 198 at 4700 rpm. Five main bearings. Hydraulic valve lifters. Carburetor: Holley 2209-2699.

(AMBASSADOR 990 SIX) Inline six. Cast iron block. Displacement: 231.9 cid. Bore and stroke: 3.75 x 3.50 inches. Compression ratio: 8.5:1. Brake horsepower: 155 at 4400 rpm. Carburetor: Carter WCD-3888S two-barrel.

(AMBASSADOR EIGHT) V-8. Overhead valves. Cast iron block. Displacement: 326.7 cid. Bore and stroke: 3.75 x 3.25 inches. Compression ratio: 8.7:1. Brake hp: 198 at 4700 rpm. Five main bearings. Hydraulic valve lifters. Carburetor: Carter WCD-3888S two-barrel.

CHASSIS FEATURES: Wheelbase: (Americans) 106 inches; (Classic/Marlin) 112 inches; (Ambassadors) 116 inches. Overall length: (Americans) 177.25 inches; (Classic station wagon) 193 inches; (other Classic/Marlin) 195 inches; (Ambassador station wagon) 197 inches. Front tread: (Ambassador/Marlin) 58.6 inches; (American) 56 inches; (Classic) 58.2 inches. Rear tread: (Ambassador/Marlin) 57.6 inches; (American) 56 inches; (Classic) 57.4 inches. Tires: (Ambassador/Marlin) 7.35 x 14; (American) 6.45 x 14; (Classic) 6.95 x 14.

OPTIONS: Air conditioner adaptor group with heavy-duty radiator and seven ampere battery ($19). All-Season air conditioner, in American ($295.85), in Classic ($312.05), in Ambassador, with power-saver fan required with V-8 ($321). 40-amp. alternator ($8.95). Appearance Group A for Americans except 440H and wagons, includes rocker panel molding, deck molding, rear fender moldings and spinner wheel discs ($55.54); same on 440H ($39.75). Appearance Group A for Classsics except not available on 770H, includes rocker panel moldings, wheel opening moldings and spinner wheel discs ($58.05); same for Classic 880 ($60.30). Appearance Group B, includes same as above except wire wheel covers: for Americans, except 440H and wagon ($92.65); for 440H ($76.95); for Classic, except 770H ($92.65); and for Classic 880 ($94.90). Back-up lights ($10.70). Heavy-duty battery ($6.50). Front disc brakes for Classic/Ambassador, includes power assist ($79.95) Slim bucket seats with console, 440 and 770 ($99.50); 440H and 770H with reclining bucket seats required ($40). Slim bucket seats with front armrest and cushion in Ambassador, with reclining seats required ($99.50). Wide bucket seats without console, 440/770 reclining seats required [standard in 440H/770H] ($59.50). Slim front bucket seats with armrest and console in Ambassador 990, requires reclining seats ($119.50); same in 990H ($20). Front bumper guards ($11.50). Front and rear bumper guards, except wagons ($23). Two-barrel carburetor with 232 cid engine ($11.55). Electric clock ($15.95). Two-tone paint on American sedans and hardtops ($16.95); on American wagons ($18.95); on Classic/Ambassador hardtops and sedans ($19.95). on Classic/Ambassador wagons ($21.95). Special paint color application, except fleet cars ($29.50). Simulated Ambassador wagon woodgrain trim ($22.20). Dowguard coolant ($5). Heavy-duty cooling system, six ($10.95); eight ($14.95). California crankcase ventilation ($5.05). L-head 196 cid engine in 220/330 ($38.55). 232 cid two-barrel engine in 220/330 with automatic required ($84.95); in 440/440H with automatic required ($49.95). in 550 only ($39.95). 327 cid V-8 four-barrel engine in Classic and Ambassador small V-8s ($81.95). Rear foam cushion ($9.95). Individually adjustable front seats ($20). Reclining front seat ($25.50). Tinted glass, in American ($27.95); in Classic/Ambassador cars ($45.50); in Classic/Ambassador wagon ($39.50). Tinted windshield only, in American ($12.95); in Classic and Ambassador ($19.95). Headrest ($12). Light Group ($19.60). Inside tilt mirror ($4.95). Outside mirror, either side ($5.30). OSRV mirror with remote-control ($11.95). Oil-bath air cleaner with L-head ($7.15). Padded panel and visors ($19.95). Padded dash panel, convertible ($17.50). Padded visors ($4.50). Power brakes, in American/Classic ($42.95), in Ambassador ($43.95). Power Saver fan ($19.60). Power steering, in American/Classic ($85.95); in Ambassador ($96.90). Power tailgate window ($31.95). American convertible power top ($49.95). Front power windows, except Americans and convertibles and hardtops ($59.50). Station wagon power front windows and power tailgate window, except Americans ($91.45). Power front and rear windows, except Americans, not available on two-door sedans or convertibles ($102.25). Wagons, except Americans, power front, rear and tailgate windows ($134.20). Radio and manual antenna in Americans ($49.50). Push-button radio and antenna in Americans ($56.50). Push-button AM radio in Classic/Ambassador ($58.50). Push-button AM/FM radio in Classic/Ambassador ($129.30). DuoCoustic rear speaker in Classic/Ambassador sedans and hardtops ($12.60). VibraTone rear speaker in Classic/Ambassador sedans and hardtops ($40.50). Heavy-duty raditor in American ($3.35). Heavy-duty radiator in Classic/Ambassador ($5.35). Airliner reclining seats in American, standard 440H, others ($25.50). Adjustable front seat in Americans, requires reclining seats ($20). Front seat belt deletion ($11 credit). Retractable front seat belts ($27.50). Retractable front and non-retractable seat belts ($26.70). Retractable seat belts front and rear ($45.80). Front and rear shocks in American ($2.70). Front and rear shocks in Classic/Ambassador ($3.85). Heavy-duty front and rear springs, in American ($5.15 cars/$6.55 wagons). Heavy-duty front and rear shocks and rear sprins in Classic/Ambassador cars ($5.05). Heavy-duty front and rear extended springs, in Classic/Ambassador ($7.30 cars/$6.55 wagons). Adjustable steering wheel in Classic/Ambassador, requires automatic transmission ($43). Custom steering wheel in 220/550 ($7.70). Side hinge tailgate, except Americans (($39.60, but standard with third seat). Station wagon third seat, except American and 550 ($85). Flash-O-Matic transmission, in American ($171.25); in Classic Six ($186.50); in Classic Eight ($193.65); in Ambassador Six ($212.35); in Ambassador Eight ($219.50). Overdrive transmission, in Americans without 232 cid engine ($105.50); in Classic Six, ($108.50); in Classic Eight or Classic with 232 cid engine ($11.35); in Ambassador ($114.95). Flash-O-Matic Shift Command transmission, in American with 232 cid engine and slim bucket seats ($186.25); in Classic with 232 cid two-barrel engine requires slim bucket seats ($201.50); in Classic V-8 requires slim bucket seats ($208.65). in Ambassador Six, requires bucket seats and console ($227.30). In Ambassador V-8, requires bucket seats and console ($234.50). Twin Stick transmission ($134.50 to $147.50). Twin Grip differential (($37.55-$42,70). Undercoating ($17.20). Wheel discs ($20.55). Wheel discs with spinners ($14.05-$34.55). Wire wheel covers with spinners ($48.70-$69.15). Windshield washer ($11.95). Electric wipers ($8.95 American/$10.95 others). Vinyl upholstery ($15-$24.50). Credit for Weather Eye heater deletion ($72-$79). Various tire options.

HISTORICAL FOOTNOTES: The full-sized Ramblers were introduced in September 1964 and the Americans appeared in dealer showrooms at the same time. Calendar year registrations of 324,669 cars were recorded. Calendar year sales were counted at 346,367 units. American Motors was the country's ninth largest auto maker this season. The 'Rambulance', a station wagon conversion for the economy class emergency vehicle market, was available again this year. A total of 246 such vehicles were sold to police and fire departments in small towns between 1960 and 1965.

1966 RAMBLER

1966 AMC, Rambler Rogue two-door hardtop coupe, V-8

229

RAMBLER AMERICAN-(SIX)-SERIES 01: The 1966 American was redesigned. It had a squared-off front end. The '330' models and the old 196.5 cid overhead valve Six were dropped. A new top-of-the-line model, called the Rambler Rogue, was introduced. Rear taillamps were larger. At the Chicago Automobile Show, in mid-season, a new 290 cid overhead valve V-8 was announced and a four-speed manual transmission was introduced.

RAMBLER I.D. DATA: VIN stamped on plate welded to right fender below hood. First symbol A=AMC. Second symbol 6=1966. Third symbol identiufies assembly plant: K=Kenosha, Wis.; B=Brampton, Ont. (Canada); Fourth symbol identifies transmission. Fifth symbol identifies body type (corresponds to fourth digit in Model Number column of charts in this catalog). Sixth symbol identifies car-line (corresponds to Model Number suffix in charts). Seventh symbol identifies series and engine. Last six symbols are sequential production number starting at 100001.

RAMBLER AMERICAN

Series Number	Body/Style Number	Body Type & Seating	Factory Price	Shipping Weight	Production Total
220 LINE					
01	6605-0	4-dr Sed-6P	2086	2574	15,940
01	6606-0	2-dr Sed-6P	2017	2554	24,440
01	6608-0	4-dr Sta Wag-6P	2369	2740	5,809
440 LINE					
01	6605-5	4-dr Sed-6P	2203	2582	14,543
01	6606-5	2-dr Sed-6P	2134	2562	5,252
01	6607-5	2-dr Conv-6P	2486	2782	2,092
01	6608-5	4-dr Sta Wag-6P	2477	2745	6,603
01	6609-5	2-dr HT Cpe-6P	2227	2610	10,255
01	6609-7	2-dr Rougue HT-5P	2370	2630	8,718

1966 AMC, Rambler Classic 770 two-door convertible, V-8

RAMBLER CLASSIC — (SIX) — SERIES 10: The 1966 Classic received a new grille, new roof and larger taillights on the same basic 1965 body. The top-level model was a two-door hardtop called the Rebel. The '660' Series designation was dropped. During the last three months of production the 290 cid V-8 took the place of the 287 cid V-8, as the smallest optional eight-cylinder engine.

RAMBLER CLASSIC

Series Number	Body/Style Number	Body Type & Seating	Factory Price	Shipping Weight	Production Total
550 LINE					
10	6615-0	4-dr Sed-6P	2238	2885	22,485
10	6616-0	2-dr Sed-6P	2189	2860	5,505
10	6618-0	4-dr Sta Wag-6P	2542	3070	9,390
770 LINE					
10	6615-5	4-dr Sed-6P	2337	2905	46,044
10	6619-5	2-dr HT Cpe-6P	2363	2935	8,736
10	6617-5	2-dr Conv-5P	2616	3070	1,806
10	6618-5	4-dr Sta Wag-6P	2629	3071	24,528
REBEL LINE					
10	6619-7	2-dr HT Cpe-5P	2523	2950	7,512

1966 AMC, Marlin two-door fastback hardtop, V-8

MARLIN — (SIX) — SERIES 50: The 1966 Marlin received a few changes. A new grille was used and many features, formerly standard, were now optional. This included power steering and brakes. The price dropped by nearly $500. And the Rambler nameplate was deleted from the rear of the car.

MARLIN

Series Number	Body/Style Number	Body Type & Seating	Factory Price	Shipping Weight	Production Total
50	6659-7	2-dr FsBk-6P	2601	3050	4,547

1966 AMC, Ambassador 990 four-door sedan, V-8

RAMBLER AMBASSADOR — (SIX) — SERIES 80: The 1966 Ambassadors had a new roof, larger and more visible taillamps and new chrome trim pieces alongside the car, on the tip of the front fenders. They took the form of a small, ribbed rectangle. The top-level Ambassador was now called the Ambassador DPL, which would soon become 'Diplomat'.

AMBASSADOR

Series Number	Body/Style Number	Body Type & Seating	Factory Price	Shipping Weight	Production Total
880 LINE					
80	6685-2	4-dr Sed-6P	2455	3006	Note 1
80	6686-2	2-dr Sed-6P	2404	2970	Note 1
80	6688-2	4-dr Sta Wag-6P	2759	3160	Note 1
990 LINE					
80	6685-5	4-dr Sed-6P	2574	3034	Note 1
80	6689-5	2-dr HT Cpe-6P	2600	3056	Note 1
80	6687-5	2-dr Conv-5P	2968	3432	(1,798)
80	6688-5	4-dr Sta Wag-6P	2880	3180	Note 1
DPL LINE					
80	6689-7	2-dr HT Cpe-5P	2756	3090	Note 1

NOTE 1: Total 1966 Rambler Ambassador Series production was 34,222 units. No breakout per body style (except for convertible) is currently available.

BASE ENGINES

Inline Six. Overhead valves. Cast iron block. Displacement: 198.8 cid. Bore and stroke: 3.75 x 3.00 inches. Compression ratio: 8.5:1. Brake hp: 128 at 4400 rpm. Seven main bearings. Hydraulic valve lifters. Carburetor: Holley one-barrel.

CHASSIS FEATURES: Wheelbase: (Americans) 106 inches; (Classic/Rebel/Marlin) 112 inches; (Ambassador) 116 inches. Overall length: (Americans) 181 inches; (Classic/Rebel/Marlin) 195 inches; (Ambassador station wagons) 200 inches; (other Ambassadors/DPL) 199 inches. Tires: (Americans) 6.45 x 14; (other station wagons) 7.35 x 14; (all other models) 6.95 x 14.

OPTIONS: Power brakes ($42). Power steering ($84). Air conditioning, Americans ($303); Classic ($319); Ambassador ($328). Power steering, Ambassador ($95). Front disc brakes ($91). Two-tone paint. Wheel disc, standard on Rogue/Rebel/DPL and Marlin. Turbo-cast wheelcovers. Wire wheelcovers with spinners. Slim band whitewall tires. Bumper guards with rubber facings. Black vinyl covered hardtop roof. Reclining seats, bucket-type standard on Rogue/Rebel/DPL. Safety headrests. Tachometer. Vinyl upholstery in station wagons. Side-hinged tailgate, Classic/Ambassador wagons only. Simulated woodgrain wagon paneling, Ambassadors only. Appearance group with wheel discs. Rocker and wheelhose moldings, standard on '990' and DPL. Rear seat foam cushions, standard on 770/Rebel/990/DPL/Marlin. Electric clock, standard in Rebel/990/DPL and Marlin. Cruise Command speed control, automatic transmission mandatory. Special black two-tone paint, for Rogue with vinyl roof. AM all-transistor radio, in Americans. Air-Guard, exhaust emissions control system. Four-Way hazard warning signals. Custom steering wheel, standard on all except '220' models. Remote control left-hand outside rearview mirror. AM/FM all transistor radio, for all except Americans. Three-speed manual transmission was standard. Automatic transmission ($187). Four-speed manual floor shift transmission was optional. American six-cylinder 232 cid 155 hp two-barrel engine ($51). All Series V-8 287 cid 198 hp two-barrel engine ($106). V-8 327 cid 250 hp two-barrel engine ($32). V-8 327 cid 270 hp four-barrel engine ($65). Positive traction rear axle was optional.

HISTORICAL FOOTNOTES: The 1966 Ramblers were introduced October 7, 1965 and the Marlin appeared in dealer showrooms around midyear. Model year production peaked at 295,897 units. Calendar year sales of 346,367 cars were recorded. R.D. Chapman, Jr. was the chief executive officer of the company this year. A total of 45,235 AMC models were made with 327 cid V-8s. Only 623 cars were built with 290 cid V-8s during the 1966 model year. The 287 cid V-8 was used in 44,300 additional units. AMC held a 3.71 percent share of the total market this year. Bucket seats were installed in 11.5 percent of all 1966 Ramblers; 27.6 percent had V-8s; 4.3 percent had disc brakes; 3.4 percent had movable steering columns; 10.9 percent had limited-slip differentials; 12.1 percent had air conditioning and only one percent had power windows.

AMC
1967-1986

In February 1962, George Romney resigned as president and chairman of American Motors Corp. (AMC) to campaign successfully for Governor of Michigan. The board of directors elected him vice-chairman and granted him a leave of absence. Richard E. Cross was named board chairman, and Roy Abernethy became president.

During the next few years, AMC invested $300 million in new advanced-designed engines, bodies and plant facilities. Romney had been the primary driving force behind AMC's small car program. When he left, Abernathy, a former Packard sales manager, began going into direct competition with the "big three" automakers.

An entirely new line of larger cars was introduced in the 1967 model year, with the Classic series becoming the Rebel. New luxury models, including convertibles, were introduced. Roy D. Chapin, Jr., son of a founder of Hudson Motor Car Co., was made chairman and chief executive officer on January 9, 1967 with William V. Luneburg as president and chief operating officer. Chapin staked out a new and bold direction for American Motors. Among the many forward steps taken to strengthen confidence in the company was a sharp reduction in delivered prices of the low-priced Rambler American models to make them more competitive with imports, which again were on the rise. Chapin pledged to introduce six new models in the next 18 months. The first was the Javelin, bowing on Sept. 26, 1967 as a 1968 model. A two-passenger sports car, the AMX, bowed in February 1968. Both cars were designed to change AMC's image, from a company that had once advertised "the only race we are interested in is the human race." Now the ads said, "We Just haven't been the same since we discovered racing."

In July 1968, AMC sold its Kelvinator appliance business to White Consolidated Industries of Cleveland, Ohio. The company could now devote its full and complete energy to the automotive business. The Hornet replaced the Rambler American in the fall of 1969 as a 1970 model. The Rambler name was discontinued. A new subcompact, the Gremlin, bowed in April 1970, as the first American subcompact.

On Feb. 5, 1970, AMC acquired Kaiser-Jeep Corp. in a transaction involving cash debentures and American Motors stock. The deal was part of planned expansion and an acquisition program. Kaiser-Jeep became the Jeep Div. of AMC and is the leading worldwide manufacturer of four-wheel-drive vehicles.

In March 1971, a wholly owned subsidiary, AM General Corp., was created by AMC. The company assumed the assets of both AMC's former U.S. government contracts and that of Kaiser-Jeep's General Products Div., whose plants are in Mishawaka, Indianapolis, and South Bend, Ind. Cruse W. Moss was named president of AM General, which made tactical wheeled vehicles for the military and delivery vehicles for the U.S. Postal Service in Studebaker's old Chippewa Ave. plant in South Bend, Ind.

In 1971, AM General announced that it planned to enter the urban transit bus field, with production in the AM General plant in Mishawaka, Ind. A deal had been arranged to manufacture a bus original designed by Canadian Flyer. In 1973, AM General began bidding successfully on contracts, including one to build buses for the Washington D.C. Mass Transit District.

1971 also marked the redesign of the Javelin and a major facelift and change of the name for the intermediates from Rebel to Matador. Also new was a station wagon version of the Hornet, called the Sportabout. It sold extremely well. A sport model of the Hornet, the SC360, that did not sell well, was also introduced.

The year 1972 brought little change to AMC's model lineup, but a gas crisis in 1972 and 1973 brought on by an Arab oil embargo increased AMC's sales of its smaller more economical models. A special model in an AMC "designer series" of cars was the Pierre Cardin Javelin. Inspired by fashion designer Pierre Cardin, it was available in 1972 and 1973. Another designer car, for 1972 only, was the Gucci Sportabout in the Hornet series.

In 1973, the Hornet series received a new body style. Its hatchback body style had many imitators after 1973.

The two-door Matador received a major facelift in 1974 making it into a sleek, new body style aimed at NASCAR racing. With its fastback body style, it came in two special models. The Matador X was a special sporty model and the Cassini Matador, inspired by fashion designer Oleg Cassini, was a luxury model. Four-door station wagon versions of the Matador received few changes and remained basically the same as the year before. The Ambassador made its last appearance in 1974. Then it was discontinued.

The big news at AMC for 1975 was the introduction of the Pacer in March 1975 at the Chicago auto show. The Pacer was AMC's first new-from-the-ground-up car to debut in a long time. It featured a unique hatchback body with larger passenger side door than driver side door, a high glass area, and a sloping hood. It had originally been intended as a front-wheel drive, rotary-engined car, but GM's discontinuance of its rotary engine program forced a return to a conventional drivetrain layout. Also for 1975, a special 'Touring' package was offered on the Sportabout, including special interior and exterior trim.

Not much changed for 1976. Matadors could get a choice of 360 cu. in. V-8s under the hood, and a stylish Barcelona package for the coupe. Gremlin offered a pair of packages to enhance its appeal; a sporty 'X' appearance option, and a more curious "Levi's" trim package. That one put simulated blue denim (actually nylon) and fake "buttons" on the seats and interior panels, and attracted quite a lot of attention. Though popular at first, Pacer sales flagged this year. People didn't seem so interested in smaller cars as the gas crunch eased.

Only Gremlin got any noticeable restyling for 1977 — along with a new four-cylinder powerplant supplied by Volkswagen. Pacer added a station wagon to the original hatchback sedan. The AMX name, abandoned a couple of years earlier, appeared again on a limited-edition Hornet model with front air dam, blackout grille, back-window louvers, and body graphics. Styling touches of that sort would become common on sporty models from just about every domestic manufacturer. High-performance 360 cu. in. V-8s were no longer offered under Matador hoods, and the 401 V-8 wasn't even available to police agencies as it had been before. Following a $46 million loss in fiscal year 1976, AMC switched to black ink on the 1977 ledger book — but as a result of strong Jeep sales, not conventional passenger cars. Even price cuts and rebates didn't help. Other companies were planning compact models, and imports gained strength each year. But, AMC couldn't seem to take hold of a major corner of the market. The agreement to purchase engines from VW/Audi was something new, and foretold the European connection that would eventually "save" AMC (temporarily) in the 1980s.

AMC took a stronger stab at the youth market in 1978, making AMX a separate model, again heavy on black accents, instead of a Hornet option. Hornet, in fact, was gone, replaced by a new and more posh (but similar) Concord. Even Gremlin turned a bit toward performance with a new GT option package that used fiberglass body components. Pacers now came with a 304 cu. in. V-8 as well as two six-cylinder choices. Matador again offered a Barcelona package for the coupe, but this would be the mid-size's final season. Corporate net earnings reached their highest peak since 1973, but again due to Jeep popularity. Negotiations with Renault were already underway. They culminated in an agreement whereby AMC would distribute the French-built Le Car (and Renault might do likewise with Jeeps). More important for the future was the announcement that the two corporations planned a joint-venture passenger car to be built in the U.S. In another agreement, AMC contracted to buy "Iron Duke" four-cylinder engines from Pontiac.

Gremlin bit the dust in 1979, replaced by a more stylish (but not dramatically different) Spirit subcompact. The performance-minded AMX switched to the shorter Spirit platform. AMC profits reached a record level as Renault bought an interest in the company for $200 million. Plans were made to begin assembly of Renault-designed cars in Kenosha as early as the 1982 model year — a target date that proved slightly premature.

V-8 engines disappeared for good the next year, as Pacer and AMX entered their final years. Eagle was the big news: the first major four-wheel drive passenger car produced in America in modern times. Riding a Concord platform, the Eagle sat three inches higher off the ground to allow for the 4WD structure. It was also the first 4WD model to be built in volume with independent front suspension, and gave AMC something to offer that no other domestic automaker had. Unfortunately, not enough customers seemed to care. In December 1980, AMC stockholders agreed to give Renault a 46 percent share of the company.

For 1981, AMC's selection shrunk to only three models: Spirit, Concord and Eagle, with Pontiac's 151 cu. in. four-cylinder engine under their hoods. The 4WD Eagle was now offered on two platforms: the original Concord size, and an SX/4 coupe (and Kammback wagon) on the smaller Spirit chassis. Both Eagles sported dark Krayton lower body treatments. Sales slipped over 8 percent, but AMC's market share hung almost steady.

Neither the compact Concord nor subcompact Spirit was doing well as 1982 began, and their days were numbered. Spirit's new five-speed overdrive manual gearbox didn't help. Switchable two/four-wheel drive was standard on Eagles, which attracted a moderate, but hardly overwhelming, following. The company's market share sunk below 2 percent, but late in the model year, production of the new Renault Alliance began in Kenosha. Early Alliance sales even helped AMC to recapture fourth place in domestic sales for the first time since 1978, beating out Volkswagen of America. Late in September, production of Spirits and the smaller Eagles moved to Canada, while big Eagles continued to be built in Wisconsin.

Only Eagles retained the Pontiac four-cylinder base engine for 1983. All final Spirits and Concords carried sixes. The new subcompact Alliance, designed in France, was revised to suit American tastes. That included electronic controls, Bendix fuel injection, and power steering and brakes. The 1.4-liter powerplants came from France. New Alliance notwithstanding, AMC managed to lose over $146 million in 1983, nearly as disastrous a figure as the 1982 loss. Early Alliances sold well, but the good news didn't last, especially as mechanical problems became evident.

A Renault-styled hatchback Encore joined the notchback Alliance for 1984, as Spirit and Concord departed. Only the big 4WD Eagle remained, powered by an AMC-built four or six. Eagle sales had looked promising at first, but sagged badly. For a change, AMC enjoyed a bit of profit this year, and could be forgiven a taste of optimism in what would prove to be a temporary respite.

The Renault-based models got a bigger engine choice for 1985, as a convertible Alliance arrived, the first AMC soft-top since the '68 Rambler Rebel. Eagle added "shift-on-the-fly" 4WD but dropped the four-cylinder engine. The modest profit for 1984 proved to be temporary, as AMC lost over $125 million this time around. Low-rate financing was tried for the first time but couldn't attract enough customers. Production was slashed several times, with workers laid off. The UAW agreed to a pay cut only after AMC threatened to cease production entirely.

Jose Dedeurwaerder, AMC's president, predicted that "the worst is over," and the loss of $91 million in 1986 wasn't as bad as previous totals. But the company's share of sales dwindled to just 1 percent, even after low-rate financing of "zero" was offered. Even American Honda earned a bigger share. Eagle sales slipped below 10,000 amid AMC's insistence that the tough 4WD would remain in production in Canada, despite rising interest in 4WD models. Chrysler's agreement to build its M-body cars at AMC's Kenosha facility helped pave the way for the Chrysler takeover (for $1 billion) a year later in August 1987. Only a few months after that, Chrysler announced imminent closing of the plant, after allegedly promising to give it another five years.

Collector interest in 1976-86 AMC models has been, to say the least, modest. But a few examples are worth noting, for curiosity value if nothing else. Gremlin 'X' is one possibility. So is the revived AMX. The last Matadors, especially with Barcelona trim, may be worth a glance. Not everyone loved Eagles when they were new, but as the pioneer of modern 4WD, they might draw modest attention one day. The SX/4 coupe, in particular, is rather attractive. AMC cars may not be exciting, but at least they're inexpensive.

1967 AMC

1967 Rambler, Rogue two-door convertible, V-8

1967 AMC, Rambler Rebel SST two-door hardtop, V-8

RAMBLER AMERICAN — (SIX) — SERIES 01: The 1967 Rambler American used the same body styling as the previous year's models, with only minor changes. New taillamps were of the same, rectangular shape, but were shorter and higher. A new side molding was used on the 440 and Rogue models. It was positioned lower on the beltline. A 343 cid V-8 with four-barrel carburetion was available as optional equipment. This was also the last season for the Rambler American convertible. Only seven convertibles were assembled with the 343 cid engine. All 1967 Ramblers and Ambassadors were marketed in six-cylinder series, with V-8 engines as options.

VIN: Starting in 1967 (but also found on some 1966 models), American Motors used a new vehicle identification system. There were 13 symbols in the Vehicle Identification Number, which was located below the hood, on the top right hand inner fender panel. The first symbol designated the manufacturer: A=American Motors. The second symbol designated the model year: 7=1967. The third symbol designated the assembly plant, B=Brampton, Canada; K=Kenosha, Wis.). The fourth symbol designated the type of transmission, as follows: S=three-speed manual on column; O=three-speed manual with overdrive; A=automatic transmission on column; C=three-speed manual with floor shift; F=four-speed manual with console and floor shift and M=four-speed manual with floor shift and no center console. The fifth symbol designated Body Style, 5=four-door sedan; 7=two-door sedan; 8=four-door station wagon and 9=two-door hardtop. The sixth symbol designated class of body: 0=220/550; 2=880; 5=440/770/990 and 7=Rogue/ SST/DPL/Marlin. The seventh symbol designated series and engine type, as follows: (American 01 Series) A=199 cid six with one-barrel carburetor; B=232 cid six with two-barrel carburetor; C=290 cid V-8 with two-barrel carburetor; D=290 cid V-8 with four-barrel carburetor; E=232 cid six with one-barrel carburetor and X=343 cid V-8 with four-barrel carburetor. (Ambassador 80 Series) M=232 cid six with two-barrel carburetor; N=290 cid V-8 with two-barrel carburetor; P=232 cid six with one-barrel carburetor; Q=343 cid V-8 with four-barrel carburetor and R=343 cid V-8 with two-barrel carburetor. (Rebel Series 10) F=232 cid six with one-barrel carburetor; G=232 cid six with two-barrel carburetor; H=290 cid V-8 with two-barrel carburetor; J=343 cid V-8 with two-barrel carburetor and K=343 cid V-8 with four-barrel carburetor. (Marlin Series 50) S=232 cid six with one-barrel carburetor; T=232 cid six with two-barrel carburetor; U=290 cid V-8 with two-barrel carburetor; V=343 cid V-8 with two-barrel carburetor and W=343 cid V-8 with four-barrel carburetor. The remaining symbols were six digits representing the sequential Serial Number and starting with 100001 for domestic cars; 700001 for Canadian cars and 10001 for export cars.

RAMBLER AMERICAN SERIES 01

Series Number	Body/Style Number	Body Type & Seating	Factory Price	Shipping Weight	Production Total
AMERICAN 220 LINE					
01	6705	4-dr Sed-6P	2142	2621	10,362
01	6706	2-dr Sed-6P	2073	2591	24,834
01	6708	4-dr Sta Wag-6P	2425	2767	2,489
AMERICAN 440 LINE					
01	6705-5	4-dr Sed-6P	2259	2613	7,523
01	6706-5	2-dr Sed-6P	2191	2586	3,317
01	6709-5	2-dr HT Cpe-6P	2283	2643	4,970
01	6708-5	4-dr Sta Wag-6P	2533	2769	4,135
AMERICAN ROGUE LINE					
01	6709-7	2-dr HT Cpe-5P	2426	2663	4,129
01	6707-7	2-dr Conv-6P	2611	2821	921

RAMBLER REBEL — (SIX) — SERIES 10: Rebel nameplates now adorned the cars that used to be called Classics, as the series' name was changed this year. The Rambler Rebel was also a totally redesigned automobile poised on a longer, 114 inch wheelbase. The top-of-the-line entry was known as the SST, which stood for Super Sport Touring (not Super Sonic Transport). New body styling featured slightly rounded body contours with a semi-fastback roof on two-door styles. In February 1967, AMC introduced three special, limited-production station wagons. The first was called the Briarcliff and 400 examples were sold only in the Eastern portion of the United States. The second was the Mariner, marketed to just 600 buyers in coastal areas. Finally, there was the Westerner, which found 500 customers in the Midwest. A new 'venturi' styled grille was one appearance highlight. The 550 models had the lowest level of trim and Rebel signatures low on the sides of the cowl, behind the front wheel opening. On 770 models a lower body molding traversed the entire length of the body, arching up over both front and rear wheelhousings. The SST was adorned with simulated air intake scoops ahead of the rear wheel openings and upper beltline accent trim, but no lower body moldings. A five year or 50,000 mile warranty covered the 1967 engines and drivetrains.

RAMBLER REBEL SERIES 10

Series Number	Body/Style Number &	Body Type Seating	Factory Price	Shipping Weight	Production Total
REBEL 550 LINE					
10	6715	4-dr Sed-6P	2319	3055	10,249
10	6716	2-dr Sed-6P	2294	3089	9,121
10	6718	4-dr Sta Wag-6P	2623	3287	6,845
REBEL 770 LINE					
10	6715-5	4-dr Sed-6P	2418	3053	24,057
10	6719-5	2-dr HT Cpe-6P	2443	3092	9,685
10	6718-5	4-dr Sta Wag-6P	2710	3288	18,240
REBEL SST LINE					
10	6719-7	2-dr HT Cpe-5P	2604	3109	15,287
10	6717-7	2-dr Conv-5P	2872	3180	1,686

1967 AMC, Marlin two-door fastback hardtop coupe, V-8

RAMBLER MARLIN — (SIX) — SERIES 50: The 1967 Marlin was longer, lower and wider and had a two-inch increase in wheelbase. The sporty Rambler entry retained its distinctive fastback roof styling and semi-elliptical side window openings. It was basically an Ambassador with a fastback roof, instead of being a streamlined Rambler Classic. American Motors hoped to increase Marlin sales by upgrading the car in this manner. It was also quite a distinctive product: a large, six-passenger sports car aimed at the family man with a 'Walter Mitty' complex. There were smoother body sides, a new rectangular gas filler door and Rally lights incorporated into the grille. Side marker lights could be seen on the trailing edge of the rear fenders, just ahead of the wraparound rear bumper ends. A full-length lower body molding helped create a slim appearance and followed the pattern seen on Rebels, arching up over both wheel housings. The rear deck area was cleaned-up a bit, by removal of the large, round medallion. Marlins (and other Ramblers) with V-8 power had V-shaped emblems at the forward edge of the front fenders. Unfortunately, the Marlin again had problems in the marketplace, as sales drooped to even lower levels.

RAMBLER MARLIN SERIES 50

Series Number	Body/Style Number	Body Type & Seating	Factory Price	Shipping Weight	Production Total
50	6759-7	2-dr FsBk Cpe-6P	2963	3342	2,545

AMBASSADOR — (SIX) — SERIES 80: The 1967 Ambassador was the top American Motors' model, but it actually shared the basic body of the new Rebel. The major difference was that the front end stretched four inches longer. It had vertically stacked quad headlights flanking a horizontal bar grille. Ambassador scripts were placed on the sides of the front fenders (behind the wheel opening) and on the left-hand edge of the hood. There was also a new stand-up hood ornament and segmented vertical taillamps. The basic 880 had low level trim and appointments with almost bare body sides. The 770 models had a full-length lower side molding similar to that described for other lines. The starring role went to the DPL, which carried a horizontal center grille divider and integral Rally lights up front. This was the final season before the curtain dropped on the Ambassador convertible.

AMBASSADOR SERIES 80

Series Number	Body/Style Number	Body Type & Seating	Factory Price	Shipping Weight	Production Total
AMBASSADOR 880 LINE					
80	6785-2	4-dr Sed-6P	2657	3279	9,772
80	6786-2	2-dr Sed-6P	2619	3310	3,623
80	6788-2	4-dr Sta Wag-6P	2962	3486	3,540
AMBASSADOR 990 LINE					
80	6785-5	4-dr Sed-6P	2776	3324	17,809
80	6789-5	2-dr HT Cpe-6P	2803	3376	6,140
80	6788-5	4-dr Sta Wag-6P	3083	3545	7,919
AMBASSADOR DPL LINE					
80	6789-7	2-dr HT Cpe-5P	2958	3394	12,552
80	6787-7	2-dr Conv-5P	3143	3434	1,260

BASE ENGINES

Inline. Six. Overhead valves. Cast iron block. Displacement: 199 cid. Bore and stroke: 3-3/4 x 3 inches. Compression ratio: 8.5:1. Brake hp: 128 at 4400 rpm. Seven main bearings. Hydraulic valve lifters. Carburetors: Carter Type RBS one or Holley one-barrel Model 1931C-3705.

Inline. Six. Overhead valves. Cast iron block. Displacement: 232 cid. Bore and stroke: 3.75 x 3.50 inches. Compression ratio: 8.5:1. Brake hp: 145 at 4400 rpm. Seven main bearings. Hydraulic valve lifters. Carburetor: Carter Type RBS onebarrel or Holley one-barrel Model 1931C-3705.

Inline Six. Overhead valves. Cast iron block. Displacement: 231.9 (232) cid. Bore and stroke: 3.75 x 3.50 inches. Compression ratio: 8.5:1. Brake hp: 145 at 4300 rpm. Seven main bearings. Hydraulic valve lifters. Carburetor: Holley one-barrel.

V-8. Overhead valves. Cast iron block. Displacement: 287.2 cid. Bore and stroke: 3.75 x 3.25 inches. Compression ratio: 8.7:1. Brake hp: 198 at 4700 rpm. Five main bearings. Hydraulic valve lifters. Carburetor: Carter WCD or Holley two-barrel.

CHASSIS FEATURES: Wheelbase: (American) 106 inches; (Rebel) 114 inches; (Ambassador/Marlin) 118 inches. Overall length: (American station wagon) 181 inches; (American passenger) 181 inches; (Rebel station wagon) 198 inches; (Rebel passenger) 197 inches; (Marlin) 201.45 inches; (Ambassador station wagon) 203 inches; (Ambassador passenger) 202.5 inches. Front tread: (American) 56 inches; (Rebel Six) 58.2 inches; (all other models) 58.6 inches. Rear tread: (American) 55 inches; (all other models) 58.5 inches. Tires: (American passenger) 6.45 x 14; (American station wagon) 6.95 x 14; (all other passenger models) 7.35 x 14; (all other station wagons) 7.75 x 14.

RAMBLER AMERICAN OPTIONS: All-transistor manual radio ($49). All-transistor push-button radio ($57). Tachometer. All-Season air conditioning ($311). Twin-Grip differential ($37). Electric washer/wipers, electric wipers mandatory with V-8 ($18). Power steering ($84). Power brakes ($42). Power tailgate window ($31). All-vinyl upholstery, standard equipment in Rogue ($24). Exterior appearance group, includes rocker moldings and wheel-covers ($77). Full wheel discs, standard on Rouge ($21). Turbo-cast wheelcovers ($61). Reclining seats for 220/440 models ($25). Reclining bucket seats with center armrest and cushion, for Rogue convertible, standard on Rogue hardtop ($98). Safety headrests ($15). Custom steering wheel for American 220, standard on other models ($8). V-8 handling package. Sports steering wheel wheel, 440 and Rogue only ($11). Black or white vinyl roof for hardtops ($75).

AMBASSADOR/MARLIN OPTIONS: Adjust-O-Tilt steering whee ($42). Cruise-Command automatic speed control ($44). Power disc brakes ($91). All-Season air conditioning ($350). Stereo system with 8-track tape ($133). Custom Trim package for DPL hardtop, includes: Morocco Brocade fabric in five colors for seats and door panels; two matching pillows and Custom nameplates ($49). Black or white vinyl roof, for hardtops and 990 sedan ($75). Two-tone paint ($19). Station wagon woodgrained exterior paneling ($100). Reclining bucket seats, standard in DPL hardtop ($142). Individually adjustable reclining seats, standard in DPL convertible ($45). Center console, with bucket seats and console shift options only ($113). Electric clock ($16). Tachometer ($48). Passenger third seat for station wagons ($112). Sports steering wheel; standard in DPL models ($16). Safety headrests ($45). AM/FM all-transistor radio ($134). Vinyl upholstery, standard in DPL convertible ($25). Full wheel discs, standard on Marlin and DPL ($21). Turbo-cast wheelcovers ($40). Light Group, standard in DPL ($16).

REBEL OPTIONS: All-Season air conditioning ($350). Adjust-O-Tilt steering wheel ($42). Cruise-Command automatic speed control ($44). Eye-level 6,000 rpm. tachometer $48). New 8-track stereo tape player, in sedans and hardtops with rear speaker ($134). Black or white vinyl roof, on hardtops and 770 sedan ($75). Two-tone paint ($26). Simulated wood-grain station wagon trim ($100). Vibra-Tone rear seat speakers for all-transistor radios ($52). Power steering ($84). Power brakes ($42). Power disc brakes for V-8 models ($91). Power tailgate window ($32). Station wagon third passenger seat ($112). Reclining, individually adjustable seats, standard in SST convertible ($45). Reclining bucket seats, standard in SST hardtop and not available in 550 model ($78). Vinyl upholstery, standard in SST convertible ($25). Solex glass, all window ($34); windshield only ($21). Sports steering wheel, in 770 and SST only ($16). Electric clock ($16). Rear foam seat cushion, standard in 770 and SST models and on third station wagon seat ($11). Wheel discs, standard on SST ($21). Turbo-cast wheelcovers ($61).

POWERTRAIN OPTIONS: Three-speed manual transmission was standard. Three-speed manual transmission with overdrive, in American Six ($109); others ($115). Automatic transmission, in American ($174); in Rebel ($186); in Ambassador/Marlin ($217). Four-speed manual transmission, with V-8 engines only ($184). Shift-Command automatic transmission with thumb-button operated floor shift, in Rogue buckets and Console ($192); in 770 and SST ($205); in Ambassador 880/Marlin ($217). Six-cylinder 232 cid 145 hp engine, in American ($39). Six-cylinder 232 cid 155 hp two-barrel engine, in American ($51); in other models ($12). V-8 290 cid 200 hp two-barrel engine, in American ($119). V-8 290 cid 225 hp four-barrel engine, in American ($32). V-8 343 cid 235 hp two-barrel engine, in Marlin/Rebel/Ambassador ($58). V-8 343 cid 280 hp four-barrel engine, in Marlin/Rebel/Ambassador ($91). 'Air Guard' exhaust emissions control system for V-8s ($45). 'Engine Mod' system for Sixes, mandatory smog-control option for California ($11). Closed crankcase ventilation system. Mandatory in California ($50). Heavy-duty clutch for V-8 with manual transmission ($5). Dual exhausts ($26).

HISTORICAL FOOTNOTES: The full-sized models were introduced Oct. 6, 1966 and the American appeared in dealer showrooms the same day. Model year production peaked at 235,522 units. Calendar year sales of 229,058 cars were recorded. R.D. Chapin, Jr. was the chief executive officer of the company this year. American Motor's financial branch, called Redisco, Inc., was sold to Chrysler this year. A new advertising agency, Wells, Rich, Greene, Inc., of New York City, was engaged by AMC's brand new management team and came up with a series of humorous, but effective ad campaigns that focused on product advantages. A total of 1.2 percent of all AMC products had four-speed manual transmissions; 16.3 percent had bucket seats, 45 percent V-8 powerplants and 11.2 percent vinyl tops.

1968 AMC

RAMBLER AMERICAN — (SIX) — SERIES 01: The 1968 Rambler American used the same body styling as the 1966 and 1967 models. There were several changes in decorative trim. A new grille featured a single horizontal strip of chrome across the insert with a Rambler nameplate at the left-hand side. Signature scripts on the sides of front fenders were moved, from in back of the headlights, to a point behind the wheel opening. Rectangular side markers were seen on both front and rear fenders. Squarish taillamps were set into the rear panel and a Rambler badge was placed near the right taillight lens. The base American 220 came as a four-door sedan or the year's only two-door sedan. This car had no side body moldings and equipment consisted of a Weather-Eye heater; front armrests; front seat foam cushions and dome or side pillar lights. The American 440 had a single, wide strip of ribbed chrome molding positioned high on the body sides. It connected the front and rear side markers. There was also a bright metal horizontal strip between the taillamps; carpeting; rear armrests; cigarette lighter; glove box lock; dual horns and Custom steering wheel. The station wagon came with all-vinyl upholstery and larger tires. The top model was the Rogue two-door hardtop which had all American 440 features, plus a larger base six-cylinder engine; special Rogue identification scripts (in place of American signatures) and higher level interior appointments. All Americans were marketed as Sixes, with V8s available as optional equipment.

VIN: Serial Numbers were located on top of the right front wheelhouse panel. The unit body data plate was riveted to the left front door, below the latch mechanism. American Motors refined the 13 symbol identification code system introduced the previous year. The third symbol was now a letter designating transmission type, instead of a number designating the assembly plant. (The transmission codes themselves were unchanged.) The fourth symbol was a number designating car line, as follows: 0=American; 1=Rebel, 3=AMX; 7=Javelin and 8=Ambassador. The fifth symbol designated body type, using the 1967 number codes. The sixth symbol designated the series or class of body, as follows: 0=American 220/Rebel 550; 2=Ambassador; 5=American 440/Rebel 770/Ambassador DPL and 7=AMX/Rogue/SST Ambassador. The seventh symbol was a letter designating engine type (without regard to car-line), as follows: A=199 cid six one-barrel; B=232 cid six one-barrel; C=232 cid six two-barrel; M=290 cid V-8 two-barrel; N=290 cid V-8 four-barrel; S=343 cid V-8 two-barrel; T=343 cid V-8 four-barrel; W=390 cid V-8 two-barrel and X=390 cid V-8 four-barrel. The following group of symbols was the sequential Serial Number and began with 100001 for cars built in Kenosha, Wis. and 700001 for cars built in Brampton, Ontario, Canada.

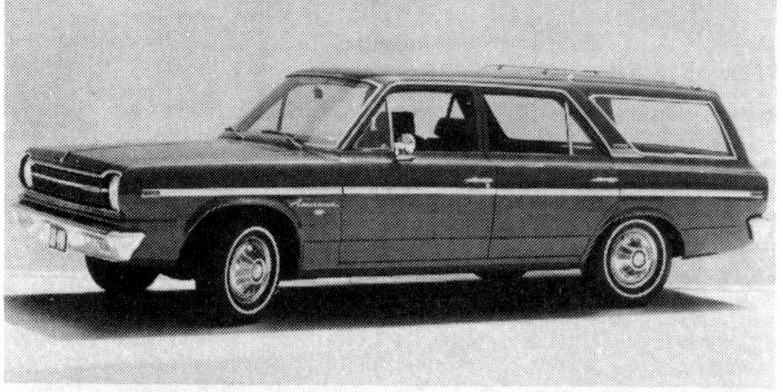

1968 AMC, Rambler American 440 four-door station wagon, V-8

RAMBLER AMERICAN SERIES 01

Series Number	Body/Style Number &	Body Type Seating	Factory Price	Shipping Weight	Production Total
AMERICAN 220 LINE					
01	6805	4-dr Sed-6P	2024	2638	16,595
01	6806	2-dr Sed-6P	1946	2604	53,824
AMERICAN 440 LINE					
01	6805-5	4-dr Sed-6P	2166	2643	11,116
01	6808-5	4-dr Sta Wag-6P	2426	2800	8,285
AMERICAN ROGUE LINE					
01	6809-7	2-dr HT Cpe-6P	2244	2678	4,549

NOTE: Prices on many American Motor cars were increased around May 1, 1968. The above prices are those in effect at the end of the model year.

RAMBLER REBEL — (SIX/V-8) — SERIES 10: Offering the only convertible left in the AMC stable, the Rebel was modestly restyled for 1968. The Ramber name was removed from the hood and the taillamps now took the form of three horizontal, curved rectangles, instead of the two large rectangles seen the year before. Square, recessed door handles were another new feature and body side moldings were eliminated. The base Rebel 550 models carried this designation below the signature scripts on the sides of front fenders and on the right-hand corner of the trunk lid's rear face. More than a dozen safety changes, enacted to satisfy government regulations, included new front side marker lamps and a preset door locking system. Regular equipment on 550s included all standard safety features; heater; front armrests (on four-doors); cigar lighter; dual headlamps; front seat foam cushions and dome or side pillar lamps. Station wagons came with a roof top travel rack and all-vinyl upholstery and convertibles featured power operated tops. Rebel 770 models had all above items, plus rear ashtrays; rear armrests; Custom steering wheel; glove box lock; dual horns and cloth and vinyl or all-vinyl seats. The 770 station wagons also had a hidden stowage compartment and vertical tailgate with power window on three-seat options. Model identification numbers were seen in the usual places and read '770'. The Rebel SST came only with V-8 power, exclusively in sport body styles, and represented the top of the line. Trim and equipment distinctions included SST lettering (below front fenderside scripts); wheelhouse trim moldings; simulated chrome air vents ahead of rear wheel opening; individually adjustable reclining seats and special interior appointments. Wheel discs were standard on the SST, as was underhood insulation. This was the final season for Rambler convertibles and the two low-production Rebel ragtops were the last that American Motors would ever offer, at least during the years covered by this catalog.

RAMBLER REBEL SERIES 10

Series Number	Body/Style Number	Body Type & Seating	Factory Price	Shipping Weight	Production Total
REBEL 550 LINE					
10	6815	4-dr Sed-6P	2443	3062	14,712
10	6817	2-dr Conv-6P	2736	3195	377
10	6818	4-dr Sta Wag-6P	2729	3301	7,427
10	6819	2-dr HT Cpe-6P	2454	3117	7,377
REBEL 770 LINE					
10	6815-5	4-dr Sed-6P	2542	3074	22,938
10	6818-5	4-dr Sta Wag-6P	2854	3306	11,375
10	6819-5	2-dr HT Cpe-6P	2556	3116	4,420
REBEL SST LINE (V-8)					
10	6817-7	2-dr Conv-6P	2999	3427	823
10	6819-7	2-dr HT Cpe-6P	2775	3348	9,876

NOTE 1: The Rebel SST came only with the 'Typhoon' 290 cid V-8 as base powerplant.
NOTE 2: See 1968 American Series note. The above prices are those in effect at the end of the model year.

1968 AMC, Javelin SST two-door hardtop coupe, V-8

1968 AMC, Ambassador SST two-door hardtop coupe, V-8

AMBASSADOR — (SIX/V-8) — SERIES 80: The 1968 Ambassador had a grille with squarer corner extensions, a grid style insert with black-out finish and a wider horizontal center divider that gave a twin slot look. Side molding treatments remained basically unchanged and consisted of a full-length chrome strip, mounted low on the body, which arched up over the wheel openings. The stand-up hood ornament was eliminated. Taillights were now divided horizontally, instead of vertically. Cars with 'Typhoon' power had V-shaped rear fender ornaments. The new paddle type AMC door handles were used. In a move that was very heavily promoted, air conditioning became standard Ambassador equipment. The base model was no longer called the 880 (although these numbers still appeared as the last three symbols in the series designation) and was simply referred to as the Ambassador. It came with All-Season air conditioning; all standard safety features; 60 ampere battery; heater; front and rear armrests; cigarette lighter; front and rear ashtrays; carpets; front foam seat cushions; dome or side pillar lights; glove box lock; dual horns and headlights (still vertically stacked) and a Custom steering wheel. Ambassador wagons had a roof rack and lockable stowage compartment, plus power tailgate window on three-seat types. The Ambassador DPL carried these three letters on a nameplate located below the side fender scripts and featured full wheel discs and upgraded appointments. Both the Ambassador and Ambassador DPL models were marketed as six-cylinder cars, with V-8 options. The flagship of the fleet was the Ambassador SST which had all DPL items, plus custom interior and exterior trim; individually adjustable reclining seats; woodgrain look panelling on dashboard; electric clock; headlights-on warning buzzer; rear foam seat cushions and a 200 hp two-barrel V-8 engine. The Ambassador SST hardtop featured a special grille treatment, with integral Rally lights, and interior courtesy lamps.

AMBASSADOR SERIES 80

Model Number	Body/Style Number	Body Type & Seating	Factory Price	Shipping Weight	Production Total
AMBASSADOR LINE					
80	6885-2	4-dr Sed-6P	2820	3193	8,788
80	6889-2	2-dr HT Cpe-6P	2842	3258	3,360
AMBASSADOR DPL LINE					
80	6885-5	4-dr Sed-6P	2920	3265	13,265
80	6889-5	2-dr HT Cpe-6P	2941	3321	3,696
80	6888-5	4-dr Sta Wag-6P	3207	3475	10,690
AMBASSADOR SST LINE (V-8)					
80	6885-7	4-dr Sed-6P	3151	3476	13,387
80	6889-7	2-dr HT-6P	3172	3530	7,686

NOTE 1: The Ambassador SST came only with the 290 cid V-8 as base powerplant.
NOTE 2: See 1968 American Series note. The above prices are those in effect at the end of the model year.

JAVELIN — (SIX/V-8): The new Javelin filled the slot vacated by the unsuccessful Marlin. It was American Motors Corporation's entry into the Pony Car market that Ford created, with the Mustang, in 1964. The car was 189 inches long and on a 109 inch wheelbase platform. Power came from either the 232 cid Six or the 343 cid V-8 in standard form. Styling characteristics included a split grille with blackout treatment and form-fitting bumper; single, square headlamp housings integrated into fenders; round parking lamps integrated into bumper (below headlights); clean-lined body with smooth-flowing lines and a semi-fastback roofline with wide, flat sail panels. The profile exhibited a 'venturi' silhouette. A full-width rear bumper; horizontal, rectangular taillights and a black-out rear panel treatment characterized the rear view. Javelin chrome signature scripts were seen in the left-hand grille insert; on the front fenders behind the wheel opening and in the center of the deck latch panel. Standard equipment included all regulation safety features; Custom steering wheel; heater, Flo-Thru ventilation; dual paint stripes along upper beltline; front armrests; cigarette lighter; front ashtray; carpeting; bucket seats with front foam cushions; compartment lights; glove box lock; dual horns; wide profile tires and, for V-8s, a performance suspension with sway bar. There was also a Javelin SST, with all the above, plus reclining front bucket seats; wood-look Sports steering wheel and door panel trims; full wheel discs and moldings for the rocker panels, side windows and hood scoop. The Javelin went on sale September 26, 1967 and by January a total of 12,390 had been sold. The company took the Javelin to Daytona Beach and other race tracks to show that it could really go. Later, the car showed promise in the new Trans-American racing series and, with full factory backing, narrowly missed unseating the championship Ford Mustang factory team. The Javelin was the last American Pony Car introduced and many enthusiasts thought it to be the best.

JAVELIN SERIES 70

Series Number	Body/Style Number	Body Type & Seating	Factory Price	Shipping Weight	Production Total
JAVELIN					
70	6879-5	2-dr FsBk-4P	2482	2826	29,097
JAVELIN SST					
70	6879-7	2-dr FsBk-4P	2587	2836	26,027

NOTE 1: The Javelin was introduced as a six-cyliner Series, with optional V-8s available at extra cost. Prices above are those in effect at the end of the model year.

1968 AMC, AMX two-door fastback sports coupe, V-8

AMX — (V-8) — SERIES 30: Round number two in the AMC revitalization program officially kicked-off on Feb. 24, 1968 with the midyear introduction of the AMX two-seater sports car at the Chicago Automobile Show. Actually, the car had made its initial public appearance about a week earlier, when it was press previewed at Daytona Raceway in Daytona Beach, Fla. The model designation stood for "American Motors Experimental" and it was the first American two-passenger, steel-bodied production type sports car to be seen since the 1955-1957 Ford Thunderbird. A kid brother to the Javelin, the AMX was built off a 97 inch wheelbase version of the same platform. Features included thin shell reclining bucket seats; carpeted interior; wood-grained steering wheel and door panel trim; the 290 cid 225 hp V-8; four-speed manual transmission; special suspension; glass-belted Goodyear tires; and four-barrel carburetor. Power options included the 343 cid V-8 and a 390 cid engine with 315 hp that turned the little sportster into a real speed demon. It looked like a short Javelin with louvered hood bulges and a non-divided grille treatment. It carried special model identification within a large ring of chrome, on the sail panels. Craig Breedlove established 106 world speed records with a 1968 AMX at Goodyear's Texas test track in February, 1968. The following month, the AMX was seen in dealer showrooms, hoping to go half as fast in the race for sales. It was, however, primarily an image car designed to exhibit AMC's new approach to design, engineering, styling and marketing. A number of replica Craig Breedlove model AMXs with Red, White and Blue paint jobs and the 290 V-8 and four-speed transmission were sold. The number of Craig Breedlove Special AMXs made is believed to have been 50 cars.

AMX SERIES 30

Model Number	Body/Style Number &	Body Type Seating	Factory Price	Shipping Weight	Production Total
30	6839-7	2-dr FsBk-2P	3245	3097	6,725

NOTE: Prices above are introductory prices. The AMX was a midyear model and its retail pricing did not change when retails for other lines increased in the spring.

BASE ENGINES

Inline. Six. Overhead valves. Cast iron block. Displacement: 199 cid. Bore and stroke: 3-3/4 x 3 inches. Compression ratio: 8.5:1. Brake hp: 128 at 4400 rpm. Seven main bearings. Hydraulic valve lifters. Carburetors: Carter Type RBS one or Holley one-barrel Model 1931C-3705.

Inline. Six. Overhead valves. Cast iron block. Displacement: 232 cid. Bore and stroke: 3.75 x 3.50 inches. Compression ratio: 8.5:1. Brake hp: 145 at 4400 rpm. Seven main bearings. Hydraulic valve lifters. Carburetor: Carter Type RBS one-barrel or Holley one-barrel Model 1931C-3705.

V-8. Overhead valves. Cast iron block. Displacement: 290 cid. Bore and stroke: 3.75 x 3.25 inches. Compression ratio: 9.0:1. Brake hp: 200 at 4600 rpm. Five main bearings. Hydraulic valve lifters. Carburetor: AMC two-barrel Model 8HM2.

AMX V-8. Overhead valves. Cast iron block. Displacement: 290 cid. Bore and stroke: 3.75 x 3.25 inches. Compression ratio: 10.0:1. Brake hp: 225 at 4700 rpm. Five main bearings. Carburetor: Carter type AFB four-barrel 4660S.

CHASSIS FEATURES: Wheelbase: (American) 106 inches; (Javelin) 109 inches; (Rebel) 114 inches; (Ambassador) 118 inches; (AMX) 97 inches. Overall length: (American) 181 inches; (Javelin) 189.2 inches; (Rebel wagon) 198 inches; (Rebel) 197 inches; (Ambassador wagon) 203 inches; (Ambassador) 202.5 inches; (AMX) 177.2 inches. Front tread: (Javelin) 57.9 inches; (AMX) 58.4 inches; (other models) See 1967 specifications. Rear tread: (Javelin and AMX) 57 inches; (other models) See 1967 specifications. Tires: (American) 6.45 x 14; (American wagon) 6.95 x 14; (Javelin Six) 6.95 x 14; (Rebel and Ambassador Six) 7.35 x 14; (Rebel/Ambassador Six wagon) 7.75 x 14; (Rebel V-8) 7.35 x 14; (Ambassador V-8) 7.75 x 14; (AMX) E70-14.

AMERICAN OPTIONS: Power brakes ($42). Power steering ($84). Air conditioning ($311). Power disc brakes, V-8 only ($97). Solex glass, all windows ($29); windshield only ($16). Front and rear bumper guards ($23); front bumper guards only ($12). Vinyl top, hardtops only ($79). Individually reclining seats ($49). Sports steering wheel ($21). Custom steering wheel in 220 models ($9). Tachometer, with V-8 only ($48). Column shift automatic transmission ($174). Four-speed manual transmission, with V-8 ($184). Vinyl seat upholstery, standard on wagon ($24). TurboCast wheelcovers ($61). Wire wheelcovers ($66). Full wheel discs, all ($21). Electric windshield wipers ($12). Wide profile high-performance tires, exchange ($64). Push-button radio and antenna ($61). Roof top station wagon travel rack ($39). Pair of headrests ($35). Special application paint colors ($28). Three-speed manual transmission was standard. Overdrive transmission was available for American. Automatic transmission was optional on all with floor control on specific models. Four-speed manual floor shift transmission, standard AMX; optional, with V-8 only, on other models. American six-cylinder 232 cid 145 hp one-barrel engine ($45). American V-8 290 cid 200 hp two-barrel engine ($119). American V-8 290 cid 225 hp four-barrel engine ($45).

REBEL OPTIONS: Power brakes ($42). Power steering ($84). Air conditioning ($356). Power disc brakes, V-8 only ($97). Solex glass, all windows ($34); windshield only ($21). Front and rear bumper guards ($23). Pair of headrests, bench seat ($35); bucket seats ($49). Two-tone paint ($32). Station wagon exterior wood-grained, 770 only ($100). SST exterior paint stripe ($14). AM push-button radio, all ($58). AM/FM push-button radio, all ($134). Black, Blue or Off-White vinyl roof, hardtop or sedan ($79). SST reclining bucket seat with center armrest ($91). Station wagon third rear-facing seat, includes power tailgate window ($112). Stereo 8-track player with two rear speakers ($134). Adjust-O-Tilt steering, with automatic only ($42). Speed control system, with automatic only ($44): Tachometer, with V-8 only ($48). SST Shift-Command console, with V-8 only ($249). Four-speed manual floor shift ($184). SST undercoating ($17). Vinyl seats, standard in convertible and wagon ($24). Sports steering wheel ($21). Turbo-cast wheelcovers; SST ($40); others ($51). Wire wheelcovers, SST ($45); other models ($66). Wheel discs, standard on SST ($45). Power windows, SST only ($ 100). Three-speed manual transmission was standard. Overdrive transmission was available for Rebel. Automatic transmission was optional on all with floor control on specific models. Four-speed manual floor shift transmission optional, with V-8 only. Rebel V-8 290 cid 200 hp two-barrel engine ($106). Rebel V-8 343 cid 235 hp two-barrel engine ($45). Rebel V-8 343 cid 280 hp four-barrel engine ($76).

AMBASSADOR OPTIONS: Power brakes ($43). Power steering ($95). Power disc brakes ($97). Twin-Grip differential ($42). Engine block heater ($18). Two-tone paint, standard color ($32). DPL Special paint with painted side panel and accent trim ($45). Simulated wood-grain side paneling, DPL only ($100). Exterior paint strip, SST only ($14). AM/FM push-button radio ($58). Vinyl top, hardtop or sedan only, three colors ($79). Reclining bucket seat with center armrest cushion ($91). Adjust-O-Tilt steering wheel ($42). Sports steering wheel ($21). Stereo 8-track with twin rear speakers ($134). Shift-Command console for V-8, SST only ($250). Four-speed manual transmission ($184). Turbo-Cast wheelcovers, SST/DPL ($40); others ($61). Wire wheelcovers, SST/DPL ($45); others ($66). Station wagon power side and tailgate windows, DPL only ($134). Power side windows, except base models ($100). Station wagon power tailgate window only ($33). Station wagon rear facing third seat with power tailgate window ($95). Tachometer, V-8 only ($48). Visibility option package ($41). Light Group option package ($22). Engine cooling option package ($16). Handling option package ($ 1 0). Two front shoulder belts, all ($23). Three-speed manual transmission was standard. Overdrive transmission was available for Ambassador. Automatic transmission was optional on all with floor control on specific models. Four-speed manual floor shift transmission, standard AMX; optional, with V-8 only, on other models. Ambassador V-8 290 cid 200 hp two-barrel engine ($16). Ambassador V-8 343 cid 235 hp two-barrel engine ($58). Ambassador V-8 343 cid 280 hp four-barrel engine ($91). Four-barrel carburetor. Positive traction rear axle. Heavy-duty clutch. NOTE: Prices for V-8s other than base 200 hp engine are in addition to basic cost of V-8 attachment.

JAVELIN OPTIONS: Power brakes ($42). Power steering ($85). Air conditioning ($356). Dual exhausts, V-8 ($21). Solex glass, all windows ($31); windshield only ($21). Headrest, with bench seat ($35); with buckets ($49). Power discs brakes, V-8 ($97). Power steering ($84). Stereo 8-track player ($195). Individually adjusting seat ($49). White vinyl seat upholstery ($20). Adjust-O-Tilt steering ($42). Shift-Command automatic transmission with floor shift and console ($269). Twin-Grip differential ($42). Vinyl roof ($85). Turbo-Cast wheelcovers, SST ($51); base model ($65). Handling package ($17). 'Go Pack' performance package includes: 343 cid V-8; dual exhausts; power disc brakes; E-70 wide profile tires; Handling Package and Rally stripes ($266). Visibility Group package ($27). FM push-button radio and antenna ($61). AM/FM push-button radio and antenna ($134). Quick-ratio manual steering ($16). Four-speed manual transmission with floor shift ($184). Wire wheelcovers, SST ($51); base Javelin ($75). Three-speed manual transmission was standard, except AMX. Overdrive transmission was available for American/Ambassador/Rebel. Automatic transmission was optional on all with floor control on specific models. Four-speed manual floor shift transmission, standard AMX; optional, with V-8 only, on other models. Four-barrel carburetor. Positive traction rear axle. Heavy-duty clutch.

NOTE: Prices for V-8s other than base 200 hp engine are in addition to basic cost of V-8 attachment.

AMX OPTIONS: Specific prices for AMX options are not available. The options list was about the same as for Javelin (at about the same prices) plus, over-the-top striping; chrome steel mag wheels and dealer accessory 'Rally Pak' gauge cluster. Three-speed manual transmission was standard, except AMX. Overdrive transmission was available for American/Ambassador/Rebel. Automatic transmission was optional on all with floor control on specific models. Four-speed manual floor shift transmission, standard AMX; optional, with V-8 only, on other models. Four-barrel carburetor. Positive traction rear axle. Heavy-duty clutch. NOTE: Prices for V-8s other than base 200 hp engine are in addition to basic cost of V-8 attachment.

HISTORICAL FOOTNOTES: The AMC models were introduced Sept. 26, 1967 and the AMX appeared in dealer showrooms during March 1968. Model year production peaked at 272,726 units. Calendar year sales of 269,334 cars were recorded. R.D. Chapin, Jr. was the chief executive officer of the company this year. For the first time since 1965, American Motors Corporation operated in the black for 1968 (although no dividends were paid). This compared to a loss of 75.8 million in 1967. The company sold its Kelvinator Division to White Consolidated Industries. Each AMC built in 1968 carried a metal dashboard plate bearing a special number (numbers 000001 to 006175 were used). This was intended to designate its rather special nature. However, the first 550 units, assembled in calendar 1967, did not have this feature.

1969 AMC

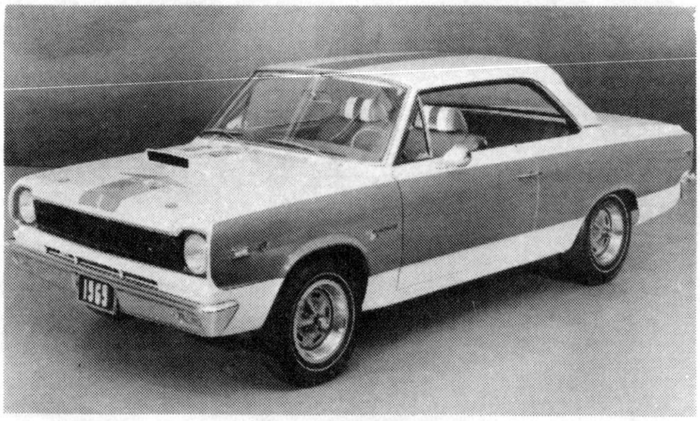

1969 AMC-Hurst, SC/Rambler two-door hardtop, V-8

RAMBLER — (SIX/V-8) — SERIES 01: Formerly the Rambler American, the 1969 base model got a shortened name. It retained its compact dimensions and overall styling in line with AMC's new policy of maintaining design continuity from year to year for its low-priced models. American nameplates were gone from the grille and a new chrome side molding was used. Some of the mechanical improvements earmarked for the more expensive AMC products were incorporated in the Rambler. They included a new accelerator cable linkage; suspended accelerator pedal; 'Clear Power 24' battery and parking lamps that remained on with headlamps. Regular equipment included all regulation safety features; front armrests; front ashtrays; heater and defroster; head restraints; front foam seat cushions; a 199 cid six and 6.45 x 14 blackwall tires. The Rambler 440 models also had rear armrests and ashtrays; cigarette lighter; glove box lock; and dual horns. The Rogue featured carpeting; 'Air Guard' system and 232 cid six. Rambler wagons had 6.95 x 14 blackwall tires. With the help of Hurst Products Corp., a special Rogue offering was built exclusively during 1969. It was called the Hurst SC/Rambler and came only with the AMX type 390 cid 315 hp V-8 engine; all-synchromesh four-speed close-ratio transmission; special Hurst four-speed shift linkage with T-handle; Sun tach mounted on steering column; dual exhaust system with special tone mufflers and chrome extensions; functional hood scoop for cold-air induction; Twin-Grip differential; 3.54:1 axle ratio; 10.5 inch diameter clutch; front power disc brakes; rear axle torque links; handling package; heavy-duty cooling system; 20:1 AMX manual steering; special application of red, white and blue exterior colors; two hood tie-downs with locking safety pins; custom teardrop racing mirrors; custom-finished grille; emblems; 14 x 6 inch mag style wheels; E70-14 Goodyear Polyglas wide-tread tires; sports steering wheel; custom upholstered headrests; all-charcoal seat upholstery; full carpeting and individually-adjustable recining seats. The retail price was "exactly" $2,998 the sales flyer said. Original programming called for a limited run of just 500 units, but supply was far outstripped by demand. Ultimately, three runs of this model were made and the total output hit 1,512 units. The first batch, or 'A' Group, had the major portion of the body sides painted red, with a blue racing stripe traveling down the middle of the body and across the roof and deck. There was also a large blue arrow, pointing towards the hood scoop. The paint code for such cars was '00.' The second, or 'B' Group of Hurst SC/Ramblers where finished more conservatively. They had a largely white exterior, with narrow red and blue stripes. These cars had either a special ('SPEC') or regular P-72 white paint code. The third group reverted to the original or 'A' style, so of the three groups, the 'A' style is predominant.

VIN: Serial Numbers were located on top of the right front wheelhouse panel. The unit body data plate was riveted to the left front door, below the latch mechanism. American Motors retained a 13 symbol identification code system. The first symbol was an A for American Motors Corp. The second symbol indicated model year, 9=1969. The third symbol indicated transmission type: S=standard column shift; O=overdrive; A=automatic, column shift; C=three-speed automatic with floor shift; and M=four-speed manual with floor shift. The fourth symbol was a number designating car line, as follows: 0=American; 1=Rebel, 3=AMX; 7=Javelin and 8=Ambassador. The fifth symbol designated body type: 5=four-door sedan; 6=two-door sedan; 8=four-door station wagon; and 9=two-door hardtop. The sixth symbol designated the series or class of body, as follows: 0=American 220/Base Rebel; 2=Base Ambassador; 5=American 440/Javelin/Ambassador DPL and 7=AMX/Rogue/SST Ambassador. The seventh symbol was a letter designating engine type (without regard to car-line), as follows: A=199 cid six one-barrel; B=232 cid six one-barrel; C=232 cid six two-barrel;

M=290 cid V-8 two-barrel; N=290 cid V-8 four-barrel; S=343 cid V-8 two-barrel; T=343 cid V-8 four-barrel; W=390 cid V-8 two-barrel and X=390 cid V-8 four-barrel. The following group of symbols was the sequential production and began with 100001 for cars built in Kenosha, Wis. and 700001 for cars built in Brampton, Ontario, Canada. Body number plate attached to left front door edge shows body serial number, Body/Style Number from second column of chart below, plus trim and paint codes.

RAMBLER SERIES 01

Series Number	Body/Style Number	Body Type & Seating	Factory Price	Shipping Weight	Production Total
RAMBLER					
01	6905	4-dr Sed-6P	2076	2638	16,234
01	6906	2-dr Sed-6P	1998	2604	51,062
RAMBLER 440					
01	6905-5	4-dr Sed-6P	2218	2643	11,957
01	6906-5	2-dr Sed-6P	2478	2800	13,233
RAMBLER ROGUE					
01	6909-7	2-dr HT Cpe-6P	2296	2296	3,543
HURST SC/RAMBLER (V-8)					
01	6909-7	2-dr HT Cpe-6P	2998	2988	1,512

NOTE 1: The SC/Rambler came only with V-8 power. A total of 1,012 A-Group editions were built. A total of 500 B-Group editions were built.

1969 AMC, Rebel SST four-door station wagon, V-8

REBEL — (SIX) — SERIES 10: The Rebel's track was increased to 60 inches for 1969. There were styling revisions including a new grille, deck lid and taillights. The outboard headlamps were housed separately in square surrounds, while the inner lamps were part of the main grille. A horizontal bar pattern insert was used in both the main opening and the outboard headlamp surrounds. The taillamps wrapped around the rear body corners. On SST models, the fake air scoops ahead of the rear wheel opening were replaced by four chrome side strips. Rebels were offered in six models; sedan, hardtop and wagon in the base series and the same three styles with SST equipment and trim. Basic equipment included standard safety features; head restraints; front armrests in two-doors; front and rear armrests in four-doors; front ashtray; cigar lighter; dual headlights; heater and defroster; front foam seat cushions; 7.35 x 14 blackwall tires; and the 145 hp 232 cid six. For all Rebels, V-8 engines were classed as optional equipment. The Rebel SST came with all the above items, plus carpets; rear armrests in two-door styles; rear ashtray; Custom steering wheel; glove box light; and dual horns. Wagons also featured a secret, lockable rear storage compartment; roof luggage rack; dual hinged tailgate; and 7.75 x 14 blackwall tires.

REBEL SERIES 10

Series Number	Body/Style Number	Body Type & Seating	Factory Price	Shipping Weight	Production Total
REBEL					
10	6915	4-dr Sed-6P	2484	3062	10,885
10	6919	2-dr HT Cpe-6P	2496	3117	5,396
10	6918	4-dr Sta Wag-6P	2817	3301	8,569
REBEL SST					
10	6915-7	4-dr Sed-6P	2584	3074	20,595
10	6919-7	2-dr HT Cpe-6P	2598	3140	5,405
10	6918-7	4-dr Sta Wag-6P	2947	3306	9,256

1969 AMC, AMX two-door fastback sports coupe, V-8

AMX — (V-8) — SERIES 30: The 1969 AMC was described as being "more racy looking than ever," but was really little changed. Its introduction as a late 1968 entry precluded major revisions for its second year. There was a new 140 mph speedometer and tachometer with a larger face. Many minor running changes evolved as the year progressed, the most obvious being the addition of a hooded dash panel cover in most 1969s. New convenience items included a passenger grab handle above the glove box and a between-the-seats package tray. Leather upholstery trims were a new option. The 290 cid four-barrel V-8 engine was the base powerplant and motors with displacements of 343 and 390 cid were available. Standard equipment included all safety items; front and rear ashtrays and armrests; cigarette lighter; collapsible spare tire; Sports steering wheel; courtesy lights; dual exhausts; carpets; Flo-Thru ventilation; glove box lock; dual horns; instrument panel gauge cluster with tachometer; rear traction bars, front head restraints; reclining bucket seats; front foam cushions; wheel discs; heavy-duty suspension; E70 x 14 fiberglass-belted blackwall tires; four-speed manual transmission with floor-shift; and 225 hp, 4-barrel V-8. Introduced as a midyear entry was the 'Big Bad AMX.' This option-created-model came in three colors and had the bumpers painted the same shade as the body. A total of 284 orange-colored Big Bad AMXS were built, as well as 195 similar models finished in blue and 283 additional cars done in green. A limited number of 52 or 53 Super Stock AMXs were made by the Hurst Corp. for AMC as special turn-key NHRA drag racing cars.

AMX SERIES 30

Series Number	Body/Style Number	Body Type & Seating	Factory Price	Shipping Weight	Production Total
AMX (V-8)					
30	6939-7	2-dr FsBk Cpe-2P	3297	3097	8,293

1969 AMC Javelin SST two-door hardtop

JAVELIN — (SIX) — SERIES 70: The 1969 Javelin had a new twin venturi grille with a round, 'bull's-eye' emblem on the left-hand side. Otherwise, it was largely unchanged. The side stripes were redesigned. There had formerly been two, narrow, parallel stripes running full-length along the beltline. Now there was a larger 'C' stripe traveling down the mid-side of the car and turning downward at the trailing edge of the front wheelhousing. This change was put into effect on Jan. 9, 1969, so both designs appeared on 1969 models. Another Javelin revision was a new trim treatment for the instrument panel in standard-level cars and extensive use of wood-grained paneling in the Javelin SST interior. Introduced as a midyear addition was the 'Mod Javelin,' which came in the same colors as the Big Bad AMX. Many Mod Javelins were marketed with the 'Craig Breedlove' options package. It included a rooftop spoiler and simulated exhaust rocker mountings. A limited number of Javelins and some Rambler 440s were built, in Germany, by Karmann. Standard equipment included the safety group; carpets; head restraints; front ashtray and armrests; cigarette lighter; Custom steering wheel; compartment lights; 'Air Guard' system; Flo-Thru ventilation; glove box lock; heater and defroster; dual horns; front bucket seats; foam front cushions; 6.95 x 14 blackwall tires; three-speed manual transmission with floor-mounted control and the 223 cid/145 hp six. Optional V-8s were available. The Javelin SST had all base items, plus a Sports steering wheel; reclining bucket seats; full wheelcovers; twin colored side stripes; and mag-styled wheels.

JAVELIN SERIES 70

Series Number	Body/Style Number	Body Type & Seating	Factory Price	Shipping Weight	Production Total
JAVELIN					
70	6979-5	2-dr FsBk Cpe-4P	2512	2826	17,389
JAVELIN SST					
70	6979-7	2-dr FsBk Cpe-4P	2633	2836	23,286

1969 AMC, Ambassador SST, four-door sedan, V-8

AMBASSADOR — (SIX/V-8) — SERIES 80: The 1969 Ambassador received a major facelift for 1969, with an all-new frontal treatment featuring horizontal, quad headlamps. The wheelbase was increased to 122 inches, a four inch gain. The track was widened to 60 inches. The new front end had a sculptured hood and a new plastic grille with twin, oblong-shaped inserts finished in black-out style. An Ambassador signature was placed at the left-hand side of the upper insert. Horizontal taillamps were set into a rear deck beauty panel. There were also notch style taillight lenses integral with the rear fender extension caps and viewable from the side of the car. They served as side marker lamps and, directly ahead of them, there was another bright metal Ambassador script. The Ambassador had the same equipment as the Rebel SST, plus standard air conditioning; front sway bar; 8.25 x 14 blackwall tires; and a 155 hp two-barrel version of the 232 cid six. The Ambassador DPL also wore wheel discs; DPL interior trim; and special, lower body exterior moldings with vinyl inserts. The Ambassador SST came standard with a clock; individually adjustable reclining seats; column-controlled Shift-Command automatic transmission; and a 200 hp, two-barrel V-8, plus all DPL equipment. There was also an SST station wagon with simulated wood-grain exterior paneling on the sides and rear of the body. New custom velour seats and stainless steel side trim were among package options for the SST. All Ambassadors had a higher capacity air conditioning system, too. In most ads of the year, an SST Ambassador was shown with a uniformed chauffer. "To make an appointment for a test ride," said the copy. "Visit your American Motors dealer. A number of them have chauffers available."

AMBASSADOR SERIES 80

Series Number	Body/Style Number	Body Type & Seating	Factory Price	Shipping Weight	Production Total
AMBASSADOR					
80	6985-2	4-dr Sed-6P	2914	3276	14,617
AMBASSADOR DPL					
80	6985-5	4-dr Sed-6P	3265	3358	12,665
80	6988-5	4-dr Sta Wag-6P	3504	3561	8,866
80	6989-5	2-dr HT Cpe-6P	3182	3403	4,504
AMBASSADOR SST (V-8)					
80	6985-7	4-dr Sed-6P	3605	3508	18,719
80	6988-7	4-dr Sta Wag-6P	3998	3732	7,825
80	6989-7	2-dr HT Cpe-6P	3622	3566	8,998

BASE ENGINES

American: Inline six. Overhead valves. Cast iron block. Displacement: 198.8 cid. Bore and stroke: 3.75 x 3.00. Compression ratio: 8.5:1. Brake hp: 128 at 4400 rpm. Seven main bearings. Hydraulic valve lifters. Carburetor: Carter RBS-4633S one-barrel.

Rogue/Javelin: Inline six. Overhead valves. Cast iron block. Displacement: 231.9 (232) cid. Bore and stroke: 3.75 x 3.50 inches. Compression ratio: 8.5:1. Brake hp: 145 at 4300 rpm. Seven main bearings. Hydraulic valve lifters. Carburetor: Carter Type RBS one-barrel Model 4631S.

Ambassador Six: Inline six. Overhead valves. Cast iron block. Displacement: 231.9 (232) cid. Bore and stroke: 3.75 x 3.50 inches. Compression ratio: 8.5: 1. Brake hp: 1 55 at 4400 rpm. Seven main bearings. Hydraulic valve lifters. Carburetor: Carter Type WCD two-barrel Model 4667S.

Base V-8: Overhead valves. Cast iron block. Displacement: 289.8 (290) cid. Bore and stroke: 3.75 x 3.28 inches. Compression ratio: 10.0:1. Brake hp: 225 at 4700 rpm. Five main bearings. Hydraulic valve lifters. Carburetor: Carter Type AFB four-barrel Model 4660S.

Optional V-8. Overhead valves. Cast iron block. Displacement: 343.1 cid. Bore and stroke: 4.08 x 3.28 inches. Compression ratio: 10.0:1. Brake hp: 280 at 4800 rpm. Five main bearings. Hydraulic valve lifters. Carburetor: Carter Type AFB four-barrel Model 4662S.

High-performance V-8. Overhead valves. Cast iron block. Displacement: 390 cid. Bore and stroke: 4.17 x 3.57 inches. Compression ratio: 10.2:1. Brake hp: 315 at 4600 rpm. Five main bearings. Hydraulic valve lifters. Carburetor: Carter Type AFB four-barrel Model 4665S.

CHASSIS FEATURES: Wheelbase: (Rambler/Rogue) 106 inches; (Rebel) 114 inches; (Javelin) 109 inches; (AMX) 97 inches; (Ambassador) 122 inches. Overall length: (Rambler/Rogue) 181 inches; (Rebel wagon) 198 inches; (Rebel) 197 inches; (Javelin) 182.2 inches; (AMX) 177.2 inches; (Ambassador wagon) 207 inches; (Ambassador) 206.5 inches. Front tread: (Rebel/Ambassador) 60 inches; (other models) See 1968. Rear tread: (Rebel/Ambassador) 60 inches; (other models) See 1968. Tires: (all models) See text.

POWERTRAIN OPTIONS: Three-speed manual transmission was standard in most models. Automatic transmission was standard in Ambassador SST with base engine. Overdrive transmission, optional in Rambler and Rebel six ($116). Automatic transmission, optional in all, ($171-223); in Ambassador SST with '343' or '390' V-8 ($22-$32). Four-speed manual floor shift transmission, in Rogue with 225 hp V-8 ($193). Javelin/close-ratio four-speed manual transmission with floor shift ($205). Rambler six-cylinder 223 cid 145 hp one-barrel engine ($45). Rebel/six-cylinder 223 cid 155 hp two-barrel engine ($16). Rogue-Javelin/V-8 290 cid 225 hp four-barrel engine ($45). Rebel-Ambassador/V-8 343 cid 280 hp four-barrel engine ($80). Rebel-Ambassador/V-8 309 cid 235 hp two-barrel engine ($52). Javelin/V-8 343 cid 280 hp four-barrel engine ($91). AMX/V-8 343 cid 280 hp four-barrel engine ($45). AMX/V-8 390 cid 315 hp four-barrel engine ($123). Javelin SST/Ambassador SST/V-8 380 cid 315 hp four-barrel engine ($168). Heavy-duty 70 ampere battery ($8). Heavy-duty battery and 55 ampere generator ($26). Heavy-duty cooling system ($53). Dual exhaust as separate V-8 option ($31). Positive traction rear axle ($42). Heavy-duty clutch, in Rambler with three-speed ($5); in Javelin with 200 hp V-8 ($11). Available rear axle gear ratios ($5).

RAMBLER CONVENIENCE OPTIONS: Power brakes ($42). Power steering ($90). Air conditioning ($324). Front and rear bumper guards, except wagon ($25). Special paint color application ($39). Two-tone paint in standard colors ($24). Tinted glass, (Javelin same price), all windows ($32); windshield ($23). push-button radio and antenna, same price all AMC models ($61). Station wagon roof-top travel rack ($39). Individually adjusting reclining bench seats ($52). Custom steering wheel, standard in Rambler 440 and Rogue ($12). Sports steering wheel, available Rambler Rogue only ($30). Automatic transmission oil cooler ($18). Column-Shift Shift-Command automatic transmission in Rambler six ($171). Column-Shift Shift-Command automatic transmission in '440' with V-8 ($190). Four-speed manual floor shift in Rogue with 225 hp V-8 ($193). Undercoating and underhood insulation pad ($21). Black or white vinyl roof, Rogue only ($79). Full-wheel discs ($21). Electric windshield wipers, required in V-8 Ramblers ($15). Air conditioning package includes Solex glass and power steering ($387). Code 56-4 Appearance Group with sill moldings and wheel discs ($39). Code 70-1 Handling Package with heavy-duty sway bar, shocks and springs ($17). Light Group, includes door switches, trunk, courtesy, glove box and other lamps ($23). Visibility Group with outside rear view remote control mirror, electric window/washer etc. ($29). Rambler sedan/hardtop, Size 6.45 x 14 two-ply whitewalls, exchange ($32). Rambler station wagon, six 6.95 x 14 two-ply whitewalls, exchange ($32). Rambler V-8, Size 6.95 x 14 two-ply whitewalls, exchange ($32).

HURST SC/RAMBLER OPTION PACKAGE: Standard equipment on the Group 'A' SC/Rambler included: AMX 390 cid V-8; four-speed all-synchromesh close-ratio transmission; special Hurst shift linkage with T-handle; Sun tach mounted on steering column; dual exhaust system with special mufflers and chrome extensions; functional hood scoop for cold-air induction; Twin-Grip differential; 10-1/2 inch diameter clutch; 3.54:1 axle ratio; front power disc brakes; rear axle torque links; handling package with heavy-duty sway bar, springs and shocks; heavy-duty radiator and cooling system; 20.0:1 manual steering ratio; special application red, white and blue exterior finish; hood lock pins; dual racing mirrors; black-out grille; special emblems on front fenders/rear panel; 14x6 inch color-keyed mag-styled wheels; five E70 x 14 Goodyear Polyglas wide-tread tires; Sports steering wheel; red, white and blue headrests; all-vinyl charcoal trim; full carpeting; individually adjustable reclining seats and more.

AMX/JAVELIN CONVENIENCE OPTIONS: Power brakes ($42). Power steering ($95). Air conditioning ($369). Rear bumper guards only ($13). Special application paint colors ($39). Console, all Javelin with Shift-Command/column shift ($53). Javelin, Instrument cluster with tachometer and 140 mph speedometer ($50). Automatic transmission oil cooler ($18). Javelin, Rally Side paint stripes, replacing pin stripes ($27). Stereo 8-track tape player, with manual radio ($195). All except Ramblers; AM/FM push-button radio and antenna ($134). Center armrest seat with cushion, bucket seats mandatory, in AMX, with four-speed ($35). Leather, upholstery trim, AMX only ($79). Quick ratio manual steering ($16). Shift-Command, column control, in Javelins except SST with '343' V-8 ($223). Shift-Command, floor control, in Javelins with 200/280 hp V-8s ($287). Four-speed close-ratio manual floor-shift, except with 200 hp V-8 ($205). Twin-Grip differential ($42). Air-Command ventilation, not available with air conditioning ($41). Wire wheelcovers, 14-inch, base Javelin ($72); Javelin SST and AMX ($51). Black, white or blue vinyl roof, all Javelins/not AMX ($100). Turbo-Cast wheelcovers, base Javelin ($67); Javelin SST and AMX ($46). 'Mag' styled wheelcovers, base Javelin only ($94). Six-inch extra-wide wheel rims ($72). Full wheel discs, all Javelins; not AMX ($21). Handling Package group ($19). AMX, higher-rate front and rear springs, 1-3/16. Heavy-Duty shocks ($19). Light Group; AMX ($20); all Javelins ($23). Electric clock plus, visibility package group ($43). Javelin E70 x 14 redline tires ($75); same AMX ($34). A desirable option is the 'Go Package' (Code 39-1), which retailed for $233.15 on cars with the '343' V-8 and $310.85 on cars with the '390' engine. It included power disc brakes; E70 redline tires; six inch wide wheel rims; handling package; Twin-Grip; heavy-duty cooling system and black, white, red, blue or silver 'over-the-top' racing stripe. The Javelin 'Go Package' (code 39-1/2) retailed for $265.50 when the '343' V-8 was ordered and $343.25 when the '390' engine was specified. It included the V-8; dual exhausts; power disc brakes; E70 wide profile redline tires; six inch wide wheel rims; handling package; and black fiberglass hood scoops.

REBEL/AMBASSADOR CONVENIENCE OPTIONS: Power brakes, Rebel ($42); Ambassador ($43). Power steering ($100). Rebel air conditioning ($376). Front and rear bumper guards, except wagons ($32). Automatic speed control, V-8/automatic ($52). SST Rebel station wagon, simulated wood-grain exterior paneling ($113). All tinted windows, in Ambassador ($39); in Rebel ($36). Automatic transmission oil cooler, standard Ambassador SST ($18). Exterior paint stripe, except SST models and DPL ($14). Power side windows, Ambassador SST/DPL cars ($105); wagons ($140). Duo-coustic rear speakers ($13). Stereo 8-track tape with manual radio, sedans/hardtops only ($134). Individually adjustable reclining seats ($58). Reclining bucket seats with front armrest and center cushion, SSTs only ($111). Station wagon third seat, includes power tailgate window ($118). Custom velour trim, Ambassador SST sedan, includes stainless trim insert ($68). All-vinyl seat upholstery, bench or individual cushion, standard in wagon ($24). Adjust-O-Tilt steering, automatic required ($45). Rebel Custom steering wheel, standard in SST ($13). Rebel SST and all Ambassadors, Sport type steering wheel ($30). Shift-Command with column control, standard Ambassador SST ($201). Shift-Command with column control, Rebel/Ambassador (except SST) with '343' V-8 ($223). Shift-Command, Ambassador SST with '343' V-8 ($22); with '390' V-8 ($37). Shift-Command with floor control, Ambassador (except SST) and Rebel SST ($280). Shift-Command with floor control, Ambassador SST with '343' V-8 ($69); with '390' V-8 ($79). Black, white or blue vinyl roof, Rebel SST except wagon ($90); Ambassador SST/DPL ($100). Wire wheelcover, SST and DPL ($51); other models ($72). Turbo-cast wheelcovers, SST and DPL ($46); other models ($67). Full wheel discs, standard SST and DPLs ($21). Visibility Group with electric clock, Ambassador SST ($29); others ($43).

HISTORICAL FOOTNOTES: The full-sized Ramblers were introduced Oct. 1, 1968 and the compact lines appeared in dealer showrooms the same day. Model year production peaked at 275,350 units. Calendar year production of 242,898 cars was recorded. R.D. Chapin was the chief executive officer of the company this year. This was the final season that the Rambler nameplate would appear. A total of 4,204,925 Ramblers were sold 1950-1969. Operations were profitable for the second year in a row, although AMC's net earnings were only $4.9 million, compared to $11.8 million one year earlier. Retail sales totaled 239,548 cars, an 11.1 percent decrease from 1968 levels. A total of 17,147 cars were made with 390 cid engines during the model year. Four and one-half percent had four-speed manual transmissions; 8.2 percent had disc brakes: 22.2 percent had bucket seats; 15.8 percent had vinyl roofs and five percent had styled steel wheels.

1970 AMC

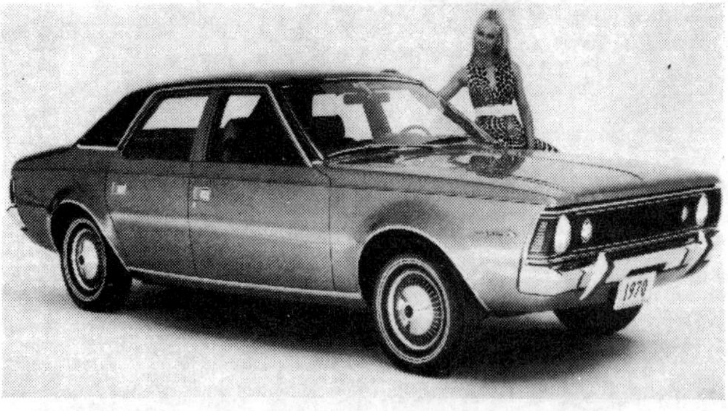

1970 AMC, Hornet SST 4-dr sedan, 6-cyl

HORNET — (SIX) — SERIES 01: In 1970, the Rambler (American) was replaced with a new car that revived an old name. This was the Hornet, which reminded some people of the days when an AMC Family member — Hudson Motors — had championed in stock car racing. The Hornet, however, was not a performance car in its basic form. It was a compact, economy model with a modern, but conventional styling theme. It was basically the Rambler/American with a major facelift and more rounded body contours. The new Hornet was offered in two body styles, two- and four-door sedan, and in base or SST trim levels.

Standard equipment, in addition to all regulations safety features, included front armrests and ashtrays, 6.45-14 blackwall tires and a 128 hp 199 cid six. The Hornet SST also had rear armrests; cigarette lighter; Custom steering wheel; colored carpets; rubber truck mat; glove box light; package tray; front foam seat cushions and a larger, 232 cid six that put out 145 hp.

VIN: Serial Numbers were located on top of the right front wheelhouse panel. The unit body data plate was riveted to the left front door, below the latch mechanism. American Motors retained a 13 symbol identification code system. The first symbol was an A for American Motors Corp. The second symbol indicated model year, 0=1970. The third symbol indicated transmission type: S=standard column shift; A=automatic, column shift; C=three-speed automatic with floor shift; and M=four-speed manual with floor shift. The fourth symbol was a number designating car line, as follows: 0=Hornet; 1=Rebel; 3=AMX; 4=Gremlin; 7=Javelin and 8=Ambassador. The fifth symbol designated body type: 5=four-door sedan; 6=two-door sedan; 8=four-door station wagon; and 9=two-door hardtop. The sixth symbol designated the series or class of body, as follows: 0=Base Hornet/Base Rebel/Gremlin; 2=Base Ambassador; 5=/Javelin/Ambassador DPL/Gremlin no. 7046-5 and 7=AMX/SST (all). The seventh symbol was a letter designating engine type (without regard to car-line), as follows: A=199 cid six; B=low-compression 199 cid six; E=232 cid six with one-barrel carburetor; F=low-compression 232 cid six with one-barrel carburetor; G=232 cid six with two-barrel carburetor; Q=low-compression 232 cid six with two-barrel carburetor; H=304 cid V-8 with two-barrel carburetor; 'I' = low-compression 304 cid V-8 with two-barrel carburetor; 'M' = low-compression 304 cid V-8 with four-barrel carburetor; 'N' = 360 cid V-8 with two-barrel carburetor; 'P' = 360 cid V-8 with four-barrel carburetor; 'S' = 390 cid V-8 with four-barrel carburetor and 'X' = 'Rebel Machine' 390 cid V-8 with four-barrel carburetor. The following group of symbols was the sequential production and began with 100001 for cars built in Kenosha, Wis. and 700001 for cars built in Brampton, Ontario, Canada. Body number plate attached to left front door edge shows body serial number, Body/Style Number from second column of chart below, plus trim and paint codes.

HORNET SERIES 01

Series Number	Body/Style Number	Body Type & Seating	Factory Price	Shipping Weight	Production Total
HORNET					
01	7005-0	4-dr Sed-6P	2072	2748	17,948
01	7006-0	2-dr Sed-6P	1994	2677	43,610
HORNET SST					
01	7005-7	4-dr Sed-6P	2221	2765	19,786
01	7006-7	2-dr Sed-6P	2144	2705	19,748

NOTE: Calendar year sales for the Hornet and Hornet SST Series were 92,458 units.

1970 AMC, Rebel 'Machine' two-door hardtop, V-8

REBEL — (SIX/V-8) — SERIES 10: The 1970 Rebel received new rear quarter panel styling and a more massive rear bumper. There were two large, horizontal-rectangular taillights with Rebel spelled out between them. There was a new vertically split and horizontally segmented grille. The Rebel SST had a bright metal molding on the front fenders, between the door handle and side marker lights. Model identification lettering was placed behind the front wheel opening and, on SST, below the rear roof pillar. There was similar Rebel lettering on the left lip of the hood. Standard Rebel features began with the regulation safety equipment (used in all 1970 AMC models), plus front and rear armrests (except two-door); front ashtray; cigarette lighter; dome or side lights; rubber trunk mat; dual headlights; front foam cushions; E78-14 fiberglass-belted blackwall tires and 145 hp 232 cid six. The Rebel SST also had rear ashtrays; glove box lock; dual horns and Custom steering wheel. Another special, Rebel-based model was developed, for AMC, by the Hurst products company. The 'Rebel Machine' was introduced at the National Hot Rod Association World Championship Drag Race, in Dallas, Texas, during October, 1969. Standard on this model were all Rebel SST features (except rear armrests and ashtrays); high-back bucket seats; Space-Saver spare tire; power front disc brakes; Ram-Air hood scoop; Handling Package; heavy-duty cooling system with Power-Flex fan; carpeting; 15 x 7 inch styled steel wheels; E60-15 fiberglass-belted tires with raised white letters; four-speed manual floor shift transmission and a 340 hp, 390 cid four-barrel V-8 with dual exhausts. The 'Machine' had the highest output motor ever used in an American Motor's product offered for public sale. Of a total of 2,326 car run, approximately the first 1,000 units were finished in white, with the lower beltline and hood done in blue. A red stripe traveled down the front fender and along the car to the deckside region. From there, the stripe crossed over the trunk and came back along the opposite body side. There were also blue and white stripes over the trunk, behind the red one and integrated into it. Later editions of the model were finished in a choice of solid colors and featured a black-out hood treatment with silver pinstriping, plus optional red, white and blue graphics that could be added on the grille and body. The 'Machine' was actually a bigger and faster car than it should have been.

REBEL SERIES 10

Series Number	Body/Style Number	Body Type & Seating	Factory Price	Shipping Weight	Production Total
REBEL					
10	7015-0	4-dr Sed-6P	2626	3129	11,725
10	7019-0	2-dr HT Cpe-6P	2660	3148	1,791
10	7018-0	4-dr Sta Wag-6P	2766	3356	8,183
REBEL SST					
10	7015-7	4-dr Sed-6P	2684	3155	13,092
10	7019-7	2-dr HT Cpe-6P	2718	3206	6,573
10	7018-7	4-dr Sta Wag-6P	3072	3375	6,846

REBEL 'MACHINE' (V-8)

10	7019-7	2-dr HT Cpe-6P	3475	3650	1,936

1970 AMC, AMX two-door fastback sports coupe, V-8

AMX — (V-8) — SERIES 30: The 1970 AMX got new rear lamps and a completely restyled front end and that was shared with Javelin performance models. The frontal treatment featured a grille that was flush with the hood and redesigned bumper housing squarish parking lamps. A horizontally divided, cross-hatched grille insert, with prominent bright vertical moldings, was used and also incorporated circular Rally lights. The restyled hood had a large Ram-Air induction scoop that took in cold air for the engine. Height was reduced about one inch, while overall length grew about two inches. Standard equipment included all items used with Javelin SST models, plus a heavy-duty 60 ampere battery; courtesy lights; rear traction bar; Space-Saver spare tire; tachometer; 140 mph speedometer; 14 x 6 inch styled steel wheels; E78-14 blackwall tires; four-speed manual floor-shift transmission and 290 hp 360 cid four-barrel V-8 with dual exhaust system. The metal dashboard plates affixed to 1970 models were numbered 014469 to 18584. This was the final year for the original type AMX, although the nameplate would be used again on performance image Javelin and Hornet-based models.

AMX SERIES 30

Series Number	Body/Style Number	Body Type & Seating	Factory Price	Shipping Weight	Production Total
AMX					
30	7039-7	2-dr FsBk Cpe-2P	3395	3126	4,116

1970 AMC, Gremlin two-door sedan, 6-cyl

GREMLIN — (SIX) — SERIES 40: Introduced Wednesday, April 1, 1970, the AMC Gremlin was hailed as the first modern U.S.-built sub-compact car. This unique entry was basically a Hornet from the trailing edge of the door forward, and there really wasn't much left after that. The rear styling had a slanted, Kammback theme. It angled from a foot, or so, behind the rear wheel opening (at the bottom) to a point on the roof that was nearly plumb with the rear wheel centerline. Model identification came from a front fender cartoon badge and model name script on the sail panel (between two angular windsplit depressions). The rear was decorated with a circular medallion set into a rectangular depression panel that arched over it. Recessed, rectangular taillamps were seen. A hatch style rear window was featured on a four-passenger version. The two available models were a two-passenger commuter and the four-passenger job with fold-down rear seat. Standard features included front armrests; front ashtray; 35 ampere alternator (55 ampere with air conditioner); dome light; exhaust emissions control system; rubber floor and trunk mats; heater and defroster; split-back front foam seat cushion; wheel trim hubcaps; dual pinstripes: 'B'-rated 6.00 x 13 blackwall Polyester tires; three-speed manual transmission with column-mounted shift controls and 128 hp 199 cid six. The four-passenger model used the rear window liftgate feature and had a foam cushioned rear seat (with folding backrest). Gremlins optionally equipped with the 232 cid six came standard with floorshift transmission controls. A selection of 30 factory-installed options or accessories, plus seven packages was provided. The Gremlin may have been undersized, but it sure wasn't under-equipped.

GREMLIN SERIES 40

Series Number	Body/Style Number	Body Type & Seating	Factory Price	Shipping Weight	Production Total
40	7046-0	2-dr Sed-2P	1879	2497	872
40	7046-5	2-dr Sed-4P	1959	2557	27,688

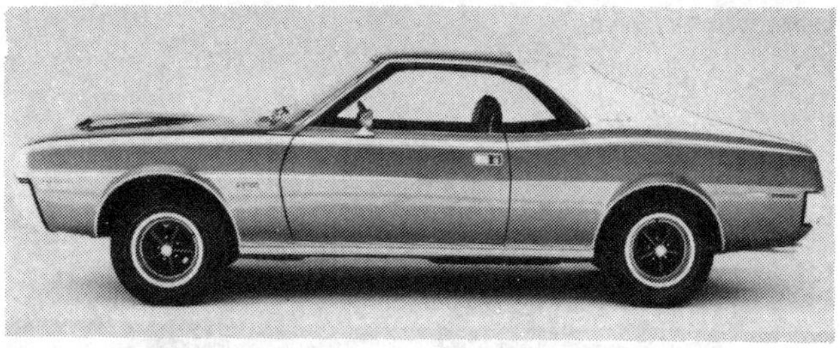

1970 AMC, Javelin SST two-door hardtop-sports coupe, V-8

JAVELIN — (SIX/V-8) — SERIES 70: The new Javelin shared its basic styling features with AMX, but retained its own twin venturi type grille without the previous 'bull's-eye' badge. The headlights were better integrated into the nose, sharing a common upper border molding with the main grille. It had the same front bumper, front parking lights and hood as the AMX and, like the two-seat 'mini-mite', was an inch lower and two inches longer. Standard equipment began with all items (except package tray) that were found on the Hornet SST. Additional features included compartment lights; dual horns; high-back bucket seats; C78-14 tires (D78-14 with V-8) and three-speed manual gearbox with shift control off the floor. The Javelin SST also had a Sports steering wheel with horn-blow rim and full wheel discs. Two-limited production Javelin SSTs were offered. The Javelin 'Trans Am' had all SST equipment (minus sill moldings and paint stripes), plus front lower and rear deck spoilers; black vinyl interior; '390' Go-Package; F70-14 glass-belted tires with raised white letters; 14 x 6 inch mag styled wheels; Space-Saver tire with regular spare wheel; AM push-button radio; tachometer and 140 mph speedometer: Visibility Group; Light Group; power steering: Twin-Grip differential; 3.91:1 axle ratio; four-speed box with Hurst floor shifter and 390 cid four-barrel V-8 with heavy-duty cooling system. These cars were replicas of the Ronnie Kaplan Trans-Am Racing Team's competition machines and were finished in a three-segment red, white and blue paint scheme created by industrial designer Brooks Stevens. Only 100 cars were built, the amount necessary to make this model eligible for the Sports Car Club of America's popular Trans Am races under 1969 'formulas.' In early 1970, the SCCA formulas were changed and so were the AMC drivers. The new rules demanded 2,500 replicas built to certain specifications. This led to the production of the 'Mark Donohue' Javelin SST. The majority of these cars had a special, thick-walled, 360 cid V-8 and all featured a unique 'duck-tail' rear spoiler with 'Mark Donohue' signature script at the right-hand side.

JAVELIN SERIES 70

Series Number	Body/Style Number	Body Type & Seating	Factory Price	Shipping Weight	Production Total
JAVELIN					
70	7079-5	2-dr FsBk Cpe-4P	2720	2845	8,496
JAVELIN SST					
70	7070-7	2-dr FsBk Cpe-4P	2848	2863	19,714
JAVELIN/SST "TRANS AM" (V-8)					
70	7079-7	2-dr FsBk Cpe-4P	3995	3340	(100)
MARK DONOHUE JAVELIN SST (V-8)					
70	7079-7	2-dr FsBk Cpe-4P	—	—	(2,501)*

NOTE 1: * The total of 19,714 Javelin SSTs includes both the TRANS AM and Mark Donohue Javelin models, as these were package models based on the Javelin SST.)
NOTE 2: Calendar year production of 28,210 units was recorded.

AMBASSADOR — (SIX/V-8) — SERIES 80: A new cross-hatched grille insert pattern characterized the 1970 Ambassador line. The rear quarter panels and bumper were also restyled. A large, rectangular taillamp crossed the back of the car. Trim distinctions between the various models followed the 1969 pattern. Standard equipment for the base Ambassador was the same as for the Rebel SST, plus 55 ampere alternator; air conditioning; F78-14 blackwall tires and 155 hp 232 cid six. The Ambassador DPL also had full wheel discs; Shift-Command automatic transmission (with base V-8) and a 210 hp 304 cid two-barrel V-8 engine. A standard extra Ambassador SST was individually adjustable reclining front seats. Ambassador DPL station wagons had all regular DPL features, plus cargo area carpets; roof-top travel rack and Dual-Swing tailgate. The Ambassador SST station wagon also had individually adjustable reclining front seats; wood-grained inside door panels a wood-grained side and rear exterior paneling.

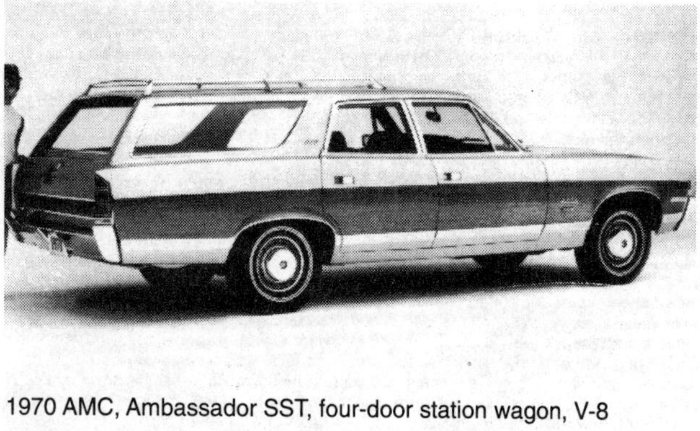

1970 AMC, Ambassador SST, four-door station wagon, V-8

AMBASSADOR SERIES 80

Series Number	Body/Style Number	Body Type & Seating	Factory Price	Shipping Weight	Production Total
AMBASSADOR					
80	7085-2	4-dr Sed-6P	3020	3328	9,565
AMBASSADOR DPL (V-8)					
80	7085-5	4-dr Sed-6P	3588	3523	6,414
80	7089-5	2-dr HT Cpe-6P	3605	3555	2,036
80	7088-5	4-dr Sta Wag-6P	3946	3817	8,270
AMBASSADOR SST (V-8)					
80	7085-7	4-dr Sed-6P	3722	3557	19,687
80	7089-7	2-dr HT Cpe-6P	3739	3606	8,255
80	7088-7	4-dr Sta Wag-6P	4122	3852	5,714

NOTE 1: The exact model year output of 1970 Ambassadors was 59,941 cars.
NOTE 2: In figures rounded-off to the nearest hundred, the model year output included the following: 3,500 base Ambassador sixes; 6,100 base Ambassador V-8s; 8,400 Ambassador DPLs (V-8 only); 27,900 Ambassador SSTs (V-8 only); 1,000 Ambassador station wagons with six-cylinder power and 13,000 Ambassador station wagons with V-8s.
NOTE 3: Calendar year production of 75,159 Ambassadors was recorded.

ENGINES

(Base six) Inline six. Overhead valves. Cast iron block. Displacement: 199 cid. Bore and stroke: 3.75 x 3 inches. Compression ratio: 8.5:1. Brake hp: 128 at 4400 rpm. Seven main bearings. Hydraulic valve lifters. Carburetor: Carter Type YF one-barrel.

(Base six) Inline six. Overhead valves. Cast iron block. Displacement: 232 cid. Bore and stroke: 3.75 x 3.50 inches. Compression ratio: 8.5:1. Brake hp: 155 at 4400 rpm. Seven main bearings. Hydraulic valve lifters. Carburetor: Carter Type WCD two-barrel.

(Base V-8) V-8. Overhead valves. Cast iron block. Displacement: 304 cid. Bore and stroke: 3.75 x 3.44 inches. Compression ratio: 9.0:1. Brake hp: 210 at 4400 rpm. Five main bearings. Hydraulic valve lifters. Carburetor: Autolite two-barrel Model 2100.

(Rebel "Machine" V-8) V-8. Overhead valves. Cast iron block. Displacement: 390 cid. Bore and stroke: 4.17 x 3.57 inches. Compression ratio: 10.0:1. Brake hp: 340 at 3600 rpm. Five main bearings. Hydraulic valve lifters. Carburetor: Autolite four-barrel Model 4300.

(AMX V-8) V-8. Overhead valves. Cast iron block. Displacement: 360 cid. Bore and stroke: 4.08 x 3.44 inches. Compression ratio: 10.0:1. Brake hp: 290 at 4800 rpm. Five main bearings. Hydraulic valve lifters. Carburetor: Autolite four-barrel Model 4300.

(JAVELIN/SST 'TRANS AM' V-8) V-8. Overhead valves. Cast iron block. Displacement: 390 cid. Bore and stroke: 4.17 x 3.57 inches. Compression ratio: 10.0:1. Brake hp: 325 at 5000 rpm. Five main bearings. Hydraulic valve lifters. Carburetor: Autolite four-barrel Model 4300.

('MARK DONOHUE' JAVELIN V-8) V-8. Overhead valves. Cast iron thick-wall block. Displacement: 360 cid. Bore and stroke: 4.08 x 3.44 inches. Compression ratio: 10.0:1. Brake hp: 290 at 4800 rpm. Five main bearings. Hydraulic valve lifters. Carburetor: Autolite four-barrel Model 4300.

CHASSIS FEATURES: Wheelbase: (Gremlin) 96 inches; (AMX) 97 inches; (Javelin) 109.9 inches; (Hornet) 108 inches; (Rebel) 114 inches; (Ambassador) 122 inches. Overall length: (Gremlin) 161-1/4 Inches; (AMX) 179 inches; (Javelin) 191 inches; (Hornet) 179.3 inches; (Rebel) 199 inches; (Wagon 198 inches); (Ambassador) 208 Inches; (Wagon 207 inches). Front tread: (Gremlin) 57.5 inches; (AMX and Javelin) 59.1 inches; (Hornet) 57.2 inches; (Rebel and Ambassador) 59.7 inches. Rear tread: (Gremlin) 57 inches; (AMX, Javelin and Hornet) 56.6 inches; (Rebel and Ambassador) 60 inches. Tires: See text.

POWERTRAIN OPTIONS: Three-speed manual transmission was standard in most AMC models. Automatic transmission was standard in Ambassador DPL and SST. Shift-Command automatic transmission was optional on all models at various prices. Four-speed manual floor shift transmission, in Javelin '360' and '390' V-8s. Close-ratio four-speed manual transmission with floor shift was standard in Rebel 'Machine' and AMX. Hornet/Gremlin six-cylinder 232 cid 145 hp one-barrel engine ($45). Base Hornet six-cylinder 232 cid 155 hp two-barrel engine ($65). Hornet SST/Rebel six-cylinder 232 cid 155 hp two-barrel engine ($19). Javelin / Ambassador / Rebel (except 'Machine') V-8 360 cid 245 hp two-barrel engine ($41). Javelin / Ambassador / Rebel (except 'Machine') V-8 360 cid 290 hp four-barrel engine ($86). Ambassador DPL/SST/Rebel SST/Javelin V-8 390 cid 325 hp four-barrel engine ($168). AMX V-8 390 cid 325 hp four-barrel engine ($1 1). Heavy-duty 70 ampere battery ($13). Axle ratios, all optional ($10). Heavy-duty cooling, standard with air ($16). Dual exhaust, as separate option ($31). Twin-Grip positive traction rear axle ($43).

NOTE: Overdrive was no longer available. Three-speed manual floor shift with Gremlin V-8s only (as standard equipment).

GREMLIN/HORNET OPTIONS: Power brakes ($43). Power steering ($96). Air conditioning ($381). Front and rear bumper guards ($25). Locking gas cap ($6). Engine block heater ($16). Gremlin roof-top luggage rack ($39). Special application paint ($39). AM push-button radio ($62). Gremlin white, black or red Rally side stripes ($25). Custom steering wheel, Gremlin ($12). Gremlin Shift-Command with column controls ($195). Wheel disc covers ($25). Electric washer/wipers, Gremlin and Hornet ($20). Tinted glass, Gremlin, all windows ($34); windshield only ($26). Electric clock, in base Hornet only ($16). Air Command ventilation system, Hornet Group only ($41). Tinted glass, Hornet all windows ($34); windshield only ($26). Vinyl insert scuff side molding, Hornet Group ($27). Hornet SST two-tone paint ($24). Hornet Group exterior body striping ($19). Front disc brakes, Hornet Group, except base series ($84). Hornet SST caranaby plaid interior trim ($78). Hornet SST, individual seats, fabric trim ($52); vinyl trim ($71). Hornet SST, bench seat, vinyl trim/standard base model ($20). Hornet base models, bench seat, cloth trim/standard SST ($13). Handling Package, Hornet Group ($23). Hornet Group Decor Package, with/Air ($34); without Air ($58). Hornet Light Group Package ($25). Hornet SST, vinyl roof ($84).

REBEL/AMBASSADOR OPTIONS: Power brakes ($43). Power steering ($105). Air conditioning, Rebel Group ($380). Cruise Control, Rebel and Ambassador, with automatic ($60). Front and rear bumper guards, except station wagons ($32). Tinted glass, all windows, Ambassador ($42); Rebel ($37). Tinted windshield, Ambassador ($32); Rebel ($30). Two-tone paint, Rebel/Ambassador sedans and hardtops ($27). Body side accent panels and moldings, wagons except Style 7088-7 ($65). Special paint, other than standard ($39). Power side windows, Ambassador DPL/SST passenger cars ($105). Power side and tailgate window, Ambassador DPL/SST wagons ($140). Power tailgate window, standard with third wagon seat ($35). AM push-button radio ($62). AM/FM push-button radio, all AMC except Hornets ($134). Eight-track stereo tape, all Ambassador, except wagons ($134). Third seat, Rebel SST and Ambassador DPL/SST wagons, with extras ($118). Vinyl bench seat, Ambassador style 70852 and base Rebel ($24). Individual seat, fabric, Ambassador (standard SST) and Rebels, except 'Machine' ($64). Velour Individual cushion interior, Ambassador SST sedan ($68). Bucket seats fabric or vinyl, Rebel/Ambassador SST hardtop ($123). Sports steering wheel with horn-blow, Ambassador, Rebel SST ($37). Tilt-O-Just steering wheel, Rebel/Ambassador with automatic ($45). Shift-Command; with floor shift, Rebel 'Machine' with '390' V-8 ($188). Black, white or blue vinyl roof, Ambassador DPL/SST, except wagons ($106). Black, white or blue vinyl roof, Rebel SST hardtop and sedan ($95). Turbo-Cast wheelcovers, Rebel (except 'Machine'); Ambassador 7085-2 ($74). Wire wheelcovers, Rebel (except 'Machine'); Ambassador Style 7085-2 ($74). Turbo-cast or wire wheelcovers, Ambassador SST or DPL ($49).

JAVELIN/AMX OPTIONS: Power brakes ($43). Power steering ($102). Air conditioning ($380). Rear bumper guards ($13). Command Air ventilation system, w/o air ($41). Center console, Javelin Group with column automatic shift only ($53). Tinted glass, Javelin/AMX prices same as Hornet. Simulated exhaust type rocker panel moldings, Javelin SST ($32). Simulated exhaust type rocker panel moldings, base Javelin ($50). AMX two-tone finish with 'black shadow' treatment ($52). Rally side stripes, solid color Javelin/AMX ($32). Power front disc brakes, V-8 engine required ($84). Eight-track stereo tape with manual radio/twin rear speakers ($195). Corduroy fabric bucket seats, Javelin SST only ($50). Leather trimmed bucket seats, AMX only ($84). Leather trimmed bucket seats, Javelin SST only ($127). quick-ratio manual steering, for racing ($16). Javelin Code 533 spoiler roof, not available with vinyl top ($33). Tachometer and 140 mph speedometer with, V-8 ($50). Shift-Command, column control, Javelin with '304' V-8 ($200). Shift-Command, column control, Javelin with '360' V-8 ($233). Shift-Command, column control, Javelin six ($195). Shift-Command, floor control, Javelin with '304' V-8 ($264). Shift-Command, floor control, Javelin with '360' V-8 ($287). Shift-Command, floor control, AMX with '390' V-8 ($118). Four-speed floor-shift, Javelins with 290/325 hp V-8 ($205). Black, white or blue vinyl roof, Javelin only ($84). Turbo-Cast wheelcovers; base Javelin ($74); Javelin SST ($49). Wire wheelcovers; base Javelin ($74); Javelin SST ($49). Styled steel wheels, 14 x 6 inch, base Javelin ($98); Javelin SST ($72). A desirable option was the code 391/2 'Go-Package', which retailed for $298.85 on the '360' AMX and $383.90 on the '390' AMX. It included one of these engines; power front disc brakes; F70-14 blackwall tires with raised white letters; Handling Package; heavy-duty cooling system and functional Ram-Air induction scoop. A desirable Javelin option was the Code 391/2 'Go Package' which retailed for $321.65 on the '360' Javelin and $409.75 on the '390' Javelin. Features included one of these engines; power front disc brakes: E70-14 redline tires; six inch wheel rims; Handling Package: AMX Ram-Air hood and dual exhausts.

HISTORICAL FOOTNOTES: The AMC line was introduced Sept. 25, 1969 and the 1970 Gremlin appeared in dealer showrooms April 1, 1970. Model year production peaked at 242,664 units. Calendar year production of 276,110 cars was recorded. R.D. Chapin, Jr. was the chief executive officer of the company this year. A total of 11,125 AMC products, built in the 1970 model year, were equipped with 390 cid V-8 engines. Twenty percent of all AMC cars had bucket seats; 14.9 percent had vinyl roofs; styled wheels were ordered by 4.6 percent of 1970 buyers and four-speed transmission installations were made in 2.6 percent of all cars. The company referred to itself as "the new American Motors" in advertisements and presented buyers with the catch line, "If you had to compete with GM, Ford and Chrysler, what would you do?"

1971 AMC

HORNET — (SIX/V-8) — SERIES 01: The 1971 Hornet received no major styling changes for 1971. Two new models were added, however, along with some springtime equipment packages. The new styles were an SST four-door Sportabout station wagon and a Hornet V-8 performance car called the SC/360 two-door sports sedan. Standard equipment on base Hornets was the same as that found on 1971 Gremlins, plus color-keyed rubber floor mats and larger, 6.45 x 14 blackwall tires. The Hornet SST models also had color-keyed carpets; cigarette lighter; Command-Air ventilation; glove box; full width package tray; Custom steering wheel; and movable rear quarter windows. The Sportabout added a carpeted cargo area; rear liftgate; cargo compartment lock; and Space-Saver spare tire. The SC/360 sports sedan had the same equipment as Hornet SST passenger models, plus front sway bar; slot style wheels; Space-Saver spare tire; D70-14 raised white letter tires; and a 245 hp 360 cid two-barrel V-8. Springtime changes included price increases for all Hornet SST V-8s and a Sportabout D/L package including color-keyed wood-grain side and rear panels; roof rack with integral air deflector; Custom wheelcovers; individual reclining seats; wood-grained instrument cluster trim; Sports rim-blow steering wheel; and D/L decals. In addition, free sun roofs were provided as part of several Hornet packages. The SC/360 stands out as the most collectable car in the group. It was designed as a low-priced performance car that could pass as a compact and, thus, side-step rising insurance rates affecting the owners and buyers of such vehicles. Its trim included special around-the-beltline decals with SC/360 call-outs at the trailing edge of rear fenders. Optional, at $199, was a 'Go-Package' that consisted of four-barrel carburetor; Ram-Air induction system; dual exhausts; Handling Package; tachometer; and Polyglas white letter tires. It included a wide hood scoop with black-out paint treatment. American Motors had programmed the model for 10,000 sales, but only 784 buyers were interested. Many expressed hopes that the new 401 cid V-8 be offered in this car, but insurance and marketing considerations precluded this. The SC/360 was sold only in 1971.

1971 AMC, Hornet SC/360 two-door sedan, V-8

VIN: Serial Numbers were located on top of the right front wheelhouse panel. The unit body data plate was riveted to the left front door, below the latch mechanism. American Motors retained a 13 symbol identification code system. The first symbol was an A for American Motors Corp. The second symbol indicated model year, 1=1971. The third symbol indicated transmission type: S=standard column shift; A=automatic, column shift; C=three-speed automatic with floor shift; F=three-speed manual with floor shift; and M=four-speed manual with floor shift. The fourth symbol was a number designating car line, as follows: H=Hornet; 1=Rebel; 3=AMX; 4=Gremlin; 7=Javelin and 8=Ambassador. The fifth symbol designated body type: 5=four-door sedan; 6=two-door sedan; 8=four-door station wagon; and 9=two-door hardtop. The sixth symbol designated the series or class of body, as follows: 0=Base Hornet/Gremlin; 1=Hornet SC/360; 2=Ambassador DPL; 5=/Javelin/Ambassador SST/Gremlin and 7=Hornet SST/Hornet Sportabout/Matador/Javelin SST; 9=Javelin AMX. The seventh symbol was a letter designating engine type (without regard to car-line), as follows: A=258 cid cid six; B=low-compression 258 cid six; E=232 cid six one-barrel carburetor; F=low-compression 232 cid six with one-barrel carburetor; H=304 cid V-8 with two-barrel carburetor; I=304 V-8 two-barrel (low-compression); 'M'= low-compression 304 cid V-8 with four-barrel carburetor; 'N' = 360 cid V-8 with two-barrel carburetor; 'P' = 360 cid V-8 with four-barrel carburetor; Z=401 cid V-8 four-barrel. The following group of symbols was the sequential production and began with 100001 for cars built in Kenosha, Wis. and 700001 for cars built in Brampton, Ontario, Canada. Body number plate attached to left front door edge shows body serial number, Body/Style Number from second column of chart below, plus trim and paint codes.

HORNET SERIES 01

Series Number	Body/Style Number	Body Type & Seating	Factory Price	Shipping Weight	Production Total
HORNET					
01	7106-0	2-dr Sed-6P	2174	2654	19,395
01	7105-0	4-dr Sed-6P	2234	2731	10,403
HORNET SST					
01	7106-7	2-dr Sed-6P	2324	2732	8,600
01	7105-7	4-dr Sed-6P	2274	2691	10,651
01	7108-7	4-dr Sta Wag-6P	2594	2827	73,471
HORNET SST SC/360 (V-8)					
01	7106-1	2-dr Spt Sed-6P	2663	3300	784

MATADOR — (SIX) SERIES 10: The former AMC Rebel became the Matador for 1971. The renamed product was significantly restyled with changes to taillamps; hood; grille; front fenders; bumper and valance panel. Wheelbase was increased to 118 inches. The Matador appearance was characterized by an integrated bumper/grille; horizontal, double-venturi grille insert and triple-rectangular taillight lenses integrated into a horizontal rear beauty panel. Body style offerings were limited to a two-door hardtop, four-door sedan and four-door station wagon marketed in a single level of trim, but with many available options packages including the high-performance 'Go Machine' group. Standard equipment was the same as for base Javelins, plus rear armrests; rear ashtray; full-back bench seat cushion in sedans and wagon/split-back in hardtop; hardtop pillar lights; color-keyed cargo area carpets in wagon (and Dual-Swing tailgate) and three-speed transmission with column shift. The Matador was marketed as a six, with V-8 options. Additional equipment, installed when a V-8 with automatic transmission was ordered, included a transmission oil cooler.

1971 AMC, Matador four-door station wagon, V-8

MATADOR SERIES 10

Series Number	Body/Style Number	Body Type & Seating	Factory Price	Shipping Weight	Production Total
10	7118-0	4-dr Sed-6P	2770	3165	5
10	7119-0	2-dr HT Cpe-6P	2799	3201	1
MATADOR					
10	7115-7	4-dr Sed-6P	3163	3437	24,918
10	7119-7	2-dr HT-6P	3129	3360	7,661
10	7118-7	4-dr Sta Wag-6P	3493	3596	10,740

1971 AMC, Gremlin 'X' two-door sedan, 6-cyl

GREMLIN — (SIX) — SERIES 40: Introduced as a midyear 1970 model, the Gremlin was unchanged for 1971 in terms of appearance. A technical revision was seen in the use of the 232 cid six as base engine and the smaller 199 cid six went out of production. There was also an attractive new Gremlin 'X' option package. It was provided only for the four-passenger model at a price of $300 with blackwall tires or $334.35 with raised white letter tires. Included were 'Spear' side stripes; black-painted grille; 14 x 6 inch slot style wheels; Space-Saver spare; Custom interior appointments; front bucket seats; special 'X' decals and D70-14 size tires (with raised white letter style considered as option-within-option). Standard equipment on all Gremlins included 35 ampere alternator; front armrests and ashtray; 50 ampere battery; dome light; exhaust emissions control; rubber floor mats; bench seats; split-back front foam seat cushion; exterior paint stripes; and 6.00 x 13 black sidewall Polyester tires. The four-passenger model also featured a glove box door; liftgate type rear window; and rear foam seat cushion with fold-down back. The 135 hp 232 cid six was base engine, with a bigger six optional.

GREMLIN SERIES 40

Series Number	Body/Style Number	Body Type & Seating	Factory Price	Shipping Weight	Production Total
40	7146-0	2-dr Sed-2P	1899	2503	2,145
40	7146-5	2-dr Sed-4P	1999	2552	74,763

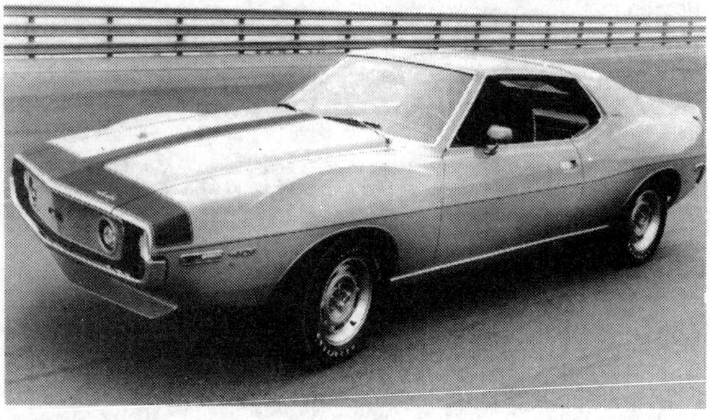

1971 AMC, Javelin AMX, two-door hardtop sports coupe, V-8

JAVELIN — (SIX/V-8) — SERIES 70: The 1971 Javelin was completely restyled. There were highly sculptured raised fenders, a twin-canopy roof with air spoiler type rear window lip, and new full-width taillamps. The interior was completely redesigned and upgraded. It featured a curved cockpit type instrument panel inspired by aircraft motifs. Three levels of trim were provided in one two-door hardtop style: base Javelin; Javelin SST and Javelin AMX. This new AMX was a four-place automobile, replacing the former two-seater. A rear-facing cowl-induction hood, flush wire mesh grille and optional front and rear spoilers were claimed as the design work of race driver Mark Donohue, who raced Javelins successfully in SCCA Trans-Am competition. Standard equipment on the Javelin was the same as for Hornets, plus Custom steering wheel; color-keyed carpets; glove box lock; dual horns: high-back bucket seats; three-speed manual transmission with floor-shift; C78-14 glass-belted tires; cigar lighter; and automatic transmission oil cooler with V-8s. The Javelin SST also had rear ashtray; rim-blow Sports steering wheel; rubber trunk mat; and wheelcovers. The AMX featured all this equipment, plus electric clock; center console (without armrest); rear deck mounted spoiler; slot style wheels; E70-14 glass-belted tires; and 245 hp 360 cid two-barrel V-8.

JAVELIN SERIES 70

Series Number	Body/Style Number	Body Type & Seating	Factory Price	Shipping Weight	Production Total
JAVELIN					
70	7179-5	2-dr FsBk Cpe-6P	2879	2887	7,105
JAVELIN SST					
70	7179-7	2-dr FsBk Cpe-6P	2999	2890	17,707
JAVELIN/AMX (V-8)					
70	7179-8	2-dr FsBk Cpe-4P	3432	3244	2,054

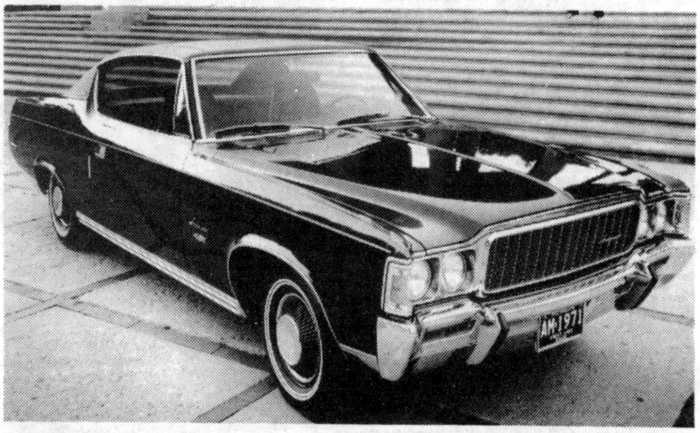

1971 AMC, Ambassador Brougham two-door hardtop, V-8

AMBASSADOR — (SIX/V-8) — SERIES 80: The 1971 Ambassador received a new, die-cast rectangular grille. There were also new front end caps incorporating side marker lights visible in face or profile view. Car-line nameplates were downgraded one notch to make it seem that each line was one level higher. The Ambassador six sedan became the DPL. Offered in SST and top-level Brougham trims were the two-door hardtop and four-door sedans and station wagon. Along with air conditioning (as before) the standard equipment list was expanded to include automatic transmission. The Ambassador six came with all Matador features, plus 55 ampere alternator: front bumper guards; All-Season air conditioning; Shift-Command transmission; and a 258 cid six. Ambassador SSTs had, in addition, full wheelcovers and a 210 hp 304 cid V-8. Ambassador Broughams also had individually adjustable reclining seats and, for wagons, wood-grained exterior panels on the sides and rear.

AMBASSADOR SERIES 80

Series Number	Body/Style Number	Body Type & Seating	Factory Price	Shipping Weight	Production Total
AMBASSADOR DPL SIX					
80	7185-2	4-dr Sed-6P	3616	3315	6,675
AMBASSADOR SST (V-8)					
80	7185-5	4-dr Sed-6P	3852	3520	5,933
80	7189-5	2-dr HT Cpe-6P	3870	3561	1,428
80	7188-5	4-dr Sta Wag-6P	4253	3815	4,465
AMBASSADOR BROUGHAM (V-8)					
80	7185-7	4-dr Sed-6P	3983	3541	13,115
80	7189-7	2-dr HT Cpe-6P	3999	3580	4,579
80	7188-7	4-dr Sta Wag-6P	4430	3862	5,479

BASE ENGINES

(Six) Inline six. Overhead valves. Cast iron block. Displacement: 232 cid. Bore and stroke: 3.75 x 3.50 inches. Compression ratio: 8.0:1. Brake hp: 135 at 4400 rpm. Seven main bearings. Hydraulic valve lifters. Carburetor: Carter Type YF one-barrel.

(SC/360 V-8) V-8. Overhead valves. Cast iron block. Displacement: 360 cid. Bore and stroke: 4.08 x 3.44 inches. Compression ratio: 8.5:1. Brake hp: 245 at 4400 rpm. Five main bearings. Hydraulic valve lifters: Carburetor: Autolite two-barrel Model 2100.

(DPL Six) Inline six. Overhead valves. Cast iron block. Displacement: 258 cid. Bore and stroke: 3.75 x 3.90 inches. Compression ratio: 8.0:1. Brake hp: 150 at 3800 rpm. Seven main bearings. Hydraulic valve lifters. Carburetor: Carter Type YF one-barrel.

(Base V-8) V-8. Overhead valves. Cast iron block. Displacement: 304 cid. Bore and stroke: 3.75 x 3.44 inches. Compression ratio: 8.4:1. Brake hp: 210 at 4400 rpm. Five main bearings. Hydraulic valve lifters. Carburetor: Autolite Model 2100 two-barrel.

POWERTRAIN OPTIONS: Automatic transmission was standard in Ambassador SST and Ambassador Brougham. Three-speed manual transmission with floor-shift was standard in the Javelin Group. Three-speed manual transmission with column-shift was standard in other models. Shift-Command with column shift, in Gremlin ($200); in Hornet six ($210); in Hornet '304' V-8 ($216); in Hornet SC/360 ($238); in Javelin/Matador six ($217); in Javelin/Matador with '304' V-8 ($223); in Ambassador SST/Brougham with '304' V-8 ($6); in Matador/Javelin with '360' V-8, except AMX ($246); in DPL with '360' V-8 ($29); in other Ambassadors with '360' V-8 ($23); in Matador with '401' V-8 ($256) and in non-DPL Ambassadors with '401' V-8 ($33). Shift Command with floor-shift and console, in Matador/Javelin with '304' V-8 and bucket seats ($279); in Brougham with '304' V-8 ($56); in non-AMX Javelin and Matador with '360' V-8 ($302); in AMX with '360' V-8 ($246); in Brougham with '360' V-8 ($79); in non-AMX Javelin and Matador with '401 'V-8 ($312); in AMX with '401 'V-8 ($256) and in '401' Ambassador Brougham ($89). Four-speed manual transmission with floorshift, available only in Matador with 'Go-Machine' package and Javelin Group with 285/330 hp V-8s ($209). (NOTE: Some transmission attachment prices increased $4-$7 at midyear). Six-cylinder 360 cid 150 hp engine, in Gremlin and Hornet, except SC/360 ($54); in Javelin and Matador, except AMX ($50). Two-barrel 360 cid 245 hp V-8 in all, except Gremlin/Hornet and except standard in AMX ($48). Four-barrel 360 cid 285 hp V-8 in Javelin AMX ($49); in Matadors, Ambassadors and other Javelins ($97). Four-barrel 401 cid 330 hp V-8 in Javelin AMX ($137); in Matador; Ambassadors and other Javelins ($137). Optional axle ratios ($12-$14). Heavy-duty 70-ampere battery ($14-$15). Twin-Grip differential in Gremlin and Hornet ($43); in all other models ($47). Dual exhausts, in all with '360' four-barrel V-8 ($31). Dual exhausts were standard with the '401' V-8. Engine block heater ($12). Cooling system including heavy-duty radiator, Power flex fan and fan shroud ($16). Cold Start package ($18-$19).

CONVENIENCE OPTIONS: Power brakes ($45-$49). Power steering ($100-$111). Air conditioning ($399). Speed control, Matador/Ambassador ($63). Front and rear bumper guards, Gremlin/Hornet ($20); Javelin/Matador ($32). Locking type gas cap, Gremlin ($6). Electric clock, in base Hornet only ($17). Electric rear defogger, Javelin/Matador/Ambassador passenger ($52). Front manual disc brakes, Javelin V-8 ($40). Engine block heater, all ($12). Heavy-duty cooling system, all ($16-$18). Tinted glass, all windows; Gremlin ($37); Hornet ($40); Matador/Javelin ($44); Ambassador ($47); Tinted windshield, Gremlin/Hornet ($30); Javelin ($32); Matador ($36); Ambassador ($38); Headlight delay system ($23). Roof luggage rack, Gremlin ($40); Matador wagon ($58). Deck luggage rack, Javelin ($35); Hornet sedan ($32); Sportabout ($47). Body side scuff molding, Gremlin/Hornet ($27); Javelin/Ambassador ($31). Two-tone paint, Hornet SST ($28); Matador/Ambassador sedan ($31); wagon ($71); Rally stripe, Javelins except AMX ($37); Gremlin ($30); Hornet (standard). Exterior paint stripe, basic Hornet — standard Hornet SST — ($10). Power front disc brakes, SC/360 and Hornet SST V-8 ($84); other V-8 ($89). Power side windows, Ambassador cars ($120); wagons with tailgate ($157). AM push-button radio, Gremlin/Hornet ($67); all others ($72). AM/FM Multiplex stereo, Javelin Group including AMX ($224). AM/FM push-button radio, Matador/Ambassador ($143). AM/radio with 8-track and two speakers, Javelin Group ($207). AM radio with 8-track and two-speakers Matador/Ambassador ($140). Leather bucket seats, Javelin SST/AMX ($84). Cordoroy fabric bucket seats, Javelin SST/AMX ($52). Serape fabric reclinable seats, Matador hardtop ($71). Center console, Javelin Group, except standard. AMX ($58). Vinyl center armrest bucket seats, Ambassador Style 7189-7 and Matador Style 7119-7 ($136). Fabric reclinable seats, Ambassador SST wagon ($96). 'Harem' fabric reclinable seats, Ambassador Brougham passenger models ($75). Third station wagon seat with power tailgate and extras ($118). Custom steering wheel, Gremlin/Hornet ($14). Rim-blow Sports steering wheel, Gremlin/Hornet ($37); Matador/Ambassador ($40). Adjust-O-Tilt steering, Javelin/Matador/Ambassador ($49). Tachometer, Hornet SC/360 only ($50). Station wagon tailgate air deflector, Gremlin ($49); others ($22). Air adjustable rear suspension, Matador/Ambassador ($42). Undercoating, Hornet SC/360 ($18). Undercoating and hood insulation, Gremlin/Hornet ($22). Black, White, Blue or Green vinyl top, Matador ($97); Ambassador ($108). Standard wheelcovers ($27-$30). Custom wheelcovers ($25-$54). Turbo-Cast or wire wheelcovers ($52-$75). Styled steel wheels, Gremlin 'X'/AMX with 'Go-Package' ($34-$37). Styled steel wheels, most other models ($99-$108); SC/360 ($46). Rear quarter vent windows, basic Hornet ($30). Electric wiper/washers, Gremlin/Hornet, except SC/360 ($21). Electric wiper/washers, Javelin/Matador/Ambassador ($22). Matador station wagon, woodgrained side panels ($117). Black, White, Blue or Green vinyl roof, Hornet, except SC/360 ($88); Javelin ($89). Calvary Twill recliner seat, Matador wagon ($96). Center armrest and cushion, Javelin without console ($51).

OPTION PACKAGES: (Matador 'Go-Machine') $373 on Matador hardtops with '360' V-8/$461.10 on Matador hardtops with '401' V-8, includes selected engine; four-barrel carburetion; dual exhausts; Handling Package; power disc brakes; E60-15 Polyglass tires with raised white letters; 15 x 7 inch styled steel wheels and Space-Saver spare tire. (Javelin/AMX 'Go Package') $410.90 on Javelin/AMX with '360' V-8/$498.95 on Javelin/AMX with '401' V-8, includes specified engine; four-barrel carburetor; dual exhausts; hood 'T' stripe decal; Rally-Pac instrumentation: Handling Package; Cowl-Air carburetor induction system; heavy duty cooling components; Twin-Grip differential; power disc brakes; E60-15 Polyglas raised white letter tires; styled-steel wheels of 15 x 7 inch size and Space-Saver spare.

HISTORICAL FOOTNOTES: The full-sized AMC models were introduced Oct. 6, 1970 and the Gremlin appeared in dealer showrooms the same date. Model year production peaked at 244,758 units. Calendar year sales of 256,963 cars were recorded. R.D. Chapin, Jr. was the chief executive officer of the company this year.

1972 AMC

1972 AMC, Hornet Sportabout 'D/L' four-door station wagon, 6-cyl

HORNET SST — (SIX/V-8) — SERIES 01: All Hornets were SST models this year. In addition, all were marketed in six and V-8 series for the first time. Physical changes were minor, consisting mainly of trim and ornamentation revisions. One was a 'silver line' treatment for the molded plastic radiator grille. There was also new taillights and an aluminum overlay panel below the trunk lid. Two model-creating options packages were available and served to upgrade the level of trim in specific attachments. For example, there was the Sportabout D/L (Deluxe) package which included color-keyed, wood-grained side and rear paneling; roof rack with air deflector; Custom wheelcovers; individually reclining seats; and 'D/L' decals. It could be had, on the station wagon only, two different ways. With vinyl upholstery, the price tag was $236.25 and with 'Scorpio' fabric upholstery the retail price was $283.55. A sporty 'X' package, priced at $118.55, included Sports steering wheel; Rally stripes; wide rocker panel moldings; slot styled wheels; C78 x 14 tires; and special 'X' emblems. The two-door sedan 'X' package included the same equipment, less wide rocker panel moldings, at the same price. A designer series Hornet Sportabout was created by famed Italian fashion genius Dr. Aldo Gucci. It featured beige seats and door panels trimmed with red and green stripes. The Gucci crest appeared on the inside door panel, front fenders and headliners. Available in just four exterior colors — Snow White; Hunter Green; Grasshopper Green and Yuca Tan — the Gucci Sportabout registered sales of 2,583 units. Standard Hornet equipment features began with all items found in Gremlins plus, rear armrest; rear ashtray; cigarette lighter; two coat hooks; color-keyed carpets; cargo mat; 16 gallon gas tank, glove box; full-width package tray; Custom steering wheel; and three-speed manual transmission with column shift. The Hornet Sportabout also had carpeting in the rear cargo area; rear liftgate; cargo compartment lock; and Space-Saver spare tire. A Space-Saver spare was also standard on any Hornet with optional styled steel wheels. The 232 cid six-cylinder engine was the base powerplant and tire sizes varied with each different engine. Size 6.45 x 14 blackwall tires came as regular equipment with sixes. However, 6.95 x 14 tires were used with sixes having an air conditioner and were standard with V-8 powered sedans and Sportabouts. An improved type of Torque-Command automatic transmission was optionally available this year. A special Hornet Rally model trim package was available in 1972 only with a special Hornet Rally stripe treatment along the lines of the 1972 SC/360 stripe treatment, on two-door Hornets. It consists of pleated vinyl bucket seats; manual disc brakes; handling package; 20:1 quick-ratio manual steering; fully-synchromesh three-speed manual transmission; three-spoke sports steering wheel (15"); and 'Rallye' emblems on rear fenders.

VIN: Serial Numbers were located on top of the right front wheelhouse panel. The unit body data plate was riveted to the left front door, below the latch mechanism. American Motors retained a 13 symbol identification code system. The first symbol was an A for American Motors Corp. The second symbol indicated model year, 2=1972. The third symbol indicated transmission type: S=standard column shift; A=automatic, column shift; C=three-speed automatic with floor shift; E=fully-synchronized three-speed manual with floor shift; F=three-speed manual with floor shift; and M=four-speed manual with floor shift. The fourth symbol was a number designating car line, as follows: 0=Hornet; 1=Rebel; 3=AMX; 4=Gremlin; 7=Javelin and 8=Ambassador. The fifth symbol designated body type: 5=four-door sedan; 6=two-door sedan; 8=four-door station wagon; and 9=two-door hardtop. The sixth symbol designated the series or class of body, as follows: 0=Base Hornet/Gremlin; 1=Hornet SC/360; 2=Ambassador DPL; 5=Ambassador SST/Gremlin and 7=Hornet SST/Javelin SST/Matador/Ambassador Brougham; 8=Javelin AMX. The seventh symbol was a letter designating engine type (without regard to car-line), as follows: A=258 cid six; B=low-compression 258 cid six; E=232 cid six one-barrel carburetor; F=low-compression 232 cid six with one-barrel carburetor; H=304 cid V-8 with two-barrel carburetor; I=304 cid V-8 two-barrel (low-compression); 'M' = low-compression 304 cid V-8 with four-barrel carburetor; 'N' = 360 cid V-8 with two-barrel carburetor; 'P' = 360 cid V-8 four-barrel carburetor; Z=401 cid V-8 four-barrel. The following group of symbols was the sequential production and began with 100001 for cars built in Kenosha, Wis. and 700001 for cars built in Brampton, Ontario, Canada. Body number plate attached to left front door edge shows body serial number, Body/Style Number from second column of chart below, plus trim and paint codes.

HORNET SST SERIES 01

Model Number	Body/Style Number	Body Type & Seating	Factory Price	Shipping Weight	Production Total
01	7206-7	2-dr Sed-6P	2199/2337	2627/2861	27,122
01	7205-7	4-dr Sed-6P	2265/2403	2691/2925	24,254
01	7208-7	4-dr Sta Wag-6P	2587/2725	2769/2998	34,065

NOTE 1: Data above slash for six/below slash for V-8.

1972 AMC, Matador two-door hardtop coupe, 6-cyl

MATADOR — (SIX/V-8) — SERIES 10: The Matador was also marketed as a six/V-8 Series for the first time this year. The 1972 Matador was fighting an identity crisis and was sometimes promoted as, "A car you probably never heard of." In attempting to give it an image, AMC redesigned the grille. Ads highlighted the Matador's use by the Los Angeles Police Department. A horizontally lined, segmented pattern grille insert with three color-keyed bars amounted to the major frontal change. The rear also had new lamps and center panel trim, while the side of the car had a dual length pinstripe treatment. Old-fashioned cable controls in the heating system were switched to a vacuum operated type. Standard equipment was the same as used on 1972 Javelins, plus a front sway bar and 19.5 gallon fuel tank. The hardtop came with two side pillar lights and split-back bench seat. Station wagons had four plastic coat hooks; cargo compartment lock; Dual-Swing tailgate; and a larger, 258 cid, base six. Tires were E78 x 14 blackwalls on passenger car models; G78 x 14 on station wagons. The base V-8 was the same as the Hornet SST's base V-8.

MATADOR SERIES 10

Model Number	Body/Style Number	Body Type & Seating	Factory Price	Shipping Weight	Production Total
10	7215-7	4-dr Sed-6P	2784/2883	3171/3355	36,899
10	7219-7	2-dr HT Cpe-6P	2818/2917	3210/3394	7,306
10	7218-7	4-dr Sta Wag-6P	3140/3239	3480/3653	10,448

NOTE 1: Data above slash for six/below slash for V-8.

1972 AMC, Gremlin 'X' two-door sedan, 6-cyl

GREMLIN — (SIX/V-8) — SERIES 40: Since it was essentially a chopped-off Hornet, it made sense to market the Gremlin the same way: as a six or V-8. The two-seater was dropped, though. Stuffing the 304 cid engine into such a small package did require some engineering modifications. Special drivetrain and suspension components were used. Standard equipment included front armrests and ashtrays; 50 ampere battery; dome light; rubber floor mats and trunk mat; glove box door; heater and defroster; foam seat cushions; split-back front seat; foldown rear seat; 21 gallon fuel tank; opening type tailgate; three-speed manual gearbox (with choice of column or floor controls); hubcaps; 37 ampere alternator; exhaust emissions controls; exterior paint stripe; and blackwall tires. Size 6.00 x 13 rubber was used with the basic six; size 6.45 x 14 with air-conditioning or V-8s. Also found on V-8 equipped models were a front sway bar and rear deck '5-Litre V-8' badge. A few new pieces of safety equipment used on all AMC products this year were three-point seat belts linked to a seat belt warning buzzer. As you can tell, the Gremlin, though small, was very well built and equipped. *POPULAR MECHANICS* magazine wrote, "The best put-together cars out of Detroit this year may come out of Wisconsin ... where American Motors makes them."

GREMLIN SERIES 40

Model Number	Body/Style Number	Body Type & Seating	Factory Price	Shipping Weight	Production Total
40	7246-5	2-dr Sed-4P	1999/2153	2494/2746	94,808

NOTE 1: Data above slash for six/below slash for V-8.
NOTE 2: Exactly 10,949 Gremlins V-8s were built in the model year.
NOTE 3: Calendar year production was 69,773 Gremlin sixes and V-8s.

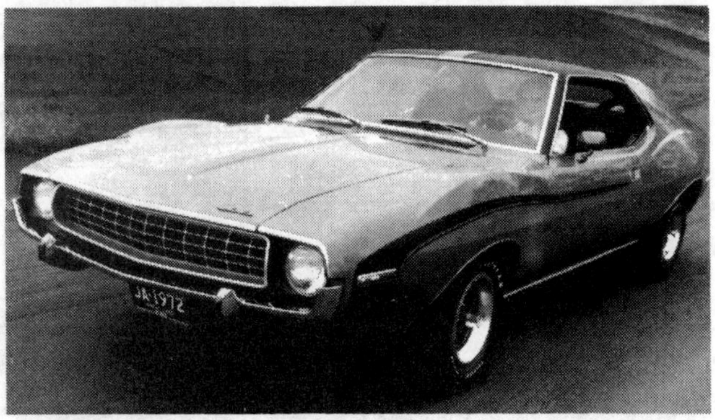

1972 AMC, Javelin SST two-door hardtop sports coupe, V-8

JAVELIN SST — (SXI/V-8) — SERIES 70: Small styling changes were the order of the year on Javelin SST. A new taillight treatment and grille were seen. The grille was a gridwork formed by three long, horizontal bright moldings and 11 shorter vertical bars. At the rear, there was full-width cross-hatch type decorative patterning in two rows on both Javelin and Javelin/AMX. The Javelin AMX, however, used a different radiator grille design, which matched the center-bulge horizontal blade pattern of 1971. For the, 1972 Javelin SST French fashion designer Pierre Cardin created the Cardin-Javelin option. It featured an interior of multi-colored pleated stripes in Chinese Red, Plum, White and Silver on a black background. Five exterior color choices were: Snow White; Stardust Silver; Diamond Blue; Trans Am Red; and Wild Plum. Experts say that a few Cardin-Javelins left the factory with midnight black finish. The crest of the House of Cardin was applied to door panels and front fenders. Production amounted to 4,152 units for the designer special. Performance, too, was part of 1972 Javelin history, as George Follomer won a second SCCA Trans-Am title for AMC this year. The standard equipment list for the Javelin SST was the same as for Hornets, with several deletions or additions. For example, no rear armrest was used. There were, however, such items as a glove box lock; dual horns; high-back bucket seats with front foam cushions; three-speed manual (full-synchromesh) transmission with floor control; Custom steering wheel; and C78-14 blackwall tires on sixes (D78-14s on V-8s). The Javelin AMX also had a Sports steering wheel; deck mounted spoiler; slot style wheels with E70-14 blackwall tires; and standard 304 cid V-8.

JAVELIN SERIES 70

Model Number	Body/Style Number	Body Type & Seating	Factory Price	Shipping Weight	Production Total
JAVELIN SST					
70	7279-7	2-dr FsBk Cpe-4P	3807/2901	2875/3118	22,964*
JAVELIN AMX (V-8)					
70	7279-8	2-dr FsBk Cpe-4P	3109	3149	3,220

NOTE 1: Prices and weights above slash are for six/below slash for V-8.
NOTE 2: Totals include 100 Javelin '401' Alabama State Police Interceptors.

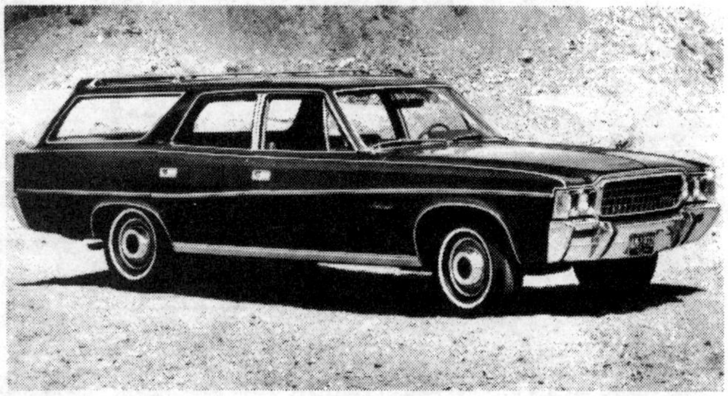

1972 AMC, Ambassador Brougham, four-door station wagon, V-8

AMBASSADOR — (V-8) — SERIES 80: The 1972 Ambassador received just a few changes, the most noticeable one being a new radiator grille. It incorporated three heavy horizontal bright blades which, together with thinner vertical blades, formed a grid of large, square openings. There was an Ambassador chrome signature script at the left-hand side. The rear view also reflected two changes: new taillights and an attractive center trim panel. Six styles were provided in two levels of trim: SST or Brougham. A six was not available. Standard equipment began with all items found in or on Matadors, plus a big 55 ampere alternator (to provide a strong electrical system with standard air conditioning); rear armrests; front and rear bumper guards on hardtops and sedans; inside hood release; power brakes; Torque Command gearbox; and the '304' V-8. The Brougham also had wheelcovers. All models, except station wagons, used E78 x 14 blackwall tires. Wagons came without the rear bumper guards, but had carpeted cargo spaces; roof-top luggage racks; and H78 x 14 tires. Brougham wagons had a tailgate air deflector and wood-grained exterior paneling. With prices beginning under $4,000, the Ambassador Eight was a good bargain.

AMBASSADOR EIGHT SERIES 80

Model Number	Body/Style Number	Body Type & Seating	Factory Price	Shipping Weight	Production Total
AMBASSADOR SST					
80	7285-5	4-dr Sed-6P	3885	3537	11,929
80	7289-5	2-dr HT Cpe-6P	3902	3579	986
80	7288-5	4-dr Sta Wag-6P	4270	3833	5256
AMBASSADOR BROUGHAM					
80	7285-7	4-dr Sed-6P	4002	3551	16,432
80	7289-7	2-dr HT Cpe-6P	4018	3581	4137
80	7288-7	4-dr Sta Wag-6P	4437	3857	5624

BASE ENGINES

NOTE: SAE net horsepower (nhp) ratings measuring output at the rear of the transmission with all accessories installed and operating are now used.

(Base six) Inline six. Overhead valves. Cast iron block. Displacement: 232 cid. Bore and stroke: 3.75 x 3.50 inches. Compression ratio: 8.0:1. SAE nhp: 100 at 3600 rpm. Seven main bearings. Hydraulic valve lifters. Carburetor: Carter Type YF one-barrel.

(Base station wagon six) Inline six. Overhead valves. Cast iron block. Displacement: 258 cid. Bore and stroke: 3.75 x 3.90 inches. Compression ratio: 8.0:1. SAE Net hp: 110 at 3500 rpm. Seven main bearings. Hydraulic valve lifters. Carburetor: Carter Type YF one-barrel.

(Base V-8) V-8. Overhead valves. Cast iron block. Displacement: 304 cid. Bore and stroke: 3.75 x 3.44 inches. Compression ratio: 8.4:1. SAE Net hp: 150 at 4400 rpm. Five main bearings. Hydraulic valve lifters. Carburetor: Autolite two-barrel Model 2100.

CHASSIS FEATURES: Wheelbase: (Gremlin) 96 inches; (Hornet) 108 inches; (Javelin) 110 inches; (Matador) 118 inches; (Ambassador) 122 inches. Overall length: (Gremlin) 161.3 inches; (Hornet) 179.3 inches; (Javelin) 191.8 inches; (Matador) 206 inches; (Ambassador) 210.8 inches. Front tread: (Gremlin and Hornet) 57.5 inches; (Javelin) 59.3 inches; (Matador and Ambassador) 59.9 inches. Rear tread: (Gremlin and Hornet) 57 inches; (Javelin, Matador and Ambassador) 60 inches.

POWERTRAIN OPTIONS: Torque-Command transmission was standard in Ambassador Series models with the base V-8 and was optional in other models at 15 different prices determined by series and engine attachment. The prices ranged from $200 for Gremlin/Hornet sixes to $257 for Matadors with the '401' V-8. In addition, Torque-Command was a $23 option for Ambassadors with the '360' V-8 and $35 extra in those with the big '401' V-8. Three-speed manual transmission with floor shift was available as an option in Gremlin/Hornet models ($32). Torque-Command with floor shift was optional in Javelins, with the '304' V-8 ($282); with the '360' V-8 ($293) and with the '401' V-8 ($305). Four-speed manual was optional in Javelins only in combination with the four-barrel '360' V-8 or '401' V-8 ($188). The big 258 cid six was optional, in Gremlin/Hornet Group ($51); in Javelin SST ($43); in Matador passenger cars ($46). The 360 cid two-barrel V-8 was optional in Hornet/Javelin Group ($42). The 360 cid four-barrel V-8 was optional in Javelin ($85); in Matador/Ambassador ($89). The 401 cid four-barrel V-8 was optional in Javelin Group ($162) and in Matador/Ambassador Group models only when F78 x 14 tires were also used ($170). Optional axle ratios, all models ($12-$14). Heavy-duty 70 ampere battery ($14-$15). Twin-Grip differential ($43-$46). Dual exhaust with '360' V-8 and four-barrel, standard with '401' V-8 ($28-$31). Heavy-duty cooling system ($16). A cowl-air induction system was included in 'Go-Package' option groups.

CONENIENCE OPTIONS: All-Season air conditioning, all models except Ambassador ($377). Gremlin/Hornet air conditioning package ($473). Front and rear bumper guards, Gremlin/Hornet ($21); Javelin ($29); Matador/Ambassador ($31). Center armrest cushion, Javelin without console ($54). Rear window defogger, Javelin ($45); Matador/Ambassador ($48). Front manual disc brakes, Gremlin/Hornet with 14 inch wheels ($47); Javelin ($47); Matador/Ambassador ($50). Engine block heater, Matador/Ambassador ($16); all others ($14). Tinted glass, all windows; Gremlin ($37); Hornet/Javelin ($40); Matador ($42); Ambassador ($49). Tinted windshield only, Gremlin/Hornet/Javelin ($30); Matador ($35). Tinted glass standard with air conditioned Ambassador. Headlights-off delay ($21). Gremlin, locking gas cap ($6). Station wagon rooftop luggage rack, Matador ($56). Sportabout, including air deflector ($61). Rear deck luggage rack, Hornet sedan ($32); Javelin SST ($32). Rally side stripes, Gremlin/Javelin SST ($33-$39). Hood 'T' Stripe AMX ($39). Color-keyed wood-grain panels, Sportabout ($95), Matador wagon ($113). Power brakes, Gremlin/Hornet/Javelin ($44) and Matador ($47). Power front disc brakes, Hornet with Rally package ($32); Hornet/Gremlin ($79); Javelin ($77); Matador ($81) and Ambassador ($50). Power steering, Gremlin/Hornet, with 14 inch wheels only. ($99); Javelin ($106) and Matador/Ambassador ($111). Power side window lifts, Matador/Ambassador only ($123). Power tailgate window, in Matador/Ambassador wagons without third seat ($35). All power windows package Ambassador wagon only ($158). AM push-button radio, Gremlin/Hornet/Javelin ($66). AM/FM Multiplex stereo with two rear speakers, Javelin ($196); Matador/Ambassador ($230). Stereo tape player in Javelin with manual radio ($190). Gremlin bench seat with Custom trim ($79). Gremlin bucket seats with Custom trim ($117). Hornet Group 'Scorpio' fabric trim with reclining seats ($109). 'Harem' fabric trim in Ambassador Brougham ($69); in Ambassador Brougham wagon, ($96). Matador/Ambassador wagon third seat, includes two safety belts, cargo mat and power tailgate window ($108). Adjustable rear air shocks, Matador/Ambassador ($40). Functional lower front spoiler, Javelin/AMX with disc brakes ($31). Quick-ratio manual steering, 20:1 ratio with Gremlin/Hornet ($11); Javelin Group, 16:1 ratio ($15). Adjust-O-Tilt steering wheel, Gremlin/Hornet with Torque-Command and Javelin ($43); Matador/Ambassador ($46). Three-spoke Sports steering wheel, Gremlin ($33); Hornet/Javelin, standard AMX, ($19); Matador/Ambassador ($20). Sun roof, Hornet two-door sedan without vinyl top, Sportabout and Gremlin ($142). Black; White; Blue; Green or Brown vinyl roof, on Hornet/Javelin ($88); on Matador passenger models ($91) and Ambassador passenger ($109). Full wheelcovers ($72-$29). Custom wheelcovers ($50-$53). Turbo-Disc wheelcovers, Hornet D/L ($25); most other models ($75-$78) and Ambassador Brougham ($50). Spoke style wheels (including Space-Saver spare on all except Matador/Ambassador), for cars with options packages that include special spoke wheel prices ($34-$50); on others ($99-$104).

OPTION PACKAGES: Gucci Sportabout package featured beige seats and door panels trimmed with red and green stripes. The Gucci crest appeared on the inside door panel, front fenders and headliners. Available in special exterior colors. Javelin Cardin bucket seats and Cardin trim, for Javelin SST only ($85). Individual reclining seat with Hornet Sportabout Gucci trim ($142). Code 391 Javelin AMX '360 Go-Package,' includes: specified engine; dual exhausts; hood 'T' stripe decal; black-out rear panel; Rally Pack instrumentation; Handling Package; Cowl-Air induction; heavy-duty cooling; Twin-Grip; power disc brakes; E60-15 polyglass raised white-letter tires; 15 x 7 inch styled steel wheels; and Space-Saver spare with regular 14 inch wheels ($428). Code 392 Javelin AMX '401 Go Package,' includes all above with 401 cid V-8 ($505). Code 633 Hornet sedan Rally Package, includes vinyl bucket seats; manual front disc brakes; Handling Package; quick 20:1 ratio manual steering; three-speed floor-shift transmission; Sports steering wheel; and 'Rally' emblems ($119). Gremlin 'X' package, includes full-length spear decal; painted grille; 14 x 6 inch slotted wheels; D70 x 14 tires; Space-Saver spare; Custom bucket seat interior; cargo region insulation; 15 inch Sports steering wheel; and interior appointments package with special decals ($285.10 w/regular tires; $319.55/with RWL tires).

HISTORICAL FOOTNOTES: The 1972 AMC models were introduced Sept. 22, 1971. Calendar year production peaked at 279,132 units. AMC set a record dollar sales total of $1.4 billion and pulled-down a $30.2 million profit. The AMC Buyer Protection Plan was an excellent 1972 sales motivation tool.

1973 AMC

1973 AMC, Hornet two-door hatchback coupe, 6-cyl

HORNET — (SIX/V-8): American Motor's popular compact, was offered in 1973, with a new hatchback model. This brought to four the number of Hornet body styles. The third 'door' combined the functions of window and trunk lid. It was top hinged and fully counter-balanced for easy opening and closing. The Hornet had new front fenders, a recessed hood with raised center crease line, and a new full-length grille. *CAR AND DRIVER* magazine called the 1973 Hornet Hatchback "the styling coupe of '73" and this was used by AMC in one of its advertisements. An interesting Levi's Jean-style interior was added as a new option on June 20, 1973. The Hornet offered plenty of interior room, good economy, and a high degree of mechanical reliability. Standard equipment and trims continued in the pattern of earlier years.

VIN: Serial Numbers were located on top of the right front wheelhouse panel. The unit body data plate was riveted to the left front door, below the latch mechanism. American Motors retained a 13 symbol identification code system. The first symbol was an A for American Motors Corp. The second symbol indicated model year, 3=1973. The third symbol indicated transmission type: S=standard column shift; A=automatic, column shift; C=three-speed automatic with floor shift; E=fully-synchronized three-speed manual with floor shift. The fourth symbol was a number designating car line, as follows: 0=Hornet; 1=Rebel; 3=AMX; 4=Gremlin; 7=Javelin and 8=Ambassador. The fifth symbol designated body type: 3=three-door hatchback; 5=four-door sedan; 6=two-door sedan; 8=four-door station wagon; and 9=two-door hardtop. The sixth symbol designated the series or class of body, as follows: 5=Ambassador SST/Gremlin; 7=Hornet SST/Javelin SST/Matador/Ambassador Brougham; 8=Javelin AMX. The seventh symbol was a letter designating engine type (without regard to car-line), as follows: A=258 cid cid six; B=low-compression 258 cid six; E=232 cid six one-barrel carburetor; F=low-compression 232 cid six with one-barrel carburetor; H=304 cid V-8 with two-barrel carburetor; I=304 cid V-8 two-barrel (low-compression); 'M' = low-compression 304 cid V-8 with four-barrel carburetor; 'N' = 360 cid V-8 with two-barrel carburetor; 'P' = 360 cid V-8 with four-barrel carburetor; Z=401 cid V-8 four-barrel. The following group of symbols was the sequential production and began with 100001 for cars built in Kenosha, Wis. and 700001 for cars built in Brampton, Ontario, Canada. Body number plate attached to left front door edge shows body serial number, Body/Style Number from second column of chart below, plus trim and paint codes.

HORNET SERIES 01

Model Number	Body/Style Number	Body Type & Seating	Factory Price	Shipping Weight	Production Total
01	7306-7	2-dr Sed-6P	2298/2436	2777/2990	23,187
01	7305-7	4-dr Sed-6P	2343/2481	2854/3067	25,452
01	7303-7	2-dr Hatch-5P	2449/2587	2818/3031	40,110
01	7308-7	4-dr Sta Wag-6P	2675/2813	2921/3134	44,719

NOTE 1: Data above slash for six/below slash for V-8.

MATADOR — (SIX/V-8) — SERIES 10: The Matador received a new grille. It consisted of four groups of slim rectangles stacked triple high. In the center of each rectangle was a short horizontal molding. There were new interior colors and fabrics, too. It was marketed with carryovers sixes and V-8s. The Matador was again a hit with the LAPD, which added more Matadors to the fleet.

1973 AMC, Matador four-door station wagon, 6-cyl

MATADOR SERIES 10

Model Number	Body/Style Number	Body Type & Seating	Factory Price	Shipping Weight	Production Total
10	7315-7	4-dr Sed-6P	2814/2853	3289/3502	33,822
10	7318-7	4-dr Sta Wag-6P	3197/3278	3627/3815	11,643
10	7319-7	2-dr H T Cpe-6P	2007/2086	3314/3527	7,067

NOTE 1: Data above slash for six/below slash for V-8.

1973 AMC, Gremlin 'X' two-door sedan, V-8

GREMLIN — (SIX/V-8): The basic Gremlin remained the same as last year's model. About the only significant changes were new safety bumpers and 6.45 x 17 standard sized tires. Equipped with the base 232 cid six, the Gremlin had lots of 'guts' and averaged about 20 mpg of gas...excellent for 1973. The Gremlin was a lot heavier than other early subcompacts and better-suited to sustained high-speed touring. It was, however, plagued by typical ills, like carburetion; cooling; squeaking; and rattles. It was also quite prone to rust. New for 1973 were several optional equipment changes. A new body side trim design had the stripe 'hopping up' behind the rear wheel opening and continuing to the rear of the fender at the higher level. It was used on the Gremlin 'X,' which also had special decals; painted grille; Custom interior appointments; bucket seats; slotted styled wheels; Space-Saver spare tire; and added cargo area sound insulation. A Levi's Gremlin package was also available, reproducing an authentic 'blue jeans' look in a spun nylon version of denim for seats, door inserts and map storage pockets. There was also orange stitching and copper rivets, just like on real jeans. Standard equipment features for 1973 were essentially unchanged.

GREMLIN SERIES 40

Model Number	Body/Style Number	Body Type & Seating	Factory Price	Shipping Weight	Production Total
40	7346-5	2-dr Sed-4P	2098/2252	2642/2867	122,844

NOTE 1: 11,672 V-8 Gremlins made in calendar 1973.
NOTE 2: Prices and weights above slash are for six/below slash for V-8.

GREMLIN SERIES 40 ENGINES

See 1972 Gremlin Series engines.

1973 AMC, Javelin two-door fastback coupe, V-8

JAVELIN — (SIX/V-8) — SERIES 70: The Javelin and Javelin AMX sports hardtops featured a new, smooth roofline. A new taillight treatment, with twin-pod lamps at each side of the car, was adopted. The Javelin AMX was unchanged otherwise, but the Javelin base models had a recessed plastic grille that was distinctive from the past. It was flush with the front of the car and incorporated rectangular Rally lights. Bucket seats were standard in both Javelin and Javelin AMX lines, along with interior packaging of the aircraft cockpit type. Both cars featured spoiler lips over the rear window, with a rear spoiler optional on AMX. Standard equipment was similar to that offered in 1972. A special model called the 'Trans Am Victory Javelin' was offered in 1973. It had a decal on the rear of the front fenders stating that the Javelin had won the SCCA Trans-AM championship for 1971 and 1972. Besides the decal, 14 inch slot-style wheels with E70-14 RWL tires and space saver spare were included at no extra charge. An advertisement featured George Follmer and Roy Woods, who had won the championship for AMC in 1972.

JAVELIN SERIES 70

Model Number	Body/Style Number	Body Type & Seating	Factory Price	Shipping Weight	Production Total
JAVELIN					
70	7379-7	2-dr FsBk Cpe-4P	2889/2983	2868/3104	25,195
JAVELIN AMX					
70	7379-8	2-dr FsBk Cpe-4P	3191	3170	5,707

1973 AMC, Ambassador Brougham two-door hardtop coupe, V-8

NOTE 1: Calendar year production of Javelins totaled 31,267 units.
NOTE 2: Data above slash for six/below slash for V-8.

AMBASSADOR — (V-8) — SERIES 80: Luxury was a key word in describing the 1973 Ambassador line of hardtop, sedan and station wagon. The SST Series was discontinued and Broughams were exclusively retained. A 304 cid V-8; automatic transmission; power steering; power front disc brakes; white sidewall tires; tinted glass; and AM radio were all standard Ambassador Brougham features. Guard-rail steel door beams, for side impact protection, were part of the construction after Jan. 1, 1973. In addition, a quieter type of seat belt warning buzzer was used. Styling features included a slightly redesigned grille with heavier vertical and horizontal moldings. The Ambassador signature was gone from the left side. On the front bumper, a black, rubber impact strip appeared. There was a hood insulation pad; left outside rearview remote-control mirror; visor vanity mirror; electric clock; and electric, variable-speed windshield wipers on every 1973 Ambassador sold. All cars were also undercoated, for protection against rust. Three optional V-8s were offered.

AMBASSADOR SERIES 80 V-8

Model Number	Body/Style Number	Body Type & Seating	Factory Price	Shipping Weight	Production Total
80	7385-7	4-dr Sed-6P	4461	3763	31,490
80	7389-7	2-dr HT Cpe-6P	4477	3774	5,534
80	7388-7	4-dr Sta Wag-6P	4861	4054	12,270

NOTE 1: Data above slash for six/below slash for V-8.

BASE ENGINES

NOTE: SAE net horsepower (nhp) ratings measuring output at the rear of the transmission with all accessories installed and operating are now used.

(Base six) Inline six. Overhead valves. Cast iron block. Displacement: 232 cid. Bore and stroke: 3.75 x 3.50 inches. Compression ratio: 8.0:1. SAE nhp: 100 at 3600 rpm. Seven main bearings. Hydraulic valve lifters. Carburetor: Carter Type YF one-barrel.

(Base station wagon six) Inline six. Overhead valves. Cast iron block. Displacement: 258 cid. Bore and stroke: 3.75 x 3.90 inches. Compression ratio: 8.0:1. SAE Net hp: 110 at 3500 rpm. Seven main bearings. Hydraulic valve lifters. Carburetor: Carter Type YF one-barrel.

(Base V-8) V-8. Overhead valves. Cast iron block. Displacement: 304 cid. Bore and stroke: 3.75 x 3.44 inches. Compression ratio: 8.4:1. SAE Net hp: 150 at 4400 rpm. Five main bearings. Hydraulic valve lifters. Carburetor: Autolite two-barrel Model 2100.

CHASSIS FEATURES: Wheelbase: (Gremlin) 96 inches; (Hornet) 108 inches; (Javelin) 110 inches; (Matador) 118 inches; (Ambassador) 122 inches. Overall length: (Gremlin) 165.5 inches; (Javelin) 192.3 inches; (Ambassador) 212.9 inches; (Matador) 208.5 inches; (Hornet) 184.9 inches. Front tread: (Gremlin) 57.5 inches; (Hornet) 56.4 inches; (Javelin) 59.3 inches. Rear tread: (Gremlin/Hornet) 57 inches; (Javelin/Ambassador/Matador) 60 inches. Tires: (Hornet) 6.95 x 14; (Matador) E78 x 14; (Ambassador) F78 x 14; (Javelin) D78 x 14 and (Gremlin) 6.45 x 14.

POWERTRAIN OPTIONS: Torque-Command automatic transmission with column control was standard in Ambassador Broughams. Three-speed manual transmission with full-synchromesh first gear and floor shift was standard in Gremlins — optional in all other models, except Ambassadors. Other available transmissions for AMC cars included Torque-Command with column control; Torque-Command with floor shift; three-speed manual with column control (standard in all except Gremlin/Javelin/Ambassador) and, in selected applications, four-speed manual with floor-mounted Hurst heavy-duty shifter. Engine choices for Gremlins were the 100 hp 232 cid six; 110 hp 258 cid six and 150 hp 304 cid V-8. These engines, plus a 175 hp 360 cid two-barrel V-8, were available in Hornets. Specific Matador models could also be ordered with a 195 hp 360 cid four-barrel V-8 or a 255 hp 401 cid four-barrel V-8. These same choices were also available for Javelin and Javelin AMX. Available in the Ambassador Brougham were the 150, 175, 195 and 255 hp engines (all V-8) with similar specifications. Optional rear axle ratios: 2.73:1; 2.87:1 and 3.54:1.

CONVENIENCE OPTIONS: Gremlin 'X' package ($285). Sportabout D/L package ($284). 'Gucci' vinyl interior, in Hornet ($142). Javelin AMX '360' Go-Package ($428). Javelin AMX '401' Go-Package ($476). Power brakes, standard in Ambassador ($44). Power disc brakes, standard in Ambassador ($79). Manual disc brakes ($47). Sunroof ($142). Station wagon third seat, includes two safety belts and power tailgate window ($108). Reclining seats, average price ($80). Bucket seats, in selected models ($131). Power steering, Gremlin ($99). Factory air conditioning, Gremlin ($377). Vinyl roof, on Hornet ($88). Factory air conditioning, Javelin/Javelin AMX ($377). AM/FM stereo, Javelin/Javelin AMX ($196). Vinyl covered top, Javelin/AMX ($88). AM/FM stereo, in Matador ($230). Vinyl covered top, on Matador, except station wagon ($91). Power windows ($123). AM/FM stereo, in Ambassador ($61). Vinyl covered top, on Ambassador ($109).

HISTORICAL FOOTNOTES: Model year output for 1973 American Motors' models was registered at exactly 320,786 cars. The company held only a 3.3 percent share of the total car business. R.D. Chapin, Jr. continued as the firm's chief executive officer this year. Introduced on the American Motor's Jeep Wagoneer line this year was the innovative 'Quadra-Trac' full-time four-wheel-drive system.

1974 AMC

1974 AMC, Hornet two-door sedan, 6-cyl

HORNET — (SIX/V-8) — SERIES 01: The styling seen on 1974 AMC Hornets was slightly revised. A new energy-absorbing front bumper looked much the same as before, but the full-width, vinyl impact strip was replaced by rubber-faced bumper guards spaced widely apart, just inboard of the grille-mounted Rally lights. The grille itself still consisted of many fine, vertical louvers, but the horizontal center bar that integrated the lamps was now finished in black. A new side trim treatment featured a thin, straight upper beltline molding that ran from the taillamps to the front fender tip and, then, down around the side marker light, with a shape paralleling the front fender edge contour. A second full-length molding ran from below the taillamp (and above the rear bumper end) to a point under the front side marker light. This molding was also straight, except in those places where it curved over the front and rear wheel openings. The Hornet nameplate was removed from the lip of the hood. Model nameplates seen alongside of the car were moved from their 1973 position (on the upper cowl sides) to a point just behind the front side marker lights. Standard equipment for the basic Hornet began with all items found on Gremlins, plus rear ashtray; color-keyed carpets; 16 gallon fuel tank; full-width package tray; Custom steering wheel; foam front bench seat; full-flow oil filter; three-speed manual column shifted transmission; 6.95 x 14 blackwall tires and, on all models except the hatchback, rear armrests. The Sportabout wagon and hatch back models also had cargo area carpeting; fold-down rear seats and a rear liftgate. Standard equipment in Hornet V-8s was a front sway bar. On cars with the optional 360 cid two-barrel V-8, a 60 ampere battery was used in place of the regular 50 ampere type.

VIN: Serial Numbers were located on top of the right front wheelhouse panel. The unit body data plate was riveted to the left front door, below the latch mechanism. American Motors retained a 13 symbol identification code system. The first symbol was an A for American Motors Corp. The second symbol indicated model year, 4=1974. The third symbol indicated transmission type: S=standard column shift; A=automatic, column shift; C=three-speed automatic with floor shift; E=fully-synchronized three-speed manual with floor shift. The fourth symbol was a number designating car line, as follows: 0=Hornet; 1=Rebel; 3=AMX; 4=Gremlin; 7=Javelin and 8=Ambassador. The fifth symbol designated body type: 3=three-door hatchback; 5=four-door sedan; 6=two-door sedan; 8=four-door station wagon; and 9=two-door hardtop. The sixth symbol designated the series or class of body, as follows: 5=Gremlin; 7=Hornet/Javelin/Matador or Ambassador Brougham; 8=Javelin AMX/Matador X/9=Matador Brougham; P=Police; T=Taxicab. The seventh symbol was a letter designating engine type (without regard to car-line), as follows: 'A' = 258 cid one-barrel six; 'E' = 232 cid one-barrel six; 'H' = 304 cid two-barrel V-8; 'N' = 360 cid two-barrel V-8; 'P' = 360 cid four-barrel V-8 and 'Z' = 401 cid four-barrel V-8. The eighth symbols indicated assembly plant, as follows: Cars coded 1-6 were built in Kenosha, Wis. and those coded 7-9 were built in Brampton, Ontario, Canada. The remaining symbols were the sequential production number and began with 100001 for cars built in Kenosha, Wis. and 700001 for cars built in Brampton, Ontario, Canada. Body number plate attached to left front door edge shows body serial number, Body/Style Number from second column of chart below, plus trim and paint codes.

HORNET SERIES 01

Series Number	Body/Style Number	Body Type & Seating	Factory Price	Shipping Weight	Production Total
01	7403-7	2-dr Hatch-5P	2849/2987	2815/3042	55,158
01	7406-7	2-dr Sed-6P	2774/2912	2774/3001	29,950
01	7405-7	4-dr Sed-6P	2824/2962	2841/3068	29,754
01	7408-7	4-dr Sta Wag-6P	3049/2987	2908/3135	71,413

NOTE 1: Data above slash for six/below slash for V-8.

1974 AMC, Matador 'X' two-door sedan, V-8

MATADOR — (SIX/V-8) — SERIES 10: AMC drastically restyled the Matador two-door coupe. Sedans and wagons had modest changes with new grilles and front/rear bumpers. A big difference in appearance was the dropping of the integrated bumper/grille for a centrally divided unit with vertical louvers, square headlamp surrounds with Argent Silver finish; round parking lamps mounted in grille (inboard of headlights) and Matador lettering on the left-hand hood lip. The front bumper was a shelf-like affair with center license plate indent flanked by rubber-faced bumper guards. Side trim on the sedans and wagons consisted of a thin, straight, three-quarter length molding. It ran from behind the front wheel opening to above the rear side marker light. There were model nameplates on the front fender, behind the wheel cutout. Standard equipment was the same as on base Javelins, plus front sway bar; full insulation package including undercoating; 19.5 gallon fuel tank; side molding vinyl inserts; Custom steering wheel; full-back bench seats; and base six or V-8 engine. The 232 cid one-barrel six or 304 cid two-barrel V-8 were standard in sedans along with E78 x 14 blackwall tires. The base station wagon also included a rubber cargo area mat; lockable hidden storage compartment; Dual-Swing tailgate; H78 x 14 tires and bigger 258 cid one-barrel base six. Vinyl inserts were not used on side moldings on wood-grained wagons. The two-door Matador coupe was unique with its long, low, fast-looking silhouette. AMC enthusiasts compare it the Jensen Interceptor. Major styling features were a long fast-sloping hood and a short rear deck. It was conceived with stock car racing in mind. After Mark Donohue captured the SCCA's Trans-Am Series championship, in 1971, AMC created a factory racing team with Donohue as driver and Roger Penske as team manager. By the time the car hit the production stage the Energy Crunch had negated the effect of performance on sales. But, 6,165 coupes with a fancy "Cassini" package were produced in 1974 and 1,817 more the following season. Matador coupes had a shorter wheelbase, than sedans and wagons. They also had some equipment differences over these models, including split-back front seats and front door light switches. There was also a special Matador 'X.' It was a full-fledged sub-model (not an option like the Gremlin 'X'), with extras including a three-spoke Sports steering wheel; wide body side stripes; hood stripes; slot styled wheels; blacked-out grille; Matador 'X' cowl nameplates; automatic transmission; and two-barrel '304' V-8. Finally, for the low-buck luxury buyer, there was the Matador Brougham coupe with all base equipment, plus black vinyl bumper nerfing strips and full wheelcovers.

MATADOR SERIES 10

Series Number	Body/Style Number	Body Type & Seating	Factory Price	Shipping Weight	Production Total
MATADOR					
10	7415-7	4-dr Sed-6P	3052/3151	3444/3659	27,608
10	7416-7	2-dr Cpe-6P	3096/3195	3459/3674	31,169
10	7416-9	2-dr Brgm Cpe-6P	3249/3348	3486/3701	21,026
10	7418-7	4-dr Sta Wag-6P	3378/3477	3769/3957	9709
MATADOR 'X' (V-8)					
10	7416-8	2-dr Cpe-6P	3699	3674	10,074

NOTE 1: Data above the slash for six/below slash for V-8.

GREMLIN — (SIX/V-8) — SERIES 40: There was a new grille for the 1974 Gremlin, but it didn't look totally fresh and new. A multitude of thin, horizontal blades filled a slightly taller opening. The molding around the entire insert was different. The side pieces had a bend instead of being straight. To completment the grille pattern, there were horizontal grooves on the headlight door/fender extension panels. They began just outside the grille surround and swept around the front body corners, with the upper grooves being interrupted by the headlamp lenses. A new bumper, of the energy-absorbing type, was used. It had a shelf-like appearance and black, rubber-faced guards. A Gremlin script was again placed on the left front face of the scoop-like hood bulge. Several new trim variations could be seen on the side of the Gremlin. The thin moldings, formerly used around the windsplits on the rear sail panels, were removed. In addition, the body side stripes were entirely redone. The overall effect was somewhat like that of a hockey stick with a pointed handle lying on its bottom edge. Changes in the rear included a thinner bumper; an AMC letter badge on the left side of the indentation panel; new, rubber-faced guards; and chrome bullet-shaped lamps surrounding the license plate (which was in a new location at the center). Standard equipment on the basic Gremlin included all regulation safety features; front armrests and ashtrays; 50 ampere battery; dome light; rubber floor and trunk mats; glove box door; heater and defroster: foam-cushioned splitback front seat; rear seat with fold-down back; 21 gallon fuel tank; opening type rear liftgate; three-speed manual transmission with floor-shift; hubcaps; 35 ampere alternator; exterior paint stripes; and either the base '232' six or '304' V-8. Standard tires were 6.45 x 14 blackwalls and V-8s also had a front suspension sway bar.

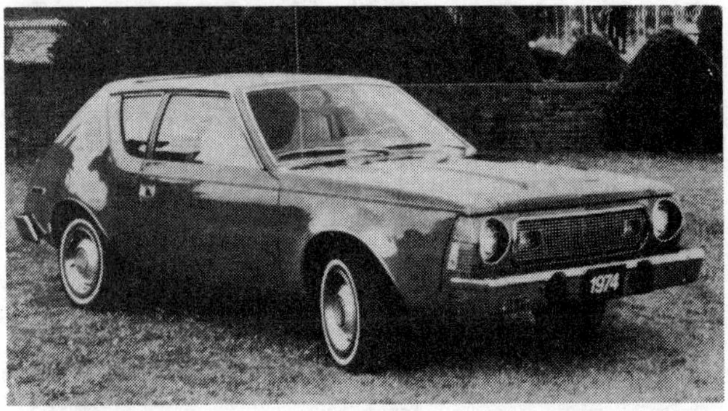

1974 AMC, Gremlin two-door sedan, 6-cyl

GREMLIN SERIES 40

Series Number	Body/Style Number	Body Type & Seating	Factory Price	Shipping Weight	Production Total
40	7446-5	2-dr Sed-4P	2481/2635	2855/3094	171,128

NOTE 1: Data above slash for six/below slash for V-8.
NOTE 2: A total of 14,137 Gremlin V-8 were produced in model year 1974.

JAVELIN — (SIX/V-8) — SERIES 70: Very few cars see major changes in their last season. The Javelin, which was about to bite the dust, followed this long tradition for 1974. About the best way to tell 1973 and 1974 base models apart is to drive them. In the '74, you'll immediately notice the new three-point lap/shoulder harness with ignition interlock. In addition, the later cars may go faster due to emissions control advances. There were Federally mandated bumper design changes, to insure the cars could meet five mph impacts. This was done with the addition of shock-absorber mountings and black rubber bumper guards. The molding around the grille insert had an inverted trapezoid shape. On the AMX, red, white and blue letters were placed in the center of the grille. Circular Rally-style parking lamps were set into larger circles creating a 'bombsight' appearance. Standard equipment included all found on base Hornets (except rear armrest), plus dual horns; foam-cushioned high-back front bucket seats; front and rear bumper guards; manual front disc brakes; rubber trunk mat; D78 x 14 blackwall tires; three-speed manual floor-shift transmission; a 232 cid six or 304 cid V-8 engine; and a front sway bar. Extra standard equipment for the AMX included a Sports steering wheel; deck mounted rear spoiler; and slotted styled wheels. Javelin AMX 'Go-Packages' were supplied, with the price depending on what kind of tires the customer ordered and which engine was installed.

JAVELIN SERIES 70

Series Number	Body/Style Number	Body Type & Seating	Factory Price	Shipping Weight	Production Total
JAVELIN					
70	7479-7	2-dr FsBk Cpe-4P	2999/3093	2875/3117	22,556
JAVELIN AMX (V-8)					
70	7479-8	2-dr FsBk Cpe-4P	3299	3184	4980

NOTE 1: Data above slash for six/below slash for V-8.
NOTE 2: Javelins and AMXs were never assembled in Canada, though some were assembled in foreign plants from U.S. made parts.

AMBASSADOR BROUGHAM — (V-8) — SERIES 80: The 1974 Ambassador received a competely new frontal treatment, which was more squared-off and designed to meet a new Federal five mile per hour barrier crash test. The grille surround was completely straight (though not flat) on top and bottom and outlined the entire grille including the dual headlights. These units had round lenses mounted in square bezels. A fine-grid pattern insert was divided, horizontally, by two thicker bright moldings forming three levels of background gridwork. There was a stand-up hood ornament and Ambassador lettering on the left-hand hood lip. The new bumper, which no longer housed the park/turn lamps, was slightly thicker at the center and had an overall shelf-like look. There was a license plate indentation at the middle, flanked by a chrome and rubber guard on each side. Vinyl nerfing strips appeared at each end, wrapping around the corners. Side trim consisted of a straight, three-quarter length strip of chrome running from behind the front wheel opening to the rear of the car; rocker panel moldings; Ambassador nameplates in back of the front wheels; and, on station wagons, redesigned wood-grained paneling positioned higher on the body sides. Two-door Ambassadors were dropped due to the Matador coupe's restyling. The two previously shared sheetmetal from the cowl back. It was felt the new coupe was not suitable for the Ambassador market. Remaining were sedans and wagons, available in only Brougham level trim. Standard equipment included all items found on base Hornets, plus a 62 ampere generator; front and rear bumper guards; inside hood release; air conditioning; full insulation package; light group; visibility group; undercoating and hood insulation; power front disc brakes; power steering; push-button AM radio; tinted glass in all windows; dual headlights; wheel opening moldings; bright metal rocker panel accents; and the '304' two-barrel V-8. Ambassador Brougham station wagons came with an in-the-floor lockable cargo compartment; Dual Swing-tailgate; durable color-keyed carpeting for cargo area; exterior wood-grained trim; rooftop travel rack and tailgate air deflector. It was the last season that the Ambassador would be offered.

AMBASSADOR BROUGHAM SERIES 80 (V-8)

Series Number	Body/Style Number	Body Type & Seating	Factory Price	Shipping Weight	Production Total
80	7485-7	4-dr Sed-6P	4559	3872	17,901
80	7488-7	4-dr Sta Wag-6P	4960	4115	7,070

NOTE 1: A total of 24,971 Ambassadors were built in calendar 1974.

BASE ENGINES

NOTE: SAE net horsepower (nhp) ratings measuring output at the rear of the transmission with all accessories installed and operating are now used.

(Base six) Inline six. Overhead valves. Cast iron block. Displacement: 232 cid. Bore and stroke: 3.75 x 3.50 inches. Compression ratio: 8.0:1. SAE nhp: 100 at 3600 rpm. Seven main bearings. Hydraulic valve lifters. Carburetor: Carter Type YF one-barrel.

(Base station wagon six) Inline six. Overhead valves. Cast iron block. Displacement: 258 cid. Bore and stroke: 3.75 x 3.90 inches. Compression ratio: 8.0:1. SAE Net hp: 110 at 3500 rpm. Seven main bearings. Hydraulic valve lifters. Carburetor: Carter Type YF one-barrel.

(Base V-8) V-8. Overhead valves. Cast iron block. Displacement: 304 cid. Bore and stroke: 3.75 x 3.44 inches. Compression ratio: 8.4:1. SAE Net hp: 150 at 4400 rpm. Five main bearings. Hydraulic valve lifters. Carburetor: Autolite two-barrel Model 2100.

CHASSIS FEATURES: Wheelbase: (Hornet) 108 inches; (Matador coupe) 114 inches; (other Matadors) 118 inches; (Gremlin) 96 inches; (Javelin) 110 inches; (Ambassador) 122 inches. Overall length: (Hornet) 187 inches; (Matador coupe) 209 inches; (other Matadors) 215.5 inches; (Matador 'X') 209.4 inches; (Gremlin) 170.3 inches; (Javelin) 195.3 inches; (Ambassador) 219.3 inches. Front tread: (Gremlin) 57.5 inches; (Hornet) 56.4 inches; (Javelin) 59.3 inches. Rear tread: (Gremlin/Hornet) 57 inches; (Javelin/Ambassador/Matador) 60 inches. Tires: (Hornet) 6.95 x 14; (Matador) E78 x 14; (Ambassador) F78 x 14; (Javelin) D78 x 14 and (Gremlin) 6.45 x 14.

POWERTRAIN OPTIONS: Torque-Command automatic transmission with column control was standard in Ambassador Broughams with the base '304' V-8. Torque-Command was also standard in the Matador 'X' with the base '304' V-8. Three-speed all-synchromesh manual transmission with floor control standard in Gremlin/Javelin groups. Three-speed all-synchromesh manual transmission with column control was standard in all other models. Torque-Command was optional in Ambassador Broughams and Matador 'X' with '360' or '401' V-8s ($13-$35). Torque-Command was optional in all other models, with prices and attachments governed by series and type of engine ($200-$257). Torque-Command with floor-shift control was optional in Hornet/Gremlin groups with prices and attachments governed by model and type of engine ($220-$251). Torque-Command with floor-shift control and center console was optional in Javelins ($280-$305); in Matadors ($291-$316) and in Matador 'X' ($59-$84) with price depending on choice of the 304, 360 or 401 V-8s; not available in sixes. Four-speed manual transmission with floor-shift control was optional in Javelins with the '232' one-barrel six or '304' two-barrel V-8 ($188). Four-speed manual transmission with Hurst floor-shifter was available in the Javelin AMX only as part of the 'Go-Package' option. A 51 ampere alternator was optional in Hornet/Gremlin/Matador groups ($13) and with air conditioning. A 62 ampere generator was optional in the Matador group ($48). An 80 ampere heavy-duty battery was optional in Ambassador/Matador groups ($21). A 10 inch heavy-duty clutch was optional in Gremlin/Hornet/Matador six ($12). A coolant recovery system was optional in Matador/Ambassador ($19). Heavy-duty cooling system, in Matador/Ambassador ($17). Dayco 'DS-7' fan belt ($5). Heavy-duty '360'-and '401' V-8, in Matador/Ambassador with heavy-duty Torque-Command only ($32). Manual low gear lock-out, with Matador V-8 and automatic transmission only (no charge). Dual exhausts were available as a separate option for the '360' V-8 and were standard with the '401' V-8. Engine option choices for the year included: 304 cid V-8 two-barrel with 150 nhp at 4200 rpm; 360 cid V-8 two-barrel with 175 nhp at 4000 rpm; 360 cid V-8 four-barrel with 195 nhp at 4400 rpm (single exhaust) and 220 nhp at 4400 rpm (dual exhausts); and 401 cid V-8 four-barrel with 235 nhp at 4600 rpm (single exhausts) and 315 nhp at 3100 rpm (dual exhausts). The 360 cid two-barrel engine was the minimum required engine size for cars sold in California. Prices for powertrain options now based on engine/transmission package (i.e. Matador X/360 V-8/Torque-Command with floor shift and console package was $71.35 above Matador base price, which included regular Torque-Command and a 304 V-8).

POPULAR CONVENIENCE OPTIONS: Air conditioning, except Ambassador ($400). Air conditioning package, in Gremlin/Hornet ($490). Matador wagon vinyl roof ($100). Gremlin 'X' hatchback package ($227). Gremlin 'X' hatchback package with Levi's trim ($298). [HORNET SEDAN] Hornet Sportabout D/L package, with vinyl trim ($284); with Custom fabric trim ($333). Hornet Sportabout 'X' package ($139). Rooftop travel rack on Matador wagon ($56). Scuff side molding, except specific models ($38). Two-tone paint, on Hornets, except Sportabout ($30); on Matador/Ambassador, except wagons ($37). Special paint application, including painted body side panels and accent moldings, on Matador wagon ($69). Rally side stripes, on Gremlins ($33); on base Javelin ($38). Hood 'T' stripes on Javelin AMX without 'Go Package' — standard with ($39). Color-keyed wood-grained exterior paneling, on Sportabout ($95); on Matador wagon, including rear panel ($113). Power brakes, except Matador/Javelin/Ambassador groups ($44). Front power disc brakes ($32-$81). Power steering ($99-$111). Power side windows, Ambassador only ($123). Power tailgate window, Matador/Ambassador two-seat wagons ($35); three-seat wagons (no charge). AM push-button radio ($66$70). AM/FM push-button radio ($179); AM/FM Multiplex radio with four speakers, in Javelin/Matador/Ambassador ($161-$230). Stereo tape player, Javelin/Ambassador ($196-$200). Domino fabric trim in Hornet hatchback ($99); in Javelin ($47); in Hornet hatchback 'X' ($50). Individually reclining seats with Venetian fabric special interior in Hornets ($109). Third seat with belts and power windows, Matador/Ambassador wagons ($108). Adjust-O-Tilt steering wheel ($43-46). Sports steering wheel, in Gremlin ($33). Aluminum trim rings and hubcaps, Gremlin/Javelin ($33). Black, white, blue, green, brown or cinnamon vinyl roof, on Matadors ($91); on Ambassadors ($109); on Hornet/Javelin roups ($88). Full wheelcovers ($27-$29). Custom wheelcovers ($23-$50). Javelin 19 x 7 inch slotted wheels ($205). Spoke-style 6 x 14 inch wheels on cars with 'D/L', 'X' or 'Go' packages ($34-$49); as a separate option, ($74-$104). Vent rear quarter windows in Gremlin/Hornet coupe ($28). DeLuxe, intermittant electric windshield wipers ($23-$24).

1974 AMC Hornet "Levis" hatchback

OPTION PACKAGES: Gremlin Custom trim, includes: Custom door and seat trim in pleated vinyl; carpeting; extra insulation; Custom steering wheel; wheel opening and drip moldings; and cargo insulation, with bench seat ($109), with bucket seat ($147). Gremlin 'X' package, includes: spear side decal; 6 x 14 inch slot wheels; Space Saver spare with base D70-14 tires; Custom trim with bucket seats: cargo insulation; carpeting; 15 inch steering wheel; special interior appointments; and decals, with base tires ($314), with RWL tires ($349), with whitewall radial tires ($298), with RWL radials ($410). Handling package ($23-$30). Hornet hatchback 'X', includes: sports steering wheel; Rally stripes; slot wheels; 'X' emblems, insulation; vinyl bucket seats; Space-Saver spare; and hidden compartment ($207). Levi's Custom trim package, includes: bucket seats with blue denim trim and Levi's buttons; special door trim; sun visors; insulation; denim litter container; blue headliner: front fender Levi's emblem; and, in Gremlin, carpets; cargo insulation and Custom steering wheel, in Gremlin ($165), in Hornet hatchback ($150), in Hornet hatchback 'X' ($101). Levi's Custom trim package, in Gremlin 'X' ($50). Rally-Pac instrumentation, in Javelin ($77). Gremlin/Hornet hatchback Rally 'X' package, includes: three-speed floor-shift or automatic; power steering; manual front disc brakes; gauges; black dash cluster; and leather Sports steering wheel, 'X' models only, with air conditioning ($100); without ($199). Sportabout D/L package, includes: Sports steering wheel; Rally stripes; slot wheels; and 'X' emblems, with vinyl trim ($264), with Custom fabric trim ($313). Designer series Cassini Matador Brougham included Custom wheelcovers with copper-colored inserts; scuff moldings; a copper-colored vinyl roof; copper grille and headlamp bezels; black-carpeted trunk compartment and tire cover; special black seat and door trim with copper buttons; black headliner; black instrument panel with copper dials and overlays; black steering wheel with copper inserts in horn rim; and copper-colored floor carpeting ($299). The upholstery in this model was quite lavish, with the individually reclining seats covered in a rich, black nylon knit fabric having a tufted look. The "Oleg Cassini of Paris, France" crest was embroidered on each front headrest and also appeared, in medallion form, on the trailing edge of front fenders below the 'dipping' feature line molding. Javelin/AMX 'Go-Package' included: (with 360 V-8) hood 'T' stripe decal; black-finished rear panel; Rally-Pac instrumentation; Handling Package; heavy-duty cooling system; Twin-Grip differential; power disc brakes; slot styled wheels; Space-Saver spare; and FR78-14 RWL tires on 14 inch wheels ($372.30). Same with E60 x 15 tires and 15 inch slot-styled wheels ($413.35). '401 Go-Package' includes the same features, but bigger V-8 ($420.50 or $461.55, respectively). AMC also offered specially-priced, factory-installed fleet options and fleet options packages.

HISTORICAL FOOTNOTES: Model year production introductions were scheduled for Sept. 15, 1973. Model year production was an all-time high of 509,496 units. Calendar year production peaked at 351,398 cars. Only 3,734 AMC models had 401 cid V-8 engines installed this year. Calendar year sales by dealer franchises in the United States totaled 355,093. Model year ended in November of 1974 instead of June 1974, to take advantage of easier 1974 emission laws. This partly accounts for the large model year production figures. R.D. Chapin, Jr. was chairman and Chief Executive Officer of AMC. A net profit of $28.6 million was made in a season that saw an energy conscious public shun the purchase of all new cars and, especially those reputed to be gas guzzlers. Luckily, AMC's traditional image did not place it deeply into this group. All 1974 AMC engines were designed to run on regular leaded, low-lead or no-lead fuels.

1975 AMC

HORNET — (SIX/V-8) — SERIES 01: Four models made up the compact Hornet Series for 1975: hatchback; Sportabout sedan/station wagon and two-door sedan. A new grille featured a bold, six segment motif and had five bright vertical division bars against blacked-out vertical louvers within each segment. The outboard divisions contained new rectangular parking lamps. There was also a slightly different look to the front bumper, which was smoother and rounder in general appearance. Six trim packages were available; the sedan offering a new D/L group with individually reclining front seats and cut-pile carpets. It also included a body side molding (between the wheel wells at mid-body height), wheelcovers and special emblems. Then, there was the Hornet 'X' group with such items as slot styled wheels; 'X' emblems and full-length Rally striping along the upper feature line. The hatchback and Sportabout could be had with the all-new 'Touring' option. Cars with extra-cost decor trim had a different side molding treatment than described in the 1974 section. Basically, the full-length lower molding was gone. Standard equipment for the base Hornet followed the same pattern described in detail for 1974 models, plus electronic ignition. Identification came from Hornet nameplates right behind the front side marker lamps and on the right rear panel. There were also AMC badges on the left-hand side of the rear panel.

1975 AMC, Hornet Sportabout four-door station wagon, 6-cyl

VIN: Serial Numbers were located on top of the right front wheelhouse panel. The unit body data plate was riveted to the left front door, below the latch mechanism. American Motors retained a 13 symbol identification code system. The first symbol was an A for American Motors Corp. The second symbol indicated model year, 5=1975. The third symbol indicated transmission type: S=standard column shift; A=automatic, column shift; C=three-speed automatic with floor shift; D=three-speed manual floor shift with overdrive; E=fully-synchronized three-speed manual with floor shift; O=three-speed manual column shift with overdrive. The fourth symbol was a number designating car line, as follows: 0=Hornet; 1=Rebel; 3=AMX; 4=Gremlin; 6=Pacer; 7=Javelin and 8=Ambassador. The fifth symbol designated body type: 3=three-door hatchback; 5=four-door sedan; 6=two-door sedan; and 8=four-door station wagon. The sixth symbol designated the series or class of body, as follows: 5=Gremlin; 7=Hornet/Matador/Pacer; P=Police; T=Taxicab. The seventh symbol was a letter designating engine type (without regard to car-line), as follows: 'A' = 258 cid one-barrel six; 'E' = 232 cid one-barrel six; 'H' = 304 cid two-barrel V-8; 'N' = 360 cid two-barrel V-8; 'P' = 360 cid four-barrel V-8 and 'Z' = 401 cid four-barrel V-8. The eighth symbols indicated assembly plant, as follows: Cars coded 1-6 were built in Kenosha, Wis. and those coded 7-9 were built in Brampton, Ontario, Canada. The remaining symbols weres the sequential production number and began with 100001 for cars built in Kenosha, Wis. and 700001 for cars built in Brampton, Ontario, Canada. Body number plate attached to left front door edge shows body serial number, Body/Style Number from second column of chart below, plus trim and paint codes.

HORNET SERIES 01

Series Number	Body/Style Number	Body Type & Seating	Factory Price	Shipping Weight	Production Total
01	7503-7	2-dr Hatch-5P	3174/3312	2839/3085	13,441
01	7505-7	4-dr Sed-6P	3124/3262	2881/3147	20,565
01	7506-7	2-dr Sed-6P	3074/3212	2815/3061	12,392
01	7508-7	4-dr Sta Wag-6P	3374/3512	3844/3878	39,563

NOTE 1: Model year output totaled 85,961 units. Of these cars 77,886 were sixes and 8,075 were V-8s.
NOTE 2: Prices and weights above slash are for six/below slash for V-8.

1975 AMC, Matador 'X' two-door coupe, V-8

MATADOR (COUPE) — (SIX/V-8) — SERIES 10: After 1974, the Matador coupe was distinct from the four-door styles. This year, American Motors emphasized the difference, by placing the two types into different series. The sedan and wagon were moved into the 80 Series slot, vacated by the Ambassador Brougham. That left the Matador coupe, by itself, in Series 10. Newly styled road wheels; front disc brakes and radial tires were standard equipment in the Matador 'X' coupe. Styling changes included a new grille with full-length horizontal bars forming a rectangular pattern. The standard engine was the 258 cid six, with three V-8s optional. 89 cars left the factory with 401 cid engines though that motor was not on the normal equipment list. Factory records indicate four 401s were Matador Coupes. The other 84 were used in Series 80 four-door sedans and wagons, probably law enforcement models. The electronic ignition system was now standard. Other regular features were the same as 1974. The Matador 'X' coupe was, technically, an option package, with a $199 price tag. Also remaining available was the Cassini Coupe of which 1,817 found buyers this season.

MATADOR (COUPE) SERIES 10

Series Number	Body/Style Number	Body Type & Seating	Factory Price	Shipping Weight	Production Total
10	7516-7	2-dr Cpe-6P	3446/3545	3562/3734	22,368

NOTE 1: Prices and weights above slash are for six/below slash for V-8.

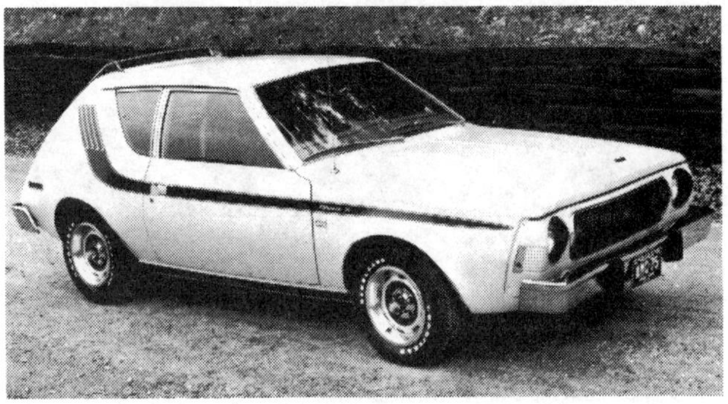

1975 AMC, Gremlin two-door Levi's sedan, 6-cyl

GREMLIN — (SIX/V-8) — SERIES 40: The Gremlin was basically unchanged for 1975. Standard equipment was the same as the previous year, plus electronic ignition system. Mechanical detail changes included a sturdier manual transmission and the optional availability of overdrive combined with six-cylinder attachments only. Also provided again, at extra-cost, were the Levis, Rally and Gremlin 'X' packages. The 'hockey stick' striping pattern was carried over on cars so-equipped. New body colors and radial-ply tires could be added. A slightly cleaner looking front bumper was used. Its upper edge had a single-bevel appearance, compared to the triple-bevel 1974 type. Also, the sail panel windsplit indentations were now in a slanted, vertical position and the flared wheel treatment was more subdued. Very close inspection would also reveal that the bumper guards had a more wedge-shaped contour. Even with such refinements, nobody had trouble spotting the Gremlin in a crowd. An interesting fact is that the price of the V-8 now came in at a dollar per pound.

GREMLIN SERIES 40

Series Number	Body/Style Number	Body Type & Seating	Factory Price	Shipping Weight	Production Total
40	7546-5	2-dr Sed-4P	2798/2952	2694/2952	56,011

NOTE 1: Model year output totaled 56,011 units. Of these 52,601 were sixes and 3,410 were V-8s.
NOTE 2: Prices and weights above slash are for the six/below slash for V-8.

PACER — (SIX) — SERIES 60: The AMC Pacer was introduced on March 1, 1975, as a midyear model. It was billed as the first wide, small car, as it was 77 inches wide but had a short 100 inch wheelbase. The Pacer used many unique features including a passenger side door that was larger than the driver's door and one of the first rack and pinion steering systems available on a U.S. built car. The car was available, for 1975, only as a two-door hatchback. It had a very short, fast-sloping hood, since is was originally designed to use GM's front-wheel-drive Wankle rotary power unit. The cancellation of that program, by GM, forced AMC to re-engineer the car on short notice. It was transformed into a rear wheel-drive piston-engined configuration. The only powerplant provided was the 232 cid six. Although its overall length was a compact-sized 171.5 inches, the Pacer's interior roominess matched or exceeded that of its full-sized contemporaries. It had a very low beltline and large expanses of glass area, giving extremely good visibility (and excellent motivation for air conditioning sales). The unconventional body featured a large rear window liftgate with dual, gas-filled cylinders for easy opening. With the rear seat folded, the cargo area expanded to nearly 30 cubic feet. Classified as a two-door sedan, buyers could order-up the 'bubbly' vehicle as a base-trim model; a sporty Pacer 'X;' or a lavish little Pacer 'D/L.' All three editions were sure to get attention in 1975 and still draw interest today.

PACER SERIES 40

Series Number	Body/Style Number	Body Type & Seating	Factory Price	Shipping Weight	Production Total
BASE PACER					
60	7566-7	2-dr Sed-6P	3299	2995	Note 1
PACER 'X' OPTION					
60	7566-7	2-dr Spt Sed-5P	3638	NA	72,158
PACER 'D/L' OPTION					
60	7566-7	2-dr DeL Sed-6P	3588	NA	(19,050)

NOTE 1: Calendar year output (Kenosha) totaled 145,528 units.
NOTE 2: Factory literature indicates that the optional reclining seats were available only as part of the Pacer D/L package. Industry records show that 26 percent of all 1975 Pacers had these seats. Based on these facts, it can be calculated that 19,050 Pacer D/L models were built during the 1975 model year. No other breakouts can be made.

MATADOR — (SIX/V-8) — SERIES 80: Moved to a higher-numbered series, the Matador four-door styles featured new hoods, grilles and bumpers. The hood was flatter and the center crease ran into the upper grille surround molding. The redesigned grille featured full-width, horizontal blades, with eight bright vertical divider moldings positioned along the protruding center section only. The horizontal blades extended right to the round headlamp lenses, making them seem better integrated into the overall appearance. The front bumper was smoother and rounder looking, eliminating the tripie-bevel look of 1974. Another change was to rectangular parking lamps. Identification came from name badges on the front fender sides, behind the wheel openings. Side trim consisted of a three-quarter length belt molding, running from a point above the name badge to the rear. Standard equipment features matched those of the previous season, plus electronic ignition. A Brougham options package was available on both models, but at two different prices. The sedan was the basis of specially-assembled taxi-cab and police car models. The latter group carried 84 of the 89 — 401 cid V-8s installed in AMC products in the 1975 model year.

MATADOR SERIES 80

Series Number	Body/Style Number	Body Type & Seating	Factory Price	Shipping Weight	Production Total
80	7585-7	4-dr Sed-6P	3452/3551	3586/3746	27,522
80	7588-7	4-dr Sta Wag-6P	3844/3943	3878/4038	9,692

NOTE 1: Data above slash for six/below slash for V-8.
NOTE 2: Model year output totaled 59.582 units built in Kenosha, Wis. Of these cars, 10,965 were sixes and 48,617 were V-8s.
NOTE 3: The six-cylinder total includes 9,390 passenger cars and 1,575 station wagons. The V-8 total includes 40,500 passenger cars and 8,117 station wagons.
NOTE 4: Calendar year output totaled 57,152 cars built at Kenosha, Wis.

BASE ENGINES

(Base six) Inline six. Overhead valves. Cast iron block. Displacement: 232 cid. Bore and stroke: 3.75 x 3.50 inches. Compression ratio: 8.0:1. SAE Net hp: 90 at 3050 rpm. Seven main bearings. Hydraulic valve lifters. Carburetor: one-barrel.

(Big six) Inline six. Overhead valves. Cast iron block. Displacement: 258 cid. Bore and stroke: 3.75 x 3.90 inches. Compression ratio: 8.0:1. SAE Net hp: 95 at 3050 rpm. Seven main bearings. Hydraulic valve lifters. Carburetor: One-barrel.

(Base V-8) Overhead valves. Cast iron block. Displacement: 304 cid. Bore and stroke: 3.75 x 3.44 inches. Compression ratio: 8.0:1. SAE Net hp: 120 at 3200 rpm. Five main bearings. Hydraulic valve lifters. Carburetor: two-barrel.

CHASSIS FEATURES: Wheelbase: (Gremlin) 96 inches-, (Pacer) 100 inches; (Hornet) 108 inches; (Matador Series 10) 114 inches; (Matador Series 80) 118 inches. Overall length: (Gremlin) 170.3 inches; (Pacer) 171.5 inches; (Hornet) 187 inches; (Matador Coupe) 209.3 inches; (Matador sedan) 216 inches; (Matador wagon) 215.5 inches. Front tread: (Pacer) 61.2 inches; (other models) See 1972 Chassis Features. Rear tread: (Pacer) 60.2 inches; (other models) See 1972 Chassis Features. Tires: (Gremlin) 6.45 x 14; (Pacer/Hornet) 6.95 x 14; (Matador coupe) ER78 x 14; (Matador sedan) FR78 x 14; (Matador wagon) HR78 x 14.

POWERTRAIN OPTIONS: Three-speed manual transmission was standard. Overdrive transmission, for six. Three-speed manual floor shift transmission. Hornet/Gremlin six-cylinder 258 cid 95 hp one-barrel engine. Pacer six-cylinder 258 cid 95 hp one-barrel engine. Hornet/Gremlin 304 cid 120 hp two-barrel engine. Matador Coupe V-8 304 cid 120 hp two-barrel engine (no charge). Matador Coupe V-8 360 cid 140 hp two-barrel engine. Matador Coupe V-8 360 cid 180 hp four-barrel engine. Matador sedan V-8 304 cid 120 hp one-barrel engine. Matador sedan V-8 360 cid 140 hp two-barrel engine. Matador sedan V-8 360 cid 180 hp four-barrel engine. Matador wagon V-8 304 cid 120 hp two-barrel engine. Matador wagon V-8 360 cid 140 hp two-barrel engine. Matador wagon V-8 360 cid 180 hp four-barrel engine. Positive traction rear axle. Heavy-duty clutch, for six.

HORNET/GREMLIN/MATADOR OPTIONS: Power steering, in Gremlin/Hornet ($119). Air conditioning, in Gremlin/Hornet group ($400). Gremlin 'X' package ($201). Gremlin Levis Custom trim package ($220). Gremlin Rally option package ($133). AM/FM stereo, all except Matador Group ($179). Styled Road wheels, Gremlin/Hornet group ($115). Hornet D/L package, on sedan ($299); on Sportabout ($293). Hornet Levis Custom trim package ($125). Hornet 'X' hatchback option package ($227). Hornet Sportabout 'X' package ($139). Sportabout rooftop travel rack ($75). Hornet Rally option package ($125). Matador air conditioning ($450). Matador AM/FM stereo system ($230). Matador AM/FM stereo with 8-track tape player ($300). Matador station wagon, third seat equipment ($121). Matador station wagon. rooftop travel rack ($59). Matador 'X' package ($199). Oleg Cassini coupe trim package ($299). Styled road wheels, on Matador group ($121). Matador Brougham package, wagon ($145); others ($105). Gremlin/Hornet group Econo-Miser package ($225). Gremlin, Custom trim package ($135). Reclining seats, as option ($75).

PACER OPTIONS: 'X' package ($339). 'D/L' package ($289). Decor package ($49). 232 CID six, one-barrel carburetor engine (standard). 258 CID six, one-barrel carburetor engine ($69). Torque-Command, Column-Shift ($239.95). Torque-Command with floor-shift, available only with bucket or reclining seats ($259.95). Three-speed manual, column shift (standard). Three-speed manual floor shift ($19.95). Three-speed manual with overdrive, column shift ($149). Twin-Grip differential ($46). Power steering ($119). Power front disc brakes ($79.35). Manual front disc brakes ($47.45). AM push-button radio ($69). AM/FM stereo radio, with four speakers ($179). Entertainment center with AM/FM stereo and tape player ($299). Hidden compartment ($29). Bucket seats ($99). Individual reclining seats, 'D/L' only. Air conditioning system ($399.95). Tinted glass, all windows ($49). Rear window defogger ($59.95). Rear window washer and wiper ($49.95). Roof rack ($49.95). Cruise-command speed control with automatic transmission only ($65). Adjust-O-Tilt steering column with automatic transmission column shift only ($49). Visibility group ($49.95). Deluxe electric windshield wipers with intermittent action ($24.95). Light group ($34.95). Door vent windows ($29.95). Sports steering wheel ($18.90). Wheel discs ($29.95). Styled road wheels ($115); with 'X' package ($50); with 'D/L' package ($85.05). Aluminum styled wheels ($200); with 'X' package ($135); with 'D/L' package ($170.05). Slot-styled wheels ('X' only). Extra quiet insulation ($29.95). Protection group ($34.95). Bumper nerfing strips ($19). Handling package, includes: heavy-duty springs, shocks and front sway bar ($29.95). Handling package with 'X' or 'D/L' package ($15). Front sway bar ($14.95). Vinyl roof ($99.95). Two-Tone paint ($49). Whitewall tires ($34.45).

HISTORICAL FOOTNOTES: The AMC models made their debut Nov. 15, 1974 and the Pacer was introduced Feb. 28, 1975. The unique 'bubble car' went on sale March 1, 1975. American Motors Corp.'s model year output hit 244,941 units. Calendar year production was recorded as 323,704 cars. Sales by U.S. dealers, for 1975 models only, were reported as 268,526 vehicles. R.D. Chapin, Jr. remained at the head of the company. Richard A. Teague, vice-president of styling gets the credit for the Pacer's unique and attractive appearance.

1975 AMC Matador four-door sedan

1976 AMC

Following the introduction of the Pacer in mid-1975, changes for the 1976 AMC model year were modest. Speedometers now reached only 90 mph. Some models had a new lockable, padded console. After January 1, brakes were enlarged to meet Federal standards. Six-cylinder engines had reshaped carburetor air passages, a new thermostat and an electric choke on some models. Since the full-sized Ambassador had been dropped for 1975, AMC stressed economy. Three-fourths of the company's vehicles rated over 20 mpg in EPA highway mileage estimates. Engines carried electronic ignition. "Safe-Command" features on all models included: energy-absorbing bumpers and steering column; front head restraints; four-way hazard warnings; lane-changer turn signals; backup and marker lights; padded sun visors and instrument panel; double-safety brake system with warning light. All models were painted in "Luster-Guard" acrylic baked enamel. Standard body colors were black; Sienna or Alpine Orange; Seaspray green; Sand tan; Firecracker red; Brilliant blue; Sunshine yellow; plus Nautical blue, Medium blue, Dark cocoa, Autumn red, Evergreen, Burnished bronze, Silver Frost or Limefire metallic.

1976 AMC Gremlin X Liftback (AMC)

GREMLIN — SERIES 40 — SIX/V-8: Introduced with considerable sales success during 1970 as the first American-made subcompact, the sawed-off Gremlin entered 1976 wearing a new grille. A horizontal crossbar stood between (and surrounded) round amber parking lights. Originally created by chopping 17 inches off the Hornet design, it remained the only domestic subcompact without a four-cylinder engine. Standard engine was a 232 cid six with three-speed column shift; a 258 six and 304 V-8 were optional. More distinctive than most small cars, with a design not universally loved, Gremlin gained refinements through its early years but few major changes in its single two-door body style. From the 'B' pillar forward, it's essentially a Hornet, but with a flat hatchback that almost looks like a real hatch. Standard equipment included 6.45 x 14 blackwall tires, foam-cushioned seats, front ashtray, folding rear seat, Weather-Eye heater/defroster, two-speed electric wipers/washers, dome light, rear lift window, front sway bar (V-8 only), rear bumper guards, front bumper nerfing strips, a 50-amp battery, and aluminum hubcaps. The sporty 'X' package, available only on Custom models, added a full-length body stripe with 'X' decal, painted lower back panel, engine-turned instrument cluster overlay, glove box decal, D70 x 14 tires on slot-style wheels, and a "Space Saver" spare tire. Available again was the unique Levi's trim package, with sporty bucket seats, stowage/litter pouches and door trim in simulated blue denim (actually spun nylon) complete with buttons; plus blue headlining and sun visors, and Levi's front fender decals. Rally stripes continued their unusual "hockey stick" design.

PACER — SERIES 60 — SIX: Great expectations greeted the wide, glassy Pacer when it appeared as a mid-1975 model. Riding a 100 in. wheelbase, it stood 77 in. wide. Billed as "the first wide small car," BUSINESS WEEK called it the "hottest car of 1975." Unique features included a passenger door nearly four inches wider than that of the driver. A surprisingly roomy four-passenger interior belies the car's compact 170 in. length. The entire rear section tied into the massive B-pillar structure, cutting both length and weight. Aerodynamic styling helped improve fuel economy. A short, sharply-sloped hood and enormous glass area give impressive visibility. But that hood was actually the result of having designed the car for a front-drive Wankel rotary engine, which never materialized from GM. Pacer was one of the first American cars to offer rack and pinion steering. A major change for 1976 was the availability of an optional 258 cid six, with one- or two-barrel carburetor, in addition to the economy 232 cid six. Air conditioning was a most desirable extra, due to the large glass area. Twin gas-filled cylinders assisted opening of the large rear lift window. The Pacer 'X' package included vinyl bucket seats with manual floor shift, woodgrain dash overlay, sports steering wheel, extra-quiet insulation, 'X' ornamentation, color-keyed body side scuff moldings, bumper nerfing strips, front sway bar, and D78 x 14 blackwall tires. The luxurious D/L package consisted of basketry fabric interior, padded steering wheel, carpeted cargo area, woodgrain-overlay dash, wheelcovers, color-keyed scuff moldings, bright license molding, nerfing strips, front sway bar, and special emblems. D/L models could also have Hyde Park fabric seat trim. Standard equipment included three-speed column shift; 6.95 x 14 tires; foam-cushioned bench seat; color-keyed carpeting and rubber cargo mat; folding second seat; body side scuff molding; Weather-Eye heater/defroster; concealed two-speed wipers/washers; dome light; 50-amp battery; aluminum hubcaps; ashtray and lighter. Body colors were identical to other models, but not including black, Sienna/Alpine orange, Dark cocoa, Limefire, or Nautical blue. Alpine white, Golden Jade metallic, Aztec copper metallic, Brandywine metallic, and Marine aqua colors were unique to Pacer.

1976 AMC Hornet four-door sedan (AMC)

HORNET — SERIES 01 — SIX/V-8: Introduced in 1970 at a cost of $40 million, the compact Hornet, new from the ground up, became one of AMC's best sellers. As in 1975, the Hornet lineup consisted of four models: a two-door hatchback (added for 1973), two- and four-door sedans, and a four-door Sportabout wagon. Appearance changed little this year, apart from thin rubber strips placed at lower bumper ends. Hornet nameplates sat behind the front side market lamps, and on the right rear panel. A 232 cid six with three-speed column shift was standard; a 258 six and a 304 V-8 were optional. Three option packages were available. Sportabouts and hatchbacks could have a "Touring Interior" with individual reclining seats in tan vinyl; matching headliner, visors and door pull straps; sports steering wheel; woodgrain dash overlay; and carpeted lower door panels. The sporty 'X' package, also for Sportabouts and hatchbacks, consisted of full-length Rally striping, slot wheels with D70 x 14 blackwall tires, 'X' ornamentation on dash and lower back panel, plus black grille accents. Sedans with the luxury D/L package had reclining front seats and trim in tan "Kasmir Knit" fabric; D/L Sportabouts came with "Potomac Stripe" fabric (or vinyl), plus woodgrain body paneling, custom wheelcovers, roof rack, and air deflector. Standard Hornet equipment included rear bumper guards, front bumper nerfing strips, Weather-Eye heater/defroster, lighter, color-keyed carpeting and mats, locking glove box, two-speed wipers/washers, front sway bar (V-8s), 50-amp battery, aluminum hubcaps, 6.95 x 14 blackwalls, and aluminum hubcaps. Hatchbacks and Sportabouts had a fold-down rear seat and rear lift window.

1976 AMC Pacer Liftback (AMC)

MATADOR (COUPE) — SERIES 10 — SIX/V-8: Starting in 1975, the stylish mid-size Matador coupe stood apart from the four-door sedan and wagon, with a different wheelbase and series number. For 1976 the coupe gained a new full-width, two-section grille that extended out to the fender tips, below the huge headlamp openings. Amber parking lights were rectangular. The Brougham package included individual reclining seats in "Hunter's Plaid" fabric, custom door trim panels, woodgrain dash overlay, full carpeting, wheel discs, hood paint stripes, bumper nerfing strips, rocker moldings, roll-down quarter windows, full-length body side scuff moldings, wheel lip and grille moldings. Going a step further, a new Barcelona luxury package (only on the Brougham) added plush, crumpled-look velvety upholstery in tan or black Knap knit to the reclining front seats, plus special wheelcovers and medallions. Barcelona buyers also got color-keyed cut pile carpeting, door pull straps and headlining, plus (with tan interior) tan grille, headlamp bezel and rear license accents. Distinctive red/yellow side striping, a unique hood ornament, hood and deck nameplates, and fender and glove box medallions identified the Barcelona. Base engine was the 258 cid six, with three V-8 options: a 304, 360 with two-barrel carburetor, and 360 with four-barrel and dual exhausts. Standard equipment was similar to Matador sedan, including front disc brakes.

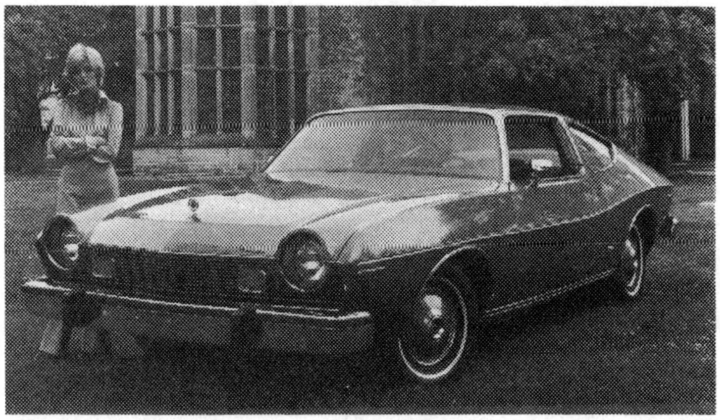

1976 AMC Matador Brougham coupe (AMC)

MATADOR (SEDAN AND WAGON) — SERIES 80 — SIX/V-8: AMC's intermediate model, little changed from 1975 when it gained a new hood, grille and bumpers, came in standard or Brougham trim. The grille has full-width horizontal blades, plus eight bright vertical divider bars along the protruding center section. Nameplates were behind the front wheel openings, just below the front of the three-quarter length belt moldings. Base engine was the 258 cid six in sedans, a 304 V-8 in wagons (and all California Matadors). Two 360 cid V-8s were optional. Only 20 Matadors held a 401 V-8, available only to law enforcement agencies. Standard equipment included a three-speed column shift transmission (automatic on wagons), foam-cushioned seats, color-keyed carpeting, front and rear ashtrays (front lighter), extra-quiet insulation package, Weather-Eye heater/defroster, day/night mirror, two-speed wipers/washers, dome light, dual-swing wagon tailgate, front sway bar, bumper guards (front only on wagon), plus manual front disc brakes on sedan (power discs on wagon). Sedans wore standard E78 x 14 blackwall tires (F78 x 14 with V-8); wagons H78 x 14. The Brougham package gave individual reclining seats in Custom Hyde Park fabric (Sof-Touch vinyl in wagons), woodgrain dash overlay, full carpeting, wheelcovers, 'Brougham' script on 'C' pillars, roof rack and tailgate air deflector (wagon), hood paint stripes, a back panel overlay (sedan), and rocker panel moldings (except two-tone or woodgrain-panel wagon).

I.D. DATA: The 13 symbol Vehicle Identification Number (VIN) was embossed on a metal plate riveted to the top left surface of the instrument panel, visible through the windshield. The first letter (A) indicated the manufacturer, American Motors. The second digit denoted the model year (6=1976). Third came a letter identifying transmission type: S=three-speed manual, column shift; O=three-speed column shift with overdrive; E=three-speed floor shift; D=three-speed floor shift with overdrive; A=column-shift automatic; C=floor-shift automatic. The fourth digit denoted the car-line (series): 0=Hornet; 1=Matador coupe; 4=Gremlin; 6=Pacer; 8=Matador sedan/wagon. Fifth digit identified body style: 3=two-door hatchback; 5=four-door sedan; 6=two-door sedan; 8=four-door station wagon. Digit six showed the model/group (body class): 3=Gremlin standard; 5=Gremlin Custom; 7=Pacer/Hornet/Matador; P=Police. The seventh letter indicated engine type: E=232-six-cylinder one-barrel; A =258-six-cylinder one-barrel; C=258-six-cylinder two-barrel; H=304 V-8 two-barrel; N=360 V-8 two-barrel; P=360 V-8 four-barrel; Z=401 V-8 four-barrel (police only). Digits 8 through 13 made up the sequential serial number, starting with 100,001 for vehicles made at Kenosha, Wisconsin and 700,001 for those manufactured at Brampton, Ontario.

NOTE: Digits 4-6 are identical to the model number. A safety sticker attached to edge of left front door shows the month and year built, plus the VIN. A unit body identification plate riveted to the edge of the left front door displays the body number, model number, trim number, paint code number, and build sequence number. A six-symbol Build Code was engraved on a machined surface of the cylinder block of six-cylinder engines, between cylinders two and three; or stamped on a metal tag attached to right bank valve cover on V-8 engines. The first digit indicates model year; second and third digits, the month of manufacture (1 = January). The fourth letter indicates engine type, and is identical to the seventh letter of the VIN. Digits five and six denote the day of manufacture. V-8 engines also have cid displacement cast into the side of the block, between the second and third freeze plugs, usually under the motor mount.

GREMLIN SERIES 40

Series Number	Body/Style Number	Body Type Seating	Factory Price	Shipping Weight	Production Total
40	7646-3	2dr Sed-4P	2889/3051	2771/3020	Note 1
40	7646-5	2dr Cus Sed-4P	2998/3160	2774/3023	Note 1

NOTE 1: Total model year production, 52,941 Gremlins (only 826 with V-8 engine).

PACER SERIES 60

60	7666-7	2dr Sed-4P	3499	3114	117,244

HORNET SERIES 01

01	7603-7	2dr Hatch-5P	3199/3344	2920/3169	Note 2
01	7605-7	4dr Sed-6P	3199/3344	2971/3220	Note 2
01	7606-7	2dr Sed-6P	3199/3344	2909/3158	Note 2
01	7608-7	4dr Sta Wag-6P	3549/3694	3040/3289	29,763

NOTE 2: Total model year production, Hornet sedans and hatchbacks, 41,814 units (41,025 with six-cylinder engine, 789 with V-8). Of the 29,763 Sportabout wagons, 26,787 had a six; 2,976 had a V-8 engine.

MATADOR (COUPE) SERIES 10

10	7616-7	2dr Cpe-6P	3621/3725	3562/3811	Note 3

MATADOR (SEDAN AND WAGON) SERIES 80

80	7685-7	4dr Sed-6P	3627/3731	3589/3838	Note 3
80	7688-7	4dr Sta Wag-6P	— /4373	— /4015	11,049

NOTE 3: Model year production of coupe and sedan totaled 30,464 units (4,993 sixes and 25,471 V-8s).

NOTE: Data above slash for six/below slash for V-8.

ENGINES: BASE SIX (Gremlin, Hornet, Pacer): Inline. OHV. Six-cylinder. Cast iron block. Displacement: 232 cid (3.8 liters). Bore & stroke: 3.75 x 3.50 in. Compression ratio: 8.0:1. Brake horsepower: 90 (SAE net) at 3050 rpm. Torque: 170 lbs.-ft. at 2000 rpm. Seven main bearings. Hydraulic valve lifters. Carburetor: one-barrel Carter YF. BASE SIX (Matador coupe/sedan); OPTIONAL (Gremlin, Hornet, Pacer): Inline. OHV. Six-cylinder. Cast iron block. Displacement: 258 cid (4.2 liters). Bore & stroke: 3.75 x 3.90 in. Compression ratio: 8.0:1. Brake horsepower: 95 at 3050 rpm. Torque: 180 lbs.-ft. at 2100 rpm. (Matador, 2000 rpm.). Seven main bearings. Hydraulic valve lifters. Carburetor: one-barrel Carter YF. OPTIONAL SIX (Pacer only): Same as 258 cid six above, but with Carter BBD two-barrel carburetor. Horsepower: 120 at 3400 rpm. Torque: 200 lbs.-ft. at 2000 rpm. BASE V-8 (Matador station wagon), OPTIONAL (Gremlin, Hornet, other Matadors): 90-degree, overhead valve V-8. Cast iron block. Displacement: 304 cid (5.0 liters). Bore & stroke: 3.75 x 3.44 in. Compression ratio: 8.4:1. Brake horsepower: 120 at 3200 rpm. Torque: 220 lbs.-ft. at 2200 rpm. Five main bearings. Hydraulic valve lifters. Carburetor: two-barrel Motorcraft 2100. OPTIONAL V-8 (Matador): 90-degree, overhead valve V-8. Cast iron block. Displacement: 360 cid (5.9 liters). Bore & stroke: 4.08 x 3.44 in. Compression ratio: 8.25:1. Brake horsepower: 140 at 3200 rpm. Torque: 260 lbs.-ft. at 1600 rpm. Five main bearings. Hydraulic valve lifters. Carburetor: two-barrel Motorcraft 2100. OPTIONAL HIGH-PERFORMANCE V-8 (Matador): Same as 360 cid V-8 above, but with four-barrel Motorcraft 4350 carburetor and dual exhausts. Horsepower: 180 at 3600 rpm. Torque: 280 lbs.-ft. at 2800 rpm. POLICE V-8 (Matador): 90-degree, overhead valve V-8. Cast iron block. Displacement: 401 cid Bore & stroke: 4.165 x 3.68 in. Compression ratio: 8.25:1. Brake horsepower: 215 at 4200 rpm. Torque: 320 lbs.-ft. at 2800 rpm. Five main bearings. Hydraulic valve lifters.

CHASSIS DATA: Wheelbase: (Gremlin) 96 in.; (Pacer) 100 in.; (Hornet) 108 in.; (Matador coupe) 114 in.; (Matador sedan/wagon) 118 in. Overall length: (Gremlin) 169.4 in.; (Pacer) 170 in.; (Hornet) 186 in.; (Matador coupe) 209.4 in.; (Matador sedan) 216 in.; (Matador wagon) 215.5 in. Height: (Gremlin) 52.3 in.; (Pacer) 52.7 in.; (Hornet) 52.2 in. except four-door sedan, 51.7 in.; (Matador coupe) 51.8 in.; (Matador sedan) 54.7 in.; (Matador wagon) 56.8 in. Front Tread: (Gremlin/Hornet) 57.5 in.; (Pacer) 61.2 in.; (Matador coupe) 59.7 in.; (Matador) 59.8 in. Rear Tread: (Gremlin/Hornet) 57.1 in.; (Pacer) 60.2 in.; (Matador) 60.0 in. Standard Tires: (Gremlin six) 6.45 x 14; (Gremlin V-8, Pacer, Hornet) 6.95 x 14; (Hornet V-8 sedan or wagon) D78 x 14; (Matador coupe/sedan, six) E78 x 14; (Matador coupe/sedan, V-8) F78 x 14; (Matador wagon) FR78 x 14 steel radial, or H78 x 14.

TECHNICAL: Three-speed manual transmission was standard: floor shift on Gremlin and Hornet hatchbacks, column shift on other models. Overdrive standard on Pacer with 258 cid six (two-barrel); optional with other Pacers and Gremlin/Hornet six. Manual transmission gear ratios: (1st) 2.99:1; (2nd) 1.75:1; (3rd) 1.00:1; (Rev) 3.17:1. Torque-Command three-speed automatic transmission optional on all models; column or floor shift selector. Standard axle ratio: (Gremlin) 2.73:1 except 304 V-8 with three-speed manual, 3.15:1 and 304 V-8 with automatic, 2.87:1; (Pacer) 2.73:1 except automatic shift and all with 258 six and two-barrel carb, 3.08:1; (Hornet) 2.73:1 except 304 V-8 with automatic, 2.87:1, and hatchback with floor shift, 3.15:1; (Matador) 2.87:1 except base six with manual shift, 3.54:1, and base six with automatic, 3.15:1. Hotchkiss drive. Steering: (Pacer) rack and pinion; (others) recirculating ball. Suspension: independent front coil springs (Pacer springs mounted between the two control arms); semi-elliptic rear leaf springs except (Matador) coil springs. Brakes: drum; front discs optional (standard on Matador). Breakerless Inductive Discharge electronic ignition. Fuel tank: (Gremlin/Matador wagon) 21 gal.; (Pacer/Hornet) 22 gal.; (Matador coupe/sedan) 24.5 gal. Uses unleaded fuel only.

DRIVETRAIN OPTIONS: 258 cid six-cylinder engine, one-barrel carb: Gremlin/Pacer/Hornet ($69); 258 cid six, two-barrel: Pacer ($99); Pacer with air conditioning ($69). 304 cid V-8, two-barrel: Gremlin ($162); Hornet ($145); Matador ($104) but standard on wagon. 360 cid V-8, two-barrel: Matador Sedan/coupe ($150); wagon ($46). 360 cid V-8, four-barrel, dual exhausts: Matador sedan/coupe ($266); wagon ($162). Three-speed floor shift: Pacer with bucket or individual reclining seats ($21). Three-speed column shift with overdrive: Pacer, Hornet Sportabout with 232 six ($157). Three-speed floor shift with overdrive: Gremlin or Hornet hatchback six ($157); Torque-Command automatic transmission, column shift: Grem./Hornet six, Pacer ($252); Gremlin V-8 ($281); Hornet V-8 ($262); Matador six ($261); Matador 304 cid V-8 ($268); Matador 360 V-8 ($281); Matador wagon ($13). Torque-Command with floor shift lever: Gremlin six, Pacer with bucket or reclining seats, and Hornet Hatchback/Sportabout six ($273); Gremlin, Hornet hatchback/Sportabout V-8 ($283); Matador coupe with 304 V-8 and bucket seats ($330); Matador coupe with 360 V-8 ($343). Optional axle ratio: Gremlin/Pacer/Hornet ($13); Matador ($14). Twin Grip differential: Gremlin/Hornet ($49); Pacer ($49); Matador ($53). Heavy-duty engine cooling system: Matador coupe ($18); others ($25); but standard with air conditioning. Heavy-duty suspension: Gremlin/Hornet ($27) including front sway bar for six-cylinder; Pacer/Matador ($32) with front sway bar on Pacer, rear on Matador. Front sway bar: Pacer ($18) but included with 'X' package. Rear sway bar: Matador ($16). 70-amp battery: Gremlin/Pacer/Hornet ($15); Matador ($16). California emission system ($50). Trailer towing package: Matador V-8 ($116); with air conditioning ($91). Highway cruising package (cruise control and 2.53:1 axle ratio with six, 2.87:1 with V-8): Gremlin with automatic transmission ($59).

GREMLIN/HORNET/MATADOR CONVENIENCE/APPEARANCE OPTIONS: 'X' package: Gremlin ($189); Hornet ($179). Hornet sedan D/L package with "Kasmir Knit" fabric ($169). Sportabout D/L package with "Potomac Stripe" fabric ($350); with vinyl ($309); with touring interior ($214). Touring interior in "Sof-Touch" vinyl: Hornet hatchback/Sportabout ($169). Brougham package: Matador sedan ($179); Matador wagon ($199); Matador coupe ($249). Barcelona package: Matador coupe with Brougham package only ($149). Levi's custom fabric trim package: Gremlin ($89). Opera windows and padded vinyl roof package: Matador coupe ($524). Interior appointment package (parcel shelf, glove box lock, lighter): Gremlin ($21). Interior decor package (appointment package, carpeted cargo area, extra-quiet insulation): Gremlin ($49). Decor package (wheelcovers and moldings): Gremlin/Hornet ($45); Gremlin 'X' ($13); Hornet 'X' or D/L ($14); Matador coupe without Brougham package ($59). Extra Quiet insulation package: Gremlin ($32); Hornet ($31). Protection group: Gremlin/Hornet ($59); Matador coupe ($69), or ($20) with Brougham package; Matador sedan/wagon ($21). Convenience group (dome/reading light, electric clock, stowage containers, dual horns): Gremlin/Hornet ($49); Matador ($52). Visibility group (remote control left mirror, visor mirror, 12 inch day/night mirror, deluxe wipers): Gremlin/Hornet ($45); Matador ($49). Remote control right mirror: Matador ($27), available with visibility group only. Power steering: Gremlin/Hornet ($125); Matador coupe ($136); Matador sedan/wagon ($137). Power front disc brakes: Gremlin/Hornet ($84); Matador ($60). Manual front disc brakes: Gremlin/Hornet ($50). Cruise control: Gremlin/Hornet ($69); Matador ($72). All Season air conditioning: Gremlin/Hornet ($425); Matador ($473). Air conditioning package (includes tinted glass and power steering): Gremlin/Hornet ($579). AM radio: Gremlin/Hornet ($75); Matador ($76). AM/FM/Stereo four-speaker radio ($199). Rear speaker for AM radio: Gremlin/Hornet ($20); Matador ($21). Eight-track tape player and AM/FM/Stereo radio: Matador ($299). Power windows: Matador Brougham sedan/wagon ($138). Power side and tailgate windows: Matador Brougham wagon ($179). Power tailgate window: Matador Brougham wagon ($41). Console: Gremlin/Hornet hatchback ($25). Hidden compartment: Gremlin ($20); Hornet hatchback ($41). Individual fabric reclining seats: Hornet ($104); Matador Brougham wagon ($31); no charge on Matador Brougham coupe. Individual vinyl reclining seats: Hornet sedan ($84); Sportabout ($63); Matador Brougham coupe/sedan ($31). Bucket seats: Hornet hatchback ($49); Matador coupe ($96), but no charge with Brougham package Fabric cushion trim, bench seat: Gremlin ($29); no charge on Hornet sedan/hatchback, Matador. Vinyl cushion trim, bucket seats: Gremlin ($49); Hornet sedan/hatchback ($21); no charge on Sportabout. Vinyl cushion trim, bench seat: Matador coupe ($31). Third seat: Matador wagon ($127). Carpeted cargo area: Gremlin Custom ($15). Deluxe wipers: Gremlin/Hornet ($26); Matador ($28). Rear defogger: Gremlin, Hornet hatchback/Sportabout ($63); Matador coupe/sedan ($73). Rear defogger, blower-type: Hornet sedan ($41). Light group: Gremlin ($26); Hornet ($30); Matador ($29). Fuel economy gauge: Gremlin/Hornet ($26); Matador ($28); but no charge with convenience group. Custom steering wheel: Gremlin ($15). Sports steering wheel (three-spoke): Gremlin ($35); Gremlin Custom, Hornet ($20); Matador ($21). Leather-wrapped steering wheel: Gremlin ($49); Gremlin Custom, Hornet ($34); Hornet hatchback/Sportabout with touring interior ($14); Matador ($35). Tilt steering wheel: Gremlin/Hornet ($52); Matador ($54). Tinted glass: Gremlin ($44); Hornet ($47); Matador ($51). Tinted windshield: Hornet ($36); Matador sedan/wagon ($42). Two-tone paint: Hornet except Sportabout ($36); Matador ($42) except wagon ($79). Special color combinations ($21) except Matador ($22). Vinyl roof: Gremlin ($74); Hornet ($93); Matador coupe/sedan ($105). Rally side stripes: Gremlin ($39); Matador coupe except Barcelona ($41). Woodgrain panels: Hornet Sportabout ($99); included with Sportabout D/L package; Matador wagon ($118). Roof rack: Gremlin ($53); Sportabout ($75); Matador wagon ($62). Locking gas cap ($6). Tailgate air deflector: Gremlin ($22); Matador wagon ($24). Inside hood release: Gremlin/Hornet ($14). Front bumper guards: Gremlin/Hornet ($15); Engine block heater: Gremlin/Hornet ($17); Matador ($18). Wheelcovers (set of four): Gremlin ($32); Hornet ($31); Matador ($34). Custom wheelcovers: Matador ($55), or ($22) with Brougham package. Styled wheels with trim rings for D-size tires: Gremlin/Hornet ($121); with Gremlin Custom 'X' or Hornet hatch/Sportabout 'X' package ($56); Sportabout D/L ($68); Matador ($127), but ($93) with Brougham or Decor package. Aluminum styled wheels for D-size tires: Gremlin/Hornet ($210); Gremlin Custom 'X' or Hornet hatch/Sportabout 'X' package ($145); Sportabout D/L package ($157); Matador ($221), but ($187) with Brougham or Decor package. Space-Saver spare tire: Gremlin or Sportabout with regular wheels ($16); no charge with special wheels/tires. Tires: 6.45 x 14 whitewall: Gremlin ($36). 6.95 x 14: Gremlin V-8 ($16). 6.95 x 14 whitewall: Gremlin V-8 ($52); Hornet ($36); D78 x 14: Hornet ($16). D78 x 14 whitewall: Gremlin with styled wheels ($68); Hornet ($52). D70 x 14 with white letters: Gremlin Custom 'X', Hornet 'X' ($48). DR78 x 14 steel radial: Gremlin ($140); Hornet ($124); Gremlin Custom 'X' or Hornet 'X' ($66). DR78 x 14 whitewall: Gremlin ($176); Hornet ($160); Gremlin Custom 'X' or Hornet 'X' ($102). DR70 x 14 radial with white letters: Gremlin ($212); Hornet ($196); Gremlin Custom 'X' or Hornet 'X' ($139). E78 x 14 whitewall: Matador ($38). E78 x 14 MSR: Matador V-8 coupe ($20). F78 x 14 whitewall: Matador ($58). F78 x 14 MSR: Matador sedan with V-8 ($20). FR78 x 14 steel radial: Matador ($134). FR78 x 14 whitewall: Matador ($172). H78 x 14 whitewall: Matador wagon ($38). HR78 x 14: Matador ($154); wagon ($114). HR78 x 14 whitewall: Matador ($192); wagon ($152).

PACER CONVENIENCE/APPEARANCE OPTIONS: 'X' package ($339). D/L package ($199); not available with 'X' package. Rally package: front console, leather-wrapped steering wheel, tachometer, electric clocks, gauges ($139). Decor package: wheelcovers and exterior moldings ($89; with D/L ($57). Extra Quiet insulation package ($35). Protection group: floor mats, rocker panel moldings, scuff panel extensions ($37). Convenience group: dome/reading light, electric clock, stowage containers, dual horns ($49); with Rally package ($34). Visibility group: remote left mirror, visor mirror, 12 in. day/night mirror, deluxe wipers ($55). Rear visibility package: rear wiper/washer, defogger ($99). Power steering ($125). Power front disc brakes ($84). Manual front disc brakes ($50). Cruise control ($69). All Season air conditioning ($425). AM radio ($75). AM/FM/Stereo four-speaker radio ($199). Rear speaker ($20). Eight-track tape player and AM/FM/Stereo radio ($299). Console ($25). Hidden compartment ($31). Individual reclining seats, D/L package: "Basketry" fabric ($63); "Hyde Park" fabric ($84). Bucket seats, "Sof-Touch" vinyl ($99); with D/L ($70). Vinyl bench cushion trim ($21). Deluxe wipers ($26). Rear defogger ($63). Rear wiper/washer ($52). Light group ($28). Fuel economy light ($18). Sports steering wheel ($20). Leather-wrapped steering wheel ($34); with 'X' package ($14). Tilt steering wheel ($52). Tinted glass ($52). Door vent windows ($30). Two-tone color ($52). Special color combinations ($21). Vinyl roof ($105). Roof rack ($52). Locking gas cap ($6). Bumper guards and nerfing strips ($53); with D/L or 'X' ($34). Engine block heater ($17). wheelcovers: set of 4 ($32). Styled wheels with trim rings for D-size tires: ($121); with 'X' package ($56); with D/L or Decor package ($89). Aluminum styled wheels for D-size tires: ($210); with 'X' ($145); with D/L or Decor package ($178). Tires: 6.95 x 14 whitewall ($36); D78 x 14 ($16); D78 x 14 WSW ($52); DR78 x 14 steel radial ($124); DR78 x 14 WSW ($160); DR70 x 14 radial with white letters ($196). Tires with 'X' package: D78 x 14 whitewall ($36); DR78 x 14 steel radial ($108); DR78 x 14 WSW ($145); DR70 x 14 radial with white letters ($181).

HISTORICAL FOOTNOTES: All 1976 AMC models debuted Sept. 24, 1975. Model year production totaled 283,275 (242,164 sixes and 41,111 V-8s). That came to 3.5 percent of the industry total, down from 3.74 percent the previous year and a healthy 5.3 percent in 1973. Calendar year production amounted to 213,606 units, well below the 323,704 total for 1975. Production halted on June 25, 1976. Calendar year sales came to 247,640, again down markedly from the previous year's 322,272. Hornets sold the best, while Matadors declined the most. While other U.S. automakers recovered from the disastrous sales slump of 1975, AMC fell short of optimistic expectations. Pacer sales, in particular, slackened after an early surge, largely the result of a weakening market for small cars. Production rose from an initial 480 Pacers per day to 800 by late 1975; but a few months later, dealers were glutted. "We were too aggressive," admitted AMC president William V. Luneberg, due to early enthusiasm. Production was slashed, and 2,700 workers laid off. As the model year opened, AMC announced the acquisition of a new plant at Richmond, Indiana to build a two-liter four- cylinder engine, whose design and tooling had been purchased from Audi for $60 million.

1977 AMC

AMC entered the 1977 model year with the same four models as before, ranging from the sub-compact Gremlin to the mid-size Matador. Only the Gremlin received significant restyling, plus a new four-cylinder engine later in the model year. All six-cylinder engines now had a "quench-head" design. Reshaped combustion chambers brought the compressed mixture closer to the spark plug. More models had a catalytic converter or Air Guard system for emission control. Coolant overflow systems were more common. Most sixes had a two-barrel carburetor. Front disc brakes became standard. Manual transmissions were fully synchronized, but the column shifter was gone; both standard three-speed and optional four-speed gearboxes had floor shifters. Standard AMC colors were Classic black, Alpine white, Sand tan, Firecracker red, Brilliant blue, Powder blue, Sunshine yellow, Lime green, Tawny orange and Sun orange; plus nine metallics: Brandywine, Silver Frost, Misty Jade, Mocha brown, Autumn red, Midnight blue, Loden green, Golden Ginger, and Captain blue. Vinyl roofs came in six "bravado grain" colors.

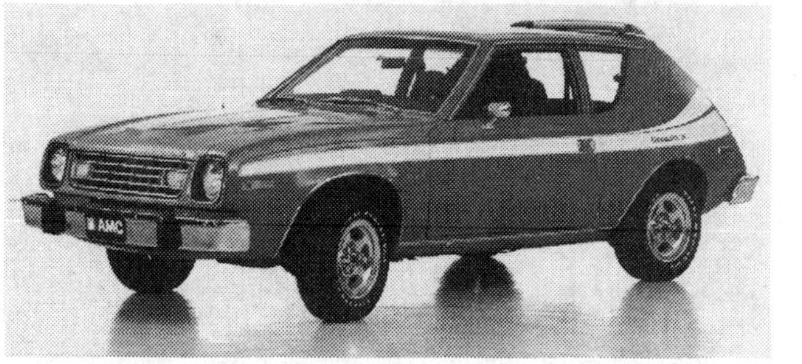

1977 AMC Gremlin X Liftback (AMC)

GREMLIN — SERIES 40 — FOUR/SIX: Gremlin entered the 1977 model year wearing a new grille, bumper and front end sheet metal, enlarged taillights, plus a bigger all-glass rear lift window. It was the first body restyle since the subcompact's 1970 debut. Though four inches shorter than before, with increased rear glass area, Gremlin's basic (and distinctive) appearance was essentially unchanged. The length loss came from reducing the car's front overhang. Rectangular parking lights were inset within the new slanted, four-row eggcrate grille. Optional front bumper guards sat farther apart than before, while horizontal amber markers stood at the forward end of the front fenders. The 'Gremlin' nameplate no longer stood on the front of the hood bulge. But the biggest change didn't come until February 1977, when the four-cylinder 121 cid engine, recently acquired from Volkswagen, was offered under Gremlin hoods. Rated 80 hp, the VW/Audi four had a belt-driven overhead cam, aluminum cross-flow head, and aluminum intake manifold. The 232 cid six remained standard, but the V-8 was gone, enhancing Gremlin's image (and gas mileage ranking) in the economy car sweepstakes. Also appearing this year was a new Borg-Warner four-speed manual transmission with floor shifter. Industry sources show that 17.6 percent of Gremlins carried a four-speed. Foam-cushioned seats were upholstered in Rally perforated vinyl. Standard equipment now included front disc brakes and three-speed floor shift, plus Weather-Eye heater/defroster, front ashtray and lighter, locking glove box, two-speed wiper/washers, dome light, color-keyed carpeting, body paint stripes, folding rear seat, rear bumper guards, and bright moldings for drip rails, wheel lips and rocker panels. Standard tires again were 6.95 x 14 blackwalls. Gremlin's sporty 'X' package used a new striping treatment, flowing at the rear in a slight curve rather than the familiar "hockey stick" shape. That package included bucket seats trimmed in Hot-Scotch plaid fabric, sports steering wheel, instrument panel overlay, 'X' decals on body stripes and glove box door, a lower back panel stripe decal, D70 x 14 tires on slot-styled wheels, and extra-quiet insulation. The unique Levi's trim package was available again this year.

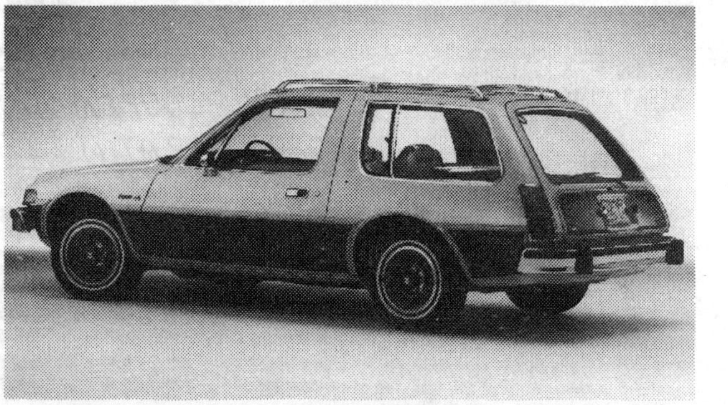

1977 AMC Pacer D/L station wagon (AMC)

PACER — SERIES 60 — SIX: Biggest news for the year-and-a-half old Pacer was the addition of a station wagon to the existing hatchback sedan. Adding cargo space to the already roomy design, it was meant to challenge Chrysler's new compact wagons. Overly optimistic production plans had to be cut back, however, since domestic small cars weren't selling well. With rear seat folded, the wide wagon, just four inches longer than the sedan, held 47.8 cu. ft. of cargo. Styling was identical to the hatchback sedan from the doors forward, but wagons had vertically-oriented three-section taillamps, unlike the sedan's horizontal units. The wagon's wide lift-up hatch reached down nearly to the bumper for easy loading. Large side windows included "flipper" vents for improved ventilation. Despite its short wheelbase, the wide (77 in.) Pacer ranked as a compact. Standard engine was again the 232 cid six, with a 258 cid six optional. Foam-cushioned bench seats were trimmed in Rally perforated vinyl on wagons, in Basketry Print fabric on hatchback sedans. Standard equipment included a three-speed floor shift, front disc brakes, 6.95 x 14 blackwall tires (D78 x 14 on the wagon), heater/defroster, two-speed wiper/washers, dome light, ashtray and lighter, built-in assist on the rear hatch, color-keyed carpeting, folding second seat, and body side scuff moldings. Wagons had a locking stowage compartment in the rear quarter panel. Pacer's 'X' sedan package included perforated vinyl bucket seats, woodgrain dash overlay, sports steering wheel, custom door trim with vinyl inserts and door pull straps, upper door moldings, DR78 x 14 tires on slot-styled wheels, plus high level ventilation and extra-quiet insulation packages. The D/L package contained individual reclining seats, custom door panels with assist straps, woodgrain dash overlay, custom steering wheel, light group, dual horns, day/night mirror, high level ventilation, wheel lip and rocker panel moldings, rear wheelhouse pads, bumper nerfing strips, extra-quiet insulation, and D78 x 14 whitewalls on styled wheels. D/L wagon bodies had woodgrain side/rear overlays.

1977 AMC Hornet D/L four-door sedan (AMC)

HORNET AND AMX — SERIES 01 — SIX/V-8: For its final year, the Hornet line was highlighted by a limited edition AMX hatchback package, bringing back a nameplate that many remembered for sporty performance a few years earlier. In addition to 'AMX' graphics between door and rear wheelhouse, and at rear, the package included a front air dam, color-coordinated bumpers, blacked-out grille, body-color rear window louvers, Euro-style brushed aluminum targa roof band, twin flat black mirrors, floor console, gauges (including tachometer), Soft-Feel steering wheel, and brushed aluminum instrument panel overlay. Flared fenders topped DR78 x 14 tires. Base Hornet engine remained the 232 six coupled to three-speed floor shift, with a 258 six or 304 V-8 optional. Sixes could get the new four-speed manual shift. AMX hatchbacks required the 258 with four-speed, or V-8 with automatic. Standard Hornet equipment included rear bumper guards, corner nerfing strips on front bumpers, color-keyed carpeting and mats, Weather-Eye heater/defroster, locking glove box, and two-speed wiper/washers. Four-door V-8 sedans and wagons wore standard D78 x 14 tires; other models, 6.95 x 14 blackwalls. All trim packages and options were upgraded. Such luxuries as vinyl bucket seats, sports steering wheel and a carpeted cargo area were standard on hatchback Hornets. Foam-cushioned bench seats were trimmed in Veracruz fabric on sedans, Rally perforated vinyl on wagons. Both 'X' and D/L packages were offered again. D/L models now included reclining front seats in Veracruz fabric on sedans (Rally perforated vinyl on wagons) plus dual horns, light group, woodgrain dash overlay, 12 in. day/night mirror, D78 x 14 whitewalls, and dual body side stripes (sedans) or woodgrain body panel overlays (wagons).

1977 AMC Matador coupe (AMC)

MATADOR (COUPE) — SERIES 10 — SIX/V-8: Unchanged externally, the new Matadors contained interior packages that had formerly been offered only in the luxury Brougham model (which was dropped this year). Engine choices remained as in 1976, but the high-performance 360 cid V-8 was gone, as was the police 401. Standard equipment now included a 258 cid six, Torque Command automatic transmission, power steering and front disc brakes, individual reclining seats in Brampton Plaid fabric, heater/defroster, color-keyed carpeting, front/rear ashtrays, front lighter, extra-quiet insulation, two-speed wiper/washers, light group, protective body side scuff moldings, wheelcovers, bumper guards, front sway bar, day/night mirror, and custom steering wheel. F78 x 14 blackwalls were standard. The luxury Barcelona package added tan nap knit fabric seats, color-keyed door trim panel panels and headlining, tan grille surround and headlight bezels, tan license plate depression and wheelcovers, plus a special hood ornament, insignias and nameplates.

1977 AMC Matador four-door sedan (AMC)

MATADOR (SEDAN AND WAGON) — SERIES 80 — SIX/V-8: As in 1976, the coupe was officially a different series than the Matador sedan and wagon. Base engine remained a 258 cid six with one-barrel carburetor, with 304 or 360 cid V-8 optional. (The 304 V-8 was standard in wagons and California Matadors). Standard equipment was similar to the coupe, but wagons wore H78 x 14 blackwalls and their seats were covered in Crush-Grain vinyl.

I.D. DATA: As before, the 13 symbol Vehicle Identification Number (VIN) was embossed on a metal plate riveted to the upper left surface of the instrument panel. Coding was the same as 1976, with the following changes. Model year code (second symbol) changed to 7=1977. Codes D, O and S=column-shift manual and overdrive transmissions (third symbol) were dropped; code M=four-speed floor shift was added. Only two digits were now used for model/group (sixth symbol): 5=Gremlin; 7=Pacer/Hornet/Matador. Engine letters (seventh symbol) changed to: G=121 four-cylinder, two-barrel; E=232 six-cylinder, one-barrel; A=258 six-cylinder, one-barrel; B=258 six-cylinder, two-barrel; H=304 V-8 two-barrel; N=360 V-8 two-barrel. In addition to month and year built, plus the VIN, the safety sticker attached to left front door now showed a safety compliance statement and consumer information (vehicle class, acceleration and passing figures, tire reserve load, and stopping distance). A federal emission control information label in the engine compartment identified the engine type and gave basic tune-up specs. The unit body identification plate was the same as 1976.

GREMLIN SERIES 40 (SIX-CYLINDER)

Series Number	Body/Style Number	Body Type & Seating	Factory Price	Shipping Weight	Production Total
40	7746-5	2dr Sed-4P	2995	2811	Note 1
40	7746-7	2dr Cus Sed-4P	3248	2824	Note 1

GREMLIN SERIES 40 (FOUR-CYLINDER)

40	7746-4	2dr Sed-4P	3248	2564	7558

NOTE 1: Total model year production, 46,171 Gremlins (38,613 six-cylinder).

PACER SERIES 60

60	7766-7	2dr Sed-4P	3649	3156	20,265
60	7768-7	2dr Sta Wag-4P	3799	3202	37,999

HORNET SERIES 01

01	7703-7	2dr Hatch-5P	3499/3662	3012/3245	11,545
01	7705-7	4dr Sed-6P	3449/3613	3035/3268	31,331
01	7706-7	2dr Sed-6P	3399/3563	2971/3204	6076
01	7708-7	4dr Sta Wag-6P	3699/3863	3100/3333	28,891

NOTE 2: Total model year production, 77,843 Hornets (73,752 with six-cylinder engine, 4091 with V-8).

MATADOR (COUPE) SERIES 10

10	7716-7	2dr Cpe-5P	4499/4619	3704/3872	6825

MATADOR (SEDAN AND WAGON) SERIES 80

80	7785-7	4dr Sed-6P	4549/4669	3713/3876	12,944
80	7788-7	4d Sta Wag-6P	—/4899	—/4104	11,078

NOTE 3: Model year production of Matador coupe and sedan totaled 19,769 units (2447 sixes and 17,322 V-8s).

NOTE: Data above slash for six-cylinder/below slash for V-8.

ENGINES: BASE FOUR (Gremlin): Inline. Overhead cam. Four-cylinder. Cast iron block; cast aluminum alloy head. Displacement: 121 cid (2.0 liters). Bore & stroke: 3.41 x 3.32 in. Compression ratio: 8.1:1. Brake horsepower: 80 at 5000 rpm. Torque: 105 lbs.-ft. at 2800 rpm. Five main bearings. Solid valve lifters. carburetor: two-barrel Holley 5210. BASE SIX (Gremlin, Hornet, Pacer): Inline. OHV. Six-cylinder. Cast iron block. Displacement: 232 cid (3.8 liters). Bore & stroke: 3.75 x 3.50 in. Compression ratio: 8.0:1. Brake horsepower: 88 at 3400 rpm. Torque: 164 lbs.-ft. at 1600 rpm. Taxable hp: 33.75. Seven main bearings. Hydraulic valve lifters. carburetor: one-barrel Carter YF. BASE SIX (Matador coupe/sedan): Inline. OHV. Six-cylinder. Cast iron block. Displacement: 258 cid (4.2 liters). Bore & stroke: 3.75 x 3.90 in. Compression ratio: 8.0:1. Brake horsepower: 98 at 3200 rpm. Torque: 193 lbs.-ft. at 1600 rpm. Seven main bearings. Hydraulic valve lifters. Carburetor: one-barrel Carter YF. OPTIONAL SIX (Gremlin, Hornet, Pacer): Same as 258 cid six above, but with Carter BBD two-barrel carburetorur. Horsepower: 114 at 3600 rpm. Torque: 192 lbs.-ft. at 2000 rpm. BASE V-8 (Matador station wagon); OPTIONAL (Hornet, other Matadors): 90-degree, overhead valve V-8. Cast iron block. Displacement: 304 cid (5.0 liters). Bore & stroke: 3.75 x 3.44 in. Compression ratio: 8.4:1. Brake horsepower: 121 at 3450 rpm. (Matador, 126 at 3600). Torque: 219 lbs.-ft. at 2000 rpm. Taxable hp: 45. Five main bearings. Hydraulic valve lifters. Carburetor: two-barrel Motorcraft 2100. OPTIONAL V-8 (Matador): 90-degree, overhead valve V-8. Cast iron block. Displacement: 360 cid (5.9 liters). Bore & stroke: 4.08 x 3.44 in. Compression ratio: 8.25:1. Brake horsepower: 129 at 3700 rpm. Torque: 245 lbs.-ft. at 1600 rpm. Taxable hp: 53.3. Five main bearings. Hydraulic valve lifters. Carburetor: two-barrel Motorcraft 2100.

CHASSIS DATA: Wheelbase: (Gremlin) 96 in., (Pacer) 100 in.; (Hornet) 108 in.; (Matador coupe) 114 in.; (Matador sedan/wagon) 118 in. Overall length: (Gremlin) 166.4 in.; (Pacer sedan) 170 in., (Pacer wagon) 174 in.; (Hornet) 186 in.; (Matador coupe) 209.4 in.; (Matador sedan) 216 in., (Matador wagon) 215.5 in. Height: (Gremlin) 52.3 in.; (Pacer) 52.7 in.; (Pacer wagon) 53.0 in.; (Hornet) 52.2 in. except four-door sedan, 52.7 in.; (Matador coupe) 51.8 in.; (Matador sedan) 54.7 in.; (Matador wagon) 56.8 in. Front Tread: (Gremlin/Hornet) 57.5 in.; (Pacer) 61.2 in.; (Matador coupe) 59.7 in.; (Matador sedan) 59.8 in. Rear Tread: (Gremlin/Hornet) 57.1 in.; (Pacer) 60.2 in.; (Matador) 60.0 in. Standard Tires: (Gremlin) 6.45 x 14; (Pacer sedan, Hornet) 6.95 x 14; (Pacer wagon; Hornet V-8 sedan/wagon) D78 x 14; (Matador coupe/sedan) F78 x 14; (Matador wagon) H78 x 14.

TECHNICAL: Three-speed, fully synchronized manual floor-shift transmission was standard on all except Matadors, which had standard automatic transmission. Four-speed floor shift optional. Three-speed manual gear ratios: (1st) 2.99:1; (2nd) 1.75:1; (Rev) 3.17:1. Four-speed ratios: (1st) 3.50:1; (2nd) 2.21:1; (3rd) 1.43:1; (Rev) 3.39:1. Torque-Command three-speed automatic transmission optional on Gremlin, Hornet and Pacer; column or floor shift selector. Three-element torque converter. Automatic gear ratios: (Low) 2.45:1; (Intermediate) 1.45:1; (High) 1.00:1; (Rev) 2.20:1. Standard axle ratios: (Gremlin/Pacer/Hornet six) 2.73:1; 2.53:1 and 3.08:1 optional. (Hornet V-8) 2.87:1; 3.15:1 optional. (Matador six) 3.15:1. (Matador V-8) 2.87:1; 3.15:1 and 3.54:1 available. Steering: (Pacer) rack and pinion; (others) recirculating ball. Suspension: independent front coil springs (Pacer springs mounted between the two control arms); semi-elliptic rear leaf springs except (Matador) coil springs. Brakes: front discs; rear drum. Breakerless Inductive Discharge electronic ignition. Fuel tank: (Gremlin and Matador wagon) 21 gallon; (Pacer/Hornet) 22 gallon; (Matador coupe/sedan) 24.5 gallon. Uses unleaded fuel only.

DRIVETRAIN OPTIONS: 258 cid six-cylinder engine, two-barrel carburetor: Gremlin/Pacer/Hornet ($79). 304 cid V-8, two-barrel: Hornet ($164); Matador ($179) but standard on wagon. 360 cid V-8, two-barrel: Matador sedan/coupe ($179); wagon ($59). Four-speed manual floor shift: Gremlin, Pacer, Hornet hatchback ($105). Torque-Command automatic transmission, column shift: Gremlin, Hornet except hatchback, Pacer ($267). Torque-Command with floor shift lever: Gremlin, Pacer Hornet hatchback/Sportabout ($289); Matador coupe with V-8 and bucket seats ($66). Twin Grip differential: Gremlin/Hornet/Pacer ($52); Matador ($56). Heavy-duty engine cooling system: Gremlin ($49); others ($27);

but standard with air conditioning. Maximum cooling system: Matador V-8 ($45); Matador V-8 with air ($18). Auxiliary automatic transmission oil cooler: Matador ($32). Heavy-duty suspension: Gremlin/Hornet ($29); Pacer/Matador ($34) with front sway bar on Pacer, rear on Matador. Front sway bar: Gremlin/Hornet/Pacer ($17) but included with radial tires. Rear sway bar: Matador ($17). Air-adjustable shock absorbers: Matador ($45); with heavy-duty suspension ($41). 70-amp battery: Gremlin/Pacer/Hornet ($16); Matador ($18). California emission system ($53). High altitude package ($15). Gremlin performance package: 258 engine, four-speed, sports steering wheel, heavy-duty suspension ($259); with floor-shift automatic ($443). Pacer performance package: same as Gremlin but including tach, clock and gauges ($349); with floor-shift automatic ($533). Hornet hatchback performance package: 258 engine, four-speed, Soft-Feel steering wheel, heavy-duty suspension ($239); with floor-shift automatic transmission ($423).

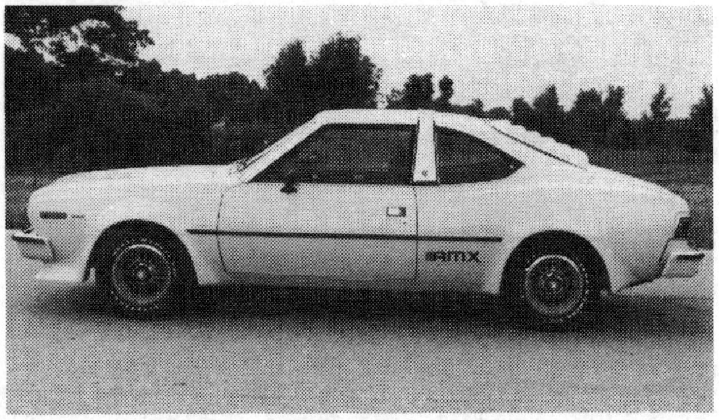

1977 AMC Hornet coupe (AMC)

GREMLIN/HORNET/MATADOR CONVENIENCE/APPEARANCE OPTIONS: AMX package: Hornet hatchback ($799). 'X' package: Gremlin ($299); Hornet hatchback/wagon ($199). Hornet D/L package: sedan ($299); wagon ($399). Barcelona package: Matador Super ($158). Levi's custom fabric trim package: Gremlin ($99); Gremlin 'X' ($50); Hornet hatchback ($49). Extra-Quiet insulation package: Gremlin/Hornet ($34); included with Gremlin 'X' package Protection group: Gremlin/Hornet ($62). Interior decor/convenience group (vanity mirror, dome/reading light, clock, stowage containers, rubber mats): Gremlin ($65); Hornet ($71); Matador ($86). Visibility group (remote control left mirror, manual right mirror, day/night mirror, deluxe wipers): Gremlin/Hornet ($67). Visibility group (dual remote mirrors, deluxe wipers): Matador ($76). Deluxe visibility group (adds rear defogger): Gremlin, Hornet hatch/wagon ($134); Hornet sedan ($111); Matador coupe ($154). Remote control left mirror: Gremlin/Hornet ($15); Matador ($16). Power steering: Gremlin/Hornet ($133). Power front disc brakes: Gremlin/Hornet ($60). Cruise control: Hornet ($60); Matador ($77). All Season air conditioning: Gremlin/Hornet ($451). Air conditioning package (included tinted glass and power steering): Gremlin/Hornet ($619); Matador ($557). AM radio: Gremlin/Hornet ($80); Matador ($81). AM/FM/Stereo four-speaker radio ($211). Rear speaker for AM radio ($21-22). Eight-track tape player and AM/FM/Stereo radio: Matador $317). Power windows: Matador sedan/wagon ($146). Power side and tailgate windows: Matador wagon ($190). Power tailgate window: Matador wagon ($44); standard in three-seat. Console: Gremlin/Hornet hatchback ($27). Hidden compartment: Hornet hatchback ($44). Individual reclining seats: Hornet sedan/wagon ($67). Bucket seats: Matador coupe ($102). Custom vinyl door panel/bucket seat trim: Gremlin ($49); NC with bench seats or 'X' package Third seat: Matador wagon ($135). Rear defogger: Gremlin, Hornet hatch/wagon ($67); Matador coupe/sedan ($78). Rear defogger, blower-type: Hornet sedan ($44). Dual horns: Gremlin/Hornet ($11). Light group: Gremlin ($28); Hornet ($33). Sports steering wheel: Gremlin/Hornet hatchback ($21); Matador ($22). Soft-Feel sports steering wheel: Gremlin, Hornet sedan/wagon ($36); Hornet hatchback or with 'X' package ($11); Matador ($37). Tilt steering wheel: Hornet sedan/wagon ($55); Matador ($57). Tinted glass: Gremlin ($47); Hornet ($50); Matador ($55). Two-tone paint: Hornet except Sportabout ($38); Matador ($45) except wagon ($84). Special color combinations ($23). Vinyl roof: Hornet ($99); Matador coupe/sedan ($111). Rally side stripes: Gremlin ($41); Matador coupe ($43). Woodgrain paneling: Hornet Sportabout except 'X' ($105); included w/Sportabout D/L package; Matador wagon ($125). Roof rack: Matador wagon ($66). Roof rack and air deflector: Hornet wagon ($80). Tailgate air deflector: Matador wagon ($25). Inside hood release: Gremlin ($15); Hornet ($25). Front bumper guards: Gremlin/Hornet ($16). Bumper nerfing strips: Matador coupe ($25). Door edge guards: Matador ($10). Engine block heater ($18-19). Wheelcovers (set of 4): Gremlin, Hornet sedan/wagon ($34). Custom wheelcovers: Gremlin, Hornet sedan/wagon ($56); Hornet hatchback, Matador ($22). Styled wheels with trim rings for D-size tires: Gremlin, Hornet sedan/wagon ($128); Hornet hatchback ($94); with 'X' or D/L package ($63). Aluminum styled wheels for D-size tires: Gremlin, Hornet sedan/wagon ($223); Hornet hatchback ($189); with 'X' or D/L ($158). Aluminum styled wheels: Matador ($198). Space-Saver spare tire: Gremlin or Sportabout with regular wheels ($17); no charge with special wheels/tires. Tires: 6.45 x 14 whitewall: Gremlin ($38). 6.95 x 14: Gremlin $17). 6.95 x 14 whitewall: Gremlin ($55); Hornet ($38). D78 x 14: Gremlin ($34); Hornet ($17). D78 x 14 WSW: Gremlin ($72); Hornet ($55); included with Hornet D/L. D70 x 14 with white letters: Gremlin, Hornet hatchback/wagon 'X' ($51). DR78 x 14 steel radial: Gremlin ($148); Hornet ($131); with Hornet D/L ($76); with 'X' package ($71). DR78 x 14 whitewall: Gremlin ($186); Hornet ($169); with Hornet sedan/wagon D/L ($114); with 'X' ($109). DR78 x 14 white-letter radial: Hornet ($208); with D/L ($153). DR70 x 14 radial with white letters: Gremlin ($225); Gremlin or Hornet hatch/wagon with 'X' ($148). F78 x 14 whitewall: Matador coupe/sedan ($40). FR78 x 14 WSW steel radial: Matador ($160). H78 x 14 WSW: Matador wagon ($40). HR78 x 14 WSW steel radial: Matador ($182); wagon ($160).

PACER CONVENIENCE/APPEARANCE OPTIONS: 'X' package: sedan ($379). D/L package: sedan ($349); wagon ($379). Decor package: body moldings ($77); included with D/L package Extra-Quiet insulation package ($37); included with 'X' and D/L pkgs. Interior decor/convenience group: lighted vanity mirror, dome/reading light, clock, stowage containers, rubber mats ($82). Visibility group: remote left mirror, right mirror, day/night mirror, deluxe wipers ($67). Deluxe visibility group: same, plus rear defogger ($134). Power steering ($133). Power disc brakes ($60). Cruise control ($73). All Season air conditioning ($451). AM radio ($80). AM/FM/Stereo four-speaker radio ($211). Rear speaker ($20). Eight-track tape player and AM/FM/Stereo radio ($317). Console ($27). Hidden compartment: sedan ($33). Individual reclining seats ($69); included with D/L package Vinyl bucket seats ($69); standard with 'X' package; no charge with D/L. Levi's custom fabric trim ($99); with 'X' or D/L ($30). Rear defogger with tinted glass ($67). Rear wiper/washer ($56). Light group ($30). Sports steering wheel ($21); included with 'X' package Soft-Feel sports steering wheel ($36); with 'X' package ($15). Tilt steering wheel ($55). Tinted glass ($20). Door vent windows ($32). Two-tone color: sedan ($55). Special color combinations ($23). Vinyl roof ($111). Roof rack ($56). Bumper guards and nerfing strips ($56). Bumper nerfing strips (#24); included with D/L package Bumper guards ($32). Engine block heater ($18). Wheelcovers: set of four ($34). Styled wheels for D-size tires: ($128); with sedan 'X' package ($63); included in D/L package Aluminum styled wheels for D-size tires: ($223); with 'X'sedan ($158); with D/L wagon ($95). Tires: 6.95 x 14 whitewall, sedan ($38); D78 x 14, sedan ($17); D78 x 14 whitewall: sedan ($55); wagon ($38); no charge with D/L. DR78 x 14 radial: sedan ($131); wagon ($114); with D/L ($76). DR78 x 14 whitewall radial: sedan ($169); wagon ($152); with D/L ($114). DR70 x 14 radial with white letters: sedan ($208); wagon ($191); with D/L ($153). Tires with sedan 'X' package: DR78 x 14 whitewall steel radial ($38); DR70 x 14 white-letter radial ($77).

HISTORICAL FOOTNOTES: All 1977 AMC models were introduced Oct. 5, 1976. Model year production: 213,125 (173,076 sixes, 32,491 V-8s and 7,558 Gremlin fours); production in U.S. only, 182,005. For calendar year 1977, U.S. production amounted to 156,994 units. Calendar year sales: 184,361 (2.0 percent of the industry total), down from 247,640 the year before. Model year sales (worldwide) amounted to 246,640, a 23 percent drop from 1976. The 1977 models began production at Kenosha on August 2. Although the Jeep business remained strong (and profitable), boosting AMC's fiscal status, passenger cars weren't selling well as the 1977 model year ended. After a $46.3 million loss in fiscal 1976, the company ended 1977 in the black to the tune of almost $8.3 million, with $2.2 billion in sales. AMC raised prices almost five percent (average) for 1977 and experienced inventory troubles, as dealers had too many '77 models on hand. This led to price cuts and rebates at the end of 1976, in an attempt to sell off the bloated stocks of Gremlins and Pacers. The intensive marketing campaign, which included such extras as free air conditioning, didn't help enough. The restyled Gremlin slipped 12 percent, Pacer sales slid 23 percent, while the Matador, nearing its last days, fell 37 percent. The production line closed periodically, for a week at a time. AMC was feeling pressured to strengthen its role as a producer of small cars, as other automakers had been introducing compact models-and many more were to come in the next few years. Although the Gremlin had been the first domestic vehicle to go head-to-head against the imports, neither it nor other AMC models were now selling well as the downsizing era began. Analysts remained puzzled as to the reason. Though common in Europe, the agreement to purchase completed four-cylinder engines from VW/Audi was new to the U.S. Those engines were delivered intact to AMC's new plant at Richmond, Indiana for hot testing. As of January 1977, there were 1,690 AMC dealer franchises. AMC's Buyer Protection Plan II now included a 24 month/24,000 mile engine/drivetrain warranty. Management received an ample shakeup. AMC president and CEO William V. Luneberg retired in May 1977, succeeded by Gerald C. Meyers. R.D. Chapin, Jr. remained as chairman, but turned over much of the responsibility to Meyers and his team.

1978 AMC

Three of the four 1977 models carried over into the 1978 model year. The Hornet name disappeared, replaced by a similar Concord compact. Instead of an option package for the Hornet, the AMX became a full-fledged model. This would be the final year for both the subcompact Gremlin and mid-size Matador. The Solid State Ignition (SSI) system introduced on some 1977 Canadian models was now standard. A new antimony battery never needed extra water. An AM/FM/stereo radio with Citizens Band and four speakers joined the option list (except on Gremlins). So did a digital clock. Standard colors were Classic black, Alpine white, Powder blue, Captain blue metallic, Midnight blue metallic, Sunshine yellow, Sand tan, Golden Ginger metallic, Mocha brown metallic, Sun orange, khaki, British bronze metallic, Loden green metallic, Quick silver metallic, Firecracker red, Autumn red metallic, and Claret metallic.

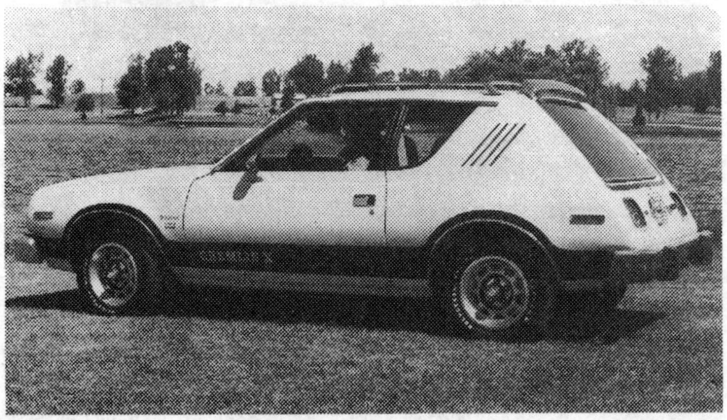

1978 AMC Gremlin X Levi's Liftback (AMC)

GREMLIN — SERIES 40 — FOUR/SIX: Following its 1977 restyle, Gremlin turned to interior refinements for the '78 model year. These included color-keyed carpeting, a custom steering wheel, and a new instrument panel with standard AM radio. Engineering improvements cut down on engine noise and exhaust manifold rattles. Thirteen colors were available on Gremlins. Air conditioning was available with the four-cylinder engine. Custom Gremlins now sported standard whitewall B78 x 14 tires with wheelcovers, plus scuff moldings. Standard equipment also included a 232 cid six-cylinder engine (or 2.0 liter four) with electronic ignition, three-speed floor shift, manual front disc brakes, heater/defroster, vinyl seat upholstery, rear bumper guards, body paint stripes, locking glove box, and rocker panel moldings. Custom models also had wheel lip moldings and a parcel shelf, plus vinyl bucket seats up front; base Gremlins had bench seats front and rear. Six-cylinder Customs could come equipped with four-speed manual floor shift. Gremlin's 'X' package, offered only on Custom models, included bucket seats and interior trim in Levi's fabric; sport steering wheel; lower body decal stripes with contrasting pinstripes and "Gremlin X" insignia on front door portion; Levi's decal; brushed aluminum instrument panel overlays; extra-quiet insulation; decals on back panel and 'C' pillar; black wiper arms, 'B' pillars and side window frames. A front sway bar and DR78 x 14 blackwall radial tires on slot styled 14 x 6 in. wheels completed the 'X' package. Early in 1978, a performance-oriented Gremlin GT hit the market, wearing fiberglass body components. Actually a $649 option package, the GT's external features included a body-colored front air dam with striping, front and rear fender flares matched to body color, black side stripes with color-keyed pinstriping, color-keyed front/rear bumper guards and nerfing strips, black grille insert, black mirrors (left one remote-controlled), hood striping, black door and quarter-window frames, and black wiper arms. GT Gremlins rode DR70 x 14 white-letter radials on spoke-style wheels with trim rings, and had a front sway bar. Bucket seats were Soft-Feel vinyl, with Levi's upholstery optional. Motoring niceties also included a sports steering wheel, gauge package, brushed aluminum instrument panel overlay, extra-quiet insulation package, and day/night mirror.

1978 AMC Pacer D/L station wagon (AMC)

PACER — SERIES 60 — : Performance-minded Pacer buyers could now choose an optional 304 cid V-8 rather than the 232 or 258 . Again offered in hatchback sedan and wagon body styles, the panoramic Pacer also gained a new hood and grille, a longer station wagon liftgate, plus improved seat and legroom up front. The new, upward-bulging eggcrate grille, dominated by horizontal bars, had two lower rows extending full-width between the headlights; two narrower upper bars protruded up into the hood area. Huge parking lights wrapped around, from headlights well into fender sides. Some of that new front-end restyling was done to squeeze in the V-8 engine. Various items that had formerly been optional now became standard, including individual reclining front seats, rear armrests, a day/night rearview mirror, electric clock, custom steering wheel, woodgrain instrument panel, and color-keyed wheelcovers. Also standard were a light group, cigarette lighter, extra-quiet insulation package, dual horns, inside hood release, locking glove box, rocker panel and wheel lip moldings, hood and fender moldings, heater/defroster, carpeting, and bumper nerfing strips. Base engine was the 232 cid with three-speed floor shift and D78 x 14 whitewall tires. As before, Pacers featured rack-and-pinion steering and an isolated suspension design, plus an oversized passenger door. Fifteen body colors were offered, including new Quick silver, khaki, British bronze, and Claret. Pacer's Sport package, available on the hatchback sedan with 258 or 304 V-8, included Soft-Feel vinyl bucket seats, a sports steering wheel, DR78 x 14 blackwall tires on slot style wheels with trim rings, and two-tone paint on lower body sides. The 'X' package was dropped.

1978 AMC Concord D/L coupe (AMC)

CONCORD — SERIES 01 — FOUR . For 1978, the compact Hornet was reworked to become a luxurious Concord, with the same 108 in. wheelbase and four body styles: two- and four-door sedan, hatchback, or station wagon. The hatchback coupe was a particularly attractive design, with huge cargo space. Sporty too, with an optional 'X' package. Styling features included rectangular headlamps and bezels, a bright V-8-section crosshatch grille with clear squarish parking lights behind the outer sections, and tri-color horizontal taillamps. All Concords had front disc brakes, front sway bar, inside hood release, rocker panel moldings, hood ornament, and full wheelcovers. Walnut-vinyl inserts accented instrument panels. Base Concords ran with a 232 cid V-8 and three-speed floor shift. Automatic shift and a 304 V-8 were optional, plus a four-cylinder engine (formerly on Gremlins alone) early in 1978. Sedans had Velveteen fabric bench seats in five colors; hatchbacks, Soft-Feel vinyl buckets. Standard equipment also included a heater/defroster, color-keyed carpeting, padded door trim panels, lighter, rear bumper guards, dual body side pinstripes, color-keyed scuff moldings, locking glove box, and C78 x 14 blackwall tires. Wagons carried a space-saver spare tire and held a locking hidden compartment. Hatchbacks had a sports steering wheel and flipper-type rear vent windows; four-doors, roll-down rear windows. Concord's Sport package, available on all models with 258 V-8 or 304 V-8, included Soft-Feel vinyl bucket seats, with individual reclining seats (vinyl or velveteen crush fabric) optional in sedans and wagons. It also included a sport steering wheel, slot-style wheels with DR78 x 14 blackwall steel-belted radials, brushed aluminum instrument panel overlays, wide rocker panel moldings, lower body side tape stripe, and extra-quiet insulation. For luxury, Concord offered the D/L package on all, but the hatchback. It featured individual reclining seats in velveteen crush fabric (Soft-Feel vinyl in wagons) plus map pockets in custom door trim panels, woodgrain instrument panel overlays, light group, dual

horns, hood pinstripes, parcel shelf, digital clock, wide rocker moldings, bumper nerfing strips front and rear, extra-quiet insulation, day/night mirror, and D78 x 14 whitewalls with color-keyed wheelcovers. D/L sedans had a landau vinyl roof and color-keyed scuff moldings, plus trunk carpeting and spare tire cover; two-doors, opera windows with silver accents. D/L wagons came with woodgrain side overlays (which could be deleted) and cargo floor skid strips. A Touring Wagon package added unique door trim panels in beige with orange/brown accents, leather-wrapped steering wheel, and wide brown/orange scuff moldings.

1978 AMC Hornet AMX coupe (AMC)

AMX HATCHBACK — SERIES 01 — . Reviving a name associated with performance a few years earlier, AMC tried once again to capture a slice of the youth/performance market with a new AMX, first introduced as a Hornet option in 1977. Built on the Concord platform, with an all new, slightly wedge-shaped front, AMX came with a load of appealing features, heavy on black. They included black fender flares front and rear, rear window louvers, an 'AMX' decal just to the rear of the doors, front air dam, painted bumpers with guards and nerfing strips, black scuff moldings, dual flat black mirrors (the left one remote-controlled), color-coordinated slot wheels-plus a blacked-out grille with round signal lights and 'AMX' emblem in the center. Inside, rivers found a floor console, Rally gauges with tachometer, map pockets in custom door trim panels, brushed aluminum instrument panel overlays, a black Soft-Feel sport steering wheel, day/night mirror, package tray, special graphics, and inside hood release. Soft-Feel vinyl bucket seats came in black, blue or beige, with Levi's trim package optional. A brushed aluminum roof band with special insignia helped make the AMX easy to spot. Five colors were offered: Alpine white, Firecracker red, Sunshine yellow, Quick silver metallic, and Classic black. That black AMX (a $49 option) had unique gold Rally striping on the roof band, front doors and fenders, plus gold stripe on black slot-styled wheels and black windshield reveal moldings. The optional decal package with Classic black body included gold/orange decals (black/orange with other body colors). A 258 cid with four-speed floor shift was standard. Automatic shift was required for the optional 304 cid V-8. Standard AMX equipment also included front disc brakes, front sway bar, and DR78 x 14 blackwall steel-belted radial tires.

1978 AMC Matador Barcelona Brougham coupe (AMC)

MATADOR (COUPE) — SERIES 10 — . For their final year, Matador coupes changed little in outside appearance but added some luxury features. The list includes power steering, power front disc brakes, automatic transmission, coolant recovery system, electric clock, dual horns, and a 12 in. day/night mirror. Individual reclining seats were upholstered in velveteen crush fabric. The light group consisted of a glove box light, engine compartment light, courtesy lights, and ashtray light. Matadors also had extra-quiet insulation, front/rear bumper guards, and a front sway bar. The coupe's Barcelona package added individual reclining seats in velveteen crush fabric with woven accent strips, plus custom door trim panels, a unique headliner, 24 oz. carpeting, headlight bezels painted in accent color, black trunk carpet, a rear sway bar, body-colored front/rear bumpers, bumper nerfing strips, landau padded vinyl roof, opera quarter windows with accents, and dual remote-control mirrors painted in body color. Barcelona medallions stood on fenders and glove box door. Two-tone combinations included Golden Ginger metallic on Sand tan, or Autumn red metallic on Claret. Standard engine was a 258 cid 6 , with 360 V-8 optional.

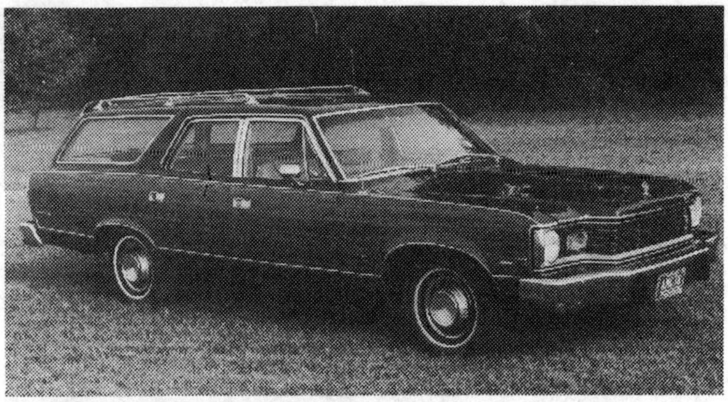

1978 AMC Matador station wagon (AMC)

MATADOR (SEDAN AND WAGON) — SERIES 80 — V-8: Like the coupe, Matador sedans and wagons changed little outside, but added the new luxury items. Station wagon seats were upholstered in crush grain vinyl. Sedans came with a 258 cid V-8, but wagons carried the 304 V-8 as standard equipment. Both had three-speed automatic transmission and door vent windows. Standard equipment was the same as the coupe, except for the coolant recovery system. Matadors came in 13 body colors, including new Quick Silver and Claret metallics, plus seven vinyl roof colors. The Barcelona package, formerly a coupe option only, was now available on sedans.

I.D. DATA: The 13 symbol Vehicle Identification Number (VIN), as before, was embossed on a metal plate riveted to the upper corner of the instrument panel, behind the left wiper pivot and 'A' pillar, visible through the windshield. Coding was the same as 1976-1977. Symbol one ('A') indicated American Motors Corp. Model year (second symbol) changed to '8' for 1978. The third symbol was a letter for transmission type: A=column-shift automatic; C=floor-shift utomatic; E=three-speed floor shift; M=four-speed floor shift. Digits indicating series (fourth symbol) were now: 0=Concord/AMX; 1=Matador coupe; 4= Gremlin; 6=Pacer; and 8=Matador sedan/wagon. The body-type digit (symbol five) changed to: 3=two-door hatchback; 5=four-door sedan; 6=two-door sedan or hatchback; 8=station wagon. Symbol V-8 indicated model/group: 4=Gremlin four-cylinder; 5=base Gremlin; 7=Pacer, Concord, Matador or Gremlin Custom; 9=AMX. Symbol seven showed engine type: A=258 six-cylinder one-barrel; C=258 six-cylinder two-barrel; E=232 six-cylinder one-barrel; G=2.0-liter four-cylinder two-barrel; H=304 V-8 two-barrel; N=360 V-8 two-barrel. The final V-8 digits were the sequential serial number: 100,001 to 699,999 for Kenosha manufacture; 700,001-999,999 for Brampton, Ontario. A non-removable Federal emission control information label in the engine compartment identified the engine family and gave basic tune-up specs. A non-removable safety sticker affixed to the edge of the left front door showed month and year built, the VIN, and a safety compliance statement, as well as consumer information (vehicle class, acceleration/passing figures, tire reserve load, and stopping distance). A unit body identification plate riveted to the edge of the left front door showed: vehicle Body Number (K=Kenosha, followed by a V-8-digit sequence number); the five-digit Model Number (model year, body, and standard appointment group); a four-digit Trim Number; Paint Code Number; and a Build Sequence Number preceded by a letter showing the production line at which the car was manufactured.

GREMLIN SERIES 40 (V-8-CYLINDER)

Series Number	Body/Style Number	Body Type & Seating	Factory Price	Shipping Weight	Production Total
40	7846-5	2dr Sed-4P	3539	2834	Note 1
40	7846-7	2dr Cus Sed-4P	3789	2822	Note 1
GREMLIN SERIES 40 (FOUR-CYLINDER)					
40	7846-4	2dr Cus Sed-4P	3789	2556	6349

NOTE 1: Total model year production, 22,104 Gremlins (15,755 V-8). I-6

PACER SERIES 60

60	7866-7	2dr Sed-4P	4048/4298	3197/3430	7411
60	7868-7	2dr Sta Wag-4P	4193/4443	3245/3478	13,820

NOTE 2: Of the 21,231 Pacers produced during the model year, 2,514 had a V-8 engine.

CONCORD SERIES 01

01	7803-7	2dr Hatch-4P	3849/4099	3051/3284	2572
01	7806-7	2dr Sed-5P	3749/3999	3029/3262	50,482
01	7805-7	4dr Sed-5P	3849/4099	3099/3332	42,126
01	7808-7	4dr Sta Wag-5P	4049/4299	3133/3366	23,573

NOTE 2: Total model year Concord production, 121,293 units (110,972 with V-8-cylinder engine, 6541 with V-8, and 3780 with a four).

AMX SERIES 01

01	7803-9	2dr Hatch-4P	4649/4899	3159/3381	2540
MATADOR (COUPE) SERIES 10					
10	7816-7	2dr Cpe-5P	4799/4989	3709/3916	2006
MATADOR (SEDAN AND WAGON) SERIES 80					
80	7885-7	4dr Sed-6P	4849/5039	3718/3921	4824
80	7888-7	4-dr Sta Wag-6P	— /5299	— /4146	3746

NOTE 3: Of the total model year production of 10,576 during the model year, only 23 came with a V-8-cylinder engine.

FACTORY PRICE AND WEIGHT NOTE: Figure before the slash in price columns (except Gremlin) is for V-8-cylinder engine, after slash for V-8 engine. ENGINES: BASE FOUR (Gremlin): Same as 1977; see 1977 specifications. BASE (Gremlin, Pacer, Concord): Inline. OHV. -cylinder. Cast iron block. Displacement: 232 cid (3.8 liters). Bore & stroke: 3.75 x 3.50 in. Compression ratio: 8.0:1. Brake horsepower: 90 at 3400 rpm. Torque: 168 lbs.-ft. at 1600 rpm. Taxable hp: 33.75. Seven main bearings. Hydraulic valve lifters. Carburetor: one-barrel Carter YF. BASE (AMX, Matador sedan/coupe); OPTIONAL (Gremlin, Pacer, Concord):Inline. OHV. cylinder. Cast iron block. Displacement: 258 cid (4.2 liters). Bore & stroke: 3.75 x 3.90 in. Compression ratio: 8.0:1. Brake horsepower: 110 at 3600 rpm. Torque: 201 lbs.-ft. at 1800 rpm. Seven main bearings. Taxable hp: 33.75. Hydraulic valve lifters. Carburetor: two-barrel Carter BBD. BASE (California models): Same as 258 cid above, but with Carter YF 7235 one-barrel carburetor. Horsepower: 100 at 3400 rpm. Torque: 200 lbs.-ft. at 1600 rpm. OPTIONAL V-8 (Pacer, Concord, AMX): 90 degree, overhead valve V-8. Cast iron block. Displacement: 304 cid (5.0 liters). Bore & stroke: 3.75 x 3.44 in. Compression ratio: 8.4:1. Brake horsepower: 130 at 3200 rpm. Torque: 238 lbs.-ft. at 2000 rpm. Taxable hp: 45. Five main bearings. Hydraulic valve lifters. Carburetor: two-barrel Motorcraft 2100. V-8 (Matador wagon); OPTIONAL (Matador coupe/sedan): 90 degree, overhead valve V-8. Cast iron block. Displacement: 360 cid (5.9 liters). Bore & stroke: 4.08 x 3.44 in. Compression ratio: 8.25:1. Brake horsepower: 140 at 3350 rpm. Torque: 278 lbs.-ft. at 2000 rpm. Taxable hp: 53.3. Five main bearings. Hydraulic valve lifters. Carburetor: two-barrel Motorcraft 2100.

CHASSIS DATA: Wheelbase: (Gremlin) 96 in.; (Pacer) 100 in.; (Concord/AMX) 108 in.; (Matador coupe) 114 in.; (Matador sed/wagon) 118 in. Overall length: (Gremlin) 166.6 in.; (Pacer sedan) 172 in.; (Pacer wagon) 177 in.; (Concord) 183.6 in.; (AMX) 186 in.; (Matador coupe) 209.9in.; (Matador sedan) 218.3 in.; (Matador wagon) 219.3 in. Height: (Gremlin) 51.5 in.; (Pacer sedan) 52.8 in.; (Pacer wagon) 53.2 in.; (Concord) 51.3-51.7 in.; (AMX) 52.2 in.; (Matador coupe) 51.6 in.; (Matador sedan) 53.9 in.; (Matador wagon) 56 in. Front Tread: (Gremlin/Concord) 57.5 in.; (Pacer) 61.4 in.; (Concord) 57.6 in.; (Matador) 59.6 in. Rear Tread: (Gremlin/Hornet) 57.1 in.; (Pacer) 60 in.; (Concord) 57.1-57.5 in. (Matador) 60.6 in. Standard Tires: (Gremlin) B78 x 14 whitewall; (Pacer) D78 x 14; (Concord) C78 x 14 except with D/L package, D78 x 14; (AMX) DR78 x 14; (Matador coupe/sedan) F78 x 14; (Matador wagon) H78 x 14.

TECHNICAL: Three-speed manual floor-shift transmission was standard on Gremlin base V-8, Concords and Pacers. Four-speed floor-shift was standard on Gremlin Custom V-8, Gremlin four, and AMX. Matadors came with Torque Command three-speed column-shift automatic; optional on others. Automatic shift gear ratios: (1st) 2.45:1; (2nd) 1.45:1; (Rev) 2.20:1. Steering: (Pacer) rack and pinion; (others) recirculating ball. Suspension: independent front coil springs (Pacer springs mounted between the two control arms); semi-elliptic rear leaf springs except (Matador) coil springs. Brakes: front discs; rear drums. Fuel tank: (Gremlin) 21 gal. except four-cylinder, 13 gal.; (Pacer) 20 gal.; Concord/AMX) 22 gal.

DRIVETRAIN OPTIONS: 258 cid V-8-cylinder engine: Gremlin/Pacer/Concord ($120). 304 cid V-8, two-barrel: Pacer/Concord/AMX ($233). 360 cid V-8, two-barrel: Matador sedan/coupe ($190). Four-speed manual floor shift: Gremlin/Pacer/Concord ($111). Torque-Command automatic transmission, column shift: Gremlin ($270); Pacer/Concord ($296). Torque-Command with floor shift lever: Gremlin ($294); Pacer/Concord/AMX ($320); Matador coupe with V-8 and bucket seats ($70). Twin Grip differential ($55-$59). Heavy-duty engine cooling system ($29-$37) but standard with air conditioning. Maximum cooling system: Matador V-8 ($48); Matador V-8 with air ($19). Auxiliary automatic transmission oil cooler: Matador ($34). Heavy-duty suspension ($31-$36); including front sway bar on Gremlin/Pacer, rear on Matador. Front sway bar: Gremlin/Pacer ($18) but included with radial tires. Rear sway bar: Matador ($18). Air-adjustable shock absorbers: Matador ($44-$48). Heavy-Duty 70-amp battery ($17-$19). California emission system ($74). High altitude package ($23).

GREMLIN/CONCORD/MATADOR CONVENIENCE/APPEARANCE OPTIONS: 'X' package: Gremlin ($249). GT package: Gremlin ($649). Concord D/L package ($200). Concord Sport package: hatchback ($289); sedan/wagon ($379). Barcelona package: Matador coupe ($849); Matador sedan ($699). Levi's trim package: AMX ($49). 'AMX' decal package (black/orange decal on hood and liftgate): AMX ($49). Extra Quiet insulation package: Gremlin/Concord/AMX ($45); included with Concord Sport and D/L package. Protection group: Gremlin ($42); Concord ($27). Interior decor/convenience group: Gremlin ($69); Concord ($79); Concord D/L ($54); AMX/Matador ($59). Gauge package: Gremlin Custom V-8 ($75); Concord ($99). Visibility group (remote control left mirror, manual right mirror, day/night mirror, intermittent wipers): Gremlin/Concord ($71). Visibility group (dual remote mirrors, deluxe wipers): Matador ($81). Remote control left mirror ($16-$17). Intermittent wipers ($30-$32). Power steering ($141-$147). Power front disc brakes ($64). Cruise control ($99); Not available on Gremlin four. Air conditioning system: Gremlin/Concord/AMX ($478). Air conditioning package (including tinted glass and power steering): Gremlin ($669); Concord/AMX ($679); Matador ($590). AM radio ($80-$81). AM/FM/Stereo four-speaker radio: Gremlin ($144); others ($224). AM/CB Radio: Gremlin ($119); others ($199) AM/FM/CB Radio ($299); not available on Gremlin. Rear speaker for AM radio ($22-23). Tape player and AM/FM/Stereo radio: Matador ($336). Digital clock: Gremlin/Concord ($25). Tachometer: Gremlin four ($49). Power windows: Matador sedan/wagon ($155). Power side and tailgate windows: Matador wagon ($202). Power tailgate window: Matador wagon ($47); standard in three-seat. Console: Gremlin/Concord ($29). Hidden compartment: Concord hatch/AMX ($31). Individual reclining seats: Matador (no charge). Vinyl bucket seats: Gremlin ($49); Concord ($69); Concord D/L and Matador coupe (NC). Third seat: Matador wagon ($143). Stowage/litter containers ($10). Rear defroster ($81-$88). Dual horns: Gremlin/Concord ($12). Light group: Concord/AMX ($30). Sports steering wheel: Gremlin Custom/Concord/Matador ($20). Soft-Feel sports steering wheel: Gremlin Custom, Concord sedan/wagon, Matador ($31); Gremlin 'X', Concord hatch or Sport package ($11). Tilt steering wheel: Concord ($64); Matador ($67). Tinted glass ($52-$64). Two-tone paint: Concord D/L ($75); Matador ($48). Delete two-tone: Matador Barcelona sedan (deduct $100). Special color combinations ($24) except AMX Classic Black ($49). Vinyl roof: Concord ($105); Matador ($118). Rally side stripes: Gremlin ($43); Matador coupe ($46). Woodgrain wagon paneling: Concord ($75); Matador ($133). Delete woodgrain from Concord D/L wagon (deduct $75). Roof rack: Gremlin ($59); Matador wagon ($84). Roof rack and air deflector: Concord wagon ($85). Tailgate air deflector: Gremlin (24); Matador ($27). Inside hood release: Gremlin ($16). Front bumper guards: Gremlin/Concord ($19). Bumper nerfing strips: Concord ($34). Side scuff molding: base Gremlin ($29). Protective innercoating ($95) except Matador ($106). Door edge guards: Matador ($11). Locking gas cap ($7). Engine block heater ($19-$20). Wheelcovers (set of four): Gremlin ($36). Custom wheelcovers ($23); Styled wheels for D-size tires ($136) except AMX, 'X' or Sport package ($71). Aluminum styled wheels ($236) except AMX, 'X' or Sport package ($171). Space-Saver spare tire: Gremlin or Concord with regular wheels ($18); no charge with special wheels/tires. Tires: B78 x 14 whitewall: Gremlin (no charge). D78 x 14: Gremlin (no charge); Concord ($18). D78 x 14 whitewall: Gremlin ($36); Concord ($63). DR78 x 14 SBR: Gremlin ($112); Concord ($139); with Concord D/L ($76); Gremlin Custom 'X' (no charge). DR78 x 14 white SBR: Gremlin ($157); Concord ($184); Concord D/L ($121); AMX, Concord Sport, Gremlin 'X' ($45). DR78 x 14 white-letter SBR: Gremlin Custom ($172); Concord ($199); Concord D/L ($136); AMX, Sport, 'X' package ($60);. DR78 x 14 white GBR: Gremlin ($36); Concord ($159); Concord Sport ($165); Concord D/L ($96). F78 x 14 whitewall: Matador coupe/sedan ($48). FR78 x 14 white SBR: Matador ($175). H78 x 14 whitewall: Matador wagon ($48). HR78 x 14 white SBR: Matador ($198); wagon ($175).

PACER CONVENIENCE AND APPEARANCE OPTIONS: Sport package ($165); not available on wagon or with/232 engine. Interior decor/convenience group ($59). Visibility group: remote left mirror, right mirror, intermittent wipers ($61). Left remote mirror ($16). Gauge package ($99). Intermittent wipers ($30). Power steering ($147). Power disc brakes ($64). Cruise control ($99). Air conditioning ($478). Air conditioning package: tinted glass and power steering ($679). AM radio ($80). AM/FM/Stereo four-speaker radio ($224). AM/CB Radio ($199). AM/FM/CB Radio ($299). Rear speaker ($22). Tape player and AM/FM/Stereo radio ($336). Console ($29). Hidden compartment:hatch ($35). Individual reclining seats: velveteen crush or reflection print fabric, or Soft-Feel vinyl (nocharge). Vinyl bucket seats (nocharge). Stowage/litter compartment ($10). Rear defroster ($81). Rear wiper/washer ($59). Sports steering wheel ($20); included with Sport package Soft-Feel sports steering wheel ($31); with Sport package ($11). Tilt steering wheel ($64). Tinted glass ($58). Door vent windows ($34). Two-tone color: hatch ($84). Special color combinations ($24). Woodgrain wagon paneling ($111). Protective inner coating ($95). Vinyl roof ($118). Roof rack ($59). Bumper guards ($38). Locking gas cap ($7). Custom wheelcovers ($23). Styled wheels ($136); with Sport hatch package ($71). Aluminum styled wheels ($236); with Sport hatch ($171). Tires: DR78 x 14 SBR ($76). DR78 x 14 white SBR ($121). DR78 x 14 white-letter SBR ($136). ER78 x 14 whitewall ($18). ER78 x 14 SBR ($94). ER78 x 14 white SBR ($139). ER78 x 14 white-letter SBR ($154). Tires with Sport hatchback package: DR78 x 14 white SBR ($45); DR78 x 14 white-letter SBR ($60); ER78 x 14 ($18); ER78 x 14 ($63); ER78 x 14 white-letter SBR ($78).

HISTORICAL FOOTNOTES: Introduced: Sept. 9, 1977 except the new Concord, in October; modified Gremlins and Concords debuted later. Model year production: (U.S.) 137,860, for 1.55 percent of the industry total; North American production for U.S. market, 175,204 units (10,129 four-cylinder, 145,467 V-8 cylinder and 19,608 V-8). Calendar year production: (U.S.) 164,351, led by Concord with 108,414 cars manufactured. Calendar year sales by U.S. dealers: 170,739 (1.8 percent of industry total). AMC's fiscal picture improved again during the 1978 model year, with net earnings at their highest point since 1973. Much of the company's strength, however, continued to stem from its Jeep subsidiary. Except for the new Concord, North American production slipped substantially in every division. Gremlin prices were cut in November 1977. Promise for the future came in the form of negotiations with Regie Nationale des Usines Renault, the Frenochargeh firm that produced the Renault. The arrangement reached between the two companies in January 1979 would permit AMC to distribute the Renault Le Car, and perhaps allow Renault to distribute AMC Jeep vehicles. Far more significant, though, was the announcement that AMC planned to produce a joint-venture Renault vehicle in the United States. Gerald C. Meyers took over as the new chairman and CEO at AMC, while W. Paul Tippett stepped into his former role as president and chief operating officer. James L. Tolley moved from PR director at Chevrolet to vice-president of public relations at AMC. Passenger car operations were consolidated at Kenosha during the year. The Milwaukee body plant turned to stamping, while the Brampton, Ontario facility converted to Jeep production. Four-cylinder engines had first been purchased intact from Volkswagenwerk. During 1977, AMC initiated a plan to assemble the engines itself, on an assembly line bought from VW. But as the 1978 model year began, AMC contracted instead to purchase "Iron Duke" fours from Pontiac.

1979 AMC

American Motors left the mid-size market by abandoning the Matador, which had been selling poorly. Meanwhile, the Gremlin was replaced by a similar, but more stylish sub-compact, the new Spirit, and the performance-oriented AMX moved to the shorter Spirit chassis. Spirit, Concord and Pacer base models included as standard equipment a high-pressure compact spare tire, manual front disc brakes, inside hood release, lighted ashtray, full wheelcovers, rear bumper guards, custom steering wheel, and color-keyed 12 oz. carpeting. A new, upscale trim level was added: the Limited. This signaled AMC's move away from austerity and toward a touch of luxury, even in economy cars. Standard colors for all models (except AMX) were Olympic white, Classic black, Quick silver metallic, Cumberland green metallic, Wedgewood blue, Starboard blue metallic, khaki, British bronze metallic, Saxon yellow, Morocco Buff, Alpaca brown metallic, Sable brown metallic, Firecracker red, Russet metallic, and Bordeaux metallic. Torque Command transmissions (supplied by Chrysler) with V-8 engines received a lockup torque converter. Over two-thirds of AMC passenger cars had air conditioning.

1979 AMC Spirit Limited Hatchback (AMC)

SPIRIT — SERIES 40 — FOUR/SIX: Gremlin's more luxurious replacement kept the same engine choices and basic design as its subcompact predecessor, but with much larger rear side windows on the two-door sedan. New quad rectangular headlamps flanked a clean four-row, horizontally ribbed, grille with center medallion. Clear, wide parking lights sat below the headlights. Horizontal amber reflectors were mounted at the forward end of the front fenders, with 'Spirit' emblems at their rear. Slightly rounded rectangular taillamps held inset back-up lights. Aluminum bumpers had black end caps. Joining that familiar sedan was a good-looking two-door liftback coupe. Front styling was similar, but the rear end held extra-wide four-section taillamps that stretched from the license plate to the outer edge of the panel. The liftback's big pop-up hatch sloped much more sharply than the sedan's back end. Basic engine was the Volkswagen/Audi 121 cid four-cylinder, with 232 and 258 cid sixes optional. Liftbacks could also have a 304 cid V-8. A four-speed floor shift transmission was standard; floor or column shift automatics optional. A three-speed manual shift was available with the 232 cid six, at reduced price. Base Spirit standard equipment included blackwall C78 x 14 tires, vinyl bucket seats, four-spoke steering wheel, lighter, locking glove box, folding rear seat, spare tire cover, dual paint stripes, plus moldings for wheel lip, drip rail, hood front edge, windshield surround and rocker panels. Liftbacks included a front sway bar to improve handling. The DL model added custom bucket seats in Caberfae corduroy fabric or Sport vinyl, walnut burl woodgrain instrument panel overlay, woodgrain steering wheel, day/night mirror, digital clock, extra-quiet insulation, dual horns, courtesy lights, package shelf, folding split rear seatback, front/rear bumper guards, and whitewall tires with color-keyed styled wheelcovers. For more luxury, Spirit's Limited included leather bucket seats, an AM radio, power door locks, power steering, power liftback release, dual remote mirrors, light and visibility groups, convenience and protection groups, tilt steering wheel, full-length console with center armrest, 18 oz. carpeting, and P195/75R14 glass-belted

whitewalls. DL and Limited liftbacks could also have the sporty GT package, which included a black full-length console, black leather-wrapped steering wheel, black instrument panel with woodgrain overlay, woodgrain door panel accents, tachometer, black bumpers with nerfing strips and guards, twin black remote mirrors, black exterior trim moldings, black grille insert and headlight bezels, and black rear venturi area. GT models carried P195/75R14 steel-belted radials on spoke style wheels; V-8s with manual shift even had a Performance Tuned exhaust sound. 'GT' emblems were placed below the 'Spirit' badges on front fenders. For handling to match the sharp looks, a GT Rally Tuned Suspension Package added a tuned front sway bar, rear sway bar, heavy-duty adjustable Gabriel "Strider" shocks, tuned strut rod bushings and rear spring iso-clamp pads, Hi-Control rear leaf springs, unique steering gears, and heavy-duty brakes.

1979 AMC Hornet AMX coupe (AMC)

AMX — SERIES 40 — SIX/V-8: AMX liftbacks entered the 1979 model year sitting on a short (96 in.) Spirit wheelbase, rather than the previous Concord platform. "Expect to be noticed," the factory declared. To make sure, every AMX had front/rear black bumpers with guards and nerfing strips, a front air dam and rear deck spoiler with accent stripes, front/rear fender flares, a new rectangular blackout grille with center 'AMX' emblem, twin remote mirrors, and Turbocast II aluminum wheels with white-letter ER60 x 14 radial tires. Eye-catching graphics included 'AMX' decals on doors and rear spoiler, plus a huge flame decal on the hood. Quad headlamps sat over clear rectangular signal/parking lights. Inside, drivers sat in Sport vinyl or Caberfae corduroy bucket seats, gripped a leather-wrapped steering wheel, watched a full gauge array with brushed aluminum instrument panel overlays, and enjoyed a center console with armrest. Standard engine was the 258 cid six (304 V-8 optional) with four-speed floor shift. A Rally Tuned suspension system helped handling, while V-8s with manual shift included a performance-tuned exhaust sound. That suspension included Gabriel "Strider" shocks, front sway bar, heavy-duty rear sway bar, and Hi-Control rear leaf springs. AMX colors were Olympic white, Classic black, Wedgewood blue, Saxon yellow, Morocco Buff, and Firecracker red.

1979 AMC Pacer Limited Liftback (AMC)

PACER — SERIES 60 — SIX/V-8: Except for an upright hood ornament above the upward-bulging grille, changes to the Pacer hatchback sedan and wagon consisted mainly of a larger (258 cid) base engine and the addition of a Limited upgrade to the basic DL model. Standard DL equipment included individual reclining seats in Caberfae corduroy or Sport vinyl, windsplit molding, front/rear bumper guards, folding rear seat back, extra-quiet insulation, day/night inside mirror, electric clock, dual horns, courtesy lights, two-speed wiper/washers, custom steering wheel with woodgrain overlays, and woodgrain-overlay instrument panel. DL models had color-keyed wheelcovers, P195/75R14 whitewall glass-belted radials, front sway bar, and color-keyed wide scuff moldings. Bucket seats uphol-stered in Caberfae corduroy fabric were also offered on DL Pacers. Beyond those features, the Limited included power steering, power windows and door locks, an AM radio, leather reclining seats with beige corduroy accents, woodgrain tilt steering wheel, dual remote-control mirrors, visibility and convenience groups, light and protection groups, folding center armrest, 18 oz. color-keyed carpeting, and color-keyed styled wheelcovers. In addition to the standard colors, Pacers were available with Misty Beige clearcoat and in six two-tone combinations. Four-speed manual shift was standard (with floor lever); column- or floor-shift automatic optional. Once again, a 304 cid V-8 was available, which required power steering and brakes. According to industry sources, only 184 Pacers had a sunroof; 347 had vinyl tops.

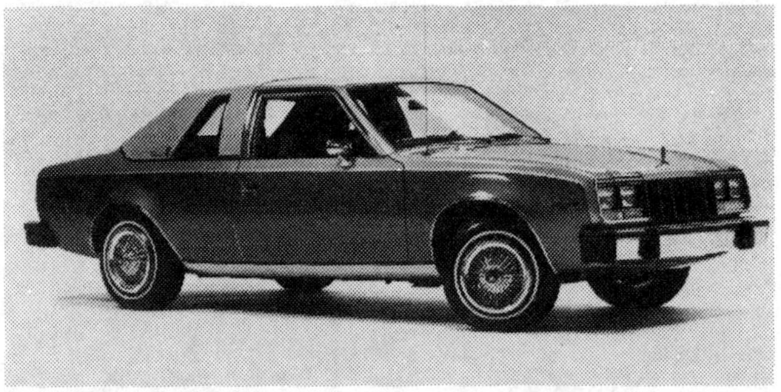

1979 AMC Concord coupe (AMC)

CONCORD — SERIES 01 — FOUR/SIX/V-8: Front-end styling changed appreciably on the compact, luxurious Concord, with wide clear parking/signal lights setting below quad rectangular headlamps. The restyled formal grille was considerably taller, with seven vertical bars. Bright aluminum bumpers had black end caps and guards. Standard engine remained the 232 cid six, but a 258 cifd six, 304 cid V-8 and 121 cid four were all optional. Four-speed manual floor shift was standard; column-shift automatic transmission optional. Concords now came in three trim levels: base, DL, and Limited (no Limited for the hatchback body). Base Concords were equipped with a front sway bar, lighted ashtray, sport vinyl notched bench seat (Striped knit fabric available with automatic transmission), color-keyed 12 oz. carpeting, full wheelcover, dual body side pinstripes, hood ornament, and moldings for drip rail, wheel lip, hood front edge, and windshield surround. Tires were D78 x 14 blackwalls. DL versions had individual reclining seats in Velveteen Crush fabric or Sport vinyl, as well as a day/night mirror, digital clock, extra-quiet insulation, dual horns, courtesy lights, underhood light, walnut burl woodgrain instrument panel overlay, custom steering wheel with woodgrain overlays, custom headliner and sun visors, package shelf, front/rear bumper guards, color-keyed wheelcovers, and whitewall tires. DL sedans and hatchbacks also sported a landau vinyl roof; wagons had woodgrain body side overlays. Genuine leather individual reclining seats with beige corduroy accents came on the plush Limiteds, as did an AM radio, power door locks, twin remote mirrors, adjustable tilt woodgrain steering wheel, light group, visibility group, convenience and protection groups, styled wheelcovers, 18 oz. carpeting, and P195/75R14 whitewall glass-belted radials (except hatchbacks).

I.D. DATA: The 13 symbol Vehicle Identification Number (VIN) again was embossed on a metal plate riveted to the top left surface of the instrument panel, visible through the windshield. Coding was the same as 1976-78. Model year code (symbol two) changed to '9' for 1979. Series codes (symbol four) now included: 0=Concord; 4=Spirit/AMX; 6=Pacer. The model/group codes (symbol six) now included: 0=base model; 5=DL; 7=Limited; and 9=AMX.

SPIRIT SERIES 40

Series Number	Body/Style Number	Body Type & Seating	Price Four/Six	Shipping Weight	Production Total
40	7943-7	2d Lift-4P	3999/4049	2545/2762	Note 1
40	7946-7	2d Sed-4P	3899/3949	2489/2706	Note 1
SPIRIT DL					
40	7943-7	2d Lift-4P	4199/4249	2635/2852	Note 1
40	7946-7	2d Sed-4P	4099/4149	2579/2796	Note 1
SPIRIT LIMITED					
40	7943-7	2d Lift-4P	5199/5249	2732/2949	Note 1
40	7946-7	2d Sed-4P	5099/5149	2676/2893	Note 1
AMX SERIES 40					
40	7943-9	2d Lift-4P	5899/6149	2899/3092	3657

NOTE 1: Total Spirit/AMX model year production, 52,714 cars (16,237 four-cylinder, 36,241 six-cylinder, 3,893 V-8).

PACER DL SERIES 60

60	7966-7	2d Hatch-4P	4699/5177	3133/3360	2863
60	7968-7	2d Sta Wag-4P	4849/5327	3170/3397	7352
PACER LIMITED					
60	7966-7	2d Hatch-4P	5699/6177	3218/3445	Note 2
60	7968-7	2d Sta Wag-4P	5849/6327	3255/3482	Note 2

NOTE 2: Production totals above include both DL and Limited models. Of the 10,215 Pacers made during the model year, 1,014 carried a V-8 engine.

CONCORD SERIES 01

01	7903-7	2d Hatch-4P	4149/4399	2888/3095	2331
01	7906-7	2d Sed-5P	4049/4299	2873/3080	40,110
01	7905-7	4d Sed-5P	4149/4399	2939/3146	40,134
01	7908-7	4d Sta Wag-5P	4349/4599	2977/3184	20,278
CONCORD DL					
01	7903-7	2d Hatch-4P	4448/4698	3003/3210	Note 3
01	7906-7	2d Sed-5P	4348/4598	2982/3189	Note 3
01	7905-7	4d Sed-5P	4448/4698	3040/3247	Note 3
01	7908-7	4d Sta Wag-5P	4648/4898	3072/3279	Note 3
CONCORD LIMITED					
01	7906-7	2d Sed-5P	5348/5598	3090/3297	Note 3
01	7905-7	4d Sed-5P	5448/5698	3146/3253	Note 3
01	7908-7	4d Sta Wag-5P	5648/5898	3177/3384	Note 3

NOTE 3: Production totals shown include base, DL and Limited models. Total Concord model year production, 102,853 (6355 four-cylinder, 91,842 six-cylinder and 4,656 V-8).
NOTE 4: Four-cylinder engine was offered on Concords at six-cylinder price.
NOTE 5: [Concord] Data above slash for four-cylinder/below slash for six-cylinder. [Other models] Data above slash for six-cylinder/below slash for V-8.

ENGINES: BASE FOUR (Spirit); OPTIONAL (Concord hatchback/sedan): Inline. Overhead cam. Four-cylinder. Cast iron block; cast aluminum alloy head. Displacement: 121 cid (2.0 liters). Bore & stroke: 3.41 x 3.32 in. Compression ratio: 8.2:1. Brake horsepower: 80 at 5000 rpm. Torque: 105 lbs.-ft. at 2800 rpm. Five main bearings. Solid valve lifters. Carburetor: two-barrel Holley 5210. BASE SIX (Concord); OPTIONAL (Spirit): Inline. OHV. Six-cylinder. Cast iron block. Displacement: 232 cid (3.8 liters). Bore & stroke: 3.75 x 3.50 in. Compression ratio: 8.0:1. Brake horsepower: 90 at 3400 rpm. Torque: 168 lbs.-ft. at 1600 rpm. Seven main bearings. Hydraulic valve lifters. Carburetor: one-barrel Carter YF. BASE SIX (AMX, Pacer); OPTIONAL (Spirit, Concord): Inline. OHV. Six-cylinder. Cast iron block. Displacement: 258 cid (4.2 liters). Bore & stroke: 3.75 x 3.90 in. Compression ratio: 8.3:1. Brake horsepower: 110 at 3200 rpm. Torque: 210 lbs.-ft. at 1800 rpm. Seven main bearings. Hydraulic valve lifters. Carburetor: two-barrel Carter BBD. OPTIONAL SIX (California models): Same as 258 cid six above, but with one-barrel Carter YF carburetor. Horsepower: 100 at 3400 rpm. Torque: 200 lbs.-ft. at 1600 rpm. Compression Ratio: 8.1:1. OPTIONAL V-8 (Spirit liftback, AMX, Pacer, Concord): 90-degree, overhead valve V-8. Cast iron block. Displacement: 304 cid (5.0 liters). Bore & stroke: 3.75 x 3.44 in. Compression ratio: 8.4:1. Brake horsepower: 125 at 3200 rpm. Torque: 220 lbs.-ft. at 2400 rpm. Five main bearings. Hydraulic valve lifters. Carburetor: two-barrel Motorcraft 2100.

CHASSIS DATA: Wheelbase: (Spirit/AMX) 96 in.; (Pacer) 100 in.; (Concord) 108 in. Overall length: (Spirit/AMX liftback) 168.5 in.; (Spirit sedan) 166.8 in.; (Pacer hatchback) 172.7 in.; (Pacer wagon) 177.7 in.; (Concord) 186 in. Height: (Spirit/AMX) 51.6 in.; (Pacer hatchback) 52.8 in.; (Pacer wagon) 53.1 in.; (Concord) 51.1-51.6 in. Front Tread: (Spirit/AMX) 58.1 in.; (Pacer) 61.2 in.; (Concord) 57.6 in. Rear Tread: (Spirit/AMX/Concord) 57.5 in.; (Pacer) 60.0 in. Standard Tires: (Spirit) C78 x 14; (AMX) ER60 x 14 Flexten belted radial OWL; (Pacer) P195/75R14 GBR. (Concord) D78 x 14 except Limited, P195/75R14 GBR.

TECHNICAL: Four-speed manual transmission with floor shift was standard. Three-speed available on Spirit. Column-shift automatic available on Spirit/Concord/Pacer. Floor-shift automatic available on Spirit/AMX/Pacer. Three-speed manual shift gear ratios: (1st) 2.99:1; (2nd) 1.75:1; (3rd) 1.00:1; (Rev) 3.17:1. Four-speed gear ratios: (1st) 3.98:1; (2nd) 2.14:1; (3rd) 1.42:1; (4th) 1.00:1; (Rev) 3.99:1. Standard axle ratio: (Spirit four) 3.08:1 with four-speed, 3.31:1 with automatic; (Spirit six) 2.53:1 except 232 with three-speed, 2.73:1; (Spirit/AMX V-8) 2.87:1 except 2.56:1 with automatic, without Twin-Grip differential; (AMX six) 2.53:1; (Pacer six) 2.53:1; (Concord four) 3.31:1 with four-speed, 3.58:1 with automatic; (Concord six) 2.53:1 except wagon with automatic, 2.73:1; (Pacer/Concord V-8) 2.56:1 exc. 2.87:1 with Twin-Grip. Steering: (Pacer) rack and pinion; (others) recirculating ball. Suspension: independent front coil springs; semi-elliptic rear leaf springs. Brakes: front disc, rear drum. Fuel tank: (Spirit four sedan) 13 gallon; (Spirit/AMX/Pacer) 21 gallon; (Concord) 22 gallon.

DRIVETRAIN OPTIONS: 232 cid six-cylinder engine: Spirit ($50). 258 cid six-cylinder engine: Spirit/Concord ($130). 304 cid V-8 ($250); not available in Spirit sedan. Three-speed manual floor shift: Spirit with 232 six (deduct $50). Torque-Command automatic transmission, column shift: Spirit six ($296); Pacer/Concord ($323). Torque-Command with floor shift: Spirit ($321); Pacer ($348); AMX ($296). Twin Grip differential ($63). Heavy-duty engine cooling system: Pacer ($31); Concord/AMC ($39) but standard with air conditioning. Maximum cooling system: Concord V-8 ($59); Concord V-8 with air ($20). Auxiliary automatic transmission oil cooler: Pacer/Concord V-8 ($36). Handling package (unique front sway bar, rear sway bar): Concord six ($30). Air shock absorbers, rear: Concord ($98). Heavy-duty shock absorbers: Pacer/Concord ($14). Heavy-duty 70 amp battery ($18); not available on Spirit; 56 amp with Pacer V-8. Cold climate group (Heavy-duty battery and alternator, engine block heater): Pacer/Concord/AMX ($96); with air conditioning or rear defroster ($38). California emission system ($78).

SPIRIT/CONCORD/AMX CONVENIENCE/APPEARANCE OPTIONS: Spirit liftback GT package ($469); with Limited ($200-$469). Spirit liftback GT Rally tuned suspension package ($99). Extra Quiet insulation package: Concord ($48). Protection group: Concord ($87); Concord DL ($21); AMX ($27). Convenience group including headlight-on buzzer, intermittent wipers, dual vanity mirrors ($75). Gauge package (including parcel shelf): Spirit six/V-8 ($104); Spirit GT ($52). Visibility group (remote control left/right mirrors, day/night mirror): Spirit/Concord ($62); DL models ($50); standard on Spirit GT. Remote control left mirror: Spirit/Concord ($17). Power steering ($152-$158); standard on Limited models. Power front disc brakes ($70). Cruise control ($104). Air conditioning system: Spirit V-8/Concord/AMX ($513). Air conditioning package (including tinted glass and power steering): Spirit/Concord/AMX ($722-$731); not available on Concord Limited. AM radio ($84); standard on Concord Limited AM/FM/Stereo four-speaker radio: Concord/AMX ($236); Concord Limited, except hatchback, ($152). AM/FM/CB Radio ($314) except Limited ($230). Rear speaker for AM radio ($24). Digital clock: Spirit/Concord ($40) but standard on Limited Tachometer: Spirit/Concord ($52); standard with Spirit GT package Power door locks: Spirit/AMX ($72); standard on Limited Power liftback release: Spirit ($30); standard on Limited Caberfae corduroy fabric bucket seats: AMX (no charge). Bench seat: Spirit sedan (no charge). Bucket seats: Spirit (no charge). Console: Spirit ($75); standard with GT. Hidden compartment: Concord hatch ($33). Rear defroster ($89). Dual horns: Spirit/Concord ($13); standard on DL and Limited Light group: Spirit ($45); Spirit DL, AMX ($35); Concord ($49); Concord DL ($39); standard on Limited models. Leather-wrapped sport steering wheel: Concord ($49) except Limited ($20). Woodgrain steering wheel: Concord DL ($29); standard on Limited Tilt steering wheel ($72); standard on Limited models. Tinted glass ($57-$60). Two-tone paint: Spirit lift ($65); Concord DL and Limited ($100). Special color combinations: ($26) except AMX. Pop-up moon roof ($178). Rally side stripes: Spirit ($46). Delete woodgrain from Concord DL wagon (deduct $75). Roof rack: Spirit sedan ($62); Concord wagon ($90). Tailgate air deflector: Spirit sedan ($26). Front bumper guards: Concord ($21); standard on DL and Limited side scuff molding: AMX ($31). Protective inner coating ($100) except Spirit. Locking gas cap ($8) except Spirit. Styled wheelcovers: Concord ($35); not available on Limited color-keyed wheelcovers: Concord DL except wagon ($45); standard on Limited. Spoke style 14 x 6 in. wheels with trim rings: Concord ($145); Concord Limited except hatchback. ($100). Turbine forged aluminum wheels: Concord ($300); Concord Limited except hatchback ($255). Tires: C78 x 14 whitewall: base Spirit ($48). D78 x 14: base Spirit ($19); D78 x 14 whitewall: Spirit ($67); Spirit DL ($19); Concord ($48). P195/75R14 GBR: Spirit ($120); Spirit DL ($72); Concord ($101); Concord DL ($53). P195/75R14 white GBR: Spirit ($168); Spirit DL ($120); Concord ($149); Concord DL ($101). P195/75R14 SBR: Spirit ($168); Spirit DL ($120); Spirit Limited (no charge); Concord ($149); Concord DL ($101); Concord Limited sedan/wagon (no charge). P195/75R14 white SBR: Spirit ($216); Spirit DL (168); Spirit Limited (48); Concord ($197); Concord DL ($149); Concord Limited sedan/wagon ($48). DR70 x 14 Flexten radial: Spirit ($238); Spirit DL ($190); Spirit Limited ($70); Concord ($219); Concord DL ($171); Concord Limited ($70). DR70 x 14 OWL Flexten radial: Spirit ($302); Spirit DL ($254); Spirit Limited ($134); Concord ($283); Concord DL ($235); Concord Limited ($134).

PACER CONVENIENCE/APPEARANCE OPTIONS: Convenience group: headlight-on buzzer, intermittent wipers, dual lighted vanity mirrors ($75); standard on Limited Visibility group: remote left and right mirrors ($50); standard on Limited Left remote mirror ($17). Gauge package ($129); not availabler with Limited or center armrest. Power steering ($158); standard on Limited. Power disc brakes ($70). Cruise control ($104). Air conditioning ($513). Air conditioning package with tinted glass and power steering ($734); not available with Limited AM radio ($84); standard with Limited AM/FM/Stereo four-speaker radio ($236); with Limited ($152). AM/FM/CB Stereo Radio ($314); with Limited ($230). Rear speaker ($24). Tape player and AM/FM/Stereo radio ($353); with Limited ($269). Center armrest ($49); standard on Limited. Hidden compartment: hatchback ($37). Caberfae corduroy bucket seats (no charge). Power door locks ($72). Power window and door locks ($194). Light group ($39); standard on Limited Rear defroster ($89). Rear wiper/washer ($62). Leather-wrapped sport steering wheel ($49); with Limited ($20). Woodgrain steering wheel ($29); including with Limited tilt steering wheel ($72); standard with Limited tinted glass ($63). Door vent windows ($36). Two-tone color: black rocker panels deleted

($65). Special color combinations ($26); not available with two-tone or Limited Misty Beige clearcoat: Limited ($60). Woodgrain wagon paneling ($117). Protective inner coating ($100). Pop-up moon roof: hatch ($178). Vinyl roof: hatch ($124). Roof rack ($62). Locking gas cap ($8). Styled wheelcovers ($35); not available on Limited spoke style 14 x 6 in. wheels ($145); with Limited ($100). Turbine forged aluminum 14 x 6 in. wheels ($300); with Limited ($255). Tires: P195/75R14 black SBR (no charge). P195/75R14 white SBR ($48).

HISTORICAL FOOTNOTES: Introduced: Sept. 19, 1978. Model year production (U.S.): 169,439 (1.8 percent of the industry total). Calendar year production: 184,636 cars, including 88,581 Concords. Calendar year sales by U.S. dealers: 162,057 (1.9 percent of industry total). Concords sold the best (85,432), Pacers the worst (only 8168). Record-breaking profits highlighted AMC's year, along with a strengthened tie to Renault, the French-owned automaker. Renault paid $200 million for an interest in AMC, carrying forward the agreement that had begun the previous year. Plans were made to begin assembling Renault autos at Kenosha for the 1982 model year, with Renault supplying the engines and transmissions. Meanwhile, Pacer sales continued to slip, due in part to marginal fuel economy, but mainly to the fact that its curious design never quite caught on. This was true even though Pacer had been, according to *AUTOMOTIVE INDUSTRIES* magazine, "widely acclaimed as a daring innovation in design." Perhaps too daring for popular tastes. As Renault became AMC's major stockholder, the company looked forward to major changes in the coming years. A three-year corrosion warranty debuted in May 1979.

1980 AMC

Major news for the 1980 model year was the appearance of the Eagle. Based on the Concord chassis, it was the first major four-wheel-drive passenger car produced in the U.S. in modern times. Otherwise, the lineup remained the same as in 1979, though this would be the final year for both Pacer and AMX. The V-8 engine was gone for good. All models had front disc brakes, standard four-speed floor shift, a high-pressure compact spare tire, inside hood release, dome light, cigar lighter, front stabilizer bar (except Spirit sedan), high level ventilation, and a parking brake warning light. To improve corrosion resistance, rust-prone regions used more galvanized materials and special coatings. One-side galvanized steel went inside hoods, deck lids, and door panels. Front fenders were plastic-lined, while petroleum wax coatings went on door panel bottoms, fender bottoms, and rear seam areas. Four-cylinder engines used a Delco High Energy Ignition system. Standard colors for all models were Cardinal red, Cameo blue, Navy blue, Saxon yellow, Cameo tan, Medium brown metallic, Classic black, and Olympic white. All models except the AMX could also be obtained in Russet, Bordeaux, Medium blue, Dark brown, Quick silver or Smoke gray metallic. One noteworthy new option: a leather-wrapped steering wheel, offered on all models. Premium sound systems could include a cassette tape player.

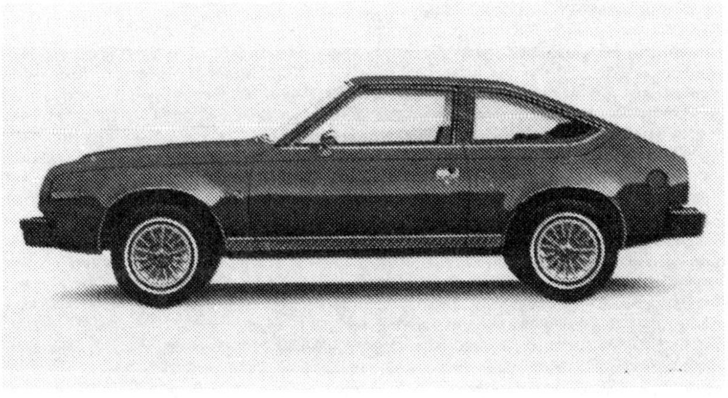

1980 AMC Spirit Liftback (AMC)

SPIRIT — SERIES 40 — FOUR/SIX: For its second year, Spirit continued with little change except for a new beltline molding on DL sedans, extending just below the front fender nameplates. The original 121 cid four departed, so base Spirits were now powered by a 2.5-liter (151 cid) four from Pontiac, with four-speed manual floor shift, and rode C78 x 14 whitewalls. A 258 six remained optional, but the V-8 was dropped. Standard equipment included manual front disc brakes, full wheelcovers, dual paint stripes, energy-absorbing bumpers, custom steering wheel, lighted ashtray, locking glove box, Sport vinyl bench seats (buckets in liftback), anodized aluminum bumpers with black end caps, and quad rectangular headlamps. In addition to the colors above, Spirits were available in Caramel. Spirit buyers could step up to a DL and get reclining bucket seats up front and split folding seat in back, with Caberfae Corduroy or Sport vinyl upholstery; an AM radio and day/night mirror; digital clock; luxury woodgrain steering wheel; woodgrain instrument panel overlay; premium cloth headliner and visors; parcel shelf; full-styled wheelcovers; front/rear bumper guards; dual horns; extra-quiet insulation; plus blackout rocker panels and wide lower body side moldings. Topping the line was the Spirit Limited, with reclining leather bucket seats and split folding rear seat, power steering, woodgrain-accent door panels, 18 oz. carpeting, console with center armrest, tilt steering wheel, power door locks and liftback release, bumper nerfing strips, protection and convenience groups, visibility group with dual remote mirrors, lights-on buzzer, and P195/75R14 whitewall glass-belted radials with wire wheelcovers. Performance fans could again get a Spirit GT liftback with tachometer and sport steering wheel inside, black bumpers and rear venturi area, blackout grille and headlight bezels, blackout moldings, black left remote mirror, dual pinstripes, and spoke styled wheels with P195/75R14 fiberglass-belted radial tires. Six-cylinder GT models with manual shift also included deep-tone exhaust. The GT package was available on base and DL Spirits. Once again, a Rally-tuned suspension package was offered for six-cylinder GT models.

AMX — SERIES 40 — SIX: The 'AMX' badge on the grille moved from the corner to the center for 1980, but other changes for the sporty performance AMC model were minor in its final outing. Base engine was the 258 cid six with four-speed manual shift. Standard gear included a GT Rally-tuned suspension system, tachometer, black left remote mirror, sports steering wheel, deep-tone exhaust (with manual shift), dual note horns, and extra-quiet insulation. An 'AMX' nameplate was on the glove box door, as well as on the grille, rear spoiler, and entry doors. Spoke style wheels held DR70 x 14 white-letter tires. As before, black accents set the tone for AMX. It had black bumpers, guards and nerfing strips; front/rear fender flares with accent stripes; front air dam; grille insert and headlight bezels; door, quarter window and rear window surround moldings; and windshield wiper arms. The color-keyed rear spoiler had accent stripes. A floor-shift automatic transmission was optional. An AMX custom interior package included reclining bucket seats in Sport vinyl or Caberfae corduroy.

PACER — SERIES 60 — SIX: Striking and innovative when it first appeared in 1975, Pacer sales never met expectations; so it was dropped after 1980. As before, the top two rows of the four-row grille bulged up into the hood area. Subdued black-out vertical grille elements were dominated by the horizontal bars. Pacer was the only AMC model that never received quad headlamps. DL Pacers came equipped with individual reclining seats in Sport vinyl, woodgrain-accent door trim panels, woodgrain instrument panel overlays, custom steering wheel with woodgrain accents, 12 oz. color-keyed full carpeting, rear seat armrests and ashtray, day/night mirror, clock, courtesy lights, key warning buzzer, locking glove box, extra-quiet insulation, and dual note horns. Tires were P195/75R14 fiberglass-belted radial whitewalls with styled wheelcovers. Pacers also had front and rear bumper guards, a squarish hood ornament, wide color-keyed body side scuff moldings, and wide rocker panel moldings. Upgrading to a Limited brought Chelsea leather reclining seats, a folding vinyl center armrest, woodgrain tilt steering wheel, and 18 oz. carpeting. Limiteds also had power steering, power windows and door locks, dual remote mirrors, an AM radio, dome/map light, glove box and under-hood lights, intermittent wipers, and lighted passenger visor mirror. In addition to the standard colors, Pacers were offered in Caramel and Misty Beige clearcoat.

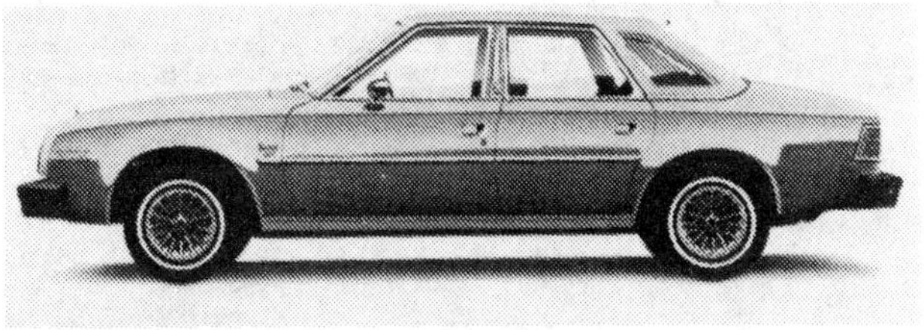

1980 AMC Concord D/L sedan (AMC)

CONCORD — SERIES 01 — FOUR/SIX: AMC's compact gained a new horizontal-bar grille for 1980, with quad rectangular headlamps and big full-width, wrap-around taillamps. 'Concord' badges sat at the side of the grille, as well as the trailing edge of the front fenders and the trunk lid. Only the front-fender amber reflectors remained. New, distinctive opera windows on DL Concords added both elegance and visibility for rear passengers. A four-cylinder 151 cid Pontiac engine was now standard, with the 258 six optional. In addition to the basic body colors, Concords came in Dark Green metallic. Added to the option list: six-way power seats, automatic-leveling air shocks, power windows, and power deck lid release. Wagons could now have a rear window wiper/washer. Base Concords rode D78 x 14 blackwall tires with full wheelcovers. Standard equipment included anodized aluminum bumpers with black end caps, dual body side pinstripes, narrow wheel lip and rocker panel moldings, bench seats in stripe knit fabric or Sport vinyl, front/rear ashtrays, hood ornament, narrow black side scuff moldings, and 12-oz. full carpeting. DL Concords added a day/night mirror, digital clock, courtesy light, dual note horns, extra-quiet insulation, front/rear bumper guards, individual reclining seats in Sport vinyl, full woodgrain instrument panel overlay, woodgrain-accented steering wheel, trunk carpeting, wide wheel lip and rocker panel moldings, 'B' pillar crests, and whitewall tires. Four-door DL sedans had rear opera windows and full vinyl roof, while two-doors featured a landau roof with stylish half-covered opera windows. Concord Limited sedans and wagons offered power steering, power door locks, tilt steering wheel, AM radio, dual remote mirrors, dome/map light, under-hood and trunk lights, lighted visor mirror, lights-on buzzer, intermittent wipers, woodgrain steering wheel, and bumper nerf strips. Their individual reclining seats were upholstered in either Chelsea leather or St. Lauren deep plush fabric. Limited tires were P195/7514 fiberglass-belted whitewall radials with wire wheelcovers.

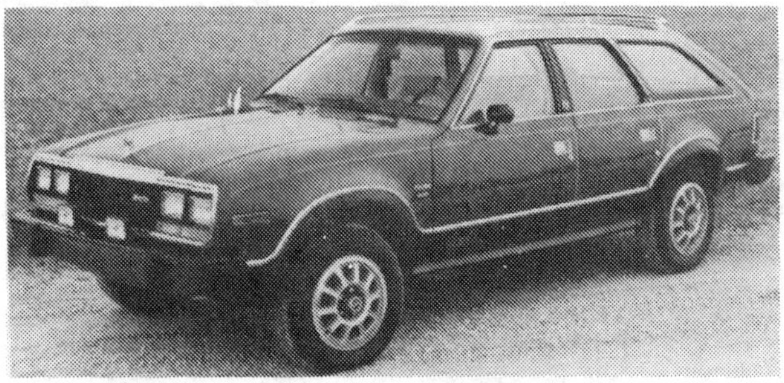

1980 AMC American Eagle station wagon (AMC)

EAGLE — SERIES 30 — SIX: Eagles arrived to the tune of a jingle, called "The Eagle Has Landed On All Fours." The Concord-based 4WD specialty vehicle was intended to combine the traction and handling of a truck with the comforts and conveniences of a luxurious passenger automobile. Measuring 109.3 in., its wheelbase was an inch longer than Concord's. The Eagle also sat three inches higher off the ground, making it easy to spot from a distance. Tires were 15-inch size. Three bodies were offered: two- and four-door sedans and a four-door wagon. Grilles had seven narrow horizontal bars and an Eagle badge in the upper corner, plus an 'AMC' emblem on the wide upper bar (at the opposite end). 'AMC' and 'Eagle' nameplates stood on the deck lid; '4 Wheel Drive' (or 'Automatic 4WD') on rear quarter panels. Distinctive opera windows were half- concealed, as on Concords. Hood ornaments were made up of twin rectangles. Eagles soon became noticed not only for their increased height and ground clearance, but for their accent-colored lower body treatment. That darker area stretched the full body length, over wide fender flares, set apart by a bright molding. Both the lower rocker sill strips and 3 in. fender flares (front and rear) were made of Krayton, a durable injection-molded plastic that was also used for bumper ends. Eagle's special bumpers had to match truck standards. A stone/gravel deflector formed the lower section of the front-end panel. AMC's Eagle was the first volume-produced 4WD with independent front suspension, allowing a lower center of gravity, smooth ride, and improved stability. The advanced 4WD system could interpret road conditions, automatically distributing power to the wheels (front or rear) that needed help most. A viscous coupling with 43 plates in the transfer case provided the necessary slippage, while it also absorbed vibration in the driveline. Standard equipment for the four-wheel-drive Eagle included a 258 cid six with two-barrel carburetor and Torque Command automatic transmission, plus power disc brakes, power steering, inside hood release, dual horns, front/rear bumper guards, electric clock, lighter, heavy-duty cooling system, and wheel opening moldings. Four-doors wore a full vinyl roof; two-doors a landau vinyl roof. Eagles rode P195/75R15B whitewall glass-belted radials with argent styled wheelcovers. Body colors were the same as Concord's. Eagle's Limited added power windows, a parcel shelf, upgraded carpeting, premium door trim, a luxury woodgrain steering wheel, visibility group, and individual reclining seats. Eagle's Sport package, available for all but the four-door sedan, included low-gloss black Krayton flares and rocker panels; black bumpers with nerfing strips; black taillamp treatment, grille insert, windshield/liftgate moldings, and remote-control twin mirrors; halogen headlamps and foglamps (except Limited), and blackwall all-weather Goodyear Tiempo steel-belted radials. Sport models wore a '4x4' silver decal on the lower door.

I.D. DATA: As in 1976-1979, the 13 symbol VIN was embossed on a metal plate riveted to the top left surface of the instrument panel, visible through the windshield. Coding was the same as before. The model year code (symbol two) changed to '0' for 1980. Under series (symbol four) a '3' code was added for the Eagle. Only two engine codes (seventh symbol) were used: B=151 four-cylinder; C=258 six-cylinder.

Series Number	Body/Style Number	Body Type & Seating	Factory Price	Shipping Weight	Production Total
SPIRIT SERIES 40					
40	8043-7	2d Lift-4P	4293/4422	2556/2758	Note 1
40	8046-7	2d Sed-4P	4193/4322	2512/2714	Note 1
SPIRIT DL					
40	8043-7	2d Lift-4P	4592/4721	2656/2854	Note 1
40	8046-7	2d Sed-4P	4492/4621	2611/2813	Note 1
SPIRIT LIMITED					
40	8043-7	2d Lift-4P	5091/5220	2675/2877	Note 1
40	8046-7	2d Sed-4P	4991/5120	2630/2832	Note 1

NOTE 1: Total production for the model year, 71,032 Spirits (37,799 four-cylinder and 33,233 six-cylinder). Model year sales: 55,392.

AMX SERIES 40					
40	8043-9	2d Lift-4P	5653	2901	N/A
PACER DL SERIES 60					
60	8066-5	2d Hatch-4P	5407	3147	405
60	8068-5	2d Sta Wag-4P	5558	3195	1341
PACER LIMITED					
60	8066-7	2d Hatch-4P	6031	3172	Note 2
60	8068-7	2d Sta Wag-4P	6182	3220	Note 2

NOTE 2: Production totals for Pacer DL also include Limited.

CONCORD SERIES 01					
01	8006-0	2d Sed-5P	4753/4882	2646/2844	27,845
01	8005-0	4d Sed-5P	4878/5007	2712/2910	35,198
01	8008-0	4d Sta Wag-5P	5078/5207	2741/2939	17,413
CONCORD DL					
01	8006-5	2d Sed-5P	5052/5181	2764/2962	Note 3
01	8005-5	4d Sed-5P	5177/5306	2834/3032	Note 3
01	8008-5	4d Sta Wag-5P	5377/5506	2855/3053	Note 3
CONCORD LIMITED					
01	8006-7	2d Sed-5P	5551/5680	2789/2987	Note 3
01	8005-7	4d Sed-5P	5676/5805	2859/3057	Note 3
01	8008-7	4d Sta Wag-5P	5876/6005	2886/3084	Note 3

NOTE 3: Production totals under base model include DL and Limited. Total model year production, 80,456 Concords (9,949 four-cylinder, 70,507 six-cylinder). Model year sales: 70,336.

EAGLE SERIES 30					
30	8036-5	2d Sed-5P	6999	3382	10,616
30	8035-5	4d Sed-5P	7249	3450	9956
30	8038-5	4d Sta Wag-5P	7549	3470	25,807
EAGLE LIMITED					
30	8036-7	2d Sed-5P	7396	3397	Note 4
30	8035-7	4d Sed-5P	7646	3465	Note 4
30	8038-7	4d Sta Wag-5P	7946	3491	Note 4

NOTE 4: Production totals shown under base Eagle include Limited. Model year Eagle sales: 34,041.

NOTE 5: [Spirit and Concord] Data above slash for four-cylinder/below slash for six-cylinder.

ENGINES: BASE FOUR (Spirit, Concord): Inline. OHV. Four-cylinder. Cast iron block and head. Displacement: 151 cid (2.5 liters). Bore & stroke: 4.0 x 3.0 in. Compression ratio: 8.2:1. Brake horsepower: 82 at 4000 rpm. Torque: 128 lbs.-ft. at 2400 rpm. Five main bearings. Hydraulic valve lifters. Carburetor: two-barrel Rochester 2SE. Manufactured by Pontiac. BASE SIX (AMX, Pacer, Eagle); OPTIONAL (Spirit, Concord): Inline. OHV. Six-cylinder. Cast iron block. Displacement: 258 cid (4.2 liters). Bore & stroke: 3.75 x 3.9 in. Compression ratio: 8.3:1. Brake horsepower: 110 at 3200 rpm. Torque: 210 lbs.-ft. at 1800 rpm. Seven main bearings. Hydraulic valve lifters. Carburetor: two-barrel Carter BBD.

CHASSIS DATA: Wheelbase: (Spirit/AMX) 96 in.; (Pacer) 100 in.; (Concord) 108 in.; (Eagle) 109.3 in. Overall length: (Spirit/AMX) 167 in.; (Pacer hatchback) 173.9 in.; (Pacer wagon) 178.8 in.; (Concord) 185 in.; (Eagle) 184 in. Height: (Spirit/AMX) 51.5 in.; (Pacer hatchback) 52.7 in.; (Pacer wagon) 53.1 in.; (Concord) 51.3-52.6 in.; (Eagle) 55-55.8 in. Front Tread: (Spirit/AMX) 58.1 in.; (Pacer) 61.2 in.; (Concord) 57.6 in.; (Eagle) 59.6 in. Rear Tread: Spirit/AMX) 57.0 in.; (Pacer) 60.0 in.; (Concord) 57.1 in.; (Eagle) 57.6 in. Standard Tires: (Spirit) C78 x 14; (AMX) DR70 x 14; (Pacer) P195/75R14; (Concord) D78 x 14; (Eagle) P195/75R15.

TECHNICAL: Four-speed floor-shift manual transmission standard; three-speed Torque Command automatic transmission optional (floor or column shift) on all models except Eagle, which had standard automatic shift. Manual shift gear ratios: Four-cylinder (1st) 3.50:1; (2nd) 2.21:1; (3rd) 1.43:1; (4th) 1.00:1; (Rev) 3.39:1. Six-cylinder (1st) 4.07:1; (2nd) 2.57:1; (3rd) 1.66:1; (4th) 1.00:1; (Rev) 3.95:1. Standard axle ratios: (Spirit/Concord four, Eagle) 3.08:1; (other models) 2.53:1. Steering: recirculating ball. Suspension: independent front coil springs; semi-elliptic rear leaf springs. Brakes: front discs, rear drum. Fuel tank: (Spirit/AMX/Pacer) 21 gal.; (Concord) 22 gal.

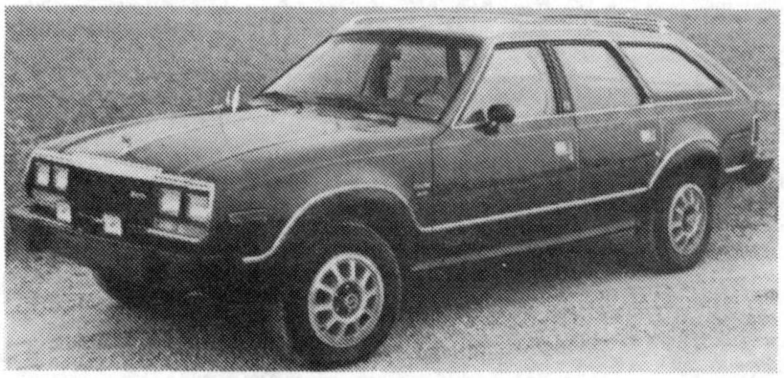

1980 AMC American Eagle station wagon (JG)

DRIVETRAIN OPTIONS: 258 cid six-cylinder engine: Spirit/Concord ($129). Torque-Command automatic transmission, column shift: Spirit ($305); Pacer/Concord ($333). Torque-Command with floor shift: Spirit/AMX ($331); Pacer ($359). Optional 3.54:1 axle ratio: Eagle ($19). Twin Grip differential ($65); not available Eagle. Heavy-duty engine cooling system (heavy-duty radiator, viscous fan): Spirit/Concord six ($41), but standard with air conditioning. Heavy-duty cooling system (heavy-duty radiator, seven-blade flex fan/shroud: Pacer ($37). Extra-duty suspension package (rear sway bar, heavy-duty shocks/springs): Eagle ($65). Handling package (unique front sway bar, rear sway bar): Spirit/Concord six except GT ($31). Front sway bar: Spirit six sedan ($20). Front suspension skid plate: Eagle ($65). Automatic load-leveling (air shocks): Concord six, Eagle ($145). Air shock absorbers, rear: Spirit ($52). Heavy-duty shock absorbers: Pacer/Concord ($15). Heavy-duty battery ($19). Cold climate group: heavy-duty battery and alternator, engine block heater ($107); with air conditioner or rear defroster ($47). California emission system ($250). Trailer towing package 'A' (to 2000 lbs.): Concord/Eagle ($85).

SPIRIT/CONCORD/AMX/EAGLE CONVENIENCE/APPEARANCE OPTIONS: Spirit liftback GT package: base/DL ($249). Spirit liftback GT Rally tuned suspension package ($109). AMX custom interior package (reclining vinyl bucket seats, custom door panels, split rear seat, day/night mirror, courtesy lights, parcel shelf): AMX with vinyl trim ($149); with fabric trim ($179). Eagle Sport package: base two-door wagon ($299). Extra-quiet insulation package: base Spirit/Concord ($50); Concord D/L wagon ($19). Protection group (stainless door edge guards, front bumpers guards and nerf strips, front mats): Spirit ($111); Spirit D/L ($67); Spirit GT, AMX ($31). Protection group with front and rear guards and mats: Concord ($114); Concord D/L ($70). Eagle protection group (stainless door edge guards, front/rear bumper nerf strips, front/rear mats): Eagle ($70) except with Sport package ($34). Convenience group including headlight-on buzzer, intermittent wipers, vanity mirror ($63). Gauge package (clock, tach, oil, amp, vacuum): Spirit six ($129) except D/L and Limited ($77); AMX GT ($75); not available base Spirit. Visibility group (remote control left/right mirrors, day/night mirror): Spirit/Concord/AMX ($64); D/L or GT ($52). Eagle visibility group (remote left/right mirrors) ($52). Remote control left mirror ($18); not available AMX. Pop-up moon roof: Spirit/AMX/Concord ($195). Power steering ($164). Power front disc brakes ($74). Cruise control ($108). Air conditioning system ($529). Air conditioning package (including tinted glass and power steering): Spirit/AMX ($752); Concord ($758). Halogen headlamps ($20). Fog lamps (dealer-installed): Eagle ($69). Light group ($37-$53). AMX hood decal ($60). AM radio ($89). AM/FM/stereo radio ($219) except Spirit D/L and Limited ($130). AM/FM/stereo radio with cassette ($335) except Spirit D/L and Limited ($246). AM/FM/CB stereo radio ($475) except Spirit D/L and Limited ($386). Premium sound system ($95). Digital clock: Spirit/Concord ($52). Tachometer: Spirit/Concord/Eagle ($54); standard with Spirit GT package. Power door locks: Spirit, AMX, Concord/Eagle two-door D/L and Limited ($75); Concord/Eagle four-door D/L and Limited ($108); not available base Spirit. Power door/window locks: Concord D/L, base Eagle ($199-$289). Power liftback release: Spirit ($32). Power decklid release: Eagle/Concord sedan and wagon ($32). Power six-way driver's seat: Concord D/L and Limited ($149). Power driver/pass. seat: Concord D/L and Limited ($249). Caberfae corduroy fabric bucket seats: Spirit D/L ($29). Vinyl bench seat: Spirit sedan (NC); Concord with striped knit fabric ($29). Vinyl reclining seats: Concord D/L (no charge). Rochelle velour stripe fabric reclining seats: Concord D/L ($29). Leather reclining seats: Concord Limited (no charge). Silver knit fabric reclining seats: Concord/Eagle Limited (deduct $100). Plaid fabric seats: Eagle ($29); not available Limited. Console: Spirit/AMX ($78). Rear defroster ($93). Rear wiper/washer: Eagle wagon ($79). Dual horns: Spirit/Concord ($14); standard on D/L and Limited. Leather-wrapped sport steering wheel: Spirit/Concord/Eagle ($51); Spirit D/L and Limited/GT, Concord Limited, AMX ($21). Woodgrain steering wheel: Concord D/L, Eagle ($30). Parcel shelf ($22). Tilt steering wheel ($75-$78). Tinted glass ($59-$65). Two-tone paint: Spirit GT ($84); Concord D/L and Limited ($103). Special color combinations: ($30) except AMX, Eagle Limited Rally side stripes: Spirit ($65). Delete woodgrain from Concord D/L and Limited or Eagle wagon (deduct $75). Roof rack: Spirit sedan ($64); Concord/Eagle wagon ($93). Tailgate air deflector: Spirit sedan ($27). Front/rear bumper guards: Spirit/Concord ($44). Side scuff molding: Spirit ($32). Locking gas cap ($9). Styled wheelcovers: Spirit/AMX/Concord ($35); not available on Limited Spoke style 14 x 6 in. wheels with trim rings: Spirit/Concord ($150) except D/L ($115); Limited ($15). Turbine forged aluminum wheels: Spirit/Concord ($310) except D/L ($275); Limited ($175); Spirit GT ($160). Turbocast II aluminum 14 x 7 in. wheels: Spirit ($350) except D/L ($315); Limited ($215); GT, AMX ($200). Wire wheelcovers: Spirit/Concord ($135) except D/L ($100). Spirit/Concord/AMX Tires: D78 x 14: Spirit (no charge). D78 x 14 whitewall: Spirit ($23); Concord ($49). P195/75R14 GBR: Spirit ($75); Concord ($101); Concord D/L and Limited ($52). P195/75R14 white GBR: Spirit ($124); Spirit GT ($50); Concord ($150); Concord D/L and Limited ($101). P195/75R14 SBR: Spirit ($125); Spirit GT ($130); Concord ($151); Concord D/L and Limited ($102). P195/75R14 white SBR: Spirit ($174); Spirit GT ($195); Concord ($200); Concord D/L and Limited ($151). DR70 x 14 Flexten radial: Spirit ($205); Concord ($231); Concord D/L and Limited ($182). DR70 x 14 OWL Flexten radial: Spirit ($270); Concord ($296); Concord D/L and Limited ($247). ER60 x 14 OWL Flexten radial: AMX with Turbocast II wheels ($53). Eagle Tires: P195/75R15 SBR Tiempo ($42). P195/75R15 white SBR Tiempo ($91) except Sport ($49).

PACER CONVENIENCE/APPEARANCE OPTIONS: Convenience group including headlight-on buzzer, intermittent wipers, right lighted vanity mirror ($63). Visibility group: remote left and right mirrors ($52). Protection group: front/rear bumper nerf strips, door edge guards, front/rear floor mats ($70). Left remote mirror ($18). Power steering ($164). Power front disc brakes ($74). Cruise control ($108). Air conditioning ($529). Air conditioning package with tinted glass and power steering ($761). AM radio ($89). AM/FM/stereo four-speaker radio ($219). AM/FM/stereo radio with eight-track ($335). AM/FM/CB stereo radio ($475). Center armrest ($51); standard on Limited Rochelle velour stripe fabric seats: D/L ($29). Vinyl sport seat trim: D/L (no cost). Power door locks ($75). Power windows with door locks ($199). Light group ($43). Rear defroster ($93). Rear wiper/washer ($79). Leather-wrapped sport steering wheel ($31); with Limited ($21). Woodgrain steering wheel ($30). Tilt steering wheel ($78). Tinted glass ($68). Door vent windows ($50). Two-tone color: black rocker panels deleted ($67). Special color combinations ($30); not available with two-tone or Limited Misty beige clearcoat: Limited ($90). Woodgrain wagon paneling ($121). Pop-up moon roof: hatch ($195). Roof rack ($78). Locking gas cap ($9). Wire wheelcovers ($100). Spoke style 14 x 6 in. wheels ($115); with Limited ($15). Turbine forged aluminum 14 x 6 in. wheels ($275); with Limited ($175). Tires: P195/75R14 white SBR ($49).

HISTORICAL FOOTNOTES: Introduced: (Eagle) Sept. 27, 1979; (others) October 11. Model year production (U.S.): 199,613, which was 2.9 percent of the industry total. Calendar year production (U.S.): 164,728 cars. Model year sales by U.S. dealers: 163,502. Calendar year sales: 149,430 (2.3 percent of total). For rating purposes, the new Eagle was declared a "four-wheel-drive automobile" rather than a passenger car. Thus, its so-so gas mileage didn't count in AMC's corporate average fuel economy (CAFE) rating, because it was classed as a light truck. And for meeting safety standards, Eagle ranked as a multi-purpose vehicle. The new Eagles were assembled in the same Kenosha production line as passenger cars, however. Early in the model year (December 1979), AMC announced that the slow-selling Pacer would be dropped to allow for increased Eagle production at the Kenosha facility. Eagle sold well, though not to AMC's 50,000-unit expectations. Eagle's 4WD had been developed by FF Developments in Britain, but made by Chrysler's New Process Gear Div. Four-wheel-drive passenger cars weren't entirely new. Britain's limited-production Jensen Interceptor had used 4WD in the 1960s. But, as the '80s decade began, Subaru's version, introduced in 1975 in wagon form, was Eagle's sole competition for the all-wheel fancier's dollars. During 1980 contract talks, AMC became the second domestic auto company (after Chrysler) to accept a member of the United Auto Workers union on its board. After record-breaking profits in the previous year, AMC ended the 1980 period with a loss of $155.7 million. Yet continued strong Jeep sales (with increased outlets worldwide) and a cash inflow from Renault, plus plans to market several French-built models in the U.S., helped keep AMC's prospects on the bright side. AMC's rustout warranty was extended to five years. Innovations: Four-wheel-drive passenger car. New microprocessor-controlled feedback carburetor system developed.

1981 AMC

With the Pacer gone, AMC's lineup dwindled to just three models. But the four-wheel drive Eagle, which appeared the year before on a modified Concord platform, added a Spirit-based version for 1981. All models came with a standard 151 cid four-cylinder engine. The optional 258 cid six was redesigned using aluminum and other lightweight materials cutting 90 pounds from its previous heft. Camshaft alterations reduced its valve overlap, allowing slower, smoother idling and more low-speed torque. Three-speed automatic transmissions now included a lockup torque converter. Galvanized steel (one-sided) outer body panels provided improved rust protection and AMC continued its five-year no-rust-through warranty, introduced the previous year. Upper deck panels were now two-sided galvanized steel. Standard 1981 colors were Classic black, Olympic white, Cameo tan, Montana blue, Moonlight blue, Autumn gold, Oriental red, and beige, plus 13 metallics (Quick silver, Medium blue, Medium or Dark brown, Copper brown, Chestnut brown, Vintage red, Deep maroon, Steel gray, blue, silver, Sherwood green, and Dark green).

1981 AMC Spirit D/L liftback coupe (AMC)

SPIRIT — SERIES 40 — FOUR/SIX: Styling changes on the subcompact Spirit included a new crossbar-style grille with emblem on lower corner, Rally stripes, altered wheelcovers, and a new selection of body colors. Power windows and radio antenna were optional for the first time. Spirits were now equipped with P185/75R14 blackwall glass-belted radial tires and had wheelcovers, front disc brakes, a lighted front ashtray and lighter, carpeting, vinyl bucket front seats, vinyl spare tire cover, and rear bumper guards. In addition to whitewalls and an AM radio, DL models offered extra-quiet insulation, dual horns, custom door panels with map pockets, a luxury woodgrain steering wheel, day/night mirror, electric clock, front bumper guards, styled wheelcovers, and a carpeted spare tire cover. A GT package with full instrumentation was available again this year.

1981 AMC Concord-Griffith AM-TC convertible (JG)

CONCORD — SERIES 01 — FOUR/SIX: For improved rust protection, all exterior Concord body panels were now galvanized steel. Glass-belted P195/75R14 blackwall radial tires with wheelcovers were standard; steel-belted tires and wire wheelcovers optional. The restyled grille used three vertical bars to accent the five horizontal bars. Opera windows were restyled. New colors and fabrics were offered. Base engine was the 151 cid four with four-speed manual floor shift. Standard equipment included a stowaway spare tire, front and rear armrests, lighter, carpeting, bench seats, folding rear wagon seat, rear bumper guards, and moldings for drip rail, wheel lip, hood front edge, windshield and rear window surrounds, rocker panels, and body side scuff area. An optional retractable cargo area cover could hide luggage in Concord wagons. A vinyl landau roof highlighted Concord's DL sedans. DL models also had stainless steel wheelcovers, individual reclining seats, a custom steering wheel, day/night mirror, cargo area skid strips, electric clock, trunk carpeting, front bumper guards, striping, dual horns, woodgrain wagon side panels, extra sound insulation, and whitewall tires. Two-doors featured opera quarter windows. The luxurious Limited added visibility and light groups as well as a luxury woodgrain steering wheel and styled wheelcovers, plus premium seat and door trim.

1981 AMC American Eagle SX/4 liftback coupe (AMC)

EAGLE SX/4 AND KAMMBACK — SERIES 50 — FOUR/SIX: Since the original four-wheel drive Eagle showed promise, AMC added a shrunken version for 1981, based on the Spirit chassis. Billed as "the sports car that doesn't always need a road," the SX/4 two-door hatchback had a sporty look, but hardly qualified as a sports car. Also in the lineup was a Kammback wagon, derived from the Spirit/Gremlin sedan design. Both rode a 97.2 in. wheelbase but carried 15 in. tires, which gave three inches more ground clearance than the Spirits. Front-end styling focused on a new 8x3 checkerboard-style grille, like the senior Eagles. An Eagle nameplate was up front, as well as at the usual front fender locations. An AMC badge sat atop the grille. 'SX/4' decals were on lower front doors, part of the wide accent-colored Krayton plastic body striping that ran from front to back, over the fender flares. A bright molding separated the two body colors. 'Four-wheel drive' emblems stood on quarter panels. Subcompact Eagles came with a 2.5-liter (151 cid) four-cylinder engine, four-speed floor shift and transfer case, the same as their bigger brothers. Three-speed automatic shift was optional, as was the 258 cid six. The smaller Eagles rode well enough to rate with ordinary passenger cars...far more smoothly than the typical off-road vehicle...it performed with reasonable liveliness and impressive gas mileage. Eagles had power front disc brakes, power steering, front sway bar, high energy ignition, 42-amp alternator, 55-380 (cold crank rating) battery, quad rectangular headlamps, 21- gallon fuel tank, and compact spare tire. Standard equipment also included P195/75R15 blackwall glass-belted radial tires with wheelcovers, vinyl front bucket seats and fold-down rear bench seat, two-speed wipers, carpeting, locking glove box, front armrests, lighted front ashtray, lighter, coat hooks, inside hood release, dome light, body pinstripes, spare tire cover, and front/rear bumper nerf strips. The DL upgrade added custom vinyl reclining bucket seats up front and a split vinyl rear seat, Alpine cloth custom headliner and visors, woodgrain instrument panel overlay, woodgrain horn cover on a custom steering wheel, day/night mirror, and digital clock. It also featured left/right remote-control chrome mirrors, chrome side marker lights, dual horns, extra-quiet insulation, and P195/75R15 whitewall glass-belted radials with argent styled wheelcovers. DL liftbacks had blackout window frames, belt moldings and door/quarter frame moldings. Off-road enthusiasts could elect an optional Sport package that included a floor shift console, parcel shelf, and vinyl sport steering wheel inside. Outside, they featured a Sport nameplate, plus low-gloss black Krayton flares and rocker panels; black bumpers with guards and nerf strips; black grille insert, moldings, taillamp treatment, and left/right remote sport mirrors. Sport models rode P195/75R15 blackwall steel-belted radial Goodyear Arriva tires with styled wheelcovers, and carried halogen headlamps and foglamps.

EAGLE — SERIES 30 — FOUR/SIX: Eagle's new checkerboard grille, with an 8x3 pattern, differed considerably from the Concord cousin. Like their new smaller companions, the big Eagles were noteworthy for their dark Krayton lower body treatment. With standard four-speed manual shift, Eagle rated 22 mpg in the EPA fuel economy rankings. Outside body panels were now one-side galvanized steel for added rust protection. Standard equipment was the same as the smaller Eagle 50, but with individual vinyl reclining front seats, dual horns, the extra-quiet insulation package, whitewall P195/75R15 glass-belted radials and argent styled wheelcovers. Eagle 30s also had front/rear armrests with woodgrain overlay, cargo area skid strips, day/night mirror, digital clock, Alpine cloth headliner and visors, locking cargo compartment, woodgrain steering wheel and instrument panel overlays, and front/rear bumper nerf strips. Two-door sedans sported a landau roof design; four-doors, a full vinyl roof. Wagon rear seats folded down. The Eagle 30 Limited added leather reclining seats, 18-oz. carpeting, luxury woodgrain steering wheel, and a parcel shelf. The Sport package, highlighted by a '4x4' silver decal on the lower door and a 'Sport' nameplate, contained a leather-wrapped sport steering wheel; low-gloss black flares and rocker panels, plus black bumpers/guards, nerf strips, hood molding, taillamp treatment, grille insert, headlamp bezels, door frames, moldings, and remote-control dual sport mirrors. Also halogen headlamps and foglamps, and P195/75R15 blackwall steel-belted Arriva radial tires.

I.D. DATA: A new 17 symbol Vehicle Identification Number (VIN) was embossed on a metal plate riveted to the upper left surface of the instrument panel, visible through the windshield. It began with a digit indicating country of manufacture: 1=U.S., 2=Canada. The second symbol identified the manufacturer: A=AMC; C=American Motors (Canada). Third symbol showed vehicle type: M=passenger car; C=multi-purpose vehicle (Eagle); E=export. Symbol four denoted engine type: B=151 cid (2.5-liter) four (from Pontiac); C=258 cid (4.2-liter) six. The fifth symbol identified transmission (and transfer case) type: M=four-speed manual floor

261

shift; H=four-speed with four-wheel drive; G=four-speed with full-time four-wheel drive; W=five-speed floor shift; N=five-speed with four-wheel drive; A=column-shift automatic; C=floor-shift automatic; and K=floor-shift automatic with four-wheel drive. The next two digits identified the line and body type: 0=Concord four-door; 06=Concord two-door; 0=Concord wagon; 35=Eagle four-door sedan; 36=Eagle two-door sedan; 38=Eagle four-door wagon; 43=Spirit two-door Liftback; 46=Spirit two-door sedan; 53=Eagle SX/4 two-door Liftback; 56=Eagle (Kammback) two-door sedan. The eighth symbol identified trim level: 0=base model; 5=DL; 7=Limited. Next came a check digit to mathematically determine validity of a car's VIN. Symbol ten was a letter indicating model year: B=1981. Symbol eleven showed manufacturing plant: K=Kenosha; B=Brampton, Ontario. Finally came a six-digit sequence number. An engine Build Code was stamped on a machined surface of the block of six-cylinder engines, between cylinders two and three; and at the rear of the engine, near the flywheel, on 151 cid fours. The fourth symbol of that code was identical to the engine code of the VIN. Symbol one is year (1= 1981); symbols 2-3, the month built (01-12); symbols 5-6, the day of the month. The VIN is also on the Federal Safety Label attached to the edge of the left door, above the door lock; or on the bottomline of a metal plate attached to upper left corner of firewall, under the hood. A Unit Body/Trim Plate on left door edge shows Body Number, Model Number, Trim Number, Paint Code, and Build Sequence Number.

SPIRIT SERIES 40

Series Number	Body/Style Number	Body Type & Seating	Factory Price	Shipping Weight	Production Total
40	8143-0	2d Lift-4P	5190/5326	2587/2716	42,252
40	8146-0	2d Sed-4P	5090/5226	2542/2671	2367

SPIRIT DL

40	8143-5	2d Lift-4P	5589/5725	2673/2802	Note 1
40	8146-5	2d Sed-4P	5489/5625	2627/2756	Note 1

NOTE 1: Production totals include base and DL models. Total model year production, 44,599 Spirits (26,075 four-cylinder and 18,524 six-cylinder). Model year sales: 38,334.

CONCORD SERIES 01

01	8106-0	2d Sed-5P	5819/5955	2672/2798	15,496
01	8105-0	4d Sed-5P	5944/6080	2738/2864	24,403
01	8108-0	4d Sta Wag-5P	6144/6280	2768/2894	15,198

1981 AMC Concord Limited coupe (AMC)

CONCORD DL

01	8106-5	2d Sed-5P	6218/6354	2767/2893	Note 2
01	8105-5	4d Sed-5P	6343/6479	2837/2963	Note 2
01	8108-5	4d Sta Wag-5P	6543/6679	2852/2978	Note 2

CONCORD LIMITED

01	8106-7	2d Sed-5P	6665/6801	2789/2915	Note 2
01	8105-7	4d Sed-5P	6790/6926	2859/2985	Note 2
01	8108-7	4d Sta Wag-5P	6990/7126	2880/3006	Note 2

NOTE 2: Production totals shown for base Concord include DL and Limited models. Total model year production, 55,097 Concords (7067 four-cylinder and 48,030 six-cylinder). Model year sales: 63,732.

1981 AMC American Eagle-Griffith convertible (JG)

EAGLE SERIES 50

50	8153-0	2d SX/4 Lift-4P	6717/6853	2967/3123	17,340
50	8156-0	2d Kamm Sed-4P	5995/6131	2919/3015	5603

EAGLE DL SERIES 50

50	8153-5	2d SX/4 Lift-4P	7119/7255	3040/3196	Note 3
50	8156-5	2d Kamm Sed-4P	6515/6651	2990/3146	Note 3

NOTE: Production totals include both base and DL models.

EAGLE SERIES 30

30	8136-5	2d Sed-5P	7847/7983	3104/3260	2378
30	8135-5	4d Sed-5P	8097/8233	3172/3328	1737
30	8138-5	4d Sta Wag-5P	8397/8533	3184/3340	10,371

EAGLE LIMITED

30	8136-7	2d Sed-5P	8244/8380	3114/3270	Note 4
30	8135-7	4d Sed-5P	8494/8630	3180/3336	Note 4
30	8138-7	4dr Sta Wag-5P	8794/8930	3198/3354	Note 4

NOTE 4: Production totals shown under base Eagle include Limited. Total model year production, 37,429 Eagles (11,344 four-cylinder and 26,085 six-cylinder). Model year sales: 42,904.

NOTE 5: Data above slash for four-cylinder/below slash for six-cylinder.

ENGINES: BASE FOUR (all models): Inline. ohv. Four-cylinder. Cast iron block and head. Displacement: 151 cid (2.5 liters). Bore & stroke: 4.0 x 3.0 in. Compression ratio: 8.24:1. Brake horsepower: 82 at 3800 rpm. Torque: 125 lbs.-ft. at 2600 rpm. Five main bearings. Hydraulic valve lifters. Carburetor: two-barrel. Rochester 2SE. OPTIONAL SIX (all models): Same as 1980 specifications (258 cid inline ohv).

CHASSIS DATA: Wheelbase: (Spirit) 96 in.; (Concord) 108 in.; (Eagle 50) 97.2 in.; (Eagle 30) 109.3 in. Overall length: (Spirit) 167 in.; (Concord) 185 in.; (Eagle 50 SX/4) 164.6 in.; (Eagle 50 Kammback) 164.4 in.; (Eagle 30) 184 in. Height: (Spirit) 51.5 in.; (Concord) 51.3-51.5 in. (Eagle 50) 55.3-55.5 in.; (Eagle 30) 55.0-55.8 in. Width: (Spirit) 72 in.; (Concord) 71 in.; (Eagle 50) 73 in.; (Eagle 30) 71.9 in. with flares. Front Tread: (Spirit) 58.1 in.; (Concord) 57.6 in.; (Eagle 50) 59.6 in. Rear Tread: (Spirit) 57 in.; (Concord) 57.1 in.; (Eagle) 57.6 in. Standard Tires: (Spirit) P185/75R14 BSW GBR; (Concord) P195/75R14 BSW GBR; (Eagle) P195/75R15 BSW GBR.

TECHNICAL: Four-speed manual floor shift standard. Gear ratios: (1st) 4.07:1; (2nd) 2.39:1; (3rd) 1.49:1; (4th) 1.00:1; (Rev) 3.95:1. Torque Command three-speed automatic transmission optional; lockup torque converter. Standard axle ratio: (Spirit/Conc. four) 3.08:1; (Spirit/Conc. six) 2.37:1; (Eagle four) 3.54:1; (Eagle six) 2.73:1. Steering: recirculating ball. Suspension: independent front coil springs; semi-elliptic rear leaf springs. Clutch dia.: 9.12 in. Brakes: front discs, rear drums; (Spirit) 10.3 in. disc; (Concord) 10.8 in. disc; (Eagle) 11 in. disc. Electronic ignition.

1981 AMC Concord-Griffith AM-TC convertible (JG)

SPIRIT/CONCORD/EAGLE DRIVETRAIN OPTIONS: 258 cid six-cylinder engine ($136). Column-shift automatic transmission: Spirit/Concord ($350). Floor-shift automatic transmission: Spirit/Eagle ($350). Optional 3.08:1 axle ratio: Eagle 30 six with automatic ($20). Twin Grip differential: Spirit/Concord ($69). Heavy-duty engine cooling system (heavy-duty radiator, viscous fan/shroud, coolant recovery system): Spirit/Concord six ($61) but standard with air cond.; Eagle four ($51). Maximum cooling system (heavy-duty radiator and viscous fan): Eagle ($62). Handling package (unique front sway bar, rear sway bar): Spirit, Concord six ($42) not available with Spirit GT. Heavy-duty shock absorbers: Spirit/Concord ($16); not available with Spirit GT Rally tuned package. Automatic load leveling (air shocks): Concord/Eagle six ($153). Extra duty suspension package (rear sway bar and heavy-duty shocks/springs): Eagle 50 ($37); Eagle 30 ($69). Extra heavy-duty suspension (heavy-duty springs, shocks, control arms/bushings): Concord ($59). Front suspension skid plate: Eagle ($69). Trailer towing package 'A' (to 2000 lbs.): Concord/Eagle 30 ($90). Trailer towing package 'B' (to 3500 lbs.): Eagle 30 ($195). heavy-duty (56/450 cold crank) battery ($20). 80-amp battery: Concord/Eagle ($32). heavy-duty alternator ($63). Cold climate group incl. heavy-duty battery/alternator, engine block heater ($113); with air conditioning or rear defroster: Spirit/Concord ($50). California emission system ($50).

CONVENIENCE/APPEARANCE OPTIONS: Spirit liftback GT package ($372); DL ($272). Spirit liftback GT Rally-tuned suspension package ($119). Eagle Sport package: Eagle 30 two-door station wagon ($314); Eagle 50 Liftback ($472); Eagle 50 DL Liftback ($367). Extra-quiet insulation package ($53). Protection group (stainless door edge guards, front/rear bumper guards and nerfing strips, front/rear floor mats): Spirit ($117); Concord ($120); DL/Limited ($71-$74); Spirit GT ($33). Eagle protection group, includes stainless door edge guards, front/rear bumper guards, front/rear floor mats ($79-$82); with Sport package. ($33-$36). Convenience group: headlight-on buzzer, intermittent wipers, right lighted vanity mirror ($67). Gauge package (clock, tach, oil, amp or volt, vacuum): Spirit, Eagle 50 ($136); Spirit GT/DL ($79-$81); Eagle 30 or DL ($81). Pop-up sunroof with tinted glass ($246). Rear spoiler: Spirit GT, Eagle Sport Liftback ($99). Light group ($46-$56). Left remote control mirror ($19) except GT; not available on Eagle 30. Left/right remote chrome mirrors ($56), except Spirit GT/DL; standard on Concord DL/Limited; not available Eagle 30. Left/right remote sport mirrors: Spirit, Eagle 50 ($56); Eagle 50 DL (no charge). Left/right electric remote chrome mirrors: Spirit/Concord ($132); Concord DL and Eagle 30 ($76-$77); Day/night mirror ($13); standard on DL/Limited. Power steering: Spirit/Concord ($173). Power front disc brakes: Spirit/Concord ($80). Electronic cruise control ($132). Air conditioning system ($531-$585). Air conditioning package including tinted glass and power steering: Spirit ($774); Concord ($833). Halogen headlamps ($40). Foglamps (dealer-installed): Eagle ($73). Dual horns: Spirit/Concord ($15). AM radio ($92). AM/FM/CB stereo four-speaker radio ($456) except Spirit DL ($364). AM/FM/stereo four-speaker radio ($192) except Spirit DL ($100). AM/FM/cassette stereo radio ($356) except Spirit DL ($264). Premium audio system including power amplifier, four hi-fi speakers, fader ($100). Power antenna ($53). Power door locks: Spirit DL, Concord DL/Limited two-door sedan, Eagle 30 two-door ($90); Spirit DL, Concord DL/Limited and Eagle 30 four-door/wagon ($130-$131). Power windows and door locks: Spirit DL, Concord DL/Limited two-door sedan, Eagle 30 two-door, 50 DL ($231); Concord DL/Limited and Eagle 30 four-door wagon ($330). Power liftback release: Spirit/Eagle 50 ($34). Power decklid release: Spirit/Concord ($50). Power six-way driver's seat: Concord DL/Limited, Eagle 30 ($157). Power six-way driver/passenger seat: Concord DL/Limited, Eagle 30 ($262). Center console with armrest: Spirit ($82). Floor front console: Eagle ($52). Parcel shelf ($24). Bench seat (striped knit fabric): Concord ($48). heavy-duty seat frame assembly for bench seat: Concord ($31). heavy-duty vinyl seat trim: Concord ($60); not available with DL or four-speed. Coventry Check seat fabric, reclining bucket seats: Spirit DL, Eagle 50 ($31). Velour stripe fabric on reclining seats: Concord DL, Eagle 30 ($58). Durham plaid fabric or Rochelle sculptured velour on reclining seats: Eagle 30 ($58). Cargo area cover: Concord/Eagle wagon ($62). Digital clock ($55); standard on DL/Limited. Rear wiper/washer ($99). Rear defroster ($102-$107). Vinyl sport steering wheel: Spirit/Eagle 50 ($34). Woodgrain steering wheel: Spirit/Concord DL, Eagle 30 ($32). Leather-wrapped sport steering wheel ($54) except GT/Limited, Eagle 50 Sport ($20-$34). Tilt steering wheel: Spirit ($79), Eagle ($82). Tinted glass ($70-$75). Two-tone paint (without pinstripes): Spirit, Concord DL/Limited ($109). Special color combinations ($32). Rally side stripes: Spirit ($79). Woodgrain paneling: Eagle wagon ($128). Delete woodgrain from Concord DL/Limited wagon (deduct $75). Side scuff moldings ($44). Roof rack: Spirit sedan, Eagle 50 ($74); Concord/Eagle 30 wagon ($98). Locking gas cap ($10). Tailgate air deflector: Spirit/Eagle 50 sedan ($29). Front/rear bumper guards ($47). Styled wheelcovers (argent): Spirit ($37) except GT; Concord ($74); Concord DL ($37); Eagle ($48). Custom wheelcovers: Concord ($37); standard on DL. Wire wheelcovers: Spirit/Concord ($142); DL ($105); not available on GT. Spoke style 14 x 6 in. wheels with trim rings: Spirit/Concord ($158) except DL ($121); Concord Limited ($16). Turbocast II aluminum 14 x 7 in. wheels: Spirit ($368) except DL ($331-$345); Spirit GT ($210); Concord Limited ($226). Aluminum 15 x 6 in. wheels: Eagle ($310-$357). Spirit/Concord Tires: P185/75R14/B white GBR: Spirit ($52). P195/75R14/B black GBR: Spirit ($25) exc. DL (no charge). P195/75R14/B white GBR: Spirit ($76); Spirit DL ($25); Concord ($52). P195/75R14/B white SBR Arriva: Spirit ($129); Spirit DL ($77); Concord ($104); Concord DL/Limited ($53). P195/70R14/B RWL polysteel radial: Spirit ($229); Spirit DL ($178); Concord ($205); Concord DL/Limited ($154). Eagle Tires: P195/75R15 white GBR: Eagle 50 ($52). P195/75R15 black SBR Arriva ($52-$53). P195/75R15 white SBR Arriva: Eagle 30 ($96); Eagle 50 ($104); Eagle DL ($96); Eagle Sport ($52). P215/65R15 OWL SBR Eagle GT: Eagle 30 ($245); Eagle Sport ($201). P215/75R15 OWL SBR Eagle GT: Eagle 50 ($254); Eagle DL ($245).

HISTORICAL FOOTNOTES: Introduced Sept. 25, 1980. Model year production: 137,125 (44,486 fours and 92,639 sixes), which came to 2.1 percent of the industry total. Calendar year production (U.S.): 109,319. Model year sales by U.S. dealers: 145,206 (including 236 leftover Pacers). Calendar year sales: 136,682 cars. Concords sold the best, followed by Eagles, then Spirits. Strengthening its tie with the French automaker, AMC marketed as a "captive import" the new Renault 18i. On Dec. 16, 1980, AMC stockholders approved the arrangement that would give Renault a 46 percent share of the corporation. Three more Renault officers joined the AMC board, making a total of five. Skyrocketing interest rates received part of the blame for sluggish sales of AMC domestic cars, which slipped 8.2 percent from 1980. Still, the company's market share was down only slightly. Operation resumed at AMC's Milwaukee plant in August 1981, after a nine-month closure. One of the two Kenosha production lines shut down to retool for manufacturing the planned Renault/AMC joint-venture front-wheel-drive subcompact model. Eagle sales edged out those of the four-wheel drive Subaru, but that flash of popularity wasn't destined to last long.

1982 AMC

Both the compact Concord and subcompact Spirit had slipped further down in sales volume, but returned for another try in 1982, along with both Eagle four-wheel drive versions. Styling changed little from 1981. The foremost technical change was the availability of a T5 five-speed overdrive manual transmission from Warner Gear, which boosted fuel economy ratings. Its overdrive fifth gear (0.76:1 or 0.86:1) was in the rear housing of the transmission. Four-speed manual and three-speed automatic transmissions were redesigned; as before, the automatics came from Chrysler. Lower final drive ratios on all models were intended to improve economy. Sixes with automatic received wider-ratio gearboxes. The optional 258 cid six-cylinder engine gained a serpentine accessory drive system for added fuel savings. One belt powered the alternator, water pump, air pump, and power steering pump. GM cars using Pontiac's "Iron Duke" four gained fuel-injection this year, but AMC's version stuck to carburetion. Front disc brakes gained low-drag calipers. New body colors were added: Topaz gold metallic, Sea blue metallic, Deep Night blue, Slate blue metallic, Jamaica beige, Mist Silver metallic, and Sun yellow. Carryover metallic colors were Deep maroon, Vintage red, Copper brown, Sherwood green, and Dark brown; plus Oriental red, Olympic white, and Classic black.

1982 AMC Spirit liftback coupe (JG)

SPIRIT — SERIES 40 — FOUR/SIX: Spirits looked the same outside, but could be purchased with the new five-speed overdrive transmission-the first domestic subcompact to offer that option. Once again, base and DL trim were available, along with the sporty GT package. Base powertrain was the 2.5-liter (151 cid) four-cylinder "Iron Duke" engine from Pontiac, with four-speed floor shift, manual front disc brakes, and P185/75R14 black glass-belted radial tires. Standard equipment also included two-speed wiper/washers, front sway bar, compact spare tire, lighter, color-keyed carpeting, inside hood release, locking glove box, dome light, vinyl bucket front seats, fold-down rear bench seat, energy-absorbing bumpers, side pinstriping, and wheelcovers. DL Spirits rode P185/75R14 whitewall glass-belted radials on argent styled wheelcovers and added reclining vinyl bucket seats, a premium split rear seat, Alpine fabric headliner and visors, AM radio, woodgrain dash overlay, day/night mirror, extra-quiet insulation package, dual horns, and a digital clock. black rocker panel moldings, chrome side marker lights, a chrome remote-control left mirror and front/rear bumper guards also marked the DL. Liftback buyers could choose a Spirit GT package, which included spoke-style 14 x 6 in. wheels with P185/75R14 blackwall glass-belted radial tires; tachometer; sport steering wheel; left-hand remote sport mirror; black body trim moldings at windshield and rear window surround, belt, drip, and 'B' pillar; black grille insert, headlight bezels, bumpers with nerfing strips and rear venturi; and 'GT' nameplates.

1982 AMC Concord sedan (JG)

CONCORD — SERIES 01 — FOUR/SIX: Base Concords again were powered by a 151 cid four with four-speed floor shift and had manual disc brakes, bench seats, a front sway bar, P195/75R14 blackwall glass-belted radial tires with wheelcovers, and a compact high-pressure spare tire. They also had the option of a five-speed overdrive floor shift, which produced a 37 mpg EPA highway rating with the four. A reworked, wider-ratio automatic transmission also helped mileage, when coupled to the optional 258 six with a lower final axle ratio. External appearance was unchanged. Standard equipment also included two-speed wiper/washers, dome light, lighter, carpeting, a hood ornament, dual body pinstripes, and black scuff moldings on body sides. Concord DL models added invididual vinyl reclining seats, molded fiberglass headliner and visors (Alpine fabric), a day/night inside mirror, dual rear ashtrays, digital electronic clock, woodgrain instrument panel overlay, dual horns, extra-quiet insulation, bumper guards front and rear, a remote-control left mirror, and custom wheelcovers with whitewall tires. As before, DL two-doors carried opera windows and a landau vinyl roof, while four-doors had a full vinyl roof and wagons sported woodgrain side panels. Moving another step up, the Limited included leather reclining seats, heavy carpeting (18-ounce in passenger area), a luxury woodgrain steering wheel, parcel shelf, chrome right-hand remote-control mirror, and wire wheelcovers with its whitewall tires.

EAGLE — SERIES 50 — FOUR/SIX: "Select Drive" let motorists switch easily between two-wheel-drive and full-time four-wheel drive on the short-wheelbase, Spirit-based Eagle. That was formerly an option. Although the 151 cid four still served as base engine, with four-speed floor shift, Eagles could also get the new five-speed manual or three-speed automatic, along with the 258 cid six-cylinder power plant. All Eagles included power steering and brakes and a front sway bar, and rode 15 in. tires with wheelcovers. 195/75R15 blackwall glass-belted radials were standard, with a compact spare tire. Eagles also came with color-keyed carpeting, custom vinyl bucket front seats (fold-down rear bench seat), front/rear bumper nerfing strips, inside hood release, dome light, locking glove box, wheelwell and rocker panel moldings, and body pinstriping. Lower door sections were painted in accent color. Eagle DL models added reclining bucket seats up front, a split rear seat, woodgrain dash overlay, day/night mirror, digital clock, Alpine fabric headliner/visors, remote-control left mirror (chrome), extra-quiet insulation, twin horns, and argent styled wheelcovers with whitewall tires. Appearing again on the SX/4 Liftback option list was an Eagle Sport package, including low-gloss black Krayton rocker panels and flares; black bumpers with guards and nerf strips; colored inserts in lower body side moldings; black grille insert, windshield, headlamp bezels, liftgate, drip/belt moldings, and taillamp treatment; a black remote-control left sport mirror, plus halogen headlamps and foglamps. Inside, Sport models had a floor shift console, parcel shelf, and vinyl sport steering wheel. They wore P195/75R15 blackwall steel-belted Arriva radial tires with styled wheelcovers.

1982 AMC American Eagle station wagon (JG)

EAGLE — SERIES 30 — FOUR/SIX: Concord-based Eagle sedans and wagons also had switch-selected two-wheel drive/four-wheel drive, with optional five-speed overdrive transmission. The 151 cid four with four-speed floor shift remained standard; 258 six optional (with automatic). Base-model big Eagles were a bit more luxurious than their smaller brothers, fitted with standard equipment that required an upgrade in the 50 Series. The list included individual reclining front seats, extra-quiet insulation, dual horns, day/night mirror, lighter, Alpine fabric headliner/visors, digital clock, vinyl door trim, left remote chrome mirror, and woodgrain dash overlay. Two-doors featured opera quarter windows with landau roof, four-doors a full vinyl roof, and wagons held a fold-down rear seat. Standard Eagles rode P195/75R15 whitewall glass-belted radials with argent styled wheelcovers. Eagle Limiteds contained leather reclining seats and heavy carpeting, plus a woodgrain steering wheel, parcel shelf, 'Limited' nameplate on front fender, and twin remote-control mirrors.

263

The Sport package included a leather-wrapped steering wheel; black bumpers and guards; black headlamp bezels, grille insert, windshield/liftgate moldings, door frames, 'B' pillars and remote left sport mirror; low-gloss black rocker panels and flares; halogen headlamps and foglamps; colored inserts in lower body side moldings, hood molding and nerf strip; a '4x4' silver decal on lower door; and steel-belted blackwall radial tires. The Sport package was available only on base Eagle two-door sedans and wagons. Two trailer towing packages were available, for light or medium loads.

I.D. DATA: The 17 symbol VIN, visible through the windshield, used the same coding as in 1981.

Series Number	Body/Style Number	Body Type & Seating	Factory Price	Shipping Weight	Production Total
SPIRIT SERIES 40					
40	8243-0	2d Lift-4P	5576/5726	2588/2687	20,063
40	8246-0	2d Sed-4P	5476/5626	2538/2637	119
SPIRIT DL					
40	8243-5	2d Lift-4P	5959/6109	2666/2765	Note 1
40	8246-5	2d Sed-4P	5859/6009	2614/2713	Note 1

NOTE 1: Production totals shown under base Spirit include DL models. Total model year production, 20,182 Spirits (9290 four-cylinder and 10,892 six-cylinder). Model year sales: 18,161.

Series Number	Body/Style Number	Body Type & Seating	Factory Price	Shipping Weight	Production Total
CONCORD SERIES 01					
01	8206-0	2d Sed-5P	5954/6104	2693/2773	6132
01	8205-0	4d Sed-5P	6254/6404	2752/2842	25,572
01	8208-0	4d Sta Wag-5P	7013/7163	2786/2876	12,106
CONCORD DL					
01	8206-5	2d Sed-5P	6716/6866	2768/2858	Note 2
01	8205-5	4d Sed-5P	6761/6911	2841/2931	Note 2
01	8208-5	4d Sta Wag-5P	7462/7612	2940/3030	Note 2
CONCORD LIMITED					
01	8206-7	2d Sed-5P	7213/7363	2790/2880	Note 2
01	8205-7	4d Sed-5P	7258/7408	2862/2952	Note 2
01	8208-7	4d Sta Wag-5P	7959/8109	2892/2982	Note 2

NOTE 2: Production totals under base Concord include DL and Limited. Total model year production, 33,693 Concords (2038 four-cylinder and 31,655 six-cylinder) including Canadian output for U.S. market. Model year sales: 36,505.

Series Number	Body/Style Number	Body Type & Seating	Factory Price	Shipping Weight	Production Total
EAGLE SX/4 SERIES 50					
50	8253-0	2d SX/4 Lift-4P	7451/7601	2972/3100	10,445
50	8256-0	2d Kamm Sed-4P	6799/6949	2933/3061	520
EAGLE 50 DL					
50	8253-5	2d SX/4 Lift-4P	7903/8053	3041/3169	Note 3
50	8256-5	2d Kamm Sed-4P	7369/7519	3000/3128	Note 3

NOTE 3: Production totals shown under base Eagle 50 include DL models. Of the 10,965 Eagle 50s manufactured, 3529 had four-cylinder engine.

Series Number	Body/Style Number	Body Type & Seating	Factory Price	Shipping Weight	Production Total
EAGLE SERIES 30					
30	8236-5	2d Sed-5P	8719/8869	3107/3235	1968
30	8235-5	4d Sed-5P	8869/9019	3172/3300	4091
30	8238-5	4d Sta Wag-5P	9566/9716	3199/3327	20,899
EAGLE 30 LIMITED					
30	8236-7	2d Sed-5P	9166/9316	3115/3243	Note 4
30	8235-7	4d Sed-5P	9316/9466	3180/3308	Note 4
30	8238-7	4d Sta Wag-5P	10013/10163	3213/3341	Note 4

NOTE 4: Production totals shown under base Eagle 30 include Eagle Limited. They include Canadian output for U.S. market. Only 6,056 Eagle 30 sedans and two wagons were made in U.S. Model year Eagle sales: 37,797.
NOTE 5: Data above slash for four-cylinder/below slash for six-cylinder.

ENGINES: BASE FOUR (all models): Inline. OHV. Four-cylinder. Cast iron block and head. Displacement: 151 cid (2.5 liters). Bore & stroke: 4.0 x 3.0 in. Compression ratio: 8.2:1. Brake horsepower: 82 at 3800 rpm. Torque: 125 lbs.-ft. at 2600 rpm. Five main bearings. Hydraulic valve lifters. Carburetor: two-barrel Rochester 2SE. except E2SE with automatic transmission. OPTIONAL SIX (all models): Inline. OHV. Six-cylinder. Cast iron block and head. Displacement: 258 cid (4.2 liters). Bore & stroke: 3.75 x 3.90 in. Compression ratio: 8.6:1. Brake horsepower: 110 at 3000 rpm. Torque: 205 lbs.-ft. at 1800 rpm. Seven main bearings. Hydraulic valve lifters. Carburetor: Carter BBD two-barrel.

CHASSIS DATA: Wheelbase: (Spirit) 96 in.; (Concord) 108 in.; (Eagle 50) 97.2 in.; (Eagle 30) 109.3 in. Overall length: (Spirit) 167 in.; (Concord) 185 in.; (Eagle 50) 164.5 in. but 166.5 in. with bumper guards; (Eagle 30) 184 in. but 186.3 in. with guards. Height: (Spirit) 51.5 in.; (Concord) 51.3-51.5 in.; (Eagle 50) 55.4 in.; (Eagle 30) 55.0-55.8 in. Front Tread: (Spirit) 58.1 in.; (Concord) 57.6 in.; (Eagle) 59.6 in.; Rear Tread: (Spirit) 57.0 in.; (Concord) 57.1 in.; (Eagle) 57.6 in.; Standard Tires: (Spirit) P185/75R14 GBR; (Concord) P195/75R14 GBR; (Eagle) P195/75R15 GBR.

TECHNICAL: Four-speed manual floor shift standard; five-speed overdrive transmission optional; three-speed automatic optional. Eagles: "Select Drive" four-wheel drive with transfer case. Four-speed manual gear ratios: (1st) 4.03:1; (2nd) 2.37:1; (3rd) 1.50:1; (4th) 1.00:1; (Rev) 3.78:1. Five-speed gear ratios: same as four-speed with additional fifth gear (0.86:1 for four-cylinder, 0.76:1 for sixes). Standard axle ratio: (four-cylinder engine) 3.08:1; (Spirit/Concord six) 2.35:1 except with auto. trans. 2.21:1; (Concord) 2.21:1; (Eagle 50 six) 2.35:1; (Eagle 30 six) 2.73:1 except with auto. 2.35:1. Steering: recirculating ball. Suspension: independent front with coil springs, upper/lower control arms, anti-roll bar; semi-elliptic rear leaf springs with "live" (rigid) rear axle. Brakes: front disc, rear drum; (Spirit/Concord) 10.8 in. disc, nine-inch drum; except Conc. wagon, 10 in. drums; (Eagle) 11 in. discs, 10 in. drums. Electronic ignition. Fuel tank: (Spirit/Eagle 50) 21 gallon; (Concord/Eagle 30) 22 gallon.

DRIVETRAIN OPTIONS: 258 cid six-cylinder engine ($150). Five-speed floor shift with overdrive ($199). Column-shift automatic transmission: Spirit/Concord ($411). Floor-shift automatic transmission: Spirit/Eagle ($411). Optional axle ratios: Concord six (2.73:1) or Eagle 30 six (3.08:1) with automatic and trailer towing package ($21). Twin Grip differential: Spirit/Concord ($75-$79). Heavy-duty engine cooling system (heavy-duty radiator, viscous fan, coolant recovery system): Spirit six, Concord ($75) but standard with Concord air conditioning; Eagle ($65). Maximum cooling system: Eagle ($68). Handling package: Spirit/Concord ($46). Heavy-duty shock absorbers: Spirit/Concord ($17); not available with Spirit GT Rally tuned package or load-leveling. Automatic load leveling: Concord six ($163). Extra duty suspension package: Eagle 30 ($75). Extra duty suspension package including rear sway bar and heavy-duty shocks: Eagle 50 ($40). Front suspension skid plate: Eagle ($75). Trailer towing package 'A': Concord/Eagle 30 ($101). Trailer towing package 'B': Eagle 50 six ($215). Heavy-duty battery ($25). Cold climate group including heavy-duty battery and engine block heater ($56). California emission system ($50).

CONVENIENCE/APPEARANCE OPTIONS: Spirit GT package: on DL liftback ($399). Spirit GT Rally tuned suspension package: liftback ($129). Eagle Sport package: Eagle 30 two-door/wagon ($333); Eagle 50 Liftback ($499); Eagle 50 DL Liftback ($394). Extra-quiet insulation package: Spirit/Concord/Eagle 50 ($59); standard on Concord DL/Limited, Eagle DL. Protection group (stainless door edge guards, bumper guards and nerfing strips, front mats): Spirit/Concord ($128); DL/Limited ($78); Spirit GT ($42). Eagle protection group including stainless door edge guards, bumper guards, floor mats ($92); with Sport package ($42). Convenience group (headlight-on buzzer, intermittent wipers, lighted vanity mirror) ($71). Gauge package (clock, tach, oil, amp or volt, vacuum): Spirit, Eagle 50 ($147); Spirit GT/DL, Eagle 30 or DL ($88). Pop-up sunroof: sedans, Eagle 50 ($279). Rear spoiler: Spirit GT, Eagle 50 Sport Liftback ($101). Light group ($59). Left remote control mirror ($30) except GT; standard on DL models; not available on Eagle 30. Right remote mirror (chrome): DL models, Eagle 30 ($31). Left/right remote chrome mirrors ($61) except Spirit GT/DL; standard on Concord Limited; not available Eagle 30. Right remote sport mirror (black): Spirit GT Liftback, Eagle Sport ($31). Left/right electric remote chrome mirrors ($142) except DL and Eagle 30 ($112); Eagle 30/Concord Limited ($81). Day/night mirror ($14); standard on DL/Limited; not available Eagle 30. Power steering: Spirit/Concord ($199). Power front disc brakes ($99). Electronic cruise control ($159). Air conditioning ($609-$679). Air conditioning package including tinted glass and power steering: Spirit/Concord ($890-$973). Halogen headlamps ($41). Halogen fog lamps: Eagle ($79). Dual horns ($16) except Eagle 30; standard on DL/Limited AM radio ($99). AM/FM/CB stereo four-speaker radio ($456) except Spirit DL ($357). AM/FM/Stereo four-speaker radio ($208) except Spirit DL ($109). AM/FM/cassette stereo radio ($356) except Spirit DL ($257). Electronically tuned AM/FM/cassette stereo radio with power amplifier and four coax speakers ($499); Spirit DL ($400). Premium audio system including power amplifier, four hi-fi speakers, fader ($115). Power antenna ($56). Power door locks: Spirit DL, Concord DL/Limited two-door sedan, Eagle DL ($106); Concord DL/Limited and Eagle 30 four-door/wagon ($152). Power door locks and windows: Spirit DL, Concord DL/Limited two-door sedan, Eagle DL ($275); Concord DL/Limited and Eagle 30 four-door/wagon ($391). Power liftback release: Spirit/Eagle 50 ($37) but including with rear spoiler. Power decklid release: Concord/Eagle sedan ($37). Power six-way driver's seat: Concord DL/Limited, Eagle 30 ($171). Power six-way driver/pass. seat: Concord DL/Limited, Eagle 30 ($281). Center console with armrest: Spirit ($89). Floor shift console: Eagle ($56). Parcel shelf ($26). Coventry Check fabric reclining bucket seats: Spirit/Eagle 50 ($32). Individual reclining seats (Castilian sculptured fabric): Concord DL, Eagle 30 ($59). Individual reclining seats (Durham plaid fabric): Eagle 30 ($59). Cargo area cover: Concord/Eagle wagon ($68). Digital clock ($59); standard on DL/Limited; not available Eagle 30. Rear wiper/washer ($119). Rear defroster ($125). Vinyl sport steering wheel: Spirit/Eagle 50 ($39) but standard with GT. Woodgrain steering wheel: DL models, Eagle 30 ($35). Leather-wrapped sport steering wheel ($58) except GT/Limited, Eagle 50 Sport ($19-$23). Tilt steering wheel ($99). Tinted glass ($82-$95). Two-tone accent color: Spirit except GT, Concord DL/Limited ($119); Special color combinations ($33). Rally stripes: Spirit ($85). Woodgrain paneling: Eagle 30 wagon ($139). Delete woodgrain from Concord DL/Limited wagon (deduct $75). Scuff moldings ($47). Roof rack: Spirit sedan/Eagle 50 ($85); Concord/Eagle 30 wagon ($105). Locking gas cap ($10). Tailgate air deflector: Spirit sedan, Eagle kammback ($32). Bumper guards ($50) but standard with DL, GT, Limited, Eagle Sport package. Styled wheelcovers (argent): Spirit ($40) except GT; standard on Spirit DL; Concord ($84); Concord DL ($43); Eagle 50 ($52). Custom wheelcovers: Concord ($41); standard on DL. Wire wheelcovers ($155); Spirit DL except GT ($115); Concord DL ($114). Spoke style 14 x 6 in. wheels with trim rings ($172) except DL ($131-$132); Concord Limited ($17). Turbocast II aluminum 14 x 7 in. wheels ($398) except DL ($357-$358); Spirit GT ($226); Concord Limited ($243). Sport aluminum 15 x 6 in. wheels: Eagle ($335-$387). Spirit/Concord Tires: P185/75R14 white GBR: Spirit ($66). P185/75R black GBR: Spirit GT sedan (no charge). P195/75R14 black GBR: Spirit ($40). P195/75R14 white GBR: Spirit ($108); Spirit DL ($40); Concord ($66). P195/75R14 white SBR Arriva: Spirit ($177); Spirit DL ($111); Concord ($137); Concord DL/Limited ($71); P205/70R14 RWL polysteel radial: Spirit ($252); Spirit DL ($186); Concord ($227); Concord DL/Limited ($161). Eagle Tires: P195/75R15 white GBR: Eagle 50 ($60). P195/75R15 black SBR Arriva: Eagle 30 ($25); Eagle 50 ($85); Eagle DL ($25). P195/75R15 white SBR Arriva: Eagle 30 ($85); Eagle 30 Sport ($60); Eagle 50 ($145); Eagle DL ($85); Eagle Sport ($60). P215/65R15 OWL SBR Eagle GT: Eagle 30 ($200); Eagle 30 Sport ($175); Eagle 50 ($260); Eagle DL ($200); Eagle Sport ($175).

HISTORICAL FOOTNOTES: Introduced Sept. 24, 1981. Model year production (U.S.): 70,898 (1.4 percent of industry total). Of that number, 14,972 were four-cylinder, 55,926 six-cylinder. Calendar year production (U.S.): 109,746 (including new Alliances for 1983). Model year sales by U.S. dealers: 99,300 (including 6,837 new Alliances), for a market share of 1.8 percent, Calendar year sales: 112,433 (2.0 percent share of industry sales). On June 15, 1982, production of the new Renault-designed, front-drive Alliance finally began at the Kenosha, Wis. plant. Its acceptance in the marketplace could signal whether AMC's passenger car operation would continue to survive. The Alliance would not debut until September, as a 1983 model. Even including its captive import Renault models (Fuego, 18i and LeCar), AMC's market share declined to well under two percent for the '82 model year. Sales of the four-wheel drive Eagle dipped too, though not nearly so badly as the other models in the lineup. Only the Renault connection, it seemed, had a reasonable chance of rescuing the ailing company. With the assistance of early Alliance sales in the fall, AMC managed to beat Volkswagen of America to recapture fourth place in the domestic rankings, for the first time since 1978. In October, the company announced a reduction in the white-collar work force, relying on attrition and early retirement as well as layoffs. An unusual 1982 agreement with the UAW allowed AMC to invest up to $2,000 from workers' paychecks in new product programs, to be repaid starting in 1985 with 10 percent interest. W. Paul Tippett Jr. was installed as chairman and CEO, replacing Gerald C. Meyers, who resigned in January 1982. In late September, all Spirit and Eagle SX/4 production moved to the plant at Brampton, Ontario. Big Eagles continued to be made in Kenosha. Concords were built at both factories. Innovations: Switchable two/four wheel drive. Built-in computer in electronic fuel feedback carburetion system to assist mechanics with swift diagnosis.

1983 AMC

Both Spirit and Concord prepared for their final outings, but American Motors entered the 1983 model year with a small French Hope: the front-drive Alliance. Mixing driver conveniences with technological sophistication, the AMC/Renault joint-venture soon would be the only two-wheel drive offering from AMC. The venerable 258 cid six gained a fuel feedback system with oxygen (knock) sensor, along with a healthy jump up to 9.2:1 compression. Gearing changed slightly to achieve better performance, in response to 1982 alterations that boosted mileage.

1983 AMC Renault Alliance sedan (AMC)

(RENAULT) ALLIANCE — SERIES 90 — FOUR: Drivetrains for the new sub-compact Alliance came from France, but cars were assembled in Wisconsin. A fuel-injected version of the 1.4-liter four-cylinder engine, as used on the imported Le Car, provided the power through a four-speed overdrive manual shift. Alliances featured rack-and-pinion steering and front drive, with fully independent (front and rear) suspension. A five-speed manual transmission was optional; also a three-speed automatic. Two- and four-door notchback sedans were offered, with standard power front disc brakes. Quad rectangular headlamps flanked a horizontal-bar grille with center emblem. Parking lights were below the bumpers. An Alliance emblem sat ahead of the front doors. MacPherson struts and coil springs made up the front suspension design, which included front and rear stabilizer bars. Alliances rode 155/80GR13 blackwall glass-belted radials with semi-styled wheels and hub covers. Body features included moldings for roof drip rail, rear window, rocker panels, body sides, and windshield surrounds. Inside were vinyl bucket seats (non-reclining), a console with lighter, fabric-covered headliner, a trip odometer, electric wipers with pulse action, and Soft-Feel steering wheel. Trunks were carpeted; hoods released from inside the car. A microcomputer monitored engine functions, sending signals to a dashboard indicator. The pedestal front seats rocked on curved tracks, adding to legroom for rear passengers. One unusual option: an infra-red door locking/unlocking device, similar to a remote-controlled garage door opener. Another: a "Systems Sentry" that warns (via lights) of low fluid levels and brake pad wear. Three upgrades were available. The L Alliance added dual accent pinstripes, a bright grille and hub covers, day/night mirror, and blacked-out rocker panels. It also carried moldings for beltline, bumper inserts and taillamps. DL models included deluxe six-way cloth bucket rocker/recliner seats, door panels with "hockey stick" armrests, a soft-hub steering wheel, extra-quiet insulation, tinted glass, dual rear ashtrays, tachometer, color-keyed remote left mirror, dual-note horn, and digital clock. The five-speed transmission was standard on DL models, which wore 175/70SR13 blackwalls with wheel trim rings. Topping the line was the Limited, with textured fabric bucket rocker/recliner seats and luxury door panels. Extras included a rear center armrest, light group, bright wheel lip moldings, luxury wheelcovers, visibility group, and halogen headlamps. Alliance body colors were: Almond beige; Olympic white; Deep Night blue; Jade Mist; Sebring red; and Sterling, Garnet, Cinnamon, Ambergiow or Diamond blue metallic clearcoat. Later in the model year an MT edition appeared, painted in special charcoal gray metallic clearcoat with MT decals and a black rear panel. Additional gear included a decklid luggage rack, right-hand remote mirror, bright instrument panel molding, body side and hood pinstriping, painted aluminum wheels, and leather-wrapped steering wheel. Inside the MT were Limited fabric rocker/recliner seats and a six-speaker, electronic-tuning stereo radio.

1983 AMC Concord sedan (JG)

SPIRIT — SERIES 40 — SIX: Not much changed on the body of the Spirit in its final year, but it gained bigger tires (P195/75R14 whitewalls), styled wheelcovers, and a push-button AM radio as standard DL fittings. The sedan was gone; only the liftback remained. Rather than the previous four, AMC's 258 cid six-cylinder engine (now with knock sensor) became the sole power plant. Four-speed manual floor shift was standard; five-speed overdrive or automatic (column or floor shift) optional. Mixing the standard goodies from the previous base and DL models, this year's Spirit came with vinyl reclining bucket seats, lighted ashtray (and lighter), front armrest, locking glove box, dome light, day/night mirror, digital clock, bumper guards, remote left mirror, styled wheelcovers, a front sway bar, and extra-quiet insulation package. Formerly an option package, the Spirit GT became a separate model this year. Performance extras included a handling package, gauge package with tachometer, and P195/75R14 SBR Arriva tires on Turbocast II aluminum wheels. Among its other goodies: a leather-wrapped sport steering wheel, black bumpers with guards and nerfing strips, black pinstripes, fog lamps, black moldings and dual remote-control sport mirrors, and center console with armrest. GT Spirits had no clock or radio as standard equipment.

CONCORD — SERIES 01 — SIX: Like the Spirit, the Concord carried a standard 258 cid six for 1983, abandoning the four-cylinder. That engine now had a new fuel feedback system and knock sensor for added efficiency. Only the twin four-door models remained: sedan and station wagon, in base or DL trim (plus a Limited wagon). DL sedans sported a full vinyl roof and opera windows. The enlarged standard equipment list included a front sway bar, front and rear ashtrays, lighter, coat hooks, color-keyed carpeting, Alpine fabric headliner and sun visors, energy-absorbing front/rear bumpers, a hood ornament, drip rail and windshield moldings, wide rocker panel moldings, and scuff belt moldings. DL and Limited equipment was similar to 1982. The Concord nameplate was on the upper corner of the grille, as well as on front fenders.

EAGLE — SERIES 50 — FOUR/SIX: Only the four-wheel-drive Eagles kept the old 151 cid (Pontiac) four-cylinder engine as base power plant, with an optional 258 six. And only the Liftback SX/4 model remained, with body graphics to prove it. Base models included a padded horn bar; DL versions a custom woodgrain steering wheel. Otherwise, equipment remained similar to the previous year. A Sport package was offered again, with halogen headlamps and fog lamps (the latter mounted above the front bumper). The package included a Sport nameplate and red or silver inserts in lower body side moldings. Other details were the same as the 1982 Sport package, with a heavy emphasis on black accents. Shorter (3.54:1) gearing with the four-cylinder engine boosted performance.

1983 AMC American Eagle station wagon (JG)

EAGLE — SERIES 30 — FOUR/SIX: Senior Eagles lost several models this year: the two-door sedan, and the Limited four-door. Standard engine was the familiar 151 four with four-speed manual shift; 258 six optional, along with five-speed overdrive gearbox or floor- shift automatic. At midyear, a new AMC-built four, measuring 150 cid, replaced the Pontiac 151. Base Eagles were well equipped, including armrests, Alpine fabric headliner/visors, digital clock, trunk and cargo area carpeting, day/night rearview mirror, dome light, locking glove compartment, a custom woodgrain steering wheel, woodgrain instrument panel, bumper nerf strips, wheel opening and rocker panel moldings, and chrome remote-control left mirror. Reclining front seats were upholstered in deluxe grain vinyl. Wagons had a fold-down rear bench seat and flip-up tailgate, plus a retractable cargo area cover. Sedans included a full vinyl roof. Whitewall P195/75R15 B glass-belted radials came with full wheelcovers. Accenting Eagle bodies were molding on the drip rail, beltline, backlight, and windshield. All Eagles had power steering and brakes. The Limited wagon held Chelsea leather reclining seats and other extras similar to the 1982 version: 18 oz. carpeting, woodgrain steering wheel, parcel shelf, and a second remote-control mirror. The Sport package was offered again, but only on the station wagon. Its contents were the same as in 1982, with red or silver inserts in the lower body side moldings, a silver '4x4' decal on the lower door, and P195/75R15 blackwall Arriva steel-belted radials.

I.D. DATA: The 17 symbol VIN, embossed on a metal plate on the top surface of the instrument panel, used the same coding as in 1981-1982; see previous section for details. The model year code changed to 'D' for 1983. Engine codes for the 151 cid four were on a pad at the right side of the block, below the cylinder head. Six-cylinder codes were on a pad between cylinders two and three. Alliance's 17 symbol VIN was similar. The first three symbols ('1AM') indicated U.S., AMC, and passenger car. The fourth symbol showed fuel injection type: D=Bendix TBI; E=Bosch multi-point. Symbol five shows transmission type: M=four-speed manual; W=five-speed; C=automatic. Digits 6-7 indicate body style: 95=four-door sedan; 95= two-door sedan. Digit eight shows trim level: 0=base model; 3=L; 6=DL; 8=Limited. Ninth is a check digit; tenth, the model year code (D=1983). In eleventh position, K=Kenosha manufacture. Finally comes a six-digit sequence number.

ALLIANCE (BASE) SERIES 90

Series Number	Body/Style Number	Body Type & Seating	Factory Price	Shipping Weight	Production Total
90	8396-0	2d Sed-5P	5595	1945	Note 1

ALLIANCE L

90	8396-3	2d Sed-5P	6020	1945	55,556
90	8395-3	4d Sed-5P	6270	1980	86,649

ALLIANCE DL

90	8396-6	2d Sed-5P	6655	1945	Note 1
90	8395-6	4d Sed-5P	6905	1980	Note 1

ALLIANCE LIMITED

90	8395-8	4d Sed-5P	7470	1980	Note 1

ALLIANCE MT

90	8396-6	2d Sed-5P	7450	N/A	Note 1
90	8395-6	4d Sed-5P	7700	N/A	Note 1

NOTE 1: Production totals shown under Alliance L include base, DL and Limited models. Model year sales: 124,687 Alliances.

SPIRIT DL SERIES 40

40	8343-5	2d Lift-4P	5995	2732	Note 2

SPIRIT GT SERIES 40

40	8343-9	2d Lift-4P	6495	2756	Note 2

NOTE 2: Total model year production, 3491. Model year sales: 6487.

CONCORD (BASE) SERIES 01

01	8305-0	4d Sed-5P	6724	2820	4433
01	8308-0	4d Sta Wag-5P	7449	2864	867

CONCORD DL

01	8305-5	4d Sed-5P	6995	2900	Note 3
01	8308-5	4d Sta Wag-5P	7730	2938	Note 3

CONCORD LIMITED

01	8308-7	4d Sta Wag-5P	8117	2990	Note 3

NOTE 3: Production totals shown under base Concord include DL and Limited models. Model year sales: 16,576 Concords.

EAGLE SX/4 SERIES 50					
50	8353-0	2d Lift-4P	7697/7852	2956/3084	2259
EAGLE SX/4 DL					
50	8353-5	2d Lift-4P	8164/8319	3025/3153	Note 4

NOTE 4: Production total includes SX/4 base and DL models.

EAGLE SERIES 30					
30	8335-5	4d Sed-5P	9162/9317	3181/3309	3093
30	8338-5	4d Sta Wag-5P	9882/10037	3201/3329	12,378
EAGLE 30 LIMITED					
30	8338-7	4d Sta Wag-5P	10343/10498	3215/3343	Note 5

NOTE 5: Station wagon production total shown includes base and Limited models. Total Eagle model year production, 17,730 (only 464 of them with four-cylinder engine). Model year sales, 31,604.
NOTE 6: Data above slash for four-cylinder/below slash for six-cylinder.

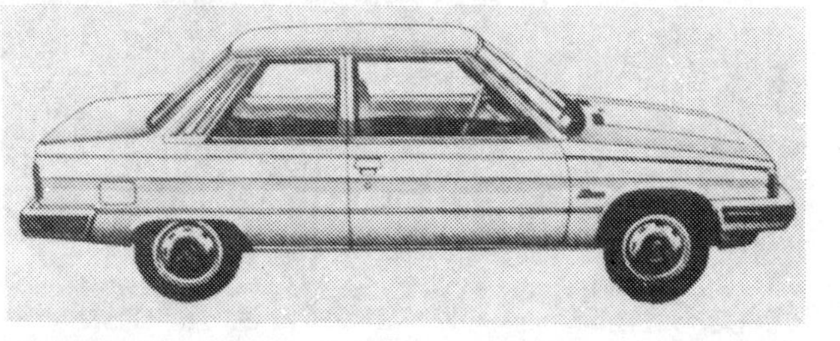

1983 AMC Alliance coupe (JG)

ENGINES: BASE FOUR (Eagle): Inline, ohv. Four-cylinder. Cast iron block and head. Displacement: 151 cid (2.5 liters). Bore & stroke: 4.0 x 3.0 in. Compression ratio: 8.2:1. Brake horsepower: 84 at 4000 rpm. Torque: 125 lbs.-ft. at 2600 rpm. Five main bearings. Hydraulic valve lifters. Carburetor: two-barrel Rochester 2SE. REPLACEMENT FOUR (Eagle): Inline, ohv. Four-cylinder. Cast iron block. Displacement: 150 cid (2.46 liters). Bore & stroke: 3.88 x 3.19 in. Compression ratio: 9.2:1. Brake horsepower: N/A. Torque: 132 lbs.-ft. at 3200 rpm. Five main bearings. Hydraulic valve lifters. Carburetor: one-barrel electronic feedback Carter YFA. BASE SIX (Spirit, Concord); OPTIONAL (Eagle): Inline, ohv. Six-cylinder. Cast iron block and head. Displacement: 258 cid (4.2 liters). Bore & stroke: 3.75 x 3.90 in. Compression ratio: 9.2:1. Brake horsepower: 110 at 3200 rpm. Torque: 210 lbs.-ft. at 1800 rpm. Seven main bearings. Hydraulic valve lifters. Carburetor: two-barrel Carter BBD. BASE FOUR (Alliance): Inline, ohv. Four-cylinder. Cast iron block; aluminum head. Transverse mounted. Displacement: 85.2 cid (1.4 liters). Bore & stroke: 2.99 x 3.03 in. Compression ratio: 8.8:1. Brake horsepower: 56 at 4200 rpm. Torque: 75 lbs.-ft. at 2500 rpm. Five main bearings. Solid valve lifters. Single-point Bendix (throttle-body) fuel-injection.

CHASSIS DATA: Wheelbase: (Alliance) 97.8 in.; (Spirit) 96 in.; (Concord) 108 in.; (Eagle 50) 97.2 in.; (Eagle 30) 109.3 in. Overall length: (Alliance) 163.8 in.; (Spirit) 167.2 in.; (Concord) 185 in.; (Eagle 50 SX/4) 164.6 in.; (Eagle 30) 183.2 in. Height: (Alliance) 54.5 in.; (Spirit) 51.5 in.; (Concord) 51-51.6 in.; (Eagle SX/4) 55 in.; (Eagle 30) 55.4 in.; (Eagle 30 wagon) 55 in. Width: (Alliance) 65.0 in.; (Spirit) 71.9 in.; (Concord) 71 in.; (Eagle SX/4) 73 in.; (Eagle 30) 72.3 in. Front Tread: (Alliance) 55.2 in.; (Spirit/Concord) 57.6 in.; (Eagle) 59.6 in. Rear Tread: (Alliance) 52.8 in.; (Spirit/Concord) 57.1 in.; (Eagle) 57.6 in. Standard Tires: (Alliance) 155/80GR13 GBR; (Spirit/Concord) P195/75R14 GBR; (Eagle) P195/75R15 GBR.

TECHNICAL: Transmission: four-speed manual floor shift standard, five-speed manual and automatic optional. Alliance transaxle: four-speed; five-speed and automatic optional. Manual transmission gear ratios (Alliance four-speed); (1st) 3.73:1; (2nd) 2.06:1; (3rd) 1.27:1; (4th) 0.90:1. (Alliance five-speed): (1st) 3.73:1; (2nd) 2.06:1; (3rd) 1.27:1; (4th) 0.90:1; (5th) 0.73:1. Spirit/Concord/Eagle (Borg-Warner T4) manual shift gear ratios: (1st) 4.03:1; (2nd) 2.37:1; (3rd) 1.50:1; (4th) 1.00:1; (Rev) 3.76:1. Borg-Warner T5 five-speed: same but 0.86:1 (four-cyl.) or 0.76:1 (six-cyl.) top gear. Standard axle ratio: (Alliance) 3.56:1 with automatic, 3.29:1 with four-speed, 3.87:1 with five-speed; (Spirit) 2.35:1; (Concord) 2.35:1 except five-speed, 2.73:1; (Eagle four) 3.54:1; (Eagle SX/4 six) 2.35:1; (Eagle 50 six) 2.73:1 except with automatic, 2.35:1. Drive: (Alliance) front; (Spirit/Concord) rear); (Eagle) 2 x 4. Clutch dia.: (Alliance) 7.1 in.; (others) 9.1 in. except six- cylinder, 9.5 in. Transverse-mounted engine (Alliance). Steering: (Alliance) rack and pinion; (others) recirculating ball. Suspension: (Spirit/Concord/Eagle) independent front coil springs with anti-roll bar, semi-elliptic rear leaf springs; (Alliance) fully independent-MacPherson strut front, twin transverse torsion bar rear, anti-roll bars. Brakes: front disc, rear drum; (Alliance) 9.4 in. disc, 8 in. drum; (Spirit/Concord) 10.8 in. disc, 9x2 in. drum; (Concord wagon) 10x1.75 in. drum; (Eagle) 11 in. disc, 10x1.75 in. drum. . Electronic ignition. Unibody construction. Fuel tank: (Alliance) 12.5 gal.; (Spirit, SX/4) 21 gal.; (Concord/Eagle) 22 gal.

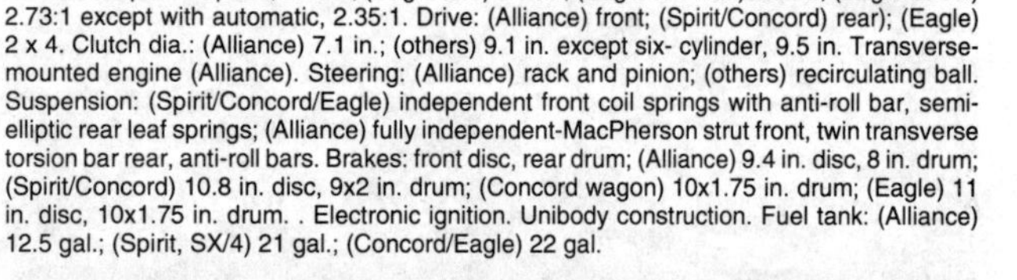

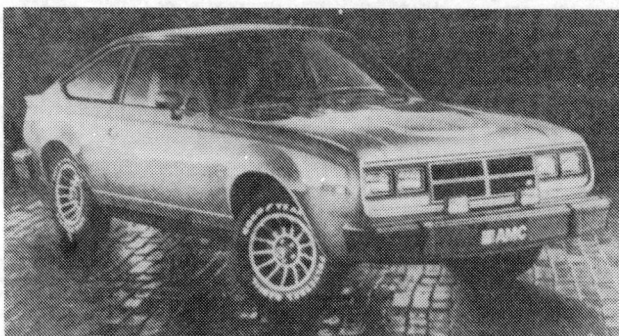

1983 AMC Spirit D/L liftback coupe (JG)

SPIRIT/CONCORD/EAGLE DRIVETRAIN OPTIONS: 258 cid six-cylinder engine: Eagle ($155). Five-speed floor shift with overdrive: Spirit/Concord ($125). Five-speed floor shift with overdrive and Select Drive: Eagle $219). Column-shift automatic transmission: Spirit/ Concord ($423). Floor-shift automatic transmission: Spirit ($423); Eagle ($437). Optional axle ratios: Concord/Eagle six (2.73:1), or Eagle 30 six (3.08:1) with automatic and trailer towing pkg. ($30). Twin Grip differential: Spirit/Concord ($82). Heavy-duty engine cooling system (heavy-duty radiator, viscous fan/shroud, coolant recovery system): Spirit/Concord ($77) but standard with air conditioning heavy-duty engine cooling (heavy-duty radiator, viscous fan): Eagle four ($67). Maximum cooling system: Eagle six ($70). Handling package (unique front sway bar; rear sway bar): Spirit/ Concord ($48). Automatic load leveling (air shocks): Concord/Eagle 30 ($169). Extra duty suspension package (special front sway bar, rear sway bar and heavy-duty shocks: Eagle SX/4 ($65); also includes heavy-duty springs, Eagle 30 ($77). Front suspension skid plate: Eagle ($77). Trailer towing package 'A' (to 2000 lbs.): Concord/Eagle 30 ($104). Trailer towing package'B' (to 3500 lbs.): Eagle 30 ($222). Heavy-duty battery ($26). Cold climate group includes heavy-duty battery and engine block heater ($58). California emission system ($65).

SPIRIT/CONCORD/EAGLE CONVENIENCE/APPEARANCE OPTIONS: Eagle Sport package: Eagle SX/4 ($516); Eagle SX/4 DL ($407); Eagle 30 wagon ($344). Extra-quiet insulation package: Concord/Eagle SX/4 ($61); standard on Concord DL/Limited Protection group (stainless door edge guards, bumper guards and nerfing strips, front mats): Spirit ($30-$40). Protection group with out nerfing strips: Concord ($132); Concord DL/Limited ($81). Eagle protection group (stainless door edge guards, bumper guards, floor mats): Eagle 30 ($75); with Sport package ($23). Eagle protection group with front mats only: SX/ 4 ($72); with Sport package ($20). Convenience group: headlight-on buzzer, intermittent wipers, right lighted vanity mirror ($73). Gauge package (clock, tach, oil, amp or volt, vacuum): Eagle SX/4 ($152); Spirit, Eagle 30, SX/4 DL ($91). Pop-up sunroof ($295). Rear spoiler (includes power liftback release): Spirit, Eagle SX/4 ($104). Light group ($61). Left remote-control mirror:Concord, Eagle SX/4 ($32); standard on DL models. Right remote mirror (chrome): DL models, Eagle 30 ($32). Left/right remote chrome mirrors: Concord, Eagle SX/4 except Sport/DL ($64). Right remote sport mirror (black): Eagle 30 Sport wagon or SX/4 Sport ($32). Left/right electric remote chrome mirrors: SX/4 ($147); DL models ($115-$116); Limited wagons ($83). Day/night mirror: Concord, SX/4 ($15). Power steering: Spirit/Concord ($212). Power front disc brakes: Spirit/Concord ($100). Electronic cruise control ($170). Air conditioning system ($670-$725). Halogen headlamps ($20). Halogen fog lamps: Eagle ($82). Dual horns: Concord, SX/4 ($17); standard on DL/Limited AM radio ($82-$83). AM/FM/CB stereo four-speaker radio ($471) except Spirit DL ($389). AM/FM/ Stereo four-speaker radio ($199) except Spirit DL ($117). AM/FM/cassette stereo radio ($329) except Spirit DL ($247). Electronically tuned AM/FM/cassette stereo radio with power amplifier and four coax speakers ($499); Spirit DL ($417). Power door locks: Spirit, SX/4 DL ($120); Concord DL/Limited and Eagle 30 ($170). Power windows and door locks: Spirit, SX/4 DL ($300); Concord DL/Limited and Eagle 30 ($425). Power liftback release: Eagle SX/4 ($40); includes with rear spoiler. Power decklid release: Concord/Eagle sedan ($40). Power six-way driver's seat: Concord DL/Limited, Eagle 30 ($189). Power six-way driver/ passenger seat: Concord DL/Limited, Eagle 30 ($302). Center console with armrest: Spirit DL ($92). Floor shift console: Eagle ($65). Parcel shelf ($27). Coventry Check fabric bucket seat trim: Spirit/Eagle SX/4 ($39). Castilian sculptured fabric seat trim: Concord DL, Eagle 30 ($67). Durham plaid fabric seat trim: Eagle 30 ($67). Digital clock: Concord, SX/4 ($61); standard on DL/Limited Rear wiper/washer ($124). Rear defroster ($135). Vinyl sport steering wheel: Spirit DL, Eagle SX/4 ($40). Woodgrain steering wheel: DL models, Eagle 30 ($36). Leather-wrapped sport steering wheel ($60) except Concord Limited wagon and Eagle 30 Limited ($24); SX/4 Sport ($20); not available Spirit DL. Tilt steering wheel ($106). Tinted glass ($95-$105). Two-tone accent color: Spirit except GT, Concord DL/Limited ($135); Special color combinations ($49). Rally stripes: Spirit DL ($88). Woodgrain paneling: Eagle 30 wagon ($144). Delete woodgrain from Concord DL/Limited wagon (deduct $75). Scuff moldings ($55). Roof rack: Concord/Eagle 30 wagon ($115). Locking gas cap ($10). Bumper guards:Concord/Eagle ($52) but standard with DL, Limited, Eagle Sport package Styled wheelcovers (argent): Concord ($87); Concord DL ($45); Eagle SX/4 ($54); not available Concord Limited Custom wheelcovers: Concord ($42); standard on DL. Wire wheelcovers: Spirit DL ($119); Concord ($160). Spoke style 14 x 6 in. wheels with trim rings: Spirit/Concord DL ($136); Concord ($178); Concord Limited wagon ($18). Turbocast II aluminum 14 x 7 in. wheels: Spirit DL ($370); Concord ($411); Concord DL/Limited wagon ($251). Sport aluminum 15 x 6 in. wheels: Eagle ($346-$400). Spirit/Concord Tires: P195/ 75R14 B white GBR: Spirit DL (no charge). P195/75R14 B black SBR Arriva: Spirit GT (no charge). P195/75R14 B white SBR Arriva: Spirit DL, Concord DL/Limited ($73); Concord ($142). P205/70R14 B RWL polysteel radial: Spirit/Concord DL/Limited ($192); Spirit GT ($188); Concord ($281). Eagle Tires: P195/75R15 B white GBR: SX/4 ($69). P195/75R15 B black SBR Arriva: Eagle 30, SX/4 DL ($20); SX/4 ($73). P195/75R15 B white SBR Arriva: SX/4 ($142); Eagle 30, SX/4 DL ($73); Sport ($69). P215/65R15 B OWL SBR Eagle GT: SX/ 4 ($269); Eagle 30, SX/4 DL ($200); Eagle Sport ($196).

ALLIANCE OPTIONS: Five-speed manual transmission, floor shift, with overdrive: L ($95). Floor-shift automatic tranmission ($420) except DL/MT/Limited ($325). heavy-duty battery ($25). Cold climate group ($36-$79). heavy-duty cooling ($67). Systems Sentry (monitors for low oil, coolant, brake fluid, disc wear, washer/power steering fluid, transaxle oil) ($125). Extra-quiet insulation package: L ($62). Protection group includes door edge guards, carpeted mats, locking gas cap ($52); not available base. Visibility group (dual remote mirrors, lighted visor mirror, intermittent wipers: L ($160); DL ($129); MT ($97). Light group ($46). Halogen headlamps: L/DL/MT ($20). Tachometer: L ($82). Power steering ($199). Speed control ($170); not available base. Air conditioning ($630). Power windows ($300-$350). Intermittent wipers: L/DL/MT ($50). Keyless entry system ($95). Power door locks ($120-$170) except base model. Rear defroster ($130). Tinted glass ($90). AM radio except MT ($82). AM/FM radio except MT ($135). AM/FM stereo radio: L/DL/Limited ($199). Electronic- tuning AM/FM stereo four-speaker radio with cassette: L/DL/Limited ($465). Speaker for left instrument panel ($28). Vinyl reclining bucket seats: base/L ($65). Vinyl rocker/reclining buckets: DL (no charge). Cloth bucket seats: L ($30). Cloth reclining bucket seats: L ($95). Leather rocker/reclining bucket seats: Limited/MT ($413). Two-tone paint: L/ DL/Limited ($160). Metallic accent paint: DL/Limited (no charge); L ($62). black leather-wrapped sport steering wheel: L/DL/Limited ($60). Luxury wheelcovers: L ($88); DL ($36). Wheel trim rings: L ($52). Aluminum wheels: L/DL/Limited ($249-$337). Tires: 155/80GR13 white GBR: base/L ($61). 175/70SR13 SBR: L ($72). 175/70SR13 white SBR: L ($132); DL/ Limited ($60). Spare tire (to replace polyspare): L/DL/MT/Limited ($35).

HISTORICAL FOOTNOTES: Introduced: Sept. 22, 1982. Model year production (U.S.): 168,726. This amounted to nearly three percent of the industry total...more than double the 1982 percentage. Of that number, 142,669 were four-cylinder, 25,057 six-cylinder. Calendar year production (U.S.): 201,993 (including new Encores for 1984). Eagle production of 6,979 units was dwarfed by the 152,581 Alliances made during 1983. But 23,012 Eagles were built in Canada. Model year sales by U.S. dealers: 183,005 (including 3,651 new Encores). Calendar year sales: 193,251 for a 2.8 percent market share, up substantially from the 1.9 percent in 1982. Designed by Renault in France, but manufactured in Kenosha, Wis., the sub-compact front-drive Alliance set the stage for other joint ventures betweeen American and foreign companies. Promoted as combining "advanced European technology with American expertise," Alliance attempted to lure buyers from both the domestic and import ranks. Its design evolved from the Renault 9, acclaimed "Car of the Year" by the European press in 1982. Tooling had begun in August 1981. Some $200 million was spent on development and production, over a 2-1/2-year period. Ample changes were made to adapt the car for American tastes, including the addition of American-designed power steering and brake systems, Bendix fuel-injection, and electronic controls. Meanwhile, the Spirit and Concord quietly disappeared as AMC focused on its joint-venture and the four-wheel drive Eagle. AMC registered a loss of $146.7 million for the year, nearly as bad as 1982, though the final quarter showed a modest profit. This seemed to point towards a better year ahead.

In July, AMC sold its defense subsidiary, AM General Corp.: next month, its lawn tractor operation, Wheel Horse Products. Corporate headquarters in Southfield, Mich. was sold too, then leased back. It was all an attempt to raise cash for product development. Production of a new 150 cid, four-cylinder engine began in February 1983. It was installed on Eagles starting at midyear.

1984 AMC

The slimmed-down AMC lineup, reduced to a single Eagle model and the Kenosha-built French sub-compact, had one major addition for 1984. A hatchback Encore, with three or five doors, joined the original twin front-drive Alliance sedans. Note: By this time, many automakers (including AMC) had begun to count the rear hatch of their small cars as a door; thus, a three-door hatchback has only two "real" doors; and a five-door has only four doors suitable for people to enter. Listings in this catalog follow the numbering used by the manufacturer.

(RENAULT) ENCORE — SERIES 90 — FOUR: Built on the same 97.8 in. wheelbase as the Alliance, the sportier Encore hatchback stood three inches shorter in overall length. It was AMC's attempt to attract youthful buyers, now that the Spirits and GTs were gone. Encores used the same 1.4-liter four-cylinder engine as the Alliance, with a standard four-speed (overdrive) manual transmission; five-speed optional. Three- and five-door versions were offered (see note above). Each had a stubby rear end with distinctive backlight and taillamp structure. Encore's emblem sat on the side of the grille, while Alliance's was at the center. Base Encores were nicely fitted inside with carpeting, fabric-covered headliner molding and visors, a day/night rearview mirror, lighter, inside hood release, console with stowage box, plus vinyl bucket front seats and 60/40 fold-down rear seat. Externally, they featured quad rectangular halogen headlamps and a horizontal-bar style black grille with bright molding, with blackout rocker panels and liftgate. Encores also had pulse wiper/washers, front and rear stabilizer bars, flip-out rear windows, power brakes, and P155/80GR13 glass-belted radials on semi-styled wheels with black hubcaps. Stored in back was a polyspare tire. The engine had electronic ignition and fuel injection. Suspension was fully independent, with MacPherson struts. Rack-and-pinion steering helped handling. Three upgrades were offered. S models held an AM radio, with dual accent pinstripes, cargo area cover, bright grille and wheelcovers; the five-door had roll-down rear windows. LS Encores moved up to 175/70SR13 steel-belted radial tires (blackwall) with luxury sport wheelcovers, plus a five-speed overdrive transmission. Other LS luxuries included tinted glass, digital clock, extra-quiet insulation, dual-note horn, oil level gauge, tachometer, "hockey stick" armrests, and rocker/recliner bucket seats in deluxe striped fabric. A chime warned of key left in ignition, headlamps-on, and seat belts unbuckled. A black rear spoiler and left remote mirror completed the LS package. Sporty GS Encores added light and visibility groups and Westchester fabric rocker/recliner buckect seats, plus black luxury sport wheelcovers, black accent stripes, black sport steering wheels, and pinstripes along hood, body side and tailgate. Five-door Encores came only in S or LS versions. A Diamond Edition Encore, added later, featured gold bumper inserts and gold aluminum wheels, plus special pinstriping on hood and body sides. Painted Olympic white or Classic black, the Diamond Encore's wing seats were upholstered in honey fabric, while its dash held an electronic-tuning AM/FM/stereo radio with cassette player.

1984 AMC Alliance sedan (JG)

(RENAULT) ALLIANCE — SERIES 90 — FOUR: After a strong start in the marketplace, Alliance changed little for its second year-though its price rose in several jumps. External niceties included blackout rocker panels and a black grille (with bright surround). Inside, a day/night mirror became standard on the base model, which wore P155/80GR13 blackwall tires. L models added dual accent pinstripes, bright grille and hub covers, and childproof rear door locks. DL rocker/recliner seats now held Lucerne fabric upholstery, while DL dashes contained low-fuel and oil level gauges. Calais fabric upholstery went on Limited rocker/recliner seats, which also offered hood pinstripes, a blackout lower back panel, and bright decklid luggage rack. Otherwise, standard equipment for each trim level was about the same as in 1983. Three colors were added and radio operation was improved. A second windshield washer outlet was added. Alliance's Diamond Edition was equipped like that of the Encore hatchback.

EAGLE — SERIES 30 — FOUR/SIX: Only the larger Eagle survived into 1984, in four-door sedan and wagon form, powered by the new AMC-manufactured four-cylinder engine. The new 150 cid (2.46 liter) four, introduced during the 1983 model run, featured a single-barrel electronic feedback carburetor. Also standard on Jeeps, it replaced the GM-built 151 cid four, which had powered Spirits and Concords as well as Eagles. The standard four-speed manual transmission now included an upshift indicator light to warn drivers when a gear change was wise. As before, the 258 cid six-cylinder engine was optional. The "live" rear axle suspension consisted of computer-selected springs and telescoping shock absorbers. Front suspension included full coil springs and stabilizer bar. Standard equipment was similar to the 1983 Eagle 30, including the familiar Krayton protective treatment on lower body sides, argent styled wheelcovers, dual pinstripes, dual horns, and a lockable wagon cargo compartment. Moldings highlighted drip and quarter windows areas, plus the belt surround. Power brakes and steering were standard; so was a hood ornament. Taillamps were large wraparound style. Leather reclining seats served as the main attraction on the Limited wagon. Eagle's Sport package, on base station wagons only, consisted of a leather-wrapped sport steering wheel, low-gloss black rocker panels and fender flares, black bumpers with guards and nerf strips, red or silver inserts in lower body side and hood moldings, black taillamp treatment, black windshield/liftgate moldings, black 'B' pillars and door frames, a left-side remote-controlled black sport mirror, and halogen headlamps. Halogen fog lamps sat above the front bumper. Sport tires were blackwall P195/75R15 steel-belted Arriva radials. A silver '4x4' decal highlighted the lower door, as did the 'Sport' nameplate.

1984 AMC Eagle four-door sedan (AMC)

I.D. DATA: Eagle's 17 symbol Vehicle Identification Number (VIN) was embossed on a metal plate riveted to the top left surface of the instrument panel, visible through the windshield. Coding was the same as 1981-1983; see 1981 for breakdown details. Engine codes (symbol four) for 1984 were: C=258-6; U=150-4. Model year code (symbol ten) changed to E for 1984. Alliance/Encore's 17-symbol VIN was also similar to 1983 coding, but with additions for the new Encore. Symbols 6-7 (body style) were now: 93=three-door liftback; 99=five-door liftback; 95=four-door sedan; 96=two-door sedan. Symbol eight (trim level) included: 0=base model; 3=Alliance L or Encore S; 6=Alliance DL or Encore LS; 8=Limited; 9=Alliance GT or Encore GS. Model year code (symbol ten) changed to E for 1984.

ENCORE (BASE) SERIES 90

Series Number	Body/Style Number	Body Type & Seating	Factory Price	Shipping Weight	Production Total
90	8493-0	3d Lift-5P	5755	1974	Note 1
ENCORE S					
90	8493-3	3d Lift-5P	6365	1985	55,343
90	8499-3	5d Lift-5P	6615	2008	32,266
ENCORE LS					
90	8493-6	3d Lift-5P	6995	2033	Note 1
90	8499-6	5d Lift-5P	7195	2059	Note 1
ENCORE GS					
90	8493-9	3d Lift-5P	7547	2043	Note 1
ENCORE DIAMOND EDITION					
90	8493-6	3d Lift-5P	7570	n.a.	n.a.
90	8499-6	5d Lift-5P	7770	n.a.	n.a.

NOTE 1: Production totals under S series include base, LS and GS Encore models. Model year sales: 72,076 Encores.

ALLIANCE (BASE) SERIES 90

90	8496-0	2d Sed-5P	5959	1934	Note 2
ALLIANCE L					
90	8496-3	2d Sed-5P	6465	1936	50,978
90	8495-3	4d Sed-5P	6715	1964	70,037
ALLIANCE DL					
90	8496-6	2d Sed-5P	7065	1975	Note 2
90	8495-6	4d Sed-5P	7365	2002	Note 2
ALLIANCE LIMITED					
90	8495-8	4d Sed-5P	8027	2019	Note 2
ALLIANCE DIAMOND EDITION					
90	8496-6	2d Sed-5P	7715	n.a.	n.a.
90	8495-6	4d Sed-5P	8015	n.a.	n.a.

NOTE 2: Production totals under L series include base, DL and Limited Alliance models. Model year sales: 105,340.

EAGLE SERIES 30

30	8435-5	4d Sed-5P	9495/9666	3189/3307	4241
30	8438-5	4d Sta Wag-5P	10225/10396	3220/3338	21,294
EAGLE LIMITED SERIES 30					
30	8438-7	4d Sta Wag-5P	10695/10866	3236/3354	Note 3

NOTE 3: Production total shown for Eagle wagon includes base and Limited models. Of the 25,535 Eagles manufactured in the model year, only 184 had a four-cylinder engine. Model year sales: 23,137.
NOTE 4: [Eagle] Data above slash for four-cyl./below slash for six-cyl.

ENGINES: BASE FOUR (Alliance/Encore): Inline. Overhead valves. Four-cylinder. Cast iron block; aluminum head. Transverse mounted. Displacement: 85.2 cid (1.4 liters). Bore & stroke: 2.99 x 3.03 in. Compression ratio: 9.0:1. Brake horsepower: 56 at 4200 rpm. Torque: 75 lbs.-ft. at 2500 rpm. Five main bearings. Solid valve lifters. Single-point Bendix (throttle-body) fuel injection. BASE FOUR (Eagle): Inline. Overhead valves. Four-cylinder. Cast iron block. Displacement: 150 cid (2.46 liter). Bore & stroke: 3.88 x 3.19 in. Compression ratio: 9.2:1. Brake horsepower: not available. Torque: 132 lbs.-ft. at 3200 rpm. Five main bearings. Hydraulic valve lifters. Carburetor: one-barrel electronic feedback Carter YFA. OPTIONAL SIX (Eagle): Same as 1983 specifications (258 cid inline Overhead valves).

CHASSIS DATA: Wheelbase: (Encore/Alliance) 97.8 in.; (Eagle) 109.3 in. Overall length: (Encore) 160.6 in.; (Alliance) 163.8 in.; (Eagle) 180.9 in. Height: (Encore/Alliance) 54.5 in.; (Eagle) 54.4 in. Width: (Alliance/Encore) 65 in.; (Eagle) 72.3 in. Front Tread: (Encore/Alliance) 55.2 in.; (Eagle) 59.6 in. Rear Tread: (Encore/Alliance) 52.8 in.; (Eagle) 57.6 in. Standard Tires: (Encore/Alliance) P155/80GR13 GBR; (Eagle) P195/75R15 GBR.

TECHNICAL: Transmission: four-speed manual standard; five-speed manual and three-speed automatic optional. Floor shift lever. Eagle four-wheel drive: selectable 2/4 wheel. Alliance manual transmission gear ratios: (1st) 3.73:1; (2nd) 2.06:1; (3rd) 1.27:1; (4th) 0.90:1; (5th) 0.73:1; (Rev) 3.54:1. Eagle manual gear ratios: (1st) 4.03:1; (2nd) 2.37:1; (3rd) 1.50:1; (4th) 1.00:1; (5th) 0.86:1 except 0.76:1 with six cylinder engine; (Rev) 3.54:1. Clutch dia.: (Alliance/Encore) 7.14 in.; (Eagle four) 9.1 in.; (Eagle six) 10.3 in. Standard axle ratio: (Alliance/Encore) 3.29:1 with four-speed; 3.87:1 with five-speed; 3.27:1 with automatic; (Eagle four) 3.54:1; (Eagle six) 2.73:1 except with towing package, 3.08:1. Steering: (Alliance/Encore) rack and pinion; (Eagle) power-assisted recirculating ball. Suspension: (Alliance/Encore) fully independent: front, MacPherson struts with lower control arms, coil springs, anti-roll bar; rear, transverse semi-torsion bars, swinging longitudinal trailing arms, anti-roll bar; (Eagle) independent front coil springs with anti-roll bar; semi-elliptic rear leaf springs. Brakes: front disc, rear drum; (Alliance) 9.4 in. discs, 8 in. drums; (Eagle) 11 in. discs, 10 in. drums. Electronic ignition. Fuel tank: (Alliance/Encore) 12.5 gallon; (Eagle) 22 gallon.

EAGLE DRIVETRAIN OPTIONS: 258 cid six-cylinder engine ($171). Five-speed floor shift with overdrive and select drive ($227). Floor-shift automatic transmission: six ($452). Optional axle ratios (2.73:1 or 3.08:1): Eagle six ($31). heavy-duty engine cooling (heavy-duty radiator, viscous fan): Eagle four ($69). Maximum cooling system: Eagle six ($72). Automatic load leveling (air shocks): six ($175). Extra duty suspension package (special front sway bar, rear sway bar, heavy-duty shocks and springs ($80). Front suspension skid plate ($80). Trailer towing package 'A' to 2000 lbs. ($108). Trailer towing package 'B' to 3500 lbs. ($230). Heavy-duty battery ($27). Cold climate group includes eavy-duty battery and engine block heater ($60). California emission system ($78).

EAGLE CONVENIENCE/APPEARANCE OPTIONS: Eagle Sport package ($356). Protection group (stainless door edge guards, bumper guards, floor mats) ($78); with Sport package ($24). Convenience group (headlight-on buzzer, intermittent wipers, right lighted vanity mirror) ($76). Gauge package (clock, tach, oil, volt, vacuum) ($94). Light group ($63). Right remote mirror: black or chrome ($33). Left/right remote chrome mirrors ($119); Limited wagon ($86); not available with Sport package Electronic cruise control ($176). Air conditioning system ($750). Halogen headlamps ($15). Halogen foglamps ($85). AM radio ($86). AM/FM/Stereo four-speaker radio ($206). AM/FM/cassette stereo radio ($340). Electronically tuned AM/FM/cassette stereo radio with power amplifier and four coax speakers ($516). Power door locks ($176). Power windows and door locks ($440). Power decklid release ($41). Power six-way driver's seat ($196). Power six-way driver/passenger seat ($313). Floor shift console ($67). Parcel shelf ($28). Fabric seat trim ($69). Rear wiper/washer: wagon ($128). Rear defroster ($140). Woodgrain steering wheel ($37) not available with Sport package Leather-wrapped sport steering wheel ($62) except Limited wagon ($25). Tilt steering wheel ($110). Tinted glass ($109). Woodgrain paneling: wagon ($149). Black scuff moldings ($57). Roof rack: wagon ($119). Locking gas cap ($10). Bumper guards ($54). Sport aluminum 15 x 6 in. wheels ($358). Tires: P195/75R15 B black SBR Arriva ($21). P195/75R15 B white SBR Arriva ($76); Sport wagon ($72). P215/65R15 B OWL SBR Eagle GT ($207); Sport wagon ($203).

ALLIANCE/ENCORE OPTIONS: Five-speed manual transmission, floor shift, without overdrive: Alliance L, Encore S ($97). Three-speed floor-shift automatic tranmission ($435) except Alliance DL/Limited, Encore LS/GS ($338). Heavy-duty battery ($27). Heavy-duty engine cooling ($69); standard with air conditioning Cold climate group (heavy-duty battery/alternator, engine coolant heater) ($38-$81). Systems Sentry (monitors for low oil, coolant, brake fluid, disc wear, washer/power steering fluid, transaxle oil) ($128); not available base models. Extra-quiet insulation package: L/S ($64). Rear black spoiler: Encore base/S ($72). Protection group (door edge guards, front/rear carpeted mats, locking gas cap) ($53); not available base models. Visibility group (dual remote mirrors, lighted visor mirror, intermittent wipers): L/S ($164); DL/LS ($133). Light group ($47); not available base models. Fog lamps ($77). Tachometer: L/S ($84). Power steering ($215). Cruise control ($174); not available base. Air conditioning ($653). Intermittent wipers ($51). Keyless entry system ($97); not available base. Power door locks ($123-$174). Power windows and door locks ($308-$359). Power liftgate lock release: Encore S/LS ($31). Digital clock ($58); not available base. Rear wiper/washer: Encore ($120). Rear defroster ($133). Tinted glass ($92). AM radio: base models ($84). AM/FM radio: base models ($147); others ($63). AM/FM four-speaker stereo radio ($120); not available base. Electronic-tuning AM/FM stereo four-speaker radio with cassette ($427); not available base. Vinyl reclining bucket seats: base/L/S ($67). Vinyl rocker/reclining bucket seats: LS/DL (no charge). Cloth bucket seats: L/S ($75). Cloth reclining bucket seats: L/S ($142). Luxury cloth rocker/reclining bucket seats: Limited ($299). Leather rocker/reclining bucket seats: DL/LS ($349); GS/Limited ($299). Metallic paint: L/S ($150); LS/GS/DL/Limited (no charge). black leather-wrapped sport steering wheel ($62); not available base. Deck lid luggage rack: Alliance L/DL ($108). Luxury wheelcovers: Alliance L ($90); DL ($37). Wheel trim rings: base/L/S ($53). Aluminum wheels: L/S ($345); DL ($292); Limited/LS/GS ($255). Tires: P155/80GR13 white GBR: base/L/S ($63). P175/70SR13 SBR: base/L/S ($74). P175/70SR13 white SBR: base/L/S ($135); others ($61). Conventional spare tire: Alliance except base ($36).

HISTORICAL FOOTNOTES: Introduced: Sept. 25, 1983. Model year production: 234,159 (208,808 four-cylinder and 25,351 Eagle six-cylinder). U.S. production totaled 208,624 cars, which amounted to more than 2.5 percent of the industry total. Calendar year production (U.S.): 192,196 Encores and Alliance made in Kenosha, Wis.; (Canada) 22,982 Eagles. Model year sales by U.S. dealers: 201,275 (including 336 leftover Concords and 386 Spirits). Calendar year sales: 190,255, for a 2.4 percent share of the market. When Eagle was introduced for 1980, its only competitor for the four-wheel drive market was Subaru. Four years later, Toyota had introduced its Tercel four-wheel drive, while Audi brought out its costly versions. Yet, AMC continued to push the Eagle, which remained the only domestic-built four-wheel drive model, despite sagging sales and rumors that it would abandon production and sell the Brampton, Ontario plant to Chrysler. As Renault/Jeep Sport vice-president R.C. Lunn insisted, the Eagle "brought a whole new dimension of functional improvement to highway driving while still retaining the off-road capabilities...many consumers coming from the two-wheel-drive segments were buying the vehicles for the security they offered for on-highway driving." (That Sport group had recently been created to promote AMC's sporty image.) For the time being, at least, Eagle remained the only domestically-built four-wheel drive passenger auto. Still, Eagle sales had declined to little more than half their 1981 level. A glance at sales figures demonstrated that the company's financial survival had to rely on the Renault front-drives. Renault now owned over 46 percent of AMC stock. The corporation showed a modest profit ($15.5 million) for the year, after a disastrous loss of $146.7 million for 1983. Even though Alliance sold strongly from the start, beating predictions by 31 percent, AMC had to halt production at Kenosha during 1984, to reduce its inventory. Encore/Alliance prices were cut late in 1984, following three price rises. Renault gradually abandoned the imported Le Car, to focus on the U.S.-built Encore. Although W. Paul Tippett Jr. remained AMC chairman, president Jose Dedeurwaerder was named CEO. Roy D. Chapin, Jr. went into retirement

1985 AMC

Renault-designed AMC models gained a new engine choice for 1985: a 1.7-liter overhead cam four, in addition to the previous 1.4-liter overhead valve four. Alliance also gained a new convertible body style. New to the option list was Keyless Entry, which used infrared waves to lock and unlock the car doors remotely.

RENAULT ENCORE — SERIES 90 — FOUR: The new 1.7-liter engine and five-speed transmission went into all Encore GS models. Others kept the 1.4-liter four and four-speed as standard, but could get the bigger engine as an option. All Encores had power disc brakes and halogen headlamps. Base Encores came with a black grille, color-keyed bumpers, carpeting, lighter, console with stowage box, cloth headliner, front courtesy lights, left-hand black remote-control mirror, day/night inside mirror, pulse wiper/washers, front/rear stabilizer bars, trip odometer, and vinyl bucket seats. Black semi-styled wheels held standard P155/80GR13 glass-belted radial tires. S Encores added an AM radio, bright grille, 60/40 split rear seat, dual pinstripes, removable carpeted cargo cover, plus moldings for bumper insert, lower liftgate and windshield surround. Stepping up a notch, the LS offered five-speed manual overdrive transmission, digital clock, rear ashtray, dual note horn, extra-quiet insulation, tachometer, black rear spoiler, bright belt moldings, and cloth-covered rocker/reclining bucket seats. Tires were P175/70SR13 steel-belted, with sport wheel-covers. This year's sporty Encore GS featured a black front air dam, black grille, light and visibility groups, black belt moldings, black sport steering wheel, and black pinstripes on the hood, liftgate and body sides. Aluminum wheels added to looks, a power liftgate release brought convenience, while a handling package helped performance and fog lamps added a practical touch.

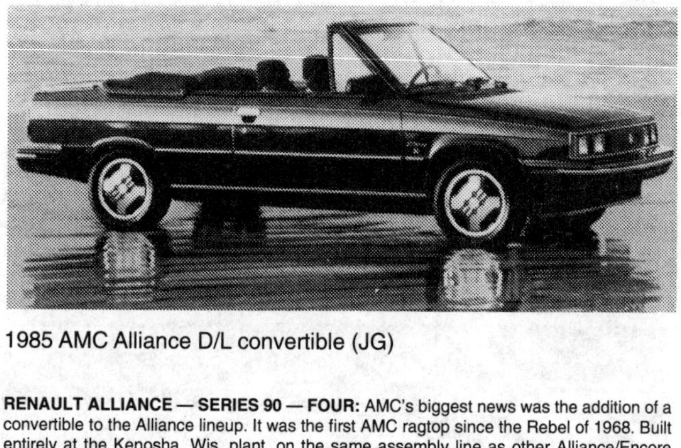

1985 AMC Alliance D/L convertible (JG)

RENAULT ALLIANCE — SERIES 90 — FOUR: AMC's biggest news was the addition of a convertible to the Alliance lineup. It was the first AMC ragtop since the Rebel of 1968. Built entirely at the Kenosha, Wis. plant, on the same assembly line as other Alliance/Encore models, the convertible adopted the new 1.7-liter Renault engine as standard power plant, with five-speed manual overdrive transmission. AMC president Jose Dedeurwaerder expected that it would "enhance the image that young, new-value buyers already have of the Alliance and Encore." With a $10,295 (minimum) base pricetag, it was promoted as the "lowest-priced domestic convertible" on the market. Three-speed automatic shift was optional, as on all models. Base and L sedans retained the smaller 1.4-liter four, but all DL Alliances also had the 1.7 four as standard. Convertibles came in two trim levels and six colors: white, beige and red, plus Mica red, Light blue, and gold metallic clearcoat. Folding tops were white or almond color. Interior trim was blue or almond. DL convertibles could also have honey or garnet interiors in vinyl or cloth. The less-costly L ragtop included a black grille and front air dam, tinted glass, heavy-duty battery, twin ashtrays, locking lighted glove box, extra-quiet insulation, dual black remote-controlled mirrors, black bumper insert moldings, roll-down quarter windows, and bright wheel trim rings. Five-speed manual transmission was standard. The power-operated top had a black inner liner and zip-out rear window. Color-keyed top boots used hidden fasteners. DL convertibles came with standard AM/FM stereo radio, digital clock, dual note horn, cloth rocker/recliner bucket seats, power steering, leather-wrapped steering wheel, door storage bins, luxury wheelcovers, and a tachometer. Both had black trim moldings and color-keyed sun visors. Base Alliance equipment was similar to the Encore liftback's, but included black-out rocker panels with bright moldings. L sedans had a bright grille and bumper insert moldings, plus an AM radio and dual accent pinstripes. DL Alliance sedans were similar to Encore's LS, with full luggage compartment trim, color-keyed mirror, bright deck lid and quarter window moldings, door storage bins, and bright wheel trim rings. They rode steel-belted P175/70SR13 radial tires. Stepping all the way up to the Limited brought buyers a rear center armrest, blackout lower back panel, light and visibility groups, deck lid luggage rack, wheel lip moldings, hood pinstripes, intermittent wipers, plus luxury cloth rocker/recliner bucket seats. Entertainment options stretched to a six-speaker electronically-tuned stereo radio with cassette player.

EAGLE — SERIES 30 — SIX: Eagles no longer had to be stationary to switch from two-wheel to four-wheel drive (and back again). "Shift-on-the-fly" capability let drivers change between the two while the car was moving. Only the 258 cid six-cylinder engine was offered this year, with standard five-speed overdrive manual transmission. Alternator capacity jumped up to 56 amps (from 42). Hoods lost their former ornament, but added a scoop effect along the top surface. Full-face radios with four speakers became standard. Other base equipment was similar to that of 1984. Eagle's Limited added a right remote chrome mirror, parcel shelf, woodgrain steering wheel and leather reclining seats, plus a new extra this year: wire wheelcovers. The Sport package was similar to 1984, with dual mirrors. Two new body colors, Medium blue and Dark blue metallic, joined Garnet, silver, Autumn brown, Almond, and Mocha Dark brown, for a total of seven choices. Interiors came in garnet, honey or almond, plus...this year...blue.

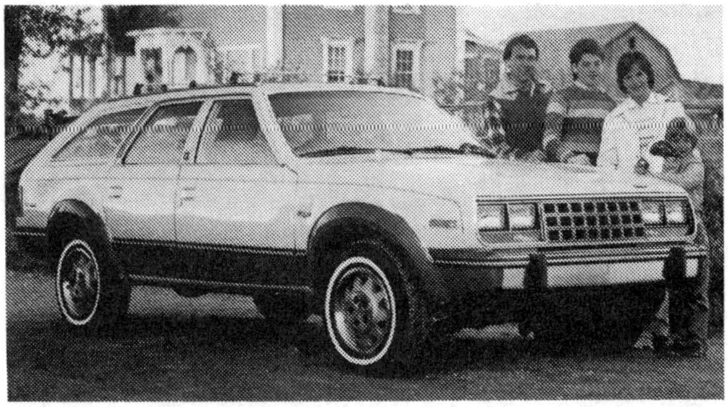

1985 AMC American Eagle station wagon (JG)

I.D. DATA: Eagle's 17 symbol Vehicle Identification Number (VIN) was again embossed on a metal plate riveted to the top left surface of the instrument panel, visible through the windshield. Coding was the same as in 1981-1984. Alliance/Encore also used a 17 symbol VIN, with coding the same as in 1984. Model year code (symbol 10) changed to 'F' for 1985. Engine codes were now: D=1.4-liter four; E=California 1.4L four; A=1.7-liter four; and C=258 cid six. The two-digit identifiers for body style (symbols 6-7) now included 97=convertible.

Series Number	Body/Style Number	Body Type & Seating	Factory Price	Shipping Weight	Production Total
ENCORE (BASE) SERIES 90					
90	8593-0	3d Lift-5P	5895	1946	Note 1
ENCORE S					
90	8593-3	3d Lift-5P	6360	1953	38,623
90	8599-3	5d Lift-5P	6610	2007	19,902
ENCORE LS					
90	8593-6	3d Lift-5P	7060	1997	Note 1
90	8599-6	5d Lift-5P	7310	2055	Note 1
ENCORE GS					
90	8593-9	3d Lift-5P	7560	2046	Note 1

NOTE 1: Production totals under S series include base, LS and GS Encore models. Model year sales: 46,923 Encores.

ALLIANCE (BASE) SERIES 90					
90	8596-0	2d Sed-5P	5995	1922	Note 2
ALLIANCE L					
90	8596-3	2d Sed-5P	6400	1929	33,617
90	8595-3	4dr Sed-5P	6650	1964	50,906
90	8597-3	2d conv-4P	10295	2153	7141
ALLIANCE DL					
90	8596-6	2d Sed-5P	7000	1962	Note 2
90	8595-6	4dr Sed-5P	7250	2001	Note 2
90	8597-6	2d conv-4P	11295	2190	Note 2
ALLIANCE LIMITED					
90	8595-8	4dr Sed-5P	7750	2190	Note 2

NOTE 2: Production totals under L series include base, DL and Limited Alliance models. Model year sales: 75,208 Alliances.
NOTE 3: Add $103 for 1.7-liter four-cylinder engine (except Encore GS, Alliance convertible and Alliance Limited, which had it as standard equipment).

EAGLE SERIES 30					
30	8535-5	4dr Sed-5P	10457	3306	2655
30	8538-5	4dr Sta Wag-5P	11217	3337	13,535
EAGLE LIMITED					
30	8538-7	4dr Sta Wag-5P	11893	3368	Note 3

NOTE 3: Production total shown for Eagle wagon includes base and Limited models. All Eagles had six-cylinder engine. Model year sales: 15,362 Eagles.

ENGINES: BASE FOUR (Alliance/Encore): Inline. Ohv. Four-cylinder. Cast iron block; aluminum head. Transverse mounted. Displacement: 85.2 cid (1.4 liters). Bore & stroke: 2.99 x 3.03 in. Compression ratio: 9.0:1. Brake horsepower: 56 at 4200 rpm. Torque: 75 lbs.-ft. at 2500 rpm. Five main bearings. Solid valve lifters. Single-point Bendix (throttle-body) fuel-injection. OPTIONAL FOUR (Alliance/Encore): Inline. Overhead cam. Four-cylinder. Cast iron block. Displacement: 105 cid (1.7 liters). Bore & stroke: 3.19 x 3.29 in. Compression ratio: 9.5:1. Brake horsepower: 77.5 at 5000 rpm. Torque: 96 lbs.-ft. at 3000 rpm. Five main bearings. Throttle-body fuel-injection (Bendix). BASE SIX (Eagle): Inline. Ohv. Six-cylinder. Cast iron block. Displacement: 258 cid (4.2 liters). Bore & stroke: 3.75 x 3.90 in. Compression ratio: 9.2:1. Brake horsepower: 110 at 3200 rpm. Torque: 210 lbs.-ft. at 1800 rpm.. Seven main bearings. Hydraulic valve lifters. Carburetor: two-barrel.

CHASSIS DATA: Wheelbase: (Encore/Alliance) 97.8 in.; (Eagle) 109.3 in. Overall length: (Encore) 160.6 in.; (Alliance) 163.8 in.; (Eagle) 180.9 in. Height: (Encore/Alliance) 54.5 in. except convertible, 53.1 in.; (Eagle) 54.4 in. Width: (Encore/Alliance) 65.0 in.; (Eagle) 72.3 in. Front Tread: (Encore/Alliance) 55.2 in.; (Eagle) 59.6 in. Rear Tread: (Encore) 52.8 in.; (Eagle) 57.6 in. Standard Tires: (Encore/Alliance) P155/80GR13 GBR; (Eagle) P195/75R15 GBR.

TECHNICAL: Transmission: (Alliance/Encore) four-speed overdrive manual shift standard; five-speed manual and three-speed automatic optional; (Eagle) five-speed overdrive manual shift standard; three-speed automatic optional. Floor shift lever. Alliance/Encore manual transmission gear ratios: (1st) 3.73:1; (2nd) 2.05:1; (3rd) 1.32:1; (4th) 0.97:1; (5th) 0.79:1; (Rev) 3.56:1. Eagle manual transmission gear ratios: (1st) 4.03:1; (2nd) 2.73:1; (3rd) 1.50:1; (4th) 1.00:1; (Rev) 3.76:1. Clutch diameter: (Alliance/Encore) 7.1 in.; (Eagle) 10.3 in. Standard axle ratio: (Alliance/Encore) 3.29:1 except 3.56:1 with five-speed transmission; (Eagle) 2.73:1. Steering: (Alliance/Encore) rack and pinion; (Eagle) recirculating ball. Suspension: (Alliance/Encore) fully independent with MacPherson front struts and stabilizer bar, trailing rear arms with transverse torsion bars and stabilizer bar; (Eagle) independent front coil springs, semi-elliptic rear leaf springs Brakes: (Alliance/Encore) 9.4 in. front discs, 8 in. rear drums; (Eagle) 11 in. front discs, 10 in. rear drums. Electronic ignition. Fuel tank: (Alliance/Encore) 12.5 gal.; (Eagle) 22 gal.

EAGLE DRIVETRAIN OPTIONS: Three-speed floor-shift automatic transmission ($366). Optional 2.73:1 or 3.08:1 axle ratio ($32). heavy-duty engine cooling ($75). Automatic load leveling ($182). Extra-duty suspension package ($83). Front suspension skid plate ($83). Trailer towing package 'A' to 2000 lbs. ($112). Trailer towing package 'B' to 3500 lbs. ($239). Heavy-duty battery ($28). Cold climate group ($62). California emission system ($81).

EAGLE CONVENIENCE/APPEARANCE OPTIONS: Eagle Sport package: base wagon ($416). Protection group: bumper guards, floot mats and door edge guards ($81); with Sport package ($25). Convenience group: headlight-on buzzer, intermittent wipers, lighted vanity mirror ($79). Gauge package ($98). Light group ($66). Right remote mirror (black or chrome): base ($34). Left/right remote chrome mirrors ($90-$124). Cruise control ($183). Air conditioning system ($781). Halogen headlamps ($16). Halogen fog lamps ($88). AM radio ($94). AM/FM/stereo radio ($235). Electronically-tuned AM/FM/cassette stereo radio ($432). Electronically-tuned AM/FM stereo radio with cassette and Dolby sound ($537). Power door locks ($183). Power windows and door locks ($458). Keyless entry system ($101). Power decklid release: sedan ($43). Power six-way driver's seat ($204). Power six-way driver/passenger seat ($326). Console ($70). Parcel shelf ($29). Cloth reclining seats: base ($72). Rear wiper/washer: wagon ($133). Rear defroster ($146). Woodgrain steering wheel ($39). Leather-wrapped sport steering wheel ($65) except Limited ($26). Tilt steering wheel ($115). Tinted glass ($113). Woodgrain paneling: wagon ($155). Special color combinations ($53). Black scuff moldings ($59). Roof rack: wagon ($124). Bumper guards ($56). Sport aluminum 15 x 6 in. wheels ($373) except Limited ($217). Wire wheelcovers ($156). Tires: P195/75R15 B black SBR Arriva ($22). P195/75R15 B white SBR Arriva ($75-$79). P215/65R15 B OWL SBR Eagle GT ($211-$215).

ALLIANCE/ENCORE OPTIONS: 1.7-liter four-cylinder engine ($103). Five-speed manual transmission, floor shift, with overdrive: Alliance L, Encore S ($100). Three-speed floor-shift automatic tranmission ($448) except Alliance L convertible, DL/Limited, Included with LS/GS ($348); not available base Encore. Handling package: S/LS 3d. ($31). Heavy-duty battery ($28). Heavy-duty cooling ($71). Cold climate group ($62-$83). Systems Sentry: monitors for low oil, coolant, brake fluid, disc wear, washer/power steering fluid, transaxle oil ($132); not available base models. Extra-quiet insulation package: L/S ($66). Rear spoiler: Encore base/S ($74). Sunshine package (sunroof, aluminum wheels and leather-wrapped steering wheel): DL/LS ($359). Pop-up sunroof ($324). Protection group ($55). Visibility group (dual remote mirrors, lighted visor mirror, intermittent wipers): L/S ($169); DL/LS ($137). Light group ($48). Fog lamps ($79). Tachometer: L/S ($87). Power steering ($221). Cruise control ($179); not available base. Air conditioning ($673). Intermittent wipers ($53). Keyless entry system ($100). Power door locks ($127-$179) Power door/window locks ($318-$370). Power liftback release: S/LS ($32). Rear defroster ($137). Rear wiper/washer: Encore ($124) except GS ($70). Tinted glass ($95). AM radio: base models ($91). AM/FM radio: base models ($160); others ($69). AM/FM stereo radio ($144); not available base. Electronic-tuning AM/FM stereo four-speaker radio with cassette ($335) except DL convertible. ($191); not availanle base. Electronic-tuning AM/FM stereo radio withcassette and Dolby ($440) except DL convertible. ($296). Digital clock: L/S ($60). Vinyl reclining bucket seats: L, S, base Alliance ($69). Vinyl rocker/reclining buckets: DL/LS (no charge). Cloth bucket seats: L/S ($77). Cloth reclining bucket seats: L/S ($146). Leather rocker/reclining bucket seats: LS/DL ($359); GS/Limited ($308). Door storage bins: L/S ($21). Metallic paint ($155) includes clearcoat. Leather-wrapped sport steering wheel ($64); not available base. Decklid luggage rack: Alliance L/DL ($111). Luxury sport wheelcovers ($93) except L convertible., DL ($38). Wheel trim rings ($55). Aluminum wheels ($216-$309). Tires: P155/80GR13 white GBR ($15). P175/70SR13 SBR ($76). P175/70SR13 white SBR ($63-$139). P185/60R14 SBR RBL: base/S/L ($155); others ($79). Spare tire (to replace polyspare): Alliance ($37).

HISTORICAL FOOTNOTES: Introduced: Oct. 1, 1984. Model year production (U.S.): 150,189 Encores and Alliances made in Kenosha, which came to 1.9 percent of the industry total; (Canada) 16,190 Eagles. Calendar year production (U.S.): 111,138 Encores and Alliances; (Canada) 11,311 Eagles. Model year sales by U.S. dealers: 137,493. Calendar year sales: 123,449, for a 1.6 percent market share. Sales dropped sharply for the model year, as consumers turned toward bigger cars-partly a result of moderate, stable gasoline prices. First-year mechanical problems also turned some prospective buyers away from AMC's French-designed duo, especially as they rose in price. After a modest profit ($15.5 million) in the previous year, AMC posted a $125.3 million loss for fiscal 1985. Alliance/Encore prices were cut in December 1984. Then, in February 1985, AMC cut the financing rate to 8.5 percent. It was the first domestic company to offer such incentives, which soon became almost normal. AMC helped pioneer extended warranties, too, with new 5-year/50,000-mile coverage on the powertrain, plus five years for rust-through. The new Alliance convertible was a joint project of AMC and American Sun Roof corporation. On a special subassembly line at Kenosha, metal tops and 'B' and 'C' pillars were sliced off two-door sedans. Then, a series of reinforcing operations strengthened the cut-down bodies before they returned to the line for paint and trim. Eagle, meanwhile, continued its downhill slide that had begun in 1981-the only year AMC beat out Subaru in four-wheel drive passenger car sales. Unlike most four-wheel drive imports, which were efficient front-drive designs with transfer mechanisms in their transaxles, Eagle was essentially a rear-drive vehicle with front-drive axle and transfer case hooked on. In short, it was a clumsy mechanism in an aged body; its design reaching back to 1969. Production fell to just 85 Eagles per day at the Brampton, Ontario factory. Just after New Year's, AMC cut production at Kenosha, laying off 600 workers. Another slash came in February, with more to follow by the start of 1986 model production in July. The June labor contract cut AMC workers' wages to a level on a par with those at GM and Ford; the agreement was accepted only after AMC took steps to close down the Kenosha plant. Innovations: "Shift-on-the-fly" four-wheel-drive. Convertible body style.

1986 AMC

All models added the high-mounted stop light required by law, and ashtrays left the standard equipment lists. Additions included gas-charged shock absorbers, a restyled instrument cluster, and larger-diameter stabilizer bars. New to the option lineup: a four-position tilt steering wheel.

ENCORE — SERIES 90 — FOUR: Base model Encores left the lineup, so S was the lowest-priced liftback. One new model was added: the top-level Electronic, whose instrument cluster held an array of hi-tech doodads. It included both digital and analog speedometers; a trip computer; tachometer; fuel gauge; oil level, temperature and pressure lights; plus a bar graph display that monitored engine functions. Elsewhere in the lineup, about the most thrilling change was an increase in alternator output from 50 to 60 amps. Only S and LS Encores were offered in five-door liftback form; the GS and Electronic were three-door (actually two passenger doors) only. The 1.4-liter four-cylinder engine was standard, 1.7-liter optional. Entry-level (S) Encores now sported black bumpers, cargo area carpeting, clearcoat paint, and an AM radio. Back seats were 60/40 split fold-down type. In addition to fancy instrumentation, Encore Electronics had cloth bucket seats, a color-keyed steering wheel, bright wheel trim rings, and a five-speed overdrive transmission. LS models lost their dual note horns, but added twin pinstripes. The top-level GS now featured dual exhausts and a Soft-Feel steering wheel, as well as the extras offered in 1985.

1986 AMC Alliance L coupe (AMC)

ALLIANCE — SERIES 90 — FOUR: The luxurious Limited Alliance left the lineup for 1986. A revised L and DL grille used dark horizontal bars over five vertical bars, with center emblem. Extended taillamps were new, headlamps smaller, interior altered. A new base four-door sedan joined the 1985 two-door. Convertibles again carried the larger 1.7-liter four, which was optional on other Alliances. Base Alliances added black bumpers and clearcoat paint to their standard equipment list. L Alliances now included a color-keyed steering wheel and bumpers, plus bright grille and Sebring red body paint. Except for semi-styled bright wheels, black belt molding and bright rocker panel and windshield moldings, the L convertible equipment was the same as in 1985. DL convertibles gained metallic paint and black body side scuff moldings with bright inserts, dual pinstripes, and a black Soft-Feel steering wheel. DL sedans lost their dual note horns, but added dual accent pinstripes.

1986 AMC American Eagle station wagon (JG)

EAGLE — SERIES 30 — SIX: Once again, Eagle was powered by the 258 six with five-speed overdrive transmission. A modest price increase was its only real change from 1985. Base models now had tinted glass (formerly an option) but lost a few minor items, including twin armrests and ashtrays. Extras on the Limited were the same as in 1985. The Sport package for base Eagle wagons was also the same as 1985, but included only the left-hand remote mirror.

I.D. DATA: Eagle's 17 symbol Vehicle Identification Number (VIN) was embossed on a metal plate riveted to the top left surface of the instrument panel, visible through the windshield. See coding details in 1981 listing. Model year (symbol 10) changed to 'G' for 1986. Alliance/Encore's VIN used the same coding as 1984-85. Model year code (10 digit) changed to 'G' for 1986. The code for the Limited ('8' in the eighth digit position) was no longer used.

ENCORE S SERIES 90

Series Number	Body/Style Number	Body Type & Seating	Factory Price	Shipping Weight	Production Total
90	8693-3	3d Lift-5P	6710	1970	12,239
90	8699-3	5d Lift-5P	6960	2003	6870

ENCORE LS

90	8693-6	3d Lift-5P	7310	1974	Note 1
90	8699-6	5d Lift-5P	7560	2007	Note 1

ENCORE ELECTRONIC

90	8693-4	3d Lift-5P	7498	1974	Note 1

ENCORE GS

90	8693-9	3d Lift-5P	7968	1977	Note 1

NOTE 1: Production totals under S series include LS, Electronic and GS Encore models. Model year sales: 17,671 Encores.

ALLIANCE (BASE) SERIES 90

90	8696-0	2d Sed-5P	5999	1923	Note 2
90	8696-5	4d Sed-5P	6199	1957	Note 2

ALLIANCE L

90	8696-3	2d Sed-5P	6510	1928	23,204
90	8695-3	4d Sed-5P	6760	1962	42,891
90	8697-3	2d conv-4P	10557	2222	2,015

ALLIANCE DL

90	8696-6	2d Sed-5P	7110	1935	Note 2
90	8695-6	4d Sed-5P	7360	1969	Note 2
90	8697-6	2d conv-4P	11557	2228	Note 2

NOTE 2: Production totals under L series include base and DL Alliance models. Model year sales: 55,603.
NOTE 3: Prices shown are for 1.4-liter engine. Add $264 for 1.7-liter four-cylinder engine (exceptept Alliance convertible, which had it as standard equipment).

EAGLE SERIES 30

30	8635-5	4d Sed-5P	10719	3307	1274
30	8638-5	4d Sta Wag-5P	11489	3341	6943

EAGLE LIMITED SERIES 30

30	8638-7	4d Sta Wag-5P	12179	3372	Note 4

NOTE 4: Production total shown for Eagle wagon includes base and Limited models. All Eagles had six-cylinder engine. Model year sales: 9020 Eagles.

1986 AMC Alliance D/L sedan (AMC)

ENGINES: BASE FOUR (Alliance/Encore): Inline. Ohv. Four-cylinder. Cast iron block; aluminum head. Transverse mounted. Displacement: 85.2 cid. (1.4 liters). Bore & stroke: 2.99 x 3.03 in. Compression ratio: 9.0:1. Brake horsepower: 56 at 4200 rpm. Torque: 75 lbs.-ft. at 2500 rpm. Five main bearings. Solid valve lifters. Single-point Bendix (throttle-body) fuel-injection. BASE FOUR (Alliance convertible, Encore Electronic/GS); OPTIONAL (other Alliance/Encore): Inline. Overhead cam. Four-cylinder. Cast iron block. Displacement: 105 cid. (1.7 liters). Bore & stroke: 3.19 x 3.29 in. Compression ratio: 9.5:1. Brake horsepower: 77.5 at 5000 rpm. Torque: 96 lbs.-ft. at 3000 rpm. Five main bearings. Bendix throttle-body fuel-injection. BASE SIX (Eagle): Same specifications as 1985.

CHASSIS DATA: Wheelbase: (Encore/Alliance) 97.8 in.; (Eagle) 109.3 in. Overall length: (Encore) 160.6 in.; (Alliance) 163.8 in.; (Eagle) 180.9 in. Height: (Encore/Alliance) 54.5 in. except convertible, 53.1 in.; (Eagle) 54.4 in. Width: (Encore/Alliance) 65.0 in.; (Eagle) 72.3 in. Front Tread: (Encore/Alliance) 55.2 in.; (Eagle) 59.6 in. Rear Tread: (Encore) 52.8 in.; (Eagle) 57.6 in. Standard Tires: (Encore/Alliance) P155/80GR13 GBR; (Eagle) P195/75R15 GBR.

TECHNICAL: Transmission: (Alliance/Encore) four-speed overdrive manual shift standard; five-speed manual and three-speed automatic optional; (Eagle) five-speed overdrive manual shift standard; three-speed automatic optional. Floor shift lever. Standard axle ratio: (Alliance/Encore) 3.29:1 except 3.56:1 with five-speed transmission; (Eagle) 2.73:1. Steering: (Alliance/Encore) rack and pinion; (Eagle) recirculating ball Suspension: (Alliance/Encore) fully independent with MacPherson front struts and stabilizer bar, trailing rear arms with transverse torsion bars and stabilizer bar; (Eagle) independent front coil springs, semi-elliptic rear leaf springs brakes: (Alliance) 9.4 in. front discs, 8 in. rear drums; (Eagle) 11 in. front discs, 10 in. rear drums. Electronic ignition. Fuel tank: (Alliance/Encore) 12.5 gal.; (Eagle) 22 gal.

EAGLE DRIVETRAIN OPTIONS: Three-speed floor-shift automatic transmission ($379). Optional 2.73:1 axle ratio ($33). Heavy-duty engine cooling ($78). Automatic load leveling ($188). Extra-duty suspension package ($86). Front suspension skid plate ($86). Trailer towing package 'A' to 2000 lbs. ($116). Trailer towing package 'B' to 3500 lbs. ($247). Heavy-duty battery ($32). Cold climate group ($64). California emissions package ($84).

EAGLE CONVENIENCE/APPEARANCE OPTIONS: Eagle Sport package: base model ($431). Protection group: front/rear bumper guards, front/rear mats, stainless steel door edge guards ($84); wagon with Sport package ($26). Convenience group: headlight-on buzzer, intermittent wipers, lighted vanity mirror ($82). Gauge package ($101). Light group ($68). Right remote mirror (black or chrome) ($35). Left/right remote chrome mirrors ($128), except Limited ($93). Cruise control ($189). Air conditioning system ($795). Halogen headlamps ($17). Halogen fog lamps ($91). AM/FM radio ($186). Electronic-tuning AM/FM/Stereo radio ($243). Electronic-tuning AM/FM/cassette stereo radio ($447). Power door locks ($189). Power windows and door locks ($474). Keyless entry system ($105). Power deck lid release: sedan ($45). Power six-way driver's seat ($211). Power six-way driver/passenger seat ($337). Floor shift console ($72). Parcel shelf ($30). Cloth seat upholstery: base ($75). Rear wiper/washer: wagon ($138). Rear defroster ($151). Woodgrain steering wheel: base ($40). Leather-wrapped sport steering wheel ($67), except Limited ($27). Tilt steering wheel ($119). Woodgrain paneling: wagon ($160). Black scuff moldings ($61). Roof rack: wagon ($128). Bumper guards ($58). Sport aluminum 15 x 6 in. wheels ($386), except Limited ($225). Wire wheelcovers ($161). Tires: P195/75R15 B black SBR Arriva ($23). P195/75R15 B white SBR Arriva ($78-$82). P215/65R15 B OWL SBR Eagle GT ($218-$223).

ALLIANCE/ENCORE OPTIONS: 1.7-liter four-cylinder engine: Alliance L/DL, Encore S/LS ($164). Five-speed manual transmission, floor shift, with overdrive: L/S ($100). Three-speed floor-shift automatic tranmission ($469) except Alliance DL/convertible, Encore S/GS/Electronic ($369). Handling package: Encore three-door, except GS ($32). Cold climate group ($85) except ($64) with air conditioning. Heavy-duty engine cooling ($73). Heavy-duty alternator and engine cooling ($107). Heavy-duty battery ($29). Extra-quiet insulation package: L/S and Electronic ($68). Sunshine package (sunroof, Soft-Feel steering wheel and 14 in. aluminum wheels): DL except convertible ($407); L/S ($541); LS ($368). Pop-up sunroof ($332). Rear spoiler: Electronic and Encore S ($76). Protection group: door edge guards and carpeted floor mats ($56). Visibility group (dual remote mirrors, intermittent wipers): L/S and Electronic ($173); DL/LS ($140); not available convertible Light group ($65). Fog lamps ($81). Tachometer: L/S ($89). Power steering ($227). Cruise control ($183); not available Alliance base. Air conditioning ($685). Digital clock: L/S/Elect. ($62). Keyless entry system ($103); not available Alliance base/convertible. Power door locks ($130-$183). Power windows and door locks: convertible ($326). Power liftgate lock release: Encore ($40). Rear wiper/washer ($127). Rear defroster ($140). Tinted glass ($99). AM radio: base Alliance ($93). AM/FM radio ($71), except base Alliance ($164). Electronic-tuning AM/FM stereo radio ($192). Electronic-tuning AM/FM stereo radio with cassette ($357), except DL convertible ($165); not available Alliance base. Electronic-tuning AM/FM stereo radio with cassette and (Dolby) Jensen Accusound ($614), except DL convertible ($422); not available Alliance base. Vinyl reclining bucket seats: base/L Alliance, Encore S/Elect. ($71). Cloth bucket seats: base/L Alliance, Encore S ($79). Cloth reclining bucket seats: L/S ($150); Electronic ($71). Cloth rocker/reclining wing back seats: DL ($105); not available convertible Sebring red paint: Alliance ($50). Metallic paint ($159); DL convertible (no charge). Black Soft-Feel sport steering wheel ($66). Tilt steering wheel: L/DL, Encore ($115). Decklid luggage rack: Alliance DL ($114). Luxury sport wheelcovers ($39) except. base/L Alliance, Encore S ($95); not available GS Wheel trim rings: base/L Alliance, S ($56). Aluminum wheels ($221-$316). Tires: P155/80GR13 white GBR ($67). P175/70SR13 SBR ($78). P175/70SR13 white SBR ($65-$143). P185/60R14 SBR RBL: DL/LS ($81); L/S/Elect. ($159); with Sunshine package (no charge). Spare tire (to replace polyspare): L/DL sedan ($65).

HISTORICAL FOOTNOTES: Introduced: Oct. 1, 1985. Model year production (U.S.): 64,873, down to only 0.8 percent of the industry total; (Total) 95,436, including 8217 Eagles. Calendar year production (U.S.): 49,435 Encores and Alliances made at Kenosha. Model year sales by U.S. dealers: 82,294. Calendar year sales: 72,849, for a market share of only 0.9 percent. AMC president Jose Dedeurwaerder believed "the worst is over" after rough times early in the model year. But, corporate fortunes and prospects continued to decline, registering a $91.3 million loss for 1986. Model year sales dropped to only one percent of the domestic market, from 1.6 the year before. Struggling to stay alive, the company even slashed financing rates all the way down to zero interest (on two-year loans, that is). But, both American Honda and Volkswagen of America beat AMC in sales and Nissan came close. By year's end, the Encore was dropped, replaced by a hatchback Alliance. Eagle sales fell below the 10,000 mark, though AMC insisted the tired old four-wheel drive would stay in production in Canada a while longer. The irony is that Eagle's failure came at a time when interest in four-wheel drive passenger cars was rising. On the positive side, Chrysler Corp. agreed to build its M-body cars at AMC's Kenosha plant, which helped pave the way for the Chrysler takeover a year later. Joseph E. Cappy became AMC president and CEO; Pierre Semerena (from Renault) was named chairman. Jose Dedeurwaerder combined duties of AMC vice-chairman with those of executive V.P. of Renault worldwide sales and marketing. Rumors flurried over the next year, until the Chrysler deal was finally concluded.

JEEP
1970-1986

American Motors purchased Jeep, from Kaiser Industries, in February 1970. At first, AMC retained such established model designations as CJ (Civilian Jeep), Wagoneer, Gladiator and Jeepster Commando. Almost immediately, the new owner began a program of dressing-up its products and streamlining marketing techniques to widen the Jeep models' appeal.

By 1971, offerings had been pared-down from 38 to 22 models. The following season, the venerable four-cylinder Jeep engine disappeared. In its place came AMC-built inline sixes and overhead valve V-8s, the top option being a 360 cid (5.9-liter) V-8 offering up to 195 hp.

The die for the future was cast in 1973. That's when the AMC-Jeep line was neatly divided into truck and non-truck models, all with names as well as numerical designations. The trucks came in single-wall Thriftside and double-wall Townside configurations. They were simply called Jeep Trucks and were no longer referred to as Gladiators. Jeep, Commando and Wagoneer labels were applied to other models. Wagoneers and four of the lighter-duty pickups had AMC's new Quadra-Trac full-time four-wheel drive setup. With this system, four-wheel drive was always in use, but each wheel could spin at its own speed making two-wheel drive unnecessary.

Optional exterior woodgrain trims, bright body side moldings and new two-tone paint treatments were made available to spruce up appearances. Color-coded instrumentation was marked with international symbols and included two new types of gauges: an ammeter and a clock. A full-width "electric shaver" type grille was seen again, too.

AMC's first all-new offering, a two-door sports utility vehicle, bowed in the 1974 Cherokee. It replaced the Commando and was much better suited to slugging it out with Broncos and Blazers in the sales wars. By this time, the line was down to a pair each of Wagoneers, Cherokees and Jeeps, plus three Townside pickups. It totaled nine models, each with a distinct identity. There was just enough variety for practically everyone and a lot less confusion and cross-over.

Calendar year registrations reflected the success that AMC's easier-to-understand marketing program was achieving. They stood at 36,354 vehicles in 1970, but leaped to 50,926 in 1972; 68,227 in 1973; and 96,835 for 1974. Model year production surpassed the magic 100,000 unit level during 1975, when some 62,000 CJs, 14,000 pickups, 17,000 Wagoneers and 16,000 Cherokees were built.

Corporately, Jeep continued to operate as a wholly owned subsidiary of AMC, with operations still based in Toledo. W.H. Jean was plant manager; J.E. MacAfee was in charge of engineering; Charles Mashigan was styling chief and E.V. Amoroso headed-up sales. John A. Conde, today a well-known automotive historian, was director of public relations.

In the later '70s, there were a number of technical changes such as adopting power disc brakes (made standard equipment in 1977) and larger six- and eight-cylinder engines. A luxurious Wagoneer Limited model-option debuted in 1979 and all other vehicles were provided with a wide choice of trim options. There was the Renegade package for CJ models and Honcho or Golden Eagle options for pickups. Cherokee S and Chief options could also be had. Such extras cost up to $1,300 in 1979. The steep prices made some extras rare, one reason they have strong appeal to collectors today.

Despite the bold prices, many high-optioned Jeeps performed well in the marketplace during this booming period for Jeep sales. Calendar year registrations went to 107,487 in 1976; 124,843 in 1977; and 168,548 in 1978. Even in off-year 1979, a season of industry-wide slack, Jeep saw 145,583 of its vehicles titled in the U.S. Then came the early '80s.

The years 1980-1982 were disastrous, not only for Jeep, but for AMC and the entire American car and truck industry. By the fall of 1983, American Motors had registered its 14th consecutive business quarter with figures written in red ink. Despite crisp new styling touches and the release of a new Scrambler model (in 1981), registrations averaged no better than 67,400 units over this three-year period.

A high note was sounded at the beginning of 1984 when the down-sized Cherokee and Wagoneer Sport Wagons were added to the AMC lineup. By this time, the company had also inked an historic agreement to produce Jeeps in China on a joint-venture basis.

A two-wheel drive version of the Cherokee was offered for the first time in 1985 and, for 1986, down-sized and modernized Commanche pickups took their place in a revitalized and expanded Jeep product stable. Entering the stage, in mid-1986 (as a 1987 model), was a totally revised type of Universal Jeep known as the Wrangler. AMC claimed that this CJ-like new model "carries the illustrious badge of Jeep's off-road durability and ruggedness, but also holds out the promise of the smoothest on-road ride yet offered in a small sport utility."

1970 JEEP

1970 AMC Jeep, J-2500 Gladiator pickup (OCW)

JEEP — 1970 SERIES — (ALL ENGINES): After American Motors purchased Kaiser-Jeep from Kaiser Industries, in February 1970, it renamed it the Jeep Corp. With the acquisition of Jeep, which had sales of over $400 million, American Motors gained entry into the four-wheel drive market, which had grown by almost 500 percent in the previous decade. Although Jeep's share of the market had fallen as other competitors had entered this field, its 20 percent share of the market was strong enough to serve as a base to re-establish Jeep as a sales leader. American Motors organized a new product development group which was assigned the long range task of developing completely new Jeep models, while making improvements in the current line. For 1970 the most significant changes to be found were a new grille, for the Gladiator truck, as well as optional two-tone color combinations.

1970 AMC Jeep, Custom Wagoneer station wagon (OCW)

I.D. DATA: The VIN is located on the left front door hinge pillar and left firewall. The VIN has 13 symbols. The first five digits indicate series and style. The sixth and seventh designate engine type. The last six digits are sequential serial numbers. The starting number varies per model.

Model	Body Type	Price	Weight	Prod. Total
J-100 — (1/4-Ton) — (4x4) — (110 in. w.b.)				
1414	4d Sta Wag	4284	3710	—
1414C	4d Sta Wag	4526	3745	—
1414X	4d Cus Sta Wag	5876	3907	—
Jeepster Commando				
8705F	2d Sta Wag	3208	2722	—
8705C	2d Rds	2917	2510	—
Jeepster				
8701	2d Conv	3822	2853	—
8702	Conv Commando	3328	2787	—
CJ-5 — (1/4-Ton) — (4x4) — (81 in. w.b.)				
8305C15	Jeep	2930	2212	—
CJ-6 — (101 in. w.b.)				
8405C15	Jeep	3026	2274	—
DJ-5 — (1/4-Ton) — (2x2) — (81 in. w.b.)				
8505C15	Jeep	2396	1872	—
Jeepster — (1/4-Ton) — (4x4) — (101 in. w.b.)				
8705H15	Pickup	3014	2659	—
Series J-2500 — (1/2-Ton) — (4x4)				
2406W17	Chassis & Cab	3361	3152	—
2406W17	Thriftside Pickup	3488	3447	—
2406W17	Townside Pickup	3516	3555	—
Series J-2600				
2406X17	Chassis & Cab	3483	3293	—
2406X17	Thriftside Pickup	3610	3588	—
2406X17	Townside Pickup	3638	3696	—
2406X17	Platform Stake	3804	3949	—
Series J-2700 — (1/2-Ton) — (4x4)				
2406Y17	Chassis & Cab	3649	3380	—
2406Y17	Thriftside Pickup	3776	3675	—
2406Y17	Townside Pickup	3804	3783	—
2406Y17	Platform Stake	3790	4036	—
Series J-3500 — (1/2-Ton) — (4x4)				
3406W17	Chassis & Cab	3381	3176	—
3406W17	Townside Pickup	3544	3604	—
Series J-3600 — (1/2-Ton) — (4x4)				
3406X17	Chassis & Cab	3505	3314	—
3406X17	Townside Pickup	3667	3742	—
3406X17	Platform Stake	3860	4018	—
Series J-3700 — (1/2-Ton) — (4x4)				
3406Y17	Chassis & Cab	3668	3401	—
3406Y17	Townside Pickup	3831	3829	—
3406Y17	Platform Stake	4024	4105	—
Series J-3800 — (1/2-Ton) — (4x4)				
3407Z19	Chassis & Cab	4320	3792	—
Series J-4500 — (1/2-Ton) — (4x4)				
3408W17	Chassis & Cab	3381	3130	—
3408W17	Townside Pickup	3544	3558	—
Series J-4600 — (3/4-Ton) — (4x4)				
3408X17	Chassis & Cab	3505	3268	—
3408X17	Townside Pickup	3668	3696	—
Series J-4700 — (3/4-Ton) — (4x4)				
3408Y17	Chassis & Cab	3668	3355	—
3408Y17	Townside Pickup	3831	3783	—

1970 AMC Jeep, XJ001 Concept Vehicle (OCW)

ENGINE (Standard CJ-5/CJ-6/DJ-5/Jeepster): Inline. F-head. Four-cylinder. Cast iron block. Bore & stroke: 3.125 x 4.375 in. Displacement: 134.2 cid. Compression ratio: 6.9:1. Brake horsepower: 72 at 4000 rpm. Torque: 114 lb.-ft. at 2000 rpm. Three main bearings. Mechanical valve lifters. Carburetor: Single one-barrel.

1970 AMC Jeep, Commando station wagon (RPZ)

1970 AMC Jeep, Jeepster interior (RPZ)

ENGINE (Optional all models except CJ-5, CJ-6, DJ-5, Jeepster): V-type. OHV. Eight-cylinder. Cast iron block. Bore & stroke: 3.8 x 3.85 in. Displacement: 350 cid. Compression ratio: 9.0:1. Brake horsepower: 230 at 4400 rpm. Max. Torque: 314 lb.-ft. at 2600 rpm. Five main bearings. Hydraulic valve lifters. Carburetor: Two-barrel.

ENGINE (Standard except CJ/DJ/Jeepster): Inline. OHV. Six-cylinder. Cast iron block. Bore & stroke: 3.75 x 3.51 in. Displacement: 232 cid. Compression ratio: 8.5:1. Brake horsepower: 145 at 4300 rpm. Torque: 215 lb.-ft. at 1600 rpm. Seven main bearings. Hydraulic valve lifters. Carburetor: Carter one-barrel model YF.

1970 AMC Jeep, CJ-5 Universal Jeep (RPZ)

ENGINE (Optional): Vee-block. Overhead valve. Six-cylinder. Cast iron block. Bore & stroke: 3.75 x 3.40 in. Displacement: 225 cid. Compression ratio: 9.0:1. Brake horsepower: 160 at 4200 rpm. Max Torque: 235 lb.-ft. at 2400 rpm. Hydraulic valve lifters. Carburetor: Two-barrel.

1970 AMC Jeep, Jeepster Commando pickups (RPZ)

CHASSIS (J-100): Wheelbase: 110. Tires: 7.75 x 15.

CHASSIS (Jeepster/Jeepster Commando): Wheelbase: 101 in. Overall length: 168.5 in. Tires: 7.35 x 15.

CHASSIS (CJ-5): Wheelbase: 81 in. Overall length: 133 in. Front tread: 48.25 in. Rear tread: 48.25 in. Tires: 6.00 x 16 in.

CHASSIS (CJ-6): Wheelbase: 101 in. Tires: 6.00 x 16 in.

CHASSIS (DJ-5): Wheelbase: 80 in. Overall length: 126 in. Front tread: 48.25 in. Rear tread: 48.25 in. Tires: 6.85 x 15 in.

CHASSIS (Jeepster): Wheelbase: 101 in. Overall length: 168.40 in. Height: 64.2 in. Front tread: 50 in. Rear tread: 50 in. Tires: 7.35 x 15 in.

CHASSIS (J-2500/J-2600/J-2700): Wheelbase: 120 in. Overall length: 193.6 in. Front tread: 63.5 in. Rear tread: 63.8 in. Tires: (J-2500) 8.25 x 15 in.; (J-2600) 7.00 x 16 in.; (J-2700) 7.50 x 16 in.

CHASSIS (J-3500/J-3600/J-3700): Wheelbase: 126 in. Tires: (J-3500) 8.25 x 15 in.; (J-3600) 7.00 x 16 in.; (J-3700) 7.50 x 16 in.

CHASSIS (J-3800/J-4500/J-4600/J-4700): Wheelbase: 132 in. Overall length: 205.6 in. Front tread: 63.9 in. Rear tread: 64.4 in. Tires: (J-3800) 7.50 x 16 in.; (J-4500) 8.25 x 15 in.; (J-4600) 7.00 x 16 in.; (J-4700) 7.50 x 15 in.

1970 AMC Jeep, CJ-5 Universal Jeep (RPZ)

TECHNICAL: Manual, synchromesh transmission. Speeds: 3F/1R. Column mounted gearshift. Dry plate clutch. (1/2-Ton Jeep models) semi-floating rear axle; (all others) full-floating rear axle. Hydraulic, four-wheel, drum brakes. Pressed steel wheels. Automatic transmission. Technical Options: Power steering. Four-speed manual transmission. Camper Package ($163).

OPTIONS: Rear bumper. Rear step bumper. AM radio. Clock. Camper Package ($163). West Coast mirror.

1970 AMC Jeep, J-100 Wagoneer station wagon (RPZ)

HISTORICAL: Introduced: Fall 1969. Calendar year sales: 30,842 (all series and models). Calendar year production: 45,805 (all series and models).

1971 JEEP

JEEP — 1971 SERIES — (ALL ENGINES): No appearance changes were made for either the Jeep CJ or J-series trucks for 1971. Early in 1971, the AMC 304 cid and 360 cid V-8 engines became optional for the J-series trucks, while the AMC 258 cid six-cylinder became their standard power plant. Of interest to modern day collectors are several special, limited-edition versions of the Jeepster issued in 1971. The first was the Hurst/Jeepster Special, which had a special air scoop hood with a tachometer on top of it. Also featured on this low-production car were Rally stripes on the cowl and tailgate, fat tires and either a Hurst Dual-Gate automatic transmission selector or manual transmission with a Hurst T-handle shifter. Also available was the SC-1 Jeepster station wagon with Butterscotch gold finish, a white top, black Rally stripes and a V-6 engine.

I.D. DATA: The VIN is located on the left front door hinge pillar and left firewall. The VIN has 13 symbols. The first five digits indicate series and style. The sixth and seventh designate engine type. The last six digits are sequential serial numbers. The starting number varies per model.

1971 AMC Jeep Commando station wagon (OCW)

Model	Body Type	Price	Weight	Prod. Total
J-100 — (1/4-Ton) — (4x4) — (110 in. w.b.)				
1414	4d Sta Wag	4447	3661	—
1414C	4d Cus Sta Wag	4526	3696	—
1414X	4d Spl Sta Wag	6114	3982	—
Jeepster Commando — (4x4)				
8705F	2d Sta Wag	3446	2722	—
8705O	2dr Rds	3197	2510	—
Jeepster Commando Six				
8705F	2dr Sta Wag	3546	2802	—
8705O	2dr Rds	3297	2590	—
87020	2dr Conv	3465	2787	—
na	2dr SC-1 Sta Wag	na	na	—
na	2dr Hurst Spl Wag	na	na	—
CJ-5 — (1/4-Ton) — (4x4)				
8305015	Jeep	2886	2112	—
CJ-6 — (1/2-Ton) — (4x4)				
8405015	Jeep	2979	2274	—
DJ-5 — (1/4-Ton) — (2x2)				
8505015	Open	2382	1872	—
Jeepster — (1/2-Ton)				
8705H15	Pickup	3291	2659	—
J-2500 — (1/2-Ton) — (4x4)				
2406W17	Chassis & Cab	3251	3125	—
2406W17	Thriftside Pickup (7-ft.)	3406	3420	—
2406W17	Townside Pickup (7-ft.)	3406	3528	—
J-3800 — (1/2-Ton) — (4x4)				
3407Z19	Chassis & Cab	4113	3792	—
3407Z19	Townside Pickup	4264	4220	—
J-4500 — (1/2-Ton) — (4x4)				
3408W17	Chassis & Cab	3281	3151	—
3408W17	Townside Pickup (8-ft.)	3443	3579	—
J-4600 — (1/2-Ton) — (4x4)				
3408X17	Chassis & Cab	3405	3289	—
3408X17	Townside Pickup (8-ft.)	3567	3717	—
J-4700 — (3/4-Ton) — (4x4)				
3408Y17	Chassis & Cab	3567	3378	—
3408Y17	Townside Pickup (8-ft.)	3729	3806	—
J-4800 — (3/4-Ton) — (4x4)				
3407Z19	Chassis & Cab	4218	3806	—
3407Z19	Townside Pickup (8-ft.)	4370	4294	—

1971 AMC Jeep, J-100 Wagoneer station wagon (OCW)

ENGINE (Standard J-100 Wagoneer): Inline. OHV. Six-cylinder. Cast iron block. Bore & stroke: 3.75 x 3.5 in. Displacement: 232 cid. Brake horsepower: 145 at 4300 rpm.
ENGINE (Optional Wagoneer V-8): Vee-block. OHV. Cast iron block. Bore & stroke: 3.8 x 3.85 in. Displacement: 350 cid. Brake horsepower: 230 at 4400 rpm.

ENGINE (Standard CJ-5/CJ-6/DJ-5/Jeepster/Commando): Inline. F-head. Four-cylinder. Cast iron block. Bore & stroke: 3.125 x 4.375 in. Displacement: 134.2 cid. Compression ratio: 6.9:1. Brake horsepower: 72 at 4000 rpm. Torque: 114 lb.-ft. at 2000 rpm. Three main bearings. Mechanical valve lifters. Carburetor: One-barrel.

1971 AMC Jeep, Commando SC-1 station wagon (OCW)

ENGINE (Optional Commando V-6): Vee-block. OHV. Bore & stroke: 3.75 x 3.4 in. Displacement: 225 cid. Brake horsepower: 160 at 4200 rpm.
ENGINE (Standard: J-2500, J-4500, J-4600, J-4700): Inline. OHV. Six-cylinder. Cast iron block. Bore & stroke: 3.75 x 3.9 in. Displacement: 258 cid. Compression ratio: 8.0:1. Brake horsepower: 110 at 3500 rpm. Seven main bearings. Hydraulic valve lifters. Carburetor: Carter one-barrel model YF.
ENGINE (Standard: J-3808, J-4800): V-type. OHV. Eight-cylinder. Cast iron block. Bore & stroke: 4.08 x 3.44 in. Displacement: 360 cid. Brake horsepower: 175. Five main bearings. Hydraulic valve lifters. Carburetor: Two-barrel.

1971 AMC Jeep, Hurst Jeepster Special (C&C)

CHASSIS: (J-100) Wheelbase: 110 in. Overall length: 183.66 in. Width: 75.60 in. Tires: 7.75 x 15.
CHASSIS: (8700) Wheelbase: 101 in. Overall length: 168.5. Tires: 7.35 x 15.
CHASSIS (Jeepster): Wheelbase: 101 in. Overall length: 168.40 in. Height: 64.2 in. Front tread: 50 in. Rear tread: 50 in. Tires: 7.35 x 15 in.
CHASSIS (J-2500): Wheelbase: 120 in. Overall length: 193.6 in. Front tread: 63.5 in. Rear tread: 63.8 in. Tires: 8.25 x 15 in.
CHASSIS (J-3800/J-4500/J-4600/J-4700/J-4800): Wheelbase: 132 in. Overall length: 205.6 in. Front tread: 63.9 in. Rear tread: 64.4 in. Tires: (J-4500) 8.25 x 15 in.; (J-4600) 7.00 x 16 in.; (J-4700) 7.50 x 16 in.; (J-4800) 7.50 x 16 in.
CHASSIS (CJ-5): Wheelbase: 81 in. Overall length: 133 in. Front tread: 48.25 in. Rear tread: 48.25 in. Tires: 6.00 x 16 in.
CHASSIS (CJ-6): Wheelbase: 101 in. Tires: 6.00 x 16 in.
CHASSIS (DJ-5): Wheelbase: 81 in. Overall length: 133 in. Tires: 6.85 x 15 in.

1971 AMC Jeep, J-100 Super Wagoneer station wagon (RPZ)

TECHNICAL: Manual, synchromesh transmission. Speeds: 3F/1R. Column mounted gearshift. Dry plate clutch. (1/2-Ton Trucks, all CJ-5/CJ-6/DJ-5) semi-floating rear axle; (all others) full-floating rear axle. Four-wheel hydraulic, drum brakes. Pressed steel wheels. Technical Options: Four-speed manual transmission. Power steering. Camper Package ($163).

1971 AMC Jeep, CJ-5 Renegade II Universal Jeep (OCW)

OPTIONS: Rear bumper. Rear step bumper. AM radio. Clock. Cigar lighter. Camper Package ($163). 327 V-8 ($220). Power steering ($135). Four-speed transmission ($105). Turbo-Hydra-Matic transmission ($280). Thriftside pickup (add $105). 225 six-cylinder engine in CJ (average $215). Four-speed transmission in CJ (average $180). CJ camper kit ($2,200). 350 cid V-8 in J-100 ($220).
HISTORICAL: Introduced: Fall 1970. Calendar year sales: 38,979 (all series). Calendar year production: 54,480 (all series). At the same time the evidence was clear that American Motors' efforts to broaden the Jeep's appeal were paying off. For the first six months of 1971, Jeep wholesale sales were up 26 percent over the same 1970 period. Overall, sales for the same period totalled 17,878 compared to 14,186 in 1970.

1972 JEEP

1972 AMC Jeep, Commando station wagon (OCW)

JEEP — 1972 SERIES — (ALL ENGINES): Numerous styling and option changes were included in the Jeep picture for 1972. The Commando got the most attention with a new front end design and interior changes. The wheelbase grew, and both the body and tread got wider.
Continuing the wave of change for 1972 at Jeep was a new line of engines for the CJ models. Although the vintage F-head four-cylinder was still offered for export, the "Dauntless" V-6 was no longer available. The standard CJ engine now was the AMC 232 cid six-cylinder with the larger 258 cid six-cylinder and 304 cid and 360 cid V-8 engines available as options. The CJ-5 and CJ-6 retained their familiar look, but now had longer wheelbases and increased length. In addition, their front and rear treads were increased by three and 1.5 inches respectively. Also adopted was a new Dana model 30 open-end front axle and a rear axle with a capacity of 3000 pounds, which was 500 pounds greater than the unit used in 1971. Use of a Dana model 20 transfer case reduced overall noise and provided a smoother shifting procedure. Appreciated by long-time Jeep fans, was the latest model's larger diameter clutch, improved heater and suspended clutch and brake pedals. Numerous new options were listed for the CJ Jeeps including a fixed rear tailgate with a rear-mounted spare, a vinyl-coated full fabric top, 15 inch wheelcovers and oil and ammeter gauges.
The Commando also had a longer wheelbase and the same engine lineup as the CJ models. Its appearance was changed due to a new front end with a stamped steel grille that enclosed both the head and parking lights. The Commando, like the CJ Jeep, also was equipped with an open-ended front axle, larger brakes and increased capacity clutch. Changes to the Commando's interior included repositioned front seats and reshaped rear wheelhousings.
The J-series trucks now could be ordered with a new 6000 GVW capacity on the 120 in. wheelbase chassis. Common to all J-trucks were larger clutches and brakes. No styling changes were made in their appearance, but interiors featured new seat trim patterns.
I.D. DATA: The VIN is located on the left front door hinge pillar and left firewall as before, but the format was changed. The VIN still has 13 symbols. The first indicates Jeep Corp. The second indicates model year. The third indicates transmission, drivetrain and assembly plant. The fourth and fifth indicate series or model. The sixth symbol identifies body style. The seventh symbol indicates model type and GVW. The eighth symbol indicates the engine. Engine codes were: E=232 cid six-cylinder; A=258 cid six-cylinder; H=304 cid V-8; N=360 cid V-8. The next five symbols are the sequential production number.

Model	Body Type	Price	Weight	Prod. Total
J-100 — (1/4-Ton) — (4 x 4)				
1414	4d Sta Wag	4398	3808	—
1414C	4d Cus Sta Wag	4640	3843	—
Jeepster Commando — (1/4-Ton) — (4 x 4)				
8705F	Sta Wag	3408	3002	—
87050	Rds	3257	2790	—
CJ-5 — (1/4-Ton) — (4x4)				
83050	Jeep	2955	2437	—
CJ-6 — (1/4-Ton) — (4x4)				
84050	Jeep	3045	2499	—
DJ-5 — (1/4-Ton) — (2x2)				
85050	Open	2475	2255	—
Commando — (1/2-Ton) — (4x4)				
8705H	Pickup	3284	2939	—
J-2500 — (1/2-Ton) — (4x4)				
2406W	Chassis & Cab	3181	3272	—
2406W	Thriftside Pickup	3328	3567	—
2406W	Townside Pickup	3328	3675	—
Series J-2600 — (1/2-Ton) — (4x4)				
2406X	Thriftside Pickup	3449	3689	—
2406X	Townside Pickup	3449	3797	—
Series J-4500 — (1/2-Ton) — (4x4)				
3408W	Chassis & Cab	3210	3298	—
3408W	Townside Pickup	3365	3726	—
Series J-4600 — (1/2-Ton) — (4x4)				
3408X	Chassis & Cab	3331	3436	—
3408X	Townside Pickup	3486	3864	—
Series J-4700 — (1/2-Ton) — (4x4)				
3408Y	Chassis & Cab	3698	3732	—
3408Y	Townside Pickup	3853	4160	—
Series J-4800 — (1/2-Ton) — (4x4)				
3407Z	Chassis & Cab	4107	4013	—
3407Z	Townside Pickup	4262	4441	—

1972 AMC Jeep, CJ-5 Universal Jeep with hardtop (OCW)

ENGINE (Standard J-100 Wagoneer): Inline. OHV. Six-cylinder. Cast iron block. Bore & stroke: 3.75 x 3.5 in. Displacement: 232 cid. Brake horsepower: 145 at 4300 rpm. Torque: 185 lb.-ft. at 1800 rpm. Seven main bearings. Hydraulic valve lifters. Carburetor: One-barrel.

ENGINE (Standard Commando V-6): Vee-block. OHV. Bore & stroke: 3.75 x 3.4 in. Displacement: 225 cid. Brake horsepower: 160 at 4200 rpm.
ENGINE (Standard CJ-5/CJ-6/DJ-5): Inline. OHV. Six-cylinder. Cast iron block. Bore & stroke: 3.75 x 3.5 in. Displacement: 232 cid. Brake horsepower: 145 at 4300 rpm. Torque: 185 lb.-ft. at 1800 rpm. Seven main bearings. Hydraulic valve lifters. Carburetor: One-barrel.
ENGINE (Standard: J-2500, J-2600, J-4500, J-4600): Inline. OHV. Six-cylinder. Cast iron block. Bore & stroke: 3.75 x 3.9 in. Displacement: 258 cid. Compression ratio: 8.0:1. Brake horsepower: 110 at 3500 rpm. Max torque: 195 lb.-ft. at 2000 rpm. Seven main bearings. Hydraulic valve lifters. Carburetor: Carter one-barrel model YF.
ENGINE (Optional J-2500, J-4500, J-4600, ($165): V-block. OHV. Eight-cylinder. Cast iron block. Bore & stroke: 3.75 x 3.44 in. Displacement: 304 cid. Compression ratio: 8.4:1. Net horsepower: 150 at 4400 rpm. Net torque: 245 lb.-ft. at 2500 rpm. Five main bearings. Hydraulic valve lifters. Carburetor: Autolite 2-barrel model 2100.
ENGINE (Standard J-4700, J-4800; Optional ($212): J-2500, J-2600, J-4500, J-4600): V-type. OHV. Eight-cylinder. Cast iron block. Bore & stroke: 4.08 x 3.44 in. Displacement: 360 cid. Compression ratio: 8.5:1. Net horsepower: 175 at 4000 rpm. Net torque: 285 lb.-ft. at 2400 rpm. Five main bearings. Hydraulic valve lifters. Carburetor: Autolite 2-barrel.

1972 AMC Jeep, J-100 Wagoneer station wagon (JAG)

CHASSIS (Wagoneer): Wheelbase: 110 in. Overall length: 183.66 in. Width: 75.6 in.
CHASSIS (Commando): Wheelbase: 104 in. Overall length: 174.5 in.
CHASSIS (CJ-5): Wheelbase: 84 in. Overall length: 138.9 in. Front tread: 51.5 in. Rear tread: 50 in. Tires: (w/232 cid six) 8.45 x 15; (w/258 cid six) H78 x 15; (with V-8) 6.00 x 16.
CHASSIS (CJ-6): Wheelbase: 104 in. Overall length: 158.9 in. Front tread: 51.5 in. Rear tread: 50 in. Tires: 7.35 x 15 in.
CHASSIS (DJ-5): Wheelbase: 84 in. Overall length: 138.9 in. Tires: 7.35 x 15 in.
CHASSIS (J-4500/J-4600/J-4700/J-4800): Wheelbase: 132 in. Overall length: 205.6 in. Front tread: 63.9 in. Rear tread: 64.4 in. Tires: (J-4500) 8.25 x 15 in.; (J-4600) 7.00 x 16 in.; (J-4700) 7.50 x 16 in.; (J-4800) 7.50 x 16 in.
CHASSIS (Commando): Wheelbase: 104 in. Overall length: 174.5 in. Front tread: 51.5 in. Rear tread: 50 in. Tires: 7.35 x 15 in.
CHASSIS (J-2500/J-2600): Wheelbase: 120 in. Overall length: 193.6 in. Front tread: 63.5 in. Rear tread: 63.8 in. Tires: (J-2500) 8.25 x 15 in.; (J-2600) 7.00 x 16 in.
TECHNICAL (J-Series): Manual, synchromesh transmission. Speeds: 3F/1R (J-4800 4F/1R). Column (floor J-4800) mounted gearshift. Dry plate clutch. (1/2-Ton Trucks, DJ-5, CJ-5, CJ-6) semi-floating rear axle; (all others) full-floating rear axle. Hydraulic drum (four-wheel) brakes. Pressed steel wheels. Technical Options: Three-speed automatic transmission. Four-speed manual transmission. Power steering. Power brakes. Trac-Lok differential. Heavy-duty cooling system. Heavy-duty alternator. Heavy-duty battery. Semi-automatic front hubs. Reserve fuel tank. Fuel tank skid plate. Heavy-duty snowplow. Winches.
TECHNICAL (CJ-5, CJ-6): Manual synchromesh transmission. Speeds: 3F/1R. Floor mounted gearshift. Dry plate clutch. Technical Options: Four-speed manual transmission. Power brakes. Power steering. Heavy-duty springs. Heavy-duty shock absorbers. Trac-Lok differential. Semi-automatic front hubs. Heavy-duty cooling system.
TECHNICAL (Commando Series): Manual synchromesh transmission. Speeds: 3F/1R. Floor mounted gearshift. Dry plate clutch. Technical Options: Three-speed automatic transmission (column shift). Four-speed (floor shift) manual transmission. Power brakes. Power steering. Heavy-duty springs. Heavy-duty shock absorbers. Heavy-duty alternator. Heavy-duty battery. Trac-Lok differential. Semi-automatic front hubs. Heavy-duty cooling system. Heavy-duty snowplow. Winches. Power take-off.
OPTIONS: Power steering ($151). Power brakes ($48). 327 cid V-8 engine ($220). Four-speed transmission ($112). Turbo-Hydramatic transmission ($280). Thriftside pickup on J2500 chassis (add $155). Townside pickup on 120 in. wheelbase chassis (add $135). Townside pickup on 132 in. wheelbase chassis (add $165). Gladiator camper on chassis and cab ($4218). Pickup body for Gladiator chassis and cab ($160).
HISTORICAL: Business was up over 30 percent in 1972, with retail sales of 51,621 Jeeps. Production was up to 71,255 units, a 34 percent boost. Deliveries for the season included 18,744 CJs, 9,115 Commandos, 14,524 Wagoneers and 9,238 trucks. For the first time, Jeep built more V-8s than other types of engines. The totals were: 134 cid four — 6,968; 232 cid L-6 9,021; 258 cid L-6 — 11,257; 304 cid V-8 — 17,696; 360 cid V-8 -18,215.

1973 JEEP

JEEP — 1973 SERIES — (ALL ENGINES): 1973 was the year that the full impact of American Motor's development plan was apparent in the design of Jeep vehicles. The J-series trucks had new double-wall side panels for the bed, a wider tailgate (operable with one hand) and a new mechanical clutch linkage that had a longer service life and required less maintenance. The interior was distinguished by a redesigned instrument panel with increased padding and easier-to-read gauges, including "direct-reading" oil and ammeter gauges. But, the big news for the J-trucks was the availability of the full-time Quadra-Trac four-wheel drive option. This system, offered on J-2500/2600/4500 and 4600 models, included the 360 cid V-8 and automatic transmission as standard equipment. An optional low range unit was available, either as a dealer-installed feature or as original equipment, direct from the factory. Quadra-Trac allowed all four wheels to operate at their own speeds, each receiving the proper portion of driving power. The key to this system was a limited-slip differential that transmitted power to the front and rear wheels.

1973 AMC Jeep, CJ-5 Universal Jeep (JAG)

The CJ Jeeps were given a new, more stylish instrument panel with a large center gauge encompassing the speedometer and the temperature and fuel gauges. Mounted to the left and right of this unit were the ammeter and oil pressure gauges. Beginning in January 1973, the Jeep "Renegade" was available. It had a standard 304 cid V-8, H78 x 16 tires on styled wheels, blacked-out hood, racing stripes, fender lip extensions, dual mirrors and visors, a custom vinyl interior, rear-mounted spare tire, plus transmission and fuel tank skid plates.

In its last year of production, the Commando was given standard upgraded tires and new axle joints.

1973 AMC Jeep, J-2500 Gladiator Townside pickup (CW)

I.D. DATA: VIN located on the left front door hinge pillar and left firewall. VIN still has 13 symbols. The first indicates Jeep Corp. The second indicates model year. The third indicates transmission, drivetrain and assembly plant. The fourth and fifth indicate series or model. The sixth symbol identifies body style. The seventh symbol indicates model type and GVW. The eighth symbol indicates the engine. Engine codes were: E=232 cid six-cylinder; A=258 cid six-cylinder; H=304 cid V-8; N=360 cid V-8; P=360 cid V-8. Z=401 cid V-8. The next five symbols are the sequential production number.

Model	Body Type	Price	Weight	Prod. Total
Jeep Wagoneer — (1/2-Ton) — (4x4)				
14	4dr Std Sta Wag	4501	3810	—
15	4dr Cus Sta Wag	4739	3850	—
Jeep Commando — (1/2-Ton) — (4x4)				
89	2dr Sta Wag	3506	3010	—
87	2dr Rds	3355	2800	—
CJ-5 — (1/4-Ton) — (4x4)				
83	Jeep	3086	2450	—
CJ-6 — (1/4-Ton) — (4x4)				
84	Jeep	3176	2510	—
DJ-5 — (1/4-Ton) — (4x2)				
85	Jeep	2606	2270	—
Commando — (1/2-Ton) — (4x4)				
88	Pickup	3382	2950	—
Series J-2500 — (1/2-Ton) — (4x4)				
25	Thriftside Pickup	3353	3570	—
25	Townside Pickup	3353	3715	—
Series J-2600 — (1/2-Ton) — (4x4)				
26	Chassis & Cab	3327	3395	—
26	Thriftside Pickup	3474	3690	—
26	Townside Pickup	3474	3835	—

Series J-4500 — (3/4-Ton) — (4x4)

45	Chassis & Cab	3235	3300	—
45	Townside Pickup	3390	3760	—

Series J-4600 — (3/4-Ton) — (4x4)

46	Chassis & Cab	3356	3435	—
46	Townside Pickup	3511	3895	—

Series J-4700 — (3/4-Ton) — (4x4)

47	Chassis & Cab	3723	3730	—

Series J-4800 — (3/4-Ton) — (4x4)

48	Chassis & Cab	4132	4015	—
48	Townside Pickup	4287	4475	—

ENGINE (Standard: Commando/CJ-5/CJ-6): Inline. OHV. Six-cylinder. Cast iron block. Bore & stroke: 3.75 x 3.5 in. Displacement: 232 cid. Compression ratio: 8.0:1. Brake horsepower: 145 at 4300 rpm. Torque: 215 lb.-ft. at 1600 rpm. Net horsepower: 100 at 3600 rpm. Torque: 185 lb.-ft. at 1800 rpm. Seven main bearings. Hydraulic valve lifters. Carburetor: Single Carter one-barrel model YF.

ENGINE (Standard: Cherokee and J-2500/J-4500/J-4600. Optional: Commando/J-4800): Inline. Six-cylinder. Cast iron block. Bore & stroke: 3.75 x 3.9 in. Displacement: 258 cid. Compression ratio: 8.0:1. Net horsepower: 110 at 3500 rpm. Net torque: 195 lb.-ft. at 2000 rpm. Seven main bearings. Hydraulic valve lifters. Carburetor: Single Carter one-barrel model YF.

1973 AMC Jeep, J-2600 Gladiator Townside pickup (JAG)

ENGINE (Optional: Commando/CJ-5/CJ-6; Standard: CJ-5/Renegade): V-type. OHV. Eight-cylinder. Cast iron block. Bore & stroke: 3.75 x 3.44 in. Displacement: 304 cid. Compression ratio: 8.4:1. Net horsepower: 150 at 4200 rpm. Net torque: 245 lb.-ft. at 2500 rpm. Five main bearings. Hydraulic valve lifters. Carburetor: Single two-barrel.

ENGINE (Standard: Wagoneer. Optional: J-2500/J-2600/J-4500/J-4700/J-4800): V-type. OHV. Eight-cylinder. Cast iron block. Bore & stroke: 4.08 x 3.44 in. Displacement: 360 cid. Compression ratio: 8.5:1. Net horsepower: 175 at 4000 rpm. Net torque: 285 lb.-ft. at 2400 rpm. Five main bearings. Hydraulic valve lifters. Carburetor: Single two-barrel.

ENGINE (Optional: J-2500/J-2600/J-4500/J-4700/J-4800): V-type. OHV. Eight-cylinder. Cast iron block. Bore & stroke: 4.08 x 3.44 in. Displacement: 360 cid. Compression ratio: 8.5:1. Net horsepower: 195 at 4400 rpm. Net torque: 295 lb.-ft. at 2900 rpm. Five main bearings. Hydraulic valve lifters. Carburetor: Single four-barrel.

1973 AMC Jeep, Super Wagoneer station wagon (JAG)

CHASSIS (Wagoneer): Wheelbase: 110 in. Overall length: 183.66 in. Width: 75.6 in.

CHASSIS (Commando): Wheelbase: 104 in. Overall length: 174.5 in. Front tread: 51.5 in. Rear tread: 50.0 in. Tires: F78-15B in.

CHASSIS (J-2500): Wheelbase: 120 in. Overall length: 193.6 in. Front tread: 63.5 in. Rear tread: 63.8 in. Tires: F78-15 in.

CHASSIS (J-2600): Wheelbase: 120 in. Overall length: 193.6 in. Front tread: 63.5 in. Rear tread: 63.8 in. Tires: 7.00 x 16 in.

CHASSIS (J-4500): Wheelbase: 132 in. Overall length: 205.6 in. Front tread: 63.9 in. Rear tread: 64.4 in. Tires: F78-15 in.

CHASSIS (J-4600): Wheelbase: 132 in. Overall length: 205.6 in. Front tread: 63.9 in. Rear tread: 64.4 in. Tires: 7.00 x 16 in.

CHASSIS (J-4700): Wheelbase: 132 in. Overall length: 205.6 in. Front tread: 63.9 in. Rear tread: 64.4 in. Tires: 7.50 x 16 in.

CHASSIS (J-4800): Wheelbase: 132 in. Overall length: 205.6 in. Front tread: 63.9 in. Rear tread: 64.4 in. Tires: 7.50 x 16 in.

CHASSIS (CJ-5): Wheelbase: 84 in. Overall length: 138.9 in. Front tread: 51.5 in. Rear tread: 50.0 in. Tires: F78-15B, four-ply.

CHASSIS (CJ-6): Wheelbase: 104 in. Overall length: 158.9 in. Front tread: 51.5 in. Rear tread: 50.0 in. Tires: F78-15B, four-ply.

1973 AMC Jeep, Wagoneer station wagon (AMC)

TECHNICAL (CJ-5, CJ-6): Manual synchromesh transmission. Speeds: 3F/1R. Floor-mounted gearshift. Semi-floating rear axle. Hydraulic drum brakes. Pressed steel wheels. Technical options: Four-speed manual transmission. Power brakes. Power steering. Heavy-duty springs. Heavy-duty shock. Variety of tire sizes. Trac-Lok differential. Semi-automatic front hubs. Heavy-duty cooling system.

TECHNICAL (Commando): Synchronized, manual transmission. Speeds: 3F/1R. Floor-mounted gearshift. Semi-floating rear axle. Technical options: Automatic Transmission. Four-speed manual transmission. Power brakes. Power steering. Heavy-duty springs. Heavy-duty shock absorbers. Heavy-duty alternator. Heavy-duty battery. Trac-Lok differential. Semi-automatic front hubs. Heavy-duty cooling system. Power take-off. Reserve fuel tank (J-Series).

TECHNICAL (J-2500/J-2600/J-4500/J-4600/J-4700/J-4800): Synchronized, manual transmission. Speeds: 3F/1R: 4F/1R. Column-mounted gearshift (J-4800: floor-mounted). Technical options: Automatic Transmission. Power brakes. Power steering. Four-speed manual transmission. Trac-Lok differential. Heavy-duty cooling system. Heavy-duty alternator. Heavy-duty battery. Power-take-off. Semi-automatic hubs.

OPTIONS: Chrome front bumper (CJ). Chrome rear bumper (CJ). AM Radio. Electric clock. Cigar lighter. Wheel covers. Full-width split front seat (Commando). Tinted glass. Air conditioning. Special Decor Group (Commando). Bucket seats with center armrest (J-Series). West Coast Mirrors (J-Series). Courtesy lights. Custom Decor Group (J-Series). Outside passenger side mirror. Dual horns (J-Series). Tonneau cover (J-Series). Rear step bumper (J-Series). Two-tone paint (J-Series). Woodgrain trim (J-Series). Safari top (Commando Roadster/CJ). Meta top (CJ). Fabric top (CJ). Front bucket seats (CJ). Rear bench seat (Commando). Power brakes ($45). 304 cid V-8 ($126-165). 258 cid six ($45). Custom Decor package ($179). Four-speed manual transmission ($107).

HISTORICAL: Introduced: Fall, 1972. Calendar year sales: (Jeep sales) 68,430. Calendar year production: (all models) 94,035.

1974 JEEP

JEEP — 1974 SERIES — (ALL ENGINES): For 1974, the Cherokee was introduced into the sports-utility field. It was a two-door station wagon with a Gladiator truck type grille. Wagoneers received a new front end treatment with the parking lamps in the grille. The Jeepster Commando line was discontinued. Renegades, which were previously available as a limited edition option, were made a regular production CJ model. In addition to the 304 cid V-8, they had special paint and graphics, a rear-mounted spare, roll bar, passenger assist rail, dual sun visors, oil and amp gauges and styled aluminum wheel rims. The Quadra-Trac four-wheel-drive system was offered on all J-series trucks with either six- or eight-cylinder engines. Other improvements included larger brakes and a shorter turning radius for the J-trucks.

1974 AMC Jeep, CJ-5 Renegade Universal Jeep (AMC)

I.D. DATA: VIN located on the left front door hinge pillar and left firewall. VIN still has 13 symbols. The first indicates Jeep Corp. The second indicates model year. The third indicates transmission, drivetrain and assembly plant. The fourth and fifth indicate series or model. The sixth symbol identifies body style. The seventh symbol indicates model type and GVW. The eigth symbol indicates the engine. Engine codes were: E=232 cid six-cylinder; A=258 cid six-cylinder; H=304 cid V-8; N=360 cid V-8; P=360 cid V-8. Z=401 cid V-8. The next five symbols are the sequential production number.

1974 AMC Jeep, J-10 Townside pickup (CW)

Model	Body Type	Price	Weight	Prod. Total
Wagoneer — (1/2-Ton) — (4x4)				
14	4dr Sta Wag	5406	4270	—
15	4dr Cus Sta Wag	5704	4290	—
Cherokee — (1/2-Ton) — (4x4)				
16	2dr Sta Wag	4161	3870	—
17	2dr "S" Sta Wag	4724	3870	—
CJ-5 — (1/4-Ton) — (4x4)				
83	Jeep	3574	2540	—
CJ-6 — (1/4-Ton) — (4x4)				
84	Jeep	3670	2600	—
J-10 — (1/2-Ton) — (4x4)				
25	Townside Pickup (SWB)	3776	3770	—
45	Townside Pickup (LWB)	3837	3820	—
J-20 — (3/4-Ton) — (4x4)				
46	Townside Pickup (LWB)	4375	4390	—

1974 AMC Jeep, Cherokee 'S' two-door station wagon (OCW)

ENGINE (Standard: Cherokee and J-10; Optional: CJ-5/CJ-6): Inline. OHV. Six-cylinder. Cast iron block. Bore & stroke: 3.85 x 3.90 in. Displacement: 258 cid. Compression ratio: 8.0:1. Net horsepower: 110 at 3500 rpm. Torque: 195 lb.-ft. at 2000 rpm. Seven main bearings. Hydraulic valve lifters. Carburetor: Carter one-barrel model YF.

ENGINE (Standard: Wagoneer and J-20; Optional: J-10): V-type. OHV. Eight-cylinder. Cast iron block. Bore & stroke: 4.08 x 3.44 in. Displacement: 360 cid. Compression ratio: 8.5:1. Net horsepower: 175 at 4000 rpm. Net torque: 285 lb.-ft. at 2400 rpm. Five main bearings. Hydraulic valve lifters. Carburetor: Single two-barrel.
ENGINE (Optional: J-10/J-20): V-type. OHV. Eight-cylinder. Cast iron block. Bore & stroke: 4.17 x 3.68 in. Displacement: 401 cid. Net horsepower: 235 at 4600 rpm. Five main bearings. Hydraulic valve lifters. Carburetor: Four-barrel.
ENGINE (Standard: CJ-5/CJ-6): Inline. OHV. Six-cylinder. Cast iron block. Bore & stroke: 3.75 x 3.50 in. Displacement: 232 cid. Compression ratio: 8.0:1. Net horsepower: 100 at 3600 rpm. Net torque: 185 lb.-ft. at 1800 rpm. Seven main bearings. Hydraulic valve lifters. Carburetor: Carter one-barrel model YF.
ENGINE (Optional: CJ-5/CJ-6; Standard: "Renegade"): V-type. OHV. Eight-cylinder. Cast iron block. Bore & stroke: 3.75 x 3.44 in. Displacement: 304 cid. Compression ratio: 8.4:1. Net horsepower: 150 at 4200 rpm. Net torque: 245 lb.-ft. at 2500 rpm. Five main bearings. Hydraulic valve lifters. Carburetor: Autolite two-barrel model 2100.

1974 AMC Jeep, J-2600 Pioneer Townside pickup (AMC)

CHASSIS (Cherokee): Wheelbase: 109 in. Overall length: 183.7 in. Tires: F78-15.
CHASSIS (Wagoneer): Wheelbase: 109 in. Overall length: 183.7 in. Tires: F78-15.
CHASSIS (CJ-5): Wheelbase: 84 in. Overall length: 138.9 in. Front tread: 51.5 in. Rear tread: 50.0 in. Tires: F78-15, four-ply.
CHASSIS (CJ-6): Wheelbase: 104 in. Overall length: 158.9 in. Front tread: 51.5 in. Rear tread: 50.0 in. Tires: F78-15, four-ply.
CHASSIS (J-10): Wheelbase: 118.7/130.7 in. Overall length: 192.5/204.5 in. Height: 69.3/69.1 in. Front tread: 63.3 in. Rear tread: 63.8 in. Tires: G78-15-15.
CHASSIS (J-20): Wheelbase: 130.7 in. Overall length: 204.5 in. Height: 70.7 in. Front tread: 63.3 in. Rear tread: 63.8 in. Tires: 8.00 x 16.5D in.
TECHNICAL (CJ-5, CJ-6, J-10, J-20): Same as 1973, except new option: Quadra-Trac full-time four-wheel-drive for J-10 and J-20 models.
OPTIONS: 258 cid six-cylinder engine in CJ-5/CJ-6 ($54). 360 cid V-8 engine in J-10 ($201). 401 cid V-8 engine, in Wagoneer ($94); in J-10 ($295); in J-20 ($94). 304 cid V-8 engine in CJ-5/CJ-6, except Renegade ($126). Four-speed transmission in CJs ($107). Renegade option package ($701). Wagoneer woodgrain trim ($179). CJ metal cab ($425). Power disc brakes for Cherokee ($65). Power brakes for CJ ($45).

1974 AMC Jeep Custom Wagoneer station wagon (AMC)

HISTORICAL: Introduced: Fall, 1973. Calendar year sales: (all Jeep models) 67,110. Calendar year production: (all models) 92,283. Model year production: 96,763. Business was up 16.6 percent. Model year output included 15,350 Wagoneers; 43,137 CJ-5s; 2,826 CJ-6s; 18,277 Cherokees; and 13,874 Jeep trucks. Jeep Corp. operated as a subsidiary of American Motors Corp., with its headquarters in Detroit. The main Jeep factory was located in Toledo, Ohio. C. Mashigan was director of design. Now famous automotive historian John A. Conde was manager of public relations. The Jeep Cherokee won the Sports Car Club of America's Pro-Rally series in 1974. There was a slight rise in popularity of six-cylinder powered Jeep vehicles in 1974, up to 49 percent of production (versus 37.8 percent in 1973).

1975 JEEP

JEEP — 1975 SERIES — (ALL ENGINES): Cherokees received sporty new trim options for 1975. At midyear, the Cherokee Chief was introduced at the Detroit Auto Show. It had a wider-than-normal track, fat tires and specific trim. The special graphics included model callouts on the rocker panels. New-for-1975 Cherokee options included Cruise Command speed control, AM/FM quadraphonic sound system and a rear window defogger. Wagoneers got electronic ignition, a new power steering system and suspension revisions. Leading the list of appearance changes for the 1975 CJ models was the availability of the Levi's vinyl front bucket and rear bench seat feature. This was standard for the Renegade model with optional for the CJ-5. The Renegade also had a newstripe format. Added to the CJ option list was a factory-installed AM radio with a weatherproof case and a fixed length whip-type antenna. The full soft top was of a new design with improved visibility and larger door openings.

The J-Series pickup trucks were available with a new Pioneer trim package consisting of woodgrain exterior trim, deep pile carpeting, pleated fabric seats, chrome front bumpers, bright exterior window moldings, deluxe door trim pads, bright wheelcovers (J-10), bright hub caps (J-20), dual horns, locking glove box, cigar lighter, woodgrain instrument cluster trim and bright armrest overlays.

1975 AMC Jeep, Cherokee 'S' station wagon (AMC)

1975 AMC Jeep, CJ-5 Levis Renegade (AMC)

I.D. DATA: VIN located on the left front door hinge pillar and left firewall. VIN still has 13 symbols. The first indicates Jeep Corp. The second indicates model year. The third indicates transmission, drivetrain and assembly plant. The fourth and fifth indicate series or model. The sixth symbol identifies body style. The seventh symbol now indicates the engine. Engine codes were: E=232 cid six-cylinder; A=258 cid six-cylinder; H=304 cid V-8; N=360 cid V-8; P=360 cid V-8. Z=401 cid V-8. The next six symbols are the sequential production number beginning at 000001.

ENGINE (Standard: Wagoneer and J-20; Optional: J-10): V-type. OHV. Eight-cylinder. Cast iron block. Bore & stroke: 4.08 x 3.44 in. Displacement: 360 cid. Compression ratio: 8.5:1. Net horsepower: 175 at 4000 rpm. Net torque: 285 lb.-ft. at 2400 rpm. Five main bearings. Hydraulic valve lifters. Carburetor: Single two-barrel.
ENGINE (Optional: J-10/J-20): V-type. OHV. Eight-cylinder. Cast iron block. Bore & stroke: 4.08 x 3.44 in. Displacement: 360 cid. Net horsepower: 195 at 4400 rpm. Five main bearings. Hydraulic valve lifters. Carburetor: Single four-barrel.
ENGINE (Optional: J-10/J-20): V-type. OHV. Eight-cylinder. Cast iron block. Bore & stroke: 4.17 x 3.68 in. Displacement: 401 cid. Net horsepower: 235 at 4600 rpm. Five main bearings. Hydraulic valve lifters. Carburetor: Single four-barrel.
ENGINE (Standard: CJ-5/CJ-6): Inline. OHV. Six-cylinder. Cast iron block. Bore & stroke: 3.75 x 3.50 in. Displacement: 232 cid. Compression ratio: 8.0:1. Net horsepower: 100 at 3600 rpm. Net torque: 185 lb.-ft. at 1800 rpm. Seven main bearings. Hydraulic valve lifters. Carburetor: Carter one-barrel model YF.
ENGINE (Standard: Renegade; Optional CJ-5/CJ-6): V-type. OHV. Eight-cylinder. Cast iron block. Bore & stroke: 3.75 x 3.44 in. Displacement: 304 cid. Compression ratio: 8.4:1. Net horsepower: 150 at 4200 rpm. Net torque: 245 lb.-ft. at 2500 rpm. Seven main bearings. Hydraulic valve lifters. Carburetor: Autolite two-barrel model 2100.
CHASSIS (Wagoneer): Wheelbase: 109 in. Overall length: 183.7 in. Tires: F78-15.
CHASSIS (Cherokee): Wheelbase: 109 in. Overall length: 183.7 in. Tires: F78-15.
CHASSIS (CJ-5): Wheelbase: 84 in. Overall length: 138.9 in. Height: 69.5 in. Front tread: 51.5 in. Rear tread: 50.0 in. Tires: F78-15B in.
CHASSIS (CJ-6): Wheelbase: 104 in. Overall length: 158.9 in. Height: 68.3 in. Front tread: 51.5 in. Rear tread: 50.0 in. Tires: F78-15B in.

1975 AMC Jeep, J-20 Townside Pioneer pickup (AMC)

1975 AMC Jeep Cherokee Chief station wagon (AMC)

Model	Body Type	Price	Weight	Prod. Total
Wagoneer — (1/2-Ton) — (4x4)				
14	4dr Std Sta Wag	6013	4240	—
15	4dr Cus Sta Wag	6246	4256	—
Cherokee — (1/2-Ton) — (4x4)				
16	2dr Sta Wag	4851	3657	—
17	2dr "S" Sta Wag	5399	3677	—
CJ-5 — (1/4-Ton) — (4x4) — (84 in. w.b.)				
83	Jeep	4099	2648	32,486
CJ-6 — (1/4-Ton) — (4x4) — (104 in. w.b.)				
84	Jeep	4195	2714	2,935
J-10 — (1/2-Ton) — (4x4) — (119/131 in. w.b.)				
25	Townside Pickup (SWB)	4228	3712	2,258
45	Townside Pickup (LWB)	4289	3770	4,721
J-20 — (3/4-Ton) — (4x4) — (131 in. w.b.)				
46	Townside Pickup	4925	4333	2,977

NOTE: See "Historical" for additional production data.

ENGINE (Standard: Cherokee and J-10; Optional: CJ-5/CJ-6): Inline. OHV. Six-cylinder. Cast iron block. Bore & stroke: 3.75 x 3.9 in. Displacement: 258 cid. Compression ratio: 8.0:1. Net horsepower: 110 at 3500 rpm. Torque: 195 lb.-ft. at 2000 rpm. Seven main bearings. Hydraulic valve lifters. Carburetor: Single Carter one-barrel. (Note: This engine was not available in California where a four-barrel version of the 360 cid V-8 was standard.)

CHASSIS (J-10 SWB): Wheelbase: 119.0 in. Overall length: 193.6 in. Height: 65.9 in. Front tread: 63.1 in. Rear tread: 64.9 in. Tires: H78-15B in.
CHASSIS (J-10/J-20 LWB): Wheelbase: 131.0 in. Overall length: 205.6 in. Height: 71.3 in. Front tread: 64.8 in. Rear tread: 66.1 in. Tires: (J-10) H78-15B; (J-20/6500-lb. GVW) 8.00 x 16.5; (J-20/7200-lb. GVW) 8.75 x 16.5 in.
TECHNICAL (CJ-5/CJ-6): Manual, synchromesh transmission. Speeds: 3F/1R. Floor-mounted gearshift. Single dry disc clutch. Semi-floating rear axle. Overall ratio: 3.73:1, (optional) 4.27:1. Manual four-wheel hydraulic drum brakes. Pressed steel, five-bolt design wheels. Technical options: Warn locking hubs. Winches (mechanical or electric). Four-speed manual transmission (with 258 cid, one-barrel engine only). Heavy-duty cooling system. Heavy-duty springs and shock absorbers (front and rear). Rear Trac-Lok differential. 70-amp battery. 62-amp alternator. Cold climate group. Draw bar. Helper springs.
TECHNICAL (Cherokee/Wagoneer/J-10/J-20): Fully synchronized. Speeds: 3F/1R. Floor-mounted gearshift. Clutch: 10.5 in., 106.75 sq. in. area, (J-20) 11.0 in., 110.96 sq. in. area. (J-10) Semi-floating rear axle; (J-20) full-floating rear axle. Overall ratio: (J-10) 3.54, 4.09:1; (J-20) 3.73:1, 4.09:1 optional. Brakes: (J-10) Hydraulic, 11 in. x 2 in.; (J-20) front power disc, 12.5 in. Wheels: (J-10) 6-bolt; (J-20) 8-bolt steel disc. Technical options: Automatic transmission Turbo Hydra-Matic. Front power disc 12.0 in. brakes (J-10). Auxiliary fuel tank. Power steering. Quadra-Trac. Rear Trak-Lok Differential (not available with Quadra-Trac). Camper Special Package. Heavy-duty battery and alternator. Four-speed transmission. Heavy-duty cooling system. Cold climate group. Heavy-duty springs and shock absorbers. Helper springs.
OPTIONS: (Wagoneer/Cherokee/CJ-5) Rear Bumperettes. Radio AM, Citizen's Band. Cigar lighter and ashtray. Swing-out tire carrier. Roll bar. Power steering, except standard in Wagoneer ($169). Four-speed transmission ($129). Power brakes. Rear seat. Renegade package ($725). Forged aluminum wheels (standard with Renegade Package). Padded instrument panel (standard with Renegade package). Passenger safety rail. Air conditioning ($421). 304 cid V-8 in CJ ($126). 360 cid V-8 engine in Cherokee and J-10 ($201). 360 cid four-barrel V-8 engine in J-10 ($295); in J-20 ($94). Tachometer. Metal top (full or half). Push bumper. Outside passenger mirror. Woodgrain wagon ($108).

OPTIONS (J-10/J-20): Rear bumper. Rear step bumper. Bumper guards with nerf strips. Radio AM, AM/FM stereo. Air conditioning. Sliding rear window. Cruise control. Power steering. Sport steering wheel. Leather-wrapped steering wheel. Tilt steering wheel. Steel-belted radial tires (not available for J-20). Tinted glass. Custom trim package. Light group. Convenience group. Wheel hub caps (J-20). Wheel covers (J-10). Outside passenger mirror. Two-tone paint. Aluminum cargo cap. CJ metal cab ($430).

1975 AMC-Jeep Wagoneer station wagon (AMC)

HISTORICAL: Introduced: Fall, 1974. Calendar year sales: (Jeep sales) 69,834. Calendar year production: (all Jeep vehicles) 105,833, (Worldwide wholesale sales) 104,936. Innovations: Introduction of Levi's seat trim. New calendar year production mark of 110,844. Model year production totaled 67,780 Jeep vehicles. This included 32,486 CJ-5s, 2,935 CJ-6s, 12,925 Cherokees, 9,296 Wagoneers, 2,258 J-2500s, 182 J-2600s, 4,721 J-4500s and 2,977 J-4600s. Use of electronic ignition system on J-Series models also started this year. In the CJ-5 series, 60.3 percent of the Jeeps had six-cylinder engines; 27 percent had AM radios and 39.7 percent had the 360 cid V-8. In the Cherokee series 71.6 percent had automatic transmission; 81 percent had power disc brakes; 73.9 percent had 360 cid or 401 cid V-8s and 14.7 percent had power steering. In the Wagoneer line, 66.8 percent had wheelcovers, 48.6 percent had an adjustable steering and 34 percent had an AM/FM stereo.

1976 JEEP

1976 AMC Jeep, CJ-5 Renegade Universal Jeep (JAG)

JEEP — 1976 SERIES — (ALL ENGINES): The major story from Jeep for 1976 was the new CJ-7 model. It was described as "the most exciting vehicle to hit the four-wheel drive market in years." With a 93.5 in. wheelbase, the CJ-7 offered more front and rear legroom, more cargo space and larger, 33.8 in. wide door openings. It was available with a one-piece removable hardtop, steel side doors and a rear liftgate. A soft top was another option. The CJ-7 made Jeep history by being the first CJ model available with an automatic transmission and Quadra-Trac. The smaller Jeep CJ-5 was also available.

Wagoneers featured a new, more rugged frame that facilitated direct steering gear mounting. The woodgrained side trim package was of a new, narrower design. The Cherokee had a new design that made extensive use of box-section side rail construction. The Cherokee Chief, with standard power disc brakes, power steering and fuel tank skid plate was a regular production model this season.

1976 AMC Jeep, CJ-7 Universal Jeep (JAG)

The Renegade package, again available for both CJ models, was upgraded to include courtesy lights under the dash, an eight-in. day/night mirror, sports steering wheel, instrument panel overlay and bright rocker panel protection molding between the front and rear wheel wells. The CJ frame was also upgraded with splayed side rails (wider in front than back) to allow for wider spacing of the rear springs. In addition, the frame had stronger cross members and an integral skid plate. Other revisions included use of longer and wider multi-springs, new shock absorbers and new hold-down mounts.

The J-series trucks also had a new frame with splayed rear side rails, hold-down mounts, springs and shocks. Its crossmembers and box section side rails were also of stronger construction. All J-trucks were equipped with an improved windshield washer system having two spray nozzles.

In January 1976 the "Honcho" package for J-10 short wheelbase models was introduced. Its primary features included five 10 x 15 in. Tracker A-T Goodyear tires with raised white lettering mounted on eight-inch slotted wheels; gold striping on the side, tailgate and fenders (with black and white accents); blue Levi's denim interior; rear step bumper; and special blue Sports steering wheel. The Honcho was offered in a choice of six exterior colors.

1976 AMC Jeep, CJ-5 Renegade Universal Jeep

I.D. DATA: VIN located on the left front door hinge pillar and left firewall. VIN still has 13 symbols. The first indicates Jeep Corp. The second indicates model year. The third indicates transmission, drivetrain and assembly plant. The fourth and fifth indicate series or model. The sixth symbol identifies body style. The seventh symbol indicates the engine. Engine codes were: E=232 cid six-cylinder; A=258 cid six-cylinder; H=304 cid V-8; N=360 cid V-8; P=360 cid V-8. Z=401 cid V-8. The next six symbols are the sequential production number beginning at 000001.

Model	Body Type	Price	Weight	Prod. Total
Wagoneer — (1/2-Ton) — (4x4)				
14	4dr Std Sta Wag	6339	4329	—
15	4dr Cus Sta Wag	6572	4345	—
Cherokee — (1/2-Ton) — (4x4)				
16	2dr Sta Wag	5258	3918	—
17	2dr Sta Wag	5806	3938	—
CJ-5 — (1/4-Ton) — (4x4) — (84 in. w.b.)				
83	Jeep	4199	2641	31,116
CJ-7 — (1/4-Ton) — (4x4) — (94 in. w.b.)				
93	Jeep	4299	2683	21,016
J-10 — (1/2-Ton) — (4x4) — (119/131 in. w.b.)				
25	Townside Pickup (SWB)	4643	3773	
45	Townside Pickup (LWB)	4704	3873	
J-20 — (3/4-Ton) — (4x4) — (131 in. w.b.)				
46	Townside Pickup	5290	4285	3,210

NOTE: See "Historical" for additional production data.

1976 AMC Jeep, CJ-7 Universal Jeep with hardtop (JAG)

ENGINE (Standard: Cherokee and J-10; Optional: CJ-5/CJ-6): Inline. OHV. Six-cylinder. Cast iron block. Bore & stroke: 3.75 x 3.9 in. Displacement: 258 cid. Compression ratio: 8.0:1. Net horsepower: 110 at 3500 rpm. Torque: 195 lb.-ft. at 2000 rpm. Seven main bearings. Hydraulic valve lifters. Carburetor: Single Carter one-barrel. (Note: This engine was not available in California where a four-barrel version of the 360 cid V-8 was standard.)
ENGINE (Standard: Wagoneer and J-20; Optional: J-10): V-type. OHV. Eight-cylinder. Cast iron block. Bore & stroke: 4.08 x 3.44 in. Displacement: 360 cid. Compression ratio: 8.5:1. Net horsepower: 175 at 4000 rpm. Net torque: 285 lb.-ft. at 2400 rpm. Five main bearings. Hydraulic valve lifters. Carburetor: Single two-barrel.

1976 AMC Jeep, J-10 Townside pickup (AMC)

ENGINE (Optional: J-10/J-20): V-type. OHV. Eight-cylinder. Cast iron block. Bore & stroke: 4.08 x 3.44 in. Displacement: 360 cid. Net horsepower: 195 at 4400 rpm. Five main bearings. Hydraulic valve lifters. Carburetor: Single four-barrel.
ENGINE (Optional: J-10/J-20): V-type. OHV. Eight-cylinder. Cast iron block. Bore & stroke: 4.17 x 3.68 in. Displacement: 401 cid. Net horsepower: 235 at 4600 rpm. Five main bearings. Hydraulic valve lifters. Carburetor: Single four-barrel.
ENGINE (Standard: CJ-5/CJ-6): Inline. OHV. Six-cylinder. Cast iron block. Bore & stroke: 3.75 x 3.50 in. Displacement: 232 cid. Compression ratio: 8.0:1. Net horsepower: 100 at 3600 rpm. Net torque: 185 lb.-ft. at 1800 rpm. Seven main bearings. Hydraulic valve lifters. Carburetor: Carter one-barrel model YF.
ENGINE (Standard: Renegade; Optional CJ-5/CJ-6): V-type. OHV. Eight-cylinder. Cast iron block. Bore & stroke: 3.75 x 3.44 in. Displacement: 304 cid. Compression ratio: 8.4:1. Net horsepower: 150 at 4200 rpm. Net torque: 245 lb.-ft. at 2500 rpm. Seven main bearings. Hydraulic valve lifters. Carburetor: Autolite two-barrel model 2100.

1976 AMC Jeep, Honcho pickup (OCW)

CHASSIS (Wagoneer): Wheelbase: 109 in. Overall length: 183.7 in. Tires: F78-15.
CHASSIS (Cherokee): Wheelbase: 109 in. Overall length: 183.7 in. Tires: F78-15.
CHASSIS (CJ-5): Wheelbase: 83.5 in. Overall length: 138.5 in. Height: 67.6 in. Front tread: 51.5 in. Rear tread: 50.0 in. Tires: F78-15B in.

CHASSIS (CJ-7): Wheelbase: 93.5 in. Overall length: 147.9 in. Height: 67.6 in. Front tread: 51.5 in. Rear tread: 50.0 in. Tires: F78-15B in.
CHASSIS (J-10): Wheelbase: 118.7 in. Overall length: 192.5 in. Front tread: (J-10) 63.3 in., (Honcho) 65.4 in. Rear tread: (J-10) 63.8 in., (Honcho) 65.8 in. Tires: H78-15B in.
CHASSIS (J-10/J-20 LWB): Wheelbase: 130.7 in. Overall length: 204.5 in. Front tread: 64.6 in. Rear tread: 65.9 in. Tires: (J-10) H78-15B; (J-20/6500-lb. GVW) 8.00 x 16.5; (J-20/7200-lb. GVW) 9.50 x 16.5 in.

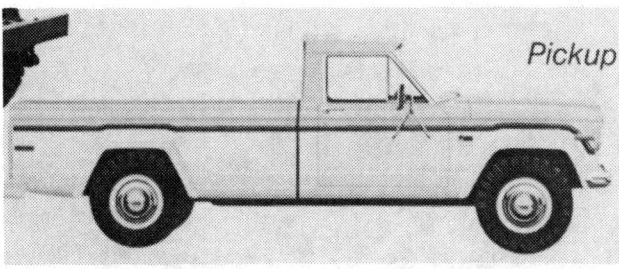

1976 AMC Jeep, J-10 Townside pickup (JAG)

TECHNICAL (Wagoneer/Cherokee/J-10/J-20): Manual, synchromesh transmission. Speeds: 3F/1R. Floor-mounted gearshift. Clutch: (J-10) 10.5 in. diameter, 106.75 sq. in., (J-20) 11.0 in. diameter, 110.96 sq. in. Rear axle: (J-10) Semi-floating; (J-20) full-floating. Overall ratio: (J-10) 3.54:1 or 4.09:1; (J-20) 3.73:1. (J-10) Four-wheel hydraulic drum brakes. (J-20) Front disc/rear drum brakes with power assist. Pressed steel wheels. Technical Options: Automatic transmission Turbo Hydra-Matic. Four-speed manual transmission. Quadra-Trac (available with automatic transmission only). 12-in. power front disc brakes (J-10). Snow Boss plow package. Winches (mechanical or 12-volt). Auxiliary fuel tank (20 gallon, for long wheelbase only, not available for California). Rear Trac-Lok differential. Camper special package. Fuel tank skid plate. Heavy-duty battery and alternator. Heavy-duty shock absorbers. Heavy-duty springs. Heavy-duty cooling system. Cold climate package. Locking hubs (not available with Quadra-Trac). Trailer Towing Package.

1976 AMC Jeep, Cherokee station wagon (JAG)

TECHNICAL (CJ-Series): Manual, synchromesh transmission. Speeds: 3F/1R. Floor-mounted gearshift. Single dry disc clutch. Semi-floating rear axle. Overall ratio: 3.54:1. Manual, hydraulic brakes. Pressed steel, five-bolt wheels. Technical Options: Automatic Transmission. Turbo Hydro-Matic transmission. Quadra-Trac (available with automatic transmission only). Four-speed manual transmission. Extra-duty suspension system. Front stabilizer bar. Power drum brakes (available with "304" V-8 only). Low range for Quadra-Trac (available for CJ-7 only). Heavy-duty cooling system. Rear Trac-Lok differential. 70-amp battery. Heavy-duty alternator. Cold climate group. Steel belted radial ply tires. Winches. Snowplow. Free-running front hubs.

1976 AMC Jeep, Wagoneer station wagon (JAG)

OPTIONS (Wagoneer/Cherokee/Trucks): 401 cid V-8 ($94). Power steering, except standard in Wagoneer ($179). Turbo-Hydramatic with Quadra-Trac, except standard on Wagoneer ($325). Four-speed transmission in Cherokee ($136). Luggage rack, Wagoneer ($80). Factory air conditioning ($480). Front bumper guards with Nerf strips. Rear step bumper. Convenience group. Light group. Custom trim package. Radios: AM/FM stereo ($209); Citizens Band; AM. Forged aluminum styled wheels (J-10 only). Bucket seats (Custom, Pioneer and Honcho only). Cruise control. Power steering. Tilt steering wheel. Sports steering wheel. Leather wrapped steering wheel. Sliding rear window. Dual low profile mirror. Aluminum cargo cap. Wheelcovers. Honcho package. Pioneer package.

1976 AMC Jeep, Custom Wagoneer station wagon (JAG)

OPTIONS (CJ-Series): Rear step bumper. Roll bar. Full soft top. Carpeting. Convenience group. Decor group. Radios: AM or Citizens Band. Sports steering wheel. Leather wrapped sports steering wheel. Rear seat. Swing-out tire carrier. Passenger grab rail. Tachometer and Rally clock. Padded instrument panel. Removable top, injection-molded (CJ-7 only). Outside passenger mirror. Renegade package.

1976 AMC Jeep, J-20 Townside pickup (131 in. w.b.)

HISTORICAL: Introduced: Fall, 1975. Calendar year sales: (all models and series) 95,506. Calendar year production: 126,125. Worldwide wholesale sales: 125,879. Model year production totaled 102,450 Jeep vehicles. This included 16,520 Wagoneers; 18,859 Cherokees; 21,016 CJ-7s; 2,431 CJ-6s (export only); 31,116 CJ-5s; 9,298 J-10s and 3,210 J-20s.

1976 AMC Jeep, Cherokee Chief two-door wagon (rear view)

1977 JEEP

1977 AMC Jeep, Honcho Townside pickup (JAG)

JEEP — 1977 SERIES — (ALL ENGINES): This was a year of changes for Jeep CJs. Probably most exciting was the midyear introduction of the "Golden Eagle" trim package for the CJ-5 and CJ-7. It included a roll bar, white-letter off-road tires and a tachometer with either the inline six or V-8. Worthwhile and noteworthy technical improvements were made. For the first time the CJ models were available with factory air conditioning. Also debuting, as a CJ option, were power front disc brakes. To faciliate their use, the CJ Jeeps had stronger front axles and wheel spindles, as well as wider wheels and tires. Also revised was the frame. It now had fully boxed side rails. All rear panels were also strengthened. The optional four-speed manual transmission had a new 6.32:1 low gear ratio. The popular Renegade package for the CJ-5 and CJ-7 models was continued. It now included new 9.00 x 15 Goodyear "Tracker" tires. More all-new options were a center console and tilt steering.

A four-door wagon was added to the Cherokee sports/utility series, which gained power front disc brakes and the 258 cid "Big Six" as standard equipment. The Cherokee two-door wagon continued and was now offered with a sporty Chief trim package and in standard and wide-stance versions. The latter featured fat raised-white-letter tires.

The most significant change made to the J-series trucks was their higher standard and optional GVW payload rating. The standard rating was increased to 6800 pounds from 6500 and the optional ratings were increased by 400 pounds to 7200 and 8000 pounds. Available once more was the Honcho package for the 119 inch wheelbase J-10. It included Levi's blue denim upholstery and door inserts, a chrome front bumper, blue sports steering wheel and Honcho graphics and lettering.

1977 AMC Jeep, Cherokee two-door station wagon (OCW)

I.D. DATA: VIN located on the left front door hinge pillar and left firewall. VIN still has 13 symbols. The first indicates Jeep Corp. The second indicates model year. The third indicates transmission, drivetrain and assembly plant. The fourth and fifth indicate series or model. The sixth symbol identifies body style. The seventh symbol indicates the engine. Engine codes were: E=232 cid six-cylinder; A=258 cid six-cylinder; H=304 cid V-8; N=360 cid V-8; P=360 cid V-8. Z=401 cid V-8. The next six symbols are the sequential production number beginning at 000001.

Model	Body Type	Price	Weight	Prod. Total
Wagoneer — (1/2-Ton) — (4x4)				
15	4dr Sta Wag	6966	4345	19,900
Cherokee — (1/2-Ton) — (4x4)				
16	2dr Sta Wag	5636	3971	—
17	2dr Wide Stance	6059	3991	—
18	4dr Sta Wag	5736	4106	—
CJ-5 — (1/4-Ton) — (4x4) — (84 in. w.b.)				
83	Jeep	4399	2659	32,996
CJ-7 — (1/4-Ton) — (4x4) — (94 in. w.b.)				
93	Jeep	4499	2701	25,414
J-10 — (1/2-Ton) — (4x4) — (119/131 in. w.b.)				
25	Townside Pickup (SWB)	4995	3826	—
45	Townside Pickup (LWB)	5059	3926	—
J-20 — (3/4-Ton) — (4x4) — (131 in. w.b.)				
46	Townside Pickup	5607	4285	3,343

NOTE: See "Historical" for additional production data.

1977 AMC Jeep, Cherokee four-door Station Wagon (OCW)

ENGINE (Standard: Cherokee and J-10; Optional: CJ-5/CJ-6): Inline. OHV. Six-cylinder. Cast iron block. Bore & stroke: 3.75 x 3.9 in. Displacement: 258 cid. Compression ratio: 8.0:1. Net horsepower: 110 at 3500 rpm. Torque: 195 lb.-ft. at 2000 rpm. Seven main bearings. Hydraulic valve lifters. Carburetor: Single Carter one-barrel. (Note: This engine was not available in California where a four-barrel version of the 360 cid V-8 was standard.)

ENGINE (Standard: Wagoneer and J-20; Optional: J-10): V-type. OHV. Eight-cylinder. Cast iron block. Bore & stroke: 4.08 x 3.44 in. Displacement: 360 cid. Compression ratio: 8.5:1. Net horsepower: 175 at 4000 rpm. Net torque: 285 lb.-ft. at 2400 rpm. Five main bearings. Hydraulic valve lifters. Carburetor: Single two-barrel.

1977 AMC Jeep, CJ-5 Universal Jeep (OCW)

ENGINE (Optional: J-10/J-20): V-type. OHV. Eight-cylinder. Cast iron block. Bore & stroke: 4.08 x 3.44 in. Displacement: 360 cid. Net horsepower: 195 at 4400 rpm. Five main bearings. Hydraulic valve lifters. Carburetor: Single four-barrel.
ENGINE (Optional: J-10/J-20): V-type. OHV. Eight-cylinder. Cast iron block. Bore & stroke: 4.17 x 3.68 in. Displacement: 401 cid. Net horsepower: 235 at 4600 rpm. Five main bearings. Hydraulic valve lifters. Carburetor: Single four-barrel.
ENGINE (Standard: CJ-5/CJ-6): Inline. OHV. Six-cylinder. Cast iron block. Bore & stroke: 3.75 x 3.50 in. Displacement: 232 cid. Compression ratio: 8.0:1. Net horsepower: 100 at 3600 rpm. Net torque: 185 lb.-ft. at 1800 rpm. Seven main bearings. Hydraulic valve lifters. Carburetor: Carter one-barrel model YF.
ENGINE (Standard: Renegade; Optional CJ-5/CJ-6): V-type. OHV. Eight-cylinder. Cast iron block. Bore & stroke: 3.75 x 3.44 in. Displacement: 304 cid. Compression ratio: 8.4:1. Net horsepower: 150 at 4200 rpm. Net torque: 245 lb.-ft. at 2500 rpm. Seven main bearings. Hydraulic valve lifters. Carburetor: Autolite two-barrel model 2100.

1977 AMC Jeep, CJ-7 Universal Jeep with hardtop (OCW)

CHASSIS (Wagoneer): Wheelbase: 109 in. Overall length: 183.7 in. Tires: F78-15.
CHASSIS (Cherokee): Wheelbase: 109 in. Overall length: 183.7 in. Tires: F78-15.
CHASSIS (CJ-5): Wheelbase: 83.5 in. Overall length: 138.5 in. Height: 67.6 in. Front tread: 51.5 in. Rear tread: 50.0 in. Tires: F78-15 in.
CHASSIS (CJ-7): Wheelbase: 93.5 in. Overall length: 147.9 in. Height: 67.6 in. Front tread: 51.5 in. Rear tread: 50.0 in. Tires: F78-15 in.
CHASSIS (J-10): Wheelbase: 118.7 in. Overall length: 192.5 in. Front tread: (J-10) 63.3 in.; (Honcho) 65.4 in. Rear tread: (J-10) 63.8 in.; (Honcho) 65.8 in. Tires: H78-15 in.
CHASSIS (J-10/J-20 LWB): Wheelbase: 130.7 in. Overall length: 204.5 in. Front tread: 64.6 in. Rear tread: 65.9 in. Tires: (J-10) F78-15; (J-20/6800 lb. GVW) 8.00 x 16.5; (J-20/ 7600-8200 lb. GVW) 9.50 x 16.5 in.

1977 AMC Jeep, J-20 Townside pickup (OCW)

TECHNICAL (Wagoneer/Cherokee/J-10/J-20): Manual, synchromesh transmission. Speeds: 3F/1R. Floor-mounted gearshift. Clutch: (J-10) 10.5 in. diameter, 106.75 sq. in., (J-20) 11.0 in. diameter, 110.96 sq. in. Rear axle: (J-10) Semi-floating; (J-20) full-floating. Overall ratio: (J-10) 3.54:1 or 4.09:1; (J-20) 3.73:1. (J-10) Four-wheel hydraulic drum brakes. (J-20) Front disc/rear drum brakes with power assist. Pressed steel wheels. Technical Options: Automatic transmission Turbo Hydra-Matic. Four-speed manual transmission. Quadra-Trac (available with automatic transmission only). 12-in. power front disc brakes (J-10). Snow Boss plow package. Winches (mechanical or 12-volt). Auxiliary fuel tank (20 gallon, for long wheelbase only, not available for California). Rear Trac-Lok differential. Camper special package. Fuel tank skid plate. Heavy-duty battery and alternator. Heavy-duty shock absorbers. Heavy-duty springs. Heavy-duty cooling system. Cold climate package. Locking hubs (not available with Quadra-Trac). Trailer Towing Package.
TECHNICAL (CJ-Series): Manual, synchromesh transmission. Speeds: 3F/1R. Floor-mounted gearshift. Single plate dry disc clutch. Semi-floating rear axle. Overall ratio: 3.54:1; 4.09:1. Manual, hydraulic four-wheel-drum brakes. Pressed steel wheels. Technical options: Automatic transmission Turbo Hydra-Matic. Quadra-Trac (available with Turbo Hydra-Matic only). Free-running front hubs ($95). Power steering ($166). Power front disc brakes ($73). Steering damper ($10). Front stabilizer bar ($27). Four-speed manual transmission (6.32:1 low gear). Extra-duty suspension system. Low-range for Quadra-Trac (CJ-7 only). Heavy-duty cooling system. Rear Trac-Lok differential. 70-amp. battery. Heavy-duty alternator. Cold climate group. Steel-belted radial ply tires. Winches. Snow plow.
OPTIONS (Wagoneer/Cherokee/Trucks): 401 cid V-8 ($94). Power steering, except standard in Wagoneer ($190). Turbo-Hydramatic with Quadra-Trac, except standard on Wagoneer ($345). Four-speed transmission in Cherokee ($144). Cherokee Chief package for Model 17 ($469). Luggage rack, Wagoneer ($85). Factory air conditioning ($509). Front bumper guards with Nerf strips. Rear step bumper. Convenience group. Light group. Custom trim package. Radios: AM/FM stereo ($222); Citizens Band; AM. Forged aluminum styled wheels (J-10 only). Bucket seats (Custom, Pioneer and Honcho only). Cruise control ($80). Power steering. Tilt steering wheel. Sports steering wheel. Leather wrapped steering wheel. Sliding rear window. Dual low profile mirror. Aluminum cargo cap. Wheelcovers. Honcho package. Pioneer package.
OPTIONS (CJ-Series): Rear step bumper. Roll bar. Full soft top ($275). Carpeting ($63). Convenience group. Decor group. AM radio ($73). Citizens Band Radio. Cigar lighter and ashtray. Renegade package, CJ-5 ($839). Outside passenger mirror. Center console ($62). Air conditioning ($499). Tachometer and Rally clock ($73). Sports steering wheel. Leather wrapped sports steering wheel. Rear seat. Swing-out tire carrier. Padded instrument panel. Injection-molded removable top (CJ-7).
HISTORICAL: Introduced: Fall, 1976. Calendar year sales: (all Jeep models and series) 115,079. Calendar year production: (Worldwide wholesales) 153,485. Model year output was 127,996 units. This included 19,900 Wagoneers; 31,308 Cherokees; 32,966 CJ-5s; 2,754 CJ-6s (export only); 25,414 CJ-7s; 12,191 J-10s and 3,343 J-20s. Innovations: Improved frame, brakes and suspension. New tires for Renegade series. Air conditioning made available for first time on CJs. "Golden Eagle" option introduced at mid-year. Historical notes: Jeep sales reached an all-time record high level in model year 1977.

1978 JEEP

1978 AMC Jeep, CJ-7 Renegade (JAG)

JEEP — 1978 SERIES — (ALL ENGINES): For 1978, Jeep Corp. stayed away from new models and made minimal design changes to the existing line. More comfort and convenience options were offered for all Jeeps. The Wagoneer got a new, midyear special edition. This Wagoneer Limited featured a luxury interior, woodgrain exterior trim and added sound insulation.
Technical and appearance changes for 1978 were limited. Continued into 1978, was the "Golden Eagle" option, which had been introduced in mid-1977. This year it was offered in a wider choice of colors and featured a large golden eagle decal on the hood and special striping on the grille, plus fender flares and gold-colored wheels. All CJ models were fitted with an improved heating system that distributed heat more efficiently, provided better fresh air ventilation and gave higher defroster temperatures. Engine efficiency was improved, due to a new ambient air intake system.
The J-series trucks were also available with the Golden Eagle package. For the J-trucks it included a chrome front bumper; rear step bumper; gold eight inch spoked wheels (with black accents); 10 x 15 inch all-terrain tires; bright window frames; pickup box-mounted roll bar (with off-road driving lights); steel grille guard; Levi's bucket seats; custom interior and an engine-turned instrument cluster. Other interior appointments were tan carpeting, sports steering wheel and bright armrest overlays. A golden eagle decal was positioned on the hood and lower door panel. Gold, green and orange striping was found on the door, cab and upper box sides.
A new trim package, the "10-4" option, was available for J-10 models on the short wheelbase chassis. It could be ordered in any of 10 colors with two-tone orange and black accent striping. A two-tone orange "10-4" decal was located on the side, just ahead of the rear

wheels. Other features included 10 x 15 inch Tracker tires, white 15 x 8 inch wheels (with red pinstriping), a bed-mounted roll bar and rear step bumper.

The J-series trucks had an additional 2.5 inch of legroom due to a modified toeboard and relocated accelerator. The J-10 GVW rating for the J-10 was moved up to 6200 pounds.

I.D. DATA: VIN located on the left front door hinge pillar and left firewall. VIN still has 13 symbols. The first indicates Jeep Corp. The second indicates model year. The third indicates transmission, drivetrain and assembly plant. The fourth and fifth indicate series or model. The sixth symbol identifies body style. The seventh symbol indicates the engine. Engine codes were: E=232 cid six-cylinder; A=258 cid six-cylinder; C=258 six-cylinder; E=232 six-cylinder; G=121 cid four-cylinder; H=304 cid V-8; N=360 cid V-8; P=360 cid V-8; Z=401 cid V-8. The next six symbols are the sequential production number beginning at 000001.

Model	Body Type	Price	Weight	Prod. Total
Wagoneer — (1/2-Ton) — (4x4)				
15	4dr Sta Wag	7695	4345	24,274
Cherokee — (1/2-Ton) — (4x4)				
16	2dr Sta Wag	6229	3971	—
17	2dr Wide Stance	6675	4084	—
18	4dr Sta Wag	6335	4106	—
CJ-5 — (1/4-Ton) — (4x4) — (84 in. w.b.)				
83	Jeep	5095	2738	37,611
CJ-7 — (1/4-Ton) — (4x4) — (94 in. w.b.)				
93	Jeep	5195	2782	38,274
J-10 — (1/4-Ton) — (4x4) — (119/131 in. w.b.)				
25	Townside Pickup (SWB)	5675	3831	—
45	Townside Pickup (LWB)	5743	3898	—
J-20 — (3/4-Ton) — (4x4) — (131 in. w.b.)				
46	Townside Pickup	6324	4269	8,085

NOTE: See "Historical" for additional production data.

ENGINE (Four): Inline. OHV. Four-cylinder. Bore & stroke: 3.41 x 3.23 in. Displacement: 121 cid. Compression: 8.2:1. Brake horsepower: 80 at 5000 rpm.

ENGINE (Standard: CJ-5/CJ-6): Inline. OHV. Six-cylinder. Cast iron block. Bore & stroke: 3.75 x 3.50 in. Displacement: 232 cid. Compression ratio: 8.0:1. Net horsepower: 100 at 3600 rpm. Net torque: 185 lb.-ft. at 1800 rpm. Seven main bearings. Hydraulic valve lifters. Carburetor: Carter one-barrel model YF.

ENGINE (Standard: Cherokee and J-10; Optional: CJ-5/CJ-6): Inline. OHV. Six-cylinder. Cast iron block. Bore & stroke: 3.75 x 3.9 in. Displacement: 258 cid. Compression ratio: 8.0:1. Net horsepower: 110 at 3500 rpm. Torque: 195 lb.-ft. at 2000 rpm. Seven main bearings. Hydraulic valve lifters. Carburetor: Single Carter one-barrel. (Note: This engine was not available in California where a four-barrel version of the 360 cid V-8 was standard.)

ENGINE (Optional: Cherokee/J-10/CJ-5/CJ-6): Inline. OHV. Six-cylinder. Cast iron block. Bore & stroke: 3.75 x 3.9 in. Displacement: 258 cid. Compression ratio: 8.0:1. Net horsepower: 120 at 3600 rpm. Torque: 195 lb.-ft. at 2000 rpm. Seven main bearings. Hydraulic valve lifters. Carburetor: Single Carter two-barrel.

ENGINE (Standard: Renegade; Optional CJ-5/CJ-6): V-type. OHV. Eight-cylinder. Cast iron block. Bore & stroke: 3.75 x 3.44 in. Displacement: 304 cid. Compression ratio: 8.4:1. Net horsepower: 150 at 4200 rpm. Net torque: 245 lb.-ft. at 2500 rpm. Seven main bearings. Hydraulic valve lifters. Carburetor: Autolite two-barrel model 2100.

ENGINE (Standard: Wagoneer and J-20; Optional: J-10): V-type. OHV. Eight-cylinder. Cast iron block. Bore & stroke: 4.08 x 3.44 in. Displacement: 360 cid. Compression ratio: 8.5:1. Net horsepower: 175 at 4000 rpm. Net torque: 285 lb.-ft. at 2400 rpm. Five main bearings. Hydraulic valve lifters. Carburetor: Single two-barrel.

ENGINE (Optional: J-10/J-20): V-type. OHV. Eight-cylinder. Cast iron block. Bore & stroke: 4.08 x 3.44 in. Displacement: 360 cid. Net horsepower: 195 at 4400 rpm. Five main bearings. Hydraulic valve lifters. Carburetor: Single four-barrel.

ENGINE (Optional: J-10/J-20): V-type. OHV. Eight-cylinder. Cast iron block. Bore & stroke: 4.17 x 3.68 in. Displacement: 401 cid. Net horsepower: 235 at 4600 rpm. Five main bearings. Hydraulic valve lifters. Carburetor: Single four-barrel.

ENGINE (Standard: CJ-5/CJ-6): Inline. OHV. Six-cylinder. Cast iron block. Bore & stroke: 3.75 x 3.50 in. Displacement: 232 cid. Compression ratio: 8.0:1. Net horsepower: 100 at 3600 rpm. Net torque: 185 lb.-ft. at 1800 rpm. Seven main bearings. Hydraulic valve lifters. Carburetor: Carter one-barrel model YF.

ENGINE (Standard: Renegade; Optional CJ-5/CJ-6): V-type. OHV. Eight-cylinder. Cast iron block. Bore & stroke: 3.75 x 3.44 in. Displacement: 304 cid. Compression ratio: 8.4:1. Net horsepower: 150 at 4200 rpm. Net torque: 245 lb.-ft. at 2500 rpm. Seven main bearings. Hydraulic valve lifters. Carburetor: Autolite two-barrel model 2100.

CHASSIS (Wagoneer): Wheelbase: 108.7 in. Overall length: 186 in. Width: 76 in. Tires: F78-15.

CHASSIS (Cherokee): Wheelbase: 108.7 in. Overall length: 186 in. Tires: F78-15.

CHASSIS (CJ-5): Wheelbase: 83.5 in. Overall length: 138.5 in. Height: 67.6 in. Front tread: 51.5 in. Rear tread: 50.0 in. Tires: H78-15 in.

CHASSIS (CJ-7): Wheelbase: 93.5 in. Overall length: 147.9 in. Height: 67.6 in. Front tread: 51.5 in. Rear tread: 50.0 in. Tires: H78-15 in.

CHASSIS (J-10): Wheelbase: 118.7 in. Overall length: 192.5 in. Front tread: (J-10) 63.3 in.; (Honcho) 65.4 in. Rear tread: (J-10) 63.8 in.; (Honcho) 65.8 in. Tires: H78-15 in.

CHASSIS (J-10/J-20 LWB): Wheelbase: 130.7 in. Overall length: 204.5 in. Front tread: 64.6 in. Rear tread: 65.9 in. Tires: (J-10) F78-15; (J-20/6800 lb. GVW) 8.00 x 16.5; (J-20/7600-8200 lb. GVW) 9.50 x 16.5 in.

CHASSIS: All chassis measurements were the same as 1977. The CJ-5 and CJ-7 had new H78-15 size tires. The J-10 had the same new size tires.

TECHNICAL (Wagoneer/Cherokee/J-10/J-20): Manual, synchromesh transmission. Speeds: 3F/1R. Floor-mounted gearshift. Clutch: (J-10) 10.5 in. diameter, 106.75 sq. in., (J-20) 11.0 in. diameter, 110.96 sq. in. Rear axle: (J-10) Semi-floating; (J-20) full-floating. Overall ratio: (J-10) 3.54:1 or 4.09:1; (J-20) 3.73:1. (J-10) Four-wheel hydraulic drum brakes. (J-20) Front disc/rear drum brakes with power assist. Pressed steel wheels. Technical Options: Automatic transmission Turbo Hydra-Matic. Four-speed manual transmission. Quadra-Trac (available with automatic transmission only). 12-in. power front disc brakes (J-10). Snow Boss plow package. Winches (mechanical or 12-volt). Auxiliary fuel tank (20 gallon, for long wheelbase only, not available for California). Rear Trac-Lok differential. Camper special package. Fuel tank skid plate. Heavy-duty battery and alternator. Heavy-duty shock absorbers. Heavy-duty springs. Heavy-duty cooling system. Cold climate package. Locking hubs (not available with Quadra-Trac). Trailer Towing Package.

TECHNICAL (CJ-Series): Manual, synchromesh transmission. Speeds: 3F/1R. Floor-mounted gearshift. Single plate dry disc clutch. Semi-floating rear axle. Overall ratio: 3.54:1, 4.09:1. Manual, hydraulic four-wheel-drum brakes. Pressed steel wheels. Technical options: Automatic transmission Turbo Hydra-Matic. Quadra-Trac (available with Turbo Hydra-Matic only). Free-running front hubs ($95). Power steering ($166). Power front disc brakes ($73). Steering damper ($10). Front stabilizer bar ($27). Four-speed manual trans-

mission (6.32:1 low gear). Extra-duty suspension system. Low-range for Quadra-Trac (CJ-7 only). Heavy-duty cooling system. Rear Trac-Lok differential. 70-amp. battery. Heavy-duty alternator. Cold climate group. Steel-belted radial ply tires. Winches. Snowplow.

OPTIONS: Most convenience and appearance options were unchanged for 1978. However, many items previously optional on the CJ models were now standard equipment. These included an ashtray and cigar lighters, passenger assist bar and passenger side exterior mirror. Standard tires were now H78-15 Suburbanite fiberglass belted types. Included in the convenience group option was an underhood light. Added to the option list for the J-series pickups was the "10-4" appearance package. New options for the J-trucks included an AM/FM Multiplex 8-track tape, 7-inch chrome-plated spoked steel wheels for J-10 models. A grille guard and a pickup bed mounted roll bar.

OPTIONS: (Wagoneer/Cherokee): Automatic transmission with Quadra-Trac, except standard in Wagoneer ($345). Wagoneer Limited trim package ($3,120).

OPTIONS (J-Series): Bumper guards AM/FM stereo/Citizen's Band ($349). AM Radio ($86). AM/Citizen's Band ($229). AM/FM stereo ($229). AM/FM stereo w/tape player ($329). Sliding rear window ($79). Tinted glass ($32). "10-4" package ($619). Levis Fabric Bucket Seats ($175). Bucket seats ($155). Roll bar ($105). Air conditioning ($557). Brush guard ($69). Pickup box cap ($324). Convenience group ($83). Cruise control ($100). Custom package ($95). Golden Eagle package, J-10 (SWB) only, with driving lights ($999); without ($974). Honcho package, J-10 (SWB) only. Light group ($46).

OPTIONS (CJ-Series): AM radio ($86). Standard soft top ($253). Levis soft top ($292). Vinyl bucket Levis seats ($73). Vinyl bench seat ($35). Rear seat ($106). Moon roof for CJ-7 ($149). Roll bar ($70). Air conditioning ($529). Metal cab for CJ-5 ($471). Removable front carpet ($44). Removable front and rear carpet ($73). Convenience group ($27). Decor group ($105). Golden Eagle package ($1249). Hardtop with doors, CJ-7 ($610). Renegade package ($799).

HISTORICAL: Introduced: Fall, 1977. Calendar year sales: (all models and series) 161,912. Model year production included 36,945 Cherokees; 24,274 Wagoneers; 37,611 CJ-5s; 743 CJ-6s (export); 38,274 CJ-7s; 9,167 J-10s and 8,085 J-20s. Innovations: Many new options released. Legroom in the Jeep pickup trucks was increased. GVW ratings of Jeep trucks increased. Historical notes: To commemorate the 25th anniversary of the CJ-5 a limited-edition "Silver Anniversary" CJ-5 was offered. A total of 3,000 copies were produced with Quick silver metallic finish, silver-toned Renegade accent striping black soft top, black vinyl bucket seats, silver accents and a special dashboard plaque.

1979 JEEP

1979 AMC Jeep, CJ-5 Renegade Universal Jeep (RPZ)

JEEP — 1979 SERIES — (ALL ENGINES): This was not a year of dramatic change for Jeep vehicles. Sales were up so much, that the outdated, ex-Willys factory in Toledo could hardly keep up. The CJ series was carried over virtually unchanged. The Renegade package featured new exterior graphics. Revisions made in the J-series trucks consisted of a new front end appearance with rectangular headlights set in a grille with a slightly protruding center section and vertical blades. The front bumper no longer had a recessed middle portion. Available for all pickups was a new "high style" box enclosure. The Honcho trim package was revised.

I.D. DATA: VIN located on the left front door hinge pillar and left firewall. VIN still has 13 symbols. The first indicates Jeep Corp. The second indicates model year. The third indicates transmission, drivetrain and assembly plant. The fourth and fifth indicate series or model. The sixth symbol identifies body style. The seventh symbol indicates the engine. Engine codes were: E=232 cid six-cylinder; A=258 cid six-cylinder; C=258 six-cylinder; E=232 six-cylinder; G=121 cid four-cylinder; H=304 cid V-8; N=360 cid V-8. The next six symbols are the sequential production number beginning at 000001.

1979 AMC Jeep, CJ-7 Renegade Universal Jeep (RPZ)

Model	Body Type	Price	Weight	Prod. Total
Wagoneer — (1/2-Ton) — (4x4)				
15	4dr Sta Wag	9065	4034	—
15	4dr Ltd Sta Wag	12,485	4181	—
Cherokee — (1/2-Ton) — (4x4)				
16	2d Sta Wag	7328	3653	—
17	2dr Wide Stance	7671	3774	—
18	4dr Sta Wag	7441	3761	—
CJ-5 — (1/4-Ton) — (4x4) — (84 in. w.b.)				
83	Jeep	5588	2623	21,308
CJ-7 — (1/4-Ton) — (4x4) — (94 in. w.b.)				
93	Jeep	5732	2666	24,580
J-10 — (1/2-Ton) — (4x4) — (119/131 in. w.b.)				
25	Townside Pickup (SWB)	6172	3693	—
45	Townside Pickup (LWB)	6245	3760	—
J-20 — (3/4-Ton) — (4x4) — (131 in. w.b.)				
46	Townside Pickup	6872	4167	10,403

NOTE: See "Historical" for additional production data.

ENGINE (Four): Inline. OHV. Four-cylinder. Bore & stroke: 3.41 x 3.23 in. Displacement: 121 cid. Compression: 8.2:1. Brake horsepower: 80 at 5000 rpm.

ENGINE (Standard: CJ-5/CJ-6): Inline. OHV. Six-cylinder. Cast iron block. Bore & stroke: 3.75 x 3.50 in. Displacement: 232 cid. Compression ratio: 8.0:1. Net horsepower: 100 at 3600 rpm. Net torque: 185 lb.-ft. at 1800 rpm. Seven main bearings. Hydraulic valve lifters. Carburetor: Carter one-barrel model YF.

ENGINE (Standard: Cherokee and J-10; Optional: CJ-5/CJ-6): Inline. OHV. Six-cylinder. Cast iron block. Bore & stroke: 3.75 x 3.9 in. Displacement: 258 cid. Compression ratio: 8.0:1. Net horsepower: 110 at 3500 rpm. Torque: 195 lb.-ft. at 2000 rpm. Seven main bearings. Hydraulic valve lifters. Carburetor: Single Carter one-barrel. (Note: This engine was not available in California where a four-barrel version of the 360 cid V-8 was standard.)

ENGINE (Optional: Cherokee/J-10/CJ-5/CJ-6): Inline. OHV. Six-cylinder. Cast iron block. Bore & stroke: 3.75 x 3.9 in. Displacement: 258 cid. Compression ratio: 8.0:1. Net horsepower: 120 at 3600 rpm. Torque: 195 lb.-ft. at 2000 rpm. Seven main bearings. Hydraulic valve lifters. Carburetor: Single Carter two-barrel.

ENGINE (Standard: Renegade; Optional CJ-5/CJ-6): V-type. OHV. Eight-cylinder. Cast iron block. Bore & stroke: 3.75 x 3.44 in. Displacement: 304 cid. Compression ratio: 8.4:1. Net horsepower: 150 at 4200 rpm. Net torque: 245 lb.-ft. at 2500 rpm. Seven main bearings. Hydraulic valve lifters. Carburetor: Autolite two-barrel model 2100.

ENGINE (Standard: Wagoneer and J-20; Optional: J-10): V-type. OHV. Eight-cylinder. Cast iron block. Bore & stroke: 4.08 x 3.44 in. Displacement: 360 cid. Compression ratio: 8.5:1. Net horsepower: 175 at 4000 rpm. Net torque: 285 lb.-ft. at 2400 rpm. Five main bearings. Hydraulic valve lifters. Carburetor: Single two-barrel.

ENGINE (Optional: J-10/J-20): V-type. OHV. Eight-cylinder. Cast iron block. Bore & stroke: 4.08 x 3.44 in. Displacement: 360 cid. Net horsepower: 195 at 4400 rpm. Five main bearings. Hydraulic valve lifters. Carburetor: Single four-barrel.

ENGINE (Optional: J-10/J-20): V-type. OHV. Eight-cylinder. Cast iron block. Bore & stroke: 4.17 x 3.68 in. Displacement: 401 cid. Net horsepower: 235 at 4600 rpm. Five main bearings. Hydraulic valve lifters. Carburetor: Single four-barrel.

ENGINE (Standard: CJ-5/CJ-6): Inline. OHV. Six-cylinder. Cast iron block. Bore & stroke: 3.75 x 3.50 in. Displacement: 232 cid. Compression ratio: 8.0:1. Net horsepower: 100 at 3600 rpm. Net torque: 185 lb.-ft. at 1800 rpm. Seven main bearings. Hydraulic valve lifters. Carburetor: Carter one-barrel model YF.

ENGINE (Standard: Renegade; Optional CJ-5/CJ-6): V-type. OHV. Eight-cylinder. Cast iron block. Bore & stroke: 3.75 x 3.44 in. Displacement: 304 cid. Compression ratio: 8.4:1. Net horsepower: 150 at 4200 rpm. Net torque: 245 lb.-ft. at 2500 rpm. Seven main bearings. Hydraulic valve lifters. Carburetor: Autolite two-barrel model 2100.

1979 AMC Jeep, Golden Eagle with hardtop (RPZ)

CHASSIS (Wagoneer): Wheelbase: 108.7 in. Overall length: 186 in. Width: 76 in. Tires: F78-15.

CHASSIS (Cherokee): Wheelbase: 108.7 in. Overall length: 186 in. Tires: F78-15.

CHASSIS (J-10 SWB): Wheelbase: 118.7 in. Overall length: 192.5 in. Height: 69.3 in. Front tread: 63.3 in. Rear tread: 63.8 in. Tires: H78-15 in.

CHASSIS (J-10/J-20 LWB): Wheelbase: 130.7 in. Overall length: 204.5 in. Height: 69.1/70.7 in. Front tread: (J-10) 63.3 in., (J-20) 64.6 in. Rear tread: (J-10) 63.8 in., (J-20) 65.9 in. Tires: (J-10) H78-15, (J-20) 8.75 x 16.5 in.

CHASSIS (CJ-5): Wheelbase: 83.5 in. Overall length: 134.8 in. Height: 67.6 in. Front tread: 51.5 in. Rear tread: 50.0 in. Tires: H78-15 in.

CHASSIS (CJ-7): Wheelbase: 93.5 in. Overall length: 144.3 in. Height: 67.6 in. Front tread: 57.5 in. Rear tread: 50.0 in. Tires: H78-15 in.

1979 AMC Jeep, Honcho Townside pickup (RPZ)

TECHNICAL (CJ-Series): Manual, synchronized transmission. Speeds: 3F/1R. Floor-mounted gearshift. Single dry disc clutch. Semi-floating rear axle. Four-wheel drum brakes. Pressed steel wheels. Options: Turbo-Hydramatic with Quadra-Trac (CJ-7 only). Four-speed manual transmission. Quadra-Trac low range. Rear Trac-Lok differential. Free-wheeling hubs. Heavy-duty cooling system. Extra-duty suspension package. Steering damper. Front stabilizer bar. Heavy-duty 70-amp. battery. Heavy-duty 63-amp. alternator. Cold Climate group.

1979 AMC Jeep, Honcho Townside pickup (CW)

TECHNICAL (Wagoneer/Cherokee/J-10/J-20): Fully synchronized manual transmission. Speeds: 3F/1R. Floor-mounted gearshift. Single dry disc clutch. (J-10) Semi-floating rear axle; (J-20) full-floating rear axle. Power front disc/rear drum brakes. Steel disc wheels. Options: Turbo-Hydramatic with Quadra-Trac. Four-speed manual transmission ($161). Quadra-Trac low range. Rear Trac-Lok differential. Free-wheeling hubs. Heavy-duty cooling system. Extra-duty suspension system. Heavy-duty shock absorbers. Front stabilizer bar. Heavy-duty 70-amp. battery. Heavy-duty 63-amp. alternator. Cold Climate group. "Snow Boss" package. 304 cid V-8 ($342).

OPTIONS (Wagoneer/Cherokee): Power steering, standard in Wagoneer ($226). 360 cid V-8, standard in Wagoneer ($378). Turbo-Hydramatic with Quadra-Trac, standard in Wagoneer ($396). AM/FM stereo ($241). AM/FM stereo with tape ($346). Cherokee "S" package ($699). Cherokee Chief package ($600). Golden Eagle package ($970).

OPTIONS (Trucks): Radios: AM, AM/FM, AM/CB, AM/FM/CB and AM/FM/Tape. Electric clock. Custom package ($126). Honcho package, SWB/J-10 ($824). Golden Eagle package, SWB/J-10 ($1224). "10-4" Package, SWB/J-10. 10-4 Package (J-10, 118.7 in. wheelbase only). Soft feel sports steering wheel. Leather-wrapped sports steering wheel. Aluminum wheels (not available J-20). 15 inch wheel covers (not available J-20). Hub caps (J-20). Convenience group. Air conditioning. Tinted glass. Tilt steering wheel. Cruise control. Light group. Low profile mirrors. Bucket seats. Rear step bumper. Sliding rear window. Cargo cap. Roll bar. Fuel tank skid plate. Front bumper guards. Brush guards. Floor mats. Spare tire lock. Rear bumperettes.

1979 AMC Jeep, Wagoneer Limited four-door wagon (RPZ)

OPTIONS (CJ Series): Renegade package ($825). Golden Eagle Package. Hardtop ($656). CJ-5 side-mounted spare and tailgate. (CJ-7) Swing-away spare tire carrier. Injection molded hardtop (CJ-7). Full metal cab (CJ-5). Removable carpet. AM radio. Moon roof (with hardtop only). Body side step. Power front disc brakes. Power steering. Tachometer and Rally clock. Tilt steering wheel. Air conditioning. Convenience group. Decor group. 15 inch wheel covers. Soft feel steering wheel (black). Soft top. Padded instrument panel. Four-speed transmission ($161). 304 cid V-8 ($342).

1979 AMC Jeep, Cherokee 'S' four-door station wagon (RPZ)

HISTORICAL: Introduced: Fall, 1978. Calendar year registrations: 145,583. Calendar year sales: (all models and series) 140,431. Calendar year production reached an all-time record of 177,575 units at the Jeep Corp. Toledo, Ohio plant. Model year output totaled 152,992 units. This included 36,195 Wagoneers; 48,813 Cherokees; 21,308 CJ-5s; 24,580 CJ-7s; 11,693 J-10s and 10,403 J-20s. Innovations: New Renegade graphics. Jeep pickups get new front end treatment. Optional "high-style" box enclosure released.

1979 AMC Jeep, Cherokee Chief two-door station wagon (RPZ)

1980 JEEP

1980 AMC Jeep, Cherokee Special Police Wagon (JAG)

JEEP — 1980 SERIES — (ALL ENGINES): American Motors' response to the dramatic upswing in gasoline prices and its inverse relationship to sales of off-road vehicle and four-wheel-drive light-duty trucks was pronounced. Throughout the Jeep line, transmissions were made more efficient and lighter in weight. Many models had freewheeling front hubs and fuel conserving driveline designs. Quadra-Trac was improved by replacement of cone clutches with viscous drives.

For the CJ models, the most dramatic news was use of the General Motors 151 cid, four-cylinder engine as standard power plant. Also new, was a standard, lightweight four-speed, close-ratio manual transmission. A new chassis design was adopted. It was both stronger and lighter than the unit it replaced. Free-wheeling front hubs were standard. Also installed on all CJ Jeeps was a roll bar. If the 258 cid six-cylinder or 304 cid V-8 were ordered for the CJ-7, they were available with a lightweight automatic transmission, part-time four-wheel drive system and freewheeling hubs.

Appearance changes were lead by a new soft top with steel doors and roll-up windows. Either CJ model was available with a new "Laredo" package. It included specially trimmed high-back bucket seats, chrome grille and accents, special striping and Wrangler radial tires.

For 1980, the J-trucks had a new lightweight, more efficient drivetrain, free-wheeling front hubs, front stabilizer bar and a high-density blow-molded fuel tank. Numerous new options were offered, including power windows and an electronic digital clock. Like the CJ models, the Jeep trucks were available with the improved Quadra-Trac, Laredo package and part-time four-wheel-drive (with either a standard four-speed manual or automatic transmission). The short-wheelbase Sportside model was available in an optional "highline" version with Honcho graphics, wooden cargo box rails and a styled roll bar.

I.D. DATA: VIN located on the left front door hinge pillar and left firewall. VIN has 13 symbols. The first indicates Jeep Corp. The second indicates model year. The third indicates transmission, drivetrain and assembly plant. The fourth and fifth indicate series or model. The sixth symbol identifies body style. The seventh symbol indicates the engine. Engine codes were: B=151 cid four-cylinder engine; C=258 cid six-cylinder engine; H=304 cid V-8; N=360 cid V-8. The next six symbols are the sequential production number beginning at 000001.

Model	Body Type	Price	Weight	Prod. Total
Wagoneer — (1/2-Ton) — (4x4)				
15	4dr Sta Wag	9732	3964	—
15	4dr Ltd Sta Wag	13,653	3990	—
Cherokee — (1/2-Ton) — (4x4)				
16	2dr Sta Wag	8180	3780	—
17	2dr Wide Stance	8823	3868	—
18	4dr Sta Wag	8380	3849	—
CJ-5 — (1/4-Ton) — (4x4) — (84 in. w.b.)				
83	Jeep	6195	2439	14,156
CJ-7 — (1/4-Ton) — (4x4) — (94 in. w.b.)				
93	Jeep	6445	2464	20,191
J-10 — (1/2-Ton) — (4x4) — (119/131 in. w.b.)				
25	Townside Pickup (SWB)	6874	3714	—
45	Townside Pickup (LWB)	6972	3776	—
J-20 — (3/4-Ton) — (4x4) — (131 in. w.b.)				
46	Townside Pickup	7837	4246	2,004

NOTE: See "Historical" for additional production data.
ENGINE (Standard: CJ): Inline. OHV. Four-cylinder. Cast iron block. Bore & stroke: 4.00 x 3.00 in. Displacement: 151 cid. Compression ratio: 8.2:1. Taxable horsepower: 21.70. Hydraulic valve lifters. Carburetor: Two-barrel.
ENGINE (Standard: J-10/CJ): Inline. OHV. Six-cylinder. Cast iron block. Bore & stroke: 3.75 x 3.90 in. Displacement: 258 cid. Compression ratio: 8.0:1. Net horsepower: 110. Taxable horsepower: 33.75. Seven main bearings. Hydraulic valve lifters. Carburetor: Single two-barrel.
ENGINE (Standard: Renegade; Optional CJ-5/CJ-6): V-type. OHV. Eight-cylinder. Cast iron block. Bore & stroke: 3.75 x 3.44 in. Displacement: 304 cid. Compression ratio: 8.4:1. Net horsepower: 150 at 4200 rpm. Net torque: 245 lb.-ft. at 2500 rpm. Seven main bearings. Hydraulic valve lifters. Carburetor: Autolite two-barrel model 2100.
ENGINE (Standard: CJ-20; Optional: J-10): Vee-block. OHV. Eight-cylinder. Cast iron block. Bore & stroke: 4.08 x 3.44 in. Displacement: 360 cid. Compression ratio: 8.25:1. Net horsepower: 175. Taxable horsepower: 53.27. Five main bearings. Hydraulic valve lifters. Carburetor: Single two-barrel.

CHASSIS (Wagoneer): Wheelbase: 108.7 in. Overall length: 186 in. Width: 76 in. Tires: F78-15.

CHASSIS (Cherokee): Wheelbase: 108.7 in. Overall length: 186 in. Tires: F78-15.

CHASSIS (CJ-5): Wheelbase: 83.5 in. Overall length: 144.3 in. Height: 67.6 in. Front tread: 51.5 in. Rear tread: 50.0 in. Tires: H78-15 in.

CHASSIS (CJ-7): Wheelbase: 93.5 in. Overall length: 153.2 in. Height: 67.6 in. Front tread: 51.5 in. Rear tread: 50.0 in. Tires: H78-15 in.

CHASSIS (J-10 SWB): Wheelbase: 118.7 in. Overall length: 192.7 in. Height: 69.3 in. Front tread: 63.3 in. Rear tread: 63.8 in. Tires: H78-15 in.

CHASSIS (J-10/J-20 LWB): Wheelbase: 130.7 in. Overall length: 204.5 in. Height: 69.1/ 70.7 in. Front tread: (J-10) 63.3 in.; (J-20) 63.3 in. Rear tread: (J-10) 63.8 in.; (J-20) 64.9 in. Tires: (J-10) H78-15; (J-20) 8.75 x 16.5 in.

CHASSIS (J-10 Sportside): Wheelbase: 118.7 in. Overall length: 196.9 in. Height: 69.1 in. Front tread: 63.3 in. Rear tread: 63.8 in. Tires: H78-15 in.

TECHNICAL (CJ-Series): Manual, synchronized transmission. Speeds: 4F/1R. Floor-mounted gearshift. Single plate dry disc clutch. Semi-floating rear axle. Overall ratio: (four-cylinder) 3.73:1, (six-cylinder) 3.07:1, (V-8) 3.07:1. Manual front disc/rear drum brakes. Steel wheels. Technical Options: Automatic transmission (CJ-7). Quadra-Trac with automatic transmission (CJ-7). Heavy-duty shock absorbers. Steering stabilizer. Automatic hubs. Air overload kit. Heavy-duty cooling system. Snowplows.

TECHNICAL (Wagoneer/Cherokee/J-Series): Fully synchronized, manual transmission. Speeds: 4F/1R. Floor-mounted gearshift. Single plate dry disc clutch. (J-10) Semi-floating rear axle; (J-20) full-floating rear axle. Front disc/rear drum brakes. Pressed steel wheels. Technical Options: Automatic transmission, column mounted. Quadra-Trac (with automatic transmission). Heavy-duty cooling system. Heavy-duty shock absorbers. Helper springs. Cruise control. Automatic hubs (not available with Quadra-Trac). Engine block heater. Extra-duty suspension package. Heavy-duty battery. Heavy-duty alternator. Cold climate group. Snow boss package.

OPTIONS (CJ-5/CJ-7): Safari soft top. Skyviewer soft top. Top boot. Sun bonnet. M3 metal cab (CJ-5 only). Cab with vertical rear door (CJ-5 only). Deluxe M3 cab (CJ-5 only). Renegade package ($899). Golden Hawk package. Laredo package. Hardtop ($676). Hand held searchlight (all models). Passenger assist handle. Ski rack. Bug screen. Automatic transmission ($333). 304 cid V-8 engine ($383). 258 cid six-cylinder engine ($129).

OPTIONS (Wagoneer/Cherokee): Power steering, included with models 15 and 17; other models ($233). Automatic transmission, standard in Wagoneer. AM/FM stereo ($245). AM/FM stereo with tape ($355). AM/FM stereo with CB radio ($495). Cruise control. Rear window defroster. Tilt steering wheel. Luggage rack. Aluminum wheels (standard on Wagoneer Limited). Wagoneer woodgrain exterior trim package ($139). Removable moon roof ($300). Power door locks. Power windows. Cherokee S package, model 16 and 18 ($784). Cherokee Chief package for model 17 ($799). Golden Eagle or Golden Hawk package for Cherokee. Cherokee Laredo package ($1,600). Custom or two-tone paint. 360 cid V-8 engine ($420). Air conditioning.

OPTIONS (J-Series Trucks): Custom package ($149). Honcho package, model 25 ($849). Laredo package, model 25 ($1,600). Sportside package, model 25 ($899). Honcho Sportside package, model 25 ($1,325). Automatic transmission ($333). 360 V-8 engine ($420).

HISTORICAL: Introduced: Fall, 1979. Calendar year registrations: 81,923. Calendar year sales: (all models and series) 77,852 including: (CJ models) 47,304, (J Series pickup) 8,656. Model year production totaled 66,520 units. This included 10,234 Wagoneers; 11,541 Cherokees; 14,156 CJ-5s; 20,191 CJ-7s; 8,394 J-10s and 2,004 J-20s. Innovations: Four-cylinder engine returns to CJ Series. New top-of-the-line Laredo package available for pickup and CJ models. Jeep dealers recorded an average of 45 sales per outlet in 1980, down from an average of 77 sales per outlet in 1979 and 93 per outlet in 1978. Jeep vehicles continued to be produced exclusively in the company's Toledo, Ohio factory. The 151 cid four, a Pontiac-built power plant, was made standard equipment in the CJs and 9,347 got it. Jeep Corp. also constructed 1,028 CJs with a four-cylinder diesel engine, but these were made strictly for export. Also, the 258 cid six was re-introduced as an economy option for the Wagoneer.

1981 JEEP

JEEP — 1981 SERIES — (ALL ENGINES): The Wagoneer continued as Jeep's six-passenger, four-wheel drive, four-door station wagon. A 258 cid (4.2-liter) six was standard (360 cid V-8 in California), as was automatic transmission with Quadra-Trac.

Cherokees continued to come in three basic models. Model 16 was a five-passenger two-door station wagon with part-time four-wheel drive and a standard 258 cid six-cylinder engine. Model 17 was the same basic vehicle in "Wide-Stance" format with fatter wheels and tires. Model 18 was the four-door Cherokee wagon.

CJ-5s represented a two/four-passenger four-wheel drive sports utility vehicle in open body or optional hardtop versions. Standard engine was the 151 cid (2.5-liter) four-cylinder linked to a four-speed manual transmission.

CJ-7s represented a longer wheelbase two/four-passenger four-wheel drive sports utility vehicle in open body or optional hardtop versions with rear-mounted or swing-away spare tire. Standard engine was again the 151 cid (2.5-liter) four-cylinder linked to a four-speed manual transmission.

The CJ-8 was a new-for-1981 sporty model, aimed at the younger or young-at-heart buyer.

CJ Jeeps for 1981 were fitted with a longer side step. CJ-7 models with the optional soft/ metal top feature had a new vent window. The Renegade package, again available for either the CJ-5 or CJ-7 Jeeps, had a new graphics package in gradations of yellow, blue or red. All CJ models had a steering damper and a front anti-roll bar, plus a new front axle assembly with gas-filled upper and lower ball joint sockets.

1981 AMC Jeep, CJ-5 Renegade Universal Jeep (OCW)

The major physical change made to the J-trucks was a new lightweight, all-plastic slotted grille. Less noticeable, but important to the J-truck's fuel economy, was a standard front air dam and the elimination of the pickups roof lip. The J-10 model was lowered by 1.25 inches, due to redesigned front and rear springs. Power steering was now standard on the J-10 models. All J-series models had low-drag brakes.

The Jeep's optional automatic transmission, which since 1980 had been the Chrysler TorqueFlite unit, now had a locking torque converter. The American Motors 258 cid six-cylinder engine was redesigned to reduce its overall weight from 535 to 445 pounds.

An important addition to the Jeep line was the Scrambler pickup, which was depicted as an import fighter.

1981 AMC Jeep, CJ-7 Laredo Univeral Jeep (OCW)

I.D. DATA: VIN located on the top left surface of the instrument panel. There are 17 symbols. The first three symbols indicate the manufacturer, make and vehicle type. The engine type is indicated by the fourth symbol: B=151 cid four-cylinder; C=258 cid six-cylinder; N=360 cid V-8. The next letter identifies the transmission type. The series and type of vehicle are identified by the sixth and seventh symbols, which are the same as the model number. The eighth character identifies the GVW rating with the next serving as the check digit. Then follows the model year and assembly point codes. The last six digits are sequential production numbers.

1981 AMC Jeep, Cherokee Chief four-door station wagon (OCW)

Model	Body Type	Price	Weight	Prod. Total
Wagoneer — (1/2-Ton) — (4x4)				
15	4dr Sta Wag	10,464	3779	—
15	4dr Ltd Sta Wag	15,164	3800	—

288

Cherokee — (1/2-Ton) — (4x4)				
16	2dr Sta Wag	9574	3699	—
17	2dr Wide Stance	9837	3748	—
18	4dr Sta Wag	10,722	3822	—
CJ-5 — (1/4-Ton) — (4x4) — (84 in. w.b.)				
85	Jeep	7240	2495	13,477
CJ-7 — (1/4-Ton) — (4x4) — (94 in. w.b.)				
87	Jeep	7490	2520	27,767
CJ-8 Scrambler — (1/4-Ton) — (4x4) — (104 in. w.b.)				
88	Jeep	7288	2650	8,355
J-10 — (1/2-Ton) — (4x4) — (119/131 in. w.b.)				
25	Townside Pickup (SWB)	7960	3702	—
26	Townside Pickup (LWB)	8056	3764	—
J-20 — (3/4-Ton) — (4x4) — (131 in. w.b.)				
27	Townside Pickup	8766	4308	1,534

NOTE: See "Historical" for additional production data.

1981 AMC Jeep, J-10 Laredo Townside pickup

ENGINE (Standard: CJ): Inline. OHV. Four-cylinder. Cast iron block. Bore & stroke: 4.00 x 3.00 in. Displacement: 151 cid. Compression ratio: 8.2:1. Taxable horsepower: 21.70. Hydraulic valve lifters. Carburetor: Two-barrel.

ENGINE (Standard: J-10/CJ): Inline. OHV. Six-cylinder. Cast iron block. Bore & stroke: 3.75 x 3.90 in. Displacement: 258 cid. Compression ratio: 8.0:1. Net horsepower: 110. Taxable horsepower: 33.75. Seven main bearings. Hydraulic valve lifters. Carburetor: Single two-barrel.

ENGINE (Standard: CJ-20; Optional: J-10): Vee-block. OHV. Eight-cylinder. Cast iron block. Bore & stroke: 4.08 x 3.44 in. Displacement: 360 cid. Compression ratio: 8.25:1. Net horsepower: 175. Taxable horsepower: 53.27. Five main bearings. Hydraulic valve lifters. Carburetor: Single two-barrel.

1981 AMC Jeep, J-10 Honcho Sportside pickup (JAG)

TECHNICAL (CJ/Scrambler): Manual, synchronized transmission. Speeds: 4F/1R. Floor-mounted gearshift. Single plate dry disc clutch. Semi-floating rear axle. Overall ratio: (four-cylinder) 3.73:1, (six-cylinder) 3.07:1, (V-8) 3.07:1. Manual front disc/rear drum brakes. Steel wheels. Technical Options: Automatic transmission (CJ-7). Quadra-Trac with automatic transmission (CJ-7). Heavy-duty shock absorbers. Steering stabilizer. Automatic hubs. Air overload kit. Heavy-duty cooling system. Snowplows.

TECHNICAL (Wagoneer/Cherokee/J-Series): Fully synchronized, manual transmission. Speeds: 4F/1R. Floor-mounted gearshift. Single plate dry disc clutch. (J-10) Semi-floating rear axle; (J-20) full-floating rear axle. Front disc/rear drum brakes. Pressed steel wheels. Technical Options: Automatic transmission, column mounted. Quadra-Trac (with automatic transmission). Heavy-duty cooling system. Heavy-duty shock absorbers. Helper springs. Cruise control. Automatic hubs (not available with Quadra-Trac). Engine block heater. Extra-duty suspension package. Heavy-duty battery. Heavy-duty alternator. Cold climate group. Snow boss package.

1981 AMC Jeep, Wagoneer Limited four-door station wagon (JAG)

CHASSIS (Wagoneer): Wheelbase: 108.7 in. Overall length: 183.5 in. Width: 75.6 in. Height: 65.9 in. GVW maximum: 6100 lbs. Cargo capacity: 95.1 cu. ft. Front overhang: 29.9 in. Rear overhang: 44.9 in. Front suspension: Leaf springs on live axle with stabilizer bar. Rear suspension: Leaf springs on live axle with tube shocks. Front brakes: power disc. Rear brakes: drum. Steering: Power-assisted; 17:1 ratio. Turning circle: 37.7 ft. Fuel: 20.3 gallons.
CHASSIS (Cherokee): Wheelbase: 108.7 in. Overall length: 183.5 in. Width: 75.6 in. Height: 66.8 in. GVW: 6100 lbs. maximum. Cargo capacity: 95.1 cu. ft. without rear seat. Front overhang: 29.9 in. Rear overhang: 44.9 in. Front suspension: Leaf springs on live axle with stabilizer bar. Rear suspension: Leaf springs on live axle with tube shocks. Front brakes: power disc. Rear brakes: drum. Steering: Power-assisted; 17:1 ratio. Turning circle: 37.7 ft. Fuel: 20.3 gallons.
CHASSIS (CJ-5): Wheelbase: 83.5 in. Overall length: 144.3 in. standard or 134.8 in. with rear-mounted spare; Width: 68.6 in. Height with open body, 67.6 in. GVW maximum: 3700 lbs. Cargo capacity: 10.2 cu. ft. without rear seat. Front overhang: 23.5 in. Rear overhang: 37.3 in. standard or 27.8 in. with rear-mounted spare. Front suspension: leaf spring on live axle with tube shocks. Rear suspension: leaf springs on live axle with tube shocks. Front brakes: manual discs. Rear brakes: drum. Steering: manual re-circulating ball, 24:1 ratio, 34.1 ft. turning circle. Fuel: 15.5 gallons.
CHASSIS (CJ-7): Wheelbase: 93.5 in. Overall length: 153.2 in. standard or 144.3 in. with rear-mounted spare; Width: 68.6 in. Height with open body, 67.6 in. GVW maximum: 4100 lbs. Cargo capacity: 13.6 cu. ft. without rear seat. Front overhang: 23.5 in. Rear overhang: 36.2 in. standard or 27.3 in. with rear-mounted spare. Front suspension: leaf spring on live axle with tube shocks. Rear suspension: leaf springs on live axle with tube shocks. Front brakes: manual discs. Rear brakes: drum. Steering: manual re-circulating ball, 24:1 ratio, 38 ft. turning circle. Fuel: 15.5 gallons.
CHASSIS (CJ-8 Scrambler): Wheelbase: 103.5 in. Overall length: 177.3 in. Height: 67.6 in. (70.5 w/hardtop). Front tread: 51.5 in. Rear tread: 50.0 in. Tires: H78-15 in.
CHASSIS (J-10 SWB): Wheelbase: 118.8 in. Overall length: 194.0 in. Height: 68.5 in. Front tread: 63.3 in. Rear tread: 63.8 in. Tires: H78-15.
CHASSIS (J-10 LWB): Wheelbase: 130.8 in. Overall length: 206 in. Height: 68.3 in. Front tread: 63.3 in. Rear tread: 63.8 in. Tires: H78-15.
CHASSIS (J-10 Sportside): Wheelbase: 118.8 in. Overall length: 194 in. Height: 69.1 in. Front tread: 63.3 in. Rear tread: 63.8 in. Tires: H78-15.
CHASSIS (J-20): Wheelbase: 130.8 in. Overall length: 206 in. Height: 70.7 in. Front tread: 64.9 in. Rear tread: 65.9 in. Tires: 8.75 x 16.5.

1981 AMC Jeep, Scrambler pickup (OCW)

OPTIONS (CJ-5): Special color combo paint ($29). Vinyl bucket seats with denim trim ($104). Optional axle ratios ($29). Center console without Laredo package ($60). Rear seat without Renegade or Laredo packages ($170). 258 cid six-cylinder engine ($136). 304 cid V-8 engine ($345). California emissions ($80). Rear Trac-Loc ($179). Laredo package soft top ($2,049). Renegade package, soft top ($945). Power front disc brakes ($79). Power steering, required with air conditioning ($206). AM radio and antenna ($90). AM/FM stereo with antenna and two speakers ($224). 63-ampere alternator, except included with Climate Group ($53). High-output battery, except included with Climate group ($42). Cold climate group without air conditioning ($104); with air conditioning ($52). Heavy-duty cooling system, included free with air conditioning ($104). Extra-heavy-duty suspension, with Laredo ($69); without Laredo ($98). Front and rear heavy-duty shocks ($30). Body side step ($24). Draw bar ($41). Roll bar accessory package ($104). Side-mount spare and removable tailgate ($25). Black or white vinyl soft top ($288). Denim vinyl soft top, blue or nutmeg color, with denim interior only ($333). Bumperettes, included in Laredo package ($28). Bumper accessory package, not available with Laredo package ($104). Front and rear floor carpet ($100). Padded dashboard, included with Renegade or Laredo packages ($52). Air conditioning, plus power steering, carpeting and six or V-8 engine required ($591). Convenience group, included with Renegade and Laredo packages ($32). Halogen fog lamps ($83).

Halogen headlamps ($48). Stowage box, not available with sunroof ($94). Tachometer and Rally clock, included in Laredo package for no extra cost ($89). Tilt steering wheel ($81). Soft-Feel steering wheel, except included in Renegade package ($47). Decor group, except included in Laredo ($132). Four 15-inch wheelcovers ($46). Five chrome-plated 15x18 inch styled steel wheels ($147).

1981 AMC Jeep, Wagoneer Brougham four-door station wagon (JAG)

OPTIONS (CJ-7): Special color combo paint ($29). Vinyl bucket seats with denim trim ($104). Optional axle ratios ($29). Center console without Laredo package ($60). Rear seat without Renegade or Laredo packages ($170). 258 cid six-cylinder engine ($136). 304 cid V-8 engine ($345). California emissions ($80). Rear Trac-Loc ($179). Laredo package soft top ($2,049). Renegade package, soft top ($945). Power front disc brakes ($79). Power steering, required with air conditioning ($206). AM radio and antenna ($90). AM/FM stereo with antenna and two speakers ($224). 63-ampere alternator, except included with Climate Group ($53). High-output battery, except included with Climate group ($42). Cold climate group without air conditioning ($104); with air conditioning ($52). Heavy-duty cooling system, included free with air conditioning ($104). Extra-heavy-duty suspension, with Laredo ($69); without Laredo ($98). Front and rear heavy-duty shocks ($30). Body side step ($24). Draw bar ($41). Roll bar accessory package ($104). Side-mount spare and removable tailgate ($25). Black or white vinyl soft top ($288). Denim vinyl soft top, blue or nutmeg color, with denim interior only ($333). Bumperettes, included in Laredo package ($28). Bumper accessory package, not available with Laredo package ($104). Front and rear floor carpet ($100). Padded dashboard, included with Renegade or Laredo packages ($52). Air conditioning, plus power steering, carpeting and six or V-8 engine required ($591). Convenience group, included with Renegade and Laredo packages ($32). Halogen fog lamps ($83). Halogen headlamps ($48). Stowage box, not available with sunroof ($94). Tachometer and Rally clock, included in Laredo package for no extra cost ($89). Tilt steering wheel ($81). Soft-Feel steering wheel, except included in Renegade package ($47). Decor group, except included in Laredo ($132). Four 15-inch wheelcovers ($46). Five chrome-plated 15x18 inch styled steel wheels ($147). Automatic transmission with part-time four-wheel drive ($350). Laredo package with hardtop ($2,520). Sunroof, rear seat mandatory ($249). Metal doors for optional soft top ($190). Black or white hardtop with doors ($710). Nutmeg hardtop with doors and bronze-toned glass ($837).

OPTIONS (Wagoneer): Two-tone paint ($184). Special non-recommended color combinations ($29). Vinyl or fabric bucket seat with center armrest ($158). Optional axle ratio ($29). 360 V-8 ($345). California emissions ($80). Automatic transmission and part-time four-wheel drive ($350). Automatic Quadra-Trac transmission ($350). Trac-Loc, except not available with Quadra-Trac ($179). Wagoneer Brougham package ($935). Snow Boss package ($1,323). Trailer towing package A ($86). Trailer towing package B ($168). Power door locks ($147). Power side windows and door locks ($372). Power tailgate window, included in Brougham package; without Brougham option ($67). Power six-way driver's bucket seat ($158). Power six-way bucket seats ($263). AM push-button radio with antenna ($90). AM/FM stereo radio with antenna and four speakers ($224). AM/FM CB stereo with antenna and four speakers ($420). AM/FM casette stereo with antenna and four speakers ($325). AM/FM 8-Track with antenna and four speakers ($320). Premium audio system, FM only ($100). 70-Amp. Heavy-duty alternator; standard with air conditioning, rear window defroster and Cold Climate group; otherwise ($53). Heavy-duty battery, except standard with Cold Climate group ($42). Cold Climate group with air conditioning or rear window defroster ($52). Auxiliary automatic transmission oil cooler ($48). Heavy-duty cooling system ($46). Extra-duty suspension ($132). Extra-heavy-duty front springs ($13). Soft Ride suspension ($44). Heavy-duty shocks ($30). Inside spare tire mount and cover ($69). Roof rack with adjustable top bars, included with Brougham; otherwise ($109). Removable sunroof ($349). Bumper guards ($72). Bumper guards and nerfing strips ($113). Retractable cargo area cover ($62). Carpeted cargo floor and insulation ($81). Extra quiet insulation, standard with Brougham; otherwise ($136). Front and rear protective floor mats, standard with Brougham package; otherwise ($21). Scuff moldings, standard with Brougham package; otherwise ($92). Air conditioning, including heavy-duty cooling and alternator ($591). All tinted glass ($75). Convenience group, standard with Brougham option; otherwise ($71). Cruise control ($132). Halogen fog lamps ($83). Light group ($73). Left- and right-hand remote-control mirrors ($132). Dual mirrors ($69). Rear window defroster ($108). Tilt steering ($81). Visibility group ($143). Leather sport steering wheel ($69). Four chrome styled wheels ($226). Four forged aluminum wheels ($339); same with Brougham option ($124). Woodgrain side panels ($146); same with Brougham option ($75).

OPTIONS (Cherokee two-door): Special non-recommended color combinations ($29). 360 V-8 ($345). California emissions ($80). Automatic transmission and part-time four-wheel drive ($350). Automatic Quadra-Trac transmission ($350). Trac-Loc, except not available with Quadra-Trac ($179). Snow Boss package ($1,323). Trailer towing package A ($86). Trailer towing package B ($168). Power tailgate window ($67). Power six-way driver's bucket seat ($158). Power six-way bucket seats ($263). AM push-button radio with antenna ($90). AM/FM stereo radio with antenna and four speakers ($224). AM/FM CB stereo with antenna and four speakers ($420). AM/FM casette stereo with antenna and four speakers ($325). AM/FM 8-Track with antenna and four speakers ($320). Premium audio system, FM only ($100). 70-Amp. Heavy-duty alternator; standard with air conditioning, rear window defroster and Cold Climate group; otherwise ($53). Heavy-duty battery, except standard with Cold Climate group ($42). Cold Climate group with air conditioning or rear window defroster ($52). Auxiliary automatic transmission oil cooler ($48). Heavy-duty cooling system ($46). Extra-duty suspension ($132). Extra-heavy-duty front springs ($13). Soft Ride suspension ($44). Heavy-duty shocks ($30). Inside spare tire mount and cover ($69). Roof rack with adjustable top bars ($109). Removable sunroof ($349). Bumper guards ($72). Bumper guards and nerfing strips ($113). Retractable cargo area cover ($62). Carpeted cargo floor and insulation ($81). Extra quiet insulation ($136). Front and rear protective floor mats otherwise ($21). Air conditioning, including heavy-duty cooling and alternator ($591). All tinted glass ($75). Convenience group ($71). Cruise control ($132). Halogen fog lamps ($83). Light group ($73). Left- and right-hand remote-control mirrors ($132). Dual mirrors ($69). Rear window defroster ($108). Tilt steering ($81). Visibility group ($143). Four styled chrome wheels included in Laredo; on Cherokee Chief ($156). Fabric bucket seats ($53). Cherokee Chief package ($839). Laredo package ($1,680). Chrome grille, standard with Laredo; other models ($53). Power door locks ($98). Power side windows and door locks ($243). Cold climate package ($104). Fixed center armrest ($67). Dual heavy-duty front shocks with heavy-duty rears ($90). Rear quarter vent window, included in Cherokee Chief/Laredo; otherwise ($79). Power sun roof, reg. Laredo ($1,460). Soft Feel steering wheel ($47). Vinyl body side moldings ($75).

OPTIONS (Cherokee four-door): Special non-recommended color combinations ($29). 360 V-8 ($345). California emissions ($80). Automatic transmission and part-time four-wheel drive ($350). Automatic Quadra-Trac transmission ($350). Trac-Loc, except not available with Quadra-Trac ($179). Snow Boss package ($1,323). Trailer towing package A ($86). Trailer towing package B ($168). Power tailgate window ($67). Power six-way driver's bucket seat ($158). Power six-way bucket seats ($263). AM push-button radio with antenna ($90). AM/FM stereo radio with antenna and four speakers ($224). AM/FM CB stereo with antenna and four speakers ($420). AM/FM casette stereo with antenna and four speakers ($325). AM/FM 8-Track with antenna and four speakers ($320). Premium audio system, FM only ($100). 70-Amp. Heavy-duty alternator; standard with air conditioning, rear window defroster and Cold Climate group; otherwise ($53). Heavy-duty battery, except standard with Cold Climate group ($42). Cold Climate group with air conditioning or rear window defroster ($52). Auxiliary automatic transmission oil cooler ($48). Heavy-duty cooling system ($46). Extra-duty suspension ($132). Extra-heavy-duty front springs ($13). Soft Ride suspension ($44). Heavy-duty shocks ($30). Inside spare tire mount and cover ($69). Roof rack with adjustable top bars ($109). Removable sunroof ($349). Bumper guards ($72). Bumper guards and nerfing strips ($113). Retractable cargo area cover ($62). Carpeted cargo floor and insulation ($81). Extra Quiet insulation ($136). Front and rear protective floor mats ($21). Air conditioning, including heavy-duty cooling and alternator ($591). All tinted glass ($75). Convenience group ($71). Cruise control ($132). Halogen fog lamps ($83). Light group ($73). Left- and right-hand remote-control mirrors ($132). Dual mirrors ($69). Rear window defroster ($108). Tilt steering ($81). Visibility group ($143). Four styled chrome wheels included in Laredo; on Cherokee Chief ($156). Fabric bucket seats ($53). Cherokee Chief package ($839). Laredo package ($1,680). Chrome grille, standard with Laredo; other models ($53). Power door locks ($138). Power side windows and door locks ($358). Cold climate package ($104). Fixed center armrest ($67). Dual heavy-duty front shocks with heavy-duty rears ($90). Rear quarter vent window, included in Cherokee Chief/ Laredo; otherwise ($79). Power sunroof, reg. Laredo ($1,460). Soft Feel steering wheel ($47). Vinyl body side moldings ($75). Forged aluminum wheels ($239).

OPTIONS (J-Series Trucks): Two-tone paint, not available with Honcho or Laredo; other models ($183). Special paint ($29). Fabric bench seat cushions included in Laredo; in other models ($24). Vinyl bucket seats, included with Laredo; in other models ($177). Fabric bucket seats, included with Honco Sportside; with regular Honcho ($177); with other models ($200). Axle ratios ($29). 360 cid V-8 ($345). California emissions system ($80). Automatic transmission with part-time four-wheel drive ($350). Quadra-Trac ($350). Trac-Loc ($179). Sportside truck package ($944). Honcho Sportside truck package ($1,392). Custom package ($157). Honcho package ($892). Laredo ($1,680). Cap for Townside box ($638). Painted rear step bumper, included in Honcho and Sportside packages ($90). Chrome rear step bumper ($151). Townside box delete ($140 credit). Brush guard instead of bumper guard, with Laredo ($38); other models ($80). 15 x 8 chrome styled wheels ($118). Chrome grille ($53).

HISTORICAL: Introduced: March 25, 1981 (Scrambler). Calendar year registrations: 58,257. Calendar year sales: (all models and series) 63,275 divided as follows: (CJ models) 30,564; (Scrambler) 7,840; (J-Series Pickups) 6,516. Model year production totaled 77,560 vehicles, including: 13,594 Wagoneers; 6,557 Cherokees; 13,447 CJ-5s; 380 CJ-6 exports; 27,767 CJ-7s; 8,335 CJ-8/Scramblers; 5,916 J-10s and 1,534 J-20s. Innovations: Renegade package revised. Improvements to CJ front suspension. Jeep trucks restyled for fuel economy savings. Power steering becomes standard and on J-10. All-new "Scrambler" model introduced. Historical notes: AMC had a total of 1,589 dealers in 1981. The company posted a loss of $160.9 million for the year. Worldwide unit sales of Jeeps totaled $104,628 million. Jeep sales per outlet continued to fall, sliding to an average of 34 units.

1982 JEEP

JEEP — 1982 SERIES — (ALL ENGINES): Like other manufacturers, Jeep Corp. was becoming more and more involved with the marketing of specialty models created by adding options to the Wagoneer, Cherokee, Jeep and Jeep truck.

In the Wagoneer line, the Custom Wagoneer came with power steering; full wheel covers; carpets; vinyl bench seats; body side moldings; a cargo mat; chrome grille and electronic ignition. The Wagoneer Brougham added cloth or deluxe vinyl upholstery; custom door trim; woodgrained dash trim; power tailgate window; styled wheels; lower tailgate molding; light group; halogen headlamps; woodgrain body side rub strips; convenience options group and visibility group. The Wagoneer Limited also included: full body side and tailgate woodgrain trim; special woodgrained luggage rack; 15-inch forged aluminum wheels; Goodyear Arriva white sidewall tires; bucket seats; woodgrained door trim; air conditioning; cruise control; tilt steering; premium AM/FM sound system; power windows and power door locks.

The Cherokee continued to be available in regular and "Wide-Wheel" two-door wagons and a four-door wagon. The base Cherokee included power steering, power brakes, four-speed manual transmission and a part-time four-wheel dive system. The Cherokee Chief added upper and lower black-out treatments; body side and tailgate striping; a blacked-out grille; Sun Valley front bucket seats and a Soft-Feel sports steering wheel. Additional equipment in the Cherokee Laredo package (available for the four-door or "Wide-Wheel" two-door) included special exterior striping; 15 x 8 chrome styled wheels with black hubs; Goodyear Wrangler radial tires (Wide-Wheel) or Goodyear Arriva white sidewall tires (four-door); bright moldings; roof rack; special folding bucket seats; Cara vinyl or Western Weave upholstery fabrics; extra carpeting; and a leather-wrapped steering wheel.The Scrambler was now delivered with a new Space-Saver spare tire mounted on the roll bar, instead of on the tailgate (as in 1981). The older arrangement, with a standard full-sized spare, was available as an option.

1982 AMC Jeep, CJ-7 Laredo Universal Jeep (RPZ)

Both the CJ-7 and J-series trucks were offered in new trim packages for 1982. The most refined CJ, the CJ-7 Limited, was depicted as the Jeep intended for "that special breed of drivers who want to blend an upgraded level of comfort and decor with their sports action on road and off." The Limited package included all base CJ-7 equipment, plus power steering and brakes; AM/FM radio with two speakers; a monochromatic paint scheme with color-keyed hardtop and wheel lip extensions; special dual-color body side striping; special grille panel; exterior Limited nameplates; and a host of other features. Of these, the most controversial was the special improved ride package, which gave the Limited a far softer ride than any other CJ in history.

Both the CJ-7 and Scrambler now had a wider front and rear tread. Respective increases were 3.4 inches and 4.6 in. All models, except the CJ-5 with the 258 cid six-cylinder engine, were available with a five-speed manual T5 transmission supplied by Warner Gear. All six-cylinder CJ Jeeps were also available with the wide-ratio, three-speed automatic transmission with a lock-up converter. CJs with manual transmissions were also available with cruise control.

The new Pioneer Townside package, available for both the J-10 and J-20, included all features found in the Custom package, plus upper side scuff molding; tailgate stripes; "Pioneer" decals; dark argent-painted grille; carpeted cab floor; front bumper guards; and full wheelcovers (J-10 only), plus several other interior and exterior features. Otherwise, changes to the J-series trucks were extremely limited. The most important drivetrain development was the availability of the five-speed gearbox.

I.D. DATA: VIN located on the top left surface of the instrument panel. There are 17 symbols. The first three symbols indicate the manufacturer, make and vehicle type. The engine type is indicated by the fourth symbol: B=151 cid four-cylinder; C=258 cid six-cylinder; N=360 cid V-8. The next letter identifies the transmission type. The series and type of vehicle are identified by the sixth and seventh symbols, which are the same as the model number. The eighth character identifies the GVW rating with the next serving as the check digit. Then follows the model year (tenth syumbol C) and assembly point codes. The last six digits are sequential production numbers.

Model	Body Type	Price	Weight	Prod. Total
Wagoneer — (1/2-Ton) — (4x4)				
15	4dr Sta Wag	11,114	3797	—
15	4dr Sta Wag Brgm	12,084	—	—
15	4dr Sta Wag Ltd	15,964	—	—
Cherokee — (1/2-Ton) — (4x4)				
16	2dr Sta Wag	9849	3692	—
17	2dr Wide Stance Wag	10,812	3741	—
18	4dr Sta Wag	11,647	3296	—
CJ-5 — (1/4-Ton) — (4x4) — (84 in. w.b.)				
85	Jeep	7515	2489	6,080
CJ-7 — (1/4-Ton) — (4x4) — (93.4 in. w.b.)				
87	Jeep	7765	2555	23,820
Scrambler — (1/2-Ton) — (4x4) — (103.4 in. w.b.)				
88	Pickup	7588	2093	7,759
J-10 — (1/2-Ton) — (4x4) — (119/131 in. w.b.)				
25	Townside Pickup (SWB)	8610	3656	—
26	Townside Pickup (LWB)	8756	3708	—
J-20 — (3/4-Ton) — (4x4) — (131 in. w.b.)				
27	Townside Pickup	9766	4270	1,418

NOTE 1: See "Historical" for additional production totals.
NOTE 2: J-10 Sportside option has 118.8 in. w.b.

ENGINE (Standard: Wagoneer/Cherokee/J-10/Optional Scrambler/CJ-5/CJ-7): Inline. OHV. Six-cylinder. Cast iron block. Bore & stroke: 3.75 x 3.90 in. Displacement: 258 cid Compression ratio: 8.3:1. Net horsepower: 110 at 3000 rpm. Net torque: 205 lbs.-ft. at 1800 rpm. Seven main bearings. Hydraulic valve lifters. Carburetor: Single two-barrel.

ENGINE (Standard: Scrambler/CJ-5/CJ-7): Inline. OHV. Four-cylinder. Cast iron block. Bore & stroke: 4.00 x 3.00 in. Displacement: 151 cid. Compression ratio: 8.2:1. Net horsepower: 82 at 4000 rpm. Torque: 125 lbs.-ft. at 2600 rpm. Hydraulic valve lifters. Carburetor: Single Two-barrel.

1982 AMC Jeep J-10 Honcho pickup (RPZ)

ENGINE (Standard: J-20; Optional: Wagoneer/Cherokee/J-10): Vee-block. OHV. Eight-cylinder. Cast iron block. Bore & stroke: 4.08 x 3.44 in. Displacement: 360 cid. Compression ratio: 8.25:1. Net horsepower: 150 at 3400 rpm. Net torque: 205 lbs.-ft. at 1500 rpm. Five main bearings. Hydraulic valve lifters. Carburetor: Single two-barrel.

CHASSIS (Wagoneer): Wheelbase: 108.7 in. Overall length: 183.5 in. Width: 75.6 in. Height: 65.9 in. GVW maximum: 6100 lbs. Cargo capacity: 95.1 cu. ft. Front overhang: 29.9 in. Rear overhang: 44.9 in. Front suspension: Leaf springs on live axle with stabilizer bar. Rear suspension: Leaf springs on live axle with tube shocks. Front brakes: power disc. Rear brakes: drum. Steering: Power-assisted; 17:1 ratio. Turning circle: 37.7 ft. Fuel: 20.3 gallons.

CHASSIS (Cherokee): Wheelbase: 108.7 in. Overall length: 183.5 in. Width: 75.6 in. Height: 66.8 in. GVW: 6100 lbs. maximum. Cargo capacity: 95.1 cu. ft. without rear seat. Front overhang: 29.9 in. Rear overhang: 44.9 in. Front suspension: Leaf springs on live axle with stabilizer bar. Rear suspension: Leaf springs on live axle with tube shocks. Front brakes: power disc. Rear brakes: drum. Steering: Power-assisted; 17:1 ratio. Turning circle: 37.7 ft. Fuel: 20.3 gallons.

CHASSIS (CJ-5): Wheelbase: 83.5 in. Overall length: 144.3 in. standard or 134.8 in. with rear-mounted spare; Width: 68.6 in. Height with open body, 67.6 in. GVW maximum: 3700 lbs. Cargo capacity: 10.2 cu. ft. without rear seat. Front overhang: 23.5 in. Rear overhang: 37.3 in. standard or 27.8 in. with rear-mounted spare. Front suspension: leaf spring on live axle with tube shocks. Rear suspension: leaf springs on live axle with tube shocks. Front brakes: manual discs. Rear brakes: drum. Steering: manual re-circulating ball, 24:1 ratio, 34.1 ft. turning circle. Fuel: 15.5 gallons.

CHASSIS (CJ-7): Wheelbase: 93.5 in. Overall length: 153.2 in. standard or 144.3 in. with rear-mounted spare; Width: 68.6 in. Height with open body, 67.6 in. GVW maximum: 4100 lbs. Cargo capacity: 13.6 cu. ft. without rear seat. Front overhang: 23.5 in. Rear overhang: 36.2 in. standard or 27.3 in. with rear-mounted spare. Front suspension: leaf spring on live axle with tube shocks. Rear suspension: leaf springs on live axle with tube shocks. Front brakes: manual discs. Rear brakes: drum. Steering: manual re-circulating ball, 24:1 ratio, 38 ft. turning circle. Fuel: 15.5 gallons.

CHASSIS (CJ-8 Scrambler): Wheelbase: 103.5 in. Overall length: 177.3 in. Height: 67.6 in. (70.5 w/hardtop). Front tread: 51.5 in. Rear tread: 50.0 in. Tires: H78-15 in.

CHASSIS (J-10 SWB): Wheelbase: 118.8 in. Overall length: 194.0 in. Height: 68.5 in. Front tread: 63.3 in. Rear tread: 63.8 in. Tires: H78-15.

CHASSIS (J-10 LWB): Wheelbase: 130.8 in. Overall length: 206 in. Height: 68.3 in. Front tread: 63.3 in. Rear tread: 63.8 in. Tires: H78-15.

CHASSIS (J-10 Sportside): Wheelbase: 118.8 in. Overall length: 194 in. Height: 69.1 in. Front tread: 63.3 in. Rear tread: 63.8 in. Tires: H78-15.

CHASSIS (J-20): Wheelbase: 130.8 in. Overall length: 206 in. Height: 70.7 in. Front tread: 64.9 in. Rear tread: 65.9 in. Tires: 8.75 x 16.5.

1982 AMC Jeep, Cherokee two-door station wagon (RPZ)

TECHNICAL: Selective synchromesh transmission. Speeds: 4F/1R. Floor-mounted gearshift. Single plate dry disc clutch. Shaft drive. Rear axle: (J-20) full-floating; (others) semi-floating. Overall drive ratio: 3.54:1. Four-wheel hydraulic front disc/rear drum power brakes. (* Power brakes optional on models 85/87/88). Steel disc wheels.

291

1982 AMC Jeep, Scrambler pickup (RPZ)

1982 AMC Jeep, Wagoneer Limited (RPZ)

1983 JEEP

JEEP — 1983 SERIES — (ALL ENGINES): The most important Jeep developments for 1983 were technical in nature. A new full-time four-wheel/two-wheel drive system, called Selec-Trac, replaced Quadra-Trac, which had been introduced in 1973. Selec-Trac was optional on J-10 pickups with either six-cylinder or V-8 engines. Selec-Trac featured a two-speed capability in the transfer case for added torque. To engage either two- or four-wheel drive, the driver of a Selec-Trac J-10 stopped the vehicle and activated a switch on the instrument panel. A safety catch was provided to avoid accidental movement of the Selec-Trac switch.

Technical refinements to the 258 cid six-cylinder engine included an increase in compression ratio to 9.2:1 from 8.6:1 and the addition of a fuel feedback system and knock sensor to improve performance and efficiency.

Appearance changes were held to a minimum. Both the Renegade package, for the CJ-5 and CJ-7, as well as the Scrambler's SR trim package, had new striping arrangements.

A new XJ model was in the works for introduction during 1983, but delays in the production startup led to the XJ line being introduced in the fall of 1983 as a new-for-1984 series.

I.D. DATA: VIN located on the top left surface of the instrument panel. There are 17 symbols. The first three symbols indicate the manufacturer, make and vehicle type. The engine type is indicated by the fourth symbol: B=151 cid (2.5 liter) four-cylinder; C=258 cid (4.2 liter) six-cylinder; N=360 cid (6.0 liter) V-8; U=150 cid (2.5 liter) four-cylinder. The next letter identifies the transmission type. The series and type of vehicle are identified by the sixth and seventh symbols, which are the same as the model number. The eighth character identifies the GVW rating with the next serving as the check digit. Then follows the model year (tenth symbol D) and assembly point codes. The last six digits are sequential production numbers.

1983 AMC Jeep, CJ-5 Renegade Universal (RPZ)

Model	Body Type	Price	Weight	Prod. Total
Wagoneer — (1/2-Ton) — (4x4)				
15	4dr Brgm Sta Wag	13,173	3869	—
15	4dr Ltd Sta Wag	16,889	—	—
Cherokee — (1/2-Ton) — (4x4)				
16	2dr Sta Wag	10,315	3764	6,186
CJ-5 — (1/4-Ton) — (4x4) — (83.4 in. w.b.)				
85	Jeep	7515	2099	3,085
CJ-7 — (1/4-Ton) — (4x4) — (93.4 in. w.b.)				
87	Jeep	6995	2595	37,673
Scrambler — (1/2-Ton) — (4x4) — (103.4 in. w.b.)				
88	Pickup	6765	2733	5,407
J-10 — (1/2-Ton) — (4x4) — (119/131 in. w.b.)				
25	Townside Pickup (SWB)	9082	3728	—
26	Townside Pickup (LWB)	9227	3790	—
J-20 — (3/4-Ton) — (4x4) — (131 in. w.b.)				
27	Townside Pickup	10,117	4336	2,740

NOTE: See "Historical" for additional production totals.

OPTIONS: 258 cid six-cylinder engine, models 85/87/88 ($150). 360 cid V-8 engine, standard w/J-20 ($351). Five-speed transmission, except J-20 ($199). Automatic transmission ($409). Automatic Quadra-Trac ($455). Air conditioning ($681). Heavy-duty alternator ($59). Optional axle ratio, without wide wheels ($33). Heavy-duty battery ($47). Power brakes, models 85/87/88 ($99). Bumper accessory package, models 85/87/88 ($113). Retractable cargo area cover ($75). Townside box cap for pickups ($684). Cold Climate group ($115). Heavy-duty cooling system ($49). Cruise control ($159). Custom package, Townside pickups ($169). Decor Group, models 85/87/88 ($86). Trac-Loc differential ($219). Doors for Scrambler/CJ-7 with soft top ($229). California emissions system ($80). Tinted glass, pickups ($48). Chrome grille, pickups ($57). Halogen headlights on CJs ($56). Honcho Package, model 25 ($949). Honcho Sportside Package, J-10 Sportside ($470). Scrambler padded instrument panel ($56). Laredo Package, model 25 ($1,749). Laredo Package for CJ, with soft top ($2,149); with hardtop ($2,599). Light group, J-10/J-20 with roll bar ($49); without roll bar ($71). CJ-7 Limited package ($2895). Pioneer package, on J-10 Townside ($599); on J-20 ($577); on Scrambler/CJ ($950). AM radio ($99). AM/FM stereo ($229). AM/FM stereo w/CB ($456). AM/FM w/cassette, pickups ($339). AM/FM ETR stereo w/cassette, pickups ($479). Premium audio system, Limited ($115). CJ Renegade package ($979). Roll bar on pickups ($135). Roll bar access package, CJs ($979). Scrambler "SL" package, with soft top ($1,999); with hardtop ($2,399). Scrambler "SR" Sport package ($799). Pickup cloth bench seat trim ($25). Cloth bucket seats, J-10 w/ Pioneer package ($181); other pickups ($205). Vinyl bucket seats, Townside pickups ($181); Scrambler including Denim trim ($106). Dual low-profile mirrors (except Scrambler), with visibility group ($26); without visibily group ($85). Dual electric remote mirrors (except Scrambler), with visibility group ($83); without visibility group ($142). Heavy-duty shocks, regular ($35); with dual fronts on J-10 ($119). Air adjustable rear shocks, except Scrambler and J-20 ($66). Snow Boss package, except Scrambler ($1,399). Scrambler spare tire lock ($9). Heavy-duty rear springs, pickups ($48). Heavy-duty front springs, J-10 ($59). Extra heavy-duty front springs, pickups ($79). Power steering on Scrambler and CJs ($229). Tachometer and rally clock, Scrambler and CJs ($96). Hardtop with doors, Scrambler ($695); CJ ($775). Scrambler soft denim top ($300). Scrambler soft vinyl top ($280). CJ soft vinyl top ($330). Scrambler tonneau cover ($122). Visibility group, except Scrambler ($152). Wheelcovers, all models ($55). Forged aluminum wheels, J-10 Townside ($399). Forged aluminum wheels, J-10 Honcho or Sportside ($200). Power side windows and door locks, J-10/J-20 ($288). Sliding rear window, J-10/J-20 ($97). Wood side rails, Scrambler ($92). Chrome styled steel wheels, models 85/87/88 ($349); J-10 ($150).

HISTORICAL: Introduced: Fall, 1981. Innovations: Wider tread for CJ and Scrambler models. New option packages released. Five-speed manual transmission available. Total calendar year production: 75,269 (includes Wagoneer and Cherokee). Calendar year registrations: 62,097 (includes Wagoneer and Cherokee). Calendar year production by model: (CJ) 37,221; (Pickups) 6,113; (Scrambler) 6,315. Calendar year sales: 67,646 (U.S. and Canada). Model year production totaled 64,965 units. This included 14,489 Wagoneers; 6,013 Cherokees; 6,080 CJ-5s; 23,820 CJ-7s; 7,759 CJ-8/Scramblers; 5,026 J-10s; and 1,418 J-20s. W. Paul Tippett, Jr. became chairman of American Motors in 1982, replacing Gerald C. Meyers. Tippett was born Dec. 27, 1932 and graduated from Wabash College, in Cincinnati, Ohio. The company posted Jeep dollar sales of $90.7 million for Jeep vehicles in calendar 1982. AMC reported a net loss of $153.5 million on its cars and trucks during this period. Texans were the leading buyers of Jeep vehicles, based on the state's registration total of 6,492 vehicles. California was second with 4,132 registrations and Pennsylvania was third with 3,568 registration. Unit sales of Jeep vehicles, for the calendar year, were 67,646 for the U.S./Canadian market and an additional 23,083 for the international market.

1983 AMC Jeep, J-10 Laredo Townside pickup (RPZ)

ENGINE (Standard: 88; Optional: 85/87): Inline. OHV. Four-cylinder. Cast iron block. Bore & stroke: 4.0 x 3.0 in. Displacement: 151 cid. Compression ratio: 9.6:1. Net horsepower: 92. Five main bearings. Hydraulic valve lifters. Carburetor: Single two-barrel.

ENGINE (Standard: 85/25/26; Optional: 88/87): Inline. OHV. Six-cylinder. Cast iron block. Bore & stroke: 3.75 x 3.90 in. Displacement: 258 cid. Compression ratio: 9.2:1. Net horsepower: 102 at 3000 rpm. Net Torque: 204 lbs.-ft. at 1650 rpm. Seven main bearings. Hydraulic valve lifters. Carburetor: Single Two-barrel.

ENGINE (Standard: 27; Optional: 25/26): Vee-block. OHV. Eight-cylinder. Cast iron block. Bore & stroke: 4.08 x 3.44 in. Displacement: 360 cid. Brake horsepower: 170 at 4000 rpm. Taxable: 280 lbs.-ft. at 2400 rpm. Five main bearings. Hydraulic valve lifters. Carburetor: Two-barrel.

CHASSIS (Wagoneer): Wheelbase: 108.7 in. Overall length: 183.5 in. Width: 75.6 in. Height: 65.9 in. GVW maximum: 6100 lbs. Cargo capacity: 95.1 cu. ft. Front overhang: 29.9 in. Rear overhang: 44.9 in. Front suspension: Leaf springs on live axle with stabilizer bar. Rear suspension: Leaf springs on live axle with tube shocks. Front brakes: power disc. Rear brakes: drum. Steering: Power-assisted; 17:1 ratio. Turning circle: 37.7 ft. Fuel: 20.3 gallons.

CHASSIS (Cherokee): Wheelbase: 108.7 in. Overall length: 183.5 in. Width: 75.6 in. Height: 66.8 in. GVW: 6100 lbs. maximum. Cargo capacity: 95.1 cu. ft. without rear seat. Front overhang: 29.9 in. Rear overhang: 44.9 in. Front suspension: Leaf springs on live axle with stabilizer bar. Rear suspension: Leaf springs on live axle with tube shocks. Front brakes: power disc. Rear brakes: drum. Steering: Power-assisted; 17:1 ratio. Turning circle: 37.7 ft. Fuel: 20.3 gallons.

CHASSIS (CJ-5): Wheelbase: 83.4 in. Overall length: 142.9 in. Width: 68.6 in. Height with open body, 69.3 in. Front tread: 52.4 in. (53.9 in. with styled wheels). Rear tread: 50.5 in. GVW maximum: 3700 lbs. Cargo capacity: 10.2 cu. ft. without rear seat. Front overhang: 23.5 in. Rear overhang: 37.3 in. standard or 27.8 in. with rear-mounted spare. Front suspension: leaf spring on live axle with tube shocks. Rear suspension: leaf springs on live axle with tube shocks. Front brakes: manual discs. Rear brakes: drum. Steering: manual re-circulating ball, 24:1 ratio, 34.1 ft. turning circle. Fuel: 15.5 gallons. Tires: G78-15.

CHASSIS (CJ-7): Wheelbase: 93.4 in. Overall length: 153.5 in. standard or 144.3 in. with rear-mounted spare; Width: 68.6 in. Front tread: 55.8 in. Rear tread: 55.1 in. Height with open body, 69.3 in. GVW maximum: 4100 lbs. Cargo capacity: 13.6 cu. ft. without rear seat. Front overhang: 23.5 in. Rear overhang: 36.2 in. standard or 27.3 in. with rear-mounted spare. Front suspension: leaf spring on live axle with tube shocks. Rear suspension: leaf springs on live axle with tube shocks. Front brakes: manual discs. Rear brakes: drum. Steering: manual re-circulating ball, 24:1 ratio, 38 ft. turning circle. Fuel: 15.5 gallons. Tires: G78-15.

CHASSIS (CJ-8 Scrambler): Wheelbase: 103.4 in. Overall length: 168.9 in. Height: 69.5 in. (70.5 w/hardtop). Front tread: 55.8 in. Rear tread: 55.1 in. Tires: G78-15 in.

CHASSIS (Model 25): Wheelbase: 118.7 in. Overall length: 192.5 in. Height: 67.4 in. Front tread: 64 in. Rear tread: 63.8 in. Tires: P225-75R15.

CHASSIS (Model 26): Wheelbase: 130.7 in. Overall length: 204.5 in. Height: 67.5 in. Front tread: 66.0 in. Rear tread: 65.8 in. Tires: P225-75R15.

CHASSIS (Model 27): Wheelbase: 130.7 in. Overall length: 204.5 in. Height: 67.6 in. Front tread: 64.6 in. Rear tread: 65.9 in. Tires: 8.75 x 16.5.

1983 AMC Jeep, Scrambler SL sport pickup (RPZ)

TECHNICAL: Selective synchromesh transmission. Speeds: 4F/1R. Floor-mounted gearshift. Single plate dry disc clutch. Shaft drive. Rear axle: (J-20) full-floating; (others) semi-floating. Ratio: 3.54:1. Four-wheel hydraulic front disc/rear drum power brakes. (* Power brakes optional on models 85/87/88). Steel disc wheels.

OPTIONS (Wagoneer/Cherokee): Automatic transmission (standard in Wagoneer); Five-speed transmission ($225). AM/FM stereo, standard in Wagoneer Limited; in other models ($199). AM/FM stereo with tape ($329). Cruise control, standard in Wagoneer Limited; in other models ($184). Rear window defroster, standard in Wagoneer Limited; in other models ($134). Tilt steering wheel, standard in Wagoneer Limited; in other models ($106). Roof rack, standard on Wagoneer Limited; on other models ($122). Aluminum wheels (standard for Wagoneer Limited). Sunroof ($389). Power sunroof ($1,637). Power door locks. Power windows. Power seats, standard in Wagoneer Limited; in other models ($290). Laredo package ($1,128). Pioneer package ($1,131). Custom or two-tone paint. 360 cid V-8 engine ($393). Air conditioning.

OPTIONS (CJ/Scrambler): Renegade package ($1,011). Laredo package ($2220). Limited package ($3,595). Scrambler "SR" Sport package ($825). Scrambler "SL" Sport package ($2,065). Soft top. Hard top. Automatic transmission. Five-speed manual transmission. 258 cid six-cylinder engine.

OPTIONS (J10/J20): Custom package ($175). Pioneer package ($619). Laredo package ($1,807). Automatic transmission. Five-speed manual transmission. 360 cid V-8 engine.

HISTORICAL: Introduced: Fall, 1982. Innovations: New Select-Trac full-time four-wheel-drive system introduced. Higher compression, fuel feedback system and knock sensor added to 258 cid six-cylinder engine. Calendar year U.S. registrations: 76,453. Calendar year sales, U.S. and Canada: 93,169. U.S. Calendar year production: (total) 75,534; (CJ) 40,758; (Pickups) 4705; (Scrambler) 5407. Worldwide Calendar year production (Toledo plant): 113,263.

Model year output totaled 64,965. This total included: 14,849 Wagoneer; 6,013 Cherokee; 6,080 CJ-5s; 23,820 CJ-7s; 7,759 CJ-8/Scramblers; 5,026 J-10s and 1,418 J-20s. Company president was Paul W. Tippett, Jr. Historical notes: Sales of four-wheel-drive Jeep vehicles in 1983 were the best since 1979 and up 29 percent from the previous year. Chairman Tippett and president Jose J. Dedeurwaerder reported, in late 1983, that a particularly encouraging sign was the fact that "Jeep production at our Toledo plant is now higher than at any time since the plant was acquired from Kaiser in 1970." Worldwide wholesale unit sales of Jeeps totaled $113,443 million, up 25 percent from 1982. Worldwide unit sales of Jeeps, in the calendar year, peaked at 93,169 vehicles in the U.S. and Canada and 20,274 more for the international market. In state-by-state registrations, California leaped to second place with 6,004 Jeeps registered for calendar 1982. Texas was still first with 6,970 registrations.

On the international scene, the May, 1983 signing of an agreement with Beijing Jeep Corp. opened the door for a joint-venture in which AMC would help build Jeeps in the People's Republic of China. Efforts to strengthen the Latin American export market were also made by AMC-Jeep.

1984 JEEP

1984 AMC Jeep, Wagoneer Limited four-door station wagon (AMC)

JEEP — 1984 SERIES — (ALL ENGINES): The all-new, down-sized Cherokee and Wagoneer Sportwagons in the XJ series bowed as a 1984 models and nearly doubled the retail sales of Jeep vehicles by the end of the year.

The Cherokee came in two-door and four-door versions, while the Wagoneer Sportwagon was a four-door only. The new bodies had a short, squarish, but neatly tailored look. Large, square headlamps sat atop the horizontally-mounted rectangular parking lamps up front. The grille featured tall, vertical bars. There was a "glassy" greenhouse and flared wheel openings. Bright bumpers with black rubber end caps were seen in front. The XJs were 21 inches shorter, six inches narrower, four inches lower and 1,000 pounds lighter than the "senior" Grand Wagoneer.

At the time, the smaller new models were the only compact sports utility vehicles offering the four-door body style and two types of four-wheel drive systems. The systems were the SelecTrac and shift-o-the-fly CommandTrac.

Now representing somewhat of a luxury station wagon, the Grand Wagoneer was again based on the old, first-seen-in-1963, body. Many equipment items that cost extra on other Jeeps were standard in the Grand Wagoneer. The Grand Wagoneer's standard equipment included automatic transmission; cruise control; rear window defroster; tilt steering wheel; luggage rack; aluminum wheels; power door locks; power windows and power seats.

The elimination of the CJ-5 from the Jeep lineup was of considerable historic importance. This was a reflection of the ever-rising popularity of the roomier CJ-7. It also reduced production complexity at Toledo. Significant to the future of Jeep vehicles was the introduction of the first four-cylinder engine ever produced by AMC. This 2.5-liter engine was the base powerplant for both the CJ-7 and the Scrambler. Unlike many other contemporary truck engines the new four-cylinder was intended specifically for use in Jeep vehicles.

Jeep's 4.2-liter (258 cid) inline six continued as base engine for the J-10 pickup. The J-20 three-quarter ton pickup continued to use the 5.9-liter (360 cid) V-8 as its sole engine choice.

GVWs for the various models were 4150-pounds for CJ-7s and Scramblers; 6200-pounds for the J-10 pickups and 7600-8400-pounds for the J-20s. The payload range for these models, in the same order, was: (CJ-7) 1227-pounds; (Scrambler) 1146-pounds; (J-10) 2156-pounds and (J-20) 3745-pounds.

1984 AMC Jeep, CJ-7 Laredo (AMC)

I.D. DATA: VIN located on the top left surface of the instrument panel. There are 17 symbols. The first three symbols indicate the manufacturer, make and vehicle type. The engine type is indicated by the fourth symbol: U=150 cid (2.5-liter) four-cylinder; W=173 cid (2.8 liter) V-6; C=258 cid (4.2 liter) six-cylinder; N=360 cid (5.9 liter) V-8. The next letter identifies the transmission type. The series and type of vehicle are identified by the sixth and seventh symbols, which are the same as the model number. The eighth character identifies the GVW rating with the next serving as the check digit. Then follows the model year (tenth symbol E) and assembly point codes. The last six digits are sequential production numbers.

Model	Body Type	Price	Weight	Prod. Total
Wagoneer Sport Wagon (L-4) — (1/2-Ton) — (4x4)				
75	4dr Sta Wag	12,444	3047	—
75	4dr Ltd Sta Wag	17,076	3222	—
Wagoneer Sport Wagon (V-6) — (1/2-Ton) — (4x4)				
75	4dr Sta Wag	12,749	3065	—
75	4dr Ltd Sta Wag	17,381	3240	—
Grand Wagoneer (V-8) — (1/2-Ton) — (4x4)				
15	4dr Sta Wag	19,306	4221	—
Cherokee (L-4) — (1/2-Ton) — (4x4)				
77	2dr Sta Wag	9995	2817	—
78	4dr Sta Wag	10,295	2979	—
Cherokee (V-6) — (1/2-Ton) — (4x4)				
77	2dr Sta Wag	10,300	2963	—
78	4dr Sta Wag	10,600	3023	—
CJ-7 — (L-4) — (1/4-Ton) — (4x4) — (93.4 in. w.b.)				
87	Jeep	7563	2598	42,644
Scrambler — (L-4) — (1/2-Ton) — (4x4) — (103.4 in. w.b.)				
88	Pickup	7563	2679	4,130
J-10 Pickup — (L-6) — (1/2-Ton) — (4x4) — (118.8/130.7 in. w.b.)				
25	Townside (SWB)	9967	3724	—
26	Townside (LWB)	10,117	3811	—
J-20 Pickup — (V-8) — (3/4-Ton) — (4x4) — (130.7 in. w.b.)				
27	Townside Pickup	11,043	4323	—

NOTE: See "Historical" for additional production data.

1984 AMC Jeep Scrambler pickup (OCW)

ENGINE (standard in Cherokee XJ/Wagoneer XJ/CJ-7/Scrambler): Inline. OHV. Four-cylinder. Cast iron block. Bore & stroke: 3.875 x 3.188 in. Displacement: 150.4 cid. Compression ratio: 9.2:1. Taxable horsepower: 24.04. Net horsepower: 86 at 3650 rpm. Net torque: 132 lbs.-ft. at 3200 rpm. Five main bearings. Hydraulic valve lifters. Carburetor: One-barrel with electronic FFB.
ENGINE (optional Wagoneer): Vee-block. OHV. Six-cylinder. Bore & stroke: 3.50 x 2.99 in. Displacement: 173 cid. Compression ratio: 8.5:1. Horsepower rating: not advertised. Carburetor: two-barrel.
ENGINE (standard in Grand Wagoneer/J-10; optional in CJ-7/Scrambler): Inline. OHV. Six-cylinder. Cast iron block. Bore & stroke: 3.75 x 3.90 in. Displacement: 258 cid. Compression ratio: 9.2:1. Taxable horsepower: 33.75. Net horsepower: 102 at 3000 rpm. Net Torque: 204 lbs.-ft. at 1650 rpm. Seven main bearings. Hydraulic valve lifters. Carburetor: Two-barrel.
ENGINE (standard in J-20; optional in Grand Wagoneer/J-10): Vee-block. OHV. Eight-cylinder. Cast iron block. Bore & stroke: 4.08 x 3.44 in. Displacement: 360 cid. Taxable horsepower: 53.27. Brake horsepower: 175. Five main bearings. Hydraulic valve lifters. Carburetor: Two-barrel.

1984 AMC Jeep, Laredo Scrambler pickup (AMC)

CHASSIS (Wagoneer XJ): Wheelbase: 101.4 in. Width: 70.5 in. Overall length: 165.3 in. Height: 64.1 in.
CHASSIS (Cherokee XJ): Wheelbase: 101.4 in. Width: 69.3 in. Overall length: 165.3 in. Height: 64.1 in.
CHASSIS (Grand Wagoneer): Wheelbase: 108.7 in. Width: 74.8 in. Overall length: 186.4 in. Height: 66.4 in. Tires: P225/75R-15
CHASSIS (CJ-7): Wheelbase: 93.4 in. Width: 65.3 in. Overall length: 153.2 in. Height: 70.9 in. Front tread: 55.8 in. Rear tread: 55.1 in. Tires: P205/75R15 in.
CHASSIS (Scrambler): Wheelbase: 103.4 in. Width: 65.3 in. Overall length: 166.2 in. Height: 70.8 in. Front tread: 55.8 in. Rear tread: 55.1 in. Tires: P205/75R15 in.
CHASSIS (J-10/Model 25): Wheelbase: 118.7 in. Width: 78.9 in. Overall length: 194 in. Height: 69 in. Front tread: 64.0 in. Rear tread: 63.8 in. Tires: P225/75R15 in.
CHASSIS (J-10/Model 26): Wheelbase: 130.7 in. Width: 78.9 in. Overall length: 206 in. Height: 69 in. Front tread: 64.0 in. Rear tread: 63.8 in. Tires: P225/75R15 in.
CHASSIS (J-20): Wheelbase: 130.7 in. Width: 78.9 in. Overall length: 206 in. Height: 70 in. Front tread: 64.6 in. Rear tread: 65.9 in. Tires: 8.25 x 16.5 in.

1984 AMC Jeep, Laredo Townside pickup (AMC)

TECHNICAL: Selective synchromesh transmission. Speeds: 4F/1R. Floor-mounted gearshift. Single dry disc clutch. Rear axle: (J-20) full-floating; (others) semi-floating. Overall ratio: 3.54:1. Power assisted hydraulic front disc/rear drum brakes. (* Power assist optional on CJ-7 and Scrambler). Steel disc wheels.
TECHNICAL (Cherokee): Selective synchromesh transmission. Speeds: 4F/1R. Floor-mounted gearshift. Single dry disc clutch. Shaft drive. Rear axle: semi-floating. Qudra-Link front suspension. Two "shift-on-the-fly" 2WD/4WD systems. UniFrame construction. Power front disc brakes/rear drum. Stabilizer bars front and rear.
TECHNICAL (Wagoneer): Three-speed automatic transmission. Speeds: 3F/1R. Column-mounted gearshift. Shaft drive. Rear axle: full-floating. Power front disc/rear drum brakes. Four-wheel drive "shift-on-the-fly" system. Selec-Trac all-road surface 2WD/4WD system optional.
OPTIONS (Wagoneer XJ/Cherokee XJ/Grand Wagoneer): Limited package ($3,595). Automatic transmission, standard in Grand Wagoneer and Limited; optional in other models. Five-speed transmission (standard in base Wagoneer XJ). AM/FM stereo, standard in Wagoneer XJ Limited; optional in other XJs. AM/FM stereo with tape, standard in Grand Wagoneer; optional in other models. Cruise control, standard in Grand Wagoneer; optional in other models ($191). Rear window defroster, standard in Grand Wagoneer and Limited; optional in other models ($140). Tilt steering wheel, standard in Grand Wagoneer and Limited; optional in other models ($110). Luggage rack, standard on Grand Wagoneer and Limited; optional on other models ($119). Aluminum wheels, standard on Grand Wagoneer and Limited; optional on other models. Power sunroof ($1,694). Power door locks, standard in Grand Wagoneer and Limited; optional in other models. Power windows, standard in Grand Wagoneer and Limited; optional in other models. Power seats, standard in Grand Wagoneer and Limited; optional in other models ($300). Cherokee Chief package ($1,310). Pioneer package ($1,000). 360 cid V-8.
OPTIONS: (CJ-7): Renegade package ($1,124). (Scrambler): "SR" Sport package ($932). Hardtop ($829). (Pickups): Pioneer package ($456). Laredo package ($2,129).
HISTORICAL: Introduced: Fall, 1983. Calendar year sales: 153,801 (all Series), includes: (CJ-7) 39,547; (Scrambler) 2846; (J-Series Pickup) 3404. Model year production 152,931 units. This included: 58,596 Cherokee XJs; 20,940 Wagoneer XJs; 42,644 CJ-7s; 4,130 Scramblers; 4,085 Jeep trucks; and 22,536 Grand Wagoneers. Innovations: First four-cylinder AMC engine introduced. The new Jeep Cherokee and Wagoneer Sport wagons were the first all-new Jeep line in 20 years. The 1984 Cherokee was honored with "4x4 of the year" awards by all three leading off-road magazines. As a result of increased Jeep production, all laid-off employees were recalled at Toledo and the Jeep work force was increased by new hires.

1985 JEEP

1985 AMC Jeep, Scrambler pickup (JAG)

JEEP — 1985 SERIES — (ALL ENGINES): The latest CJ-7, available in base, Renegade and Laredo levels, had new, optional fold and tumble rear seats in place of the older fixed-back type. Both the Laredo and Renegade CJ-7 Jeeps featured new exterior tape stripe patterns, three new exterior colors and one new interior color. High-back bucket seats were now standard.

Highlights of the 1985 Scrambler included replacement of the SR Sport version by the Renegade and the introduction of the Scrambler Laredo, which filled the position previously occupied by the soft- and hard-top SL Sport. Both the Renegade and Laredo had new exterior and interior decor, a new interior color and new soft- and hardtop colors. All Scramblers, including the base model, had high-back bucket seats.

Again available was the down-sized Cherokee XJ two-door sport utility wagons, with an inline four-cylinder engine standard and an overhead cam four-cylinder diesel engine optional. The four-door Wagoneer XJ was back again, too, offering the 2.5 liter inline four or 2.8 liter V-6 below its hood.

Customers who wanted a larger vehicle were not left out. They could order the Grand Wagoneer station wagon, which had the 4.2 liter "Big Six" standard and offered the 5.9 liter V-8 at extra-cost.

J-series trucks were unchanged for 1985 as the company geared up to produce a compact Commanche pickup for 1986. However, there was a new two-wheel drive Cherokee released at midyear as a 1985-1/2 model. It was aimed at giving buyers in the Sun Belt states a Jeep pickup that cost $1,200 less than the four-wheel drive model.

1985 AMC Jeep, CJ-7 Universal Jeep (JAG)

I.D. DATA: VIN located on the top left surface of the instrument panel. There are 17 symbols. The first three symbols indicate the manufacturer, make and vehicle type. The engine type is indicated by the fourth symbol: B=126 cid (2.1 liter) OHC-4 turbo diesel; U=150 cid (2.5-liter) four-cylinder; W=173 cid (2.8 liter) V-6; C=258 cid (4.2 liter) six-cylinder; N=360 cid (5.9 liter) V-8. The next letter identifies the transmission type. The series and type of vehicle are identified by the sixth and seventh symbols, which are the same as the model number. The eighth character identifies the GVW rating with the next serving as the check digit. Then follows the model year (tenth symbol F) and assembly point codes. The last six digits are sequential production numbers.

1985 AMC Jeep, CJ-7 Renegade (AMC)

Model	Body Type	Price	Weight	Prod. Total
Wagoneer Sport Wagon — (L-4) — (1/2-Ton) — (4x4)				
75	4d Sta Wag	13,255	3,063	—
75	4d Ltd Sta Wag	17,953	3,222	—
Wagoneer Sport Wagon — (V-6) — (1/2-Ton) — (4x4)				
75	4dr Sta Wag	13,604	3,106	—
75	4dr Ltd Sta Wag	18,302	3,265	—
Grand Wagoneer — (L-6) — (1/2-Ton) — (4x4)				
15	4dr Sta Wag	20,462	4,228	9,010
Cherokee Sport Wagon — (L-4) — (1/2-Ton) — (4x2)				
73	2dr Sta Wag	9195	2777	—
74	4dr Sta Wag	9766	2828	—
Cherokee Sport Wagon — (L-4) — (1/2-Ton) — (4x4)				
77	2dr Sta Wag	10,405	2923	—
78	4dr Sta Wag	10,976	2984	—
Cherokee Sport Wagon — (V-6) — (1/2-Ton) — (4x2)				
73	2dr Sta Wag	9544	2852	—
74	4dr Sta Wag	10,115	2903	—
Cherokee Sport Wagon — (V-6) — (1/2-Ton) — (4x4)				
77	2dr Sta Wag	10,754	2998	—
78	4dr Sta Wag	11,325	3059	—
CJ-7 — (1/4-Ton) — (4x4) — (93.5 in. w.b.)				
87	Jeep	7282	2601	21,770
Scrambler — (1/2-Ton) — (4x4) — (103.5 in. w.b.)				
88	Jeep	7282	2701	1,050
J-10 Pickup — (1/2-Ton) — (4x4) — (131 in. w.b.)				
26	Pickup	10,311	3799	—
J-20 Pickup — (3/4-Ton) — (4x4) — (131 in. w.b.)				
27	Pickup	11,275	4353	—

NOTE: See historical for additional production data.

ENGINE (Optional Cherokee XJ/Wagoneer XJ): Inline. Overhead cam. OHV. Turbocharged. Diesel. Four-cylinder. Bore & stroke: 2.99 x 3.56 in. Displacement: 126 cid. Compression: 21.5:1. Fuel system: Indirect injection. Brake hp: not advertised. Taxable hp: 18.34.

ENGINE (Standard: CJ-7/Scrambler/Cherokee XJ/Wagoneer XJ): Inline. OHV. Four-cylinder. Cast iron block. Bore & stroke: 3.88 x 3.19 in. Displacement: 150 cid. Compression ratio: 9.2:1. Taxable horsepower: 24.04. Net horsepower: 86 at 3650 rpm. Net torque: 132 lbs.-ft. at 3200 rpm. Five main bearings. Hydraulic valve lifters. Carburetor: One-barrel.

ENGINE (Optional Wagoneer/Cherokee): Vee-block. OHV. Six-cylinder. Bore & stroke: 3.50 x 2.99 in. Displacement: 173 cid. Compression ratio: 8.5:1. Horsepower rating: not advertised. Carburetor: two-barrel.

ENGINE (Standard: Grand Wagoneer/J-10; Optional: CJ-7/Scrambler): Inline. OHV. Six-cylinder. Cast iron block. Bore & stroke: 3.75 x 3.90 in. Displacement: 258 cid. Compression ratio: 9.2:1. Taxable horsepower: 33.75. Net horsepower: 102 at 3000 rpm. Net torque: 204 lbs.-ft. at 1650 rpm. Seven main bearings. Hydraulic valve lifters. Carburetor: Two-barrel.

ENGINE (Standard: J-20/Optional: J-10/Grand Wagoneer): Vee-block. OHV. Eight-cylinder. Cast iron block. Bore & stroke: 4.08 x 3.44 in. Displacement: 360 cid. Compression ratio: 8.25:1. Taxable horsepower: 53.27. Five main bearings. Hydraulic valve lifters. Carburetor: Two-barrel.

1985 AMC Jeep, Cherokee Pioneer four-door station wagon (JAG)

CHASSIS (CJ-7): Wheelbase: 93.4 in. Overall length: 153.2 in. Height: (hardtop) 71.0 in.; (open) 69.1 in. Front tread: 55.8 in. Rear tread: 55.1 in. Tires: P205/75R15 in. "Arriva", steel-belted.

CHASSIS (Wagoneer XJ): Wheelbase: 101.4 in. Width: 70.5 in. Overall length: 165.3 in. Height: 64.1 in.

CHASSIS (Cherokee XJ): Wheelbase: 101.4 in. Width: 69.3 in. Overall length: 165.3 in. Height: 64.1 in.

CHASSIS (Grand Wagoneer): Wheelbase: 108.7 in. Width: 74.8 in. Overall length: 186.4 in. Height: 66.4 in. Tires: P225/75R-15

CHASSIS (CJ-7): Wheelbase: 93.4 in. Width: 65.3 in. Overall length: 153.2 in. Height: 70.9 in. Front tread: 55.8 in. Rear tread: 55.1 in. Tires: P205/75R15 in.

CHASSIS (Scrambler): Wheelbase: 103.4 in. Width: 65.3 in. Overall length: 166.2 in. Height: 70.8 in. Front tread: 55.8 in. Rear tread: 55.1 in. Tires: P205/75R15 in.

CHASSIS (J-10/Model 25): Wheelbase: 118.7 in. Width: 78.9 in. Overall length: 194 in. Height: 69 in. Front tread: 64.0 in. Rear tread: 63.8 in. Tires: P225/75R15 in.

CHASSIS (J-10/Model 26): Wheelbase: 130.7 in. Width: 78.9 in. Overall length: 206 in. Height: 69 in. Front tread: 64.0 in. Rear tread: 63.8 in. Tires: P225/75R15 in.

CHASSIS (J-20): Wheelbase: 130.7 in. Width: 78.9 in. Overall length: 206 in. Height: 70 in. Front tread: 64.6 in. Rear tread: 65.9 in. Tires: 8.25 x 16.5 in.

TECHNICAL (CJ-7/Scrambler): Selective synchromesh transmission. Speeds: 4F/1R. Floor-shift. Single dry disc clutch. Shaft drive. Semi-floating rear axle. Overall ratio: 3.54:1. Manual front disc/rear drum brakes. 15 x 6 in., five-bolt pressed steel wheels.

TECHNICAL (J-10): Selective synchromesh transmission. Speeds: 4F/1R. Floor-mounted gearshift. Single dry disc clutch. Shaft drive. Rear axle: semi-floating. Overall ratio: 2.73:1. Power front disc/rear drum brakes. 15 x 6 in. pressed steel wheels.

TECHNICAL (J-20): Automatic transmission. Speeds: 3F/1R. Column-mounted gearshift. Shaft drive. Rear axle: full-floating. Overall ratio: 3.73:1. Power front disc/rear drum brakes. 16.5 x 6 in. pressed steel wheels.

TECHNICAL (Cherokee): Selective synchromesh transmission. Speeds: 4F/1R. Floor-mounted gearshift. Single dry disc clutch. Shaft drive. Rear axle: semi-floating. Quadra-Link front suspension. Two "shift-on-the-fly" 2WD/4WD systems. UniFrame construction. Power front disc brakes and rear drum brakes. Stabilizer bars front and rear.

TECHNICAL (Wagoneer): Three-speed automatic transmission. Speeds: 3F/1R. Column-mounted gearshift. Shaft drive. Rear axle: full-floating. Power front disc/rear drum brakes. Four-wheel drive "shift-on-the-fly" system. Selec-Trac all-road surface 2WD/4WD system optional.

OPTIONS: (CJ-7/Scrambler): Renegade package. Laredo package. Variable-ratio power steering (required with air conditioning). Five-speed manual transmission with overdrive. Power front disc brakes. Cold Climate Group. Heavy-duty alternator. Heavy-duty battery. Heavy-duty cooling system. Coolant recovery system. Automatic/part-time four-wheel-drive (not available with four-cylinder) Rear Trac-Lok differential. Heavy-duty shock absorbers. Extra-duty suspension package. Soft ride suspension (not available for base models). Front chrome bumper (Scrambler only, standard Laredo). Chrome rear step bumper (Scrambler only, standard Laredo). Painted rear step bumper (Scrambler only, not available Laredo). Bumper accessory package (not available for Laredo). Rear bumperettes (standard Laredo). Doors for soft and fiberglass tops. Outside passenger side mirror. Body Side step. Full soft top. Vinyl soft tops. Radios: AM/AM-FM Stereo/AM-FM Stereo/Cassette player (require factory hard or soft top). Hardtop. Air conditioning (not available with four-cylinder) Cruise Control. Fog lamps (clear lens). Halogen headlamps. Extra Quiet Insulation (hardtop only). Styled steel chrome wheels. Styled steel painted wheels. Bumper accessory package (not available Laredo). Carpeting (front and rear) (standard Laredo). Center Console (standard Laredo). Convenience Group (base model only, standard all others). Decor Group (standard Renegade, not available for Laredo). Roll Bar Accessory package. Fold and tumble rear seat (standard on Renegade and Laredo) CJ-7 only. Soft-feel sport steering wheel (standard on Renegade, not available for Laredo). Leather-wrapped steering wheel (standard on Laredo). Rear storage box. Tachometer and Rally clock (standard on Laredo). Tilt steering wheel. Various wheel and tire combinations.

OPTIONS (Pickups): Heavy-duty alternator. Heavy-duty battery. Cold Climate Group. Heavy-duty cooling system. Heavy-duty GVW options: 8400 GVW (J-20); 6200 GVW (J-10). Rear Trac-Lok differential. Heavy-duty shock absorber. Extra heavy-duty springs. Radios: AM/AM-FM Stereo/Electronically-tuned AM/FM with cassette. Automatic transmission. 5.9-liter V-8 engine. Pioneer package ($475). White styled wheels. Visibility Group. Tilt steering wheel (not available with manual transmission). Light Group. Insulation package. Protective floor mats. Grain vinyl bench seat (included in Pioneer package). Soft feel sport steering wheel. Leather-wrapped steering wheel. Chrome front grille. Dual low-profile exterior mirrors. Sliding rear window. Air conditioning. Convenience Group.

OPTIONS (Wagoneer/Cherokee): Stereo tape. Cruise control (*). Tilt steering wheel (*). Custom wheelcovers (*). Power sunroof. Sunroof. Power door locks (*). Power windows (*). Power seat (*). Chief package (Cherokee). Laredo package (Cherokee). Pioneer package (Cherokee). Turbo diesel engine. Six-cylinder engine (Standard in Grand Wagoneer). Power steering. Air conditioning.

HISTORICAL: Introduced: Fall, 1984. Calendar year production: (all Jeep vehicles including AM General) 236,277. Includes: (CJ-Series) 46,553. (J-Series Pickup) 1,953. Model year production of AMC Jeep light-duty trucks (excluding AM General) was 92,310 units. This includes 48,540 Chrokee XJs; 10,620 Wagoneer XJs; 21,720 CJ-7s; 1,050 Scramblers; 9,010 Grand Wagoneers; 1,320 J-10/J-20 pickups. Innovations: New Scrambler "Renegade" model introduced. CJ-7 gets new fold-and-tumble rear seat option. New trim packages. Historical notes: Paul W. Tippett continued as chairman of Renault/AMC/Jeep.

1986-1/2 AMC Jeep, Wrangler Laredo (OCW)

JEEP — 1986 SERIES — (ALL ENGINES): An historic milestone in light-duty truck history was the end of production of the CJ Universal Jeep during model year 1986. At the start of the season, the 1986 sales brochure gave little hint that the CJ series would be phased-out. "CJ has always been special," said the promotional copy. "And for 1986 it makes more sense when you consider its long list of standard features, its exceptional fuel economy and its affordable price." The end came in Jan. 1986, closing 40 years of civilian production during which some 1.6 million units were built. Naturally, there were no significant changes in the 1986 CJ-7 model.

Its replacement, seen as early as February, at the Chicago Auto Show, carried a May 13, 1986 showroom release date. It was known as the Wrangler (or Jeep YJ in Canada). Although the basic Jeep CJ silhouette was apparent in the Wrangler, the forward half of the vehicle, including hood, fender lines and grille, were more contemporary in appearance. Characteristics included a seven-slot, horizontally "veed" grille, square headlamps and parking lamps, a raised hood center panel, modern instrumentation and doors. It had a 93.4 in. wheelbase, 152-153 in. overall length, 66 in. width and 58 in. front/rear tread. The Wrangler was merchandised as a 1987 model and came in base, Sport Decor and Laredo trim levels. Standard equipment included a fuel-injected version of AMC's 2.5-liter four-cylinder engine, five-speed manual transmission; power brakes; P215/75R-15 tires; high-back front bucket seats; fold-and-tumble rear seat; mini front carpet mat, column-mounted wiper/washer controls; padded roll bar with side bars extending to windshield frame; swing-away rear tailgate; soft top with metal half-doors and tinted windshield.

New, and promoted as a 1986 model, was a compact Comanche pickup truck. This slab-sided, sporty pickup was designed to meet the Japanese imports head-on. It had a grille with 10 vertical segments between its rectangular headlamps and a sleek cab with trim panels on the rear corners. A Jeep UniFrame with 120 in. wheelbase used a Quadra-Link front suspension and Hotchkiss rear layout with semi-elliptical multi-leaf springs. Standard equipment included a throttle-body-injected 2.5-liter four, five-speed manual transmission, power assisted front disc/rear drum brakes and P195/75R15 steel radial tires. With the 4 x 4 models, Jeep's "Command-Trac" shift-on-the-fly system was used.

The Cherokee XJ continued as AMC's compact sports utility wagon with two-door and four-door versions available. The 1985-1/2 two-wheel drive version of the two-door, which had earned seven percent of total 1985 Cherokee sales, was back again. It and the four-wheel drive four-door model used the 2.5 liter L-4 with throttle-body-injection as standard engine. The 2.1 liter turbo diesel and 2.8 liter V-6 were optional.

The Wagoneer XJ four-door, four-wheel drive sport utility wagon offered the same power-train choices as the Cherokee.

For 1986, the Grand Wagoneer had a distinctive new front end with twin horizontal openings and vertically arranged dual headlamps. Power steering; power brakes; a mini console; carpets; gages; body and hood stripes; wheel trim rings; and P205/75R15 radial whitewalls were standard. A Wagoneer Limited package added air conditioning; heavy-duty electrical system; bumper guards; cruise control; power windows; power door locks; AM/FM radio; roof rack; woodgrain body sides; Michelin tires and more.

"An enormous capacity for work" was promised for 1986 buyers of the J-10/J-20 series of standard pickups. These 4x4 only models had few appearance changes, other than new graphics treatments. They again had the Townside look set off by rectangular headlamps and a 13 slot "electric shaver" grille. Base engines were the 4.2-liter six in the half-ton J-10 and the 5.9-liter V-8 in the 3/4-ton J-20.

I.D. DATA: VIN located on the top left surface of the instrument panel. There are 17 symbols. The first three symbols indicate the manufacturer, make and vehicle type. The engine type is indicated by the fourth symbol: B=126 cid (2.1 liter) OHC-4 turbo diesel; H=150 cid (2.5-liter) four-cylinder TBI engine; W=173 cid (2.8 liter) V-6; C=258 cid (4.2 liter) six-cylinder; N=360 cid (5.9 liter) V-8. The next letter identifies the transmission type. The series and type of vehicle are identified by the sixth and seventh symbols, which are the same as the model number. The eighth character identifies the GVW rating with the next serving as the check digit. Then follows the model year (tenth symbol G) and assembly point codes. The last six digits are sequential production numbers.

Model	Body Type	Price	Weight	Prod. Total
1985 AMC Jeep, CJ-7 Renegade (AMC)				
Model	Body Type	Price	Weight	Prod. Total
Wagoneer Sport Wagon — (L4/TBI) — (1/2-Ton) — (4x4)				
75	4d Sta Wag	13,360	3,039	—
75	4d Ltd Sta Wag	18,600	3,234	—
Wagoneer Sport Wagon — (V-6) — (1/2-Ton) — (4x4)				
75	4dr Sta Wag	14,607	3,104	—
75	4dr Ltd Sta Wag	19,037	3,299	—
Grand Wagoneer — (L-6) — (1/2-Ton) — (4x4)				
15	4dr Sta Wag	21,350	4,252	-16,252
Cherokee Sport Wagon — (L4/TBI) — (1/2-Ton) — (4x2)				
73	2dr Sta Wag	9335	2751	—
74	4dr Sta Wag	9950	2802	—
Cherokee Sport Wagon — (L4/TBI) — (1/2-Ton) — (4x4)				
77	2dr Sta Wag	10,695	2917	—
78	4dr Sta Wag	11,320	2968	—
Cherokee Sport Wagon — (V-6) — (1/2-Ton) — (4x2)				
73	2dr Sta Wag	9772	2847	—
74	4dr Sta Wag	10,387	2898	—
Cherokee Sport Wagon — (V-6) — (1/2-Ton) — (4x4)				
77	2dr Sta Wag	11,132	3013	—
78	4dr Sta Wag	11,757	3064	—
CJ-7 — (1/4-Ton) — (4x4) — (93.5 in. w.b.)				
87	Jeep	7500	2596	25,929
Wrangler — (1/4-Ton) — (4-cyl) — (93.4 in. w.b.)				
81	Jeep	8396	—	—
Wrangler — (1/4-Ton) — (V-6) — (93.4 in. w.b.)				
81	Jeep	9899	—	—
Comanche — (1/2-Ton) — (4x2) — (120 in. w.b.)				
66	Pickup	7049	2931	—
Comanche — (1/2-Ton) — (4x4) — (120 in. w.b.)				
65	Pickup	8699	3098	—
J-10 — (1/2-Ton) — (4x4) — (131 in. w.b.)				
26	Pickup	10,870	3808	—
J-20 — (1/2-Ton) — (4x4) — (131 in. w.b.)				
27	Pickup	12,160	4388	—

NOTE: See historical for additional production data.

1986 AMC Jeep, Commanche pickup (JAG)

ENGINE (Turbo Diesel; Optional: 65/66): Inline. OHV. Four-cylinder. Cast iron block. Bore & stroke: 2.99 x 3.5 in. Displacement: 126 cid/2.1L. Taxable horsepower: 18.34. Hydraulic valve lifters. Turbo charged.
ENGINE (Standard: Wrangler/66; 65; 87): Inline. OHV. Four-cylinder. Cast iron block. Bore & stroke: 3.88 x 3.19 in. Displacement: 150 cid/2.5L. Compression ratio: 9.2:1. Brake horsepower: 117 at 5000 rpm. Taxable horsepower: 24.04. Hydraulic valve lifters. Throttle Body Injection (EFI). Torque (Compression): 135 lbs.-ft. at 3500 rpm.
ENGINE (Optional: 65/66): Vee-block. Six-cylinder. Cast iron block. Bore & stroke: 3.5 x 2.99 in. Displacement: 173 cid/2.8L. Brake horsepower: 115 at 4800 rpm. Taxable horsepower: 29.45. Hydraulic valve lifters. Two-barrel carburetor. Torque (Compression) 150 lbs.-ft. at 2100 rpm.
ENGINE (Standard: J-10; Optional: 66/65/Wrangler): Inline. OHV. Six-cylinder. Cast iron block. Bore & stroke: 3.75 x 3.9 in. Displacement: 258 cid/4.2L. Compression ratio: 9.2:1. Brake horsepower: 112 at 3000 rpm. Taxable horsepower: 33.75. Hydraulic valve lifters. Two-barrel carburetor. Torque (Compression) 210 lbs.-ft. at 3000 rpm.
ENGINE (Standard: 27; Optional: 26): Vee-block. Eight-cylinder. Cast iron block. Bore & stroke: 4.08 x 3.44 in. Displacement: 360 cid/5.9L. Taxable horsepower: 53.27. Hydraulic valve lifters. Four-barrel carburetor.

1986 AMC Jeep, Grand Wagoneer limousine (OCW)

CHASSIS (CJ-7): Wheelbase: 93.5 in. Overall length: 153.2 in. Height: (hardtop) 71.0 in. Front tread: 55.8 in. Rear tread: 55.1 in. Tires: P205/75R-15 in.
CHASSIS (Scrambler): Wheelbase: 103.4 in. Width: 65.3 in. Overall length: 166.2 in. Height: 70.8 in. Front tread: 55.8 in. Rear tread: 55.1 in. Tires: P205/75R15 in.
CHASSIS (Wrangler): Wheelbase: 93.4 in. Overall length: 152 in. Height: (hardtop) 69.3 in. Front tread: 58 in. Rear tread: 58 in. Tires: P215/75R-15 in.
NOTE: Comanche and Wrangler 4x4s use P225/75R-15 in. tires.
CHASSIS (Wagoneer XJ): Wheelbase: 101.4 in. Width: 70.5 in. Overall length: 165.3 in. Height: 64.1 in.
CHASSIS (Cherokee XJ): Wheelbase: 101.4 in. Width: 69.3 in. Overall length: 165.3 in. Height: 64.1 in.
CHASSIS (Grand Wagoneer): Wheelbase: 108.7 in. Width: 74.8 in. Overall length: 186.4 in. Height: 66.4 in. Tires: P225/75R-15
CHASSIS (J-10): Wheelbase: 131 in. Overall length: 206 in. Height: 70.7 in. Front tread: 63.3 in. Rear tread: 63.8 in. Tires: P225/75R-15 in.
CHASSIS (J-20): Wheelbase: 131 in. Overall length: 206 in. Height: 70.7 in. Front tread: 63.3 in. Rear tread: 63.8 in. Tires: P225/75R-15 in.
TECHNICAL (Wrangler): Selective synchromesh transmission. Speeds: (Standard) 5F/1R. Floor controls. Single dry disc clutch. Shaft drive. Semi-floating rear axle. Overall drive ratio: (Standard) 4.11.1. Brakes: Disc front/drum rear with power assist. Wheels: 15 x 7.0 disc wheels.
NOTE: For Standard 4x2 model w/2.5L four.

TECHNICAL (CJ-7/Scrambler): Selective synchromesh transmission. Speeds: 4F/1R. Floor-shift. Single dry disc clutch. Shaft drive. Semi-floating rear axle. Overall ratio: 3.54:1. Manual front disc/rear drum brakes. 15 x 6 in., five-bolt pressed steel wheels.
TECHNICAL (J-10): Selective synchromesh transmission. Speeds: 4F/1R. Floor-mounted gearshift. Single dry disc clutch. Shaft drive. Rear axle: semi-floating. Overall ratio: 2.73:1. Power front disc/rear drum brakes. 15 x 6 in. pressed steel wheels.
TECHNICAL (J-20): Automatic transmission. Speeds: 3F/1R. Column-mounted gearshift. Shaft drive. Rear axle: full-floating. Overall ratio: 3.73:1. Power front disc/rear drum brakes. 16.5 x 6 in. pressed steel wheels.
TECHNICAL (Cherokee): Selective synchromesh transmission. Speeds: 4F/1R. Floor-mounted gearshift. Single dry disc clutch. Shaft drive. Rear axle: semi-floating. Qudra-Link front suspension. Two "shift-on-the-fly" 2WD/4WD systems. UniFrame construction. Power front disc brakes/rear drum. Stabilizer bars front and rear.
TECHNICAL (Wagoneer): Three-speed automatic transmission. Speeds: 3F/1R. Column-mounted gearshift. Shaft drive. Rear axle: full-floating. Power front disc/rear drum brakes. Four-wheel drive "shift-on-the-fly" system. Selec-Trac all-road surface 2WD/4WD system optional.
OPTIONAL: Chrysler A-3 automatic transmission available only with six-cylinder engine.
TECHNICAL (Comanche Pickup): Selective synchromesh transmission. Speeds: 5F/1R. Floor controls. Single dry disc clutch. Shaft drive. Hotchkiss solid rear axle. Overall drive ratio: 3.73:1. Brakes: Power assisted front disc/rear drum. Wheels: 15 x 6 in. five-bolt pressed steel wheels.
NOTE: For Standard 4x2 model w/base engine.
OPTIONS (Wrangler): Carbureted 4.2L six-cylinder engine. Three-speed automatic transmission. Hardtop (standard with Laredo). Tilt steering. (Sport Decor Group): includes, most standard features, plus AM/FM monaural radio; black side cowl carpet; special Wrangler hood decals; special Wrangler lower body side stripes; P215/75R15 Goodyear all-terrain Wrangler tires; conventional size spare with lock and convenience group. (Laredo Hardtop Group): includes richer interior trim; AM/FM monaural radio; Buffalo-grain vinyl upholstery; front and rear carpeting; center console; extra-quiet insulation; leather-wrapped Sport steering wheel; special door trim panels and map pockets; chrome front bumper; rear bumperettes; grille panel; headlamp bezels; tow hooks; color-keyed wheel flares; full-length mud guards; integrated steps; deep tinted glass; OSRV door mirrors; bumper accessory package; special hood and body side stripes; convenience group; 15 x 7 in. aluminum wheels; P215/75R-15 Goodyear Wrangler RWL radial tires, spare tire and matching aluminum spare wheel. Rear trac-lok differential. Air conditioning. Extra-quiet insulation. Full-carpets. Halogen fog lamps. Power steering. Cruise Control (6-cyl. only). Leather-wrapped Sport steering wheel. Electric rear window defogger (hardtop). Heavy-duty suspension. Heavy-duty cooling. Aluminum wheels. Off-Road equipment package. Conventional spare tire. Metallic exterior paints.
OPTIONS: (CJ-7/Scrambler): Renegade package. Laredo package. Variable-ratio power steering (required with air conditioning. Five-speed manual transmission with overdrive. Power front disc brakes. Cold Climate Group. Heavy-duty alternator. Heavy-duty battery. Heavy-duty cooling system. Coolant recovery system. Automatic/part-time four-wheel-drive (not available with four-cylinder) Rear Trac-Lok differential. Heavy-duty shock absorbers. Extra-duty suspension package. Soft ride suspension (not available for base models). Front chrome bumper (Scrambler only, standard Laredo). Chrome rear step bumper (Scrambler only, standard Laredo). Painted rear step bumper (Scrambler only, not available Laredo). Bumper accessory package (not available for Laredo). Rear bumperettes (standard Laredo). Doors for soft and fiberglass tops. Outside passenger side mirror. Body Side step. Full soft top. Vinyl soft tops. Radios: AM/AM-FM Stereo/AM-FM Stereo/Cassette player (require factory hard or soft top). Hardtop. Air conditioning (not available with four-cylinder) Cruise Control. Fog lamps (clear lens). Halogen headlamps. Extra Quiet Insulation (hardtop only). Styled steel chrome wheels. Styled steel painted wheels. Bumper accessory package (not available Laredo). Carpeting (front and rear) (standard Laredo). Center Console (standard Laredo). Convenience Group (base model only, standard all others). Decor Group (standard Renegade, not available for Laredo). Roll Bar Accessory package. Fold and tumble rear seat (standard on Renegade and Laredo CJ-7 only. Soft-feel sport steering wheel (standard on Renegade, not available for Laredo). Leather-wrapped steering wheel (standard on Laredo). Rear storage box. Tachometer and Rally clock (standard on Laredo). Tilt steering wheel. Various wheel and tire combinations.
OPTIONS (Wagoneer/Cherokee): Stereo tape. Cruise control (*). Tilt steering wheel (*). Custom wheelcovers (*). Power sunroof. Sunroof. Power door locks (*). Power windows (*). Power seat (*). Chief package (Cherokee). Laredo package (Cherokee). Pioneer package (Cherokee). Turbo diesel engine. Six-cylinder engine (Standard in Grand Wagoneer). Power steering. Air conditioning.
OPTIONS (Comanche): "X" package. "XLS" package. 2.8L six-cylinder engine. 2.1L turbo diesel engine. 2205-pound payload package. Power steering w/17.5:1 ratio. P225/75R15 radial tires. 24-gallon fuel tank. Automatic transmission. Command-Trac 4x4 system (Standard on 4x4 models). Selec-Trac 4x4 system (optional on 4x4 models). 4:11.1 rear axle ratio.
HISTORICAL: Introduced: Fall, 1985; (Wrangler) May, 1986. Innovations: New Comanche mini-pickup series. New Wrangler replaces CJ-Series in Jan., 1986 (considered a 1987 model). Model year production (excluding AM General products) totaled 243,406. This included: 25,929 CJ-7s; 122,968 Cherokee XJs; 13,716 Wagoneer XJs; 45,219 Comanche pickups; 1,657 J-10/J-20 pickups; 17,665 Grand Wagoneers and 16,252 of the 1987 Wranglers. A California Jeep dealer launched an unsuccessful nationwide publicity campaign to "Save the Jeep CJ." Joseph E. Cappy became president of AMC during 1986. Cappy referred to the new Comanche as "AMC's first state-of-the-art-product in the two-wheel-drive market, which accounts for 75 percent of the light truck market." Other names considered for the new compact pickup were Renegade, Commando, Wrangler and Honcho. AMC also announced that a total of five years had been spent in development of the 1987 Wrangler as an up-to-date replacement for the famous Jeep CJ.

HOW TO USE THIS AMC PRICE GUIDE

On the following pages is an **AMC PRICE GUIDE**. The worth of an old car is a "ballpark" estimate at best. The estimates contained in this book are based upon national and regional data compiled by the editors of *Old Cars News & Marketplace* and *Old Cars Price Guide*. These data include actual bids and prices at collector car auctions and sales, classified and display advertising of such vehicles, verified reports of private sales and input from experts.

Price estimates are listed for cars in six different states of condition. These conditions (1-6) are illustrated and explained in the **VEHICLE CONDITION SCALE** on the following pages. Values are for complete vehicles — not parts cars — except as noted. Modified car values are not included, but can be estimated by figuring the cost of restoring the subject vehicle to original condition and adjusting the figures shown here accordingly.

Appearing below is a section of a chart taken from the 1985 AMC Jeep section to illustrate the following elements:

A. MAKE The make of vehicle, or marque name, appears in large, boldface type at the beginning of each value section.

B. DESCRIPTION The extreme left-hand column indicates vehicle year, model name, body type, engine configuration and, in some cases, wheelbase.

C. CONDITION CODE The six columns to the right are headed by the numbers one through six (1-6) which correspond to the conditions described in the **VEHICLE CONDITION SCALE** on the following page.

D. PRICE The price estimates, in dollars, appear below their respective condition code headings and across from the vehicle descriptions.

A. MAKE ——— **AMC JEEP**

1985	6	5	4	3	2	1
Wagoneer, 4-cyl, 101.4" wb						
4d Sta Wag	350	750	1200	2350	4900	7000
4d Ltd Sta Wag	350	750	1300	2450	5250	7500
Wagoneer, 8-cyl, 101.4" wb						
4d Sta Wag	350	750	1300	2450	5250	7500
4d Ltd Sta Wag	350	800	1450	2750	5600	8000
Grand Wagoneer, 108.7" wb						
4d Sta Wag	350	750	1200	2350	4900	7000
Cherokee, 4-cyl, 101.4" wb						
2d Sta Wag 2WD	200	650	1050	2250	4200	6000
4d Sta Wag 2WD	200	675	1050	2250	4350	6200
2d Sta Wag 4WD	350	700	1150	2300	4550	6500
4d Sta Wag 4WD	350	725	1150	2300	4700	6700
Cherokee, 6-cyl, 101.4" wb						
2d Sta Wag 2WD	200	675	1050	2250	4350	6200
4d Sta Wag 2WD	350	700	1100	2300	4500	6400
2d Sta Wag 4WD	350	725	1150	2300	4700	6700
4d Sta Wag 4WD	350	725	1200	2350	4850	6900

B. DESCRIPTION / C. CONDITION CODE / D. PRICE

VEHICLE CONDITION SCALE

1) EXCELLENT: Restored to current maxiumum professional standards of quality in every area, or perfect original with components operating and appearing as new. A 95-plus point show vehicle that is not driven.

2) FINE: Well-restored, or a combination of superior restoration and excellent original. Also, an *extremely* well-maintained original showing very minimal wear.

3) VERY GOOD: Completely operable original or "older restoration" showing wear. Also, a good amateur restoration, all presentable and serviceable inside and out. Plus, combinations of well-done restoration and good operable components or a partially restored vehicle with all parts necessary to complete and/or valuable NOS parts.

4) GOOD: A driveable vehicle needing no or only minor work to be functional. Also, a deteriorated restoration or a very poor amateur restoration. All components may need restoration to be "excellent," but the vehicle is mostly useable "as is."

5) RESTORABLE: Needs *complete* restoration of body, chassis and interior. May or may not be running, but isn't weathered, wrecked or stripped to the point of being useful only for parts.

6) PARTS VEHICLE: May or may not be running, but is weathered, wrecked and/or stripped to the point of being useful primarily for parts.

PRICING

AMC

	6	5	4	3	2	1
1958-1959						
American DeLuxe, 6-cyl.						
2 dr Sed	200	550	900	2000	3600	5200
Sta Wag (1959 only)	200	550	900	2100	3700	5300
American Super, 6-cyl.						
2 dr Sed	200	550	900	2100	3700	5300
Sta Wag (1959 only)	200	550	900	2150	3800	5400
Rambler DeLuxe, 6-cyl.						
4 dr Sed	200	550	900	2000	3600	5200
Sta Wag	200	500	850	1850	3350	4900
Rambler Super, 6-cyl.						
4 dr Sed	200	500	850	1850	3350	4900
4 dr HdTp	200	500	850	1900	3500	5000
Sta Wag	200	550	900	2150	3800	5400
Rambler Custom, 6-cyl.						
4 dr Sed	200	600	1000	2200	4000	5700
4 dr HdTp	200	650	1000	2200	4150	5900
Sta Wag	200	600	950	2150	3850	5500
Rebel Super V-8						
4 dr Sed DeL (1958 only)	200	600	1000	2200	4000	5700
4 dr Sed	200	650	1000	2200	4100	5800
Sta Wag	200	650	1000	2200	4150	5900
Rebel Custom, V-8						
4 dr Sed	200	650	1000	2200	4150	5900
4 dr HdTp	200	650	1050	2250	4200	6000
Sta Wag	200	650	1050	2250	4200	6000
Ambassador Super, V-8						
4 dr Sed	200	600	1000	2200	4000	5700
Sta Wag	200	650	1000	2200	4100	5800
Ambassador Custom, V-8						
4 dr Sed	200	650	1000	2200	4100	5800
4 dr Hdtp	200	650	1000	2200	4150	5900
Sta Wag	200	650	1000	2200	4150	5900
HdTp Sta Wag	200	675	1050	2250	4300	6100
1960						
American DeLuxe, 6-cyl.						
2 dr Sed	200	500	850	1900	3500	5000
4 dr Sed	200	500	850	1850	3350	4900
Sta Wag	200	500	850	1950	3600	5100
American Super, 6-cyl.						
2 dr Sed	200	500	850	1900	3500	5000
4 dr Sed	200	500	850	1950	3600	5100
Sta Wag	200	550	900	2000	3600	5200
American Custom, 6-cyl.						
2 dr Sed	200	500	850	1900	3500	5000
4 dr Sed	200	500	850	1850	3350	4900
Sta Wag	200	500	850	1950	3600	5100
Rambler DeLuxe, 6-cyl.						
4 dr Sed	200	500	850	1850	3350	4900
Sta Wag	200	500	850	1850	3350	4900
Rambler Super, 6-cyl.						
4 dr Sed	200	500	850	1900	3500	5000
6P Sta Wag	200	500	850	1900	3500	5000
8P Sta Wag	200	500	850	1950	3600	5100
Rambler Custom, 6-cyl.						
4 dr Sed	200	500	850	1950	3600	5100
4 dr HdTp	200	550	900	2000	3600	5200
6P Sta Wag	200	500	850	1950	3600	5100
8P Sta Wag	200	550	900	2000	3600	5200
Rebel Super, V-8						
Sed	200	550	900	2100	3700	5300
6P Sta Wag	200	550	900	2000	3600	5200
8P Sta Wag	200	550	900	2100	3700	5300
Rebel Custom, V-8						
Sed	200	550	900	2150	3800	5400
4 dr HdTp	200	600	950	2150	3850	5500
6P Sta Wag	200	550	900	2150	3800	5400
8P Sta Wag	200	600	950	2150	3850	5500
Ambassador Super, V-8						
Sed	200	600	950	2200	3900	5600
6P Sta Wag	200	600	950	2150	3850	5500
8P Sta Wag	200	600	950	2200	3900	5600
Ambassador Custom, V-8						
Sed	200	600	1000	2200	4000	5700
4 dr HdTp	200	650	1000	2200	4100	5800
6P Sta Wag	200	600	1000	2200	4000	5700
HdTp Sta Wag	200	650	1050	2250	4200	6000
8P Sta Wag	200	650	1000	2200	4150	5900
1961						
American						
DeL Sed	150	400	750	1600	3100	4400
2 dr DeL Sed	150	400	750	1650	3150	4500
4 dr DeL Sta Wag	150	450	750	1700	3200	4600
2 dr DeL Sta Wag	150	400	750	1650	3150	4500
4 dr Sup Sed	150	400	750	1650	3150	4500
2 dr Sup Sed	150	450	750	1700	3200	4600
4 dr Sup Sta Wag	150	450	800	1750	3250	4700
2 dr Sup Sta Wag	150	450	750	1700	3200	4600
4 dr Cus Sed	150	450	750	1700	3200	4600
2 dr Cus Sed	150	450	800	1750	3250	4700
Cus Conv	200	650	1050	2250	4200	6000
4 dr Cus Sta Wag	150	450	800	1750	3250	4700
2 dr Cus Sta Wag	150	450	800	1800	3300	4800
400 Sed	150	450	800	1750	3250	4700
400 Conv	200	675	1050	2250	4350	6200
Rambler Classic						
DeL Sed	150	400	750	1650	3150	4500
DeL Sta Wag	150	450	750	1700	3200	4600
Sup Sed	150	450	750	1700	3200	4600
Sup Sta Wag	150	450	800	1750	3250	4700
Cus Sed	150	450	800	1750	3250	4700
Cus Sta Wag	150	450	800	1800	3300	4800
400 Sed	150	450	800	1800	3300	4800

NOTE: Add 5 percent for V-8.

Ambassador						
DeL Sed	150	450	750	1700	3200	4600
Sup Sed	150	450	800	1750	3250	4700
5 dr Sup Sta Wag	150	450	800	1800	3300	4800
4 dr Sup Sta Wag	150	450	800	1750	3250	4700
Cus Sed	150	450	800	1800	3300	4800
5 dr Cus Sta Wag	200	500	850	1900	3500	5000
4 dr Cus Sta Wag	200	500	850	1850	3350	4900
400 Sed	200	500	850	1850	3350	4900
1962						
American						
DeL Sed	150	350	750	1350	2800	4000
2 dr DeL Sed	150	350	750	1350	2800	4000
4 dr DeL Sta Wag	150	350	750	1450	2900	4100
2 dr DeL Sta Wag	150	350	750	1350	2800	4000
Cus Sed	150	350	750	1450	2900	4100
2 dr Cus Sed	150	350	750	1450	2900	4100
4 dr Cus Sta Wag	150	350	750	1450	3000	4200
2 dr Cus Sta Wag	150	350	750	1450	2900	4100
4 dr 400	150	350	750	1450	2900	4100
2 dr 400	150	350	750	1450	3000	4200
400 Conv	200	675	1050	2250	4350	6200
400 Sta Wag	150	400	750	1650	3150	4500
Classic						
DeL Sed	150	350	750	1350	2800	4000
2 dr DeL	150	350	750	1450	2900	4100
DeL Sta Wag	150	350	750	1450	3000	4200
Cus Sed	150	400	750	1550	3050	4300
2 dr Cus	150	400	750	1600	3100	4400
4 dr Cus Sta Wag	150	400	750	1550	3050	4300
5 dr Cus Sta Wag	150	400	750	1600	3100	4400
400 Sed	150	400	750	1600	3100	4400
2 dr 400	150	400	750	1650	3150	4500
400 Sta Wag	150	450	750	1700	3200	4600

NOTE: Add 5 percent for V-8.

Ambassador						
4 dr Cus Sed	150	350	750	1450	3000	4200
2 dr Cus Sed	150	400	750	1550	3050	4300
Cus Sta Wag	150	450	800	1750	3250	4700
4 dr 400 Sed	150	400	750	1650	3150	4500
2 dr 400 Sed	150	450	750	1700	3200	4600
4 dr 400 Sta Wag	150	450	800	1750	3250	4700
5 dr 400 Sta Wag	150	450	800	1800	3300	4800
1963						
American						
4 dr 220 Sed	125	250	700	1150	2500	3600
2 dr 220 Sed	150	300	700	1250	2600	3700
220 Bus Sed	125	250	700	1150	2450	3500
220 Bus Wag	150	300	700	1250	2600	3700
2 dr 220 Sta Wag	125	250	700	1150	2500	3600
330 Sed	150	300	700	1250	2600	3700
2 dr 330 Sed	125	250	700	1150	2500	3600
330 Sta Wag	150	300	700	1250	2600	3700
330 2 dr Wag	150	300	700	1250	2650	3800
440 Sed	150	300	750	1350	2700	3900
2 dr 440 Sed	150	350	750	1350	2800	4000
440 HdTp	150	400	750	1550	3050	4300
440-H HdTp	200	500	850	1900	3500	5000
440 Conv	200	600	950	2150	3850	5500
440 Sta Wag	150	300	750	1350	2700	3900
Classic						
550 Sed	125	250	700	1150	2450	3500
2 dr 550 Sed	125	250	700	1150	2500	3600
550 Sta Wag	125	250	700	1150	2450	3500
660 Sed	125	250	700	1150	2450	3500
2 dr 660 Sed	125	250	700	1150	2500	3600
660 Sta Wag	150	300	700	1250	2600	3700
770 Sed	150	300	750	1350	2700	3900
2 dr 770 Sed	150	300	700	1250	2650	3800
770 Sta Wag	150	300	700	1250	2600	3700

NOTE: Add 5 percent for V-8 models.

Ambassador						
800 Sed	150	300	700	1250	2650	3800
2 dr 800 Sed	150	300	750	1350	2700	3900
800 Sta Wag	150	350	750	1350	2800	4000
880 Sed	150	300	750	1350	2700	3900
2 dr 880 Sed	150	350	750	1350	2800	4000
880 Sta Wag	150	350	750	1450	2900	4100
990 Sed	150	300	750	1350	2800	4000
2 dr 990 Sed	150	350	750	1450	2900	4100
5 dr 990 Sta Wag	150	350	750	1450	3000	4200
4 dr 990 Sta Wag	150	350	750	1350	2800	4000
1964						
American						
220 Sed	125	250	700	1150	2500	3600
2 dr 220	150	300	700	1250	2600	3700
220 Sta Wag	150	300	700	1250	2650	3800
330 Sed	150	300	700	1250	2650	3800
2 dr 330	150	300	750	1350	2700	3900
330 Sta Wag	150	300	750	1350	2700	3900
440 Sed	150	300	700	1250	2650	3800
440 HdTp	150	400	750	1550	3050	4300
440-H HdTp	200	500	850	1900	3500	5000
Conv	200	600	950	2150	3850	5500

303

Model	6	5	4	3	2	1
Classic						
550 Sed	125	250	700	1150	2450	3500
2 dr 550	125	250	700	1150	2500	3600
550 Sta Wag	150	300	700	1250	2600	3700
660 Sed	125	250	700	1150	2500	3600
2 dr 660	150	300	700	1250	2600	3700
660 Sta Wag	150	300	700	1250	2650	3800
770 Sed	150	300	700	1250	2600	3700
2 dr 770	150	300	700	1250	2650	3800
770 Hdtp	150	400	750	1550	3050	4300
770 Typhoon	200	600	950	2150	3850	5500
770 Sta Wag	150	300	700	1250	2650	3800
NOTE: Add 5 percent for V-8 models.						
Ambassador						
Sed	150	400	750	1550	3050	4300
HdTp	150	450	800	1800	3300	4800
990H	200	500	850	1900	3500	5000
Sta Wag	150	350	750	1350	2800	4000
1965						
American						
220 Sed	150	300	700	1250	2600	3700
2 dr 220	150	300	700	1250	2650	3800
220 Sta Wag	150	300	700	1250	2650	3800
330 Sed	150	300	700	1250	2650	3800
2 dr 330	150	350	750	1350	2800	4000
330 Sta Wag	150	350	750	1450	2900	4100
440 Sed	150	350	750	1350	2800	4000
440 HdTp	200	500	850	1900	3500	5000
440-H HdTp	200	550	900	2000	3600	5200
Conv	200	600	950	2200	3900	5600
Classic						
550 Sed	125	250	700	1150	2500	3600
2 dr 550	150	300	700	1250	2600	3700
550 Sta Wag	150	300	700	1250	2600	3700
660 Sed	150	300	750	1350	2700	3900
2 dr 660	150	350	750	1350	2800	4000
660 Sta Wag	150	350	750	1450	2900	4100
770 Sed	150	300	750	1350	2700	3900
770 HdTp	150	350	750	1450	3000	4200
770-H HdTp	200	550	900	2100	3700	5300
770 Conv	200	600	1000	2200	4000	5700
770 Sta Wag	150	350	750	1350	2800	4000
NOTE: Add 5 percent for V-8 models.						
Marlin						
FstBk	200	600	950	2150	3850	5500
Ambassador						
880 Sed	150	350	750	1350	2800	4000
2 dr 880	150	350	750	1450	2900	4100
880 Sta Wag	150	350	750	1450	3000	4200
990 Sed	150	350	750	1450	2900	4100
990 HdTp	150	400	750	1550	3050	4300
990-H HdTp	200	500	850	1900	3500	5000
Conv	200	600	950	2150	3850	5500
Sta Wag	150	350	750	1350	2800	4000
Marlin, V-8						
FstBk	200	550	900	2000	3600	5200
1966						
American						
220 Sed	125	250	700	1150	2450	3500
220 2 dr Sed	125	250	700	1150	2500	3600
220 Wag	150	300	700	1250	2600	3700
440 Sed	150	300	700	1250	2650	3800
440 2 dr Sed	150	300	750	1350	2700	3900
440 Conv	150	450	800	1800	3300	4800
440 Wag	150	300	700	1250	2600	3700
440 HdTp	150	350	750	1350	2800	4000
Rogue	150	400	750	1650	3150	4500
Classic						
550 Sed	125	250	700	1150	2500	3600
550 2 dr Sed	125	250	700	1150	2500	3600
550 Sta Wag	150	300	700	1250	2600	3700
770 Sed	150	300	700	1250	2650	3800
770 HdTp	150	350	750	1450	3000	4200
770 Conv	200	600	950	2150	3850	5500
770 Sta Wag	150	300	700	1250	2600	3700
Rebel						
2 dr HdTp	200	500	850	1900	3500	5000
Marlin						
FsBk Cpe	200	600	950	2150	3850	5500
Ambassador						
880 Sed	150	300	750	1350	2700	3900
880 2 dr Sed	150	350	750	1350	2800	4000
880 Sta Wag	150	350	750	1450	3000	4200
990 Sed	150	350	750	1450	2900	4100
990 HdTp	150	400	750	1650	3150	4500
990 Conv	200	650	1050	2250	4200	6000
990 Sta Wag	125	250	700	1150	2450	3500
DPL (Diplomat)						
DPL HdTp	200	500	850	1900	3500	5000
1967						
American 220						
Sed	125	250	700	1150	2450	3500
2 dr Sed	125	250	700	1150	2450	3500
Sta Wag	125	250	700	1150	2450	3500
American 440						
Sed	125	250	700	1150	2500	3600
2 dr Sed	125	250	700	1150	2500	3600
HdTp	150	350	750	1350	2800	4000
Sta Wag	125	250	700	1150	2450	3500
American Rogue						
HdTp	200	600	950	2200	3900	5600
Conv	350	700	1150	2300	4550	6500
Rebel 550						
Sed	125	250	700	1150	2450	3500
2 dr Sed	125	250	700	1150	2450	3500
Sta Wag	125	250	700	1150	2450	3500
Rebel 770						
Sed	125	250	700	1150	2500	3600
HdTp	150	350	750	1350	2800	4000
Sta Wag	125	250	700	1150	2450	3500
Rebel SST						
HdTp	150	350	750	1450	3000	4200
Conv	200	650	1050	2250	4200	6000
Rambler Marlin						
FsBk Cpe	200	600	950	2150	3850	5500
Ambassador 880						
Sed	125	250	700	1150	2500	3600
2 dr Sed	125	250	700	1150	2500	3600
Sta Wag	150	300	700	1250	2600	3700
Ambassador 990						
Sed	150	350	750	1350	2800	4000
HdTp	150	450	800	1800	3300	4800
Sta Wag	150	350	750	1450	2900	4100
Ambassador DPL						
HdTp	200	500	850	1950	3600	5100
Conv	350	700	1150	2300	4550	6500
1968						
American 220						
Sed	150	300	700	1250	2600	3700
2 dr Sed	150	300	700	1250	2600	3700
American 440						
Sed	150	300	700	1250	2650	3800
Sta Wag	150	300	700	1250	2600	3700
Rogue						
HdTp	200	650	1050	2250	4200	6000
Rebel 550						
Sed	150	300	700	1250	2600	3700
Conv	150	450	800	1750	3250	4700
Sta Wag	125	250	700	1150	2450	3500
HdTp	150	350	750	1450	3000	4200
Rebel 770						
Sed	150	300	700	1250	2600	3700
Sta Wag	125	250	700	1150	2500	3600
HdTp	150	400	750	1600	3100	4400
Rebel SST						
Conv	200	650	1050	2250	4200	6000
HdTp	150	450	750	1700	3200	4600
Ambassador						
Sed	150	300	700	1250	2650	3800
HdTp	150	400	750	1650	3150	4500
Ambassador DPL						
Sed	150	350	750	1350	2800	4000
HdTp	150	450	800	1750	3250	4700
Sta Wag	150	350	750	1350	2800	4000
Ambassador SST						
Sed	150	350	750	1350	2800	4000
HdTp	200	500	850	1900	3500	5000
Javelin						
FsBk	350	750	1200	2350	4900	7000
Javelin SST						
FsBk	350	900	1550	3050	5900	8500
NOTE: Add 20 percent for GO pkg. Add 30 percent for Big Bad pkg.						
AMX						
FsBk	550	1700	2800	5600	9800	14,000
NOTE: Add 25 percent for Craig Breedlove Edit.						
1969						
Rambler						
Sed	125	250	700	1150	2500	3600
2 dr Sed	125	250	700	1150	2500	3600
Rambler 440						
Sed	150	300	700	1250	2600	3700
2 dr Sed	150	300	700	1250	2600	3700
Rambler Rogue						
HdTp	200	650	1050	2250	4200	6000
Rambler Hurst S/C						
HdTp	500	1550	2600	5200	9100	13,000
Rebel						
Sed	125	250	700	1150	2450	3500
HdTp	150	350	750	1350	2800	4000
Sta Wag	125	250	700	1150	2500	3600
Rebel SST						
Sed	150	300	700	1250	2600	3700
HdTp	150	350	750	1450	3000	4200
Sta Wag	150	300	700	1250	2600	3700
AMX						
FsBk Cpe	550	1700	2800	5600	9800	14,000
NOTE: Add 25 percent for Big Bad Pkg.						
Javelin						
FsBk Cpe	350	750	1200	2350	4900	7000
Javelin SST						
FsBk Cpe	350	900	1550	3050	5900	8500
NOTE: Add 20 percent for GO Pkg. Add 30 percent for Big Bad Pkg.						
Ambassador						
Sed	150	300	700	1250	2650	3800
Ambassador DPL						
Sed	150	350	750	1350	2800	4000
Sta Wag	150	350	750	1350	2800	4000
HdTp	150	350	750	1450	3000	4200
Ambassador SST						
Sed	150	300	700	1250	2650	3800
HdTp	150	400	750	1550	3050	4300
1970						
Hornet						
Sed	125	250	700	1150	2450	3500
2 dr Sed	125	250	700	1150	2450	3500
Hornet SST						
Sed	125	250	700	1150	2500	3600
2 dr Sed	125	250	700	1150	2500	3600
Rebel						
Sed	150	300	700	1250	2600	3700
HdTp	150	400	750	1650	3150	4500
Sta Wag	150	350	750	1450	2900	4100
Rebel SST						
Sed	150	300	700	1250	2650	3800
HdTp	200	650	1050	2250	4200	6000
Sta Wag	150	300	700	1250	2600	3700
Rebel 'Machine'						
HdTp	450	1500	2500	5000	8800	12,500
AMX						
FsBk Cpe	550	1700	2800	5600	9800	14,000
Gremlin						
2 dr Comm	150	300	700	1250	2650	3800
2 dr Sed	150	300	750	1350	2700	3900
Javelin						
FsBk Cpe	350	750	1300	2450	5250	7500
Javelin SST						
FsBk Cpe	450	1000	1600	3300	6250	8900
NOTE: Add 20 percent for GO pkg. Add 30 percent for Big Bad pkg.						
'Trans Am'						
FsBk Cpe	400	1200	2000	3950	7000	10,000
'Mark Donahue'						
FsBk Cpe	450	1100	1700	3650	6650	9500

	6	5	4	3	2	1
Ambassador						
Sed	150	300	700	1250	2650	3800
Ambassador DPL						
Sed	150	300	750	1350	2700	3900
HdTp	150	350	750	1350	2800	4000
Sta Wag	150	300	700	1250	2650	3800
Ambassador SST						
Sed	150	350	750	1350	2800	4000
HdTp	150	350	750	1450	2900	4100
Sta Wag	150	300	750	1350	2700	3900
1971						
Gremlin						
2 dr Sed	125	250	700	1150	2450	3500
Sed	125	250	700	1150	2450	3500
Hornet						
2 dr Sed	125	250	700	1150	2500	3600
Sed	125	250	700	1150	2500	3600
Hornet SST						
2 dr Sed	125	250	700	1150	2450	3500
Sed	125	250	700	1150	2450	3500
Hornet SC/360						
HdTp	350	800	1450	2750	5600	8000
Javelin						
HdTp	150	400	750	1650	3150	4500
SST HdTp	200	600	950	2150	3850	5500
NOTE: Add 10 percent for 401 V-8.						
Javelin AMX						
HdTp	350	750	1300	2450	5250	7500
NOTE: Add 15 percent for GO Pkg.						
Matador						
Sed	125	250	700	1150	2450	3500
HdTp	150	300	700	1250	2600	3700
Sta Wag	125	250	700	1150	2500	3600
Ambassador DPL						
Sed	125	250	700	1150	2500	3600
Ambassador SST						
Sed	150	300	700	1250	2600	3700
HdTp	150	300	750	1350	2700	3900
Sta Wag	150	300	700	1250	2650	3800
NOTE: Add 10 percent to Ambassador SST for Broughams.						
1972						
Hornet SST						
2 dr Sed	125	250	700	1150	2450	3500
Sed	125	250	700	1150	2500	3600
Sta Wag	150	300	700	1250	2600	3700
Gucci	150	400	750	1650	3150	4500
DeL Wag	150	300	700	1250	2650	3800
'X' Wag	150	300	700	1250	2600	3700
Matador						
Sed	150	300	700	1250	2600	3700
HdTp	150	300	750	1350	2700	3900
Sta Wag	150	300	700	1250	2650	3800
Gremlin						
2 dr Sed	125	250	700	1150	2500	3600
'X' Sed	150	400	750	1650	3150	4500
Javelin						
SST	150	400	750	1650	3150	4500
AMX	200	650	1050	2250	4200	6000
Go '360'	350	750	1200	2350	4900	7000
Go '401'	350	800	1350	2700	5500	7900
Cardin	350	725	1200	2350	4800	6800
NOTE: Add 20 percent for 401 V-8.						
Add 25 percent for 401 Police Special V-8.						
Add 30 percent for GO Pkg.						
Ambassador SST						
Sed	150	300	700	1250	2600	3700
HdTp	150	300	750	1350	2700	3900
Sta Wag	150	300	700	1250	2650	3800
Ambassador Brougham						
NOTE: Add 10 percent to SST prices for Brougham.						
Gremlin V8						
2 dr	150	350	750	1450	2900	4100
Hornet V8						
2 dr	150	300	750	1350	2700	3900
4 dr	150	300	700	1250	2650	3800
2 dr Hatchback	150	350	750	1350	2800	4000
Sta Wag	150	300	750	1350	2700	3900
AMX V8						
2 dr HdTp	350	850	1500	2900	5700	8200
NOTE: Add 15 percent for GO Pkg.						
Matador V8						
4 dr Sed	150	300	700	1250	2600	3700
2 dr HdTp	150	300	700	1250	2650	3800
Sta Wag	150	300	700	1250	2600	3700
Ambassador Brgm V8						
4 dr Sed	150	300	700	1250	2650	3800
2 dr HdTp	125	200	600	1100	2300	3300
Sta Wag	150	300	700	1250	2650	3800
1973						
Gremlin V8						
2 dr	150	400	750	1650	3150	4500
Hornet V8						
2 dr	150	300	750	1350	2700	3900
4 dr	150	300	700	1250	2650	3800
2 dr Hatchback	150	350	750	1350	2800	4000
Sta Wag	150	300	750	1350	2700	3900
Javelin V8						
2 dr HdTp	200	500	850	1900	3500	5000
AMX V8						
2 dr HdTp	350	800	1450	2750	5600	8000
Matador V8						
4 dr Sed	150	300	700	1250	2600	3700
2 dr HdTp	150	300	700	1250	2650	3800
Sta Wag	150	300	700	1250	2600	3700
Ambassador Brgm V8						
4 dr Sed	150	300	700	1250	2650	3800
2 dr HdTp	150	300	750	1350	2700	3900
Sta Wag	150	300	700	1250	2650	3800
1974						
Gremlin V8						
2 dr Sed	150	400	750	1650	3150	4500
Hornet						
Sed	125	250	700	1150	2400	3400
2 dr Sed	125	250	700	1150	2450	3500
Hatch	125	250	700	1150	2500	3600
Sta Wag	125	250	700	1150	2450	3500
Javelin						
FsBk	150	350	750	1450	3000	4200

	6	5	4	3	2	1
Javelin AMX						
FsBk	350	700	1150	2300	4550	6500
Matador						
Sed	125	200	600	1100	2250	3200
2 dr Sed	125	250	700	1150	2500	3600
Sta Wag	125	200	600	1100	2300	3300
Matador Brougham						
Cpe	150	300	700	1250	2600	3700
Matador 'X'						
Cpe	150	300	700	1250	2650	3800
Ambassador Brougham						
Sed	125	200	600	1100	2300	3300
Sta Wag	125	250	700	1150	2400	3400
NOTE: Add 10 percent for Oleg Cassini coupe.						
Add 12 percent for 'Go-Package'.						
1975						
Gremlin						
2 dr Sed	150	300	700	1250	2650	3800
Hornet						
Sed	125	250	700	1150	2450	3500
2 dr Sed	125	250	700	1150	2400	3400
Hatch	125	250	700	1150	2450	3500
Sta Wag	125	250	700	1150	2450	3500
Pacer						
2 dr Sed	150	300	750	1350	2700	3900
Matador						
Sed	125	250	700	1150	2400	3400
Cpe	125	250	700	1150	2500	3600
Sta Wag	125	250	700	1150	2450	3500
1976						
Gremlin, V-8						
2 dr Sed	125	250	700	1150	2450	3500
Cus 2 dr Sed	150	300	700	1250	2650	3800
Hornet, V-8						
4 dr Sed	125	200	600	1100	2200	3100
2 dr Sed	100	175	525	1050	2100	3000
2 dr Hatch	125	200	600	1100	2250	3200
4 dr Sptabt	125	200	600	1100	2300	3300
Pacer, 6-cyl.						
2 dr Sed	125	250	700	1150	2450	3500
Matador, V-8						
4 dr Sed	100	175	525	1050	2100	3000
Cpe	125	200	600	1100	2250	3200
Sta Wag	125	200	600	1100	2200	3100
NOTE: Deduct 5 percent for 6 cylinder.						
1977						
Gremlin, V-8						
2 dr Sed	150	300	700	1250	2600	3700
Cus 2 dr Sed	150	300	700	1250	2650	3800
Hornet, V-8						
4 dr Sed	125	200	600	1100	2250	3200
2 dr Sed	125	200	600	1100	2200	3100
2 dr Hatch	125	200	600	1100	2300	3300
Sta Wag	125	250	700	1150	2400	3400
Pacer, 6-cyl.						
2 dr Sed	125	250	700	1150	2500	3600
Sta Wag	150	300	700	1250	2600	3700
Matador, V-8						
4 dr Sed	125	200	600	1100	2200	3100
Cpe	125	200	600	1100	2300	3300
Sta Wag	125	200	600	1100	2250	3200
NOTE: Deduct 5 percent for 6 cylinder.						
Add 10 percent for AMX package.						
1978						
Gremlin						
2 dr Sed	125	250	700	1150	2400	3400
Cus 2 dr Sed	125	250	700	1150	2450	3500
Concord						
4 dr Sed	100	175	525	1050	2050	2900
2 dr Sed	100	175	525	1050	1950	2800
2 dr Hatch	100	175	525	1050	2100	3000
Sta Wag	125	200	600	1100	2200	3100
Pacer						
2 dr Hatch	125	200	600	1100	2300	3300
Sta Wag	125	250	700	1150	2400	3400
AMX						
2 dr Hatch	125	250	700	1150	2500	3600
Matador						
4 dr Sed	100	175	525	1050	1950	2800
Cpe	100	175	525	1050	2100	3000
Sta Wag	100	175	525	1050	2050	2900
1979						
Spirit, 6-cyl.						
2 dr Hatch	125	250	700	1150	2450	3500
2 dr Sed	125	250	700	1150	2400	3400
Spirit DL, 6-cyl.						
2 dr Hatch	125	250	700	1150	2500	3600
2 dr Sed	125	250	700	1150	2450	3500
Spirit Limited, 6-cyl.						
2 dr Hatch	150	300	700	1250	2600	3700
2 dr Sed	125	250	700	1150	2500	3600
NOTE: Deduct 5 percent for 4-cyl.						
Concord, V-8						
4 dr Sed	125	200	600	1100	2200	3100
2 dr Sed	100	175	525	1050	2100	3000
2 dr Hatch	125	200	600	1100	2250	3200
4 dr Sta Wag	125	200	600	1100	2250	3200
Concord DL, V-8						
4 dr Sed	125	200	600	1100	2250	3200
2 dr Sed	125	200	600	1100	2200	3100
2 dr Hatch	125	200	600	1100	2300	3300
4 dr Sta Wag	125	200	600	1100	2300	3300
Concord Limited, V-8						
4 dr Sed	125	200	600	1100	2300	3300
2 dr Sed	125	200	600	1100	2250	3200
4 dr Sta Wag	125	250	700	1150	2400	3400
NOTE: Deduct 5 percent for 6-cyl.						
Pacer DL, V-8						
2 dr Hatch	125	250	700	1150	2400	3400
2 dr Sta Wag	125	250	700	1150	2450	3500
Pacer Limited, V-8						
2 dr Hatch	125	250	700	1150	2450	3500
2 dr Sta Wag	125	250	700	1150	2500	3600
NOTE: Deduct 5 percent for 6-cyl.						
AMX, V-8						
2 dr Hatch	150	300	700	1250	2600	3700
NOTE: Deduct 7 percent for 6-cyl.						

	6	5	4	3	2	1
1980						
Spirit, 6-cyl.						
2 dr Hatch	150	350	750	1350	2800	4000
2 dr Cpe	150	300	750	1350	2700	3900
2 dr Hatch DL	150	350	750	1450	2900	4100
2 dr Cpe DL	150	350	750	1350	2800	4000
2 dr Hatch Limited	150	400	750	1550	3050	4300
2 dr Cpe Limited	150	350	750	1450	3000	4200
NOTE: Deduct 10 percent for 4-cyl.						
Concord, 6-cyl.						
4 dr Sed	125	250	700	1150	2500	3600
2 dr Cpe	125	250	700	1150	2450	3500
4 dr Sta Wag	150	300	700	1250	2600	3700
4 dr Sed DL	150	300	700	1250	2600	3700
2 dr Cpe DL	125	250	700	1150	2500	3600
4 dr Sta Wag DL	150	300	700	1250	2650	3800
4 dr Sed Limited	150	300	750	1350	2700	3900
2 dr Cpe Limited	150	300	700	1250	2650	3800
4 dr Sta Wag Limited	150	300	750	1350	2700	3900
Pacer, 6-cyl.						
2 dr Hatch DL	125	250	700	1150	2500	3600
2 dr Sta Wag DL	150	300	700	1250	2600	3700
2 dr Hatch Limited	150	300	700	1250	2650	3800
2 dr Sta Wag Limited	150	300	750	1350	2700	3900
AMX, 6-cyl.						
2 dr Hatch	150	350	750	1450	3000	4200
Eagle 4WD, 6-cyl.						
4 dr Sed	200	500	850	1900	3500	5000
2 dr Cpe	200	500	850	1850	3350	4900
4 dr Sta Wag	200	550	900	2000	3600	5200
4 dr Sed Limited	200	550	900	2000	3600	5200
2 dr Cpe Limited	200	500	850	1950	3600	5100
4 dr Sta Wag Limited	200	550	900	2150	3800	5400
1981						
Spirit, 4-cyl.						
2 dr Hatch	150	300	700	1250	2600	3700
2 dr Cpe	125	250	700	1150	2500	3600
2 dr Hatch DL	150	300	750	1350	2700	3900
2 dr Cpe DL	150	300	700	1250	2650	3800
Spirit, 6-cyl.						
2 dr Hatch	150	350	750	1450	2900	4100
2 dr Cpe	150	350	750	1350	2800	4000
2 dr Hatch DL	150	400	750	1550	3050	4300
2 dr Cpe DL	150	350	750	1450	3000	4200
Concord, 6-cyl.						
4 dr Sed	150	300	700	1250	2600	3700
2 dr Cpe	125	250	700	1150	2500	3600
4 dr Sta Wag	150	300	700	1250	2650	3800
4 dr Sed DL	150	300	700	1250	2650	3800
2 dr Cpe DL	150	300	700	1250	2600	3700
4 dr Sta Wag DL	150	300	750	1350	2700	3900
4 dr Sed Limited	150	300	750	1350	2700	3900
2 dr Cpe Limited	150	300	700	1250	2650	3800
4 dr Sta Wag Limited	150	350	750	1350	2800	4000
NOTE: Deduct 12 percent for 4-cyl.						
Eagle 50 4WD, 4-cyl.						
2 dr Hatch SX4	150	450	800	1800	3300	4800
2 dr Hatchback	150	450	800	1750	3250	4700
2 dr Hatch SX4 DL	200	500	850	1900	3500	5000
2 dr Hatchback DL	200	500	850	1850	3350	4900
Eagle 50 4WD, 6-cyl.						
2 dr Hatch SX4	200	550	900	2000	3600	5200
2 dr Hatchback	200	500	850	1950	3600	5100
2 dr Hatch SX4 DL	200	550	900	2150	3800	5400
2 dr Hatchback DL	200	550	900	2100	3700	5300
1982						
Spirit, 6-cyl.						
2 dr Hatch	150	350	750	1450	3000	4200
2 dr Cpe	150	350	750	1450	2900	4100
2 dr Hatch DL	150	400	750	1600	3100	4400
2 dr Cpe DL	150	400	750	1550	3050	4300
NOTE: Deduct 10 percent for 4-cyl.						
Concord, 6-cyl.						
4 dr Sed	150	300	700	1250	2650	3800
2 dr Cpe	150	300	700	1250	2600	3700
4 dr Sta Wag	150	300	750	1350	2700	3900
4 dr Sed DL	150	300	750	1350	2700	3900
2 dr Cpe DL	150	300	700	1250	2650	3800
4 dr Sta Wag DL	150	350	750	1350	2800	4000
4 dr Sed Limited	150	350	750	1350	2800	4000
2 dr Cpe Limited	150	300	750	1350	2700	3900
4 dr Sta Wag Limited	150	350	750	1450	2900	4100
NOTE: Deduct 12 percent for 4-cyl.						
Eagle 50 4WD, 4-cyl.						
2 dr Hatch SX4	200	500	850	1850	3350	4900
2 dr Hatchback	150	450	800	1800	3300	4800
2 dr Hatch SX4 DL	200	500	850	1950	3600	5100
2 dr Hatchback DL	200	500	850	1900	3500	5000
Eagle 50 4WD, 6-cyl.						
2 dr Hatch SX4	200	550	900	2100	3700	5300
2 dr Hatchback	200	550	900	2000	3600	5200
2 dr Hatch SX4 DL	200	600	950	2150	3850	5500
2 dr Hatchback DL	200	550	900	2150	3800	5400
Eagle 30 4WD, 4-cyl.						
4 dr Sed	150	450	800	1750	3250	4700
2 dr Cpe	150	450	750	1700	3200	4600
4 dr Sta Wag	150	450	800	1800	3300	4800
4 dr Sed Limited	150	450	800	1800	3300	4800
2 dr Cpe Limited	150	450	800	1750	3250	4700
4 dr Sta Wag Limited	200	500	850	1900	3500	5000
Eagle 30 4WD, 6-cyl.						
4 dr Sed	200	500	850	1950	3600	5100
2 dr Cpe	200	500	850	1900	3500	5000
4 dr Sta Wag	200	550	900	2100	3700	5300
4 dr Sed Limited	200	550	900	2100	3700	5300
2 dr Cpe Limited	200	550	900	2000	3600	5200
4 dr Sta Wag Limited	200	600	950	2150	3850	5500
1983						
Spirit, 6-cyl.						
2 dr Hatch DL	150	400	750	1550	3050	4300
2 dr Hatch GT	150	400	750	1600	3100	4400
Concord, 6-cyl.						
4 dr Sed	150	300	750	1350	2700	3900
4 dr Sta Wag	150	350	750	1350	2800	4000
4 dr Sed DL	150	350	750	1350	2800	4000
4 dr Sta Wag DL	150	350	750	1450	2900	4100
4 dr Sta Wag Limited	150	400	750	1550	3050	4300
Alliance, 4-cyl.						
2 dr Sed	125	250	700	1150	2500	3600
4 dr Sed L	150	300	700	1250	2600	3700
2 dr Sed L	150	300	700	1250	2600	3700
4 dr Sed DL	150	300	700	1250	2650	3800
2 dr Sed DL	150	300	700	1250	2650	3800
4 dr Sed Limited	150	300	750	1350	2700	3900
Eagle 50 4WD, 4-cyl.						
2 dr Hatch SX4	200	500	850	1900	3500	5000
2 dr Hatch SX4 DL	200	550	900	2000	3600	5200
Eagle 50 4WD, 6-cyl.						
2 dr Hatch SX4	200	550	900	2150	3800	5400
2 dr Hatch SX4 DL	200	600	950	2200	3900	5600
Eagle 30 4WD, 4-cyl.						
4 dr Sed	150	450	800	1800	3300	4800
4 dr Sta Wag	200	500	850	1900	3500	5000
4 dr Sta Wag Limited	200	550	900	2000	3600	5200
Eagle 30 4WD, 6-cyl.						
4 dr Sed	200	550	900	2000	3600	5200
4 dr Sta Wag	200	550	900	2150	3800	5400
4 dr Sta Wag Limited	200	600	950	2200	3900	5600
1984						
Alliance, 4-cyl.						
2 dr	150	300	700	1250	2600	3700
L						
4 dr	150	300	700	1250	2650	3800
2 dr	150	300	700	1250	2650	3800
DL						
4 dr	150	300	750	1350	2700	3900
2 dr	150	300	750	1350	2700	3900
Limited						
4 dr	150	350	750	1350	2800	4000
Renault Encore, 4-cyl.						
2 dr Liftback	125	250	700	1150	2400	3400
S						
2 dr Liftback	125	250	700	1150	2450	3500
4 dr Liftback	125	250	700	1150	2450	3500
LS						
2 dr Liftback	125	250	700	1150	2500	3600
4 dr Liftback	125	250	700	1150	2500	3600
GS						
2 dr Liftback	150	300	700	1250	2600	3700
Eagle 4WD, 4-cyl.						
4 dr Sed	200	500	850	1850	3350	4900
4 dr Sta Wag	200	500	850	1950	3600	5100
4 dr Sta Wag Limited	200	550	900	2100	3700	5300
Eagle 4WD, 6-cyl.						
4 dr Sed	200	550	900	2100	3700	5300
4 dr Sta Wag	200	600	950	2150	3850	5500
4 dr Sta Wag Limited	200	600	1000	2200	4000	5700
1985 - 1986						
Alliance (Renault)						
2 dr Sed	100	150	450	1000	1800	2600
4 dr Sed L	100	175	525	1050	1950	2800
2 dr Sed L	100	175	525	1050	2050	2900
Conv L	150	300	700	1250	2600	3700
4 dr Sed DL	125	200	600	1100	2250	3200
2 dr Sed DL	125	250	700	1150	2450	3500
Conv DL	150	350	750	1450	2900	4100
Limited 4 dr Sed	150	300	750	1350	2700	3900
Eagle 4WD						
4 dr Sed	200	550	900	2150	3800	5400
4 dr Sta Wag	200	600	950	2200	3900	5600
Limited 4 dr Sta Wag	200	650	1000	2200	4100	5800

METROPOLITAN

1954						
Series E, (Nash), 4-cyl., 85" wb, 42 hp						
HdTp	200	550	900	2000	3600	5200
Conv	200	600	1000	2200	4000	5700
1955						
Series A & B, (Nash), 4-cyl., 85" wb, 42 hp						
HdTp	200	550	900	2000	3600	5200
Conv	200	600	1000	2200	4000	5700
Series A & B, (Hudson), 4-cyl., 85" wb, 42 hp						
HdTp	200	550	900	2000	3600	5200
Conv	200	600	1000	2200	4000	5700
1956						
Series 1500, (Nash), 4-cyl., 85" wb, 52 hp						
HdTp	200	550	900	2000	3600	5200
Conv	200	600	1000	2200	4000	5700
Series A, (Nash), 4-cyl., 85" wb, 42 hp						
HdTp	150	450	800	1800	3300	4800
Conv	200	550	900	2150	3800	5400
Series 1500, (Hudson), 4-cyl., 85" wb, 52 hp						
HdTp	200	550	900	2000	3600	5200
Conv	200	600	1000	2200	4000	5700
Series A, (Hudson), 4-cyl., 85" wb, 42 hp						
HdTp	150	450	800	1800	3300	4800
Conv	200	550	900	2150	3800	5400
1957						
Series 1500, (Nash), 4-cyl., 85" wb, 52 hp						
HdTp	200	550	900	2000	3600	5200
Conv	200	600	1000	2200	4000	5700
Series A-85, (Nash), 4-cyl., 85" wb, 42 hp						
HdTp	200	500	850	1850	3350	4900
Conv	200	600	950	2150	3850	5500
1958						
Series 1500, (AMC), 4-cyl., 85" wb, 55 hp						
HdTp	200	550	900	2000	3600	5200
Conv	200	600	1000	2200	4000	5700
1959						
Series 1500, (AMC), 4-cyl., 85" wb, 55 hp						
HdTp	200	600	1000	2200	4000	5700
Conv	350	700	1150	2300	4550	6500
1960						
Series 1500, (AMC), 4-cyl., 85" wb, 55 hp						
HdTp	200	600	1000	2200	4000	5700
Conv	350	700	1150	2300	4550	6500
1961						
Series 1500, (AMC), 4-cyl., 85" wb, 55 hp						
HdTp	200	600	1000	2200	4000	5700
Conv	350	700	1150	2300	4550	6500

1962

	6	5	4	3	2	1
Series 1500, (AMC), 4-cyl., 85" wb, 55 hp						
HdTp	200	600	1000	2200	4000	5700
Conv	350	700	1150	2300	4550	6500

AMC - JEEP

1970

	6	5	4	3	2	1
Model J-100, 110" wb						
4d Sta Wag	200	650	1000	2200	4100	5800
4d Cus Sta Wag	200	650	1000	2200	4150	5900
Model J-100, 101" wb						
4d Cus Sta Wag	200	600	950	2200	3900	5600
Jeepster Commando, 101" wb						
Sta Wag	200	650	1000	2200	4150	5900
Rds	350	725	1150	2300	4700	6700
Jeepster						
Conv	350	725	1200	2350	4800	6800
Conv Commando	350	725	1200	2350	4850	6900
CJ-5, 1/4-Ton, 81" wb						
Jeep	400	1200	2000	3950	7000	10,000
CJ-6, 101" wb						
Jeep	450	1100	1700	3650	6650	9500
DJ-5, 1/4-Ton, 81" wb						
Jeep	200	650	1050	2250	4200	6000
Jeepster, 1/4-Ton, 101" wb						
PU	200	650	1050	2250	4200	6000
Wagoneer V-8						
4d Cus Sta Wag	200	675	1050	2250	4350	6200
NOTE: Deduct 10 percent for 6-cyl.						
Series J-2500						
Thriftside PU	200	500	850	1850	3350	4900
Townside PU	200	500	850	1900	3500	5000
Series J-2600						
Thriftside PU	150	450	800	1750	3250	4700
Townside PU	150	450	800	1800	3300	4800
Platform Stake	150	400	750	1650	3150	4500
Series J-2700, 3/4-Ton						
Thriftside PU	150	350	750	1450	3000	4200
Townside PU	150	400	750	1550	3050	4300
Platform Stake	150	350	750	1450	2900	4100
Series J-3500, 1/2-Ton						
Townside PU	150	350	750	1450	3000	4200
Series J-3600, 1/2-Ton						
Townside PU	150	350	750	1450	3000	4200
Platform Stake	150	350	750	1350	2800	4000
Series J-3700, 3/4-Ton						
Townside PU	150	350	750	1450	2900	4100
Platform Stake	150	300	750	1350	2700	3900
Series J-4500						
Townside PU	150	350	750	1350	2800	4000
Series J-4600						
Townside PU	150	300	750	1350	2700	3900
Series J-4700						
Townside PU						

1971

	6	5	4	3	2	1
Model J-100, 110" wb						
4d Sta Wag	200	600	1000	2200	4000	5700
4d Cus Sta Wag	200	650	1000	2200	4150	5900
4d Spl Sta Wag	200	650	1000	2200	4100	5800
Jeepster Commando, 101" wb						
Sta Wag	200	650	1000	2200	4150	5900
Rds	350	725	1150	2300	4700	6700
Jeepster Commando Six						
Sta Wag	200	675	1050	2250	4300	6100
Rds	350	725	1200	2350	4800	6800
Conv	350	725	1200	2350	4850	6900
CJ-5, 1/4-Ton						
Jeep	400	1200	2000	3950	7000	10,000
CJ-6, 1/2-Ton						
Jeep	450	1100	1700	3650	6650	9500
DJ-5, 1/4-Ton						
Open	350	900	1550	3050	5900	8500
Jeepster, 1/2-Ton						
PU	200	650	1050	2250	4200	6000
Wagoneer V-8						
4d Sta Wag Cus	200	650	1000	2200	4150	5900
NOTE: Deduct 10 percent for 6-cyl.						
Series J-2500						
Thriftside PU	200	500	850	1850	3350	4900
Townside PU	150	450	800	1800	3300	4800
Series J-3800						
Townside PU	150	450	800	1750	3250	4700
Series J-4500						
Townside PU	150	450	750	1700	3200	4600
Series J-4600						
Townside PU	150	400	750	1650	3150	4500
Series J-4700						
Townside PU	150	400	750	1600	3100	4400
Series J-4800						
Townside PU	150	400	750	1550	3050	4300

1972

	6	5	4	3	2	1
Wagoneer, 6-cyl.						
4d Sta Wag	200	650	1000	2200	4100	5800
4d Cus Sta Wag	200	650	1050	2250	4200	6000
Commando, 1/2-Ton, 6-cyl, 101" wb						
4d Sta Wag	200	675	1050	2250	4300	6100
Rds	350	725	1200	2350	4800	6800
CJ-5, 1/4-Ton						
Jeep	400	1200	2000	3950	7000	10,000
CJ-6, 1/4-Ton						
Jeep	450	1100	1700	3650	6650	9500
DJ-5, 1/4-Ton, 2WD						
Open	200	650	1050	2250	4200	6000
Series J-2500, 1/2-Ton						
Thriftside PU	200	500	850	1850	3350	4900
Townside PU	150	450	800	1800	3300	4800
Series J-2600, 1/2-Ton						
Thriftside PU	150	450	800	1800	3300	4800
Townside PU	150	450	800	1750	3250	4700
Series J-4500, 3/4-Ton						
Townside PU	150	450	750	1700	3200	4600
Series J-4600, 3/4-Ton						
Townside PU	150	400	750	1650	3150	4500
Series J-4700, 3/4-Ton						
Townside PU	150	400	750	1600	3100	4400
Series J-4800, 3/4-Ton						
Townside PU	150	400	750	1550	3050	4300

1973

	6	5	4	3	2	1
Wagoneer, 6-cyl, 110" wb						
4d Sta Wag	200	650	1000	2200	4100	5800
4d Cus Sta Wag	200	650	1050	2250	4200	6000
Jeep Commando, 6-cyl						
2d Sta Wag	200	600	1000	2200	4000	5700
Rds	350	725	1150	2300	4700	6700
CJ-5, 1/4-Ton						
Jeep	400	1200	2000	3950	7000	10,000
CJ-6, 1/4-Ton						
Jeep	450	1100	1700	3650	6650	9500
Commando, 1/2-Ton, 104" wb						
PU	200	550	900	2100	3700	5300
Series J-2500, 1/2-Ton						
Thriftside PU	200	500	850	1900	3500	5000
Townside PU	200	500	850	1950	3600	5100
Series J-2600, 1/2-Ton						
Thriftside PU	200	500	850	1850	3350	4900
Townside PU	200	500	850	1900	3500	5000
Series J-4500, 3/4-Ton						
Townside PU	150	450	800	1800	3300	4800
Series J-4600, 3/4-Ton						
Townside PU	150	450	800	1750	3250	4700
Series J-4800, 3/4-Ton						
Townside PU	150	450	750	1700	3200	4600

1974

	6	5	4	3	2	1
Wagoneer, V-8, 109" wb						
4d Sta Wag	200	650	1000	2200	4150	5900
4d Cus Sta Wag	200	675	1050	2250	4300	6100
Cherokee, 6-cyl, 109" wb						
2d Sta Wag	200	650	1000	2200	4100	5800
2d "S" Sta Wag	200	650	1000	2200	4150	5900
CJ-5, 1/4-Ton, 84" wb						
Jeep	400	1200	2000	3950	7000	10,000
CJ-6, 1/4-Ton, 104" wb						
Jeep	450	1100	1700	3650	6650	9500
Series J-10, 1/2-Ton, 110"-131" wb						
Townside PU, SWB	150	450	800	1800	3300	4800
Townside PU, LWB	150	450	800	1750	3250	4700
Series J-20, 3/4-Ton						
Townside PU, LWB	150	450	750	1700	3200	4600

1975

	6	5	4	3	2	1
Wagoneer, V-8						
4d Sta Wag	200	650	1000	2200	4150	5900
4d Cus Sta Wag	200	675	1050	2250	4300	6100
Cherokee, 6-cyl						
2d Sta Wag	200	650	1000	2200	4100	5800
2d "S" Sta Wag	200	650	1000	2200	4150	5900
CJ-5, 1/4-Ton, 84" wb						
Jeep	400	1200	2000	3950	7000	10,000
CJ-6, 1/4-Ton, 104" wb						
Jeep	450	1100	1700	3650	6650	9500
Series J-10, 1/2-Ton, 119" or 131" wb						
Townside PU, SWB	150	450	800	1800	3300	4800
Townside PU (LWB)	150	450	800	1750	3250	4700
Series J-20, 3/4-Ton, 131" wb						
Townside PU	150	450	750	1700	3200	4600

1976

	6	5	4	3	2	1
Wagoneer, V-8						
4d Sta Wag	200	650	1000	2200	4150	5900
4d Cus Sta Wag	200	675	1050	2250	4350	6200
Cherokee, 6-cyl.						
2d Sta Wag	200	650	1000	2200	4100	5800
2d "S" Sta Wag	200	650	1000	2200	4150	5900
CJ-5, 1/4-Ton, 84" wb						
Jeep	400	1200	2000	3950	7000	10,000
CJ-7, 1/4-Ton, 94" wb						
Jeep	450	1100	1700	3650	6650	9500
Series J-10, 1/2-Ton, 119" or 131" wb						
Townside PU, SWB	150	450	800	1750	3250	4700
Townside PU, LWB	150	450	750	1700	3200	4600
Series J-20, 3/4-Ton, 131" wb						
Townside PU, LWB	150	400	750	1650	3150	4500

1977

	6	5	4	3	2	1
Wagoneer, V-8						
4d Sta Wag	200	675	1050	2250	4300	6100
Cherokee, 6-cyl						
2d Sta Wag	200	650	1000	2200	4100	5800
2d "S" Sta Wag	200	650	1000	2200	4150	5900
4d Sta Wag	200	650	1000	2200	4100	5800
CJ-5, 1/4-Ton, 84" wb						
Jeep	450	1050	1650	3500	6400	9200
CJ-7, 1/4-Ton, 94" wb						
Jeep	450	1000	1650	3350	6300	9000
Series J-10, 1/2-Ton, 119" or 131" wb						
Townside PU, SWB	150	450	750	1700	3200	4600
Townside PU, LWB	150	400	750	1650	3150	4500
Series J-20, 3/4-Ton, 131" wb						
Townside PU	150	400	750	1600	3100	4400

1978

	6	5	4	3	2	1
Wagoneer, 108.7" wb						
4d Sta Wag	200	675	1050	2250	4300	6100
Cherokee, 6-cyl						
2d Sta Wag	200	650	1000	2200	4100	5800
2d "S" Sta Wag	200	650	1000	2200	4150	5900
4d Sta Wag	200	650	1000	2200	4100	5800
CJ-5, 1/4-Ton, 84" wb						
Jeep	450	1050	1650	3500	6400	9200
CJ-7, 1/4-Ton, 94" wb						
Jeep	450	1000	1650	3350	6300	9000
Series J-10, 1/2-Ton, 119" or 131" wb						
Townside PU, SWB	150	450	750	1700	3200	4600
Townside PU, LWB	150	400	750	1650	3150	4500
Series J-20, 3/4-Ton, 131" wb						
Townside PU	150	400	750	1600	3100	4400

1979

	6	5	4	3	2	1
Wagoneer, V-8, 108.7" wb						
4d Sta Wag	200	650	1050	2250	4200	6000
4d Ltd Sta Wag	200	675	1050	2250	4350	6200
Cherokee, 6-cyl						
2d Sta Wag	200	650	1000	2200	4100	5800
2d "S" Sta Wag	200	650	1000	2200	4150	5900
4d Sta Wag	200	650	1000	2200	4100	5800

	6	5	4	3	2	1
CJ-5, 1/4-Ton, 84" wb						
Jeep	450	1000	1600	3300	6250	8900
CJ-7, 1/4-Ton, 94" wb						
Jeep	350	900	1550	3050	5900	8500
Series J-10, 1/2-Ton, 119" or 131" wb						
Townside PU, SWB	150	450	750	1700	3200	4600
Townside PU, LWB	150	400	750	1650	3150	4500
Series J-20, 3/4-Ton, 131" wb						
Townside PU	150	400	750	1600	3100	4400
1980						
Wagoneer, 6-cyl, 108.7" wb						
4d Sta Wag	200	650	1050	2250	4200	6000
4d Ltd Sta Wag	200	675	1050	2250	4350	6200
Cherokee, 6-cyl						
2d Sta Wag	200	600	950	2150	3850	5500
2d "S" Sta Wag	200	600	1000	2200	4000	5700
4d Sta Wag	200	650	1000	2200	4150	5900
CJ-5, 1/4-Ton, 84" wb						
Jeep	350	900	1550	3050	5900	8500
CJ-7, 1/4-Ton, 94" wb						
Jeep	350	800	1450	2750	5600	8000
Series J-10, 1/2-Ton, 119" or 131" wb						
Townside PU, SWB	150	450	750	1700	3200	4600
Townside PU, LWB	150	400	750	1650	3150	4500
Series J-20, 3/4-Ton, 131" wb						
Townside PU	150	400	750	1600	3100	4400
1981						
Wagoneer, 108.7" wb						
4d Sta Wag	200	650	1050	2250	4200	6000
4d Brgm Sta Wag	200	675	1050	2250	4350	6200
4d Ltd Sta Wag	350	700	1100	2300	4500	6400
Cherokee						
2d Sta Wag	200	500	850	1850	3350	4900
2d Sta Wag, Wide Wheels	200	500	850	1900	3500	5000
4d Sta Wag	200	500	850	1950	3600	5100
Scrambler, 1/2-Ton, 104" wb						
PU	125	250	700	1150	2450	3500
CJ-5, 1/4-Ton, 84" wb						
Jeep	150	350	750	1350	2800	4000
CJ-7, 1/4-Ton, 94" wb						
Jeep	150	350	750	1450	3000	4200
Series J-10, 1/2-Ton, 119" or 131" wb						
Townside PU, SWB	150	400	750	1650	3150	4500
Townside PU, LWB	150	400	750	1600	3100	4400
Series J-20, 3/4-Ton, 131" wb						
Townside PU	150	400	750	1550	3050	4300
1982						
Wagoneer, 6-cyl						
4d Sta Wag	200	650	1000	2200	4150	5900
4d Brgm Sta Wag	200	675	1100	2250	4400	6300
4d Ltd Sta Wag	350	725	1150	2300	4700	6700
Cherokee, 6-cyl						
2d Sta Wag	200	500	850	1850	3350	4900
4d Sta Wag	200	500	850	1950	3600	5100
Scrambler, 1/2-Ton, 103.4" wb						
PU	125	250	700	1150	2450	3500
CJ-5, 1/4-Ton, 84" wb						
Jeep	350	800	1450	2750	5600	8000
CJ-7, 1/4-Ton, 93.4" wb						
Jeep	350	750	1300	2450	5250	7500
Series J-10, 1/2-Ton, 119" or 131" wb						
Townside PU, SWB	150	400	750	1650	3150	4500
Townside PU, LWB	150	400	750	1600	3100	4400
Series J-20, 3/4-Ton, 131" wb						
Townside PU	150	400	750	1550	3050	4300
Series J-10, 1/2-Ton, 119" wb						
Sportside PU	150	450	750	1700	3200	4600
1983						
Wagoneer, 6-cyl						
4d Brgm Sta Wag	350	750	1200	2350	4900	7000
4d Ltd Sta Wag	350	750	1300	2450	5250	7500
Cherokee, 6-cyl						
2d Sta Wag	200	650	1050	2250	4200	6000
Scrambler, 1/2-Ton, 103.4" wb						
PU	125	250	700	1150	2500	3600
CJ-5, 1/4-Ton, 83.4" wb						
Jeep	350	800	1450	2750	5600	8000
CJ-7, 1/4-Ton, 93.4" wb						
Townside Pickup (4WD)						
Jeep	350	750	1300	2450	5250	7500
Series J-10, 1/2-Ton, 119" or 131" wb						
Townside PU, SWB	150	400	750	1600	3100	4400
Townside PU, LWB	150	400	750	1550	3050	4300
Series J-10, 1/2-Ton						
Sportside PU	150	400	750	1650	3150	4500
Series J-20, 3/4-Ton, 131" wb						
Townside PU	150	350	750	1450	3000	4200
1984						
Wagoneer, 4-cyl						
4d Sta Wag	350	750	1250	2400	5050	7200
4d Ltd Sta Wag	350	750	1200	2350	4900	7000
Wagoneer, 6-cyl						
4d Sta Wag	350	750	1300	2400	5200	7400
4d Ltd Sta Wag	350	750	1300	2450	5250	7500
Grand Wagoneer, V-8						
4d Sta Wag	350	750	1350	2650	5450	7800
Cherokee, 4-cyl						
2d Sta Wag	350	750	1250	2350	5000	7100
4d Sta Wag	350	750	1200	2350	4900	7000
Cherokee, 6-cyl						
2d Sta Wag	350	750	1250	2400	5100	7300
4d Sta Wag	350	750	1250	2400	5050	7200
Scrambler, 1/2-Ton, 103.4" wb						
PU	150	400	750	1600	3100	4400
CJ-7, 1/4-Ton, 93.4" wb						
Jeep	350	750	1350	2650	5450	7800
Series J-10, 1/2-Ton, 119" or 131" wb						
Townside PU	200	550	900	2000	3600	5200
Series J-20, 3/4-Ton, 131" wb						
Townside PU	200	550	900	2100	3700	5300
1985						
Wagoneer, 4-cyl, 101.4" wb						
4d Sta Wag	350	750	1200	2350	4900	7000
4d Ltd Sta Wag	350	750	1300	2450	5250	7500
Wagoneer, 8-cyl, 101.4" wb						
4d Sta Wag	350	750	1300	2450	5250	7500
4d Ltd Sta Wag	350	800	1450	2750	5600	8000

	6	5	4	3	2	1
Grand Wagoneer, 108.7" wb						
4d Sta Wag	350	750	1200	2350	4900	7000
Cherokee, 4-cyl, 101.4" wb						
2d Sta Wag 2WD	200	650	1050	2250	4200	6000
4d Sta Wag 2WD	200	675	1050	2250	4350	6200
2d Sta Wag 4WD	350	700	1150	2300	4550	6500
4d Sta Wag 4WD	350	725	1150	2300	4700	6700
Cherokee, 6-cyl, 101.4" wb						
2d Sta Wag 2WD	200	675	1050	2250	4350	6200
4d Sta Wag 2WD	350	700	1100	2300	4500	6400
2d Sta Wag 4WD	350	725	1150	2300	4700	6700
4d Sta Wag 4WD	350	725	1200	2350	4850	6900
Scrambler, 1/2-Ton, 103.5" wb						
PU	200	675	1050	2250	4350	6200
CJ-7, 1/4-Ton, 93.5" wb						
Jeep	350	800	1450	2750	5600	8000
Series J-10, 1/2-Ton, 131" wb, 4WD						
Townside PU	200	650	1050	2250	4200	6000
Series J-20, 3/4-Ton, 131" wb, 4WD						
Townside PU	350	700	1150	2300	4550	6500
1986						
Wagoneer						
4d Sta Wag	350	900	1550	3050	5900	8500
4d Ltd Sta Wag	450	950	1600	3250	6150	8800
4d Grand Sta Wag	450	1000	1650	3350	6300	9000
Cherokee						
2d Sta Wag 2WD	350	750	1200	2350	4900	7000
4d Sta Wag 2WD	350	750	1300	2450	5250	7500
2d Sta Wag 4WD	350	750	1300	2450	5250	7500
4d Sta Wag 4WD	350	800	1450	2750	5600	8000
Wrangler, 1/4-Ton, 93.4" wb						
Jeep 2WD						
Comanche, 120" wb						
PU	350	700	1100	2300	4500	6400
CJ-7, 1/4-Ton, 93.5" wb						
Jeep	350	900	1550	3050	5900	8500
Series J-10, 131" wb, 4WD						
Townside PU	350	750	1200	2350	4900	7000
Series J-20, 131" wb, 4WD						
Townside PU	350	750	1300	2450	5250	7500
1987						
Wrangler, 4WD, 93.5" wb						
Jeep	450	1050	1650	3500	6400	9200
"S" Jeep	350	900	1550	3050	5900	8500
Comanche - 113" or 120" wb						
PU SBx	350	725	1200	2350	4800	6800
PU LBx	350	725	1200	2350	4850	6900
J-10, 131" wb						
PU	350	750	1300	2450	5250	7500
J-20, 131" wb						
PU	350	800	1450	2750	5600	8000

HUDSON

	6	5	4	3	2	1
1909						
Model 20, 4-cyl.						
Rds	550	1800	3000	6000	10,500	15,000
1910						
Model 20, 4-cyl.						
Rds	550	1700	2800	5600	9800	14,000
Tr	550	1700	2800	5600	9800	14,000
1911						
Model 33, 4-cyl.						
Rds	450	1500	2500	5000	8800	12,500
Tor Rds	500	1600	2700	5400	9500	13,500
Pony Ton	500	1600	2700	5400	9500	13,500
Tr	450	1450	2400	4800	8400	12,000
1912						
Model 33, 4-cyl.						
Rds	550	1800	3000	6000	10,500	15,000
Tor Rds	550	1700	2800	5600	9800	14,000
Tr	800	2500	4200	8400	14,700	21,000
Cpe	650	2000	3350	6700	11,700	16,700
Limo	700	2150	3600	7200	12,600	18,000
1913						
Model 37, 4-cyl.						
Rds	600	1850	3100	6200	10,900	15,500
Tor Rds	600	1900	3200	6400	11,200	16,000
Tr	650	2050	3400	6800	11,900	17,000
Cpe	550	1700	2800	5600	9800	14,000
Limo	650	2000	3350	6700	11,700	16,700
Model 54, 6-cyl.						
2P Rds	600	1850	3100	6200	10,900	15,500
5P Rds	600	1900	3200	6400	11,200	16,000
Tor Rds	600	2000	3300	6600	11,600	16,500
Tr	650	2050	3400	6800	11,900	17,000
7P Tr	650	2100	3500	7000	12,300	17,500
Cpe	600	1850	3100	6200	10,900	15,500
Limo	650	2050	3400	6800	11,900	17,000
1914						
Model 40, 6-cyl.						
Rbt	500	1550	2600	5200	9100	13,000
Tr	550	1700	2800	5600	9800	14,000
Cabr	500	1600	2700	5400	9500	13,500
Model 54, 6-cyl.						
7P Tr	600	1850	3100	6200	10,900	15,500
1915						
Model 40, 6-cyl.						
Rds	600	1900	3200	6400	11,200	16,000
Phae	600	2000	3300	6600	11,600	16,500
Tr	550	1800	3000	6000	10,500	15,000
Cabr	450	1500	2500	5000	8800	12,500
Cpe	400	1250	2100	4200	7400	10,500
Limo	450	1400	2300	4600	8100	11,500
Lan Limo	450	1450	2400	4800	8400	12,000
Model 54, 6-cyl.						
Phae	650	2100	3500	7000	12,300	17,500
7P Tr	650	2050	3400	6800	11,900	17,000
Sed	450	1450	2400	4800	8400	12,000
Limo	550	1700	2800	5600	9800	14,000
1916						
Super Six, 6-cyl.						
Rds	550	1800	3000	6000	10,500	15,000
Cabr	600	1850	3100	6200	10,900	15,500

	6	5	4	3	2	1
Phae	600	1900	3200	6400	11,200	16,000
Tr Sed	400	1300	2200	4400	7700	11,000
TwnC	400	1350	2250	4500	7900	11,300
Model 54, 6-cyl.						
7P Phae	550	1800	3000	6000	10,500	15,000
1917						
Super Six, 6-cyl.						
Rds	450	1500	2500	5000	8800	12,500
Cabr	500	1550	2600	5200	9100	13,000
7P Phae	450	1500	2500	5000	8800	12,500
Tr Sed	400	1200	2000	3950	7000	10,000
TwnC	450	1500	2500	5000	8800	12,500
Twn Lan	500	1550	2600	5200	9100	13,000
Limo Lan	450	1500	2500	5000	8800	12,500
1918						
Super Six, 6-cyl.						
Rds	500	1550	2600	5200	9100	13,000
Cabr	500	1600	2700	5400	9500	13,500
4P Phae	450	1450	2400	4800	8400	12,000
5P Phae	450	1500	2500	5000	8800	12,500
4P Cpe	450	1000	1650	3350	6300	9000
Tr Sed	450	1100	1700	3650	6650	9500
Sed	450	1100	1700	3650	6650	9500
Tr Limo	450	1450	2400	4800	8400	12,000
TwnC	450	1450	2400	4800	8400	12,000
Limo	450	1450	2400	4800	8400	12,000
Twn Limo	450	1500	2500	5000	8800	12,500
Limo Lan	450	1450	2400	4800	8400	12,000
F F Lan	450	1500	2500	5000	8800	12,500
1919						
Super Six Series O, 6-cyl.						
Cabr	450	1450	2400	4800	8400	12,000
4P Phae	450	1500	2500	5000	8800	12,500
7P Phae	500	1550	2600	5200	9100	13,000
4P Cpe	350	900	1550	3050	5900	8500
Sed	350	750	1300	2450	5250	7500
Tr Limo	450	1100	1700	3650	6650	9500
TwnC	400	1300	2200	4400	7700	11,000
Twn Lan	450	1400	2300	4600	8100	11,500
Limo Lan	450	1450	2400	4800	8400	12,000
1920						
Super Six Series 10-12, 6-cyl.						
4P Phae	450	1500	2500	5000	8800	12,500
7P Phae	450	1450	2400	4800	8400	12,000
Cabr	500	1550	2600	5200	9100	13,000
Cpe	350	900	1550	3050	5900	8500
Sed	350	750	1300	2450	5250	7500
Tr Limo	450	1000	1650	3350	6300	9000
Limo	450	1100	1700	3650	6650	9500
1921						
Super Six, 6-cyl.						
4P Phae	450	1500	2500	5000	8800	12,500
7P Phae	450	1450	2400	4800	8400	12,000
Cabr	500	1550	2600	5200	9100	13,000
4P Cpe	350	900	1550	3050	5900	8500
Sed	350	800	1450	2750	5600	8000
Tr Limo	450	1000	1650	3350	6300	9000
Limo	450	1100	1700	3650	6650	9500
1922						
Super Six, 6-cyl.						
Spds	500	1550	2600	5200	9100	13,000
Phae	500	1600	2700	5400	9500	13,500
Cabr	550	1700	2800	5600	9800	14,000
Cpe	350	750	1350	2600	5400	7700
2 dr Sed	350	750	1300	2450	5250	7500
Sed	350	750	1250	2350	5000	7100
Tr Limo	450	1000	1650	3350	6300	9000
Limo	450	1100	1700	3650	6650	9500
1923						
Super Six, 6-cyl.						
Spds	550	1700	2800	5600	9800	14,000
Phae	500	1550	2600	5200	9100	13,000
2 dr Sed	350	750	1300	2450	5250	7500
Sed	350	800	1450	2750	5600	8000
7P Sed	350	900	1550	3050	5900	8500
Cpe	450	1000	1650	3350	6300	9000
1924						
Super Six, 6-cyl.						
Spds	500	1600	2700	5400	9500	13,500
Phae	500	1550	2600	5200	9100	13,000
2 dr Sed	350	750	1300	2450	5250	7500
Sed	350	850	1500	2800	5650	8100
7P Sed	350	900	1550	3050	5900	8500
1925						
Super Six, 6-cyl.						
Spds	550	1700	2800	5600	9800	14,000
Phae	500	1600	2700	5400	9500	13,500
2 dr Sed	350	750	1300	2450	5250	7500
Brgm	450	1000	1650	3350	6300	9000
Sed	350	750	1300	2450	5250	7500
7P Sed	350	800	1450	2750	5600	8000
1926						
Super Six, 6-cyl.						
Phae	700	2150	3600	7200	12,600	18,000
2 dr Sed	350	900	1550	3050	5900	8500
Brgm	400	1250	2100	4200	7400	10,500
7P Sed	450	1000	1650	3350	6300	9000
1927						
Standard Six, 6-cyl.						
Phae	700	2200	3700	7400	13,000	18,500
2 dr Sed	350	850	1500	2900	5700	8200
Spec 2 dr Sed	350	900	1550	3050	5900	8500
Brgm	450	1100	1700	3650	6650	9500
7P Sed	450	1100	1700	3650	6650	9500
Super Six						
Cus Rds	1250	3950	6600	13,200	23,100	33,000
Cus Phae	1300	4100	6800	13,600	23,800	34,000
2 dr Sed	450	1100	1700	3650	6650	9500
Sed	400	1300	2200	4400	7700	11,000
Cus Brgm	600	1900	3200	6400	11,200	16,000
Cus Sed	650	2050	3400	6800	11,900	17,000
1928						
First Series, 6-cyl., (Start June, 1927)						
Std 2 dr Sed	350	850	1500	2900	5700	8200
Std Sed	350	900	1550	3000	5850	8400

	6	5	4	3	2	1
2 dr Sed	350	950	1600	3200	6050	8700
Sed	450	1100	1700	3650	6650	9500
Rds	750	2400	4000	8000	14,000	20,000
Cus Phae	850	2750	4600	9200	16,100	23,000
Cus Brgm	550	1800	3000	6000	10,500	15,000
Cus Sed	600	1900	3200	6400	11,200	16,000
Second Series, 6-cyl., (Start Jan. 1928)						
2 dr Sed	450	1000	1650	3350	6300	9000
Sed	450	1100	1700	3650	6650	9500
RS Cpe	450	1450	2400	4800	8400	12,000
Rds	750	2400	4000	8000	14,000	20,000
Ewb Sed	350	950	1600	3200	6050	8700
Lan Sed	450	1000	1650	3350	6300	9000
Vic	450	1050	1650	3500	6400	9200
7P Sed	450	1100	1700	3650	6650	9500
1929						
Series Greater Hudson, 6-cyl., 122" wb						
RS Rds	1200	3850	6400	12,800	22,400	32,000
Phae	1300	4100	6800	13,600	23,800	34,000
Cpe	500	1550	2600	5200	9100	13,000
2 dr Sed	400	1200	2000	3950	7000	10,000
Conv	1100	3500	5800	11,600	20,300	29,000
Vic	400	1250	2100	4200	7400	10,500
Sed	400	1200	2000	3950	7000	10,000
Twn Sed	400	1250	2100	4200	7400	10,500
Lan Sed	400	1250	2100	4200	7400	10,500
Series Greater Hudson, 6-cyl., 139" wb						
Spt Sed	700	2150	3600	7200	12,600	18,000
7P Sed	750	2400	4000	8000	14,000	20,000
Limo	850	2650	4400	8800	15,400	22,000
DC Phae	1600	5150	8600	17,200	30,100	43,000
1930						
Great Eight, 8-cyl., 119" wb						
Rds	1400	4450	7400	14,800	25,900	37,000
Phae	1450	4700	7800	15,600	27,300	39,000
RS Cpe	700	2150	3600	7200	12,600	18,000
2 dr Sed	400	1200	2000	3950	7000	10,000
Sed	400	1250	2100	4200	7400	10,500
Conv Sed	1500	4800	8000	16,000	28,000	40,000
Great Eight, 8-cyl., 126" wb						
Phae	1600	5050	8400	16,800	29,400	42,000
Tr Sed	400	1300	2200	4400	7700	11,000
7P Sed	500	1550	2600	5200	9100	13,000
Brgm	400	1250	2100	4200	7400	10,500
1931						
Greater Eight, 8-cyl., 119" wb						
Rds	1600	5050	8400	16,800	29,400	42,000
Phae	1650	5300	8800	17,600	30,800	44,000
Cpe	400	1300	2200	4400	7700	11,000
Special Cpe	450	1450	2400	4800	8400	12,000
RS Cpe	450	1400	2300	4600	8100	11,500
2 dr Sed	450	1000	1650	3350	6300	9000
Sed	450	1100	1700	3650	6650	9500
Twn Sed	450	950	1600	3250	6150	8800
Great Eight, l.w.b., 8-cyl., 126" wb						
Spt Phae	1750	5500	9200	18,400	32,200	46,000
Brgm	450	1450	2400	4800	8400	12,000
Fam Sed	450	1400	2300	4600	8100	11,500
7P Sed	450	1450	2400	4800	8400	12,000
Clb Sed	450	1400	2300	4600	8100	11,500
Tr Sed	400	1200	2000	3950	7000	10,000
Special Sed	400	1300	2200	4400	7700	11,000
1932						
(Standard) Greater, 8-cyl., 119" wb						
2P Cpe	450	1100	1700	3650	6650	9500
4P Cpe	400	1200	2000	3950	7000	10,000
Spec Cpe	450	1450	2400	4800	8400	12,000
Conv	1100	3500	5800	11,600	20,300	29,000
2 dr Sed	400	1200	2000	3950	7000	10,000
5P Sed	400	1250	2100	4200	7400	10,500
Twn Sed	400	1200	2000	3950	7000	10,000
(Sterling) Series, 8-cyl., 132" wb						
Spec Sed	400	1300	2200	4400	7700	11,000
Sub	400	1250	2100	4200	7400	10,500
Major Series, 8-cyl., 132" wb						
Phae	1050	3350	5600	11,200	19,600	28,000
Tr Sed	450	1400	2350	4700	8300	11,800
Clb Sed	450	1500	2500	5000	8800	12,500
Brgm	500	1600	2700	5400	9500	13,500
7P Sed	450	1400	2300	4600	8100	11,500
1933						
Pacemaker Super Six, 6-cyl., 113" wb						
Conv	850	2650	4400	8800	15,400	22,000
Phae	850	2750	4600	9200	16,100	23,000
Bus Cpe	350	750	1300	2450	5250	7500
RS Cpe	350	800	1450	2750	5600	8000
2 dr Sed	350	750	1200	2350	4900	7000
Sed	350	750	1300	2450	5250	7500
Pacemaker Standard, 8-cyl., 119" wb						
Conv	900	2900	4800	9600	16,800	24,000
RS Cpe	450	1000	1650	3350	6300	9000
2 dr Sed	350	750	1300	2450	5250	7500
Sed	450	1100	1700	3650	6650	9500
Pacemaker Major, 8-cyl., 132" wb						
Phae	1000	3100	5200	10,400	18,200	26,000
Tr Sed	350	800	1450	2750	5600	8000
Brgm	350	850	1500	2900	5700	8200
Clb Sed	350	900	1550	3050	5900	8500
7P Sed	350	850	1500	2950	5800	8300
1934						
Special, 8-cyl., 116" wb						
Conv	1000	3250	5400	10,800	18,900	27,000
Bus Cpe	350	750	1250	2400	5050	7200
Cpe	350	750	1300	2450	5250	7500
RS Cpe	350	800	1450	2750	5600	8000
Comp Vic	350	750	1350	2600	5400	7700
2 dr Sed	350	750	1300	2450	5250	7500
Sed	350	750	1200	2350	4900	7000
Comp Sed	350	800	1450	2750	5600	8000
DeLuxe Series, 8-cyl., 116" wb						
2P Cpe	350	750	1300	2400	5200	7400
RS Cpe	350	800	1350	2700	5500	7900
Comp Vic	350	900	1550	3050	5900	8500
2 dr Sed	350	750	1350	2650	5450	7800
Sed	350	750	1250	2400	5050	7200
Comp Sed	350	750	1300	2450	5250	7500

	6	5	4	3	2	1
Challenger Series, 8-cyl., 116" wb						
2P Cpe	350	750	1300	2450	5250	7500
RS Cpe	350	800	1450	2750	5600	8000
Conv	1100	3500	5800	11,600	20,300	29,000
2 dr Sed	350	750	1300	2450	5250	7500
Sed	350	750	1350	2600	5400	7700
Major Series, 8-cyl., 123" wb						
(Special)						
Tr Sed	350	900	1550	3050	5900	8500
Comp Trs	350	950	1600	3200	6050	8700
(DeLuxe)						
Clb Sed	450	950	1600	3250	6150	8800
Brgm	350	900	1550	3100	6000	8600
Comp Clb Sed	350	900	1550	3050	5900	8500
1935						
Big Six, 6-cyl., 116" wb						
Conv	1150	3600	6000	12,000	21,000	30,000
Cpe	350	900	1550	3050	5900	8500
RS Cpe	450	1000	1650	3350	6300	9000
Tr Brgm	350	750	1200	2350	4900	7000
2 dr Sed	350	700	1150	2300	4550	6500
Sed	350	750	1200	2350	4900	7000
Sub Sed	350	750	1300	2400	5200	7400
Eight Special, 8-cyl., 117" wb						
Conv	1150	3700	6200	12,400	21,700	31,000
Cpe	350	750	1350	2600	5400	7700
RS Cpe	350	900	1550	3050	5900	8500
Tr Brgm	350	750	1250	2400	5050	7200
2 dr Sed	350	750	1250	2350	5000	7100
Sed	350	750	1300	2500	5300	7600
Sub Sed	350	750	1350	2600	5400	7700
Eight DeLuxe						
Eight Special, 8-cyl., 124" wb						
Brgm	350	750	1300	2500	5300	7600
Tr Brgm	350	750	1350	2600	5400	7700
Clb Sed	350	750	1200	2350	4900	7000
Sub Sed	350	750	1300	2450	5250	7500
Eight DeLuxe, 8-cyl., 117" wb						
2P Cpe	350	750	1350	2650	5450	7800
RS Cpe	350	900	1550	3100	6000	8600
Conv	1200	3850	6400	12,800	22,400	32,000
Tr Brgm	350	750	1250	2400	5100	7300
2 dr Sed	350	750	1250	2400	5050	7200
4 dr Sed	350	750	1350	2600	5400	7700
Sub Sed	350	750	1350	2650	5450	7800
Eight Custom, 8-cyl., 124" wb						
Brgm	350	750	1350	2600	5400	7700
Tr Brgm	350	750	1350	2650	5450	7800
Sed	350	750	1300	2450	5250	7500
Sub Sed	350	750	1350	2650	5450	7800
Late Special, 8-cyl., 124" wb						
Brgm	350	750	1250	2400	5050	7200
Tr Brgm	350	750	1250	2400	5100	7300
Club Sed	350	750	1250	2350	5000	7100
Sub Sed	350	750	1350	2650	5450	7800
Late DeLuxe, 8-cyl., 124" wb						
Brgm	350	750	1250	2400	5100	7300
Tr Brgm	350	750	1300	2400	5200	7400
Club Sed	350	750	1250	2400	5050	7200
Sub Sed	350	800	1350	2700	5500	7900
1936						
Custom Six, 6-cyl., 120" wb						
Conv	1100	3500	5800	11,600	20,300	29,000
Cpe	350	750	1300	2450	5250	7500
RS Cpe	350	800	1450	2750	5600	8000
Brgm	350	750	1200	2350	4900	7000
Tr Brgm	350	750	1250	2350	5000	7100
Sed	350	750	1200	2350	4900	7000
Tr Sed	350	750	1300	2450	5250	7500
DeLuxe Eight, Series 64, 8-cyl., 120" wb						
Conv	1200	3850	6400	12,800	22,400	32,000
Cpe	350	750	1350	2600	5400	7700
RS Cpe	350	900	1550	3000	5850	8400
Brgm	350	750	1250	2400	5050	7200
Tr Brgm	350	750	1250	2400	5050	7200
DeLuxe Eight, Series 66, 8-cyl., 127" wb						
Sed	350	750	1350	2600	5400	7700
Tr Sed	350	800	1450	2750	5600	8000
Custom Eight, Series 65, 120" wb						
2P Cpe	350	750	1350	2650	5450	7800
RS Cpe	350	900	1550	3050	5900	8500
Conv	1200	3850	6400	12,800	22,400	32,000
Brgm	350	750	1250	2400	5100	7300
Tr Brgm	350	750	1300	2400	5200	7400
Custom Eight, Series 67, 127" wb						
Sed	350	750	1300	2500	5300	7600
Tr Sed	350	750	1350	2600	5400	7700
1937						
Custom Six, Series 73, 6-cyl., 122" wb						
Conv	1150	3700	6200	12,400	21,700	31,000
Conv Brgm	1200	3850	6400	12,800	22,400	32,000
Bus Cpe	350	700	1150	2300	4550	6500
3P Cpe	350	750	1200	2350	4900	7000
Vic Cpe	350	750	1300	2450	5250	7500
2 dr Brgm	350	700	1150	2300	4550	6500
2 dr Tr Brgm	350	725	1150	2300	4700	6700
Sed	350	750	1200	2350	4900	7000
Tr Sed	350	750	1250	2350	5000	7100
DeLuxe Eight, Series 74, 8-cyl., 122" wb						
Cpe	350	800	1450	2750	5600	8000
Vic Cpe	350	900	1550	3050	5900	8500
Conv	1150	3700	6200	12,400	21,700	31,000
2 dr Brgm	350	750	1300	2500	5300	7600
2 dr Tr Brgm	350	750	1350	2600	5400	7700
Sed	350	750	1350	2600	5400	7700
Tr Sed	350	750	1350	2650	5450	7800
Conv Brgm	1050	3350	5600	11,200	19,600	28,000
DeLuxe Eight, Series 76, 8-cyl., 129" wb						
Sed	450	1000	1650	3350	6300	9000
Tr Sed	450	1100	1700	3650	6650	9500
Custom Eight, Series 75, 8-cyl., 122" wb						
Cpe	350	800	1450	2750	5600	8000
Vic Cpe	350	850	1500	2900	5700	8200
Conv Cpe	1200	3850	6400	12,800	22,400	32,000
2 dr Brgm	350	750	1350	2650	5450	7800
2 dr Tr Brgm	350	800	1450	2750	5600	8000
Sed	350	750	1350	2650	5450	7800
Tr Sed	350	800	1350	2700	5500	7900
Conv Brgm	1250	3950	6600	13,200	23,100	33,000
Custom Eight, Series 77, 8-cyl., 129" wb						
Sed	350	800	1450	2750	5600	8000
Tr Sed	350	850	1500	2900	5700	8200
1938						
Standard Series 89, 6-cyl., 112" wb						
Conv	1150	3700	6200	12,400	21,700	31,000
Conv Brgm	1200	3850	6400	12,800	22,400	32,000
3P Cpe	200	550	900	2150	3800	5400
Vic Cpe	350	700	1100	2300	4500	6400
Brgm	350	700	1100	2300	4500	6400
Tr Brgm	350	700	1150	2300	4550	6500
Sed	350	700	1150	2300	4600	6600
Tr Sed	350	725	1150	2300	4700	6700
Utility Series 89, 6-cyl., 112" wb						
Cpe	200	500	850	1850	3350	4900
2 dr Sed	150	450	800	1800	3300	4800
2 dr Tr Sed	200	500	850	1850	3350	4900
DeLuxe Series 89, 6-cyl., 112" wb						
Conv	1150	3600	6000	12,000	21,000	30,000
Conv Brgm	1150	3700	6200	12,400	21,700	31,000
3P Cpe	350	750	1250	2400	5050	7200
Vic Cpe	350	750	1300	2450	5250	7500
Brgm	350	725	1200	2350	4850	6900
Tr Brgm	350	700	1150	2300	4600	6600
Sed	350	725	1150	2300	4700	6700
Tr Sed	350	725	1200	2350	4800	6800
Custom Series 83, 6-cyl., 122" wb						
Conv	1150	3700	6200	12,400	21,700	31,000
Conv Brgm	1200	3850	6400	12,800	22,400	32,000
3P Cpe	350	800	1450	2750	5600	8000
Vic Cpe	350	900	1550	3050	5900	8500
Brgm	350	750	1300	2400	5200	7400
Tr Brgm	350	750	1250	2400	5050	7200
Sed	350	725	1200	2350	4800	6800
Tr Sed	350	725	1200	2350	4850	6900
DeLuxe Series 84, 8-cyl., 122" wb						
Conv	1150	3700	6200	12,400	21,700	31,000
Conv Brgm	1200	3850	6400	12,800	22,400	32,000
3P Cpe	350	800	1450	2750	5600	8000
Vic Cpe	350	900	1550	3050	5900	8500
Brgm	350	800	1350	2700	5500	7900
Tr Brgm	350	725	1200	2350	4850	6900
Tr Sed	350	750	1200	2350	4900	7000
Custom Series 85, 8-cyl., 122" wb						
3P Cpe	350	900	1550	3050	5900	8500
Vic Cpe	450	1000	1650	3350	6300	9000
Brgm	350	900	1550	3000	5850	8400
Tr Brgm	450	1000	1600	3300	6250	8900
Sed	350	750	1250	2350	5000	7100
Tr Sed	350	750	1250	2400	5100	7300
Country Club Series 87, 8-cyl., 129" wb						
Sed	350	750	1250	2350	5000	7100
Tr Sed	350	750	1250	2400	5100	7300
1939						
DeLuxe Series 112, 6-cyl., 112" wb						
Conv	1150	3600	6000	12,000	21,000	30,000
Conv Brgm	1150	3700	6200	12,400	21,700	31,000
Trav Cpe	350	750	1300	2400	5200	7400
Utl Cpe	350	750	1250	2400	5100	7300
3P Cpe	350	750	1300	2400	5200	7400
Vic Cpe	350	750	1300	2500	5300	7600
2 dr Utl Sed	350	700	1150	2300	4550	6500
Tr Brgm	350	725	1200	2350	4850	6900
Tr Sed	350	750	1200	2350	4900	7000
Sta Wag	450	1000	1600	3300	6250	8900
Pacemaker Series 91, 6-cyl., 118" wb						
3P Cpe	350	750	1300	2450	5250	7500
Vic Cpe	350	750	1350	2600	5400	7700
Tr Brgm	350	750	1250	2350	5000	7100
Tr Sed	350	750	1250	2350	5000	7100
Series 92, 6-cyl., 118" wb						
Conv	1200	3850	6400	12,800	22,400	32,000
Conv Brgm	1250	3950	6600	13,200	23,100	33,000
3P Cpe	350	750	1300	2500	5300	7600
Vic Cpe	350	750	1350	2650	5450	7800
Tr Brgm	350	750	1250	2400	5100	7300
Tr Sed	350	750	1250	2400	5050	7200
Country Club Series 93, 6-cyl., 122" wb						
Conv	1250	3950	6600	13,200	23,100	33,000
Conv Brgm	1300	4100	6800	13,600	23,800	34,000
3P Cpe	350	750	1350	2650	5450	7800
Vic Cpe	350	800	1450	2750	5600	8000
Tr Brgm	350	750	1300	2450	5250	7500
Tr Sed	350	750	1250	2400	5100	7300
Big Boy Series 96, 6-cyl., 129" wb						
6P Sed	350	750	1300	2500	5300	7600
7P Sed	350	800	1350	2700	5500	7900
Country Club Series 95, 8-cyl., 122" wb						
Conv	1300	4100	6800	13,600	23,800	34,000
Conv Brgm	1300	4200	7000	14,000	24,500	35,000
3P Cpe	350	800	1450	2750	5600	8000
Vic Cpe	350	850	1500	2900	5700	8200
Tr Brgm	350	800	1350	2700	5500	7900
Tr Sed	350	750	1300	2400	5200	7400
Custom Series 97, 8-cyl., 129" wb						
5P Tr Sed	350	800	1350	2700	5500	7900
7P Sed	350	900	1550	3000	5850	8400
1940						
Traveler Series 40-T, 6-cyl., 113" wb						
Cpe	350	750	1300	2400	5200	7400
Vic Cpe	350	800	1350	2700	5500	7900
2 dr Tr Sed	350	725	1200	2350	4850	6900
4 dr Tr Sed	350	725	1200	2350	4850	6900
DeLuxe Series, 40-P, 6-cyl., 113" wb						
Conv 6 Pass	1000	3100	5200	10,400	18,200	26,000
Cpe	350	750	1300	2500	5300	7600
Vic Cpe	350	800	1450	2750	5600	8000
2 dr Tr Sed	350	750	1200	2350	4900	7000
4 dr Sed	350	750	1250	2350	5000	7100
Super Series 41, 6-cyl., 118" wb						
Conv 5 Pass	1000	3250	5400	10,800	18,900	27,000
Conv 6 Pass	1050	3350	5600	11,200	19,600	28,000
Cpe	350	750	1350	2650	5450	7800
Vic Cpe	350	850	1500	2800	5650	8100
2 dr Tr Sed	350	750	1250	2350	5000	7100

	6	5	4	3	2	1
4 dr Tr Sed	350	750	1250	2400	5050	7200
Country Club Series 43, 6-cyl., 125" wb						
6P Sed	350	750	1250	2400	5050	7200
7P Sed	350	750	1300	2400	5200	7400
Series 44, 8-cyl., 118" wb						
Conv 5 Pass	1050	3350	5600	11,200	19,600	28,000
Conv 6 Pass	1100	3500	5800	11,600	20,300	29,000
Cpe	350	800	1450	2750	5600	8000
Vic Cpe	350	900	1550	3000	5850	8400
2 dr Tr Sed	350	750	1250	2400	5050	7200
4 dr Tr Sed	350	750	1250	2400	5100	7300
DeLuxe Series 45, 8-cyl., 118" wb						
2 dr Tr Sed	350	750	1250	2400	5100	7300
4 dr Tr Sed	350	750	1300	2400	5200	7400
Country Club Eight Series 47, 8-cyl., 125" wb						
Tr Sed	350	750	1300	2500	5300	7600
7P Sed	350	800	1350	2700	5500	7900
Big Boy Series 48, 6-cyl., 125" wb						
C-A Sed	350	750	1300	2450	5250	7500
7P Sed	350	750	1300	2500	5300	7600

1941

	6	5	4	3	2	1
Utility Series 10-C, 6-cyl., 116" wb						
Cpe	350	750	1200	2350	4900	7000
2 dr Sed	350	725	1150	2300	4700	6700
Traveler Series 10-T, 6-cyl., 116" wb						
Cpe	350	750	1250	2350	5000	7100
Clb Cpe	350	750	1250	2400	5050	7200
2 dr Sed	350	750	1200	2350	4900	7000
4 dr Sed	350	750	1200	2350	4900	7000
DeLuxe Series 10-P, 6-cyl., 116" wb						
Conv	1000	3250	5400	10,800	18,900	27,000
Cpe	350	725	1200	2350	4800	6800
Clb Cpe	350	725	1200	2350	4850	6900
2 dr Sed	350	725	1150	2300	4700	6700
4 dr Sed	350	725	1200	2350	4800	6800
Super Series 11, 6-cyl., 121" wb						
Conv	1100	3500	5800	11,600	20,300	29,000
Cpe	350	725	1200	2350	4850	6900
Clb Cpe	350	750	1250	2350	5000	7100
2 dr Sed	350	700	1150	2300	4600	6600
4 dr Sed	350	725	1150	2300	4700	6700
Sta Wag	450	1500	2500	5000	8800	12,500
Commodore Series 12, 6-cyl., 121" wb						
Conv	1150	3600	6000	12,000	21,000	30,000
Cpe	350	800	1450	2750	5600	8000
Clb Cpe	350	850	1500	2900	5700	8200
2 dr Sed	350	750	1350	2600	5400	7700
Sed	350	750	1350	2650	5450	7800
Commodore Series 14, 8-cyl., 121" wb						
Conv	1150	3700	6200	12,400	21,700	31,000
Cpe	350	850	1500	2900	5700	8200
Clb Cpe	350	900	1550	3000	5850	8400
2 dr Sed	350	800	1450	2750	5600	8000
Sed	350	850	1500	2800	5650	8100
Sta Wag	550	1700	2800	5600	9800	14,000
Commodore Custom Series 15, 8-cyl., 121" wb						
Cpe	450	1000	1650	3350	6300	9000
Clb Cpe	450	1050	1700	3600	6600	9400
Commodore Custom Series 17, 8-cyl., 128" wb						
Sed	350	900	1550	3000	5850	8400
7P Sed	450	1000	1650	3350	6300	9000
Big Boy Series 18, 6-cyl., 128" wb						
C-A Sed	350	850	1500	2900	5700	8200
7P Sed	350	900	1550	3000	5850	8400

1942

	6	5	4	3	2	1
Traveler Series 20-T, 6-cyl., 116" wb						
Cpe	350	750	1250	2400	5050	7200
Clb Cpe	350	750	1300	2400	5200	7400
2 dr Sed	350	750	1250	2350	5000	7100
4 dr Sed	350	750	1250	2400	5050	7200
DeLuxe Series 20-P, 6-cyl., 116" wb						
Conv	950	3000	5000	10,000	17,500	25,000
Cpe	350	750	1300	2400	5200	7400
Clb Cpe	350	800	1350	2700	5500	7900
2 dr Sed	350	750	1250	2400	5050	7200
4 dr Sed	350	750	1300	2400	5200	7400
Super Series 21, 6-cyl., 121" wb						
Conv	1000	3100	5200	10,400	18,200	26,000
Cpe	350	750	1350	2600	5400	7700
Clb Cpe	350	850	1500	2800	5650	8100
2 dr Sed	350	750	1250	2400	5100	7300
Sed	350	750	1300	2500	5300	7600
Sta Wag	450	1450	2400	4800	8400	12,000
Commodore Series 22, 6-cyl., 121" wb						
Conv	1050	3350	5600	11,200	19,600	28,000
Cpe	350	900	1550	3050	5900	8500
Clb Cpe	450	950	1600	3250	6150	8800
2 dr Sed	350	900	1550	3000	5850	8400
Sed	350	900	1550	3050	5900	8500
Commodore Series 24, 8-cyl., 121" wb						
Conv	1150	3600	6000	12,000	21,000	30,000
Cpe	450	950	1600	3250	6150	8800
Clb Cpe	450	1000	1650	3350	6300	9000
2 dr Sed	350	900	1550	3100	6000	8600
Sed	350	950	1600	3200	6050	8700
Commodore Custom Series 25, 8-cyl., 121" wb						
Clb Cpe	450	1000	1650	3350	6300	9000
Commodore Series 27, 8-cyl., 128" wb						
Sed	350	950	1600	3200	6050	8700

1946-1947

	6	5	4	3	2	1
Super Series, 6-cyl., 121" wb						
Conv	850	2750	4600	9200	16,100	23,000
Cpe	350	800	1350	2700	5500	7900
Clb Cpe	350	800	1450	2750	5600	8000
2 dr Sed	350	700	1150	2300	4600	6600
Sed	350	725	1200	2350	4850	6900
Commodore Series, 6-cyl., 121" wb						
Clb Cpe	350	900	1550	3100	6000	8600
Sed	350	800	1450	2750	5600	8000
Super Series, 8-cyl., 121" wb						
Clb Cpe	350	950	1600	3200	6050	8700
Sed	350	850	1500	2900	5700	8200
Commodore Series, 8-cyl., 121" wb						
Conv	1050	3350	5600	11,200	19,600	28,000
Clb Cpe	450	1050	1700	3550	6500	9300
Sed	450	950	1600	3250	6150	8800

1948-1949

	6	5	4	3	2	1
Super Series, 6-cyl., 124" wb						
Sed	450	1050	1700	3550	6500	9300
Conv	950	3000	5000	10,000	17,500	25,000
Cpe	450	1100	1700	3650	6650	9500
Clb Cpe	450	1150	1800	3800	6800	9700
2 dr Sed	450	1050	1650	3500	6400	9200
Commodore Series, 6-cyl., 124" wb						
Conv	1050	3350	5600	11,200	19,600	28,000
Clb Cpe	450	1150	1900	3850	6850	9800
Sed	450	1100	1700	3650	6650	9500
Super Series, 8-cyl., 124" wb						
Clb Cpe	450	1150	1900	3900	6900	9900
2 dr Sed (1949 only)	450	1050	1700	3550	6500	9300
Sed	350	750	1300	2500	5300	7600
Commodore Series, 8-cyl., 124" wb						
Conv	1150	3600	6000	12,000	21,000	30,000
Clb Cpe	400	1300	2200	4400	7700	11,000
Sed	400	1200	2000	3950	7000	10,000

1950

	6	5	4	3	2	1
Pacemaker Series 500, 6-cyl., 119" wb						
Conv	950	3000	5000	10,000	17,500	25,000
Bus Cpe	450	1050	1700	3600	6600	9400
Clb Cpe	450	1150	1800	3800	6800	9700
2 dr Sed	450	1050	1700	3550	6500	9300
Sed	450	1000	1650	3400	6350	9100
DeLuxe Series 50A, 6-cyl., 119" wb						
Conv	1000	3100	5200	10,400	18,200	26,000
Clb Cpe	450	1150	1900	3850	6850	9800
2 dr Sed	450	1150	1800	3800	6800	9700
Sed	450	1100	1700	3650	6650	9500
Super Six Series 501, 6-cyl., 124" wb						
Conv	1000	3250	5400	10,800	18,900	27,000
Clb Cpe	450	1150	1900	3900	6900	9900
2 dr Sed	450	1150	1900	3850	6850	9800
Sed	450	1100	1800	3700	6700	9600
Commodore Series 502, 6-cyl., 124" wb						
Conv	1050	3350	5600	11,200	19,600	28,000
Clb Cpe	400	1300	2200	4400	7700	11,000
Sed	400	1200	2000	3950	7000	10,000
Super Series 503, 8-cyl., 124" wb						
Clb Cpe	450	1400	2300	4600	8100	11,500
2 dr Sed	400	1300	2200	4400	7700	11,000
Sed	400	1250	2100	4200	7400	10,500
Commodore Series 504, 8-cyl., 124" wb						
Conv	1150	3600	6000	12,000	21,000	30,000
Clb Cpe	450	1450	2400	4800	8400	12,000
Sed	400	1300	2200	4400	7700	11,000

1951

	6	5	4	3	2	1
Pacemaker Custom Series 4A, 6-cyl., 119" wb						
Conv	900	2900	4800	9600	16,800	24,000
Cpe	450	1150	1900	3900	6900	9900
Clb Cpe	400	1200	2000	4000	7100	10,100
2 dr Sed	450	1150	1800	3800	6800	9700
Sed	450	1150	1900	3850	6850	9800
Super Custom Series 5A, 6-cyl., 124" wb						
Conv	950	3000	5000	10,000	17,500	25,000
Clb Cpe	400	1250	2100	4200	7400	10,500
2 dr Sed	450	1150	1900	3850	6850	9800
Sed	400	1200	2000	3950	7000	10,000
Hlywd	450	1450	2400	4800	8400	12,000
Commodore Custom Series 6A, 6-cyl., 124" wb						
Conv	1000	3100	5200	10,400	18,200	26,000
Clb Cpe	400	1250	2100	4200	7300	10,400
Sed	400	1200	2000	4000	7100	10,100
Hlywd	500	1550	2600	5200	9100	13,000
Hornet Series 7A, 6-cyl., 124" wb						
Conv	1100	3500	5800	11,600	20,300	29,000
Clb Cpe	400	1250	2100	4200	7400	10,600
Sed	400	1250	2100	4200	7300	10,400
Hlywd	550	1700	2800	5600	9800	14,000
Commodore Custom Series 8A, 8-cyl., 124" wb						
Conv	1150	3600	6000	12,000	21,000	30,000
Clb Cpe	400	1300	2200	4400	7700	11,000
Sed	400	1250	2100	4200	7400	10,600
Hlywd	550	1800	3000	6000	10,500	15,000

1952

	6	5	4	3	2	1
Pacemaker Series 4B, 6-cyl., 119" wb						
Cpe	450	1150	1800	3800	6800	9700
Clb Cpe	450	1150	1900	3850	6850	9800
2 dr Sed	450	1100	1700	3650	6650	9500
Sed	450	1100	1800	3700	6700	9600
Wasp Series 5B, 6-cyl., 119" wb						
Conv	950	3000	5000	10,000	17,500	25,000
Hlywd	400	1300	2200	4400	7700	11,000
Clb Cpe	400	1200	2000	3950	7000	10,000
2 dr Sed	450	1100	1800	3700	6700	9600
Sed	450	1150	1800	3800	6800	9700
Commodore Series 6B, 6-cyl., 124" wb						
Conv	1000	3100	5200	10,400	18,200	26,000
Hlywd	500	1550	2600	5200	9100	13,000
Clb Cpe	400	1200	2050	4100	7100	10,200
Sed	400	1200	2000	3950	7000	10,000
Hornet Series 7B, 6-cyl., 124" wb						
Conv	1000	3250	5400	10,800	18,900	27,000
Hlywd	550	1700	2800	5600	9800	14,000
Clb Cpe	400	1250	2100	4200	7300	10,400
Sed	400	1200	2000	4000	7100	10,100
Commodore Series 8B, 8-cyl., 124" wb						
Conv	1050	3350	5600	11,200	19,600	28,000
Hlywd	550	1800	3000	6000	10,500	15,000
Clb Cpe	400	1250	2100	4200	7300	10,400
Sed	400	1200	2000	4000	7100	10,100

1953

	6	5	4	3	2	1
Jet Series 1C, 6-cyl., 105" wb						
4 dr Sed	350	750	1200	2350	4900	7000
Super Jet Series 2C, 6-cyl., 105" wb						
2 dr Clb Sed	350	750	1300	2500	5300	7600
4 dr Sed	350	750	1300	2450	5250	7500
Wasp Series 4C, 6-cyl., 119" wb						
Clb Cpe	450	950	1600	3250	6150	8800
2 dr Sed	350	900	1550	3050	5900	8500
Sed	350	900	1550	3100	6000	8600
Super Wasp Series 5C, 6-cyl., 119" wb						
Conv	900	2900	4800	9600	16,800	24,000
Hlywd	400	1300	2200	4400	7700	11,000
Clb Cpe	450	1000	1650	3350	6300	9000

	6	5	4	3	2	1
2 dr Sed	350	900	1550	3100	6000	8600
4 dr Sed	350	950	1600	3200	6050	8700
Hornet Series 7C, 6-cyl., 124" wb						
Conv	1000	3100	5200	10,400	18,200	26,000
Clb Cpe	400	1250	2100	4200	7400	10,500
Sed	400	1200	2000	3950	7000	10,000
Hlywd	500	1550	2600	5200	9100	13,000

1954

	6	5	4	3	2	1
Jet Series 1D, 6-cyl., 105" wb						
2 dr Utl Sed	350	750	1200	2350	4900	7000
2 dr Clb Sed	350	750	1250	2400	5050	7200
4 dr Sed	350	750	1250	2350	5000	7100
Super Jet Series 2D, 6-cyl., 105" wb						
2 dr Clb Sed	350	750	1300	2450	5250	7500
4 dr Sed	350	750	1300	2500	5300	7600
Jet Liner Series 3D, 6-cyl., 105" wb						
2 dr Clb Sed	350	750	1350	2600	5400	7700
4 dr Sed	350	800	1450	2750	5600	8000
Wasp Series 4D, 6-cyl., 119" wb						
Clb Cpe	350	750	1300	2450	5250	7500
Clb Sed	350	750	1250	2400	5100	7300
Sed	350	750	1300	2400	5200	7400
Super Wasp Series 5D, 6-cyl., 119" wb						
Conv	850	2750	4600	9200	16,100	23,000
Hlywd	450	1450	2400	4800	8400	12,000
Clb Cpe	350	900	1550	3050	5900	8500
Clb Sed	350	750	1350	2650	5450	7800
Sed	350	800	1450	2750	5600	8000
Hornet Special Series 6D, 6-cyl., 124" wb						
Clb Cpe	400	1200	2000	3950	7000	10,000
Clb Sed	450	1000	1650	3400	6350	9100
Sed	450	1050	1700	3600	6600	9400
Hornet Series 7D, 6-cyl., 124" wb						
Brgm Conv	1000	3100	5200	10,400	18,200	26,000
Hlywd	550	1700	2800	5600	9800	14,000
Clb Cpe	400	1300	2200	4400	7700	11,000
Sed	450	1100	1800	3700	6700	9600
Italia, 6-cyl.						
2 dr	1050	3350	5600	11,200	19,600	28,000

1955

	6	5	4	3	2	1
Super Wasp, 6-cyl., 114" wb						
Sed	350	700	1150	2300	4550	6500
Custom Wasp, 6-cyl., 114" wb						
Hlywd	400	1300	2200	4400	7700	11,000
Sed	350	700	1150	2300	4600	6600
Hornet Super, 6-cyl., 121" wb						
Sed	350	725	1200	2350	4850	6900
Hornet Custom, 6-cyl., 121" wb						
Hlywd	450	1450	2400	4800	8400	12,000
Sed	350	750	1300	2450	5250	7500
Italia, 6-cyl.						
2 dr	1050	3350	5600	11,200	19,600	28,000

NOTE: Add 5 percent for V-8.
For Hudson Rambler prices see AMC.

1956

	6	5	4	3	2	1
Super Wasp, 6-cyl., 114" wb						
Sed	350	700	1150	2300	4550	6500
Super Hornet, 6-cyl., 121" wb						
Sed	350	750	1200	2350	4900	7000
Custom Hornet, 6-cyl., 121" wb						
Hlywd	450	1400	2300	4600	8100	11,500
Sed	350	800	1450	2750	5600	8000
Hornet Super Special, 8-cyl., 114" wb						
Hlywd	450	1500	2500	5000	8800	12,500
Sed	350	850	1500	2900	5700	8200
Hornet Custom, 8-cyl., 121" wb						
Hlywd	500	1600	2700	5400	9500	13,500
Sed	350	900	1550	3050	5900	8500

NOTE: For Hudson Rambler prices see AMC.

1957

	6	5	4	3	2	1
Hornet Super, 8-cyl., 121" wb						
Hlywd	450	1450	2400	4800	8400	12,000
Sed	450	1100	1700	3650	6650	9500
Hornet Custom, 8-cyl., 121" wb						
Hlywd	550	1700	2800	5600	9800	14,000
Sed	400	1250	2100	4200	7400	10,500

NOTE: For Hudson Rambler prices see AMC.

HUDSON

1929

	6	5	4	3	2	1
Dover Series						
Canopy Exp	450	950	1600	3250	6150	8800
Screenside Dly	350	900	1550	3050	5900	8500
Panel Dly	450	1000	1650	3350	6300	9000
Flareboard PU	450	1100	1700	3650	6650	9500
Bed Rail PU	450	1450	2400	4800	8400	12,000
Sed Dly	400	1200	2000	3950	7000	10,000
Mail Truck w/sl. doors	750	2400	4000	8000	14,000	20,000

NOTE: The Dover mail truck in Harrah's 1986 auction sold for $26,000, but the Harrah sale tended to inflate values based on the fact that William F. Harrah owned the vehicles.

1930

	6	5	4	3	2	1
Essex Commercial Car Series						
PU	450	1100	1700	3650	6650	9500
Canopy Exp	350	850	1500	2900	5700	8200
Screenside Exp	350	900	1550	3050	5900	8500
Panel Exp	450	1000	1650	3350	6300	9000
Sed Dly	400	1200	2000	3950	7000	10,000

NOTE: Add 30 percent for original dealer side rail (service truck) pickup conversions.

1931

	6	5	4	3	2	1
Essex Commercial Car Series						
PU	450	1100	1700	3650	6650	9500
Canopy Exp	350	850	1500	2900	5700	8200
Screenside Exp	350	900	1550	3050	5900	8500
Panel Dly	450	1000	1650	3350	6300	9000
Sed Dly	400	1200	2000	3950	7000	10,000

NOTE: Add 30 percent for original dealer side rail (service truck) pickup conversions.

1933

	6	5	4	3	2	1
Essex-Terraplane Series						
PU Exp	350	850	1500	2900	5700	8200
Canopy Dly	350	750	1350	2600	5400	7700
Screenside Dly	350	800	1450	2750	5600	8000
Panel Dly	350	850	1500	2950	5800	8300
DeL Panel Dly	350	900	1550	3050	5900	8500
Sed Dly	450	1000	1650	3350	6300	9000
Mail Dly Van	550	1800	3000	6000	10,500	15,000

1934

	6	5	4	3	2	1
Terraplane Series						
Cab PU	350	900	1550	3050	5900	8500
Sed Dly	450	1000	1650	3350	6300	9000
Utl Coach	200	650	1050	2250	4200	6000
Com Sed Taxicab	350	750	1200	2350	4900	7000
Cantrell Sta Wag	500	1550	2600	5200	9100	13,000
Cotton Sta Wag	450	1450	2400	4800	8400	12,000

1935

	6	5	4	3	2	1
Terraplane Series GU						
Cab PU	350	900	1550	3050	5900	8500
Sed Dly	450	1000	1650	3350	6300	9000
Utl Coach	200	650	1050	2250	4200	6000
Com Sta Wag	500	1550	2600	5200	9100	13,000
Taxicab	350	750	1200	2350	4900	7000

1936

	6	5	4	3	2	1
Terraplane Series 61						
Cab PU	350	800	1450	2750	5600	8000
Cus Panel Dly	350	900	1550	3050	5900	8500
Utl Coach	200	600	950	2150	3850	5500
Cus Sta Wag	450	1450	2400	4800	8400	12,000
Taxicab	200	650	1050	2250	4200	6000

1937

	6	5	4	3	2	1
Terraplane Series 70 - (1/2 Ton)						
Utl Coach	200	550	900	2150	3800	5400
Utl Cpe PU	350	750	1350	2600	5400	7700
Terraplane Series 70 - (3/4 Ton)						
Cab PU	350	750	1300	2450	5250	7500
Panel Dly	350	750	1350	2650	5450	7800
Sta Wag	450	1400	2300	4600	8100	11,500
"Big Boy" Series 78 - (3/4 Ton)						
Cab PU	350	750	1200	2350	4900	7000
Cus Panel Dly	350	750	1300	2450	5250	7500
Big Boy Taxicab	200	650	1050	2250	4200	6000

1938

	6	5	4	3	2	1
Hudson-Terraplane Series 80						
Cab PU	350	750	1200	2350	4900	7000
Cus Panel Dly	350	800	1450	2750	5600	8000
Utl Cpe	200	650	1000	2200	4100	5800
Utl Coach	350	700	1100	2300	4500	6400
Utl Tr Coach	200	600	950	2150	3850	5500
Sta Wag	500	1550	2600	5200	9100	13,000
Hudson "Big Boy" Series 88						
Cab PU	350	800	1450	2750	5600	8000
Cus Panel Dly	350	900	1550	3050	5900	8500
Hudson 112 Series 89						
Cab PU	350	750	1300	2450	5250	7500
Panel Dly	350	800	1450	2750	5600	8000
Utl Cpe	200	600	950	2150	3850	5500
Utl Coach	200	650	1050	2250	4200	6000
Utl Tr Coach	350	700	1100	2300	4500	6400

1939

	6	5	4	3	2	1
Hudson 112 Series						
PU	350	750	1300	2450	5250	7500
Cus Panel	350	800	1450	2750	5600	8000
(Business Cars)						
Utl Coach	200	650	1050	2250	4200	6000
Utl Cpe	350	700	1150	2300	4550	6500
Sta Wag	500	1550	2600	5200	9100	13,000
Hudson "Big Boy" Series						
PU	350	800	1450	2750	5600	8000
Cus Panel	350	900	1550	3050	5900	8500
Taxicab (86 HP)	450	1000	1650	3350	6300	9000
7P Partition Taxicab	350	700	1150	2300	4550	6500
Hudson Pacemaker Series						
Cus Panel	450	950	1600	3250	6150	8800

1940

	6	5	4	3	2	1
Hudson Six Series						
PU	350	700	1150	2300	4550	6500
Panel Dly	350	750	1300	2450	5250	7500
Traveler Line						
Utl Coach	200	600	950	2150	3850	5500
Utl Cpe	200	650	1050	2250	4200	6000
Sta Wag	500	1550	2600	5200	9100	13,000
Taxicab	200	600	950	2150	3850	5500
"Big Boy" Series						
PU	350	900	1550	3050	5900	8500
Panel Dly	350	800	1450	2750	5600	8000
9P Carryall Sed	200	650	1050	2250	4200	6000
7P Sed	200	600	950	2150	3850	5500

1941

	6	5	4	3	2	1
Hudson Six Series						
PU	350	900	1550	3050	5900	8500
All-Purpose Dly	350	900	1550	3050	5900	8500
Traveler Line						
Utl Cpe	200	650	1050	2250	4200	6000
Utl Coach	200	600	950	2150	3850	5500
"Big Boy" Series						
PU	350	900	1550	3050	5900	8500
9P Carryall Sed	200	650	1050	2250	4200	6000
Taxicab	200	600	950	2150	3850	5500

1942

	6	5	4	3	2	1
Traveler Series						
Utl Cpe	200	650	1050	2250	4200	6000
Utl Coach	200	600	950	2150	3850	5500
Hudson Six Series						
PU	350	900	1550	3050	5900	8500
Hudson "Big Boy" Series						
PU	350	750	1300	2450	5250	7500

1946

	6	5	4	3	2	1
Cab Pickup Series						
Cab PU	450	1000	1650	3350	6300	9000

1947

	6	5	4	3	2	1
Series 178						
PU	350	750	1200	2350	4900	7000

LAFAYETTE

1920-1921
Model 134
8-cyl, 132" wb, 90 hp, V-8

	6	5	4	3	2	1
Sed	650	2050	3400	6800	11,900	17,000
Cpe	750	2400	4000	8000	14,000	20,000
Tr	1200	3850	6400	12,800	22,400	32,000
Torpedo	1200	3850	6400	12,800	22,400	32,000
Limo	800	2500	4200	8400	14,700	21,000

1922
Model 134
8-cyl, 132" wb, 90 hp, V-8

Sed	650	2050	3400	6800	11,900	17,000
Cpe	750	2400	4000	8000	14,000	20,000
Rds	1250	3950	6600	13,200	23,100	33,000
Tr	1200	3850	6400	12,800	22,400	32,000
Torpedo	1200	3850	6400	12,800	22,400	32,000
Limo	800	2500	4200	8400	14,700	21,000

1923
Model 134
8-cyl, 132" wb, 100 hp

Sed	700	2150	3600	7200	12,600	18,000
Cpe	750	2400	4000	8000	14,000	20,000
Rds	1300	4100	6800	13,600	23,800	34,000
Tr	1250	3950	6600	13,200	23,100	33,000
Torpedo	1200	3850	6400	12,800	22,400	32,000
Limo	800	2500	4200	8400	14,700	21,000
Vestibule Sed	850	2650	4400	8800	15,400	22,000

1924
Model 134
8-cyl, 132" wb, 100 hp

Sed	700	2150	3600	7200	12,600	18,000
Cpe	750	2400	4000	8000	14,000	20,000
Rds	1300	4100	6800	13,600	23,800	34,000
Tr	1250	3950	6600	13,200	23,100	33,000
Torpedo	1200	3850	6400	12,800	22,400	32,000
Limo	800	2500	4200	8400	14,700	21,000
Imperial Limo	850	2750	4600	9200	16,100	23,000

NASH

1918
Series 680, 6-cyl.

	6	5	4	3	2	1
7P Tr	500	1550	2600	5200	9100	13,000
5P Tr	500	1550	2600	5200	9100	13,000
4P Rds	550	1700	2800	5600	9800	14,000
7P Tr	500	1550	2600	5200	9100	13,000
Sed	450	1000	1650	3350	6300	9000
Cpe	450	1050	1650	3500	6400	9200

1919
Series 680, 6-cyl.

Rds	450	1450	2400	4800	8400	12,000
Spt	500	1550	2600	5200	9100	13,000
5P Tr	450	1450	2400	4800	8400	12,000
7P Tr	500	1550	2600	5200	9100	13,000
4P Rds	500	1550	2600	5200	9100	13,000
Sed	350	800	1450	2750	5600	8000
Cpe	350	900	1550	3050	5900	8500

1920
Series 680, 6-cyl.

5P Tr	500	1550	2600	5200	9100	13,000
Rds	450	1450	2400	4800	8400	12,000
7P Tr	550	1700	2800	5600	9800	14,000
Cpe	350	900	1550	3050	5900	8500
Sed	350	800	1450	2750	5600	8000
Spt	350	950	1600	3200	6050	8700

1921
Series 680, 6-cyl.

5P Tr	400	1300	2200	4400	7700	11,000
Rds	450	1450	2400	4800	8400	12,000
Spt	500	1550	2600	5200	9100	13,000
Tr	450	1450	2400	4800	8400	12,000
Cpe	350	900	1550	3050	5900	8500
Sed	350	750	1200	2350	4900	7000

Series 40, 4-cyl.

Tr	400	1200	2000	3950	7000	10,000
Rds	400	1250	2100	4200	7400	10,500
Cpe	350	725	1200	2350	4800	6800
Sed	350	700	1150	2300	4550	6500
Cabr	400	1200	2000	3950	7000	10,000

1922
Series 680, 6-cyl.

5P Tr	400	1300	2200	4400	7700	11,000
7P Tr	450	1450	2400	4800	8400	12,000
7P Sed	350	800	1450	2750	5600	8000
Cpe	350	750	1300	2450	5250	7500
Rds	450	1500	2500	5000	8800	12,500
Spt	500	1550	2600	5200	9100	13,000
5P Sed	350	750	1200	2350	4900	7000

Series 40, 4-cyl.

Tr	400	1250	2100	4200	7400	10,500
Rds	400	1300	2200	4400	7700	11,000
Cpe	350	750	1200	2350	4900	7000
Sed	350	700	1150	2300	4550	6500
Cabr	400	1200	2000	3950	7000	10,000
Ca'ole	350	750	1350	2650	5450	7800

1923
Series 690, 6-cyl., 121" wb

Rds	450	1450	2400	4800	8400	12,000
Tr	500	1550	2600	5200	9100	13,000
Spt	550	1700	2800	5600	9800	14,000
Sed	350	750	1350	2650	5450	7800
Cpe	350	800	1450	2750	5600	8000

Series 690, 6-cyl., 127" wb

Tr	500	1550	2600	5200	9100	13,000
Sed	350	800	1450	2750	5600	8000
Cpe	350	900	1550	3050	5900	8500

Series 40, 4-cyl.

Tr	400	1300	2200	4400	7700	11,000
Rds	450	1450	2400	4800	8400	12,000
Spt	500	1550	2600	5200	9100	13,000
Ca'ole	500	1600	2700	5400	9500	13,500
Sed	350	750	1250	2400	5050	7200

1924
Series 690, 6-cyl., 121" wb

	6	5	4	3	2	1
Rds	450	1450	2400	4800	8400	12,000
Tr	400	1300	2200	4400	7700	11,000
Spec Del	350	700	1150	2300	4550	6500
Cpe	350	700	1150	2300	4550	6500
Spec Sed	200	650	1050	2250	4200	6000

Series 690, 6-cyl., 127" wb

7P Tr	500	1550	2600	5200	9100	13,000
7P Sed	350	750	1200	2350	4900	7000
Vic	350	725	1200	2350	4850	6900

4 cyl.

Tr	500	1550	2600	5200	9100	13,000
Rds	500	1600	2700	5400	9500	13,500
Cab	500	1650	2750	5500	9600	13,700
5P Sed	350	700	1150	2300	4550	6500
Sed	200	650	1050	2250	4200	6000
Spt	550	1700	2800	5600	9800	14,000
Cpe	350	700	1100	2300	4500	6400

1925
Advanced models, 6-cyl.

Tr	450	1500	2500	5000	8700	12,400
7P Tr	500	1600	2700	5400	9400	13,400
4 dr Sed	350	750	1250	2350	5000	7100
Vic Cpe	350	900	1550	3000	5850	8400
7P Sed	350	750	1300	2400	5200	7400
Rds	550	1750	2900	5800	10,100	14,400
Cpe	350	750	1300	2400	5200	7400
2 dr Sed	350	700	1100	2300	4500	6400

Special models, 6-cyl.

Tr	450	1500	2500	5000	8700	12,400
4 dr Sed	200	650	1000	2200	4150	5900
Rds	500	1600	2700	5400	9400	13,400
2 dr Sed	200	650	1000	2200	4150	5900

Light six, (Ajax), 6-cyl.

Tr	400	1250	2100	4200	7300	10,400
Sed	200	550	900	2150	3800	5400

1926
Advanced models, 6-cyl.

5P Tr	550	1750	2900	5800	10,100	14,400
7P Tr	600	1850	3100	6200	10,800	15,400
2 dr Sed	350	725	1200	2350	4850	6900
4 dr Sed	350	750	1250	2400	5050	7200
7P Sed	350	750	1300	2400	5200	7400
4 dr Cpe	350	750	1250	2400	5050	7200
Rds	600	1850	3100	6200	10,800	15,400
Vic Cpe	350	900	1550	3000	5850	8400

Special models, 6-cyl.

Rds	500	1650	2800	5600	9700	13,900
2 dr Sed	350	700	1100	2300	4500	6400
7P Sed	350	750	1250	2350	5000	7100
Cpe	350	725	1200	2350	4850	6900
4 dr Sed	350	750	1200	2350	4900	7000
Spec Rds	550	1750	2900	5800	10,100	14,400

Light Six (formerly Ajax)

Tr	400	1250	2100	4200	7400	10,600
Sed	200	550	900	2150	3800	5400

1927
Standard, 6-cyl.

Tr	400	1300	2200	4400	7600	10,900
Cpe	350	750	1250	2350	5000	7100
2 dr Sed	200	675	1050	2250	4350	6200
4 dr Sed	200	675	1100	2250	4400	6300
DeL Sed	350	700	1100	2300	4500	6400

(Begin September 1926)

Rds	550	1800	3000	6000	10,400	14,900
Tr	500	1600	2700	5400	9400	13,400
Cpe	450	1050	1650	3500	6400	9200
2 dr Sed	350	900	1550	3100	6000	8600
4 dr Sed	450	1000	1650	3350	6300	9000

(Begin January 1927)

Cav Sed	450	1050	1650	3500	6400	9200
4 dr Sed	450	1000	1650	3350	6300	9000
RS Cab	550	1750	2900	5800	10,100	14,400
RS Rds	600	1850	3100	6200	10,800	15,400

Advanced, 6-cyl.

(Begin August 1926)

Rds	600	1850	3100	6200	10,800	15,400
5P Tr	600	1900	3200	6400	11,100	15,900
7P Tr	600	1950	3300	6600	11,500	16,400
Cpe	450	1150	1900	3900	6900	9900
Vic	450	1050	1700	3600	6600	9400
2 dr Sed	350	850	1500	2900	5700	8200
4 dr Sed	350	900	1550	3000	5850	8400
7P Sed	450	950	1600	3250	6150	8800

(Begin January 1927)

RS Cpe	400	1250	2100	4200	7300	10,400
Spec Sed	450	950	1600	3250	6150	8800
Amb Sed	450	1050	1650	3500	6400	9200

1928
Standard, 6-cyl.

Tr	600	1850	3100	6200	10,800	15,400
Cpe	350	750	1250	2350	5000	7100
Conv Cabr	600	1900	3200	6400	11,100	15,900
2 dr Sed	350	725	1200	2350	4800	6800
4 dr Sed	350	750	1250	2400	5050	7200
Lan Sed	350	750	1250	2400	5050	7200

Special, 6-cyl.

Tr	650	2100	3500	7000	12,200	17,400
RS Rds	650	2100	3500	7000	12,200	17,400
Cpe	450	1050	1700	3600	6600	9400
Conv Cabr	600	1950	3300	6600	11,500	16,400
Vic	350	950	1600	3200	6050	8700
2 dr Sed	350	900	1550	3100	6000	8600
4 dr Sed	450	950	1600	3250	6150	8800
4 dr Cpe	450	1050	1650	3500	6400	9200

Advanced, 6-cyl.

Spt Tr	750	2350	3900	7800	13,600	19,400
Tr	700	2200	3700	7400	12,900	18,400
RS Rds	700	2200	3700	7400	12,900	18,400
Cpe	450	1050	1700	3600	6600	9400
Vic	450	1100	1700	3650	6650	9500
2 dr Sed	350	850	1500	2950	5800	8300
4 dr Sed	350	900	1550	3100	6000	8600
4 dr Cpe	450	1150	1900	3900	6900	9900
7P Sed	450	1050	1700	3600	6600	9400

1929

Standard, 6-cyl.	6	5	4	3	2	1
Sed	350	750	1250	2400	5050	7200
Tr	550	1750	2900	5800	10,100	14,400
Cabr	500	1600	2700	5400	9400	13,400
2 dr Sed	350	750	1250	2400	5050	7200
2P Cpe	350	750	1250	2350	5000	7100
4P Cpe	350	750	1300	2400	5200	7400
Lan Sed	350	725	1200	2350	4850	6900
Special, 6-cyl.						
2 dr Sed	350	750	1300	2400	5200	7400
2P Cpe	350	800	1350	2700	5500	7900
4P Cpe	350	900	1550	3000	5850	8400
Rds	750	2350	3900	7800	13,600	19,400
Sed	350	800	1350	2700	5500	7900
Cabr	700	2200	3700	7400	12,900	18,400
Vic	350	900	1550	3000	5850	8400
Advanced, 6-cyl.						
Cpe	450	1100	1700	3650	6650	9500
Cabr	750	2350	3900	7800	13,600	19,400
2 dr Sed	350	900	1550	3000	5850	8400
7P Sed	450	1050	1700	3600	6600	9400
Amb Sed	450	1150	1900	3900	6900	9900
4 dr Sed	350	950	1600	3200	6050	8700

1930

Single, 6-cyl.	6	5	4	3	2	1
Rds	700	2200	3700	7400	12,900	18,400
Tr	600	1850	3100	6200	10,800	15,400
2P Cpe	350	700	1100	2300	4500	6400
2 dr Sed	350	750	1200	2350	4900	7000
4P Cpe	350	900	1550	3000	5850	8400
Cabr	600	1850	3100	6200	10,800	15,400
4 dr Sed	350	750	1250	2350	5000	7100
DeL Sed	350	750	1250	2400	5100	7300
Lan'let	350	800	1350	2700	5500	7900
Twin-Ign, 6-cyl.						
Rds	850	2750	4600	9200	16,100	23,000
7P Tr	950	3000	5000	10,000	17,500	25,000
5P Tr	900	2900	4800	9600	16,800	24,000
2P Cpe	350	800	1350	2700	5500	7900
4P Cpe	350	900	1550	3000	5850	8400
2 dr Sed	350	800	1350	2700	5500	7900
Cabr	800	2600	4300	8600	15,100	21,500
Vic	450	1500	2500	5000	8700	12,400
4 dr Sed	350	900	1550	3000	5850	8400
7P Sed	350	950	1600	3200	6050	8700
Twin-Ign, 8-cyl.						
2 dr Sed	350	900	1550	3000	5850	8400
2P Cpe	450	1050	1700	3600	6600	9400
4P Cpe	400	1250	2100	4200	7300	10,400
Vic	600	1950	3300	6600	11,500	16,400
Cabr	1350	4300	7200	14,400	25,200	36,000
Sed	450	1000	1600	3300	6250	8900
Amb Sed	400	1250	2100	4200	7300	10,400
7P Sed	400	1250	2100	4200	7300	10,400
7P Limo	450	1350	2300	4600	8000	11,400

1931

Series 660, 6-cyl.	6	5	4	3	2	1
5P Tr	550	1750	2900	5800	10,100	14,400
2P Cpe	350	725	1200	2350	4850	6900
4P Cpe	350	750	1300	2400	5200	7400
2 dr Sed	350	725	1200	2350	4850	6900
4 dr Sed	350	750	1300	2400	5200	7400
Series 870, 8-cyl.						
2P Cpe	450	1000	1600	3300	6250	8900
4P Cpe	350	900	1550	3000	5850	8400
Conv Sed	1700	5400	9000	18,000	31,500	45,000
2 dr Sed	350	800	1350	2700	5500	7900
Spec Sed	350	900	1550	3000	5850	8400
Series 880 - Twin-Ign, 8-cyl.						
2P Cpe	450	1050	1700	3600	6600	9400
4P Cpe	450	1150	1900	3900	6900	9900
Conv Sed	1800	5750	9600	19,200	33,600	48,000
Sed	450	1050	1700	3600	6600	9400
Twn Sed	450	1150	1900	3900	6900	9900
Series 890 - Twin-Ign, 8-cyl.						
7P Tr	1500	4800	8000	16,000	28,000	40,000
2P Cpe	800	2500	4200	8400	14,700	21,000
4P Cpe	850	2650	4400	8800	15,400	22,000
Cabr	1800	5750	9600	19,200	33,600	48,000
Vic	550	1800	3000	6000	10,500	15,000
2 dr Sed	550	1800	3000	6000	10,500	15,000
Amb Sed	600	1900	3200	6400	11,200	16,000
7P Sed	650	2050	3400	6800	11,900	17,000
7P Limo	700	2300	3800	7600	13,300	19,000

1932

Series 960, 6-cyl.	6	5	4	3	2	1
5P Tr	1150	3700	6200	12,400	21,700	31,000
2P Cpe	450	1000	1650	3350	6300	9000
4P Cpe	450	1100	1700	3650	6650	9500
2 dr Sed	350	750	1200	2350	4900	7000
4 dr Sed	350	800	1450	2750	5600	8000
Series 970, 8-cyl., 116.5" wb						
2P Cpe	400	1300	2200	4400	7700	11,000
4P Cpe	450	1450	2400	4800	8400	12,000
Conv Sed	1800	5750	9600	19,200	33,600	48,000
2 dr Sed	450	1100	1700	3650	6650	9500
Spec Sed	400	1200	2000	3950	7000	10,000
Series 980 - Twin-Ign, 8-cyl., 121" wb						
2P Cpe	900	2900	4800	9600	16,800	24,000
4P Cpe	950	3000	5000	10,000	17,500	25,000
Conv Sed	1950	6250	10,400	20,800	36,400	52,000
Sed	800	2500	4200	8400	14,700	21,000
Twn Sed	850	2650	4400	8800	15,400	22,000
Series 990 - Twin-Ign, 8-cyl., 124"-133" wb						
7P Tr	1800	5750	9600	19,200	33,600	48,000
2P Cpe	1000	3100	5200	10,400	18,200	26,000
4P Cpe	1000	3250	5400	10,800	18,900	27,000
Cabr	1900	6000	10,000	20,000	35,000	50,000
Vic	1000	3100	5200	10,400	18,200	26,000
2 dr Sed	850	2650	4400	8800	15,400	22,000
Spec Sed	900	2900	4800	9600	16,800	24,000
Amb Sed	950	3000	5000	10,000	17,500	25,000
7P Sed	900	2900	4800	9600	16,800	24,000
Limo	1100	3500	5800	11,600	20,300	29,000

1933

Standard Series	6	5	4	3	2	1
Rds	900	2900	4800	9600	16,800	24,000
2P Cpe	350	750	1300	2450	5250	7500
4P Cpe	450	1000	1650	3350	6300	9000
4 dr Sed	350	800	1450	2750	5600	8000
Twn Sed	350	900	1550	3050	5900	8500
Special Series, 8-cyl.						
Rds	1050	3350	5600	11,200	19,600	28,000
2P Cpe	450	1000	1650	3350	6300	9000
4P Cpe	450	1100	1700	3650	6650	9500
4 dr Sed	450	1000	1650	3350	6300	9000
Conv Sed	1500	4800	8000	16,000	28,000	40,000
Twn Sed	450	1050	1700	3550	6500	9300
Advanced Series, 8-cyl.						
Cabr	1150	3600	6000	12,000	21,000	30,000
2P Cpe	450	1150	1800	3800	6800	9700
4P Cpe	450	1100	1700	3650	6650	9500
4 dr Sed	450	1050	1650	3500	6400	9200
Conv Sed	1750	5650	9400	18,800	32,900	47,000
Vic	450	1450	2400	4800	8400	12,000
Ambassador Series, 8-cyl.						
Cabr	1600	5050	8400	16,800	29,400	42,000
Cpe	500	1550	2600	5200	9100	13,000
4 dr Sed	450	1450	2400	4800	8400	12,000
Conv Sed	2050	6600	11,000	22,000	38,500	55,000
Vic	1000	3250	5400	10,800	18,900	27,000
142" Brgm	900	2900	4800	9600	16,800	24,000
142" Sed	850	2650	4400	8800	15,400	22,000
142" Limo	1050	3350	5600	11,200	19,600	28,000

1934

Big Six, 6-cyl.	6	5	4	3	2	1
Bus Cpe	350	750	1300	2450	5250	7500
Cpe	350	900	1550	3050	5900	8500
Brgm	350	800	1450	2750	5600	8000
2 dr Sed	350	750	1200	2350	4900	7000
Twn Sed	350	750	1300	2450	5250	7500
Tr Sed	350	750	1300	2400	5200	7400
Advanced, 8-cyl.						
Bus Cpe	350	800	1450	2750	5600	8000
Cpe	350	950	1600	3200	6050	8700
Brgm	450	1000	1650	3350	6300	9000
2 dr Sed	350	750	1300	2450	5250	7500
Twn Sed	350	900	1550	3050	5900	8500
Tr Sed	350	800	1450	2750	5600	8000
Ambassador, 8-cyl.						
Brgm	350	900	1550	3050	5900	8500
2 dr Sed	350	800	1450	2750	5600	8000
Tr Sed	350	850	1500	2900	5700	8200
7P Sed	450	1000	1650	3350	6300	9000
Limo	450	1450	2400	4800	8400	12,000
Lafayette, 6-cyl.						
2 dr Sed	350	700	1150	2300	4550	6500
Twn Sed	350	700	1150	2300	4600	6600
Brgm	350	725	1200	2350	4800	6800
Spec Cpe	350	800	1450	2750	5600	8000
Spec 4P Cpe	350	900	1550	3050	5900	8500
Spec Tr Sed	350	750	1200	2350	4900	7000
Spec Sed	350	750	1250	2400	5050	7200
Brgm	350	750	1300	2450	5250	7500

1935

Lafayette, 6-cyl.	6	5	4	3	2	1
Bus Cpe	350	725	1200	2350	4850	6900
2 dr Sed	350	725	1150	2300	4700	6700
Brgm	350	750	1200	2350	4900	7000
Tr Sed	350	725	1200	2350	4800	6800
Twn Sed	350	725	1200	2350	4850	6900
Spec Cpe	350	900	1550	3050	5900	8500
Spec 6W Sed	350	750	1300	2450	5250	7500
6W Brgm	350	750	1300	2500	5300	7600
Advanced, 6-cyl.						
Vic	350	800	1350	2700	5500	7900
6W Sed	350	725	1200	2350	4850	6900
Advanced, 8-cyl.						
Vic	350	850	1500	2800	5650	8100
6W Sed	350	750	1250	2400	5050	7200
Ambassador, 8-cyl.						
Vic	450	1100	1700	3650	6650	9500
6W Sed	350	900	1550	3050	5900	8500

1936

Lafayette, 6-cyl.	6	5	4	3	2	1
Bus Cpe	350	700	1150	2300	4550	6500
Cpe	350	725	1200	2350	4850	6900
Cabr	600	2000	3300	6600	11,600	16,500
Sed	350	700	1150	2300	4550	6500
Vic	350	700	1150	2300	4600	6600
Tr Sed	200	650	1000	2200	4100	5800
400 Series, 6-cyl.						
Bus Cpe	200	650	1000	2200	4150	5900
Cpe	350	700	1150	2300	4550	6500
Vic	350	725	1200	2350	4850	6900
Tr Vic	350	725	1150	2300	4700	6700
Sed	350	700	1150	2300	4600	6600
Tr Sed	350	700	1150	2300	4550	6500
Spec Bus Cpe	350	700	1150	2300	4550	6500
Spec Cpe	350	700	1150	2300	4600	6600
Spec Spt Cabr	800	2500	4200	8400	14,700	21,000
Spec Vic	350	725	1200	2350	4850	6900
Spec Tr Vic	350	725	1200	2350	4850	6900
Spec Sed	350	700	1150	2300	4600	6600
Spec Tr Sed	350	700	1150	2300	4550	6500
Ambassador Series, 6-cyl.						
Vic	350	750	1300	2450	5250	7500
Tr Sed	350	700	1150	2300	4550	6500
Ambassador Series, 8-cyl.						
Tr Sed	350	750	1300	2450	5250	7500

1937

Lafayette 400, 6-cyl.	6	5	4	3	2	1
Bus Cpe	200	600	950	2150	3850	5500
Cpe	200	600	950	2200	3900	5600
A-P Cpe	200	600	950	2150	3850	5500
Cabr	650	2050	3400	6800	11,900	17,000
Vic Sed	200	600	950	2150	3850	5500
Tr Sed	200	600	950	2200	3900	5600
Ambassador, 6-cyl.						
Bus Cpe	350	700	1150	2300	4550	6500
Cpe	350	725	1150	2300	4700	6700
A-P Cpe	200	675	1050	2250	4300	6100

	6	5	4	3	2	1
Cabr	700	2150	3600	7200	12,600	18,000
Vic Sed	200	650	1000	2200	4150	5900
Tr Sed	350	700	1150	2300	4550	6500
Ambassador, 8-cyl.						
Bus Cpe	350	725	1150	2300	4700	6700
Cpe	350	725	1200	2350	4850	6900
A-P Cpe	350	725	1150	2300	4700	6700
Cabr	800	2500	4200	8400	14,700	21,000
Vic Sed	350	700	1150	2300	4550	6500
Tr Sed	350	750	1300	2450	5250	7500
1938						
Lafayette						
Master, 6-cyl.						
Bus Cpe	200	500	850	1850	3350	4900
Vic	200	500	850	1950	3600	5100
Tr Sed	200	500	850	1900	3500	5000
DeLuxe, 6-cyl.						
Bus Cpe	200	600	950	2150	3850	5500
A-P Cpe	200	650	1050	2250	4200	6000
Cabr	600	1850	3100	6200	10,900	15,500
Vic	200	600	950	2150	3850	5500
Tr Sed	200	600	950	2200	3900	5600
Ambassador, 6-cyl.						
Bus Cpe	350	700	1150	2300	4550	6500
A-P Cpe	350	750	1200	2350	4900	7000
Cabr	500	1650	2800	5600	9700	13,900
Vic	200	650	1000	2200	4150	5900
Tr Sed	350	700	1150	2300	4550	6500
Ambassador, 8-cyl.						
Bus Cpe	350	725	1150	2300	4700	6700
A-P Cpe	350	725	1200	2350	4800	6800
Cabr	750	2400	4000	8000	14,000	20,000
Vic	200	675	1100	2250	4400	6300
Tr Sed	350	725	1200	2350	4850	6900
1939						
Lafayette, 6-cyl.						
(Add 10 percent for DeLuxe)						
Bus Cpe	200	600	950	2150	3850	5500
2 dr Sed	200	500	850	1900	3500	5000
4 dr Sed	200	550	900	2100	3700	5300
Tr Sed	200	500	850	1950	3600	5100
A-P Cpe	150	450	800	1750	3250	4700
A-P Cabr	700	2200	3700	7400	13,000	18,500
Tr Sed	200	500	850	1950	3600	5100
Ambassador, 6-cyl.						
Bus Cpe	350	750	1300	2500	5300	7600
A-P Cpe	350	800	1350	2700	5500	7900
A-P Cabr	850	2650	4400	8800	15,400	22,000
2 dr Sed	350	725	1150	2300	4700	6700
4 dr Sed	350	725	1200	2350	4800	6800
Tr Sed	350	725	1200	2350	4850	6900
Ambassador, 8-cyl.						
Bus Cpe	350	800	1350	2700	5500	7900
A-P Cpe	350	850	1500	2800	5650	8100
A-P Cabr	1100	3500	5800	11,600	20,300	29,000
2 dr Sed	350	750	1300	2450	5250	7500
4 dr Sed	350	750	1300	2450	5250	7500
Tr Sed	350	750	1300	2450	5250	7500
1940						
DeLuxe Lafayette, 6-cyl.						
Bus Cpe	350	700	1150	2300	4550	6500
A-P Cpe	350	700	1150	2300	4600	6600
A-P Cabr	1000	3100	5200	10,400	18,200	26,000
2 dr FsBk	200	675	1050	2250	4350	6200
4 dr FsBk	350	700	1150	2300	4550	6500
4 dr Trk Sed	200	675	1100	2250	4400	6300
Ambassador, 6-cyl.						
Bus Cpe	350	750	1300	2500	5300	7600
A-P Cpe	350	750	1350	2650	5450	7800
A-P Cabr	1200	3850	6400	12,800	22,400	32,000
2 dr FsBk	350	750	1300	2500	5300	7600
4 dr FsBk	350	750	1350	2600	5400	7700
4 dr Trk Sed	350	750	1350	2650	5450	7800
Ambassador, 8-cyl.						
Bus Cpe	350	900	1550	3050	5900	8500
A-P Cpe	350	900	1550	3100	6000	8600
A-P Cabr	1300	4100	6800	13,600	23,800	34,000
2 dr FsBk	350	800	1350	2700	5500	7900
4 dr FsBk	350	800	1450	2750	5600	8000
4 dr Trk Sed	350	850	1500	2800	5650	8100
1941						
Ambassador 600, 6-cyl.						
Bus Cpe	200	650	1000	2200	4150	5900
2 dr FsBk	200	600	950	2200	3900	5600
4 dr FsBk	200	600	1000	2200	4000	5700
DeL Bus Cpe	200	675	1050	2250	4300	6100
DeL Brgm	350	700	1150	2300	4550	6500
DeL 2 dr FsBk	200	650	1050	2250	4200	6000
DeL 4 dr FsBk	200	650	1000	2200	4150	5900
4 dr Tr Sed	200	675	1050	2250	4300	6100
Ambassador, 6-cyl.						
Bus Cpe	350	750	1300	2500	5300	7600
Spec Bus Cpe	350	750	1300	2450	5250	7500
A-P Cabr	1050	3350	5600	11,200	19,600	28,000
2 dr Brgm	350	750	1350	2650	5450	7800
4 dr Spec Sed	350	750	1300	2450	5250	7500
4 dr Spec FsBk	350	750	1300	2450	5250	7500
4 dr DeL FsBk	350	750	1300	2500	5300	7600
4 dr Tr Sed	350	800	1350	2700	5500	7900
Ambassador, 8-cyl.						
A-P Cabr	1150	3700	6200	12,400	21,700	31,000
2 dr DeL Brgm	350	850	1500	2800	5650	8100
4 dr Spec FsBk	350	800	1350	2700	5500	7900
4 dr DeL FsBk	350	800	1450	2750	5600	8000
4 dr Tr Sed	350	900	1550	3050	5900	8500
1942						
Ambassador 600, 6-cyl.						
2 dr Bus Cpe	200	675	1050	2250	4350	6200
2 dr Brgm	350	700	1150	2300	4600	6600
2 dr SS	350	700	1150	2300	4550	6500
4 dr SS	350	725	1150	2300	4700	6700
4 dr Tr Sed	350	700	1150	2300	4550	6500
Ambassador, 6-cyl.						
Bus Cpe	350	750	1300	2450	5250	7500
2 dr Brgm	350	750	1350	2650	5450	7800
2 dr SS	350	750	1300	2500	5300	7600

	6	5	4	3	2	1
4 dr SS	350	750	1350	2600	5400	7700
4 dr Tr Sed	350	750	1350	2650	5450	7800
Ambassador, 8-cyl.						
Bus Cpe	350	800	1350	2700	5500	7900
2 dr Brgm	350	800	1450	2750	5600	8000
2 dr SS	350	750	1350	2650	5450	7800
4 dr SS	350	800	1350	2700	5500	7900
4 dr Tr Sed	350	800	1450	2750	5600	8000
1946						
600, 6-cyl.						
2 dr Brgm	200	500	850	1950	3600	5100
4 dr Sed	200	500	850	1900	3500	5000
4 dr Trk Sed	200	550	900	2100	3700	5300
Ambassador, 6-cyl.						
2 dr Brgm	350	750	1300	2450	5250	7500
4 dr Sed	350	750	1300	2450	5250	7500
4 dr Trk Sed	350	750	1300	2500	5300	7600
4 dr Sub Sed	850	2750	4600	9200	16,100	23,000
1947						
600, 6-cyl.						
2 dr Brgm	200	550	900	2000	3600	5200
4 dr Sed	200	500	850	1950	3600	5100
4 dr Trk Sed	200	600	950	2150	3850	5500
Ambassador, 6-cyl.						
2 dr Brgm	350	750	1300	2450	5250	7500
4 dr Sed	350	750	1300	2500	5300	7600
4 dr Trk Sed	350	750	1350	2600	5400	7700
4 dr Sub Sed	850	2750	4600	9200	16,100	23,000
1948						
600, 6-cyl.						
DeL Bus Cpe	200	600	950	2200	3900	5600
4 dr Super Sed	200	600	950	2150	3850	5500
4 dr Super Trk Sed	200	600	950	2150	3850	5500
2 dr Super Brgm	200	600	1000	2200	4000	5700
4 dr Cus Sed	200	650	1000	2200	4150	5900
4 dr Cus Trk Sed	200	675	1050	2250	4300	6100
2 dr Cus Brgm	200	650	1050	2250	4200	6000
Ambassador, 6-cyl.						
4 dr Sed	350	750	1250	2400	5050	7200
4 dr Trk Sed	350	750	1250	2400	5100	7300
2 dr Brgm	350	750	1250	2350	5000	7100
4 dr Sub Sed	900	2900	4800	9600	16,800	24,000
Custom Ambassador, 6-cyl.						
4 dr Sed	350	750	1300	2450	5250	7500
4 dr Trk Sed	350	750	1300	2500	5300	7600
2 dr Brgm	350	750	1300	2450	5250	7500
2 dr Cabr	950	3000	5000	10,000	17,500	25,000
1949						
600 Super, 6-cyl.						
4 dr Sed	200	600	950	2200	3900	5600
2 dr Sed	200	600	950	2200	3900	5600
2 dr Brgm	200	600	950	2200	3900	5600
600 Super Special, 6-cyl.						
4 dr Sed	200	675	1050	2250	4300	6100
2 dr Sed	200	650	1000	2200	4150	5900
2 dr Brgm	200	600	1000	2200	4000	5700
600 Custom, 6-cyl.						
4 dr Sed	200	650	1000	2200	4100	5800
2 dr Sed	200	600	1000	2200	4000	5700
2 dr Brgm	200	650	1000	2200	4100	5800
Ambassador Super, 6-cyl.						
4 dr Sed	350	750	1200	2350	4900	7000
2 dr Sed	350	725	1200	2350	4850	6900
2 dr Brgm	350	750	1200	2350	4900	7000
Ambassador Super Special, 6-cyl.						
4 dr Sed	350	750	1250	2350	5000	7100
2 dr Sed	350	750	1200	2350	4900	7000
2 dr Brgm	350	750	1250	2350	5000	7100
Ambassador Custom, 6-cyl.						
4 dr Sed	350	750	1250	2400	5050	7200
2 dr Sed	350	750	1250	2350	5000	7100
2 dr Brgm	350	750	1250	2400	5050	7200
1950						
Rambler Custom, 6-cyl.						
2 dr Conv Lan	350	700	1150	2300	4550	6500
2 dr Sta Wag	200	600	950	2150	3850	5500
Nash Super Statesman, 6-cyl.						
2 dr DeL Cpe	200	675	1050	2250	4300	6100
4 dr Sed	200	600	950	2200	3900	5600
2 dr Sed	200	600	950	2200	3900	5600
2 dr Clb Cpe	350	700	1150	2300	4600	6600
Nash Custom Statesman, 6-cyl.						
4 dr Sed	200	600	950	2200	3900	5600
2 dr Sed	200	600	1000	2200	4000	5700
2 dr Clb Cpe	200	650	1000	2200	4100	5800
Ambassador, 6-cyl.						
4 dr Sed	350	725	1150	2300	4700	6700
2 dr Sed	350	725	1200	2350	4800	6800
2 dr Clb Cpe	350	725	1200	2350	4850	6900
Ambassador Custom, 6-cyl.						
4 dr Sed	350	725	1200	2350	4800	6800
2 dr Sed	350	725	1200	2350	4850	6900
2 dr Clb Cpe	350	750	1200	2350	4900	7000
1951						
Rambler, 6-cyl.						
2 dr Utl Wag	200	600	950	2200	3900	5600
2 dr Cus Clb Sed	200	600	1000	2200	4000	5700
2 dr Cus Conv Sed	350	750	1200	2350	4900	7000
2 dr Cus Sta Wag	200	600	1000	2200	4000	5700
Nash Statesman, 6-cyl.						
2 dr DeL Bus Cpe	200	600	1000	2200	4000	5700
4 dr Sup Sed	200	650	1000	2200	4100	5800
2 dr Sup	200	600	1000	2200	4000	5700
2 dr Sup Cpe	200	650	1000	2200	4100	5800
2 dr Cus Cpe	200	675	1050	2250	4300	6100
2 dr Cus	200	650	1000	2200	4150	5900
Ambassador, 6-cyl.						
4 dr Sup Sed	350	725	1200	2350	4850	6900
2 dr Sup	350	725	1200	2350	4800	6800
2 dr Sup Cpe	350	725	1200	2350	4850	6900
4 dr Cus Sed	350	750	1250	2350	5000	7100
2 dr Cus	350	750	1250	2400	5050	7200
2 dr Cus Cpe	350	750	1250	2400	5100	7300
Nash-Healy						
Spt Car	1200	3850	6400	12,800	22,400	32,000

1952-1953
Rambler, 6-cyl.

	6	5	4	3	2	1
2 dr Utl Wag	200	600	950	2150	3850	5500
2 dr Cus Clb Sed	200	600	1000	2200	4000	5700
2 dr Cus Conv Sed	350	750	1300	2450	5250	7500
2 dr Cus Sta Wag	200	600	950	2150	3850	5500

Nash Statesman, 6-cyl.
(Add 10 percent for Custom)

2 dr Sed	200	600	1000	2200	4000	5700
4 dr Sed	200	600	950	2200	3900	5600
2 dr Cus Ctry Clb	350	750	1300	2450	5250	7500

Ambassador, 6-cyl.
(Add 10 percent for Custom)

2 dr Sed	350	750	1250	2350	5000	7100
4 dr Sed	350	750	1200	2350	4900	7000
2 dr Cus Ctry Clb	450	1100	1700	3650	6650	9500

Nash-Healey

Cpe	1350	4300	7200	14,400	25,200	36,000
Spt Car	1500	4800	8000	16,000	28,000	40,000

1954
Rambler, 6-cyl.

2 dr DeL Clb Sed	200	600	950	2150	3850	5500
2 dr Sup Clb Sed	200	600	950	2200	3900	5600
2 dr Ctry Clb Sed	200	600	1000	2200	4000	5700
2 dr Utl Wag	200	600	1000	2200	4000	5700
4 dr Sup Sed (108")	200	600	950	2150	3850	5500
2 dr Cus Ctry Clb	350	750	1200	2350	4900	7000
2 dr Cus Conv	350	800	1450	2750	5600	8000
2 dr Cus Sta Wag	200	600	950	2150	3850	5500
4 dr Cus Sed (108")	200	600	950	2150	3850	5500
2 dr Cus Wag (108")	200	600	1000	2200	4000	5700

Nash Statesman, 6-cyl.

4 dr Sup Sed	200	650	1050	2250	4200	6000
2 dr Sup Sed	200	675	1050	2250	4300	6100
4 dr Cus Sed	200	675	1050	2250	4350	6200
2 dr Cus Ctry Clb	450	1000	1650	3350	6300	9000

Nash Ambassador, 6-cyl.
(Add 5 percent for LeMans option)

4 dr Sup Sed	350	750	1250	2400	5050	7200
2 dr Sup Sed	350	750	1250	2400	5100	7300
4 dr Cus Sed	350	750	1300	2450	5250	7500
2 dr Cus Ctry Clb	400	1300	2200	4400	7700	11,000

Nash-Healey

Cpe	1400	4500	7500	15,000	26,300	37,500
Spt Car	1600	5050	8400	16,800	29,400	42,000

1955
Rambler, 6-cyl.

2 dr DeL Clb Sed	200	600	950	2200	3900	5600
2 dr DeL Bus Sed	200	600	950	2150	3850	5500
4 dr DeL Sed (108")	200	600	950	2150	3850	5500
2 dr Sup Clb Sed	200	600	950	2200	3900	5600
2 dr Utl Wag	200	600	1000	2200	4000	5700
4 dr Sup Sed (108")	200	600	950	2200	3900	5600
4 dr Sup Crs Ctry (108")	200	600	1000	2200	4000	5700
2 dr Cus Ctry Clb	350	725	1200	2350	4850	6900
4 dr Cus Sed (108")	200	600	1000	2200	4000	5700
4 dr Cus Crs Ctry (108")	200	650	1000	2200	4100	5800

Nash Statesman, 6-cyl.

2 dr Sup Sed	200	600	950	2150	3850	5500
2 dr Cus Sed	200	600	950	2200	3900	5600
2 dr Cus Ctry Clb	350	750	1300	2450	5250	7500

Nash Ambassador, 6-cyl.

4 dr Sup Sed	350	700	1150	2300	4600	6600
4 dr Cus Sed	350	725	1150	2300	4700	6700
2 dr Cus Ctry Clb	450	1450	2400	4800	8400	12,000

Nash Ambassador, 8-cyl.

4 dr Sup Sed	350	725	1150	2300	4700	6700
4 dr Cus Sed	350	750	1200	2350	4900	7000
2 dr Cus Ctry Clb	500	1550	2600	5200	9100	13,000

1956
Rambler, 6-cyl.

4 dr DeL Sed	200	500	850	1850	3350	4900
4 dr Sup Sed	200	500	850	1900	3500	5000
4 dr Sup Crs Ctry	200	650	1000	2200	4100	5800
4 dr Cus Sed	200	500	850	1850	3350	4900
4 dr Cus HdTp	200	600	950	2200	3900	5600
4 dr Cus Crs Ctry	200	675	1100	2250	4400	6300
4 dr HdTp Wag	350	725	1150	2300	4700	6700

Nash Statesman, 6-cyl.

4 dr Sup Sed	200	550	900	2150	3800	5400

Nash Ambassador, 6-cyl.

4 dr Sup Sed	350	700	1150	2300	4600	6600

Nash Ambassador, 8-cyl.

4 dr Sup Sed	350	725	1150	2300	4700	6700
4 dr Cus Sed	350	725	1200	2350	4800	6800
2 dr Cus HdTp	500	1550	2600	5200	9100	13,000

1957
Rambler, 6-cyl.

4 dr DeL Sed	200	500	850	1900	3500	5000
4 dr Sup Sed	200	550	900	2000	3600	5200
4 dr Sup HdTp	200	600	950	2200	3900	5600
4 dr Sup Crs Ctry	200	650	1000	2200	4100	5800
4 dr Cus Sed	200	500	850	1950	3600	5100
4 dr Cus Crs Ctry	200	650	1000	2200	4150	5900

Rambler, 8-cyl.

4 dr Sup Sed	200	550	900	2000	3600	5200
4 dr Sup Crs Ctry Wag	200	650	1000	2200	4150	5900
4 dr Cus Sed	200	550	900	2100	3700	5300
4 dr Cus HdTp	200	650	1050	2250	4200	6000
4 dr Cus Crs Ctry Wag	200	675	1050	2250	4300	6100
4 dr Cus Crs Ctry HdTp	350	700	1150	2300	4600	6600

Rebel, 8-cyl.

4 dr HdTp	450	1000	1650	3350	6300	9000

Nash Ambassador, 8-cyl.

4 dr Sup Sed	350	725	1150	2300	4700	6700
4 dr Sup HdTp	350	800	1450	2750	5600	8000
4 dr Cus Sed	350	750	1200	2350	4900	7000
2 dr Cus HdTp	500	1600	2700	5400	9500	13,500

RAMBLER

1902
One cylinder, 4 hp

	6	5	4	3	2	1
2P Rbt	850	2650	4400	8800	15,400	22,000

1903
One cylinder, 6 hp

	6	5	4	3	2	1
2/4P Lt Tr	800	2500	4200	8400	14,700	21,000

1904
Model E, 1-cyl., 7 hp, 78" wb

Rbt	650	2050	3400	6800	11,900	17,000

Model G, 1-cyl., 7 hp, 81" wb

Rbt	650	2100	3500	7000	12,300	17,500

Model H, 1-cyl., 7 hp, 81" wb

Tonneau	650	2100	3500	7000	12,300	17,500

Model J, 2-cyl., 16 hp, 84" wb

Rbt	700	2150	3600	7200	12,600	18,000

Model K, 2-cyl., 16 hp, 84" wb

Tonneau	700	2150	3600	7200	12,600	18,000

Model L, 2-cyl., 16 hp, 84" wb

Canopy Ton	700	2200	3700	7400	13,000	18,500

1905
Model G, 1-cyl., 8 hp, 81" wb

Rbt	650	2050	3400	6800	11,900	17,000

Model H, 1-cyl., 8 hp, 81" wb

Tr	650	2050	3400	6800	11,900	17,000

Type One, 2-cyl., 18 hp, 90" wb

Tr	700	2150	3600	7200	12,600	18,000

Type Two, 2-cyl., 20 hp, 100" wb

Surrey	700	2300	3800	7600	13,300	19,000
Limo	850	2650	4400	8800	15,400	22,000

1906
Model 17, 2-cyl., 10/12 hp, 88" wb

2P Rbt	600	1900	3200	6400	11,200	16,000

Type One, 2-cyl., 18/20 hp, 90" wb

5P Surrey	650	2050	3400	6800	11,900	17,000

Type Two, 2-cyl., 20 hp, 100" wb

5P Surrey	650	2100	3500	7000	12,300	17,500

Type Three, 2-cyl., 18/20 hp, 96" wb

5P Surrey	700	2150	3600	7200	12,600	18,000

Model 14, 4-cyl., 25 hp, 106" wb

5P Tr	700	2200	3700	7400	13,000	18,500

Model 15, 4-cyl., 35/40 hp, 112" wb

5P Tr	700	2300	3800	7600	13,300	19,000

Model 16, 4-cyl., 35/40 hp, 112" wb

5P Limo	750	2350	3900	7800	13,700	19,500

1907
Model 27, 2-cyl., 14/16 hp, 90" wb

2P Rbt	600	1900	3200	6400	11,200	16,000

Model 22, 2-cyl., 20/22 hp, 100" wb

2P Rbt	600	2000	3300	6600	11,600	16,500

Model 21, 2-cyl., 20/22 hp, 100" wb

5P Tr	650	2050	3400	6800	11,900	17,000

Model 24, 4-cyl., 25/30 hp, 108" wb

5P Tr	650	2100	3500	7000	12,300	17,500

Model 25, 4-cyl., 35/40 hp, 112" wb

5P Tr	700	2150	3600	7200	12,600	18,000

1908
Model 31, 2-cyl., 22 hp, 106" wb

Det Tonneau	700	2150	3600	7200	12,600	18,000

Model 34, 4-cyl., 32 hp, 112" wb

3P Rds	700	2200	3700	7400	13,000	18,500
5P Tr	700	2300	3800	7600	13,300	19,000

1909
Model 47, 2-cyl., 22 hp, 106" wb

2P Rbt	700	2150	3600	7200	12,600	18,000

Model 41, 2-cyl., 22 hp, 106" wb

5P Tr	700	2300	3800	7600	13,300	19,000

Model 44, 4-cyl., 34 hp, 112" wb

5P Tr	750	2400	4000	8000	14,000	20,000
4P C.C. Tr	750	2400	4000	8000	14,000	20,000

Model 45, 4-cyl., 45 hp, 123" wb

7P Tr	1000	3100	5200	10,400	18,200	26,000
4P C.C. Tr	1000	3100	5200	10,400	18,200	26,000
3P Rds	950	3000	5000	10,000	17,500	25,000

1910
Model 53, 4-cyl., 34 hp, 109" wb

Tr	900	2900	4800	9600	16,800	24,000

Model 54, 4-cyl., 45 hp, 117" wb

Tr	1000	3100	5200	10,400	18,200	26,000

Model 55, 4-cyl., 45 hp, 123" wb

Tr	1050	3350	5600	11,200	19,600	28,000
Limo	700	2300	3800	7600	13,300	19,000

1911
Model 63, 4-cyl., 34 hp, 112" wb

Tr	950	3000	5000	10,000	17,500	25,000
Rds	900	2900	4800	9600	16,800	24,000
Cpe	600	1900	3200	6400	11,200	16,000
Twn Car	700	2150	3600	7200	12,600	18,000

Model 64, 4-cyl., 34 hp, 120" wb

Tr	1000	3250	5400	10,800	18,900	27,000
Toy Ton	1000	3250	5400	10,800	18,900	27,000
Lan'let	700	2300	3800	7600	13,300	19,000

Model 65, 4-cyl., 34 hp, 128" wb

Tr	1050	3350	5600	11,200	19,600	28,000
Toy Ton	1050	3350	5600	11,200	19,600	28,000
Limo	750	2400	4000	8000	14,000	20,000

1912
Four, 38 hp, 120" wb

5P CrCtry Tr	1050	3350	5600	11,200	19,600	28,000
4P Sub Ctry Club	1050	3350	5600	11,200	19,600	28,000
2P Rds	1000	3250	5400	10,800	18,900	27,000
4P Sed	600	1900	3200	6400	11,200	16,000
7P Gotham Limo	700	2300	3800	7600	13,300	19,000

Four, 50 hp, 120" wb

Ctry Club	1200	3850	6400	12,800	22,400	32,000
Valkyrie	1200	3850	6400	12,800	22,400	32,000

Four, 50 hp, 128" wb

Morraine Tr	1300	4200	7000	14,000	24,500	35,000
Metropolitan	1350	4300	7200	14,400	25,200	36,000
Greyhound	1350	4300	7200	14,400	25,200	36,000
Knickerbocker	1450	4700	7800	15,600	27,300	39,000

1913
Four, 42 hp, 120" wb

2/3P CrCtry Rds	1150	3600	6000	12,000	21,000	30,000
4/5P CrCtry Tr	1150	3700	6200	12,400	21,700	31,000
4P Inside Drive Cpe	700	2300	3800	7600	13,300	19,000
7P Gotham Limo	850	2650	4400	8800	15,400	22,000

JEFFERY

1914

	6	5	4	3	2	1
Four, 40 hp, 116" wb						
5P Tr	600	1900	3200	6400	11,200	16,000
5P Sed	400	1300	2200	4400	7700	11,000
Four, 27 hp, 120" wb						
2P Rds	700	2150	3600	7200	12,600	18,000
4P/5P/7P Tr	650	2050	3400	6800	11,900	17,000
Six, 48 hp, 128" wb						
5P Tr	750	2400	4000	8000	14,000	20,000
6P Tr	700	2300	3800	7600	13,300	19,000
7P Limo	600	1900	3200	6400	11,200	16,000

1915

	6	5	4	3	2	1
Four, 40 hp, 116" wb						
5P Tr	700	2300	3800	7600	13,300	19,000
2P Rds	750	2400	4000	8000	14,000	20,000
2P All-Weather	800	2500	4200	8400	14,700	21,000
7P Limo	650	2050	3400	6800	11,900	17,000
4P Sed	500	1550	2600	5200	9100	13,000
Chesterfield Six, 48 hp, 122" wb						
5P Tr	800	2500	4200	8400	14,700	21,000

	6	5	4	3	2	1
2P Rds	850	2650	4400	8800	15,400	22,000
2P All-Weather	850	2750	4600	9200	16,100	23,000

1916

	6	5	4	3	2	1
Four, 40 hp, 116" wb						
7P Tr	850	2650	4400	8800	15,400	22,000
5P Tr	850	2750	4600	9200	16,100	23,000
7P Sed	500	1550	2600	5200	9100	13,000
5P Sed	450	1450	2400	4800	8400	12,000
3P Rds	900	2900	4800	9600	16,800	24,000
Chesterfield Six, 48 hp, 122" wb						
5P Tr	1000	3250	5400	10,800	18,900	27,000

1917

	6	5	4	3	2	1
Model 472, 4-cyl., 40 hp, 116" wb						
7P Tr	800	2500	4200	8400	14,700	21,000
2P Rds	850	2650	4400	8800	15,400	22,000
7P Sed	450	1450	2400	4800	8400	12,000
Model 671, 6-cyl., 48 hp, 125" wb						
7P Tr	850	2750	4600	9200	16,100	23,000
3P Rds	900	2900	4800	9600	16,800	24,000
5P Sed	500	1550	2600	5200	9100	13,000

For Collector Car Enthusiasts...

Standard Guide to Military Vehicles, 1940-1965
* Includes jeeps, half-tracks, tanks, and more
* Historical information & technical specifications
* Pricing in 6 grades for collectors & veterans
ONLY $29.95

Standard Guide to American Muscle Cars, 1949-1992
* High performance cars from the late '40s through '92
* Highlights classic muscle cars from the '60s & '70s
* 500 photos, specifications, production information
ONLY $19.95

American Volunteer Fire Trucks
* A pictorial history of the rigs used across America
* Over 500 models are featured in photos & text
* A must for firefighters & fans of these vehicles
ONLY $16.95

American Motors, the Last Independent
* An in-depth historical view of American Motors Corp.
* Fully illustrated & well told by Patrick R. Foster
* Includes Hudson, Nash, Rambler, Jeep and today's cars
ONLY $19.95

The Fabulous '50s - The Cars, The Culture
* A photographic history of the cars we loved in the '50s
* Over 500 automotive photos, advertisements & billboards
* Lively captions depict the cultural trivia of this era
ONLY $14.95

Standard Catalog of 4x4's, 1945-1993
* A comprehensive guide to American four-wheel-drive vehicles
* Includes trucks, vans, sports sedans and sports cars
* Technical specifications, production totals, I.D. numbers
ONLY $24.95

1993 Standard Guide to Cars & Prices, 5th Edition
* 130,000 current values for vehicles built from 1901-1985
* Complete listings of models from America & around the world
* Includes 20 current "Bargain Buys" for under $7,000 each
ONLY $15.95

Standard Catalog of Imported Cars, 1946-1990
* This unique book profiles hundreds of models worldwide
* Complete with technical specifications, histories & photos
* Current values based on the I-to-6 condition scale
ONLY $24.95

Antique Car Wrecks
* Adapted from OLD CARS WEEKLY's "Wreck of the Week" column
* Includes more than 400 photos, 80% previously unpublished
* Driving safety is reinforced by this photographic history
ONLY $14.95

Police Cars: A Photographic History
* An up close and personal look at the cars against crime
* 300 pages of early paddy wagons to 1990's Corvette Cops
* A must for historical fans of our men and women in blue
ONLY $14.95

MASTERCARD & VISA CUSTOMERS, CALL TOLL-FREE TO ORDER...
(800) 258-0929
Monday - Friday 6:30 am-8:00 pm, Saturday 8:00 am-2:00 pm Central Standard Time

Catalogs For Car Collectors...

Standard Catalog of American Cars, 1805-1942
* 5,000 auto builders' makes and models from 1805-1942
* 4,500 photos present visual details to help restorers
ONLY $45.00

Standard Catalog of American Cars, 1946-1975
* Presenting more than 1,000 vehicle listings from 1946-1975
* Over 1,500 photographs aid in vehicle identification
ONLY $27.95

Standard Catalog of American Cars, 1976-1986
* Presents thousands of cars manufactured in America
* Helps pinpoint tomorrow's collector cars today
ONLY $19.95

Standard Catalog of American Light-Duty Trucks
* All new 2nd edition presents 500 truck listings, 1896-1986
* ID data, serial numbers, codes, specs, current pricing
ONLY $29.95

Standard Catalog of Ford
* Profiles every Ford make and model built from 1903-1990
* More than 500 photographs aid in identification
* ID data, serial numbers, codes, specs, current pricing
ONLY $19.95

Standard Catalog of Chevrolet
* Profiles every Chevy model manufactured from 1912-1990
* Over 500 photos bring restorers visual aid during projects
* ID data, serial numbers, codes, specs, current pricing
ONLY $19.95

Standard Catalog of Chrysler
* Profiles each Chrysler make and model in detail, from 1924-1990
* Presents I-to-6 conditional pricing through 1983 models
* ID data, serial numbers, codes, specs, production totals
ONLY $19.95

Standard Catalog of Buick
* Profiles every Buick model & make crafted from 1903-1990
* Chassis specs, body types, shipping weights, current pricing
* Fascinating stories, historical perspectives, photo profiles
ONLY $18.95

Standard Catalog of Cadillac
* All the makes that made Cadillac famous, from 1903-1990
* Photographic perspectives, codes, specs, ID and serial numbers
* Includes I-to-6 conditional pricing for current value comparisons
ONLY $18.95

Standard Catalog of American Motors
* Presents every model in the AMC family from 1903-1987
* Hudson, Nash, Metropolitan, Rambler, AMC/Jeep, AMX
* Factory prices, today's values, technical specifications
ONLY $19.95

Standard Guide to Automotive Restoration
* Matt Joseph's technical, hands-on guide for all restorers
* Complete system-by-system instructions for all makes & models
* More than 400 detailed photos aid in correct application
ONLY $24.95

MASTERCARD & VISA CUSTOMERS, CALL TOLL-FREE TO ORDER...
(800) 258-0929
Monday - Friday 6:30 am-8:00 pm, Saturday 8:00 am-2:00 pm Central Standard Time

MARQUE YOUR CHOICE!

OLD CARS
WEEKLY NEWS & MARKETPLACE
New Subscriber Offer

52 Weekly Issues...Only **$19.50**

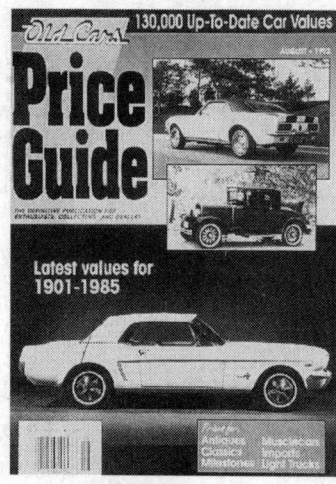

OLD CARS PRICE GUIDE

6 Bi-Monthly Issues...Only **$16.95**

CALL TOLL-FREE TO ORDER
(800) 258-0929 Dept. AEU3
Mon.-Fri., 6:30 a.m. - 8 p.m. • Sat., 8 a.m.-2 p.m., CST

KRAUSE PUBLICATIONS
IOLA, WISCONSIN 54990-0001